Lecture Notes in Computer Science   8558

Commenced Publication in 1973
Founding and Former Series Editors:
Gerhard Goos, Juris Hartmanis, and Jan van Leeuwen

Gerwin Klein   Ruben Gamboa (Eds.)

Interactive
Theorem Proving

5th International Conference, ITP 2014
Held as Part of the Vienna Summer of Logic, VSL 2014
Vienna, Austria, July 14-17, 2014
Proceedings

 Springer

Volume Editors

Gerwin Klein
NICTA
Sydney, NSW, Australia
E-mail: gerwin.klein@nicta.com.au

Ruben Gamboa
University of Wyoming
Laramie, WY, USA
E-mail: ruben@uwyo.edu

ISSN 0302-9743   e-ISSN 1611-3349
ISBN 978-3-319-08969-0   e-ISBN 978-3-319-08970-6
DOI 10.1007/978-3-319-08970-6
Springer Cham Heidelberg New York Dordrecht London

Library of Congress Control Number: 2014942571

LNCS Sublibrary: SL 1 – Theoretical Computer Science and General Issues

Typesetting: Camera-ready by author, data conversion by Scientific Publishing Services, Chennai, India

Printed on acid-free paper

Springer is part of Springer Science+Business Media (www.springer.com)

logic n. **1** the science of reasoning.
– ORIGIN from Greek *logikē tekhnē*
'art of reason'.

Foreword

In the summer of 2014, Vienna hosted the largest scientific conference in the history of logic. The Vienna Summer of Logic (VSL, http://vsl2014.at) consisted of twelve large conferences and 82 workshops, attracting more than 2000 researchers from all over the world. This unique event was organized by the Kurt Gödel Society and took place at Vienna University of Technology during July 9 to 24, 2014, under the auspices of the Federal President of the Republic of Austria, Dr. Heinz Fischer.

The conferences and workshops dealt with the main theme, logic, from three important angles: logic in computer science, mathematical logic, and logic in artificial intelligence. They naturally gave rise to respective streams gathering the following meetings:

Logic in Computer Science / Federated Logic Conference (FLoC)

- 26th International Conference on Computer Aided Verification (CAV)
- 27th IEEE Computer Security Foundations Symposium (CSF)
- 30th International Conference on Logic Programming (ICLP)
- 7th International Joint Conference on Automated Reasoning (IJCAR)
- 5th Conference on Interactive Theorem Proving (ITP)
- Joint meeting of the 23rd EACSL Annual Conference on Computer Science Logic (CSL) and the 29th ACM/IEEE Symposium on Logic in Computer Science (LICS)
- 25th International Conference on Rewriting Techniques and Applications (RTA) joint with the 12th International Conference on Typed Lambda Calculi and Applications (TLCA)
- 17th International Conference on Theory and Applications of Satisfiability Testing (SAT)
- 76 FLoC Workshops
- FLoC Olympic Games (System Competitions)

Mathematical Logic

- Logic Colloquium 2014 (LC)
- Logic, Algebra and Truth Degrees 2014 (LATD)
- Compositional Meaning in Logic (GeTFun 2.0)
- The Infinity Workshop (INFINITY)
- Workshop on Logic and Games (LG)
- Kurt Gödel Fellowship Competition

Logic in Artificial Intelligence

- 14th International Conference on Principles of Knowledge Representation and Reasoning (KR)
- 27th International Workshop on Description Logics (DL)
- 15th International Workshop on Non-Monotonic Reasoning (NMR)
- 6th International Workshop on Knowledge Representation for Health Care 2014 (KR4HC)

The VSL keynote talks which were directed to all participants were given by Franz Baader (Technische Universität Dresden), Edmund Clarke (Carnegie Mellon University), Christos Papadimitriou (University of California, Berkeley) and Alex Wilkie (University of Manchester); Dana Scott (Carnegie Mellon University) spoke in the opening session. Since the Vienna Summer of Logic contained more than a hundred invited talks, it would not be feasible to list them here.

The program of the Vienna Summer of Logic was very rich, including not only scientific talks, poster sessions and panels, but also two distinctive events. One was the award ceremony of the Kurt Gödel Research Prize Fellowship Competition, in which the Kurt Gödel Society awarded three research fellowship prizes endowed with 100.000 Euro each to the winners. This was the third edition of the competition, themed Logical Mind: Connecting Foundations and Technology this year.

The 1st FLoC Olympic Games formed the other distinctive event and were hosted by the Federated Logic Conference (FLoC) 2014. Intended as a new FLoC element, the Games brought together 12 established logic solver competitions by different research communities. In addition to the competitions, the Olympic Games facilitated the exchange of expertise between communities, and increased the visibility and impact of state-of-the-art solver technology. The winners in the competition categories were honored with Kurt Gödel medals at the FLoC Olympic Games award ceremonies.

Organizing an event like the Vienna Summer of Logic was a challenge. We are indebted to numerous people whose enormous efforts were essential in making this vision become reality. With so many colleagues and friends working with us, we are unable to list them individually here. Nevertheless, as representatives of the three streams of VSL, we would like to particularly express our gratitude to all people who helped to make this event a success: the sponsors and the Honorary Committee; the Organization Committee and the local organizers; the conference and workshop chairs and Program Committee members; the reviewers and authors; and of course all speakers and participants of the many conferences, workshops and competitions.

The Vienna Summer of Logic continues a great legacy of scientific thought that started in Ancient Greece and flourished in the city of Gödel, Wittgenstein and the Vienna Circle. The heroes of our intellectual past shaped the scientific world-view and changed our understanding of science. Owing to their achievements, logic has permeated a wide range of disciplines, including computer science, mathematics, artificial intelligence, philosophy, linguistics, and many more. Logic is everywhere – or in the language of Aristotle, πάντα πλήρη λογικῆς τέχνης.

July 2014

Matthias Baaz
Thomas Eiter
Helmut Veith

Preface

This volume contains the papers presented at ITP 2014: 5th International Conference on Interactive Theorem Proving held during July 13–16, 2014 in Vienna, as part of the Federated Logic Conference (FLoC, July 10–24, 2014), which was part of the Vienna Summer of Logic (July 9–24, 2014).

ITP 2014 was the 5th conference on Interactive Theorem Proving and related areas, ranging from theoretical foundations to implementation aspects and applications in program verification, security, and formalization of mathematics. The inaugural meeting of ITP was held during 11–14 July 2010 in Edinburgh, Scotland, as part of the Federated Logic Conference (FLoC, 9–21 July 2010). ITP is the evolution of the TPHOLs conference series to the broad field of interactive theorem proving. TPHOLs meetings took place every year from 1988 until 2009.

There were 59 submissions to ITP 2014, each of which was reviewed by at least three Program Committee members. The Committee decided to accept 35 papers, 4 of which were rough diamonds. The program also included invited talks by Anna Slobodova, Rod Chapman, and Peter Sewell.

We would like to thank the co-chairs of the Vienna Summer of Logic (Matthias Baaz, Thomas Eiter, Helmut Veith), the FLoC general chair (Moshe Vardi), all the other members of the VSL and FLoC Organizing Committee, and David Pichardie, who was kind enough to take care of ITP's satellite workshops. We gratefully acknowledge the support of VSL's partners and sponsors, including TU Wien, IST Austria, Universität Wien, Stadt Wien, Austrian Airlines, the Federal Ministry of Science Research and Economy, the University of Manchester, Cateringkultur, Vienna Convention Bureau, Elsevier's Artificial Intelligence, European Association for Computer Science Logic, Blacklane Limousines, Fraunhofer, OrbiTeam, Rigorous Systems Engineering, and College Publications. We also gratefully acknowledge the support of FLoC's sponsors: Microsoft Research, ARM, and NEC. And we extend a special thanks to ITP's sponsors: NICTA, and HappyJack Software.

Next year's conference will be held in Nanjing, China. The site was chosen by the ITP Steering Committee in consultation with the broader ITP community.

May 2014

Gerwin Klein
Ruben Gamboa

Organization

Program Committee

Jeremy Avigad	Carnegie Mellon University, USA
Lennart Beringer	Princeton University, USA
Yves Bertot	Inria, France
Thierry Coquand	Chalmers University, Sweden
Amy Felty	University of Ottawa, Canada
Ruben Gamboa	University of Wyoming, USA
Georges Gonthier	Microsoft Research, USA
Elsa Gunter	University of Illinois at Urbana-Champaign, USA
John Harrison	Intel Corporation, UK
Matt Kaufmann	University of Texas at Austin, USA
Gerwin Klein	NICTA and UNSW, Australia
Alexander Krauss	Technische Universität München, Germany
Ramana Kumar	University of Cambridge, UK
Joe Leslie-Hurd	Intel Corporation, UK
Assia Mahboubi	Inria - École polytechnique, France
Panagiotis Manolios	Northeastern University, USA
Magnus O. Myreen	University of Cambridge, UK
Tobias Nipkow	Technische Universität München, Germany
Michael Norrish	NICTA and ANU, Australia
Sam Owre	SRI International, USA
Christine Paulin-Mohring	Université Paris-Sud, France
Lawrence Paulson	University of Cambridge, UK
David Pichardie	Inria Rennes - Bretagne Atlantique, France
Lee Pike	Galois Inc., USA
Jose-Luis Ruiz-Reina	University of Seville, Spain
Julien Schmaltz	Open University of the Netherlands, The Netherlands
Bas Spitters	Radboud University Nijmegen, The Netherlands
Sofiene Tahar	Concordia University, Canada
René Thiemann	University of Innsbruck, Austria
Laurent Théry	Inria, France
Christian Urban	King's College London, UK
Tjark Weber	Uppsala University, Sweden
Makarius Wenzel	Université Paris-Sud 11, France

Additional Reviewers

Aravantinos, Vincent
Baelde, David
Blanchette, Jasmin Christian
Cachera, David
Chamarthi, Harsh Raju
Demange, Delphine
Dunchev, Cvetan
Helali, Ghassen
Hölzl, Johannes
Jain, Mitesh
Joosten, Sebastiaan
Kaliszyk, Cezary
Keller, Chantal
Khan-Afshar, Sanaz
Kiniry, Joe
Mahmoud, Mohamed Yousri
Martin-Mateos, Francisco-Jesus

Matichuk, Daniel
Murray, Toby
Rager, David
Ravitch, Tristan
Rubio, Julio
Sewell, Thomas
Siddique, Umair
Tankink, Carst
Tassi, Enrico
Tomb, Aaron
Urbain, Xavier
Verbeek, Freek
Wang, Yuting
Winkler, Sarah
Winwood, Simon
Ziliani, Beta

Abstracts of Invited Talks

Microcode Verification – Another Piece of the Microprocessor Verification Puzzle[*]

Jared Davis, Anna Slobodova, and Sol Swords

Centaur Technology, Inc.

{jared,anna,sswords}@centtech.com

Abstract. Despite significant progress in formal hardware verification in the past decade, little has been published on the verification of microcode. Microcode is the heart of every microprocessor and is one of the most complex parts of the design: it is tightly connected to the huge machine state, written in an assembly-like language that has no support for data or control structures, and has little documentation and changing semantics. At the same time it plays a crucial role in the way the processor works.

We describe the method of formal microcode verification we have developed for an x86-64 microprocessor designed at Centaur Technology. While the previous work on high and low level code verification is based on an unverified abstract machine model, our approach is tightly connected with our effort to verify the register-transfer level implementation of the hardware. The same microoperation specifications developed to verify implementation of the execution units are used to define operational semantics for the microcode verification.

While the techniques used in the described verification effort are not inherently new, to our knowledge, our effort is the first interconnection of hardware and microcode verification in context of an industrial size design. Both our hardware and microcode verifications are done within the same verification framework.

Are We There Yet? 20 Years of Industrial Theorem Proving with SPARK[*]

Roderick Chapman and Florian Schanda

Altran UK Limited, 22 St Lawrence Street, Bath BA1 1AN, United Kingdom
roderick.chapman@gmail.com, florian.schanda@altran.com

Abstract. This paper presents a retrospective of our experiences with applying theorem proving to the verification of SPARK programs, both in terms of projects and the technical evolution of the language and tools over the years.

---

[*] The full versions of these papers are available within this book.

Retrofitting Rigour

Peter Sewell

University of Cambridge
Peter.Sewell@cl.cam.ac.uk,
http://www.cl.cam.ac.uk/users/pes20/

Abstract. We rely on an enormous number of software components, which continue to be developed —as they have been since the 1940s— by methods based on a test-and-debug cycle and informal-prose specifications. Replacing this legacy infrastructure and development practices wholesale is not (and may never be) feasible, so if we want to improve system quality by using mathematically rigorous methods, we need a strategy to do so incrementally. A first step is to focus on the more stable extant interfaces: processor architectures, programming languages, and key APIs, file formats, and protocols. By characterising these with new precise abstractions, we can start to retrofit rigour into the system, using those (1) to improve existing components by testing, (2) as interfaces for replacement formally verified components, and (3) simply to gain understanding of what the existing behavioural interfaces are. Many issues arise in building such abstractions and in establishing confidence in them. To validate them against existing systems, they must be made executable, not necessarily in the conventional sense but as test oracles, to check whether observed behaviour of the actual system is admitted by the specification. The appropriate treatment of loose specification is key here, and varies from case to case. To validate them against the intent of the designers, they must be made broadly comprehensible, by careful structuring, annotation, and presentation. And to support formal reasoning by as broad a community as possible, they must be portable into a variety of reasoning tools. Moreover, the scale of real-world specifications adds further engineering concerns. This talk will discuss the issues of loose specification and some lessons learnt from developing and using our Ott and Lem lightweight specification tools, which are aimed at supporting all these pre-proving activities (and which target multiple theorem provers).

Acknowledgements. Ott and Lem have been developed in joint work with Francesco Zappa Nardelli, Scott Owens, Dominic Mulligan, Thomas Tuerk, Kathryn Gray, and Thomas Williams, and with feedback from many other users. I acknowledge funding from EPSRC grants EP/H005633 (Leadership Fellowship) and EP/K008528 (REMS Programme Grant).

Table of Contents

Rough Diamonds

Microcode Verification – Another Piece of the Microprocessor Verification Puzzle

Jared Davis, Anna Slobodova, and Sol Swords

Centaur Technology, Inc.
{jared,anna,sswords}@centtech.com

Abstract. Despite significant progress in formal hardware verification in the past decade, little has been published on the verification of microcode. Microcode is the heart of every microprocessor and is one of the most complex parts of the design: it is tightly connected to the huge machine state, written in an assembly-like language that has no support for data or control structures, and has little documentation and changing semantics. At the same time it plays a crucial role in the way the processor works.

We describe the method of formal microcode verification we have developed for an x86-64 microprocessor designed at Centaur Technology. While the previous work on high and low level code verification is based on an unverified abstract machine model, our approach is tightly connected with our effort to verify the register-transfer level implementation of the hardware. The same microoperation specifications developed to verify implementation of teh execution units are used to define operational semantics for the microcode verification.

While the techniques used in the described verification effort are not inherently new, to our knowledge, our effort is the first interconnection of hardware and microcode verification in context of an industrial size design. Both our hardware and microcode verifications are done within the same verification framework.

1 Introduction

Microprocessor design is a complex effort that takes hundreds of man-years. Verification of the microprocessor design remains the bottleneck of the design process. It consumes an increasing amount of resources and deploys more and more sophisticated methods including high-performance simulators and formal technology. There are many aspects to verifying the correctness of a microprocessor based system. In this paper, we will discuss only *functional verification*. Most of the papers about microprocessor verification are solely concerned with the verification of hardware. We will focus on the verification of *microcode* which is the heart and soul of a microprocessor.

While the external interface to a microprocessor is mandated by its Instruction Set Architecture (ISA), its internal behavior is governed by some processor-specific microarchitecture. For instance, contemporary x86 processors externally

G. Klein and R. Gamboa (Eds.): ITP 2014, LNAI 8558, pp. 1–16, 2014.

support a wealth of instructions, sizes, and modes that take thousands of pages to describe (see Intel®64 and IA32 Architecture Software Developer Manuals). For performance reasons, implementations of modern x86 processors have a *frontend* that translates x86 instructions into simpler microoperations (uops), and a *backend* for executing these uops (Figure 1). Microcode (ucode) bridges the external world of Complex Instruction Set Computing and the internal world of Reduced Instruction Set Computing.

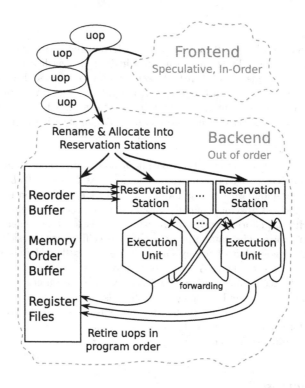

Fig. 1. Processor Backend: uops from the front-end are renamed, placed into the reorder buffer, and given to the reservation station of the appropriate execution unit. Each uop executes once its unit and operands become available. Results are forwarded among the execution units and also sent back to the reorder buffer, where they remain until retirement.

While simple and common x86 instructions are often translated into a single uop, more complex or obscure operations are implemented as ucode programs that are stored in a microcode ROM. Microcode programs are responsible for many complex features that a processor provides, e.g., they are used to implement transcendental functions, hardware virtualization, processor initialization, security features, and so on. Accordingly, their correctness is critical.

Unfortunately, there are many challenges to verifying microcode programs. Microcode verification can be seen as an instance of hardware/software co-verification, with all of the associated challenges. Whether using formal or testing-based methods, validation involves understanding both the micro-architecture and the microcode program, neither of which is easy.

Microcode is a very primitive, low-level language without even basic control constructs or data structures. At the same time, ucode programs are designed for efficiency rather than verification. During the design effort, not only are ucode programs frequently updated, but the very microcode language is extended with new operations and features. Even as the end of an effort nears and the hardware design is frozen, ucode programs continue to change—indeed, ucode patches become the preferred way of fixing bugs and adding or removing features.

The formal verification team at Centaur Technology applies formal methods to problems in various stages of the design process, including equivalence checking of transistor level and Register-Transfer Level (RTL) designs. In the area of the RTL verification, we have applied symbolic simulation supported by SAT-based and BDD-based technology to verify execution of individual microoperations in assigned execution units [1–3]. All of this verification has been carried out within the ACL2 system [4].

This paper presents our approach to formally verifying microcode routines for a new x86-64 processor in development at Centaur Technology. Our methods draw inspiration from the work of many published sources, but our work differs from each of these works in one or more aspects listed below:

i Our target is microcode – a language below the ISA level;
ii Our verification is done on an industrial scale design – an implementation of a fully x86-64 compatible microprocessor. In addition, it is done on a live project that undergoes continuous changes on the specification and implementation levels.
iii Our formal model of the microarchitecture is based on the specifications used in the RTL proofs. To our knowledge this is the first such interconnection of hardware and ucode verification done on a microarchitecture of such complexity.

Section 2 describes our formal ACL2 model of the processor's microarchitectural state and uop execution semantics. Our model can be run as a high-speed microcode simulator (around 250k uops/sec), and is also designed to achieve good reasoning performance in the theorem prover.

Section 3 gives a sketch of our approach to verifying microoperation sequences and loops, and how those can be composed to achieve correctness theorems about parts of code that constitutes subroutines. The sequential composition of the blocks is based on exploiting the power of the simplification engine within a theorem prover.

Section 4 describes the degree to which our abstract machine model has been proven to correspond to the actual hardware implementations. Parts of our model are contrived, but significant parts are directly based on specification functions that have a mechanically proven correspondence to the Verilog modules of our processor.

Section 5 summarizes related work. Finally, Section 6 concludes the paper with comments about our future work.

2 Microcode Modeling

Microcode originates as a text program. Figure 2 shows an example of a microcode program.

```
clr_pram:
 MVIG.S64    g0, 0;        g0 = 0
clr_loop:
 STORE_PRAM g0, ADR, 0;   PRAM[ADR] = g0
 ADDIG.S64   ADR, ADR, 8; ADR = ADR + 8
 NLOOPE.S64 g8, 1, ret;   g8--; if !g8 goto ret
 JMP_ALL     clr_loop;    goto clr_loop
ret:
 JLINK       ;             return
```

Fig. 2. A microcode routine that zeroes an area of PRAM memory. Here g0 and g8 are 64-bit registers. ADR is an alias to the 64-bit register g9. Labels like clr_pram, clr_loop and ret are resolved into ROM addresses by the assembler.

Figure 3 shows the relation between the model and hardware. An assembler translates microcode program into a binary image which is stored on the chip in a ROM. When executing microcode, the microtranslator unit fetches instructions from the ROM and translates them into backend uops that are then executed.

Our model consists of four parts described in more details below:

- **Microcode** - a constant representing the entire ROM image.
- **Microcode translator** - a function that maps ROM instructions into backend uops, plus a ucode sequencing instruction that determines the control flow after the uops' execution.
- **State** - a data structure representing the microarchitectural state.
- **Operational semantics** for backend uops, defining their effects on the state.

To build the microcode ROM, the source code files are collected and processed by a microcode assembler. The assembler converts each instruction into its binary encoding, producing a binary image that captures every ucode routine. When the processor is manufactured, this image is embedded into its ucode ROM. We extended the microcode assembler to produce, besides the ROM image and other debugging and statistical information it already generated, a Lisp/ACL2

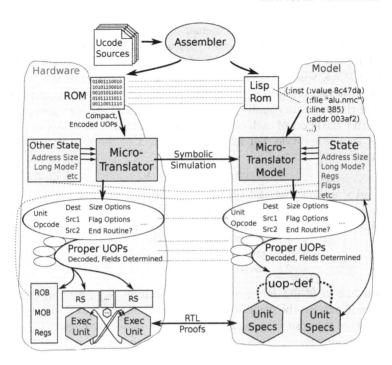

Fig. 3. Connections between our model and the processor implementation

model of the ucode listing which can be conveniently loaded into our verification framework. This way, our model of the microcode program stays up-to-date as microcode routines or the assembler itself are changed.

The microcode translator functionality is derived by means of symbolic simulation from the RTL design of the microtranslator unit. A mapping from inputs representing a ROM instruction to outputs representing a sequence of backend uops and control flow information is represented by a set of (Boolean) formulas. For execution speed, we precompute a specialized version of these formulas for each instruction in the ROM and store them into a hash table. Some instructions' translations depend on the machine state (e.g. the current mode of operation), so the specialization does not always yield a final, concrete result, but only simpler formulas. In those cases, we finish the translation at execution time. To execute a single ROM instruction, we look up that instruction's specialized translation formula and evaluate that formula by substituting in any relevant state bits. The obtained sequence of uops and the control information is interpreted with respect to defined semantics which determines the next state, including the value of the program counter. Automatically deriving the micro-translator functionality from the actual RTL keeps this part of the model up-to-date.

The machine state is represented as a tuple containing essential machine variables such as the program counter, the stack, various sets of registers (e.g., the x86 general-purpose registers are in a field called *gregs* and the SSE media registers are in *mregs*) and scratch memories. Parts of the state correspond to registers defined by the x86 architecture, but much of it is specific to the concrete micro-architecture of the project. A machine state may be a *running, halting,* or *divergent* state, where a divergent state is used to represent the result of a program that never terminates. These three types of states are distinguished by the value of the program counter; a natural number indicates a running state.

We describe the effect of each uop as a transformation of the state, in the usual interpreter semantics style. For instance, the effect of the XADD uop is modeled by the function

$$xadd\text{-}def(uop, s) \rightarrow s',$$

where s and s' are current and transformed states, resp., and uop represents a particular instance of a microoperation with all relevant information – operands, operand size, flag information, unit where the uop is executed, *etc.*:

dest g9	src width	64	opcode	XADD
src1 g8	dest width	64	exec unit	int
src2 8	write flags?	no		

A function like *xadd-def* interprets this information, e.g., for the instruction above it would extract the value of gregs[8] and interpret it as 64-bit value, add it to src2, store the result in gregs[9], and not update any flags.

Semantic functions for all types of uops are combined into a universal uop definition:

$$\mu op\text{-}def(uop, s) \stackrel{\text{def}}{=}$$
$$\begin{cases} xadd\text{-}def(uop, s) & \text{if uop.type is XADD} \\ xsub\text{-}def(uop, s) & \text{if uop.type is XSUB} \\ \ldots & \ldots \\ error(s) & \text{otherwise} \end{cases}$$

To model the execution of consecutive steps from a state s, we use an approach by Ray and Moore [5,6]. We first define the function $run(n, s)$ which returns the new state after executing n steps:

$$run(n, s) \stackrel{\text{def}}{=} \begin{cases} s & \text{if } n = 0 \\ run(n - 1, step(s)) & \text{otherwise} \end{cases}$$

Note that *step* executes one ROM instruction that can consist of several uops. Its definition and connection to our hardware verification proofs are explained in more detail in Section 4.

Finally, we can define $run^* : s \rightarrow s$, which runs the machine until it enters a halting state. If the program does not terminate, then logically it returns the divergent state $\bot$ (whereas the actual execution of run^* would never terminate). This avoids the need to explicitly determine how many steps a program takes and allows us to pursue partial correctness results without proving termination (see *run-measure* in Section 3).

$$run^*(s) \stackrel{\text{def}}{=} \begin{cases} s & \text{if } halting(s) \\ run^*(step(s)) & \text{if } \neg divergent(s) \\ \bot & \text{otherwise} \end{cases}$$

The next section will explain how we reason about this model. Then, in Section 4, we will explain the relationship between this model and the hardware design.

3   Microcode Verification

Before describing our verification methodology, we need to explain our verification objective. Unlike higher level languages, microcode does not have any nice control structures like for loops, while loops, if-then-else constructs, etc, and there is no such entity as a main program in microcode. It consists of a sequence of micro operations residing in a ROM. Figure 2 shows an example of a microcode snippet. Microoperations can move values between registers and to/from scratch memory, and can manipulate values by means of arithmetic and logical operations. Loops are implemented with conditional and unconditional jumps. Microcode is written for efficiency and would not please the eye of any programmer.

Our objective in microcode verification is to characterize the effect of executing microcode from an entry point to an exit point on the machine state. In order to do this, we incrementally verify blocks of code and compose the theorems into theorems about larger blocks. We have defined a macro *def-uc-block* that supports verification of a general block of code, and a macro *def-uc-loop* that supports reasoning about loops.

3.1   Def-uc-block

We use the *def-uc-block* macro to specify blocks of straight-line code and to compose together previously defined blocks (including loops). The user specifies:

1. *start-pc* as an initial value for the program counter;
2. *block-precondition* as a state predicate;
3. *run-block*, a function that executes the machine model until the end of block is reached. This may be a simple application of the universal *run* function for a given number of steps, or as a combination of *run* and applications of previously defined blocks.
4. *block-specification* that describes the machine state after execution of this block (the post-state). This definition is in terms of updates to the start-state. As a consequence, those parts of the state that remain unchanged are left out of the description of the change. While we need to keep track of the changing values in some registers and memories, other parts of the state are used as temporary storage and become irrelevant for the final result. To avoid precisely characterizing these don't care values, we copy them from the actual post-state (produced by *run-block*) into the specification state, making their equivalence trivial. This is known as *wormhole abstraction* [7]).

For the code in Figure 2, the verification could start with the definition of *def-uc-block* with the arguments described in Figure 4.

> name: *clr_loop_last*
> pc: *get-label(clr_loop)*
> run: *run(4, s)*
>
> precondition: Let $adr = s.gregs[9]$, $cnt = s.gregs[8]$ in
> $addr\text{-}ok(adr) \wedge (cnt = 1) \wedge$
> $\neg empty(s.retstack)$
>
> specification: Let $adr = s.gregs[9]$, $val = s.gregs[0]$ in {
> $s = pram\text{-}store(val, adr, s)$;
> $s.gregs[8] = 0$;
> $s.gregs[9] = adr + 8$;
>   return $stack\text{-}pop(s)$;
>   }

Fig. 4. *def-uc-block* example (last run through the loop): *get-label* translates a label into initial value of the program counter. Running the block takes 4 steps. In the precondition, *addr-ok* identifies valid address to the PRAM memory. While g9 is used as a pointer to memory, g8 is a counter that controls the loop. The loop terminates when the counter clears. The specification describes the state update caused by the last run through the loop: the value of g0 is stored in the PRAM at address specified by g9, g8 is decremented, g9 is incremented by the size of the written entry, and the program counter is set to the value on the top of the return stack.

The expansion of the macro defines all the functions above and automatically proves some theorems about them. For instance, the *run-block* function has to satisfy following properties:

R1: $run^*(run\text{-}block(s)) = run^*(s)$
  applying run^* to the post-state brings us to the same state as applying run^* to the start state.
R2: $halting(s) \implies run\text{-}block(s) = s$
  run-block will not advance from a halting state.
R3: $divergent(s) \implies run\text{-}block(s) = \bot$
  Whenever we get into a divergent state, we converge into the $\bot$ state.
R4: $\neg divergent(run^*(s)) \implies \neg divergent(run\text{-}block(s))$
  If run^* terminates, then *run-block* terminates.
R5: *run-block* makes progress in the termination of run^*:

$$\neg halting(s) \wedge \neg divergent(run\text{-}block(s))$$
$$\implies$$
$$run\text{-}measure(run\text{-}block(s)) < run\text{-}measure(s)$$

where *run-measure* is a non-executable function whose value is the minimum number of steps needed to bring the machine to a halting state, if that exists, and zero otherwise. It is defined as a Skolem witnessing function using the ACL2 feature *defchoose*.

R6: *run-measure(run-block(s))* ≤ *run-measure(s)*
A weaker monotonic condition.

For the *block-precondition* predicate we have an option to do a simple vacuity check. It exploits a symbolic simulator [8] that converts an ACL2 object that is defined over a finite domain into a symbolic object encoded as an And-Inverter Graph. Finding a state that satisfies the precondition is thus transformed into satisfiability of a Boolean formula, which we then translate into CNF and solve using an off-the-shelf SAT solver [9].

The main result of the *def-uc-block* expansion is the correctness theorem:

Theorem 1 (block-correct).

$$s.pc = start\text{-}pc \land block\text{-}precondition(s)$$
$$\Longrightarrow$$
$$run\text{-}block(s) = block\text{-}specification(s, run\text{-}block(s))$$

Theorem *block-correct* is the crucial point of the verification. We have two distinct methods for proving this theorem for each block.

- We can use bit-level symbolic execution [8], which computes a Boolean formula representing the correctness condition and attempts to solve it using a SAT solver. This is preferred for short blocks whose correctness proofs do not depend on much mathematics. This method is largely automatic (though it can be tuned with rules that determine how to process certain functions), and in many cases can either quickly prove the desired theorem or produce a counterexample showing a difference between the spec and the actual behavior of the routine. However, this method suffers from capacity limitations and is also difficult to debug in cases where a proof times out or otherwise fails.
- We can use ACL2's native proof engines, together with a litany of hints and rules that optimize its behavior on this sort of problem. E.g., we instruct the prover to only open the definition of *run* if the program counter of the state can be determined.

Both methods support composition of blocks, and both also support wormhole abstraction, obviating the need to specify and spend proof effort on don't-care fields of the state.

3.2 Def-uc-loop

Although there is no explicit loop construct in the microcode, loops do appear in the code in various forms. Macro *def-uc-loop* supports their verification. Through the arguments to this macro, the user specifies:

1. *start-pc*, an initial value for the program counter
2. *loop-precondition*, a starting state predicate
3. *loop-specification*, the loop's effect on the machine state

4. *measure*, a term used in the proof of termination of the loop (e.g, value of a register that serves as a counter).
5. *done*, a condition (state predicate) that is satisfied upon entering the last execution of the loop.
6. *run-last*, run function for the execution of the last time through the loop.
7. *run-next*, run function for the execution of any but the last time trough the loop.

def-uc-loop also supports proving partial correctness for loops that may not terminate; in these cases, the measure may be omitted.

Execution of a *def-uc-loop* is usually preceded by two executions of *def-uc-block* that specify the effect of executing one round of the loop. In particular, the two cases describe the *run-last* block (executed under the precondition $done(s) \wedge loop\text{-}precondition(s)$) and the *run-next* block (under the precondition $\neg done(s) \wedge loop\text{-}precondition(s)$). Figure 5 shows an example of *def-uc-loop* arguments for the code on Figure 2.

Expansion of the macro defines a function *run-loop* that repeatedly executes *run-next* until the *done* condition first holds, then finishes by executing *run-last*. Properties R1–R6 are proved for *run-loop*, and the correctness theorem has exactly the same form as the correctness theorem for *def-uc-block*:

Theorem 2 (loop-correct).

$$s.pc = start\text{-}pc \wedge loop\text{-}precondition(s)$$
$$\implies$$
$$run\text{-}loop(s) = loop\text{-}specification(s, run\text{-}loop(s))$$

This theorem is proved using induction defined by the scheme of the *run-loop* function and the two block-correct theorems for the *run-next* and *run-last* functions. The proof may be done either using ACL2's built-in proof engines or by applying our bit-level proof engine separately to the base case and induction step.

4   Hardware Connection

We would like our microcode model to be useful both for ad-hoc testing of microcode routines and for carrying out formal proofs of correctness about these routines. The closer the model is to the actual processor, the stronger the results of tests and proofs. Figure 3 shows the corresponding parts of our model and the processor implementation. Dark blue parts were proved to match dark orange parts of the hardware model.

We derive our instruction listing from the same microcode assembler that also produces the content of ROM, and we model the microcode decoder by effectively simulating the RTL of Micro-translator, as described in Section 2. Thus, our model of the translation from the text microcode into uops has a strong connection to the real design of the processor's frontend.

name:	clr_loop
pc:	$get\text{-}label(clr\_loop)$
measure:	$s.gregs[8]$
done:	$s.gregs[8] = 1$
run-last:	$run\text{-}clr\text{-}loop\text{-}last(s)$
run-next:	$run\text{-}clr\text{-}loop\text{-}next(s)$

precondition: Let $adr = s.gregs[9]$, $cnt = s.gregs[8]$ in
$\qquad addr\text{-}ok(adr \cdot (cnt - 1)) \wedge (cnt > 0) \wedge$
$\qquad \neg empty(s.retstack)$

specification: Let $adr = s.gregs[9]$, $val = s.gregs[0]$,
$\qquad cnt = s.gregs[8]$, $idx = adr \div 8$ in {
$\qquad s = clr\text{-}pram\text{-}k(idx, idx + cnt, val, s)$;
$\qquad s.gregs[8] = 0$;
$\qquad s.gregs[9] = adr + 8 \cdot cnt$;
$\qquad$ return $stack\text{-}pop(s)$;
$\qquad$ }

Fig. 5. *def-uc-loop* example: The measure (the value of g8) will decrease with each run through the loop. The precondition assures that the last address to which we write is within a boundary and that the starting value of g8 is positive, assuring termination. The *run* function is composed from two previously defined run functions *(run-clr_loop_last s)* and *(run-clr_loop_but_last s)*. The specification describes the state upon termination of the loop: *clr-pram-k(start, end, val, s)* copies value from g0 throughout $PRAM[start : end - 1]$; g8 is set to 0; g9 is set to point at the address followed by the last written address; and the program counter is set to the value from the top of the return stack.

As for the backend, our model is a significant abstraction of the actual processor, which is depicted in Figure 1. For instance, we abstract away out-of-order execution of the micro operations. Consequently, things that appear very simple in our model, say, "get the current value in register g0," are actually quite complicated, involving, e.g., the register aliasing table, the reservation stations, forwarding, the reorder buffer, *etc.* This said, significant parts of our model *do* have a strong connection to the real hardware design. In previous work [1–3] we described how we have developed an RTL-level verification framework within ACL2, and used it to prove that our execution units for integer, media, and floating point instructions implement desired operations. This previous work means that, for many uops, we have a specification function, written in ACL2, that functionally matches the execution of the uop in a particular unit. Thanks to regression proofs, we can be quite confident that these specifications remain up-to-date.

Now we can sketch how the *step* function is defined. It takes a state of our abstract machine and returns the state of the machine after executing the ROM instruction pointed to by the current PC (*s.pc*). The extraction of the ROM instruction (*get-rom-inst*) is a simple lookup in the constant *ucode* – a list representing the content of the ROM. It will then lookup the pre-computed formula for the result of running the micro-translator unit on this instruction. The hash table lookup returns a sequence of uops and additional sequencer information in the form of symbolic formulas (sym_uops, sym_seq). These formulas are pretty simple, depending on a few variables whose values can be extracted from the current state. $exec\_uops$ executes the sequence of uops by repeatedly applying $\mu op\text{-}def$ (see Section 2) one by one. $\mu op\text{-}def$ is directly connected to the proofs of hardware. In case of a conditional jump instruction, it also decides whether the branch is taken. The function $next\text{-}pc$ will determine the value of the next PC which completes the execution of one instruction.

$$
\begin{aligned}
&step : s \to s' = \{ \\
&\quad inst = get\text{-}rom\text{-}inst(s.pc, *ucode*) \\
&\quad (sym\_uops, sym\_seq) = lookup\text{-}uxlator(inst) \\
&\quad (uops, useq) = eval(s, sym\_uops, sym\_seq) \\
&\quad (branch\_taken, s) = exec\text{-}uops(uops, s) \\
&\quad s.pc = next\text{-}pc(s, useq, branch\_taken) \\
&\}
\end{aligned}
$$

It is important to note that our model is defined in extensible way. It allows us to relatively seamlessly move the boundary between the parts that are validated and those that are contrived.

5   Related Work

Our work builds upon countless ideas and advancements in microprocessor and machine code verification published over decades. Our contributions are in combining these advances and using them in an industrial setting, and in connecting methods for software and hardware verification under one unifying framework.

Operational semantics as a formalization of the meaning of programs was introduced in the 60s by McCarthy [10]. Early applications can be found in a technical report by van Wijngaarden et al. describing ALGOL68 [11]. Since then, structural operational semantics has been extensively used for mechanical verification of complex programs using various theorem provers: ACL2 and its predecessor NQTHM [12–14]; Isabelle/HOL [15]; and PVS [16]. Smith and Dill used operational semantics along with domain specific simplifications using a SAT solver and ACL2 for automatic equivalence checking of object code implementations of block ciphers [17]. More recent attempts to formalize operational semantics of complex ISAs come from Goel and Hunt [18] for x86, and Fox et al. for ARM [19]. All these papers model languages on the ISA level or above, and their operational semantics is not supported by any further verification. Wilding et al. [20] made use of the executability of ACL2 functions to validate their model by extensive testing against the hardware.

Many papers have discussed methodology for verifying the correctness of a microprocessor's microarchitecture with respect to its ISA [21, 22]. While these defined crucial concepts and methods for bridging the two different abstraction levels, they did not go beyond theoretical models of small to moderate size. Even the more comprehensive machine verification project described by Hunt [23], which includes some simple microcode verification connected to top-down hardware verification, does not have the complexity and dimensions of industrial scale designs.

Some work has been published by researchers from Intel® on verification of backward compatibility of microcode [24, 25]. The idea is based on creating symbolic execution paths, storing them in a database and using them either for testing or for checking assertions. This work differs from ours in several aspects. First, it is not connected to hardware verification. Their operational semantics of microcode is defined through a translation to an intermediate language with predefined operational semantics. There is no direct connection of this semantics to what is actually implemented in the hardware. Second, their approach uses SAT/SMT, while we are using mostly theorem prover and symbolic simulator that is built-in and verified within the prover. Finally, the verification objectives of our work are very different: while we compare the effect of running a microcode routine to a fully or partially defined specification that can be written on a high abstraction level, their objective is to compare the behavior of two microcode routines for backward compatibility.

Since the beginning of the computing era, the correctness of programs has been on the minds of great computer scientists like Floyd [26], Hoare [27], Manna [28], *etc.* The first papers concerned with program verification were based on assertions, but at that point, researchers weren't equipped with high-level mechanized proof systems. Matthews *et al.* [29] merged the idea of operational semantics with assertion verification. Their work is closest to our verification approach. Both our approach and that of Matthews *et al.* decompose the program into blocks separated by cutpoints. The difference is that Matthews *et al.* use the inductive invariant approach: a set of cutpoint/assertion pairs is defined and the goal is to prove a global invariant of the form:

$$\forall i \ (pc(s) = cutpoint_i) \Rightarrow assertion_i(s).$$

In our approach, we characterize (fully or partially) the effect of running each block (a sequence of operations between two cutpoints) on the state, then sequentially compose blocks together to build up a characterization of the effect of a full microcode routine on the state. These two approaches have been shown to be logically equivalent (in sufficiently expressive logic) [5]. Wormhole abstraction has been introduced in [7]. The target of [7, 29] is a slightly higher level of machine code. The main difference to our work is that their operational semantics remains unverified.

Last year, a paper by Alex Horn, Michael Tautschig and others [30] demonstrated validation of firmware and hardware interfaces on several interesting examples. Their work differs from ours in several aspects: Their methods require both hardware and software to be described in the same language (C or SystemC); the hardware is much simpler than a microprocessor; and the main method used for verification is model-checking.

6   Conclusion and Future Work

We presented an approach to microcode verification that is tightly connected to ongoing hardware verification. Since our RTL and microcode proofs are done within the same system, we are able to reuse the functions specifying hardware behavior to model microoperations. While the microcode model is far from being completely verified, our methodology has the flexibility to move the boundary between proved and contrived parts of the model as we achieve more of its validation.

We tried our approach on several microcode routines. One routine, that was a representative of arithmetic operation routines, was a 54-instruction microcode routine that performs unsigned integer division of a 128-bit dividend by a 64-bit divisor, storing the 64-bit quotient and remainder. We proved the correctness of this routine using 11 *def-uc-block* forms. Of these, five specified the behavior of low-level code blocks, and the rest primarily composed these sub-blocks together according to the control flow.

Another example was a verification of one of the critical algorithms that run as a part of the machine bring-up process. The algorithm deals with *decompression* of strings. The processor reads a an input that is a concatenation of compressed strings of variable lengths and places it in scratch memory. The main routine runs in a loop where one round identifies the beginning of the next compressed substring and converts it back to uncompressed form. The beginning and the type of a compressed substring is identified by its header. The decompression algorithm is implemented in about 800 lines of code. Its logical structure contains several loops (including nested loops) and many subroutines. A handful of designated registers keep track of pointers to scratch memory, positions within strings, counters, and currently processed strings, and many more are used as temporary value holders. The correctness proof of the decompression uses about 50 *def-uc-block/ def-uc-loop* structures. Since the algorithm and its implementation were frequently changing (adding new compression types, changing storage location, etc.), it was important to have automated support for the detection of these changes.

The main difference between our work and work done in academia is that we have to deal with a real contemporary industrial design that is not only very complex and sparsely documented, but also is constantly changing: both the hardware model and microcode are constantly updated. This requires automation to detect the changes and (wherever possible) make the relevant adjustments to proofs. Our automation includes regularly building the design model and running a regression suite for both hardware and microcode proofs.

Even though our microcode verification methodology is based on the powerful rewriting capabilities of the ACL2 theorem proving system [4], the approach could be applied using other rewriting systems as well.

In the future we would like to expand our verification effort to cover more of the critical microcode, e.g. in security-related areas. We also would like to pursue a more systematic approach to the verification of ISA instructions, connecting our specifications to a formal model of x86 such as the one one developed by Goel and Hunt [18].

Acknowledgement. We would like to thank Warren Hunt for comments on the first draft of this paper.

References

1. Hunt Jr., W.A., Swords, S.: Centaur Technology media unit verification. In: Bouajjani, A., Maler, O. (eds.) CAV 2009. LNCS, vol. 5643, pp. 353–367. Springer, Heidelberg (2009)
2. Hunt Jr., W.A., Swords, S., Davis, J., Slobodova, A.: Use of Formal Verification at Centaur Technology. In: Hardin, D. (ed.) Design and Verification of Microprocessor Systems for High-Assurance Applications, pp. 65–88. Springer (2010)
3. Slobodova, A., Davis, J., Swords, S., Hunt Jr., W.: A flexible formal verification framework for industrial scale validation. In: Proceedings of the 9th IEEE/ACM International Conference on Formal Methods and Models for Codesign (MEMOCODE), Cambridge, UK, pp. 89–97. IEEE/ACM (July 2011)
4. Kaufmann, M., Moore, J.S., Boyer, R.S.: ACL2 version 6.1 (2013), http://www.cs.utexas.edu/~moore/acl2/
5. Ray, S., Moore, J.S.: Proof styles in operational semantics. In: Hu, A.J., Martin, A.K. (eds.) FMCAD 2004. LNCS, vol. 3312, pp. 67–81. Springer, Heidelberg (2004)
6. Moore, J.S.: Proving theorems about Java and the JVM with ACL2. In: Models, Algebras and Logic of Engineering Software, pp. 227–290 (2003)
7. Hardin, D.S., Smith, E.W., Young, W.D.: A robust machine code proof framework for highly secure applications. In: Proceedings of the Sixth International Workshop on the ACL2 Theorem Prover and its Applications, pp. 11–20. ACM (2006)
8. Swords, S., Davis, J.: Bit-blasting ACL2 theorems. In: ACL2 2011. Electronic Proceedings in Theoretical Computer Science, vol. 70, pp. 84–102 (2011)
9. Davis, J., Swords, S.: Verified AIG algorithms in ACL2. In: Proceedings of ACL2 Workshop (2013)
10. McCarthy, J.: Towards a mathematical Scioence of computation. In: Information Processing Congress, vol. 62, pp. 21–28. North-Holland (1962)
11. van Wijngaarden, A., Mailloux, B., Peck, J., Koster, C., Sintzoff, M., Lindsey, C., Meertens, L., Fisker, R.G.: Revised report on the algorithmic language ALGOL 68 (1968)
12. Boyer, R., Moore, J.: Mechanized formal reasoning about programs and computing machines. In: Automated Reasoning and its Applications: Essays in Honor of Larry Woss, pp. 141–176 (1996)
13. Greeve, D., Wilding, M., Hardin, D.: High-speed, analyzable simulators. In: Kaufmann, M., Moore, J.S., Manolios, P. (eds.) Computer-Aided Reasoning: ACL2 Case Studies, pp. 89–106. Kluwer Academic Publishers (2000)
14. Yu, Y.: Automated proofs of object code for a widely used microprocessor. PhD. Thesis (1992)
15. Strecker, M.: Formal verification of a Java compiler in Isabelle. In: Voronkov, A. (ed.) CADE 2002. LNCS (LNAI), vol. 2392, pp. 63–77. Springer, Heidelberg (2002)
16. Hamon, G., Rushby, J.: An operational semantics for stateflow. In: Wermelinger, M., Margaria-Steffen, T. (eds.) FASE 2004. LNCS, vol. 2984, pp. 229–243. Springer, Heidelberg (2004)
17. Smith, E., Dill, D.: Automatic formal verification of Block Cipher implementations. In: Cimatti, A., Jones, R. (eds.) Proceedings of the Conference on Formal Methods in Computer-Aided Design (FMCAD), pp. 45–51. IEEE/ACM (2008)

18. Goel, S., Hunt Jr., W.A.: Automated code proofs on a formal model of the X86. In: Cohen, E., Rybalchenko, A. (eds.) VSTTE 2013. LNCS, vol. 8164, pp. 222–241. Springer, Heidelberg (2013)
19. Fox, A., Myreen, M.O.: A trustworthy monadic formalization of the ARMv7 instruction set architecture. In: Kaufmann, M., Paulson, L.C. (eds.) ITP 2010. LNCS, vol. 6172, pp. 243–258. Springer, Heidelberg (2010)
20. Wilding, M., Greeve, D., Richards, R., Hardin, D.: Formal verification of partition management of the AAMP7G microprocessor. In: Hardin, D. (ed.) Design and Verification of Microprocessor Systems for High-Assurance Applications, pp. 175–192. Springer (2010)
21. Cyrluk, D.: Microprocessor verification in pvs. A methodology and simple example. (February 1994), http://www.csl.sri.com/papers/csl-93-12/
22. Sawada, J., Hunt Jr., W.: Verification of FM9801: An out-of-order microprocessor model with speculative execution, exceptions, and program-modifying capability. J. of Formal Methods in System Design 20(2), 187–222 (2002)
23. Hunt Jr., W.A.: FM8501: A Verified Microprocessor. LNCS, vol. 795. Springer, Heidelberg (1994)
24. Arons, T., Elster, E., Fix, L., Mador-Haim, S., Mishaeli, M., Shalev, J., Singerman, E., Tiemeyer, A., Vardi, M.Y., Zuck, L.D.: Formal verification of backward compatibility of microcode. In: Etessami, K., Rajamani, S.K. (eds.) CAV 2005. LNCS, vol. 3576, pp. 185–198. Springer, Heidelberg (2005)
25. Franzén, A., Cimatti, A., Nadel, A., Sebastiani, R., Shalev, J.: Applying SMT in symbolic execution of microcode. In: Proceedings of the 2010 Conference on Formal Methods in Computer-Aided Design (FMCAD), Austin, TX, pp. 121–128, FMCAD Inc (2010)
26. Floyd, R.: Assigning meanings to programs. In: Mathematical Aspects of Computer Science, Proceedings of Symposia in Applied Mathematics, vol. XIX, pp. 19–32. American Mathematical Society (1967)
27. Hoare, C.: An axiomatic basis to computer programming. Communications of the ACM 12, 576–583 (1969)
28. Manna, Z.: The correctness of programs. Journal of Computer and System Sciences 3, 119–127 (1969)
29. Matthews, J., Moore, J.S., Ray, S., Vroon, D.: Verification Condition Generation Via Theorem Proving. In: Hermann, M., Voronkov, A. (eds.) LPAR 2006. LNCS (LNAI), vol. 4246, pp. 362–376. Springer, Heidelberg (2006)
30. Horn, A., Tautschnig, M., Val, C., Liang, L., Mehlham, T., Grundy, J., Kroening, D.: Formal co-validation of low-level hardware/software interfaces. In: Jobstman, B., Ray, S. (eds.) Proceedings of the Formal Methods in Computer-Aided Design (FMCAD), pp. 121–128. ACM/IEEE (2013)

Are We There Yet? 20 Years of Industrial Theorem Proving with SPARK

Roderick Chapman and Florian Schanda

Altran UK Limited, 22 St Lawrence Street, Bath BA1 1AN, United Kingdom
roderick.chapman@gmail.com, florian.schanda@altran.com

Abstract. This paper presents a retrospective of our experiences with applying theorem proving to the verification of SPARK programs, both in terms of projects and the technical evolution of the language and tools over the years.

Keywords: Formal Methods, Program Verification, Theorem Proving, SPARK.

1 Introduction

This paper reflects on our experience with proving properties of programs written in SPARK[2] - a programming language and verification toolset that we have designed, maintained, sold, and used with some success for nearly 20 years.

2 Projects and Technologies

The following sections present a retrospective on the use of theorem proving in SPARK, roughly alternating between technical developments in the language and tools and the experiences of various projects, coming from both our own experience and that of a selection of external SPARK users in industry.

2.1 Early Days - 1987ish

It all started in about 1987. Our predecessors at the University of Southampton and then PVL had designed and implemented a Hoare-logic based verification system for a subset of Pascal[9]. Their next goal seemed almost absurdly bold - to design a programming language and verification system that would offer sound verification for non-trivial programs, but be scalable and rich enough to write "real world" embedded critical systems. The base language chosen was Ada83, and through judicious subsetting, semantic strengthening, and the addition of contracts, SPARK (the "SPADE Ada Ratiocinative Kernel" - we kid you not) was born, with a first language definition appearing in March 1987.

The highest priority design goal was the provision of soundness in all forms of verification. This led to a need for a completely unambiguous dynamic semantics,

G. Klein and R. Gamboa (Eds.): ITP 2014, LNAI 8558, pp. 17–26, 2014.

so that the results of verification would be reliable for all compilers and target machines, and so that all valid SPARK programs would be valid Ada programs with the same meaning and also, at a stroke, removing the need for us to produce a "special" compiler for SPARK.

The toolset consisted of three main tools

- The Examiner. This consists of a standard compiler-like "front-end", followed by various analyses that check the language subset, aliasing rules, data- and information-flow analysis, then (finally) generation of verification conditions (VCs) in a language called FDL - a typed first-order logic that has a relatively simple mapping from SPARK. The VCs generated include those for partial correctness with respect to preconditions, postconditions and loop-invariants, but also (lots of) VCs for "type safety" such as the freedom from buffer overflow, arithmetic overflow, division by zero and so on.
- The Checker. This is an interactive proof assistant for FDL. It emerged from of one of the team's PhD research[19]. Written in PROLOG.
- The Simplifier[1]. This is an heuristic, pattern-matching theorem prover, based on the same core inference engine as the Checker. It started out as a way of literally "Simplifying" VCs (a bit) before the real fun could start with the Checker, but as we will see, it grew substantially in scope and power over the years.

2.2   SHOLIS Project

Our first attempt at serious proof of a non-trivial system came with the SHOLIS project in 1995, by which time PVL and the SPARK technology had been acquired by Praxis (now Altran UK). This was the first effort to meet the requirements of the (then) rather onerous Interim version of the UK's Def-Stan 00-55 for critical software at Integrity Level "SIL4".

SHOLIS is a system that assists Naval crew with the safe operation of helicopters at sea, advising on safety limits (for example incident wind vector on the flight deck, and ship's roll and pitch) for particular operations such as landing, in-air refuelling, crew transfer and so on.

The SHOLIS software[16] comprised about 27 kloc (logical) of SPARK code, 54 kloc of information-flow contracts, and 29 kloc of proof contracts, plus some tiny fragments of assembly language to support CPU start-up and system boot. There was no operating system and no COTS libraries of any kind - a significant simplification at this level of integrity, and a programming model that SPARK was explicitly designed to support.

The SHOLIS code generated nearly 9000 VCs, of which 3100 were for functional and safety properties, and 5900 for type safety. Of the 9000 total, 6800 (75.5 %) were proven automatically by the Simplifier, with the remaining 2200 being "finished off" using the interactive Checker.

---

[1] Not to be confused with Greg Nelson's better-known Simplify prover.

The experience was painful. Computing resource was scarce; one UNIX server (shared by the whole company) was used for all the proof work. Simplification times for a single subprogram were measured in hours or days.

A key output was the identification of the need to support state abstraction and refinement in SPARK proofs. This mechanism is used in SPARK to control the volume of states that are visible, and hence has a direct impact on the complexity and size of contracts and the "postcondition explosion" problem. This was implemented (too late for SHOLIS), but had a major impact on later projects.

2.3   C130J Project

The Lockheed-Martin C130J is the most recent generation of the enormously successful "Hercules" military transport aircraft. The Mission Computer application software is written in SPARK, and was subject to a large verification effort in the UK as part of the acquisition of the aircraft by the UK RAF.

Verification consisted of full-blown SPARK analysis, including verification of partial correctness for most critical functions with respect to the system's functional specification, which was expressed using the "Parnas Tables" [20] notation.

Unusually, the "proof" component of the work was performed in the UK in the late 1990s *after* the formal testing (to meet the objectives of DO-178B Level A) of the Mission Computer.

Results from the development phase of the project are best reported in [18] while the results of the later proof work (and comparisons of the SPARK code with other systems and programming languages) can be found in [13].

2.4   Improving the VCG and Simplifier

Both the SHOLIS and C130J projects led to a serious analysis of how we could improve the completeness of the Simplifier. Analysis identified a number of key areas where improvement was sorely needed:

- Tactics for unwrapping and instantiation of universally quantified conclusions, especially those that commonly arise from arrays in SPARK.
- Modular (aka "unsigned") arithmetic - very common in low-level device driver code, ring buffers, cryptographic algorithms and so on.
- Tracking the worst-case ranges of integer expressions.

These were implemented in 2002. We also spent effort on improving the VC Generator (VCG) itself - it turns out it's only too easy for the VCG to produce a VC that omits some vital hypothesis so that *no* prover could prove it, no matter how clever. Ever more detailed and precise modelling of the language semantics inside the VCG continues to this day.

Finally, we noted one other factor that critically impacted the usefulness of the proof system - the "proof friendliness" of the code under analysis. This seemed to correlate with common software engineering guidance - simplicity, low

information-flow coupling, and proper use of abstraction to control the name-
and state-space of any one subprogram. In short - we were learning how to write
provable programs, so we started to set a goal for projects to "hit" a particular
level of automatic proof (e.g. 95 % of VCs discharge automatically), making the
proof a design-level challenge rather than a retrospective slog.

2.5   Tokeneer Project

Tokeneer is an NSA-funded demonstrator of high-security software engineering.
We were given a clean-sheet to work from, so the project deployed various forms
of formal methods, including a system specification and security properties in Z,
and implementation in SPARK[1].

This was the first time we had attempted to prove non-trivial security properties of a software system with SPARK. Owing to the budget, the system was
small (only about 10 kloc logical, producing 2623 VCs), but critically, 2513 of
those were proven automatically (95.8 %), with only 43 left to the Checker and
67 discharged by review (i.e. we looked at them really hard).

Unusually (and some years later, in 2008) the NSA granted a licence that
effectively allowed a fully "open source" release of the entire Tokeneer project
archive. It remains the focus of various research efforts[28].

2.6   Speeding Up and Going FLOSS

With the emergence of cheap multi-core CPUs, we implemented an obvious
improvement to the Simplifier in 2007 - the ability to run the prover on several
VCs at once in parallel. It turns out the VCG generates *lots* of small, simple VCs
which are completely independent of one another, so are ripe for parallelization.
This almost wasn't a conscious design goal in 1990ish, but came as a pleasantly
surprising benefit.

The results are dramatic. On a modern (2013-era) quad-core machine, the
entire SHOLIS software can be re-proven from scratch in about 11 minutes,
with only 440 undischarged VCs (compare with weeks and 2200 undischarged
VCs in the original project).

Secondly, in 2009 we took the dramatic step, in partnership with AdaCore,
to "go open source", with the entire toolset moving to a FLOSS development
model and GPL licence, as a way of promoting interest, teaching and research
with the language.

2.7   User-Defined Rules

In response to customer demand, we implemented an approach for users to
write and "insert" additional rules or lemmas into the Simplifier to "help" it
with particularly tricky VCs, or in areas where its basic reasoning power proved
insufficient.

This approach opens an obvious soundness worry (users can write non-sensical
rules), but does offer an attractive middle-ground between the onerous use of the

Checker and the rather relaxed idea of just "eyeballing" the undischarged VCs to see if they look OK. Additionally, a single well-written rule can be written once but used thousands of times by the Simplifier, so the effort to get the rules right should pay off. Finally, the Checker be used to verify the soundness of a user-defined rule from first principles if required.

2.8 iFACTS Project - Scaling Up

Starting in 2006, the implementation of the NATS iFACTS system is the most ambitious SPARK project to date.

iFACTS augments the tools available to en-route air-traffic controllers in the UK. In particular, it supplies electronic flight-strip management, trajectory prediction, and medium-term conflict detection for the UK's en-route airspace, giving controllers a substantially improved ability to plan ahead and predict conflicts in a sector[21].

The project has a formal functional specification (again expressed mostly in Z), and the majority of the code (about 250 kloc logical lines of code, as counted by GNATMetric) is implemented in SPARK. Figure 1 illustrates that the proportion of SPARK contracts is minor compared to the bulk of the executable code and comments (and whitespace).

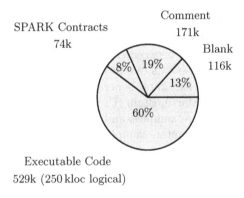

Fig. 1. Project size in physical lines of code (counted with `wc -l`)

Proof concentrates on type-safety, but not functional correctness, since the system has stringent requirements for reliability and availability - in short, the software must be proven "crash proof".

The current operational build produces 152927 VCs, of which 151026 (98.76 %) are proven entirely automatically by the Simplifier alone. User-defined rules are used for another 1701 VCs, with only 200 proved "by review". This remains the highest "hit rate" for automatic proof with SPARK that we have ever encountered.

The proof is reconstructed daily (and overnight), and developers are under standing orders that all code changes *must* prove OK before code can be committed into the CM system. Developers who "break the proof" receive a terse notification the following morning.

2.9   Reaching Out - SMT and Counterexamples

Licensing SPARK under the GPL has also made it much easier to collaborate with academia, and we have seen two useful improvements for SPARK 2005 as a result of this.

Alternative Provers. The existing automatic theorem prover for SPARK was good (98.76 % on well written code), but due to its nature had difficulties discharging certain VCs. Paul Jackson from the University of Edinburgh has written Victor[15], a translator and prover driver to allow SMT solvers to be used with SPARK. As SMT solvers are fundamentally different from rewrite systems such as the Simplifier, they are able to easily discharge many of the VCs the Simplifier cannot deal with.

For some projects, such as SPARKSkein, using a modern SMT solver allowed 100 % of all VCs to be discharged automatically. Victor is now shipped with SPARK.

Counterexamples. The existing SPARK proof tools were all geared towards proving VCs. As a consequence it was difficult (and thus time-consuming, and thus expensive) to distinguish between true VCs that could not be proved due to prover limitations and false VCs due to specification or programming errors.

Riposte [24] is the counter-example generator for SPARK that was developed under a joint project with Martin Brain (University of Bath, now Oxford). This tool provides helpful counter-examples, and can even be used as a proof tool as it is sound, so the lack of a counter-example guarantees none exist (although the tool is incomplete, so sometimes a counter-example might be generated where none exists).

2.10   SPARKSkein Project - Fast, Formal, Nonlinear

In 2010, we implemented the Skein[25] hash algorithm in SPARK. The goal was to show that a "low-level" cryptographic algorithm like Skein could be implemented in SPARK, be subject to a complete proof of type safety, be readable and "obviously correct" with respect to the Skein mathematical specification, and also be as fast or faster than the reference implementation offered by the designers, which is implemented in C.

The proofs of type safety turned out to be quite tricky. Firstly, finding the correct loop invariants proved difficult, and this was compounded by the plethora of modular types and non-linear arithmetic in the VC structures. Of the 367 VCs, 23 required use of the Checker to complete the proof - not bad but these still required a substantial effort to complete. Full results from the project are reported in [7]. All sources, proofs and tests have been released under GPL and can be obtained here[26].

One final unexpected side-effect was the discovery of a subtle corner-case bug in the designers' C implementation (an arithmetic overflow, which leads to a loop iterating zero times, which leads to an undefined output).

2.11   CacheSimp - Speeding Up Even More

In an industrial context, verification time is not just a number, it has a significant qualitative effect on how verification tools are used and thus on project management. If verification takes 4 hours, then a developer has to organise their working day around this activity. If the same activity can be done in 30 minutes or less then this has a significant and positive impact on how the tools are used (and often mistakes are found much earlier).

Figure 2 shows that a single proof run of iFACTS takes around 3 hours (green line with squares) on a fast desktop computer, regardless of how big the actual change is (blue bars). We implemented a very simple caching system [6] in around 250 lines of code using memcached [12], where each invocation of the proof tools first checks if the result is already known. As the memcached server was run on a separate computer accessible to all developers, this system was both incremental and distributed, leading to an average 29-fold speedup (red line with circles) and a strong correlation to the size of the change.

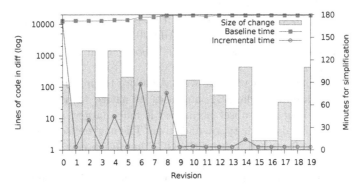

Fig. 2. Results showing the effects of using a simple caching system

2.12   Reaching Out - Interactive Provers

One particular SPARK user, secunet, kept pushing the boundaries of what was reasonable to achieve using automated verification, and what was expressible using the simple first order logic constructs available in SPARK annotations. Stefan Berghofer (from secunet) implemented a plug-in for Isabelle/HOL that allows one to express much richer properties using SPARK proof functions [4] and complete the proof in Isabelle. They have kindly released their work under a free software license, including a fully verified big-number library which they have used to implement RSA. Given the recent OpenSSL "Heartbleed"[17] bug, this importance of this contribution should not be underestimated.

2.13   Muen Project

This project[27] span off from the work of secunet. Muen is a FLOSS separtion-kernel for the x86_64 architecture, but unusually, almost the entire kernel is written in SPARK - something we might have considered impossible some years ago. The kernel code is subject to an automated proof of type-safety. We look forward to further results from this group.

3   Future Trends

This final section reflects on two topics - the future of SPARK, and the role that theorem-proving evidence can play in the wider context of regulation and acceptance of critical software.

3.1   Technologies and Languages - SPARK 2014

Despite many positive experiences, SPARK and the use of the proof tools remain a challenge for many customers - the "adoption hurdle" is often perceived as too high. Secondly, it became clear to us that improving the Simplifier had become a game of diminishing returns, and it was time to move to more modern proof technologies. Finally, the arrival of Ada 2012 brought contracts into the mainstream Ada syntax, so it was time to "reboot" both the language design and underlying technologies.

Since 2012, we have been working with AdaCore to produce SPARK 2014. The language is based on Ada 2012, and uses its native "aspect" notation for all contracts. The language subset permitted is much larger, including variant records, generic units, dynamic types, and so on. The new toolset is based on the full GNAT Pro Ada front-end (part of the GCC family), a new information-flow analysis engine (based on analysis of program-dependence graphs[11,14]), and a new proof system that uses the Why3 language and VCG[5] and modern SMT solvers such as Alt-Ergo[8] and CVC4[3]. The new tools also bring a significant improvement in the area of floating point verification: where the old tools used an unsound model using real numbers, the new tools employ a sound encoding; current versions use an axiomatised rounding function. We feel that a transition to the upcoming SMTLIB floating-point standard will bring many benefits, in particular preliminary experiments (with the University of Oxford) using SMT solvers implementing ACDL [10] have shown promise.

SPARK 2014 also (perhaps unusually) takes the step of unifying the dynamic (i.e. run-time) and static (i.e. for proof) semantics of contracts, so that they can be proved, or tested at run-time (as in, say, Eiffel), or both - providing some interesting possibilities for mixing verification styles and/or mixing languages (i.e. programs partly written in SPARK, Ada, C, or anything else) in a single program.

Both the "Pro" (supported) and "GPL" (free, unsupported) versions of the SPARK 2014 toolset will be available before this conference.

3.2 Assurance and Acceptance for Critical Systems

Over the years, both customers and regulators have taken a variety of stances on the use of strong static analysis and theorem proving in critical software. Some regulators remain sceptical, perhaps owing to the novelty of the idea or the perceived unreliability (i.e. unsoundness) of common "bug finding" style static analysis tools.

The future looks bright, though, in the aerospace with the advent of DO-178C[22] and its formal methods supplement DO-333[23] which explicitly allows "formal methods" as a combination of an unambiguous language and analysis methods which can be shown to be sound. Additionally, DO-178C allows later verification activities (e.g. testing) to be reduced or eliminated if it can be argued that formal analytical approaches have met the required verification objective(s). This supports (we hope) a strong economic incentive for the adoption of more formal and static approaches.

We look forward to the day when software will be delivered with its proofs, which can be re-generated at will by the customer or regulator, perhaps even by diverse verification tools. That would surely move us towards a claim to being a true engineering discipline.

References

1. Barnes, J., Chapman, R., Johnson, R., Widmaier, J., Cooper, D., Everett, B.: Engineering the Tokeneer enclave protection software. In: 1st IEEE International Symposium on Secure Software Engineering (March 2006)
2. Barnes, J.: SPARK: The Proven Approach to High Integrity Software. Altran Praxis (2012)
3. Barrett, C., Conway, C.L., Deters, M., Hadarean, L., Jovanović, D., King, T., Reynolds, A., Tinelli, C.: CVC4. In: Gopalakrishnan, G., Qadeer, S. (eds.) CAV 2011. LNCS, vol. 6806, pp. 171–177. Springer, Heidelberg (2011)
4. Berghofer, S.: Verification of dependable software using SPARK and isabelle. In: SSV, pp. 15–31 (2011)
5. Bobot, F., Filliâtre, J.C., Marché, C., Paskevich, A.: Why3: Shepherd your herd of provers. In: Boogie 2011: First International Workshop on Intermediate Verification Languages, Wrocław, Poland, pp. 53–64 (August 2011), http://proval.lri.fr/publications/boogie11final.pdf
6. Brain, M., Schanda, F.: A lightweight technique for distributed and incremental program verification. In: Joshi, R., Müller, P., Podelski, A. (eds.) VSTTE 2012. LNCS, vol. 7152, pp. 114–129. Springer, Heidelberg (2012)
7. Chapman, R., Botcazou, E., Wallenburg, A.: SPARKSkein: A formal and fast reference implementation of skein. In: Simao, A., Morgan, C. (eds.) SBMF 2011. LNCS, vol. 7021, pp. 16–27. Springer, Heidelberg (2011)

8. Conchon, S., Contejean, E., Kanig, J.: Ergo: A theorem prover for polymorphic first-order logic modulo theories (2006), http://ergo.lri.fr/papers/ergo.ps
9. Cullyer, W., Goodenough, S., Wichmann, B.: The choice of computer languages for use in safety-critical systems. Software Engineering Journal 6(2), 51–58 (1991)
10. D'Silva, V., Haller, L., Kroening, D.: Abstract conflict driven learning. In: POPL, pp. 143–154 (2013)
11. Ferrante, J., Ottenstein, K.J., Warren, J.D.: The program dependence graph and its use in optimization. ACM Trans. Program. Lang. Syst. 9(3), 319–349 (1987), http://doi.acm.org/10.1145/24039.24041
12. Fitzpatrick, B., et al.: Memcached - a distributed memory object caching system (2003), http://memcached.org
13. German, A.: Software static code analysis lessons learned. Crosstalk 16(11) (2003)
14. Horwitz, S., Reps, T., Binkley, D.: Interprocedural slicing using dependence graphs. In: Proceedings of the ACM SIGPLAN 1988 Conference on Programming Language Design and Implementation, PLDI 1988, pp. 35–46. ACM, New York (1988), http://doi.acm.org/10.1145/53990.53994
15. Jackson, P.B., Passmore, G.O.: Proving SPARK verification conditions with smt solvers. Technical Report, University of Edinburgh (2009)
16. King, S., Hammond, J., Chapman, R., Pryor, A.: Is proof more cost-effective than testing? IEEE Transactions on Software Engineering 26(8), 675–686 (2000)
17. Mehta, N.: Cve-2014-0160 (April 2014)
18. Middleton, P., Sutton, J.: Lean Software Strategies: Proven Techniques for Managers and Developers. Productivity Press (2005)
19. O'Neill, I.: Logic Programming Tools and Techniques for Imperative Program Verification. Ph.D. thesis, University of Southampton (1987)
20. Parnas, D.L., Madey, J., Iglewski, M.: Precise documentation of well-structured programs. IEEE Transactions on Software Engineering 20(12), 948–976 (1994)
21. Rolfe, M.: How technology is transforming air traffic management, http://nats.aero/blog/2013/07/how-technology-is-transforming-air-traffic-management
22. RTCA: DO-178C: Software considerations in airborne systems and equipment certification (2011)
23. RTCA: DO-333: Formal methods supplement to do-178c and do-278a (2011)
24. Schanda, F., Brain, M.: Using answer set programming in the development of verified software. In: ICLP (Technical Communications), pp. 72–85 (2012)
25. Schneier, B., Ferguson, N., Lucks, S., Whiting, D., Bellare, M., Kohno, T., Walker, J., Callas, J.: The skein hash function family. Submission to NIST (Round 3) (2010)
26. http://www.skein-hash.info
27. http://muen.codelabs.ch
28. Woodcock, J., Aydal, E.G., Chapman, R.: The Tokeneer Experiments. In: Reflections on the Work of CAR Hoare, pp. 405–430. Springer (2010)

Towards a Formally Verified Proof Assistant

Abhishek Anand and Vincent Rahli

Cornell University, Ithaca, NY, USA

Abstract. This paper presents a formalization of Nuprl's metatheory in Coq. It includes a nominal-style definition of the Nuprl language, its reduction rules, a coinductive computational equivalence, and a Curry-style type system where a type is defined as a *Partial Equivalence Relation* (PER) à la Allen. This type system includes Martin-Löf dependent types, a hierarchy of universes, inductive types and partial types. We then prove that the typehood rules of Nuprl are valid w.r.t. this PER semantics and hence reduce Nuprl's consistency to Coq's consistency.

1 Introduction

Trustworthiness of Proof Assistants. In order to trust a proof checked by a proof assistant, we have to trust many aspects of both its theory and implementation. Typically, the core of a proof assistant consists of a proof checking machinery which ensures that proof terms are indeed proofs of the corresponding statements. In constructive type theories such as the ones implemented in Agda [10], Coq [9], and Nuprl [4], this is accomplished with typechecking rules, which are derived from a semantic model, e.g., a computational model based on an applied λ-calculus. Parts of these theories have been formally described in various documents [37,7,3,15,24,36].

This is not a completely satisfactory state of affairs because: (1) It is possible to overlook inconsistencies between the different parts formalized on paper; and (2) Mistakes are possible in these large proofs (often spanning hundreds of pages) which are never carried out in full details. For example, we at least once added an inconsistent rule to Nuprl even after extensive discussions regarding its validity. A Bug that lead to a proof of False was found in Agda's typechecker[1]. Recently, the Propositional Extensionality axiom was found to be inconsistent with Coq[2]. However, consistency of Propositional Extensionality is a straightforward consequence of Werner's proof-irrelevant semantics [37] of the Prop universe. Werner allows only structurally recursive definitions while Coq's termination analyzer seems to be more permissive. A similar bug was discovered in Agda[3].

Fortunately, proof assistants have matured enough [26,19] that we can consider formalizing proof assistants in themselves [8], or in other proof assistants.

---

[1] https://lists.chalmers.se/pipermail/agda/2012/003700.html
[2] See https://sympa.inria.fr/sympa/arc/coq-club/2013-12/msg00119.html for more details. We do not use this axiom in our development.
[3] https://lists.chalmers.se/pipermail/agda/2014/006252.html

G. Klein and R. Gamboa (Eds.): ITP 2014, LNAI 8558, pp. 27–44, 2014.
© Springer International Publishing Switzerland 2014

By Tarski's undefinability theorem, these rich logics cannot formalize their own semantics. However, because Martin-Löf's type theories are stratified into cumulative universes, it is plausible that a universe can be modeled in a higher universe of the same theory. Even better, universes of two equally powerful theories can be interleaved so that universes up to a level i of one of the theories could be modeled using at most universes up to level $i + 1$ of the other theory, and vice-versa. The latter approach seems likely to catch more mistakes if the two theories and their implementations are not too closely correlated.

Also, some type theories are proof-theoretically stronger than others. For example, Agda supports inductive-recursive definitions and can likely prove Nuprl and the predicative fragment of Coq consistent. Among other things, this paper also illustrates how one can define Nuprl's entire hierarchy of universes in Agda.

Advantages of a Mechanized Metatheory. A mechanized metatheory can guide and accelerate innovation. Currently, developers of proof assistants are reluctant to make minor improvements to their proof assistants. Many questions have to be answered. Is the change going to make the theory inconsistent? Is it going to break existing developments? If so, is there a way to automatically transform developments of the old theory to the new one while preserving the semantics? A mechanized metatheory can be used to confidently answer these questions. Moreover, we would no longer have to sacrifice efficiency for simplicity.

Mechanized Formalization of Nuprl. Therefore, this paper tackles the task of formalizing Nuprl in Coq. We chose Coq over other constructive proof assistants [10] because of its powerful tactic mechanism, and over other non-constructive proof assistants [32] because of the convenience of extracting programs from proofs for free. The two theories differ in several ways. While Coq (like Agda) is based on an intensional type theory with intrinsic typing, Nuprl is based on an extensional one with extrinsic typing. Also, over the years Nuprl has been extended with several expressive types not found in Coq, e.g., quotient types [4]; refinement types [4]; intersection and union types [24]; partial types [15]; and recursive types [28]. Partial types enable principled reasoning about partial functions (functions that may not converge on all inputs). This is better than the approach of Agda[4] and Idris [11] where one can disable termination checking and allow oneself to accidentally prove a contradiction.

Following Allen [3] we implement W types (which can be used to encode inductive types [2] in an extensional type theory like Nuprl) instead of Mendler's recursive types, therefore making the entire Nuprl metatheory predicatively justifiable as evidenced by our illustrative Agda model in Sec. 5. Mendler's recursive types can construct the least fixpoint of an arbitrary monotone endofunction (not necessarily strictly positive) in a Nuprl universe. Formalizing this notion seems to require an impredicative metatheory. Our current system includes a hierarchy of universes, function types, partial types, and we also extend Allen's work to include parametrized W types (similar to parametrized inductive types [33]).

Our formalization of Nuprl's metatheory in Coq proceeds as follows in three steps: (1) We define an inductive type of Nuprl terms. This definition is

---

[4] http://wiki.portal.chalmers.se/agda/pmwiki.php?n=Main.AgdaVsCoq

parametrized by a collection of operators and makes it possible to add new constructs without changing the core definitions. We then define substitution and α-equality and prove several of their properties. (2) We define Nuprl's lazy computation system and a coinductive computational approximation relation that was defined and proved to be a congruence by Howe [21]. We formalize this proof as well as the domain theoretic properties that were used by Crary to justify some typehood rules about partial types [15]. We then define a computational equivalence relation [21] which plays a key role in the definition of our type system. (3) Following Allen's approach, we define types as PERs. This definition determines which closed terms denote equal types and which closed terms are equal members of types. Finally, we define Nuprl's sequents and prove the validity of many inference rules. We also show that using induction-recursion in Agda results in a more intuitive and simple definition of Nuprl's type system.

We describe below the key details of each of these steps. More details can be found in our technical report or in our publicly available code [5]. A key aspect of this formalization is that it gives us a *verified trusted core* of Nuprl. Although one can use Nuprl's tactics to prove typehood properties[5], these tactics can only use the above mentioned

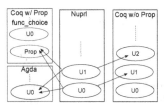

rules in the end. Moreover, this work tackles the task of formally describing in a unified framework all the extensions and changes that were made to Nuprl's type theory since CTT86 [14] and since Allen's PER semantics [3]. The core of this work is a formalization of [21,3,15,24]. Unlike previous works, we pin down three precise metatheories that can model (parts of) Nuprl. This is best illustrated by the figure above. An arrow from a Nuprl universe A to some universe B means that the PERs of types in A can be defined as relations in the universe B. As expected of large mechanized formalizations like ours, we found at least one minor mistake in one of these extensions. With the help of the original author, we were able to fix this proof in our formalization.

2   Uniform Term Model and Computation System

We use a nominal approach (bound variables have names) to define Nuprl terms. This definition closely matches the way terms are represented in Nuprl's current implementation. It is also very close to definitions used in paper descriptions of Nuprl [21]. Many alternative approaches have been discussed in the literature. See [6] for a survey. However, our choice avoided the overhead of translating the paper definitions about Nuprl to some other style of variable bindings.

We often show colored code[6]: blue is used for inductive types, dark red for constructors of inductive types, green for defined constants, functions and lemmas, red for keywords and some notations, and purple for variables.

---

[5] Nuprl has never had a typechecker. It relies on customizable tactics to do typechecking.

[6] Some of the colored items are hyperlinked to the place they are defined, either in this document, in the standard library, or in our publicly available code.

Inductive NVar : Set := | nvar : nat → NVar. Inductive Opid : Set := | Can : CanonicalOp → Opid | NCan : NonCanonicalOp → Opid.	Inductive NTerm : Set := | vterm: NVar → NTerm | oterm: Opid → list BTerm → NTerm with BTerm : Set := | bterm: (list NVar) → NTerm → BTerm.

Fig. 1. Uniform Term Model

Fig. 1 defines NTerm, the type of Nuprl terms. Variable bindings are made explicit by the simultaneously defined BTerm type. bterm takes a list of variables lv and a term nt and constructs a bound term. Intuitively, a variable that is free in nt gets bound to its first occurrence in lv, if any. The rest of our term definition is parametrized by a collection of operators Opid. We divide our Opids into two groups, CanonicalOps and NonCanonicalOps (see [5, Sec. 2.1] for their definitions). This distinction is only relevant for defining the computation system and its properties. Intuitively, an operator constructs a term from a list BTerm. For example, oterm (Can NLambda) [bterm [nvar 0] (vterm (nvar 0))] represents a λ-term of the form $\lambda x.x$. Nuprl has a lazy computation system and any NTerm of the form (oterm (Can _) _) is already in canonical form and is called a value.

Not all the members of NTerm are well-formed: (nt_wf t) asserts that t is well-formed. For example, a well-formed λ-term must have exactly one bound term as an argument. Moreover, that bound term must have exactly one bound variable. Fig. 2 compactly describes the syntax of an illustrative subset of the language that we formalized. There, v ranges over values. There is a member of CanonicalOp for each clause of v. For example, the constructor Nint of type $\mathbb{Z} \to$ CanonicalOp corresponds to the first clause that denotes integers. In Fig. 2, vt ranges over the values that represent types. A term t is either a variable, a value, or a non-canonical term represented as (oterm (NCan _) _) in our term model. These have arguments (marked in boxes) that are said to be *principal*. As mentioned below, principal arguments are subterms that have to be evaluated to a canonical form before checking whether the term itself is a redex or not.

We next define our simultaneous substitution function (lsubst) and α-equality (alpha_eq and alpha_eq_bterm). As expected of a nominal approach to variable bindings, we spent several weeks proving their properties that were required to formalize Nuprl. One advantage is that these and many other definitions and proofs [5, Sec. 2.2] apply to any instantiation of Opid in which equality is decidable. This considerably simplifies the process of extending the language.

We formalize Nuprl's computation system by defining a one step computation function [5, Sec. 3.1] on NTerm. When evaluating a non-canonical term, it first checks whether one of the principal arguments is non-canonical. If so, it recursively evaluates it. The interesting cases are when all the principal arguments are canonical. Fig. 2 compactly describes these cases for an illustrative subset of the formalized system.

$v ::= vt$	(type)	$\mathtt{inl}(t)$	(left injection)	$\mathtt{Ax}$	(axiom)
$\mid \underline{i}$	(integer)	$\mathtt{inr}(t)$	(right injection)	$\langle t_1, t_2 \rangle$	(pair)
$\mid \lambda x.t$	(lambda)	$\mathtt{sup}(t_1, t_2)$	(supremum)		

$vt ::= \mathbb{Z}$	(integer type)	$x{:}t_1 \to t_2$	(function type)
$\mid x : t_1 \times t_2$	(product type)	$t_2 = t \in t_1$	(equality type)
$\mid \mathbf{Base}$	(base type)	$\mid \bar{t}$	(partial type)
$\mid \cup x : t_1.t_2$	(union type)	$\cap x{:}t_1.t_2$	(intersection type)
$\mid t_1 // t_2$	(quotient type)	$t_1 + t_2$	(disjoint union type)
$\mid \{x : t_1 \mid t_2\}$	(set type)	$\mathtt{W}(x{:}t_1, t_2)$	(W type)

$t ::= x$	(variable)	$\mid v$	(value)
$\mid \boxed{t_1}\, t_2$	(application)	$\mid \mathtt{fix}(\boxed{t})$	(fixpoint)
$\mid \mathtt{let}\ x := \boxed{t_1}\ \mathtt{in}\ t_2$	(call-by-value)	$\mid \mathtt{let}\ x, y = \boxed{t_1}\ \mathtt{in}\ t_2$	(spread)
$\mid \mathtt{case}\ \boxed{t_1}\ \mathtt{of}\ \mathtt{inl}(x) \Rightarrow t_2 \mid \mathtt{inr}(y) \Rightarrow t_3$	(decide)		
$\mid \mathtt{if}\ \boxed{t_1} =_{\mathbb{Z}} \boxed{t_2}\ \mathtt{then}\ t_3\ \mathtt{else}\ t_4$	(integer equality)		

$(\lambda x.F)\ a$	$\to\ F[x \backslash a]$	$\mathtt{fix}(v)$	$\to\ v\ \mathtt{fix}(v)$
$\mathtt{let}\ x, y = \langle t_1, t_2\rangle\ \mathtt{in}\ F$	$\to\ F[x \backslash t_1; y \backslash t_2]$	$\mathtt{let}\ x := v\ \mathtt{in}\ t$	$\to\ t[x \backslash v]$
$\mathtt{case}\ \mathtt{inl}(t)\ \mathtt{of}\ \mathtt{inl}(x) \Rightarrow F \mid \mathtt{inr}(y) \Rightarrow G$	$\to\ F[x \backslash t]$		
$\mathtt{case}\ \mathtt{inr}(t)\ \mathtt{of}\ \mathtt{inl}(x) \Rightarrow F \mid \mathtt{inr}(y) \Rightarrow G$	$\to\ G[y \backslash t]$		
$\mathtt{if}\ \underline{i}_1 =_{\mathbb{Z}} \underline{i}_2\ \mathtt{then}\ t_1\ \mathtt{else}\ t_2$		$\to\ t_1,\ \text{if}\ \underline{i}_1 = \underline{i}_2$	
$\mathtt{if}\ \underline{i}_1 =_{\mathbb{Z}} \underline{i}_2\ \mathtt{then}\ t_1\ \mathtt{else}\ t_2$		$\to\ t_2,\ \text{if}\ \underline{i}_1 \neq \underline{i}_2$	

Fig. 2. Syntax (top) and operational semantics (bottom) of Nuprl

Definition olift $(R : \mathsf{NTerm} \to \mathsf{NTerm} \to [\mathsf{univ}])\ (x\ y :\mathsf{NTerm}) : [\mathsf{univ}] :=$
    $\mathsf{nt\_wf}\ x \times \mathsf{nt\_wf}\ y \times$
    $\forall\ sub:\ \text{Substitution},\ \mathsf{wf\_sub}\ sub \to\ \text{programs}\ [\mathsf{lsubst}\ x\ sub,\ \mathsf{lsubst}\ y\ sub]$
                    $\to R\ (\mathsf{lsubst}\ x\ sub)\ (\mathsf{lsubst}\ y\ sub).$

Definition blift $(R : \mathsf{NTerm} \to \mathsf{NTerm} \to [\mathsf{univ}])\ (bt1\ bt2: \mathsf{BTerm}): [\mathsf{univ}] :=$
    $\{lv:\ (\text{list NVar}) \times \{nt1, nt2 : \mathsf{NTerm} \times R\ nt1\ nt2$
        $\times \mathsf{alpha\_eq\_bterm}\ bt1\ (\mathsf{bterm}\ lv\ nt1) \times \mathsf{alpha\_eq\_bterm}\ bt2\ (\mathsf{bterm}\ lv\ nt2)\ \}\}.$

Definition lblift $(R: \mathsf{NTerm} \to \mathsf{NTerm} \to [\mathsf{univ}])\ (l\ r: \text{list BTerm}): [\mathsf{univ}] :=$
    $\text{length}\ l = \text{length}\ r \times \forall\ n : \mathsf{nat},\ n < \text{length}\ l \to \mathsf{blift}\ R\ (l[n])\ (r[n]).$

Fig. 3. Lifting operations. The notation $\{\_ : \_ \times\ \_\}$ denotes sigma types (sigT)

3   Computational Approximation and Equivalence

When we define the type system in Sec. 6, we want it to respect many
computational relations. For example, most type systems satisfy the subject
reduction property. In Agda, Coq and Nuprl, if t reduces to t', and t is in some
type T, then t and t' are equal in T. In addition, it is useful to have our types
respect a *congruence* that contains the computation relation. For example,
Coq has a notion of definitional equality which is a congruence. In Nuprl, we have

---

CoInductive approx : (*tl tr* :NTerm) : [univ] :=
| approx_fold: close_compute approx *tl tr* → approx *tl tr*.

CoInductive approx_aux (*R* : NTerm → NTerm → [univ]) (*tl tr*: NTerm): [univ] :=
| approx_fold: close_compute (approx_aux *R* \2/ *R*) *tl tr* → approx_aux *R* *tl tr*.

Definition approx := approx_aux (fun _ _ : NTerm ⇒ False).

Fig. 4. Computational approximation

a computation equivalence ∼ [21], which is a coinductively defined congruence
that permits more powerful equational reasoning. For example, all diverging
programs are equivalent under ∼. The same holds for all programs that generate
an infinite stream of zeroes. Howe first defines a preorder approx on closed terms,
proves that it is a congruence and finally defines *t1* ∼ *t2* as approx *t1 t2* ×
approx *t2 t1*. We first define olift, blift, lblift in Fig. 3. These will be used to lift a
binary relation on closed NTerms to one on terms that are not necessarily closed,
to BTerms, and to lists of BTerms, respectively. Note that programs *l* asserts that
every member of *l* is both well-formed and closed; wf_sub *sub* asserts that the
range of the substitution *sub* consists of well-formed terms; *t* ⇓ *tv* asserts that
t converges to the value *tv* [5, Sec. 3.1]. *t* ⇓ asserts that *t* converges to some
value. Although the notation [univ] currently stands for Type, most of our
development works unchanged if we change it to Prop.

One can think of approx as the greatest fixpoint of the following operator on
binary relations:

Definition close_compute (*R*: NTerm → NTerm → [univ]) (*tl tr* : NTerm): [univ]:=
programs [*tl*, *tr*] × ∀ (*c* : CanonicalOp) (*tls* : list BTerm),
(*tl* ⇓ oterm (Can *c*) *tls*)
→ {*trs* : list BTerm × (*tr* ⇓ oterm (Can *c*) *trs*)× lblift (olift *R*) *tls trs* }.

One could now directly define approx as at the top of Fig. 4, where \2/ denotes
disjunction of binary relations. However, this approach would require using the
cofix tactic of Coq for proving the properties of approx. Unfortunately, cofix
does very conservative productivity checking [23] and often rejects our legitimate
proofs. So we use parametrized coinduction [23] to define it in a slightly indirect
way (the next two items in Fig. 4). With this technique, we only need to use
cofix once to prove a "coinduction-principle" [5, Sec. 3.2] for approx and use
that principle for the coinductive proofs about approx. Howe then proves that
(olift approx) (abbreviated as approx_open) is a congruence [21]:

Theorem approx_open_congruence : ∀ (*o* : Opid) (*lbt1 lbt2* : list BTerm),
lblift approx_open *lbt1 lbt2*
→ nt_wf (oterm *o lbt2*) → approx_open (oterm *o lbt1*) (oterm *o lbt2*).

The proof is not easy. He first defines another relation approx_star, which contains approx_open and which is a congruence by definition. Then he proves that
approx_star implies approx_open. This proof assumes that all Opids satisfy a condition called *extensionality*. We formalize his proof and prove that all the Opids

T type	iff $T \equiv T$	$T_1 \equiv T_2$ if $T_1 \Downarrow T_1' \wedge T_2 \Downarrow T_2' \wedge T_1' \equiv T_2'$
$t \in T$	iff $t \equiv t \in T$	$t_1 \equiv t_2 \in T$ if $t_1 \Downarrow t_1' \wedge t_2 \Downarrow t_2' \wedge T \Downarrow T' \wedge t_1' \equiv t_2' \in T'$

$$t_1 \equiv t_2 \in \texttt{Base} \qquad \text{iff } t_1 \sim t_2$$

$$\texttt{Ax} \equiv \texttt{Ax} \in (a = b \in A) \quad \text{iff } (a = b \in A) \text{ type } \wedge a \equiv b \in A$$

$$t_1 \equiv t_2 \in \overline{A} \qquad \text{iff } (\overline{A}) \text{ type } \wedge (t_1 \Downarrow \Longleftrightarrow t_2 \Downarrow) \wedge (t_1 \Downarrow \Rightarrow t_1 \equiv t_2 \in A)$$

$$f_1 \equiv f_2 \in x{:}A \rightarrow B \qquad \text{iff } (x{:}A \rightarrow B) \text{ type}$$
$$\wedge \forall a_1, a_2.\ a_1 \equiv a_2 \in A \Rightarrow f_1(a_1) \equiv f_2(a_2) \in B[x \backslash a_1]$$

$$\sup(a_1, f_1) \equiv \sup(a_2, f_2) \in \texttt{W}(x{:}A, B)$$
$$\text{iff } (\texttt{W}(x{:}A, B)) \text{ type } \wedge a_1 \equiv a_2 \in A$$
$$\wedge \forall b_1, b_2.\ b_1 \equiv b_2 \in B[x \backslash a_1] \Rightarrow f_1(b_1) \equiv f_2(b_2) \in \texttt{W}(A{:}x, B)$$

$$\texttt{Base} \equiv \texttt{Base}$$

$$(a_1 = a_2 \in A) \equiv (b_1 = b_2 \in B) \quad \text{iff } A \equiv B$$
$$\wedge (a_1 \equiv b_1 \in A \vee a_1 \sim b_1) \wedge (a_2 \equiv b_2 \in A \vee a_2 \sim b_2)$$

$$\overline{A} \equiv \overline{B} \qquad \text{iff } A \equiv B \wedge (\forall a.\ a \in A \Rightarrow a \Downarrow)$$

$$x_1{:}A_1 \rightarrow B_1 \equiv x_2{:}A_2 \rightarrow B_2 \qquad \text{iff } A_1 \equiv A_2$$
$$\wedge \forall a_1, a_2.\ a_1 \equiv a_2 \in A_1 \Rightarrow B_1[x_1 \backslash a_1] \equiv B_2[x_2 \backslash a_2]$$

$$\texttt{W}(x_1{:}A_1, B_1) \equiv \texttt{W}(x_2{:}A_2, B_2) \qquad \text{iff } A_1 \equiv A_2$$
$$\wedge \forall a_1, a_2.\ a_1 \equiv a_2 \in A_1 \Rightarrow B_1[x_1 \backslash a_1] \equiv B_2[x_2 \backslash a_2]$$

Fig. 5. Informal definition of a core part of a Nuprl universe

of the current Nuprl system are extensional. Hence, we obtain that approx_open, approx and $\sim$ are congruences [5, Sec. 3.2.1].

Domain Theoretic Properties. The preorder approx has interesting domain theoretic properties [15] such as compactness and the least upper bound principle. Let $\bot$ be $\texttt{fix}(\lambda x.x)$. It is the least element (up to $\sim$) w.r.t. approx, i.e., for any closed term t, approx $\bot$ t. The least upper bound principle says that for any terms G and f, $G(\texttt{fix}(f))$ is the least upper bound of the (approx) chain $G(f^n(\bot))$ for $n \in \mathbb{N}$. Compactness says that if $G(\texttt{fix}(f))$ converges, then there exists a natural number n such that $G(f^n(\bot))$ converges. We formalized proofs of both these properties [5, Sec. 3.3]. Crary used compactness to justify his fixpoint induction principle. It provides an important way for proving that a term of the form $\texttt{fix}(f)$ is in a partial type [5, Sec. 5.2]. We have used the least upper bound principle to justify some untyped computational equivalences that are useful for automatic optimization of Nuprl extracts [34].

4 PER Semantics

We now define Nuprl's type system. Several semantics have been proposed for Nuprl over the years, such as: Allen's PER semantics [3,15]; Mendler's adaptation of Allen's PER semantics [28]; and Howe's set-theoretic semantics [22]. In this paper, we use Allen's PER semantics because it can be defined predicatively. Also, the various additions made to Nuprl over the years have been validated using this semantics. Allen's method determines which closed terms are types and which closed terms are equal members of types, therefore, defining types as PERs. A PER is symmetric and transitive, but not necessarily reflexive. Partiality is required because the domain of these relations is the entire collection of closed terms, and not just the members of the type. We say that t is a member of type T when the PER definition of T relates t and t.

Fig. 5 informally presents a core part of Nuprl's type system à la Crary, which can be made formal using induction-recursion (where equality in W types has to be defined inductively). We write $T_1 \equiv T_2$ to mean that T_1 and T_2 are equal types, and $a \equiv b \in T$ to mean that a and b are equal in type T. Allen designed his semantics to work with systems that have an extensional type equality (meaning that two types T and S are equal if for all t and s, $t \equiv s \in T$ iff $t \equiv s \in S$) and suggests a way to deal with type systems that have a more intensional type equality, such as Nuprl. For example, two true equality types such as $0 = 0 \in \mathbb{N}$ and $1 = 1 \in \mathbb{N}$ are not equal types in Nuprl even though they have the same PER. Also, note that type equality in Nuprl is coarser than computational equivalence ($\sim$). For example $(\lambda x.((x + 1) - 1) = \lambda x.x \in \mathbb{N} \to \mathbb{N})$ and $(\lambda x.x = \lambda x.x \in \mathbb{N} \to \mathbb{N})$ are equal equality types, but the two terms are not computationally equivalent because $\lambda x.((x + 1) - 1)$ gets stuck when applied to a term that is not a number while $\lambda x.x$ does not. Crary [15] provides a formal account of the adaptation of Allen's semantics to deal with systems that have non-fully extensional type equality (he adds a few type constructors and leaves off the W types). Following Crary, our type system defines which terms denote equal types instead of simply defining which terms denote types as in Allen's semantics.

As mentioned by Allen, simple induction mechanisms such as the one of Coq are not enough to provide a straightforward definition of Nuprl's semantics, where one would define typehood and member equality by mutual induction [3, Sec. 4.2]. The problem is that in the inductive clause that defines the dependent function types, the equality predicate occurs at a non-strictly-positive position. Allen suggests that the definition should however be valid because it is "half-positive". This is what induction-recursion, as implemented in Agda, achieves [17,18]. Instead of making that induction-recursion formal, the approach taken by Allen was to define ternary relations between types and equalities.

This trick of translating a mutually inductive-recursive definition to a single inductive definition has been formally studied by Capretta [13]. He observes that this translation is problematic when the return type of the function mentions a predicative universe. Indeed, we experienced the same problem while formalizing Allen's definition in a precise metatheory like Coq. In particular, we can only predicatively formalize a finite number of universes of Nuprl in Coq. This is not surprising given the results of Setzer that intensional and extensional versions of various dependent type theories have the same proof theoretic strength [35].

We will first explain how induction-recursion can be used to define the type system in an intuitive way. Although the inductive-recursive definition is easier to understand and lets us predicatively prove Nuprl's consistency, Agda lacks a tactic language, which is critical to automate many otherwise tedious proofs. Therefore, we chose Coq over Agda and used Allen's trick to define Nuprl's type

```
mutual
    data equalType (n : ℕ) (iUnivs : Vec PER n)        equalInType : {T1 T2 : NTerm} {n : ℕ}
        (T1 T2 : NTerm) : Set where                        (iUnivs : Vec PER n)
                                                           (teq : equalType n iUnivs T1 T2)
    PINT : { _ : T1 ⇓ mk_Int}                              → PER
          { _ : T2 ⇓ mk_Int}
          → (equalType n iUnivs T1 T2)                 equalInType iUnivs PINT   t t' =
    PFUN : {A1 B1 A2 B2 : NTerm} {v1 v2 : NVar}        ∑ ℤ (λ n
          { _ : T1 ⇓ (mk_Fun A1 v1 B1)}                    → (t ⇓ (mk_int n))
          { _ : T2 ⇓ (mk_Fun A2 v2 B2)}                    × t' ⇓ (mk_int n))
          (pA : equalType n iUnivs A1 A2)
          (pB : (a1 a2 : NTerm)                        equalInType iUnivs (PFUN pA pB) t t' =
             (pa : equalInType iUnivs pA a1 a2)            (a1 a2 : NTerm)
             → equalType n iUnivs                          (pa : equalInType iUnivs pA a1 a2)
             (subst B1 v1 a1)                              → equalInType   iUnivs
             (subst B2 v2 a2))                                (pB a1 a2 pa)
          →  (equalType n iUnivs T1 T2)                      (mk_apply t a1)
    PUNIV : (m : Fin n)                                      (mk_apply t' a2)
          { _ : T1 ⇓ (mk_Univ (toℕ m))}
          { _ : T2 ⇓ (mk_Univ (toℕ m))}               equalInType iUnivs (PUNIV m) T1 T2 =
          → (equalType n iUnivs T1 T2)                 lookup m iUnivs T1 T2
```

Fig. 6. Agda Inductive-Recursive definition

system. At first, this purely inductive definition in Sec. 6 might seem overly complicated. However, it can be understood as applying Capretta's general recipe [13] to the inductive-recursive definition.

5   An Inductive-Recursive Approach

Crary first presents an intuitive definition of Nuprl's type system in the style of Fig. 5 and asserts that it is not a valid inductive definition. Then he uses Allen's trick to convert it to a purely inductive definition [15, page 51]. Using induction-recursion [18], this section shows that the definitions in Fig. 5 are meaningful. Moreover, we show how to define the entire predicative hierarchy of universes of Nuprl in the first universe of Agda's predicative hierarchy. This is not surprising, given that induction-recursion is known to increase the proof theoretic strength of type theories [18]. As mentioned above, because Agda does not have a tactic machinery necessary for our proof automation, this definition is only for illustrative purposes. We define universes having only integers and dependent function types ([5, Sec. 4.1] has more types, e.g., W types).

In Fig. 6, where , where PER is NTerm→NTerm→Set, equalType inductively defines which types are equal and equalInType recursively defines which terms are equal in a type[7]. These definitions refer to each other and are simultaneously defined using the mutual keyword. Both definitions are parametrized by a number

---

[7] For brevity, we ignore the issue of closedness of terms here.

n and $iUnivs$, a vector of PERs of size n. The idea is that we have already defined the first n universes by now and the PER defined by (equalType n $iUnivs$) will serve as the equality of types in the next universe. The mth member (where $m < n$) of $iUnivs$ is the already constructed PER that determines which two terms denote equal types in the mth universe. Given an evidence that $T1$ and $T2$ are equal types in this universe, equalInType returns the PER of this type. Note that equalInType is structurally recursive on its argument teq. Note also that equalInType occurs at a negative position in the PFUN clause, but this is allowed since equalInType is defined by structural recursion and not induction.

6 An Inductive Approach Based on Allen's Semantics

6.1 Metatheory

Now, we return to Coq, where induction-recursion is not yet implemented. As mentioned above, all the Coq definitions presented so far would typecheck either in Prop or Type. This is not true about the definition of the type system. As mentioned above, we define two metatheories of Nuprl in Coq. One uses its predicative Type hierarchy. This metatheory uses $n + 1$ Coq universes to model n Nuprl universes. Because universe polymorphism is still under development in Coq, this currently requires that we duplicate the code at each level, which is impractical. Hence, we have only verified this translation for $n = 3$, and we will not discuss that metatheory further [5, Sec. 4.3].

The other metatheory uses Coq's Prop impredicative universe with the FunctionalChoice_on axiom[8] ([5, Sec. 4.2.3] explains why this axiom is needed). In this metatheory, we can model all of Nuprl's universes. Also, it allows us to justify some principles that a classical mathematician might wish to have. For example, in the Prop model, using the law of excluded middle for members of Prop (known to be consistent with Coq[8]), following Crary [15] we have proved [5, Sec. 5.2] that the following weak version of the law of excluded middle is consistent with Nuprl: $\forall P : \mathbb{U}_i. \downarrow(P + (P \rightarrow \text{Void}))$. Because the computational content of the disjoint union is erased (using the squashing operator $\downarrow$), one cannot use this to construct a magical decider of all propositions.

6.2 Type Definitions

This section illustrates our method by defining base, equality, partial, function, and W types. As mentioned above, types are defined as PERs on closed terms. A CTerm is a pair of an NTerm and a proof that it is closed. Let per stand for CTerm $\rightarrow$ CTerm $\rightarrow$ Prop. A *type system* is defined below as a *candidate type system* that satisfies some properties such as symmetry and transitivity of the PERs, where a candidate type system is an element of the type cts, which we define as CTerm $\rightarrow$ CTerm $\rightarrow$ per$\rightarrow$ Prop. Given a cts c, c $T1$ $T2$ eq asserts

---

[8] http://coq.inria.fr/cocorico/CoqAndAxioms

that $T1$ and $T2$ are equal types in the type system c and eq is the PER that determines which terms are equal in these types.

We now define the PER constructors (of the form per_TyCon) for each type constructor TyCon of Nuprl's type theory. Intuitively, each per_TyCon is a monotonic operator that takes a candidate type system ts and returns a new candidate type system where all the types compute to terms of the form (TyCon $T1 \ldots Tn$) where $T1, \ldots, Tn$ are types of ts.

In the definitions below we use $\{\_:\_,\_\}$ for propositional existential types (i.e., ex from the standard library). Also, we use Nuprl term constructors of the form mkc_TyCon. We omit their definitions as they should be obvious. Finally, for readability we sometimes mix Coq and informal mathematical notations.

Base. The values of type Base (suggested by Howe [21]) are closed terms and its equality is computational equivalence. In the following definition, ts is not used because Base does not have any arguments:

Definition per_base $(ts : \mathrm{cts})$ $T1$ $T2$ $(eq : \mathrm{per}) : \mathrm{Prop} :=$
  $T1 \Downarrow \mathsf{Base} \times T2 \Downarrow \mathsf{Base} \times \forall\, t\ t',\, eq\ t\ t' \Leftrightarrow t \sim t'.$

Equality. Unlike Coq, Nuprl has primitive equality types which reflect the metatheoretical PERs as propositions that users can reason about. Note that Uniqueness of Identity Proofs, aka UIP, holds for Nuprl, i.e., Ax is the unique canonical inhabitant of equality types.

Definition per_eq $(ts : \mathrm{cts})$ $T1$ $T2$ $(eq : \mathrm{per}) : \mathrm{Prop} :=$
  $\{A,\ B,\ a1,\ a2,\ b1,\ b2 : \mathrm{CTerm}\,,\ \{eqa : \mathrm{per}$
  $,\ T1 \Downarrow (\mathrm{mkc\_equality}\ a1\ a2\ A) \times T2 \Downarrow (\mathrm{mkc\_equality}\ b1\ b2\ B)$
  $\times\ ts\ A\ B\ eqa \times (eqa\ \backslash 2/\ \sim)\ a1\ b1 \times (eqa\ \backslash 2/\ \sim)\ a2\ b2$
  $\times\ (\forall\, t\ t',\, eq\ t\ t' \Leftrightarrow (t \Downarrow \mathsf{Ax} \times t' \Downarrow \mathsf{Ax} \times eqa\ a1\ a2))\ \}\}.$

This definition differs from the one present in earlier Nuprl versions, where $(eqa\ \backslash 2/\ \sim)$ was simply eqa. This means that $t \in T$ is now a type when T is a type and t is in Base. In earlier versions of Nuprl [15] (as well as in other type theories [36]) membership was a non-negatable proposition, i.e., $t \in T$ was not a proposition unless it was true. This change allows us to reason in the theory about a wider range of properties that could previously only be stated in the metatheory. For example, we can now define the *subtype* type $A \sqsubseteq B$ as $\lambda x.x \in A \to B$ because $\lambda x.x$ is in Base. Sec. 6.3 shows how we had to change the definition of a type system in order to cope with this modification.

Partial Type. Given a type T that has only converging terms, we form the partial type $\overline{T}$ (see Fig. 5). Equal members of $\overline{T}$ have the same convergence behaviour, and if either one converges, they are equal in T.

Definition per_partial $(ts : \mathrm{cts})$ $T1$ $T2$ $(eq : \mathrm{per}) : \mathrm{Prop} :=$
  $\{A1,\ A2 : \mathrm{CTerm}\,,\ \{eqa : \mathrm{per}$
  $,\ T1 \Downarrow (\mathrm{mkc\_partial}\ A1) \times T2 \Downarrow (\mathrm{mkc\_partial}\ A2)$
  $\times\ ts\ A1\ A2\ eqa \times (\forall\, a,\, eqa\ a\ a \to a\Downarrow)$
  $\times\ (\forall\, t\ t',\, eq\ t\ t' \Leftrightarrow ((t\Downarrow \Leftrightarrow t'\Downarrow) \times (t\Downarrow \to eqa\ t\ t')))\ \}\}.$

Type Family. Allen [3] introduces the concept of type families to define dependent types such as function types and W types. A type family $TyCon$ is defined as a family of types B parametrized by a domain A. In the following definition per-fam(*eqa*) stands for $(\forall\ (a\ a'\ :\ \text{CTerm})\ (p\ :\ eqa\ a\ a'),\ \text{per})$, which is the type of PERs of type families over a domain with PER *eqa*; and CVTerm l is the type of terms with free variables contained in l.

```
Definition type_family TyCon (ts : cts) T1 T2 eqa (eqb : per-fam(eqa)) : Prop:=
    {A, A' : CTerm , {v, v' : NVar , {B : CVTerm [v] , {B' : CVTerm [v'] ,
      T1 ⇓ (TyCon A v B) × T2 ⇓ (TyCon A' v' B')
      × ts A A' eqa
      × (∀ a a', ∀ e : eqa a a', ts (B[v\a]) (B'[v'\a']) (eqb a a' e))}}}}.
```

Equalities of type families in our formalization (such as *eqb* above) are five place relations, while they are simply three place relations in Allen's and Crary's formalizations. This is due to the fact that conceptually $\forall\ (a'\ :\ \text{CTerm})$ and $\forall\ (p\ :\ eqa\ a\ a')$ could be turned into intersection types because *eqb* only depends on the fact that the types are inhabited and does not make use of the inhabitants.

Dependent Function. Dependent function types are defined so that functional extensionality is a trivial consequence.

```
Definition per_func (ts : cts) T1 T2 (eq : per) : Prop :=
    {eqa : per, {eqb : per-fam(eqa)
      , type_family mkc_function ts T1 T2 eqa eqb
      × (∀ t t', eq t t' ⇔
          (∀ a a' (e : eqa a a'), eqb a a' e (mkc_apply t a) (mkc_apply t' a')))}}.
```

W. We define W types, by first inductively defining their PERs called weq [3]:

```
Inductive weq (eqa : per) (eqb : per-fam(eqa)) (t t' : CTerm) : Prop :=
| weq_cons :
    ∀ (a f a' f' : CTerm) (e : eqa a a'),
    t ⇓ (mkc_sup a f)
    → t' ⇓ (mkc_sup a' f')
    → (∀ b b', eqb a a' e b b' → weq eqa eqb (mkc_apply f b) (mkc_apply f' b'))
    → weq eqa eqb t t'.
```

```
Definition per_w (ts : cts) T1 T2 (eq : per) : Prop :=
    {eqa : per, {eqb : per-fam(eqa) ,
      type_family mkc_w ts T1 T2 eqa eqb × (∀ t t', eq t t' ⇔ weq eqa eqb t t')}}.
```

Our technical report [5, Sec. 4.2] extends W types to parametrized W types and uses them to define parametrized inductive types (e.g., vectors).

6.3 Universes and Nuprl's Type System

Universes. As Allen [3] and Crary [15] did, we now define Nuprl's universes of types, and finally Nuprl's type system. First, we inductively define a close operator on candidate type systems. Given a candidate type system *cts*, this operator builds another candidate type system from *ts* that is closed w.r.t. the type constructors defined above:

```
Inductive close (ts : cts) (T T' : CTerm) (eq : per) : Prop :=
  | CL_init : ts T T' eq → close ts T T' eq
  | CL_base : per_base (close ts) T T' eq → close ts T T' eq
  | CL_eq : per_eq (close ts) T T' eq → close ts T T' eq
  | CL_partial : per_partial (close ts) T T' eq → close ts T T' eq
```

| CL_func : per_func (close ts) T T' eq → close ts T T' eq
| CL_w : per_w (close ts) T T' eq → close ts T T' eq.

We define $\mathbb{U}(i)$, the Nuprl universe type at level i, by recursion on $i \in \mathbb{N}$:

```
Fixpoint univi (i : nat) (T T' : CTerm) (eq : per) : Prop :=
  match i with
  | 0 ⇒ False
  | S n ⇒ (T ⇓ (U(n)) × T' ⇓ (U(n))
          × ∀ A A', eq A A' ⇔ {eqa : per, close (univi n) A A' eqa})
        {+} univi n T T' eq    end.
```

Finally, we define univ, the collection of all universes, and the Nuprl type system as follows:

```
Definition univ (T T' : CTerm) (eq : per) := {i : nat , univi i T T' eq}.
Definition nuprl := close univ.
```

We can now define $t_1 \equiv t_2 \in T$ as $\{eq : per , \text{nuprl } T\ T\ eq \times eq\ t_1\ t_2\}$ and $T \equiv T'$ as $\{eq : per , \text{nuprl } T\ T'\ eq\}$.

Type System. Let us now prove that nuprl is a *type system*, i.e., that it is a candidate type system ts that satisfies the following properties [3,15]:

1. *uniquely valued*: $\forall$ T T' eq eq', ts T T' eq → ts T T' eq' → $eq \Leftarrow_2\Rightarrow eq'$.
2. *equality respecting*: $\forall$ T T' eq eq', ts T T' eq → $eq \Leftarrow_2\Rightarrow eq'$ → ts T T' eq'.
3. *type symmetric*: $\forall$ eq, symmetric (fun T T' ⇒ ts T T' eq).
4. *type transitive*: $\forall$ eq, transitive (fun T T' ⇒ ts T T' eq).
5. *type value respecting*: $\forall$ T T' eq, ts T T eq → $T \sim T'$ → ts T T' eq.
6. *term symmetric*: $\forall$ T eq, ts T T eq → symmetric eq.
7. *term transitive*: $\forall$ T eq, ts T T eq → transitive eq.
8. *term value respecting*: $\forall$ T eq, ts T T eq → $\forall$ t t', eq t t → $t \sim t'$ → eq t t'.

A type system uniquely defines the PERs of its types. The last six properties state that $\_\equiv\_\in\_$ and $\_\equiv\_$ are PERs that respect computational equivalence. This definition differs from Crary's [15] as follows: (1) We added condition 2 because Allen and Crary consider equivalent PERs to be equal and we decided not to add the propositional and function extensionality axioms; (2) We strengthened conditions 5 and 8 by replacing $\Downarrow$ with $\sim$. This seemed necessary to obtain a strong enough induction hypothesis when proving that our new definition of per_eq preserves the type system properties [5, Sec. 4.2]. These properties and the congruence of $\sim$ allow us to do computation in any context in sequents.

Finally, the following lemma corresponds to Crary's Lemma 4.13 [15]: For all $i \in \mathbb{N}$, univi i is a type system; and the following theorem corresponds to Crary's Lemma 4.14 [15]: univ and nuprl are type systems.

6.4   Sequents and Rules

In Nuprl, one reasons about the nuprl type system types using a sequent calculus, which is a collection of rules that captures many properties of the nuprl type system and its types. For example, for each type we have introduction and elimination rules. This calculus can be extended as required by adding more types and/or rules. This section presents the syntax and semantics of Nuprl's sequents and rules. We then prove that these rules are valid w.r.t. the above semantics, and therefore that Nuprl is consistent. This paper provides a safe way to add new rules by allowing one to formally prove their validity, which is a difficult task without the help of a proof assistant[9].

Syntax of Sequents and Rules. Sequents are of the form $h_1, \ldots, h_n \vdash T \lfloor \text{ext } t \rfloor$, where t is the *extract/evidence* of T, and where an hypothesis h is either of the form $x : A$ (non-hidden) or of the form $[x : A]$ (hidden). Such a sequent states that T is a type and t is a member of T. A rule is a pair of a sequent and a list of sequents, which we write as $(S_1 \wedge \cdots \wedge S_n) \Rightarrow S$. To understand the necessity of hidden hypotheses, let us consider the following intersection introduction rule:

$$H, [x : A] \vdash B[x] \lfloor \text{ext } e \rfloor \ \wedge \ H \vdash A = A \in \mathbb{U}_i \ \Rightarrow \ H \vdash \cap a{:}A.B[a] \lfloor \text{ext } e \rfloor$$

This rule says that to prove that $\cap a{:}A.B[a]$ is true with extract e, one has to prove that $B[x]$ is true with extract e, assuming that x is of type A. The meaning of intersection types requires that the extract e be the same for all values of A, and is therefore called the *uniform evidence* of $\cap a{:}A.B[a]$. The fact that x is hidden means that it cannot occur free in e (but can occur free in B). The same mechanism is required to state the rules for, e.g., subset types or quotient types.

Semantics of Sequents and Rules. Several definitions for the truth of sequents occur in the Nuprl literature [14,15,24]. Among these, Kopylov [24]'s definition was the simplest. We provide here an even simpler definition and we have proved in Coq that all these definitions are equivalent [5, Sec. 5.1]. The semantics we present uses a notion of *pointwise functionality* [15, Sec. 4.2.1], which says that each type in a true sequent must respect the equalities of the types on which it depends. This is captured by formula 1 below for the hypotheses of a sequent, and by formula 2 for its conclusion. For the purpose of this discussion, let us ignore the possibility that some hypotheses can be hidden.

Let H be a list of hypotheses of the form $x_1 : T_1, \ldots, x_n : T_n$, let s_1 be a substitution of the form $(x_1 \mapsto t_1, \ldots, x_n \mapsto t_n)$, and let s_2 be a substitution of the form $(x_1 \mapsto u_1, \ldots, x_n \mapsto u_n)$.

Similarity. Similarity lifts the notion of equality in a type (i.e., $\_{=}\_{\in}\_$) to lists of hypotheses. We say that s_1 and s_2 are similar in H, and write $s_1 {\equiv} s_2 {\in} H$, if for all $i \in \{1, \ldots, n\}$, $t_i {\equiv} u_i {\in} T_i[x_1 \backslash t_1; \cdots ; x_{i-1} \backslash t_{i-1}]$. Let $s {\in} H$ be $s {\equiv} s {\in} H$.

Equal Hypotheses. The following notion of equality lifts the notion of equality between types (i.e., $\_{\equiv}\_$) to lists of hypotheses. We say that the hypotheses H are equal w.r.t. s_1 and s_2, and write $s_1(H) {\equiv} s_2(H)$, if for all $i \in \{1, \ldots, n\}$, $T_i[x_1 \backslash t_1; \cdots ; x_{i-1} \backslash t_{i-1}] {\equiv} T_i[x_1 \backslash u_1; \cdots ; x_{i-1} \backslash u_{i-1}]$.

Hypotheses Functionality. We say that the hypotheses H are pointwise functional w.r.t. the substitution s, and write $H@s$ if

$$\forall s'. \ s {\equiv} s' {\in} H \Rightarrow s(H) {\equiv} s'(H) \tag{1}$$

---

[9] Howe [22] writes: "Because of this complexity, many of the Nuprl rules have not been completely verified, and there is a strong barrier to extending and modifying the theory."

Truth of Sequents. We say that a sequent of the form $H \vdash T \lfloor \text{ext } t \rfloor$ is true if

$$\forall s_1, s_2.\ s_1 \equiv s_2 \in H \land H@s_1 \Rightarrow T[s_1] \equiv T[s_2] \land t[s_1] \equiv t[s_2] \in T[s_1] \qquad (2)$$

In addition the free variables of t have to be non-hidden in H.

Validity of Rules. A rule of the form $(S_1 \land \cdots \land S_n) \Rightarrow S$ is valid if assuming that the sequents $S_1, \ldots, S_n$ are true then the sequent S is also true.

Consistency. Using the framework described in this paper we have currently verified over 70 rules, including the usual introduction and elimination rules to reason about the core type system presented above in Sec. 6.2 [5, Sec. 5.2].

A Nuprl proof is a tree of sequents where each node corresponds to the application of a rule. Because we have proved that the above mentioned rules are correct, using the definition of the validity of a rule, and by induction on the size of the tree, this means that the sequent at the root of the tree is true w.r.t. the above PER semantics. Hence, a proof of False, for any meaningful definition of False, i.e., a type with an empty PER such as $(0 = 1 \in \mathbb{Z})$, would mean that the PER is in fact non-empty, which leads to a contradiction.

Building a Trusted Core of Nuprl. Using our proofs that the Nuprl rules are correct, and the definition of the validity of rules, we can then build a verified proof refiner (rule interpreter) for Nuprl. Our technical report [5, Sec. 5.3] illustrates this by presenting a Ltac based refiner in Coq that allows one to prove Nuprl lemmas. These proofs are straightforward translations of the corresponding Nuprl proofs and we leave for future work the automation of this translation. An appealing use of such a tool is that it can then be used as Nuprl's trusted core which checks that proofs are correct, i.e., if the translation typechecks in Coq, this means that the Nuprl proof is correct.

7   Related Work

Perhaps the closest work to ours is that of Barras [7]. He formalizes Werner's [37] set theoretic semantics for a fragment of Coq in Coq by first axiomatizing the required set theory in Coq. While this fragment has the Peano numbers, inductive types are missing. Werner's semantics assumes the existence of inaccessible cardinals to give denotations to the predicative universes of Coq as sets. In earlier work, Barras and Werner [8] provide a deep embedding of a fragment of Coq that excludes universes and inductive types. They thereby obtain a certified typechecker for this fragment.

Similarly, Harrison [20] verified: (1) a weaker version of HOL Light's kernel (with no axiom of infinity) in HOL Light (a proof assistant based on classical logic); and (2) HOL Light in a stronger variant of HOL light (by adding an axiom that provides a larger universe of sets). Myreen et al. are extending this work to build a verified implementation of HOL Light [31] using their verified ML-like programming language called CakeML [25]. They fully verified CakeML in HOL4 down to machine code. Similarly, Myreen and Davis formally proved the soundness of the Milawa theorem prover [30] (an ACL2-like theorem prover) which runs on top of their verified Lisp runtime called Jitawa [29]. Both these projects go further by verifying the implementations of the provers down to machine code. Also, Nuprl's logic is different from those of HOL and Milawa, e.g., HOL does not support dependent types and Milawa's logic is a first-order logic of total, recursive functions with induction.

Also, Buisse and Dybjer [12] partially formalize a categorical model of an intensional Martin-Löf type theory in Agda.

Uses of induction-recursion to define shallow embeddings of hierarchies of Martin-Löf universes have often been described in the literature [17,27]. However, because we have a deep embedding, our inductive-recursive definition is parametrized over the already defined terms. This deep-embedding approach is required for our goal of extracting an independent correct by construction proof assistant. Also, the extensionality of Nuprl complicates our definitions a bit. For example, we have to define equality of types instead of just typehood. Danielsson [16] uses induction-recursion to define a deep embedding of a dependently typed language that does not have universes and inductive types.

8   Future Work and Acknowledgments

As future work, we want to formalize a tactic language and build a user (Emacs/NetBeans) interface for our verified Nuprl version, based on the code extracted from our Coq development. Also, we plan to add a way to directly write inductive definitions (possibly parametrized and/or mutual) and have a formally verified and transparent translation to our formalized parametrized W types. Finally, we plan to formalize a typechecker for a large part of Nuprl.

We thank the Coq and Agda mailing lists' members for helping us with various issues while using these tools. We thank Mark Bickford, Robert L. Constable, David Guaspari, and Evan Moran for their useful comments as well as Jason Gross from whom we learned that Agda allows inductive-recursive definitions.

References

1. Blazy, S., Paulin-Mohring, C., Pichardie, D. (eds.): ITP 2013. LNCS, vol. 7998. Springer, Heidelberg (2013)
2. Abbott, M., Altenkirch, T., Ghani, N.: Representing nested inductive types using W-types. In: Díaz, J., Karhumäki, J., Lepistö, A., Sannella, D. (eds.) ICALP 2004. LNCS, vol. 3142, pp. 59–71. Springer, Heidelberg (2004)
3. Allen, S.F.: A Non-Type-Theoretic Semantics for Type-Theoretic Language. PhD thesis, Cornell University (1987)
4. Allen, S.F., Bickford, M., Constable, R.L., Eaton, R., Kreitz, C., Lorigo, L., Moran, E.: Innovations in computational type theory using Nuprl. J. Applied Logic 4(4), 428–469 (2006), http://www.nuprl.org/
5. Anand, A., Rahli, V.: Towards a formally verified proof assistant. Technical report, Cornell University (2014), http://www.nuprl.org/html/Nuprl2Coq/
6. Aydemir, B.E., Charguéraud, A., Pierce, B.C., Pollack, R., Weirich, S.: Engineering formal metatheory. In: POPL 2008, pp. 3–15. ACM (2008)

7. Barras, B.: Sets in Coq, Coq in sets. Journal of Formalized Reasoning 3(1), 29–48 (2010)
8. Barras, B., Werner, B.: Coq in Coq. Technical report, INRIA Rocquencourt (1997)
9. Bertot, Y., Casteran, P.: Interactive Theorem Proving and Program Development. Springer (2004), http://coq.inria.fr/
10. Bove, A., Dybjer, P., Norell, U.: A brief overview of Agda – A functional language with dependent types. In: Berghofer, S., Nipkow, T., Urban, C., Wenzel, M. (eds.) TPHOLs 2009. LNCS, vol. 5674, pp. 73–78. Springer, Heidelberg (2009), http://wiki.portal.chalmers.se/agda/pmwiki.php
11. Brady, E.: Idris —: systems programming meets full dependent types. In: 5th ACM Workshop Programming Languages meets Program Verification, PLPV 2011, pp. 43–54. ACM (2011)
12. Buisse, A., Dybjer, P.: Towards formalizing categorical models of type theory in type theory. Electr. Notes Theor. Comput. Sci. 196, 137–151 (2008)
13. Capretta, V.: A polymorphic representation of induction-recursion (2004), http://www.cs.ru.nl/~venanzio/publications/induction_recursion.ps
14. Constable, R.L., Allen, S.F., Bromley, H.M., Cleaveland, W.R., Cremer, J.F., Harper, R.W., Howe, D.J., Knoblock, T.B., Mendler, N.P., Panangaden, P., Sasaki, J.T., Smith, S.F.: Implementing mathematics with the Nuprl proof development system. Prentice-Hall, Inc., Upper Saddle River (1986)
15. Crary, K.: Type-Theoretic Methodology for Practical Programming Languages. PhD thesis, Cornell University, Ithaca, NY (August 1998)
16. Danielsson, N.A.: A formalisation of a dependently typed language as an inductive-recursive family. In: Altenkirch, T., McBride, C. (eds.) TYPES 2006. LNCS, vol. 4502, pp. 93–109. Springer, Heidelberg (2007)
17. Dybjer, P.: A general formulation of simultaneous inductive-recursive definitions in type theory. J. Symb. Log. 65(2), 525–549 (2000)
18. Dybjer, P., Setzer, A.: Induction-recursion and initial algebras. Ann. Pure Appl. Logic 124(1-3), 1–47 (2003)
19. Gonthier, G., Asperti, A., Avigad, J., Bertot, Y., Cohen, C., Garillot, F., Le Roux, S., Mahboubi, A., O'Connor, R., Biha, S.O., Pasca, I., Rideau, L., Solovyev, A., Tassi, E., Théry, L.: A machine-checked proof of the odd order theorem. In: ITP 2013, [1], pp. 163–179
20. Harrison, J.: Towards Self-verification of HOL Light. In: Furbach, U., Shankar, N. (eds.) IJCAR 2006. LNCS (LNAI), vol. 4130, pp. 177–191. Springer, Heidelberg (2006)
21. Howe, D.J.: Equality in lazy computation systems. In: Proceedings of Fourth IEEE Symposium on Logic in Computer Science, pp. 198–203. IEEE Computer Society (1989)
22. Howe, D.J.: Semantic foundations for embedding HOL in Nuprl. In: Wirsing, M., Nivat, M. (eds.) AMAST 1996. LNCS, vol. 1101, pp. 85–101. Springer, Heidelberg (1996)
23. Hur, C.-K., Neis, G., Dreyer, D., Vafeiadis, V.: The power of parameterization in coinductive proof. In: POPL 2013, pp. 193–206. ACM (2013)
24. Kopylov, A.: Type Theoretical Foundations for Data Structures, Classes, and Objects. PhD thesis, Cornell University, Ithaca, NY (2004)
25. Kumar, R., Myreen, M.O., Norrish, M., Owens, S.: CakeML: a verified implementation of ML. In: POPL 2014, pp. 179–192. ACM (2014)
26. Leroy, X.: Formal certification of a compiler back-end or: programming a compiler with a proof assistant. In: POPL 2006, pp. 42–54. ACM (2006)

27. McBride, C.: Hier soir, an OTT hierarchy (2011),
    http://sneezy.cs.nott.ac.uk/epilogue/?p=1098
28. Mendler, P.F.: Inductive Definition in Type Theory. PhD thesis, Cornell University,
    Ithaca, NY (1988)
29. Myreen, M.O., Davis, J.: A verified runtime for a verified theorem prover. In:
    van Eekelen, M., Geuvers, H., Schmaltz, J., Wiedijk, F. (eds.) ITP 2011. LNCS,
    vol. 6898, pp. 265–280. Springer, Heidelberg (2011)
30. Myreen, M.O., Davis, J.: The reflective milawa theorem prover is sound (Down to
    the machine code that runs it). In: Klein, G., Gamboa, R. (eds.) ITP 2014. LNCS
    (LNAI), vol. 8558, pp. 413–428. Springer, Heidelberg (2014)
31. Myreen, M.O., Owens, S., Kumar, R.: Steps towards verified implementations of
    hol light. In: ITP 2013 [1], pp. 490–495
32. Nipkow, T., Paulson, L.C., Wenzel, M.: Isabelle/HOL. LNCS, vol. 2283. Springer,
    Heidelberg (2002)
33. Paulin-Mohring, C.: Inductive definitions in the system Coq - rules and properties.
    In: Bezem, M., Groote, J.F. (eds.) TLCA 1993. LNCS, vol. 664, pp. 328–345.
    Springer, Heidelberg (1993)
34. Rahli, V., Bickford, M., Anand, A.: Formal program optimization in Nuprl using
    computational equivalence and partial types. In: ITP 2013, [1], pp. 261–278
35. Setzer, A.: Proof theoretical strength of Martin-Löf Type Theory with W-type and
    one universe. PhD thesis, Ludwig Maximilian University of Munich (1993)
36. I.A.S. The Univalent Foundations Program. Homotopy Type Theory: Univalent
    Foundations of Mathematics. Univalent Foundations (2013)
37. Werner, B.: Sets in types, types in sets. In: Ito, T., Abadi, M. (eds.) TACS 1997.
    LNCS, vol. 1281, pp. 530–546. Springer, Heidelberg (1997)

Implicational Rewriting Tactics in HOL

Vincent Aravantinos[1] and Sofiène Tahar[2]

[1] Software & Systems Engineering, Fortiss GmbH,
Guerickestraße 25, 80805, Munich, Germany
vincent.aravantinos@fortiss.org
http://www.fortiss.org/en
[2] Electrical and Computer Engineering Dept., Concordia University,
1455 De Maisonneuve Blvd. W., Montreal, Canada
tahar@ece.concordia.ca
http://hvg.ece.concordia.ca

Abstract. Reducing the distance between informal and formal proofs in interactive theorem proving is a long-standing matter. An approach to this general topic is to increase automation in theorem provers: indeed, automation turns many small formal steps into one big step. In spite of the usual automation methods, there are still many situations where the user has to provide some information manually, whereas this information could be derived from the context. In this paper, we characterize some very common use cases where such situations happen, and identify some general patterns behind them. We then provide solutions to deal with these situations automatically, which we implemented as HOL Light and HOL4 tactics. We find these tactics to be extremely useful in practice, both for their automation and for the feedback they provide to the user.

1 Introduction

Interactive theorem proving has well-known benefits: it allows to build a formal proof with the help of a computer ensuring the proof is correct. It does not have the restrictions of automated theorem proving since it can appeal to the user's creativity through interaction. But this interaction is also the shortcoming of interactive theorem proving: the user is forced to make explicit some formal steps that are obvious to a human. Thus, automation of these steps, when possible, relieves the user from many tedious manipulations. These complex manipulations are one of the essential reasons why interactive theorem proving is not so popular in practical applications. Thus automation is a key ingredient to bring formal reasoning to a wider audience by making it closer to intuitive reasoning.

In interactive theorem proving, the main tool for automation of reasoning is *decision or semi-decision procedures* [18,25]: e.g., for propositional reasoning [10], linear arithmetic reasoning [8], reasoning modulo various theories [5], or even first-order [12] or higher-order reasoning [16]. These have been an independent subject of research for many years with several of the corresponding progresses being transfered to interactive theorem proving, either by direct implementation or by call to external tools [24]. But decision procedures are useful only to conclude goals: the user must resort to interaction with the theorem prover if the goal is too complex to be proven by a decision procedure (which is the case of most goals). (S)he still has access however to another kind of automation: rewriting.

G. Klein and R. Gamboa (Eds.): ITP 2014, LNAI 8558, pp. 45–60, 2014.

Rewriting enables automation of equality reasoning: given a theorem stating an equality $t = u$ and a term v which contains a subterm matching t, rewriting replaces this subterm by the corresponding instantiation of u (we refer to [3,21] for details). An extremely useful generalization is *conditional rewriting* [9] (often called "simplification" in the HOL Light [14] and HOL4 [29] communities): given a theorem whose statement has the form $p \Rightarrow l = r$, conditional rewriting replaces, in a term t, any subterm matching l, *if the condition p can be proven automatically for the corresponding instantiation*. (Conditional) rewriting may not prove goals as complex as the ones proven by some decision procedure, but it can be used at any step of a proof, even if it does not terminate this proof: it simply allows to *make progress* in the proof, which is an extremely useful feature in an interactive context. However, (conditional) rewriting still requires regularly some user explicit input could be automated as we show now.

Example 1. Consider the following goal (for clarity, we use mathematical notations instead of HOL Light ASCII text to represent mathematical expressions): $\forall x, y, z.$ prime $y \wedge xy > z \wedge x - y = 5 \wedge x^2 - y^2 = z^2 \wedge 0 < z \wedge 0 < y \Rightarrow \frac{x}{x}y = y$ where prime y indicates that y is a prime number.

Assume that the immediate objective of the user is to rewrite $\frac{x}{x}$ into 1. To do so, (s)he calls the rewriting tactic with the theorem $\vdash \forall x.\ x \neq 0 \Rightarrow \frac{x}{x} = 1$. This of course does not work since this theorem is not purely equational: one must here use *conditional* rewriting in order to get rid of the condition $x \neq 0$. However which theorems should be provided here in order to prove $x \neq 0$ automatically? In this goal, this follows from the fact that $x.y > z$ and $0 < z$: therefore $x.y > 0$; thus $x.y \neq 0$, and hence $x \neq 0$. This proof is not mathematically difficult, however it requires a lot of thought from the user before being able to come up with the tactic call which will accomplish the intended action, since (s)he basically needs to mentally build a formal proof. Once this is done, one can apply conditional rewriting with the following theorems: $\vdash \forall x.\ x \neq 0 \Rightarrow \frac{x}{x} = 1$, $\vdash \forall x,y.\ x > y \Leftrightarrow y < x$, $\vdash \forall x,y.\ x < y \wedge y < z \Rightarrow x < z$, $\vdash \forall x.\ 0 < x \Rightarrow 0 \neq x$, $\vdash \forall x,y.\ x.y < 0 \wedge 0 < y \Rightarrow 0 < x$.

The whole process is extremely tedious. Even more embarrassing: this process is about building mentally a formal proof whereas helping such a task is precisely what an interactive theorem prover is made for! Therefore, there can of course be mistakes in this proof, or omission of some intermediate theorems when calling the tactic. In addition, this is a situation where the user is forced to interrupt the flow of his/her proof in order to adapt this proof to the tool at use: that is precisely the sort of situation leading a user to conclude that interactive theorem proving is counter-intuitive, tedious to use, and therefore to maybe give up on using it. Finally, notice that the simplifier might also simply not be able to deal with the sort of reasoning involved in the proof. In all these cases, *nothing* happens: the tactic does not apply any change to the goal and the user is left with no clue which of the above flaws is the reason for the lack of progress.

To avoid this, a simpler and more frequently used approach is to simply assert $x \neq 0$, and prove it as an independent subgoal. This allows to get rid of all the above flaws: proving this condition is done under the control of the theorem prover, which helps the user with the goal and tactic mechanism, thus providing some useful feedback while avoiding any mistake in the formal proof. In addition, one can use complex reasoning that is out of reach for the conditional rewriter of HOL Light or HOL4. However, this approach forces the user to write *manually* the subgoal $x \neq 0$. Manually writing explicit information in a script is extremely fragile with respect to proof change: if ever x is renamed in the original goal, then the proof script has to be updated; similarly if an earlier modification changes the situation into the exact same one, but with a non-trivial expression instead of x; in other cases, some change of prior definitions, or the removal of some assumptions might all as well lead to necessary updates of the subgoal. Finally, all these updates are even more tedious when the subgoal is big (also entailing more potential typing errors), which happens frequently in real-life situations.[1]

In this paper we target precisely this sort of problem. We characterize frequent situations presenting similar problems and provide solutions to them. This includes the problem mentioned in the above example and others, more or less frequent where the user also has to explicitly write some information which could be derived automatically by the theorem prover. Note that the forms of reasoning which we address in this paper are actually simple: they are much simpler than any decision procedures. But it is precisely because they are simple, that it is necessary and useful to automatize them.

The rest of the paper is organized as follows: Section 2 presents our solution to the problem of Example 1. Section 3 describes the underlying algorithm. Section 4 proposes several refinements to our solution. Section 5 presents solutions to other situations presenting the same sort of proof-engineering defects. Finally Section 6 discusses the related work and Section 7 concludes the paper.

This work is entirely implemented in HOL Light and HOL4. The HOL Light version has been integrated in the official distribution of HOL Light and the sources for HOL4 can be publicly found at [1]. Both implementations come with a manual providing technical details to use the tactics.

2 Implicational Rewriting

This section deals precisely with the situation presented in Example 1. Terms and substitutions are defined as usual, with $t\sigma$ denoting the application of the substitution σ to the term t. For terms t, u, v, the notation $t[u/v]$ denotes the term obtained from t by replacing all occurrences of v by u. A *formula* is any term of type boolean. For a formula of the form $\phi \wedge \psi$ (resp. $\neg\phi$, $\forall x.\phi$) we say that ϕ and ψ (resp. ϕ, resp. ϕ) are *direct subformulas* of the formula, and similarly for the other connectives and quantifier. The *subformulas* of a formula are defined by transitive closure of the relation "is a direct subformula". An *atomic subformula* is a subformula whose head symbol is not a logical connective or quantifier. Given a formula ϕ, a subformula ψ occurs *positively* (resp. *negatively*) in ϕ if it occurs in the scope of an even (resp. odd) number of negations or implication premisses[2]. The fact of occurring positively or negatively is called the *polarity* of the subformula.

Consider again the Example 1. In this example, implicational rewriting consists in replacing in the conclusion the atom $x \neq 0 \wedge \frac{x}{x}y = y$ (and only this atom, not the top formula) by $1.y = y$, i.e., $\frac{x}{x}$ is replaced by 1, and the conjunction with $x \neq 0$ is added to the atom. In case this atom was occurring negatively in the goal (e.g., the same goal but with $\frac{x}{x}y \neq y$) then $\frac{x}{x}y \neq y$ would have been replaced by $\neg(x \neq 0 \Rightarrow \frac{x}{x}y = y)$. Formally, this is generalized as follows:

---

[1] We do not claim that *every* explicitly-written subgoal is a bad practice w.r.t. proof engineering: many subgoals provide important high-level information which is out of reach for automation. However, the problem here is that *this subgoal could be automatically generated*. So the user should not be left with the burden of writing it.

[2] For simplicity, we do not consider formulas with equivalences, even though it is easily handled in practice, e.g., by rewriting $\phi \Leftrightarrow \psi$ into $(\phi \Rightarrow \psi) \wedge (\psi \Rightarrow \phi)$.

Definition 1 (Implicational Rewriting). *Consider a goal with conclusion c. Then, we call* implicational rewriting *by th any tactic replacing one or more atomic subformula A of c by:*

$$p\sigma_1 \wedge \cdots \wedge p\sigma_k \vartriangleright A[r\sigma_1/l\sigma_1]\ldots[r\sigma_k/l\sigma_k] \tag{1}$$

where $\sigma_1, \ldots, \sigma_k$ *are the matching substitutions of some subterms of A matching l and where* $\vartriangleright = \wedge$ *(resp.* $\Rightarrow$*) if A occurs positively (resp. negatively) in c.*

Note that the definition leaves a lot of freedom about the strategy to use for the rewrite: not all subterms need to be rewritten, and we do not specify which subterms should be rewritten in priority. In the following, $\mathcal{IR}_{th}(c)$ denotes some function implementing the specification of Definition 1.

The first property of $\mathcal{IR}$ is that *it replaces indeed some term matching l by the corresponding instantiation of r*. But the tactic would be useless if it was not sound:

Theorem 1 (Soundness). *For every goal of conclusion c and every theorem th, it holds that* $\mathcal{IR}_{th}(c) \Rightarrow c$.

Contrarily to (conditional) rewriting, the resulting goal only entails the initial goal, but is not equivalent to it: this can be seen both as a weakness (one needs to backtrack if the result becomes not provable anymore) or as a strength (many more possible inferences are accessible for reasoning). Note that the case distinction about the polarity in Definition 1 is capital for this theorem to hold.

An essential property of implicational rewriting, as opposed to conditional rewriting, is that it provides *feedback* to the user about the condition that is required to be proven: where the user is left with a simply unchanged goal when conditional rewriting cannot prove the side condition, implicational rewriting provides instead the precise condition instantiation which has to be proven. In addition, since this condition now appears in the goal, the the theorem prover can be used to prove it, exactly as if the condition had been stated explicitly as a subgoal by the user.

Note that there exists an easier solution to the problem presented in Example 1: given a goal having a subterm matching l, replace $l\sigma$ by $r\sigma$ and introduce automatically a new *subgoal* stating $p\sigma$. This is the approach of [30] in HOL Light, or of the `force` function of ACL2. It is also very similar to [15] in HOL4: the only difference with the latter is that $p\sigma$ is added as a conjunction *on top of the overall formula* (and not at the level of the atom as implicational rewriting does) instead of being stated as a separate subgoal. Similar tactics are also available in Coq and Isabelle. This approach will be called *dependent rewriting* in the following, according to [15]. Table 1 sums up the different approaches (with only one rewriting for presentational reasons), where $g \rightsquigarrow g'$ denotes a tactic turning a goal g into g', an expression $\phi[t]$ in g means that t occurs in ϕ, then the expression $\phi[t']$ in g' denotes ϕ in which this occurrence is replaced by t'.

So the major difference between implicational and dependent rewriting is the fact that the latter applies *deeply*. We argue now that this is not a cosmetic feature but actually has a high impact on the *compositionality* of the tactic.

Example 2. Consider a goal $g : \forall x, y.\ P\ x\ y \Rightarrow \frac{x}{x} * y = y$, where $P\ x\ y$ is a big expression entailing in particular that $x \neq 0$. With implicational rewriting, using the theorem $\vdash \forall x.\ x \neq 0 \Rightarrow \frac{x}{x} = 1$, we obtain immediately the goal: $\forall x, y.\ P\ x\ y \Rightarrow x \neq 0 \wedge 1 * y = y$. Instead, with *dependent* rewriting, the tactic will try to replace the goal by $g' : x \neq 0 \wedge \forall x, y.\ P\ x\ y \Rightarrow 1 * y = y$. But this does not work since $x \neq 0$ is not in the scope of $\forall x$. Therefore $g' \not\Rightarrow g$ and the tactic application is not valid. Consequently, with dependent rewriting, one must first

Table 1. Definitions of the different sorts of rewriting

	Definition	Condition
Rewriting	$c[l\sigma] \rightsquigarrow c[r\sigma]$	if $\vdash l = r$
Conditional rewriting	$c[l\sigma] \rightsquigarrow c[r\sigma]$	if $\vdash p \Rightarrow l = r$ and $\vdash p$
Dependent rewriting	$c[l\sigma] \rightsquigarrow p\sigma \wedge c[r\sigma]$	if $\vdash p \Rightarrow l = r$
Implicational rewriting	$c[A[l\sigma]] \rightsquigarrow c[p\sigma \wedge / \Rightarrow A[r\sigma]]$	if A occurs pos./neg. in c

remove the quantifiers using the adequate tactics (GEN_TAC in HOL4 and HOL Light), and then only can apply dependent rewriting. We then obtain the goal $x \neq 0 \wedge (P\ x\ y \Rightarrow 1 * y = y)$. However, this is still not satisfying because the new goal is not provable: indeed $x \neq 0$ derives from $P\ x\ y$, but since $x \neq 0$ is not "in the scope" of $P\ x\ y$, it cannot be proven. Therefore, even though the tactic is valid, it is of no help for the proof. Consequently, one must first discharge the hypothesis $P\ x\ y$ in the assumptions before applying the dependent rewrite, yielding finally the goal $x \neq 0 \wedge 1.y = y$.

As the example demonstrates, a lot of book-keeping manipulations are necessary with dependent rewriting but not with implicational rewriting. But it gets even worse when one starts to try proving $x \neq 0$: if P is complex, it will also itself probably require some rewrites, however this is not possible in a simple way since $P\ x\ y$ is now in the assumptions. So a first option is to put the assumption back in the goal (tedious when goals have many assumptions), which actually amounts to manually doing what implicational rewriting does automatically.

A second option is to rethink the flow of the proof and give up on using dependent rewriting. This is what happens most commonly in practice: one will try, from the initial goal, to find a proof of $x \neq 0$ from $P\ x\ y$, apply the corresponding tactics, and finally use conditional rewriting. Note in addition that the proof of $x \neq 0$ must be done in forward reasoning since $x \neq 0$ does not appear in the goal: this makes the process even harder since interactive provers do not emphasize this type of reasoning. This is unless the user decides to set manually the subgoal $x \neq 0$ with the flaws already mentioned in Example 1. As we can see, all the intended benefits of using dependent rewriting are lost whatever is the chosen option and we get back precisely to the situation that was described in Example 1: the user is forced to rethink his/her proof against his/her original intuition; (s)he cannot use automation and has to input data manually, with all the proof-engineering problems already mentioned.

Even though this is a toy example, it is representative of an *extremely frequent* situation when using dependent rewrite. Actually this tactic seems to be seldom used in HOL4, maybe showing that these flaws prevent it from being useful in practice. Table 3 sums up the advantages of the different approaches.

3   Implementation

Let th be a theorem of the form $\vdash \forall x_1, \ldots, x_k.\ p \Rightarrow l = r$. In this section, we provide an implementation of implicational rewriting by th, called IR. This implementation requires the following steps:

1. go through the atoms of the goal, keeping track of their polarity;
2. for each atom, go through its subterms;
3. for each subterm t_i matching l with substitution σ_i, replace it by $r\sigma_i$ while keeping track of the matching substitution σ_i;

Table 2. Pros & cons of the different sorts of rewriting

	replaces $l\sigma$ by $r\sigma$	conditional equations	no need to prove the condition	compositionality
Rewriting	✓			
Cond. rewr.	✓	✓		
Dep. rewr.	✓	✓	✓	
Imp. rewr.	✓	✓	✓	✓

4. reconstitute the atom with the replaced subterms;
5. add the conjunction $p\sigma_1 \wedge p\sigma_2 \wedge \ldots$;
6. reconstitute the complete goal.

In addition, to obtain a valid tactic, the process should not just generate a formula ϕ, but also a *proof* that this formula entails the conclusion of the initial goal c. To do so, we actually generate a theorem $\vdash \phi \Rightarrow c$ (which holds if the implementation satisfies the specification of Definition 1, by Theorem 1).

Steps 2, 3, and 4 are implemented by a function IRC_{th} (for Implicational Rewriting Conversion) which takes an atom A as input and returns a theorem $p\sigma_1, \ldots, p\sigma_k \vdash A = A[r\sigma_1/l\sigma_1] \ldots [r\sigma_k/l\sigma_k]$[3]: IRC_{th} is a recursive function defined by case analysis on the structure of A:

$$\text{IRC}_{th}(t) \stackrel{\text{def}}{=} \begin{cases} \dfrac{\vdash p \Rightarrow l = r}{p\sigma \vdash l\sigma = r\sigma} & \begin{array}{l} \text{if } t \text{ has the form } l\sigma \\ \text{for some substitution } \sigma \end{array} \\[2ex] \dfrac{}{\vdash t = t} & \text{if } t \text{ is a variable or a constant} \\[2ex] \dfrac{\Gamma \vdash u = v}{\Gamma \vdash t = \lambda x.v} & \begin{array}{l} \text{if } t \text{ has the form } \lambda x.u \\ \text{and } \text{IRC}_{th}(u) = \Gamma \vdash u = v \end{array} \\[2ex] \dfrac{\Gamma_1 \vdash t_1 = u_1 \quad \Gamma_2 \vdash t_2 = u_2}{\Gamma_1 \cup \Gamma_2 \vdash t_1 t_2 = u_1 u_2} & \begin{array}{l} \text{if } t \text{ has the form } t_1 t_2, \\ \text{IRC}_{th}(t_1) = \Gamma_1 \vdash t_1 = u_1 \\ \text{and } \text{IRC}_{th}(t_2) = \Gamma_2 \vdash t_2 = u_2 \end{array} \end{cases}$$

We impose that when the first rule and another can be applied, the first rule always has priority, and, if necessary, that th is renamed in order to avoid captures (i.e., so as not to contain any variable which is bound in the rewritten term). Since IRC_{th} must return a theorem, we present not only the resulting theorem but also the proof that allows to obtain it: for instance, the third rule means that $\text{IRC}_{th}(\lambda x.u)$ calls first $\text{IRC}_{th}(u)$; this recursive call must have the form $\Gamma \vdash u = v$, from which can thus be deduced $\Gamma \vdash \lambda x.u = \lambda x.v$. Note that, in the end, the function only returns this latter theorem, i.e., the result of the inference and not the inference itself, contrarily to what the definition of IRCmight suggest: this is done only for explanation purposes. The provided inferences are intended to give a clue to the reader about the way to obtain the result, but do not correspond to some precise inference rule: however all of them can be easily implemented using functions provided by both HOL Light and HOL4.

IRC_{th} can actually be seen as a *usual* rewriting by th. In practice, we therefore make use of *conversions* [23] to implement these steps (i.e., functions which, given a term t returns a theorem of the form $\vdash t = u$).

---

[3] Note that, in HOL, equality among booleans is just the same as equivalence.

Step 5 is then easily obtained from this result by propositional reasoning. The corresponding function is called AIR (for Atomic Implicational Rewriting) and exists both in a positive form AIR^+ and in a negative form AIR^-: it is up to the function which calls AIR to determine the adequate form according to the context (positive or negative atom). [4]

$$\text{AIR}^+_{th}(A) \stackrel{\text{def}}{=} \frac{\Gamma \vdash A = A'}{\vdash (\bigwedge_{\phi \in \Gamma} \phi) \wedge A' \Rightarrow A} \qquad \text{AIR}^-_{th}(A) \stackrel{\text{def}}{=} \frac{\Gamma \vdash A = A'}{\vdash A \Rightarrow ((\bigwedge_{\phi \in \Gamma} \phi) \Rightarrow A')}$$

where $\Gamma \vdash A = A'$ be the result of $\text{IRC}_{th}(A)$.

Finally, steps 1 and 6 are achieved by $\text{IR}^\pi_{th}(\phi)$, where ϕ is the intended conclusion of the goal to be implicationally rewritten and $\pi \in \{+, -\}$ is a polarity:

$$\text{IR}^\pi_{th}(\phi) \stackrel{\text{def}}{=} \begin{cases} \text{AIR}^\pi(\phi) & \text{if } \phi \text{ is atomic} \\[2ex] \dfrac{\vdash \phi_1 \Rightarrow \psi_1}{\vdash \neg\psi_1 \Rightarrow \neg\phi_1} & \begin{array}{l} \text{if } \pi = + \text{ (resp. } -), \\ \phi \text{ is a negation } \neg\phi_1 \text{ (resp. } \neg\psi_1), \\ \text{and } \text{IR}^{\overline{\pi}}_{th}(\phi_1)(\text{resp. } \psi_1) = \vdash \phi_1 \Rightarrow \psi_1 \end{array} \\[3ex] \dfrac{\vdash \psi_1 \Rightarrow \phi_1 \quad \vdash \psi_2 \Rightarrow \phi_2}{\vdash \psi_1 \wedge \psi_2 \Rightarrow \phi_1 \wedge \phi_2} & \begin{array}{l} \text{if } \pi = + \text{ (resp. } -), \\ \phi \text{ is a conjunction } \phi_1 \wedge \phi_2 \text{ (resp. } \psi_1 \wedge \psi_2), \\ \text{IR}^\pi_{th}(\phi_1)(\text{resp. } \psi_1) = \vdash \psi_1 \Rightarrow \phi_1, \\ \text{and } \text{IR}^\pi_{th}(\phi_2)(\text{resp. } \psi_2) = \vdash \psi_2 \Rightarrow \phi_2 \end{array} \\[3ex] \dfrac{\vdash \psi_1 \Rightarrow \phi_1}{\vdash \forall\psi_1 \Rightarrow \forall\phi_1} & \begin{array}{l} \text{if } \pi = + \text{ (resp. } -), \\ \phi \text{ is a quantifed formula } \forall\phi_1 \text{ (resp. } \forall\psi_1), \\ \text{and } \text{IR}^\pi_{th}(\phi_1)(\text{resp. } \psi_1) = \vdash \psi_1 \Rightarrow \phi_1 \end{array} \\[3ex] \dfrac{\vdash \phi_1 \Rightarrow \psi_1 \quad \vdash \psi_2 \Rightarrow \phi_2}{\vdash (\psi_1 \Rightarrow \psi_2) \Rightarrow (\phi_1 \Rightarrow \phi_2)} & \begin{array}{l} \text{if } \pi = + \text{ (resp. } -), \\ \phi \text{ is an implic. } \phi_1 \Rightarrow \phi_2 \text{ (resp. } \psi_1 \Rightarrow \psi_2), \\ \text{IR}^{\overline{\pi}}_{th}(\phi_1)(\text{resp. } \psi_1) = \vdash \phi_1 \Rightarrow \psi_1, \\ \text{and } \text{IR}^\pi_{th}(\phi_2)(\text{resp. } \psi_2) = \vdash \psi_2 \Rightarrow \phi_2 \end{array} \end{cases}$$

Disjunction is handled like conjunction, simply replacing $\vee$ by $\wedge$, and exist. quantification like univ. quantification, replacing $\forall$ by $\exists$.

where $\overline{\pi}$ is defined as $\overline{+} \stackrel{\text{def}}{=} -$ and $\overline{-} \stackrel{\text{def}}{=} +$. As in the definition of IRC, not only we give the conclusion of the resulting theorem but also the (big-step) inference rule used to derive this theorem from the recursive calls. We call IRC "implicational conversions" because the rules are very similar to conversions, except that implication is used instead of equality.

At the top-level, only the positive polarity is used: in the end, IR^+_{th} returns a theorem of the form $\vdash \phi' \Rightarrow \phi$ where ϕ' is the implicationally rewritten version of ϕ. So, given a goal of conclusion c, one can call IR^+_c and apply the Modus Ponens tactic – i.e., the tactic which, given a goal of conclusion c and a theorem $c' \Rightarrow c$, turns c into c' – to the result. We can prove that this tactic indeed implements implicational rewriting:

---

[4] In practice, a distinction has to be made between the assumptions introduced by IRC and the assumptions that come from the original goal. This is easy to achieve but not presented for readability.

Theorem 2. *For every theorem th and every formula ϕ, the tactic consisting in applying Modus Ponens to* $\mathrm{IR}_{th}^{+}(\phi)$ *implements implicational rewriting by th.*

4   Refinements

In this section, we investigate a few refinements of implicational rewriting.

4.1   Theorems Introducing Variables

A usual problem with rewriting is how to handle theorems that introduce new variables when applying the rewrite, i.e., theorems of the form $\vdash l = r$, where r contains variables not occurring in l. Consider for instance the rewriting of the term $y + 0$ by the theorem $\vdash 0 = x - x$: since x does not appear in $y + 0$, what sense would it have to replace 0 by $x - x$? And what if the original term indeed contains x, e.g., if we rewrite $x + 0$ instead of $y + 0$? Shall the rewrite replace 0 by $x - x$ or rename x to avoid a possibly unintended capture? This problem is usually considered to be rare enough that it is not worth considering: since x does not occur in the original term, the user simply should avoid rewriting with these theorems. However, it can make sense to apply such a rewrite, e.g., it can be useful to replace 0 by $x - x$ if, in the context of the proof, one wants to apply a theorem about substractions.

With implicational rewriting, this situation is even more frequent since new variables can also come from the condition of the theorem:

Example 3. Consider for example that one wants to prove $\forall s, \cdot, x, y, z, t.\ P \Rightarrow ((x \cdot y) \cdot z) \cdot t = x \cdot (y \cdot (z \cdot t))$, where P is an irrelevant premise. We assume that P entails in particular the predicate $group\,(s, \cdot)$, which states the usual group axioms for the operation $\cdot$ over s. A first step is to (implicationally) rewrite the goal with the associativity theorem $\vdash \forall g, op, x, y, z.\ group\,(g, op) \wedge x \in g \wedge y \in g \wedge z \in g \Rightarrow op\,(op\,x\,y)\,z = op\,(x(op\,y\,z))$. This yields: $\forall s, \cdot, x, y, z, t.\ P \Rightarrow group\,(g, \cdot) \wedge x \in g \wedge y \in g \wedge z \in g \wedge (x \cdot (y \cdot z)) \cdot t = x \cdot (y \cdot (z \cdot t))$. Here, the new variable g has been introduced in the goal whereas it does not have any meaning there. As a consequence, the goal is not provable anymore.

What happens is that matching $op\,(op\,x\,y)\,z$ (the l.h.s. of the associativiy theorem) indeed provides instantiations for op, x, y and z, but it does not provide any instantiation for g. But in practice, the user usually knows the instantiation of g, or will find it out later on in the course of the proof. Therefore, a satisfying solution would apply the rewrite but would still leave to the user the possibility to instantiate g manually. We achieve this by detecting automatically variables that are introduced by the rewrite, then applying the rewrite, and finally *quantifying existentially over the introduced variables*. In the above example, one would obtain the goal: $\forall s, \cdot, x, y, z, t.\ P \Rightarrow \exists g.\ group\,(g, \cdot) \wedge x \in g \wedge y \in g \wedge z \in g \wedge (x \cdot (y \cdot z)) \cdot t = x \cdot (y \cdot (z \cdot t))$. In our first example, $y + 0$ would be replaced by $\exists x.\ y + (x - x)$. This solution allows to maintain the provability of the goal, while preserving the advantages of (implicational) rewriting.

Formally, this consists simply in replacing the expression (1) of Definition 1 by $\exists x_1, \ldots, x_k.\ p\tau\sigma_1 \wedge \cdots \wedge p\tau\sigma_k \triangleright A[r\tau\sigma_1/l\sigma_1] \ldots [r\tau\sigma_k/l\sigma_k]$, where $x_1, \ldots, x_k$ denote all the variables introduced by the theorem $\vdash p\tau \Rightarrow l\tau = r\tau$ (i.e., formally, variables occurring in $r\tau$ and $p\tau$ but not in $l\tau$) and where τ is a renaming substitution used to avoid potential captures between the variables of the goal and the variables that occur in r and p but not in l. The proof of Theorem 1 still applies: only the base case of the induction slightly changes.

This refinement is implemented by modifying IRC so that the introduced variables are renamed to avoid possible captures; then by modifying AIR so that the function adds the required existential quantifications over these newly introduced variables. Note that the quantification's scope is only the premise (resp. conclusion) of the resulting theorem in the positive (resp. negative) case.

4.2   Preprocessing and Postprocessing

Preprocessing the input theorems to allow more theorems than just those of the form $\forall x_1, \ldots, x_k. \, p \Rightarrow l = r$ is a simple way to allow many improvements. For instance, purely equational theorems $\forall x_1, \ldots, x_k. \, l = r$ can be turned into $\forall x_1, \ldots, x_k. \, \top \Rightarrow l = r$, which entails that implicational rewrite can also be used as a substitute to standard rewriting.

In addition, some further preprocessing allows to accept theorems of the form $p \Rightarrow c$ (resp. $p \Rightarrow \neg c$) by turning them into $p \Rightarrow c = \top$ (resp. $p \Rightarrow c = \bot$). Note that this is commonly used in HOL4 and HOL Light for usual rewriting. For implicational rewriting, it also means that it can be used as a substitute for the (matching) Modus Ponens tactic, i.e., the tactic which, given a theorem of the form $\forall x_1, \ldots, x_k. \, p \Rightarrow c$ and a goal of the form $c\sigma$ for some substitution σ, generates the goal $p\sigma$. Indeed, in such a situation, implicational rewriting turns the goal $c\sigma$ into $\top$ (thanks to the described preprocessing) and adds the conjunct $p\sigma$. We thus obtain the goal $p\sigma \wedge \top$ which is equivalent to what is obtained by the matching Modus Ponens. However, implicational rewriting is then even more powerful than this tactic since it is able to apply this sort of reasoning *deeply* (the lack of this feature is a common criticism).

However, this latter solution is a little bit unsatisfying since it yields $p\sigma \wedge \top$ instead of the expected $p\sigma$. Of course it is trivial to get rid of $\top$ here, but it would obviously be better to achieve this automatically. This is easily done by maintaining a set of basic rewriting theorems containing commonly used propositional facts like $\forall p. \, p \wedge \top \Leftrightarrow p$, or $\top \wedge \top \Leftrightarrow \top$. Again, similar solutions are used for rewriting in HOL4 and HOL Light. In order to do this rewriting on the fly, implicational rewriting must be adapted to be able to rewrite with several theorems. Since the current definitions already handle multiple rewrites with the same theorem, it is trivial to extend them to deal with several theorems simply by considering a set of theorems instead of just one.

Finally, theorems of the form $\forall x_1, \ldots, x_k. \, p \Rightarrow \forall y_1, \ldots, y_n. \, l = r$ (i.e., additional quantifiers appear before the equation) can also be handled at no cost (all definitions adapt trivially).

4.3   Taking the Context into Account

Example 4. Consider the goal $\forall x, y. \, x \neq 0 \Rightarrow \frac{x}{x} * y = y$. Applying implicational rewriting with $\vdash \forall x. \, x \neq 0 \Rightarrow \frac{x}{x} = 1$ yields the goal $\forall x, y. \, x \neq 0 \Rightarrow x \neq 0 \wedge 1 * y = y$. The context obviously entails the inner $x \neq 0$, but one still needs additional manual reasoning to obtain $\forall x, y. \, x \neq 0 \Rightarrow 1 * y = y$.

Therefore a further refinement is to modify implicational rewriting so that the context is taken into account. This can be handled by adding the contextual hypotheses to the set of usable theorems, while going down the theorem. This means that, e.g., in the positive case of the implicational rule of IR's definition (fourth case), the formula ϕ_1 is added to the set of theorems that can be used by the recursive call to compute ψ_2 from ϕ_2. Similar treatments can be applied in the negative case and for conjunction. This is overall similar to what is done in usual rewriting to handle the context, see, e.g., [23] for a sketch of these ideas. However these additions require a particular care in order, in particular, to avoid recomputing the same contexts several times, see. Many solutions exist, see, e.g., the source codes of HOL4, HOL Light or Isabelle. Note that this allows to get rid of many of the introduced conditions automatically, exactly like conditional rewrite does. Therefore implicational rewriting can also often be used as a replacement for conditional rewriting.

In the end, all these refinements (including efficient use of the context) are implemented in one single tactic called IMP_REWRITE_TAC: this tactic takes a list of theorems and applies implicational rewriting with all of them repeatedly until no more application is possible. Note that all these refinements are not just small user-friendly improvements: they also improve again the compositionality of the tactic by allowing to chain seamlessly implicational rewrites of many theorems. Since, as shown in the above refinements, this tactic subsumes rewriting, conditional rewriting, and Modus Ponens tactics, it integrates very well in the usual tactic-style proving workflow: one can use just one tactic to cover all the other cases, plus the ones that were not covered before. In practice, this combination happens to be *extremely powerful*: many proof steps can be turned into only one call to this tactic with several theorems. For instance, the tactic has been extensively used for months to develop in particular the work presented in [19,20,27]. In particular, the library presented in [20] was completely rewritten using implicational rewriting, which showed a dramatic reduction of its code size. In addition the time taken to prove new theorems was also much reduced due to the relevant feedback provided by the tactic.

The main improvements we foresee for the above refinement process are regarding performance. This could probably benefit from all the optimizations that already exist for usual rewriting, e.g., [22]. Another easy possible refinement would be to allow implicational rewriting in the assumptions instead of the conclusion of the goal. This should be easily achieved simply by considering the reverse implication.

5   Other Interaction-Intensive Situations

In this section, we tackle the automation of situations that present the same loopholes as the ones that motivated our development of implicational rewriting, i.e., situations where the user has to input manually some information that could be computed automatically by the theorem prover, leading to fragility of proof scripts and tediousness of the interaction.

5.1   Contextual Existential Instantiation

Consider a slight variation of Example 3:

Example 5. Let be the goal $\forall s, \cdot, x, y, z, t.\ group\,(s, \cdot) \Rightarrow ((x{\cdot}y){\cdot}z){\cdot}t = x{\cdot}(y{\cdot}(z{\cdot}t))$. Applying implicational rewriting now yields: $\forall s, \cdot, x, y, z, t.\ group\,(s, \cdot) \Rightarrow \exists sg.\ group\,(sg, \cdot) \wedge (x \cdot (y \cdot z)) \cdot t = x \cdot (y \cdot (z \cdot t))$. The usual solution is then to strip the quantifiers and discharge $group\,(s, \cdot)$ to obtain $\exists sg.\ group\,(sg, \cdot) \wedge (x \cdot (y \cdot z)) \cdot t = x \cdot (y \cdot (z \cdot t))$ as a conclusion. Then one can provide explicitly the witness s for sg in order to obtain: $group\,(s, \cdot) \wedge (x \cdot (y \cdot z)) \cdot t = x \cdot (y \cdot (z \cdot t))$.

Here, the user must provide explicitly a witness. In this example, the witness is just the one character s, but in other situations it can be big complex terms. Therefore, the same reproaches can be applied to this situation as those that gave raise to implicational rewriting (fragility of the script, tediousness for the user, etc.). However, once again, it is a situation where the context could be used by the theorem prover to find the witness by itself: it is easy to go through all the assumptions and find an atom which matches an atom of the conclusion (e.g., in the above example, $group\,(s, \cdot)$ matches $group\,(sg, \cdot)$).

One can even instantiate existential quantifications that appear *deep* in the goal instead of just as the top connective. Then, not only the assumptions, but also the context of the formula can be used. This was implemented in a tactic called HINT_EXISTS_TAC, of which a preliminary version has already been integrated into HOL4. As shown in the above example, this sort of situation can also happen when using implicational rewriting, therefore this tactic is also integrated transparently in IMP_REWRITE_TAC.

Note that the underlying algorithm shares some common structure with the one of implicational rewriting. More precisely, IR can actually be reused only changing the base call to AIR. For space reasons, we refer to the source code in [1] for the details of the implementation.

5.2 Cases Rewrite

Implicational rewriting can be seen as a situation where the user has a at his/her disposal a theorem $\vdash \forall x_1, \ldots, x_k.\ p \Rightarrow c$ *and is ready to accept whatever it takes* to make use of the information provided by c. Since this cannot be done at no cost, implicational rewriting accepts to do it, at the condition to add the necessary instantiation of p. When one uses implicational rewriting, one makes the underlying assumption that (s)he will have the ability to prove p later on.

But, sometimes, we do not want to make such a strong assumption; instead, we want to split the proof by considering both what happens if p holds and what happens if p does not hold. In such cases, the usual solution is to explictly use a case-split tactic to consider two branches of the proof: one where p holds, and one where $\neg p$ holds. But, once again, the user has to explicitly state some information which is possibly verbose, fragile and tedious, when the prover could do the same automatically by retrieving the relevant information from the theorem $\vdash \forall x_1, \ldots, x_k.\ p \Rightarrow c$.

This yields *cases rewriting* which requires also a theorem of the form $\vdash \forall x_1, \ldots, x_k.\ p \Rightarrow l = r$ (or $\vdash \forall x_1, \ldots, x_k.\ p \Rightarrow c$ with preprocessing) and looks for an atom A with a subterm matching l (say with substitution σ). However, unlike implicational rewriting, it does not replace A by $p\sigma \wedge A[r\sigma/l\sigma]$ or $p\sigma \Rightarrow A[r\sigma/l\sigma]$, but rather by $(p\sigma \Rightarrow A[r\sigma/l\sigma]) \wedge (\neg p\sigma \Rightarrow A)$. This was implemented in the tactic CASES_REWRITE_TAC. We refer to the manual of [1] for more details.

5.3 Target Rewrite

Example 6 ([2]). Consider the goal $\forall n, m.\ SUC\ n \leq SUC\ m \Rightarrow n \leq m$. Assume we already proved that the predecessor function is monotonic: $\vdash \forall n, m.\ n \leq m \Rightarrow PRE\ n \leq PRE\ m$ and that the predecessor is the left inverse of the successor: $\vdash \forall n.\ PRE\,(SUC\ n) = n$. A natural proof would start by replacing $n \leq m$ by $PRE\,(SUC\ n) \leq PRE\,(SUC\ m)$. But rewriting with $\vdash \forall n.\ n = PRE\,(SUC\ n)$ will obviously not terminate. So we should rewrite the goal only once (as allowed by the HOL4 and HOL Light tactic ONCE_REWRITE_TAC). But this rewrites everywhere, yielding: $\forall n, m.\ PRE\ (SUC\ (SUC\ n)) \le PRE\ (SUC\ (SUC\ m))$ $\Rightarrow PRE\ (SUC\ n) \le PRE\ (SUC\ m)$. Instead, we have to use a special tactic allowing to state precisely that we want to rewrite only in the conclusion of the implication, and only once (e.g., GEN_REWRITE_TAC in HOL Light and HOL4), or more elaborate solutions like [13] in Coq). This requires to provide explicitly to the prover which part of the goal has to be rewritten: so, exactly as in the previous situations, the user has to provide explicitly some information which is very unintuitive and very dependent on the current shape of the goal. Thus, once again, this solution is both fragile and tedious to the user.

In this example, the important information is actually not the *location* of the rewrite, but the *objective* that the user has in mind. And this objective is to rewrite the goal *in order to use* the theorem $\vdash \forall n, m.\ n \le m \Rightarrow PRE\ n \le PRE\ m$. Generally, when one wants to apply a precisely located rewrite with a theorem, the underlying objective is to get a goal which allows the use of *another* theorem. So we define *target rewriting* which takes *two theorems* as input: the one used for the rewrite (called *supporting* theorem) and the one that we intend to use after the rewrite (called *target* theorem). Very naively, the tactic simply explores all the possible rewrites of the goal using the supporting theorem until one of these rewrites yields a term which can be rewritten by the target theorem.

Example 7. In Example 6, the supporting theorem is $\vdash \forall n.\ n = PRE\ (SUC\ n)$ and the target theorem is $\vdash \forall n, m.\ n \le m \Rightarrow PRE\ n \le PRE\ m$. The list of all possible 1-step rewrites explored by the tactic is the following:

1 $\forall n, m.\ PRE\ (SUC\ (SUC\ n)) \le SUC\ m \Rightarrow n \le m$
2 $\forall n, m.\ SUC\ n \le PRE\ (SUC\ (SUC\ m)) \Rightarrow n \le m$
3 $\forall n, m.\ SUC\ n \le SUC\ m \Rightarrow PRE\ (SUC\ n) \le m$
4 $\forall n, m.\ SUC\ n \le SUC\ m \Rightarrow n \le PRE\ (SUC\ m)$

Then the list of possible 2-step rewrites is enumerated as follows:

1 $\forall n, m.\ PRE\ (SUC\ (SUC\ n)) \le PRE\ (SUC\ (SUC\ m)) \Rightarrow n \le m$
2 $\forall n, m.\ PRE\ (SUC\ (SUC\ n)) \le SUC\ m \Rightarrow PRE\ (SUC\ n) \le m$
3 $\forall n, m.\ PRE\ (SUC\ (SUC\ n)) \le SUC\ m \Rightarrow n \le PRE\ (SUC\ m)$
4 $\forall n, m.\ SUC\ n \le PRE\ (SUC\ (SUC\ m)) \Rightarrow PRE\ (SUC\ n) \le m$
5 $\forall n, m.\ SUC\ n \le PRE\ (SUC\ (SUC\ m)) \Rightarrow n \le PRE\ (SUC\ m)$
6 $\forall n, m.\ SUC\ n \le SUC\ m \Rightarrow PRE\ (SUC\ n) \le PRE\ (SUC\ m)$

Here the tactic stops because the last rewrite allows to apply the target theorem.

Because of the exhaustive enumeration, this tactic can of course be extremely costly. Still, in the numerous practical cases where its execution time is reasonable, it is tremendously helpful since the user does not have to provide explicit information anymore: the only additional effort is to step back and look a step ahead in the intended proof to know which theorem shall be used afterwards. Many concrete situations happen to match this use case pattern. For instance, associativity-commutativity (AC) rewriting [28] is a particular case of it:

Example 8. Consider the goal $a + (b * c + -a) + a = b * c + a$ [13]. A standard proof of this goal is to first rearrange the innermost addition into $a + (-a + b * c) + a = b * c + a$ by carefully using the commutativity of addition and then using the theorem $\vdash \forall x\ y.\ x + (-x + y) = y$ to conclude. Instead, one can just use target rewriting with the AC of addition as the supporting theorem and $\vdash \forall x\ y.\ x + (-x + y) = y$ as the target theorem.

We called this tactic TARGET_REWRITE_TAC. It is actually able to take several supporting theorems as input (though, of course, more supporting theorems means more possibilities to explore, and thus a bigger execution time). In addition, it actually does not work with rewriting but with implicational rewriting, which allows to use implicational theorems as supporting theorems.

6 Related Work

In proof theory, applying reasoning *deep* in a goal is the precise focus of deep inference [7]. At first sight, it can hardly be said that implicational rewriting shares more than a conceptual relation to deep inference though, but some of the benefits can be seen as similar since in both cases the fact of reasoning deep allows to get rid of some "bureaucratic" manipulations. If considering sets of clauses instead of arbitrary formulas, implicational rewriting turns out to be very close to *superposition* [4]. Since formulas are normalized, superposition does not need to consider the polarity of the formula where it applies, therefore it actually corresponds only to the negative case of implicational rewriting. Studying this connection in detail to improve our implementation is part of future work.

As already mentioned, the implementation of IR can be seen as a particular form of rewriting where implication is used instead of equality. This is very similar to the "consequence conversions" in HOL4 [33], and more generally can be seen as a particular case of rewriting with preorders [17,32], where the used preorders are both $\Rightarrow$ and $\Leftarrow$ used in an interleaved way. Taking the context into account builds on top of several implementations serving similar ideas. Conceptually, one can find many connections with "window inference" [26].

Target rewriting is very close to the "smart matching" tactic of Matita [2]: this tactic uses as supporting theorems the whole database of already proven equational theorems and the target theorem is provided explicitly like in our approach. It uses a sophisticated implementation of superposition instead of our naive approach which just enumerates all possible rewrites, thus making it much more efficient. However it uses only equational theorems as supporting theorems, whereas target rewriting also accepts implicational theorems: this was particularly useful in, e.g., [20], where most theorems are prefixed by assumptions. In addition, smart matching only tries to match the top goal, whereas target rewriting also works deeply, with the same advantages as for implicational rewriting: no need to use book-keeping tactics, and therefore more compositionality. This is made possible precisely because we use implicational rewriting instead of matching Modus Ponens as is done for smart matching. However, this has of course a much bigger impact on the performance. The perfect solution would probably lie in between: using superposition to make target rewriting more efficient, or extending smart matching with ideas of target rewriting.

Conceptually, both smart matching and target rewriting are connected to *deduction modulo* [11] since they are essentially about making a distinction between the "important" steps of a proof and the ones that are just "glue" between the important steps: in our context, this glue is the use of supporting theorems; in the context of smart application, it is the use of the available equational knowledge base; and in deduction modulo, it is "calculation" steps as opposed to deduction steps. However, to the best of our knowledge, there is no tool similar to target rewriting or smart matching making use of deduction modulo.

Finally, as explained earlier, AC-rewriting can be seen as a special case of target rewriting. Many works have been devoted to this, e.g., in HOL90 [28], or more recently in Coq [6].

The advantage of these works is that they are of course more efficient, since they deal with a very special case. But they are much less general, and therefore not as useful as target rewriting.

7 Conclusion

We presented in this paper some tactics to reduce human interaction in interactive theorem provers. Their objective is not to provide the automation of intricate reasoning that the user could not achieve by him/herself, but rather to assist him/her in some quite simple but frequent reasoning tasks. The most important of our tactics is implicational rewriting, whose core idea was presented in detail. We argued how a big advantage of it is that it allows for a better composionality thus making it extremely useful in practice. We presented an implementation of implicational rewriting as well as some refinements improving further its usefulness. Finally we covered a few other tactics pursuing similar objectives of reducing the tediousness of human interaction and the fragility of proof scripts. The objective of this latter aspect is also to improve human interaction, but in a longer term perspective: when proof scripts are robust to change, they make the development of theories easier, and thus improve the user experience.

In practice, many proofs are surprisingly sequences of calls to IMP_REWRITE_ TAC interleaved with a few calls to TARGET_REWRITE_TAC, and only rarely other tactics. Of course, in cases where human creativity is really required, some subgoals or lemmas are set, but this is to be expected when reasoning in higher-order logic. Apart from these cases, one can observe that these tactics serve the purpose they were designed for: many reasoning tasks which are simple for a human become simple with the theorem prover. *In the end, the only reproach which can be made to this approach is that removing most explicitly set subgoals reduces the readability of the proof scripts.* We argue instead that the purpose of a proof script is not to mimic usual mathematical proofs: the information that can be found automatically should be found out by the machine. Tools like Proviola [31] can still be used in order to provide some valuable feedback to the user.

Acknowledgements. The author thanks L. Liu, M. Y. Mahmoud and G. Helali from Concordia University for their feedback after extensive use of this work, as well as N. Peltier from the Laboratory of Informatics of Grenoble, and the contributors of the hol-info mailing list for fruitful discussions.

References

1. Aravantinos, V.: Implicational Conversions for HOL Light and HOL4 (2013), https://github.com/aravantv/impconv, https://github.com/aravantv/impconv/HOL4-impconv (respectively)
2. Asperti, A., Tassi, E.: Smart Matching. In: Autexier, S., Calmet, J., Delahaye, D., Ion, P.D.F., Rideau, L., Rioboo, R., Sexton, A.P. (eds.) AISC/Calculemus/MKM 2010. LNCS (LNAI), vol. 6167, pp. 263–277. Springer, Heidelberg (2010)
3. Baader, F., Nipkow, T.: Term Rewriting and All That. Cambridge University Press (1999)
4. Bachmair, L., Ganzinger, H.: Rewrite-Based Equational Theorem Proving with Selection and Simplification. Journal of Logical Computation 4(3), 217–247 (1994)
5. Barrett, C.W., Sebastiani, R., Seshia, S.A., Tinelli, C.: Satisfiability Modulo Theories. In: Handbook of Satisfiability. Frontiers in Artificial Intelligence and Applications, vol. 185, pp. 825–885. IOS Press (2009)

6. Braibant, T., Pous, D.: Tactics for Reasoning Modulo AC in Coq. In: Jouannaud, J.-P., Shao, Z. (eds.) CPP 2011. LNCS, vol. 7086, pp. 167–182. Springer, Heidelberg (2011)

7. Brünnler, K., Tiu, A.F.: A Local System for Classical Logic. In: Nieuwenhuis, R., Voronkov, A. (eds.) LPAR 2001. LNCS (LNAI), vol. 2250, pp. 347–361. Springer, Heidelberg (2001)

8. Cooper, D.C.: Theorem Proving in Arithmetic Without Multiplication. Machine Intelligence 7, 91–99 (1972)

9. Brand, D., Darringer, J., Joyner, W.: Completeness of Conditional Reductions. Research Report RC-7404, IBM (1978)

10. Davis, M., Logemann, G., Loveland, D.W.: A Machine Program for Theorem-Proving. Communications of the ACM 5(7), 394–397 (1962)

11. Dowek, G., Hardin, T., Kirchner, C.: Theorem Proving Modulo. Journal of Automated Reasoning 31(1), 33–72 (2003)

12. Fitting, M.: First-Order Logic and Automated Theorem Proving. Texts and Monographs in Computer Science. Springer (1990)

13. Gonthier, G., Tassi, E.: A Language of Patterns for Subterm Selection. In: Beringer, L., Felty, A. (eds.) ITP 2012. LNCS, vol. 7406, pp. 361–376. Springer, Heidelberg (2012)

14. Harrison, J.: The HOL Light System Reference (2013), http://www.cl.cam.ac.uk/~jrh13/hol-light/reference.html

15. Homeier, P.V.: HOL4 source code (2002), http://ww.src/1/dep_rewrite.sml

16. Huet, G.P.: A Mechanization of Type Theory. In: International Joint Conference on Artificial Intelligence, pp. 139–146. William Kaufmann (1973)

17. Inverardi, P.: Rewriting for preorder relations. In: Lindenstrauss, N., Dershowitz, N. (eds.) CTRS 1994. LNCS, vol. 968, pp. 223–234. Springer, Heidelberg (1995)

18. Kroening, D., Strichman, O.: Decision Procedures: An Algorithmic Point of View. Texts in Theoretical Computer Science. An EATCS Series. Springer (2008)

19. Liu, L., Hasan, O., Aravantinos, V., Tahar, S.: Formal Reasoning about Classified Markov Chains in HOL. In: Blazy, S., Paulin-Mohring, C., Pichardie, D. (eds.) ITP 2013. LNCS, vol. 7998, pp. 295–310. Springer, Heidelberg (2013)

20. Mahmoud, M.Y., Aravantinos, V., Tahar, S.: Formalization of Infinite Dimension Linear Spaces with Application to Quantum Theory. In: Brat, G., Rungta, N., Venet, A. (eds.) NFM 2013. LNCS, vol. 7871, pp. 413–427. Springer, Heidelberg (2013)

21. Mayr, R., Nipkow, T.: Higher-Order Rewrite Systems and Their Confluence. Theoretical Computer Science 192(1), 3–29 (1998)

22. Norrish, M.: Rewriting Conversions Implemented with Continuations. Journal of Automated Reasoning 43(3), 305–336 (2009)

23. Paulson, L.C.: A Higher-Order Implementation of Rewriting. Science of Computer Programming 3(2), 119–149 (1983)

24. Paulson, L.C., Blanchette, J.C.: Three Years of Experience with Sledgehammer, a Practical Link Between Automatic and Interactive Theorem Provers. In: IWIL@LPAR. EPiC Series, vol. 2, pp. 1–11 (2010)

25. Robinson, J.A., Voronkov, A. (eds.): Handbook of Automated Reasoning. Elsevier and MIT Press (2001)

26. Robinson, P.J., Staples, J.: Formalizing a Hierarchical Structure of Practical Mathematical Reasoning. Journal of Logical Computation 3(1), 47–61 (1993)

27. Siddique, U., Aravantinos, V., Tahar, S.: Formal Stability Analysis of Optical Resonators. In: Brat, G., Rungta, N., Venet, A. (eds.) NFM 2013. LNCS, vol. 7871, pp. 368–382. Springer, Heidelberg (2013)

28. Slind, K.: AC Unification in HOL90. In: Joyce, J.J., Seger, C.-J.H. (eds.) HUG 1993. LNCS, vol. 780, pp. 436–449. Springer, Heidelberg (1994)

29. Slind, K., Norrish, M.: A Brief Overview of HOL4. In: Mohamed, O.A., Muñoz, C., Tahar, S. (eds.) TPHOLs 2008. LNCS, vol. 5170, pp. 28–32. Springer, Heidelberg (2008)
30. Solovyev, A.: SSReflect/HOL Light manual. Flyspeck project (2012)
31. Tankink, C., Geuvers, H., McKinna, J., Wiedijk, F.: Proviola: A Tool for Proof Re-animation. In: Autexier, S., Calmet, J., Delahaye, D., Ion, P.D.F., Rideau, L., Rioboo, R., Sexton, A.P. (eds.) AISC/Calculemus/MKM 2010. LNCS (LNAI), vol. 6167, pp. 440–454. Springer, Heidelberg (2010)
32. Türk, T.: HOL4 source code (2006), http://src/simp/src/congLib.sml
33. Türk, T.: HOL4 source code (2008), http://src/1/ConseqConv.sml

A Heuristic Prover for Real Inequalities

Jeremy Avigad, Robert Y. Lewis, and Cody Roux

Carnegie Mellon University, Pittsburgh, PA 15213, USA

Abstract. We describe a general method for verifying inequalities between real-valued expressions, especially the kinds of straightforward inferences that arise in interactive theorem proving. In contrast to approaches that aim to be complete with respect to a particular language or class of formulas, our method establishes claims that require heterogeneous forms of reasoning, relying on a Nelson-Oppen-style architecture in which special-purpose modules collaborate and share information. The framework is thus modular and extensible. A prototype implementation shows that the method is promising, complementing techniques that are used by contemporary interactive provers.

1 Introduction

Comparing measurements is fundamental to the sciences, and so it is not surprising that ordering, bounding, and optimizing real-valued expressions is central to mathematics. A host of computational methods have been developed to support such reasoning, using symbolic or numeric methods, or both. For example, there are well-developed methods of determining the satisfiability or unsatisfiability of linear inequalities [31,32], polynomial inequalities [6], nonlinear inequalities involving functions that can be approximated numerically [17], [25], and inequalities involving convex functions [9]. The "satisfiability modulo theories" framework [5], [27], provides one way of integrating such methods with ordinary logical reasoning and proof search; integration with resolution theorem proving methods has also been explored [1], [30]. Interactive theorem provers like Isabelle [28] and HOL Light [19] now incorporate various such methods, either constructing correctness proofs along the way, or reconstructing them from appropriate certificates. (For a small sample, see [8], [10], [20], [23].) Such systems provide powerful tools to support interactive theorem proving. But, frustratingly, they often fail when it comes to fairly routine calculations, leaving users to carry out explicit calculations painstakingly by hand. Consider, for example, the following valid implication:

$$0 < x < y, \ u < v \ \Rightarrow \ 2u + exp(1 + x + x^4) < 2v + exp(1 + y + y^4)$$

The inference is not contained in linear arithmetic or even the theory of real-closed fields. The inference is tight, so symbolic or numeric approximations to the exponential function are of no use. Backchaining using monotonicity properties of addition, multiplication, and exponentiation might suggest reducing the goal

G. Klein and R. Gamboa (Eds.): ITP 2014, LNAI 8558, pp. 61–76, 2014.
© Springer International Publishing Switzerland 2014

to subgoals $2u < 2v$ and $exp(1 + x + x^4) < exp(1 + y + y^4)$, but this introduces some unsettling nondeterminism. After all, one could just as well reduce the goal to

- $2u < exp(1 + y + y^4)$ and $exp(1 + x + x^4) < 2v$, or
- $2u + exp(1 + x + x^4) < 2v$ and $0 < exp(1 + y + y^4)$, or even
- $2u < 2v + 7$ and $exp(1 + x + x^4) < exp(1 + y + y^4) - 7$.

And yet, the inference is entirely straightforward. With the hypothesis $u < v$ in mind, you probably noticed right away that the terms $2u$ and $2v$ can be compared; similarly, the comparison between x and y leads to comparisons between x^4 and y^4, then $1 + x + x^4$ and $1 + y + y^4$, and so on.

The method we propose is based on such heuristically guided forward reasoning, using properties of addition, multiplication, and the function symbols involved. As is common for resolution theorem proving, we try to establish the theorem above by negating the conclusion and deriving a contradiction. We then proceed as follows:

- Put all terms involved into a canonical normal form. This enables us to recognize terms that are the same up to a scalar multiple, and up to associativity and commutativity of addition and multiplication.
- Iteratively call specialized modules to learn new comparisons between subterms, and add these new comparisons to a common "blackboard" structure, which can be accessed by all modules.

The theorem is verified when any given module derives a contradiction using this common information. The procedure fails, on the other hand, when none of the modules can learn anything new. We will see in Section 6 that the method is far from complete, and may not even terminate. On the other hand, it is flexible and extensible, and easily verifies a number of inferences that are not obtained using more principled methods. As a result, it provides a useful complement to more conventional approaches.

We have thus far designed and implemented modules to learn comparisons from the additive and multiplicative structure of terms, as well as a module to instantiate axioms involving general functions. The additive and multiplicative modules have two different implementations, with different characteristic strengths and weaknesses. The first uses a natural but naive Fourier-Motzkin elimination, and the second uses more refined geometric techniques. Our prototype implementation, written in Python, is available online:

https://github.com/avigad/polya

We have named the system "Polya," after George Pólya, in recognition of his work on inequalities as well as his thoughtful studies of heuristic methods in mathematics (e.g. [18], [29]).

The general idea of deriving inequalities by putting terms in a normal form and combining specialized modules is found in Avigad and Friedman [3], which examines what happens when the additive and multiplicative fragments of real arithmetic are combined. This is analogous to the situation handled by SMT solvers, with the added twist that the languages in question share more than just the equality symbol. Avigad and Friedman show that the universal fragment remains decidable even if both theories include multiplication by rational constants, while the full first-order theory is undecidable. The former decidability result, however, is entirely impractical, for reasons discussed there. Rather, it is the general framework for combining decision procedures and the use of canonical normal forms that we make use of here.

2   The Framework

2.1   Terms and Canonical Forms

We wish to consider terms, such as $3(5x + 3y + 4xy)^2 f(u + v)^{-1}$, that are built up from variables and rational constants using addition, multiplication, integer powers, and function application. To account for the associativity of addition and multiplication, we view sums and products as multi-arity rather than binary operations. We account for commutativity by imposing an arbitrary ordering on terms, and ordering the arguments accordingly.

Importantly, we would also like to easily identify the relationship between terms t and t' where $t = c \cdot t'$, for a nonzero rational constant c. For example, we would like to keep track of the fact that $4y + 2x$ is twice $x + 2y$. Towards that end, we distinguish between "terms" and "scaled terms": a scaled term is just an expression of the form $c \cdot t$, where t is a term and c is a rational constant. We refer to "scaled terms" as "s-terms" for brevity.

Definition 1. *We define the set of* terms $\mathcal{T}$ *and* s-terms $\mathcal{S}$ *by mutual recursion:*

$$t, t_i \in \mathcal{T} \quad := \quad 1 \mid x \mid \textstyle\sum_i s_i \mid \prod_i t_i^{n_i} \mid f(s_1, \ldots, s_n)$$
$$s, s_i \in \mathcal{S} \quad := \quad c \cdot t.$$

Here x ranges over a set of variables, *f ranges over a set of* function symbols, *$c \in \mathbb{Q}$, and $n_i \in \mathbb{Z}$.*

Thus we view $3(5x + 3y + 4xy)^2 f(u + v)^{-1}$ as an s-term of the form $3 \cdot t$, where t is the product $t_1^2 t_2^{-1}$, t_1 is a sum of three s-terms, and t_2 is the result of applying f to the single s-term $1 \cdot (u + v)$.

Note that there is an ambiguity, in that we can also view the coefficient 3 as the s-term $3 \cdot 1$. This ambiguity will be eliminated when we define a notion of *normal form* for terms. The notion extends to s-terms: an s-term is in normal form when it is of the form $c \cdot t$, where t is a term in normal form. (In the special case where $c = 0$, we require t to be the term 1.) We also refer to terms in normal form as *canonical*, and similarly for s-terms.

To define the notion of normal form for terms, we fix an ordering $\prec$ on variables and function symbols, and extend that to an ordering on terms and s-terms. For example, we can arbitrarily set the term 1 to be minimal in the ordering, then variables, then products, then sums, and finally function applications, recursively using lexicographic ordering on the list of arguments (and the function symbol) within the latter three categories. The set of terms in normal form is then defined inductively as follows:

- $1, x, y, z, \ldots$ are terms in normal form.
- $\sum_{i=1\ldots n} c_i \cdot t_i$ is in normal form provided $c_1 = 1$, each t_i is in normal form, and $t_1 \prec t_2 \prec \ldots \prec t_n$.
- $\prod_i t_i^{n_i}$ is in normal form provided each t_i is in normal form, and $1 \neq t_1 \prec t_2 \prec \ldots \prec t_n$.
- $f(s_1, \ldots, s_n)$ is in normal form if each s_i is.

The details are spelled out in Avigad and Friedman [3]. That paper provides an explicit first-order theory, T, expressing commutativity and associativity of addition and multiplication, distributivity of constants over sums, and so on, such that the following two properties hold:

1. For every term t, there is a unique s-term s in canonical form, such that T proves $t = s$.
2. Two terms t_1 and t_2 have the same canonical normal form if and only if T proves $t_1 = t_2$.

For example, the term $3(5x + 3y + 4xy)^2 f(u + v)^{-1}$ is expressed canonically as $75 \cdot (x + (3/5) \cdot y + (4/5) \cdot xy)^2 f(u + v)^{-1}$, where the constant in the additive term $5x + 3y + 4xy$ has been factored so that the result is in normal form.

The two clauses above provide an axiomatic characterization of what it means for terms to have the same canonical form. As discussed in Section 7, extending the reach of our methods requires extending the notion of a canonical form to include additional common operations.

2.2   The Blackboard

We now turn to the blackboard architecture, which allows modules to share information in a common language. To the addition module, multiplication is a black box; thus it can only make sense of additive information in the shared pool of knowledge. Conversely, the multiplication module cannot make sense of addition. But both modules can make sense of information in the form $t_1 < c \cdot t_2$, where t_1 and t_2 are subterms occurring in the problem. The blackboard enables modules to communicate facts of this shape.

When the user asserts a comparison $t > 0$ to the blackboard, t is first put in canonical form, and names $t_0, t_1, t_2, \ldots$ are introduced for each subterm. It is convenient to assume that t_0 denotes the canonical term 1. Given the example in the last section, the method could go on to define

$$t_1 := x, \quad t_2 := y, \quad t_3 := t_1 t_2, \quad t_4 := t_1 + (3/5) \cdot t_2 + (4/5) \cdot t_3,$$
$$t_5 := u, \quad t_6 := v, \quad t_7 := t_5 + t_6, \quad t_8 = f(t_7), \quad t_9 := t_4^2 t_8^{-1}$$

In that case, $75 \cdot t_9$ represents $3(5x+3y+4xy)^2 f(u+v)^{-1}$. Any subterm common to more than one term is represented by the same name. Separating terms in this way ensures that each module can focus on only those definitions that are meaningful to it, and otherwise treat subterms as uninterpreted constants.

Now any comparison $s \bowtie s'$ between canonical s-terms, where $\bowtie$ denotes any of $<, \leq, >, \geq, =$, or $\neq$, translates to a comparison $c_i t_i \bowtie c_j t_j$, where t_i and t_j name canonical terms. But this, in turn, can always be expressed in one of the following ways:

- $t_i \bowtie 0$ or $t_j \bowtie 0$, or
- $t_i \bowtie c \cdot t_j$, where $c \neq 0$ and $i < j$.

The blackboard therefore maintains the following data:

- a defining equation for each t_i, and
- comparisons between named terms, as above.

Note that this means that, *a priori*, modules can only look for and report comparisons between terms that have been "declared" to the blackboard. This is a central feature of our method: the search is deliberately constrained to focus on a small number of terms of interest. The architecture is flexible enough, however, that modules can heuristically expand that list of terms at any point in the search. For example, our addition and multiplication modules do not consider distributivity of multiplication over addition, beyond multiplication of rational scalars. But if a term $x(y + z)$ appears in the problem, a module could heuristically add the identity $x(y + z) = xy + xz$, adding names for the new terms as needed.

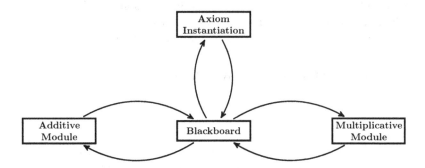

Fig. 1. The Blackboard Architecture

To verify an implication, the user asserts the hypotheses to the blackboard, together with the negation of the conclusion. Individual modules then take turns learning new comparisons from the data, and asserting them to the blackboard as well, until a contradiction is obtained, or no further conclusions can be drawn. The setup is illustrated by Figure 1. Notice that this is essentially the Nelson-Oppen architecture [5], [27], in which (disjoint) theories communicate by means of a shared logical symbol, typically equality. Here, the shared language is instead assumed to contain the list of comparisons $<, \leq, >, \geq, =, \neq$, and multiplication by rational constants.

Now suppose a module asserts an inequality like $t_3 < 4t_5$ to the blackboard. It is the task of the central blackboard module to check whether the assertion provides new information, and, if so, to update its database accordingly. The task is not entirely straightforward: for example, the blackboard may already contain the inequality $t_3 < 2t_5$, but absent sign information on t_3 or t_5, this does not imply $t_3 < 4t_5$, nor does the converse hold. However, if the blackboard includes the inequalities $t_3 < 2t_5$ and $t_3 \leq 7t_5$, the new assertion is redundant. If, instead, the blackboard includes the inequalities $t_3 < 2t_5$ and $t_3 \leq 3t_5$, the new inequality should replace the second of these. A moment's reflection shows that at most two such inequalities need to be stored for each pair t_i and t_j (geometrically, each represents a half-plane through the origin), but comparisons between t_i or t_j and 0 should be counted among these.

There are additional subtleties: a weak inequality such as $t_3 \leq 4t_5$ paired with a disequality $t_3 \neq 4t_5$ results in a strong inequality; a pair of weak inequalities $t_3 \leq 4t_5$ and $t_3 \geq 4t_5$ should be replaced by an equality; and, conversely, a new equality can subsume previously known inequalities.

3   Fourier-Motzkin

The Fourier-Motzkin algorithm is a quantifier-elimination procedure for the theory of the structure $\langle \mathbb{R}, 0, +, < \rangle$, that is, the real numbers as an additive ordered group. Nothing changes essentially if we add to the language of that theory the constant 1 and scalar multiplication by c, for each rational c. Here we see that the method can be used to infer comparisons between variables from additive data, and that this can be transported to the multiplicative setting as well.

3.1   The Fourier-Motzkin Additive Module

The Fourier-Motzkin additive module begins with the comparisons $t_i \bowtie c \cdot t_j$ stored in the blackboard, where $\bowtie$ is one of $\leq, <, \geq, >, =$ (disequalities are not used). It also makes use of comparisons $t_i \bowtie 0$, and all definitions $t_i = \sum_j c_j t_{k_j}$ in which the right-hand side is a sum. The goal is to learn new comparisons of the form $t_i \bowtie c \cdot t_j$ or $t_i \bowtie 0$. The idea is simple: to learn comparisons between t_i and t_j, we need only eliminate all the other variables. For example, suppose, after substituting equations, we have the following three inequalities:

$$3t_1 + 2t_2 - t_3 > 0$$
$$4t_1 + t_2 + t_3 \geq 0$$
$$2t_1 - t_2 - 2t_3 \geq 0$$

Eliminating t_3 from the first two equations we obtain $7t_1 + 3t_2 > 0$, from which we can conclude $t_1 > (-3/7)t_2$. Eliminating t_3 from the last two equations we obtain $10t_1 + t_2 \geq 0$, from which we can conclude $t_1 \geq (-1/10)t_2$. More generally, eliminating all the variables other than t_i and t_j gives the projection of the convex region determined by the constraints onto the i, j plane, which determines the strongest comparisons for t_i and t_j that are implied by the data.

Constants can be represented using the special variable $t_0 = 1$, which can be treated as any other variable. Thus eliminating all variables except for t_i and t_0 yields all comparisons between t_i and a constant.

The additive module simply carries out the elimination for each pair i, j. In general, Fourier-Motzkin elimination can require doubly-exponential time in the number of variables. With a bit of cleverness, one can use previous eliminations to save some work, but for a problem with n subterms, one is still left with $O(n^2)$-many instances of Fourier-Motzkin with up to n variables in each. It is interesting to note that for the examples described in Section 6, the algorithm performs reasonably well. In Section 4, however, we describe a more efficient approach.

3.2   The Fourier-Motzkin Multiplicative Module

The Fourier-Motzkin multiplication module works analogously: given comparisons $t_i \bowtie c \cdot t_j$ or $t_i \bowtie 0$ and definitions of the form $t_i = \prod_j t_{k_j}^{n_j}$, the module aims to learn comparisons of the first two forms. The use of Fourier-Motzkin here is based on the observation that the structure $\langle \mathbb{R}, 0, +, < \rangle$ is isomorphic to the structure $\langle \mathbb{R}^+, 1, \times, < \rangle$ under the map $x \mapsto e^x$. With some translation, the usual procedure works to eliminate variables in the multiplicative setting as well. In the multiplicative setting, however, several new issues arise.

First, the multiplicative module only makes use of terms t_i which are known to be strictly positive or strictly negative. The multiplicative module thus executes a preprocessing stage which tries to infer new sign information from the available data.

Second, the inequalities that are handled by the multiplicative module are different from those handled by the additive module, in that terms can have a rational coefficient. For example, we may have an inequality $3t_2^2 t_5 > 1$; here, the multiplicative constant 3 would correspond to an additive term of $\log 3$ in the additive procedure. This difference makes it difficult to share code between the additive and multiplicative modules, but the rational coefficients are easy to handle.

Finally, the multiplicative elimination may produce information that cannot be asserted directly to the blackboard, such as a comparison $t_i^2 < 3t_j^2$ or $t_i^3 < 2t_j^2$. In that case, we have to pay careful attention to the signs of t_i and t_j and their relation to ± 1 to determine which facts of the form $t_i \bowtie c \cdot t_j$ can be inferred. We compute exact roots of rational numbers when possible, so a comparison $t_i^2 < 9t_j^2$ translates to $t_i < 3t_j$ when t_i and t_j are known to be positive. As a last resort, faced with a comparison like $t_i^2 < 2t_j^2$, we use a rational approximation of $\sqrt{2}$ to try to salvage useful information.

4   Geometric Methods

Although the Fourier-Motzkin modules perform reasonably well on small problems, they are unlikely to scale well. The problem is that many of the inequalities that are produced when a single variable is eliminated are redundant, or subsumed by the others. Thus, by the end of the elimination, the algorithm may be left with hundreds or thousands of comparisons of the form $t_i \bowtie c \cdot t_j$, for different values of c. Some optimizations are possible, such as using simplex based methods (e.g. [14]) to filter out some of the redundancies. In this section, however, we show how methods of computational geometry can be used to address the problem more directly. On many problems in our test suite, performance is roughly the same. But on some problems of moderate complexity we have found our implementation of the geometric approach to be faster than the Fourier-Motzkin approach by a factor of five, with the most notable improvement occurring in problems where the ratio of learned comparisons to number of variables is high.

4.1   The Geometric Additive Module

Geometric methods provide an alternative perspective on the task of eliminating variables. A linear inequality $c \le \sum_{i=1}^{k} c_i \cdot t_i$ determines a half-space in $\mathbb{R}^{k+1}$; when $c = 0$, as in the homogenized inequalities in our current problem, the defining hyperplane of the half-space contains the origin. A set of n homogeneous inequalities determines an unbounded pyramidal polyhedron in $\mathbb{R}^k$ with vertex at the origin, called a "polyhedral cone." (Equalities, represented as $(k-1)$-dimensional hyperplanes, simply reduce the dimension of the polyhedron.) The points inside this polyhedron represent solutions to the inequalities. The problem of determining the strongest comparisons between t_i and t_j then reduces to finding the maximal and minimal ratios of the i-th and j-th coordinates of points inside the polyhedron.

We use the following well-known theorem of computational geometry (see [33, Section 1.1]):

Theorem 1. *A subset $P \subseteq \mathbb{R}^k$ is a sum of a convex hull of a finite point set and a conical combination of vectors (a $\mathcal{V}$-polyhedron) if and only if it is an intersection of closed half-spaces (an $\mathcal{H}$-polyhedron).*

A description of a $\mathcal{V}$-polyhedron is said to be a $\mathcal{V}$-*representation* of the polyhedron, and similarly for $\mathcal{H}$-polyhedrons; there are a number of effective methods to convert between representations.

The comparisons and additive equalities stored in the central blackboard essentially describe an $\mathcal{H}$-representation of a polyhedron. After constructing the corresponding $\mathcal{V}$-representation, it is easy to pick out the implied comparisons as follows. For every pair of variables t_i and t_j, project the set of vertices to the $t_i t_j$ plane by setting all the other coordinates to 0. If there is anything to be learned, all (nonzero) vertices must fall in the same halfplane; find the two outermost points (as in Figure 2b) and compute their slopes to the origin. These

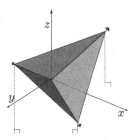
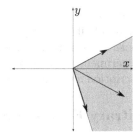

(a) A polyhedral cone in $\mathbb{R}^3$, defined by three half-spaces

(b) Projected to the xy plane, the polyhedron implies $x \geq 2y$ and $x \geq -\frac{1}{3}y$

Fig. 2. Variable elimination by geometric projection

slopes determine the coefficients c in two comparisons $t_i \bowtie c \cdot t_j$, and the relative position of the two vertices determine the inequality symbols in place of $\bowtie$.

We chose to use Avis' *lrs* implementation of the reverse-search algorithm [4] to carry out the geometric computations. Vertex enumeration algorithms typically assume convexity of the polyhedron: that is, all inequalities are taken to be weak. As it is essential for us to distinguish between $>$ and $\geq$, we use a trick taken from Dutertre and de Moura [14, Section 5]. Namely, given a set of strict inequalities $\{0 < \sum_{i=1}^{k} c_i^m \cdot t_i : 0 \leq m \leq n\}$, we introduce a new variable δ with constraints $0 \leq \delta$ and $\{\delta \leq \sum_{i=1}^{k} c_i^m \cdot t_i : 0 \leq m \leq n\}$, and generate the corresponding polyhedron. If, in the vertex representation, every vertex has a zero δ-coordinate, then the inequalities are only satisfiable when $\delta = 0$, which implies that the system with strict inequalities is unsatisfiable. Otherwise, a comparison $t_i \bowtie c \cdot t_j$ is strict if and only if every vertex on the hyperplane $t_i = c \cdot t_j$ has a zero δ coordinate, and weak otherwise.

4.2   The Geometric Multiplicative Module

As with the Fourier-Motzkin method, multiplicative comparisons $1 \leq \prod_{i=1}^{k} t_i^{e_i}$ can be handled in a similar manner, by restricting to terms with known sign information and taking logarithms. Once again, there is a crucial difference from the additive setting: taking the logarithm of a comparison $c \cdot t_i \cdot t_j^{-1} \bowtie 1$ with $c \neq 1$, one is left with an irrational constant $\log c$, and the standard computational methods for vertex enumerations cannot perform exact computations with these terms.

To handle this situation we introduce new variables to represent the logarithms of the prime numbers occurring in these constant terms. Let $p_1, \ldots, p_l$ represent the prime factors of all constant coefficients in such a problem, and for each $1 \leq i \leq l$, let q_i be a variable representing $\log p_i$. We can then rewrite each $c \cdot t_i \cdot t_j^{-1} \bowtie 1$ as $p_1^{d_0} \cdot \ldots \cdot p_l^{d_l} \cdot t_i \cdot t_j^{-1} \bowtie 1$. Taking logarithms of all such inequalities produces a set of additive inequalities in $k + l$ variables.

In order to find the strongest comparisons between t_i and t_j, we can no longer project to the $t_i t_j$ plane, but instead look at the $t_i t_j q_1 \ldots q_l$ hyperplane. The simple arithmetical comparisons to find the two strongest comparisons are no longer applicable; we face the harder problem of converting the vertex representation of a polyhedron to a half-space representation. This problem is dual to the conversion in the opposite direction, and the same computational packages are equipped to solve it. Experimentally, we have found Fukuda's *cdd* implementation of Motzkin's double description method [15] to be faster than *lrs* for this procedure.

5   Arbitrary Function Symbols

The inferences captured by the addition and multiplication modules constitute a fragment of the theory of real-closed fields, roughly, that theory "minus" the distributivity of multiplication over addition [3]. Recall, however, that we have also included arbitrary function symbols in the language. An advantage to our framework is that we do not have to treat function terms as uninterpreted constants; rather, we can seamlessly add modules that (partially) interpret these symbols and learn relevant inequalities concerning them.

To start with, a user may wish to add axioms asserting that a particular function f is nonnegative, monotone, or convex. For example, the following axiom expresses that f is nondecreasing:

$$\forall x, y. \; x \le y \to f(x) \le f(y)$$

Given such an axiom, Polya's function module searches for useful instantiations during the course of a search, and may thus learn useful information.

Specifically, given a list of universal axioms in variables $v_1 \ldots v_n$, the instantiation module searches for relevant assignments $v_i \mapsto c_i \cdot t_{j_i}$, where each c_i is a constant and each t_{j_i} is a subterm in the given problem. Each axiom is then instantiated with these assignments, and added to the central blackboard as a set of disjunctive clauses. As the search progresses, elements of these clauses are refuted; when only one remains, it is added to the blackboard, as a new piece of information available to all the modules.

The task is a variant of the classic matching problem, but there are at least three aspects of our framework that present complications. First, given that we consider terms with a rational scalar multiplicative constant, the algorithm has to determine those values. So, in the example above, x and y can be instantiated to an s-term $c \cdot t_i$ for any c, when such an instantiation provides useful information. Second, we need to take into account the associativity and commutativity of operations like addition and multiplication, so, for example, a term $f(x + y)$ can be unified with a term $f(t_i + t_j + t_k)$ found in the blackboard in multiple ways. Finally, although the framework is built around the idea of restricting attention to subterms occurring in the original problem, at times it is useful to consider new terms. For example, given the axiom

$$\forall x, y. \; f(x + y) \le f(x) + f(y) \; ,$$

it is clearly a good idea to instantiate x and y to t_i and t_j, respectively, whenever $f(t_i + t_j)$, $f(t_i)$, and $f(t_j)$ all appear in the blackboard, even if the term $f(t_i) + f(t_j)$ does not.

In short, we wish to avoid difficult calculations of rational constants, expensive matching up to associativity and commutativity (see e.g. [11]), and unrestrained creation of new terms, while at the same time making use of potentially useful instantiations. The solution we adopted is to use function terms to trigger and constrain the matching process, an idea commonly used by SMT solvers [12] [27]. Given a universal axiom $(\forall v_1 \ldots v_n)F(v_1, \ldots, v_n)$, F is first converted into clausal normal form, and each clause F' is treated separately. We take the *trigger set* of F' to be the set of all functional subterms contained in F. A straightforward unification procedure finds all assignments that map each trigger element to a (constant multiple of a) problem term, and these assignments are used to instantiate the full clause F'. The instantiated clause is asserted to the central blackboard, which checks for satisfied and falsified literals.

For u a term containing unification variables $\{v_i\}$ and σ an assignment mapping $v_i \mapsto c_i \cdot t_{j_i}$, the problem of matching $\sigma(u)$ to a problem term t_j is nontrivial: the matching must be done modulo equalities stored in the blackboard. For example, if $t_1 = t_2 + t_3$, $t_4 = 2t_3 - t_5$, and $t_6 = f(t_1 + t_4)$, then given the assignment $\{v_1 \mapsto t_2 - t_5, v_2 \mapsto 3t_3\}$, the term $u = f(v_1 + v_2)$ should be matched to t_6. We thus combine a standard unification algorithm, which suggests candidate assignments to the variables occurring in an axiom, with Gaussian elimination over additive and multiplicative equations, to find the relevant matching substitutions.

The function module thus provides a general and natural scheme for incorporating axioms. It is flexible enough to integrate other methods for matching axioms and triggering the creation of new problem terms. We expect to achieve better performance for special function symbols, however, by refining the notion of canonical normal form. For example, we can handle the associative-commutative function $\min(c_1 \cdot t_1, \ldots, c_n \cdot t_n)$ by sorting the arguments according to the ordering on terms and scaling so that $c_1 = \pm 1$. Similarly, we can assume that in any subterm $|c \cdot t|$, we have $c = 1$. We intend to explore use for normal forms for arbitrary powers and logarithms, making use of ideas from [26]. Moreover, special-purpose modules can be used to contribute more refined information and inferences. For example, we expect that such a special-purpose module is needed to effectively manage the additive and multiplicative nature of the exponentiation, e.g. under the identity $exp(\sum_i c_i \cdot t_i) = \prod_i exp(t_i)^{c_i}$.

6 Examples

The current distribution of Polya includes a number of examples that are designed to illustrate the method's strengths, as well as some of its weaknesses. For comparison, we verified a number of these examples in Isabelle, trying to use Isabelle's automated tools as much as possible. These include "auto," an internal tableau theorem prover which also invokes a simplifier and arithmetic reasoning methods, and Sledgehammer [24], [8], which heuristically selects a body of facts from the local context and background library, and exports it to various provers. We also sent some of the inferences directly to the SMT solver Z3 [13].

To start with, Polya handles inferences involving linear real inequalities, which are verified automatically by many interactive theorem proving systems. It can also handle purely multiplicative inequalities such as

$$0 < u < v < 1,\ 2 \leq x \leq y \Rightarrow 2u^2 x < vy^2, \tag{1}$$

which are not often handled automatically. It can solve problems that combine the two, like these:

$$x > 1 \Rightarrow (1 + y^2)x > 1 + y^2 \tag{2}$$

$$0 < x < 1 \Rightarrow 1/(1 - x) > 1/(1 - x^2) \tag{3}$$

$$0 < u,\ u < v,\ 0 < z,\ z + 1 < w \Rightarrow (u + v + z)^3 < (u + v + w)^5 \tag{4}$$

It also handles inferences that combine such reasoning with axiomatic properties of functions, such as:

$$(\forall x.\ f(x) \leq 1),\ u < v,\ 0 < w \Rightarrow u + w \cdot f(x) < v + w \tag{5}$$

$$(\forall x, y.\ x \leq y \to f(x) \leq f(y)),\ u < v,\ x < y \Rightarrow u + f(x) < v + f(y) \tag{6}$$

Isabelle's auto and Sledgehammer fail on all of these but (5) and (6), which are proved by resolution theorem provers. Sledgehammer can verify more complicated variants of (5) and (6) by sending them to Z3, but fails on only slightly altered examples, such as:

$$(\forall x.\ f(x) \leq 2),\ u < v,\ 0 < w \Rightarrow u + w \cdot (f(x) - 1) < v + w \tag{7}$$

$$(\forall x, y.\ x \leq y \to f(x) \leq f(y)),\ u < v,\ 1 < v,\ x \leq y \Rightarrow$$
$$u + f(x) \leq v^2 + f(y) \tag{8}$$

$$(\forall x, y.\ x \leq y \to f(x) \leq f(y)),\ u < v,\ 1 < w,\ 2 < s,$$
$$(w + s)/3 < v,\ x \leq y \Rightarrow u + f(x) \leq v^2 + f(y) \tag{9}$$

Z3 gets most of these when called directly, but also fails on (8) and (9). Moreover, when handling nonlinear equations, Z3 "flattens" polynomials, which makes a problem like (4) extremely difficult. It takes Z3 a couple of minutes when the exponents 3 and 5 in that problem are replaced by 9 and 19, respectively. Polya verifies all of these problems in a fraction of a second, and is insensitive to the exponents in (4). It is also unfazed if any of the variables above are replaced by more complex terms.

Polya has no problem with examples such as

$$0 < x < y,\ u < v \Rightarrow 2u + exp(1 + x + x^4) < 2v + exp(1 + y + y^4), \tag{10}$$

mentioned in the introduction. Sledgehammer verifies this using resolution, and slightly more complicated examples by calling Z3 with the monotonicity of *exp*. Sledgehammer restricts Z3 to linear arithmetic so that it can reconstruct proofs in Isabelle, so to verify (10) it provides Z3 with the monotonicity of the power function as well. When called directly on this problem with this same information, however, Z3 resorts to nonlinear mode, and fails.

Sledgehammer fails on an example that arose in connection with a formalization of the Prime Number Theorem, discussed in [2]:

$$0 \le n, \ n < (K/2)x, \ 0 < C, \ 0 < \varepsilon < 1 \ \Rightarrow \ \left(1 + \frac{\varepsilon}{3(C+3)}\right) \cdot n < Kx \quad (11)$$

Z3 verifies it when called directly. Sledgehammer also fails on these [3]:

$$0 < x < y \ \Rightarrow \ (1 + x^2)/(2 + y)^{17} < (1 + y^2)/(2 + x)^{10} \quad (12)$$

$$(\forall x, y. \ x < y \rightarrow exp(x) < exp(y)),$$
$$0 < x < y \ \Rightarrow \ (1 + x^2)/(2 + exp(y)) \ge (2 + y^2)/(1 + exp(x)) \ . \quad (13)$$

Z3 gets (12) but not (13). Neither Sledgehammer nor Z3 get these:

$$(\forall x, y. \ f(x + y) = f(x)f(y)), \ a > 2, \ b > 2 \ \Rightarrow \ f(a + b) > 4 \quad (14)$$

$$(\forall x, y. \ f(x + y) = f(x)f(y)), \ a + b > 2, \ c + d > 2 \ \Rightarrow \ f(a + b + c + d) > 4 \quad (15)$$

Polya verifies all of the above easily.

Let us consider two examples that have come up in recent Isabelle formalizations by the first author. Billingsley [7, page 334] shows that if f is any function from a measure space to the real numbers, the set of continuity points of f is Borel. Formalizing the proof involved verifying the following inequality:

$$i \ge 0, \ |f(y) - f(x)| < 1/(2(i + 1)),$$
$$|f(z) - f(y)| < 1/(2(i + 1)) \ \Rightarrow \ |f(x) - f(y)| < 1/(i + 1) \ . \quad (16)$$

Sledgehammer and Z3 fail on this, while Polya verifies it given only the triangle inequality for the absolute value.

The second example involves the construction of a sequence $f(m)$ in an interval (a, b) with the property that for every $m > 0$, $f(m) < a + (b - a)/m$. The proof required showing that $f(m)$ approaches a from the right, in the sense that for every $x > a$, $f(m) < x$ for m sufficiently large. A little calculation shows that $m \ge (b - a)/(x - a)$ is sufficient. We can implicitly restrict the domain of f to the integers by considering only arguments $\lceil m \rceil$; thus the required inference is

$$(\forall m. \ m > 0 \rightarrow f(\lceil m \rceil) < a + (b - a)/\lceil m \rceil),$$
$$a < b, \ x > a, \ m \ge (b - a)/(x - a) \ \Rightarrow \ f(\lceil m \rceil) < x \ . \quad (17)$$

Sledgehammer and Z3 do not capture this inference, and the Isabelle formalization was tedious. Polya verifies it immediately using only the information that $\lceil x \rceil \ge x$ for every x.

When restricted to problems involving linear arithmetic and axioms for function symbols, the behavior of Z3 and Polya is similar, although Z3 is vastly more efficient. As the examples above show, Polya's advantages show up in problems that combine multiplicative properties with either linear arithmetic or axioms. In particular, Z3 procedures for handling nonlinear problems do not incorporate axioms for function symbols.

Of course, Polya fails on wide classes of problems where other methods succeed. It is much less efficient than the best linear solvers, for example, and so should not be expected to scale to large industrial problems. Because the multiplicative module only takes advantage of equations where the signs of all terms are known, Polya fails disappointingly on the trivial inference

$$x > 0, \ y < z \ \Rightarrow \ xy < xz \ . \tag{18}$$

But the problem is easily solved given a mechanism for splitting on the signs of y and z (see the discussion in the next section). Another shortcoming, in contrast to methods which begin by flattening polynomials, is that Polya does not, *a priori*, make use of distributivity at all, beyond the distributivity of multiplication by a rational constant over addition. Any reasonable theorem prover for the theory of real closed fields can easily establish

$$x^2 + 2x + 1 \geq 0, \tag{19}$$

which can also be obtained simply by writing the left-hand side as $(x+1)^2$. But, as pointed out by Avigad and Friedman [3], the method implemented by Polya is, in fact, nonterminating on this example.

We also tried a number of these problems with MetiTarski [1], which combines resolution theorem proving with procedures for real-closed fields as well as symbolic approximations to transcendental function. We found that MetiTarski does well on problems in the language of real-closed fields, but not with axioms for interpreted functions, nor with the examples with *exp* above.

For problems like these, time constraints are not a serious problem. Polya solves a suite of 51 problems, including all the ones above, in about 2 seconds on an ordinary desktop using the polytope packages, and in about 5.5 seconds using Fourier-Motzkin. Test files for Isabelle, Z3, and MetiTarski, as well as more precise benchmark results, can be found in the distribution.

7   Conclusions and Future Work

One advantage of the method described here is that it should not be difficult to generate proof certificates that can be verified independently and used to construct formal derivations within client theorem provers. For procedures using real closed fields, this is much more difficult; see [23] [20].

Another interesting heuristic method, implemented in ACL2, is described in [21]. We have not carried out a detailed comparison, but the method is considerably different from ours. (For example, it flattens polynomial terms.)

We envision numerous extensions to our method. One possibility is to implement case splitting and conflict-driven clause learning (CDCL) search, as do contemporary SMT solvers. For example, recall that the multiplicative routines only work insofar as the signs of subterms are known. It is often advantageous, therefore, to split on the signs on subterms. More generally, we need to implement mechanisms for backtracking, and also make the addition and multiplication modules incremental.

There are many ways our implementation could be optimized, and, of course, we would gain efficiency by moving from Python to a compiled language like C++. We find it encouraging, however, that even our unoptimized prototype performs well on interesting examples. It seems to us to be more important, therefore, to explore extensions of these methods, and try to capture wider classes of inequalities. This includes reasoning with exponents and logarithms; reasoning about the integers as a subset of the reals; reasoning about common functions, such as trigonometric functions; and heuristically allowing other natural moves in the search, such as flattening or factoring polynomials, when helpful. We would also like to handle second-order operators like integrals and sums, and integrate better with external theorem proving methods.

We emphasize again that this method is not designed to replace conventional methods for proving linear and nonlinear inequalities, which are typically much more powerful and efficient in their intended domains of application. Rather, our method is intended to complement these, capturing natural but heterogeneous patterns of reasoning that would otherwise fall through the cracks. What makes the method so promising is that it is open-ended and extensible. Additional experimentation is needed to determine how well the method scales and where the hard limitations lie.

Acknowledgment. We are grateful to Leonardo de Moura and the anonymous referees for helpful corrections, information, and suggestions.

References

1. Akbarpour, B., Paulson, L.C.: MetiTarski: An Automatic Prover for the Elementary Functions. In: Autexier, S., Campbell, J., Rubio, J., Sorge, V., Suzuki, M., Wiedijk, F. (eds.) AISC/Calculemus/MKM 2008. LNCS (LNAI), vol. 5144, pp. 217–231. Springer, Heidelberg (2008)
2. Avigad, J., Donnelly, K., Gray, D., Raff, P.: A formally verified proof of the prime number theorem. ACM Trans. Comput. Logic 9(1), 2 (2007)
3. Avigad, J., Friedman, H.: Combining decision procedures for the reals. Log. Methods Comput. Sci. 2(4), 4:4, 42 (2006)
4. Avis, D.: Living with lrs. In: Akiyama, J., Kano, M., Urabe, M. (eds.) JCDCG 1998. LNCS, vol. 1763, pp. 47–56. Springer, Heidelberg (2000)
5. Barrett, C., Sebastiani, R., Seshia, S.A., Tinelli, C.: Satisability modulo theories. In: Biere, A., et al. (eds.) Handbook of Satisability, pp. 825–885. IOS Press (2008)
6. Basu, S., Pollack, R., Roy, M.: Algorithms in real algebraic geometry. Springer (2003)
7. Billingsley, P.: Probability and measure, 3rd edn. John Wiley & Sons Inc. (1995)
8. Blanchette, J.C., Böhme, S., Paulson, L.C.: Extending Sledgehammer with SMT Solvers. In: Bjørner, N., Sofronie-Stokkermans, V. (eds.) CADE 2011. LNCS (LNAI), vol. 6803, pp. 116–130. Springer, Heidelberg (2011)
9. Boyd, S., Vandenberghe, L.: Convex optimization. Cambridge University Press (2004)
10. Chaieb, A., Nipkow, T.: Proof synthesis and reflection for linear arithmetic. J. Autom. Reasoning 41(1), 33–59 (2008)
11. Contejean, E.: A Certified AC Matching Algorithm. In: van Oostrom, V. (ed.) RTA 2004. LNCS, vol. 3091, pp. 70–84. Springer, Heidelberg (2004)

12. de Moura, L., Bjørner, N.: Efficient E-Matching for SMT Solvers. In: Pfenning, F. (ed.) CADE 2007. LNCS (LNAI), vol. 4603, pp. 183–198. Springer, Heidelberg (2007)
13. de Moura, L., Bjørner, N.: Z3: An Efficient SMT Solver. In: Ramakrishnan, C.R., Rehof, J. (eds.) TACAS 2008. LNCS, vol. 4963, pp. 337–340. Springer, Heidelberg (2008)
14. Dutertre, B., de Moura, L.: A fast linear-arithmetic solver for DPLL(T). In: Ball, T., Jones, R.B. (eds.) CAV 2006. LNCS, vol. 4144, pp. 81–94. Springer, Heidelberg (2006)
15. Fukuda, K., Prodon, A.: Double description method revisited. In: Deza, M., Manoussakis, I., Euler, R. (eds.) CCS 1995. LNCS, vol. 1120, pp. 91–111. Springer, Heidelberg (1996)
16. Garling, D.J.H.: Inequalities: a journey into linear analysis. Cambridge University Press, Cambridge (2007)
17. Gao, S., Avigad, J., Clarke, E.M.: Delta-complete decision procedures for satisfiability over the reals. In: Gramlich, B., et al. (eds.) IJCAR, pp. 286–300 (2012)
18. Hardy, G.H., Littlewood, J.E., Pólya, G.: Inequalities. Cambridge University Press, Cambridge (1988), Reprint of the 1952 edition
19. Harrison, J.: HOL light: a tutorial introduction. In: Srivas, M., Camilleri, A. (eds.) FMCAD 1996. LNCS, vol. 1166, pp. 265–269. Springer, Heidelberg (1996)
20. Harrison, J.: Verifying Nonlinear Real Formulas Via Sums of Squares. In: Schneider, K., Brandt, J. (eds.) TPHOLs 2007. LNCS, vol. 4732, pp. 102–118. Springer, Heidelberg (2007)
21. Hunt Jr., W.A., Krug, R.B., Moore, J.: Linear and nonlinear arithmetic in ACL2. In: Geist, D., Tronci, E. (eds.) CHARME 2003. LNCS, vol. 2860, pp. 319–333. Springer, Heidelberg (2003)
22. Jones, C.N., Kerrigan, E.C., Maciejowski, J.M.: Equality set projection: A new algorithm for the projection of polytopes in halfspace representation. Technical report, Department of Engineering, University of Cambridge (March 2004)
23. McLaughlin, S., Harrison, J.: A proof-producing decision procedure for real arithmetic. In: Nieuwenhuis, R. (ed.) CADE 2005. LNCS (LNAI), vol. 3632, pp. 295–314. Springer, Heidelberg (2005)
24. Meng, J., Paulson, L.: Lightweight relevance filtering for machine-generated resolution problems. J. Applied Logic 7(1), 41–57 (2009)
25. Moore, R., Kearfott, R., Cloud, M.: Introduction to interval analysis. Society for Industrial and Applied Mathematics (SIAM) (2009)
26. Moses, J.: Algebraic simplification: A guide for the perplexed. Communications of the ACM 14, 527–537 (1971)
27. Nelson, G., Oppen, D.: Simplification by cooperating decision procedures. ACM Transactions of Programming Languages and Systems 1, 245–257 (1979)
28. Nipkow, T., Paulson, L.C., Wenzel, M.T.: Isabelle/HOL. LNCS, vol. 2283. Springer, Heidelberg (2002)
29. Polya, G.: How to solve it. Princeton University Press, Princeton (1945)
30. Prevosto, V., Waldmann, U.: SPASS+T. In: Sutcliffe, G., et al. (eds.) ESCoR: Empirically Successful Computerized Reasoning 2006. CEUR Workshop Proceedings, pp. 18–33 (2006)
31. Pugh, W.: The omega test: a fast and practical integer programming algorithm for dependence analysis. Communications of the ACM 8, 4–13 (1992)
32. Schrijver, A.: Theory of linear and integer programming. John Wiley & Sons (1986)
33. Ziegler, G.: Lectures on polytopes. Springer (1995)

A Formal Library for Elliptic Curves in the Coq Proof Assistant

Evmorfia-Iro Bartzia[1] and Pierre-Yves Strub[2]

[1] INRIA Paris-Rocquencourt, France
iro.bartzia@inria.fr
[2] IMDEA Software Institute, Spain
pierre-yves@strub.nu

Abstract. A preliminary step towards the verification of elliptic curve cryptographic algorithms is the development of formal libraries with the corresponding mathematical theory. In this paper we present a formalization of elliptic curves theory, in the SSReflect extension of the Coq proof assistant. Our central contribution is a library containing many of the objects and core properties related to elliptic curve theory. We demonstrate the applicability of our library by formally proving a non-trivial property of elliptic curves: the existence of an isomorphism between a curve and its Picard group of divisors.

1 Introduction

The design of cryptographic algorithms is a complicated task. Besides functional correctness, cryptographic algorithms need to achieve contradictory goals such as efficiency and side channel resistance. Faulty implementations of algorithms may endanger security [4]. This is why formal assurance about their correctness is essential. Our motivation is to develop libraries that allow the formal verification of asymmetric cryptographic algorithms. As of today, the work on formal verification of security protocols has been assuming that the cryptographic libraries correctly implement all algorithms [2]. The first step towards the formal verification of cryptographic algorithms is the development of libraries that formally express the corresponding mathematical theory. In this paper we present a formal library for elementary elliptic curve theory that will enable formal analysis of elliptic-curve algorithms.

Elliptic curves have been used since the 19th century to approach a wide range of problems such as the fast factorization of integers and the search for congruent numbers. In the 20th century, researchers have regained interest in elliptic curves because of their applications in cryptography, first suggested in 1985 independently by Neal Koblitz [14] and Victor Miller [15]. Their use in cryptography relies principally on the existence of a group law that is a good candidate for public key cryptography, as its Discrete Logarithm Problem is hard relatively to the size of the parameters used. Elliptic curves also allow the definition of digital signatures and of new cryptographic primitives, such as identity-based encryption [17], based on bilinear (Weil and Tate) pairings.

G. Klein and R. Gamboa (Eds.): ITP 2014, LNAI 8558, pp. 77–92, 2014.
© Springer International Publishing Switzerland 2014

The mathematics of elliptic curves used in cryptography start from defining the group law and continue to theory from algebraic geometry [9].

Because our formalization involves algebraic structures such as rings and groups, polynomials, rational functions and matrices, we use the SSReflect extension [11] of the Coq proof-assistant [19] and its mathematical components library [1]. The Coq development can be found on the second author website (http://pierre-yves.strub.nu/).

Contributions. This paper presents an attempt to formalize non-trivial objects of algebraic geometry such as elliptic curves, rational functions and divisors. Our library is designed in such a way that will enable formal proofs of functional correctness of elliptic-curve algorithms. We validate the applicability of our theory by formally proving the Picard theorem, i.e. that an elliptic curve is structurally equivalent with its Picard group of divisors. Our formalization follows an elementary proof from Guillot [12] and Charlap [5].

Paper Outline. In sections 2 to 4, we present a formal proof of the following proposition, referred later as the *Picard theorem*:

> *The set of points of an elliptic curve together with its operation is isomorphic to its Picard group of divisors.*

We first define the two structures - namely the elliptic curve (Section 2) and the Picard group (Section 3) - and then prove that there exists a group isomorphism between them. In contrast to the definition of an elliptic curve, which goes smoothly, the definition of the Picard group involves several steps and forms the main matter of this paper. By construction, the Picard group of divisors is equipped with a group structure. In Section 4, we prove that the two structures are isomorphic. By transport of structure, the set of points of an elliptic curve together with its operation forms a group. In Section 5 and 6, we discuss related and future work.

2   Formalizing Elliptic Curves

An elliptic curve is a special case of a projective algebraic curve that can be defined as follows:

Definition 1. *Let $\mathbb{K}$ be a field. Using an appropriate choice of coordinates, an elliptic curve $\mathcal{E}$ is a plane cubic algebraic curve $\mathcal{E}(x, y)$ defined by an equation of the form:*

$$\mathcal{E}: y^2 + a_1 xy + a_3 y = x^3 + a_2 x^2 + a_4 x + a_6$$

where the a_i's are in $\mathbb{K}$ and the curve has no singular point (i.e. no cusps or self-intersections). The set of points, written $\mathcal{E}(\mathbb{K})$, is formed by the solutions (x, y) of $\mathcal{E}$ augmented by a distinguished point $\mathcal{O}$ (called point at infinity*):*

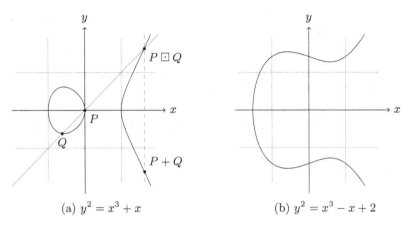

(a) $y^2 = x^3 + x$ (b) $y^2 = x^3 - x + 2$

Fig. 1. Catalog of Elliptic Curves Graphs

$$\mathcal{E}(\mathbb{K}) = \{(x, y) \in \mathbb{K} \mid \mathcal{E}(x, y)\} \cup \{\mathcal{O}\}$$

Figure 1 provides graphical representations of such curves in the real plane.

When the characteristic of $\mathbb{K}$ is different from 2 and 3, the equation $\mathcal{E}(x, y)$ can be simplified into its *Weierstrass* form:

$$y^2 = x^3 + ax + b.$$

Moreover, such a curve does not present any singularity if $\Delta(a, b) = 4a^3 + 27b^2$ — the curve's discriminant — is not equal to 0. Our work lies in this setting.

The parametric type `ec` represents the points on a specific curve. It is parameterized by a `K : ecuFieldType` — the type of fields with characteristic not in $\{2, 3\}$ — and a `E : ecuType` — a record that packs the curve parameters a and b along with a proof that $\Delta(a, b) \neq 0$. An inhabitant of the type `ec` is a point of the projective plane (represented by the type `point`), along with a proof that the point is on the curve.

```
Record ecuType := { A : K; B : K; _ : 4 * A^3 + 27 * B^2 != 0 }.
Inductive point := EC_Inf | EC_In of K & K.
Notation "(x, y)" := (EC_In x y).
Definition oncurve (p : point) :=
  if p is (x, y) then y^2 == x^3 + A * x + B else true.
Inductive ec : Type := EC p of oncurve p.
```

The points of an elliptic curve can be equipped with a structure of an abelian group. We give here a geometrical construction of the law. Let P and Q be points on the curve $\mathcal{E}$ and l be the line that goes through P and Q (or that is tangent to the curve at P if $P = Q$). By the Bezout theorem, counting multiplicities, l intersects $\mathcal{E}$ at a third point, denoted by $P \square Q$. The sum $P + Q$ is the opposite of $P \square Q$, obtained by taking the symmetric of $P \square Q$ with respect to the x axis. Figure 1 highlights this construction. To sum up:

1. $\mathcal{O}$ is defined to be the neutral element: $\forall P.\ P + \mathcal{O} = \mathcal{O} + P = P$,
2. the opposite of a point (x_P, y_P) (resp. $\mathcal{O}$) is $(x_P, -y_P)$ (resp. $\mathcal{O}$), and
3. if three points are collinear, their sum is equal to $\mathcal{O}$.

This geometrical definition can be translated into an algebraic setting, obtaining polynomial formulas for the definition of the law. Having such polynomial formulas leads to the following definitions:

```
Definition neg (p : point) :=
  if p is (x, y) then (x, -y) else EC_Inf.

Definition add (p1 p2 : point) :=
  let p1 := if oncurve p1 then p1 else EC_Inf in
  let p2 := if oncurve p2 then p2 else EC_Inf in
    match p1, p2 with
    | EC_Inf, _ => p2 | _, EC_Inf => p1
    | (x1, y1), (x2, y2) =>
      if x1 == x2 then ... else
        let s := (y2 - y1) / (x2 - x1) in
        let xs := s^2 - x1 - x2 in
          (xs, - s * (xs - x1) - y1)
  end.
```

Note that these definitions do not directly work with points on the curve, but instead on points of the projective plane (points that do not lie on the curve are projected to $\mathcal{O}$). We then prove that these operations are internal to the curve and lift them to $\mathcal{E}$:

```
Lemma add0 (p q : point): oncurve (add p q).
Definition addec (p1 p2 : ec) : ec := EC p1 p2 (add0 p1 p2).
```

We link back this algebraic definition to its geometrical interpretation. First, we define a function line: given two points P, Q on the curve, it returns the equation $ux + vy + c = 0$ of the line (PQ) intersecting the curve at P and Q (resp. the equation of the tangent to the curve at P if $P = Q$). We then show that, if (PQ) is not parallel to the y axis (i.e. is not intersecting the curve at $\mathcal{O}$), then (PQ) is intersecting $\mathcal{E}$ exactly at P, Q and $-(P + Q) = P \square Q$ as defined algebraically. This proof mainly relies on Vieta's formulas that relate the coefficients of a polynomial to sums and products of its roots. Although only a specific instance of Vieta's formulas is needed, we formalized the general ones:

Lemma 1 (Vieta's formulas). *For any polynomial $p = \sum_{i \le n} a_i X^i$ with roots $x_1, \ldots, x_n$, over an algebraically closed field, we have:*

$$\forall k.\ \sigma_k(x_1, \ldots, x_n) = (-1)^k \cdot \frac{a_{n-k}}{a_n}$$

where σ_k is the kth-elementary symmetrical polynomial.

3   The Picard Group of Divisors

From now on, let $\mathcal{E}$ be a smooth elliptic curve with equation $y^2 = x^3 + ax + b$ over the field $\mathbb{K}$. We assume that $\mathbb{K}$ is not of characteristic 2, nor 3. Related to this curve, we assume two Coq parameters K : ecuFieldType and E : ecuType K. We now move to the construction of the *Picard group* $\text{Pic}(\mathcal{E})$. This construction is split into several steps:

1. We start by constructing two objects: the field of rational functions $\mathbb{K}(\mathcal{E})$ over $\mathcal{E}$ and the group of $\mathcal{E}$-divisors $\text{Div}(\mathcal{E})$, i.e. the set of formal sums over the points of $\mathcal{E}$. From $\text{Div}(\mathcal{E})$ we construct $\text{Div}^0(\mathcal{E})$ which is the subgroup of zero-degree divisors.

2. We attach to each rational function $f \in \mathbb{K}(\mathcal{E})$ a divisor $\text{Div}(f)$ (called *principal divisor*) that characterizes f up to a scalar multiplication. This allows us to define the subgroup $\text{Prin}(\mathcal{E})$ of $\text{Div}(\mathcal{E})$, namely the *group of principal divisors*. The quotient group $\text{Div}^0(\mathcal{E})/\text{Prin}(\mathcal{E})$ forms the *Picard group*.

3.1   The Field of Rational Functions $\mathbb{K}(\mathcal{E})$

We denote the ring of bivariate polynomials over $\mathbb{K}$ by $\mathbb{K}[x, y]$.

Definition 2. *The ring $\mathbb{K}[\mathcal{E}]$ of polynomials over the curve is defined as the quotient ring of $\mathbb{K}[x, y]$ by the prime ideal $\langle y^2 - (x^3 + ax + b) \rangle$. The field $\mathbb{K}(\mathcal{E})$ is defined as the field of fractions of the integral domain $\mathbb{K}[\mathcal{E}]$.*

In other words, $\mathbb{K}[\mathcal{E}]$ is defined as the quotient of $\mathbb{K}[x, y]$ by the following equivalence relation $\sim$:

$$p \sim q \text{ if and only if } \exists k \in \mathbb{K}[x, y] \text{ such that } p - q = k(y^2 - x^3 - ax - b).$$

Since the polynomials y^2 and $x^3 + ax + b$ are identified in $\mathbb{K}[\mathcal{E}]$, we can associate, to any equivalence class of $\mathbb{K}[\mathcal{E}]$, a canonical representative of the form $p_1 y + p_2$ ($p_1, p_2 \in \mathbb{K}[x]$), obtained by iteratively substituting y^2 by $x^3 + ax + b$ in any element of the equivalence class. As such, instead of going through the path of formalizing ideals and ring quotients, we give a direct representation of $\mathbb{K}[\mathcal{E}]$ solely based on {poly K}, the type of univariate polynomials over K:

```
Inductive ecring := ECRing of {poly K} * {poly K}.
Notation "[ecp p1 *Y + p2]" := (ECRing p1 p2).
Coercion ecring_val (p : ecring) := let: ECRing p := p in p.
```

The type ecring is simply a copy of {poly K} * {poly K}, an element ([ecp p1 *Y + p2] : ecring) representing the class of the polynomial $p_1 y + p_2 \in \mathbb{K}[\mathcal{E}]$. We explicitly define the addition and multiplication, that are compatible with the one induced by the ring quotient, on the canonical representatives.

For instance:

$$(p_1 y + p_2)(q_1 y + q_2) = p_1 q_1 y^2 + (p_1 q_2 + q_1 p_2)y + p_2 q_2$$
$$= (p_1 q_2 + q_1 p_2)y + (p_1 q_1(x^3 + ax + b) + p_2 q_2)$$

leads to:

```
Notation XPoly := 'X^3 + A *: 'X + B.
Definition dotp (p q : ecring) := p.2 * q.2 + (p.1 * q.1) * Xpoly.
Definition mul (p q : ecring) := [ecp p.1*q.2 + p.2*q.1 *Y + dotp p q].
```

where .1 and .2 resp. stand for the first and second projections.

The set $\mathbb{K}[\mathcal{E}]$, as a ring quotient by a prime ideal, is an integral domain. As such, we are able to equip the type `ecring` with an `integralDomain` structure, proving all the required axioms of the structure. We can then use the `fraction` [6] library to built the type {`fraction ecring`} representing $\mathbb{K}(\mathcal{E})$, the field of fractions over $\mathbb{K}[\mathcal{E}]$.

3.2   Order and Evaluation of Rational Functions

In complex analysis, the *zeros* and *poles* of functions, and their *order* of vanishing are notions related to analytic functions and their Laurent expansion; while in abstract algebra, they refer to algebraic varieties and discrete valuation rings [9]. For our formalization, we follow the more elementary definitions given in [12]. More precisely, the evaluation of a function $f \in \mathbb{K}(\mathcal{E})$ at a point $P = (x_P, y_P) \in \mathcal{E}$ is defined as follows:

Definition 3. *A rational function $f \in \mathbb{K}(\mathcal{E})$ is said to be regular at $P = (x_P, y_P)$ if there exists a representative g/h of f such that $h(x_P, y_P) \neq 0$. If f is regular at P, the evaluation of f at P is the value $f(P) = \frac{g(x_P, y_P)}{h(x_P, y_P)}$, which is independent of the representative of f. If f is not regular at P, then P is called a pole of f and the evaluation of f at P is defined as $f(P) = \infty$.*

However, such a definition cannot be formalized as-is. Instead, we rely on the following extra notions allowing us to decompose any rational function in some canonical representative:

Definition 4. *A function $u \in \mathbb{K}(\mathcal{E})$ is called a uniformizer at $P \in \mathcal{E}(\mathbb{K})$ if i) $u(P) = 0$, and ii) every non-zero function $f \in \mathbb{K}(\mathcal{E})$ can be written in the form $f = u^v g$ with $g(P) \neq 0, \infty$ and $v \in \mathbb{Z}$.*
The exponent v is independent from the choice of the uniformizer and is called the order of f at P, a quantity denoted by $\mathrm{ord}_f(P)$.

Lemma 2. *There exists a uniformizer for every point on the curve.*

To get an intuition of the previous definitions, one can make a parallel with the notion of multiplicity for roots of univariate polynomials or with the notion of zeros and poles in $\mathbb{K}(x)$, the field of plain rational functions.

For instance, let us first consider the ring of polynomials $\mathbb{K}[x]$. Let p be a polynomial in $\mathbb{K}[x]$ and r be an element of $\mathbb{K}$. We can factorize p as $p = (x-r)^m q$ such that $m \in \mathbb{N}$ and $q(r) \neq 0$. The exponent m is the multiplicity of p at r. The multiplicity of r is 1 for the polynomial factor $(x-r)$. Evaluation and multiplicity are closely related: r is a root of p iff $m > 0$.

In an analogous way, we can consider the field of fractions $\mathbb{K}(\mathcal{E})$. Let P be in $\mathcal{E}$ and f in $\mathbb{K}(\mathcal{E})$. Then, one can always write f in the form $f = u^v g$ with $v \in \mathbb{Z}$ uniquely defined and P neither a zero nor a pole of g $(g(P) \neq 0, \infty)$. The exponent v is the order of f at P. (Here, the function u corresponds to the polynomial factor $(x - r)$ for univariate polynomials) If $v > 0$ then P is a zero for f, and if $v < 0$ then P is a pole for f.

As said, the given definition of evaluation is not constructive. However, the proof of Lemma 2 is constructive and gives all the necessary material to define these notions. Let P be a point on the curve. For every $f \in \mathbb{K}[\mathcal{E}]$ (of type `ecring`) we explicitly give the decomposition $f = u_P^v(n/d)$ such that $n(P), d(P) \neq 0$, and u_P is a fixed rational function depending solely on P:

```
Definition unifun (P : point) : {fraction ecring} :=
  match P with
  | (x, y) => if y == 0 then [ecp 1 *Y + 0] else [ecp 0 *Y + ('X - x)]
  | EC_Inf => [ecp 0 *Y + X] / [ecp 1 *Y + 0]
  end.

Definition poly_order (f : ecring) (P : point) :=
  match P with
  | EC_Inf => let d := (degree f).-1 in
      (-d, ('X^d * f, [ecp 1 *Y + 0]^d)).
  | (x, y) => ...
```

and then prove that the decomposition is correct and unique:

```
Definition uniok (f u : fraction ecring) (p : point) o (n d : ecring) :=
  match p with
  | (x, y) => [&& f == u^o * (n // d), n.[x, y] != 0 & d.[x, y] != 0]
  | EC_Inf => ...

Lemma poly_order_correct :
  forall (f : ecring) (p : point), f != 0 -> oncurve p ->
    let: (o, (g1, g2)) := poly_order f p in
      uniok (unifun p) f p o g1 g2.

Lemma uniok_uniq:
  forall f p, f != 0 -> oncurve p ->
    forall o1 o2 n1 n2 d1 d2,
         uniok (unifun p) f p o1 n1 d1
      -> uniok (unifun p) f p o2 n2 d2
      -> (o1 == o2) && (n1 // d1 == n2 // d2).
```

We then lift these definitions to the quotient {fraction ecring}, and prove that all the lifted functions are stable by taking the quotient, allowing us to lift all the proved properties over $\mathbb{K}[\mathcal{E}]$ to $\mathbb{K}(\mathcal{E})$ (i.e. from ecring to {fraction ecring}). For instance, the order on {fraction ecring} is defined as:

```
Definition orderf (f : {ratio ecring}) p : int :=
  if \n_f == 0 then 0 else (poly_order \n_f p).1 - (poly_order \d_f p).1.

Definition order (f : {fraction ecring}) p := orderf (repr f) p.
```

We can then formalize Definition 3 by a simple case analysis over the *order*, relying on the decomposition of rational functions we have just formalized:

```
Definition eval (f : {fraction ecring}) p :=
  match p, order f p with
  | _, Posz _.+1 => 0
  | _, Negz _ => [inf]
  | (x, y), Posz 0 => (decomp f ecp).1.[x,y] / (decomp f ecp).2.[x,y]
  | EC_Inf, _ => ...
  end.
```

Due to the lack of space, we cannot give much details on the whole formalization of valuation theory, and move to the key lemma of this section:

Lemma 3. *A rational function $f \in \mathbb{K}(\mathcal{E})$ has a finite number of poles and zeros. Moreover, assuming that $\mathbb{K}$ is algebraically closed, $\sum_{P \in \mathcal{E}}(\mathrm{ord}_P(f)) = 0$.*

This lemma will be central when moving to the construction of the isomorphism between an elliptic curve and its Picard group.

3.3   Principal Divisors

From now on, we assume that $\mathbb{K}$ is algebraically closed.

Principal divisors are introduced as a tool for describing the zeros and poles of rational functions on an elliptic curve:

Definition 5 (Principal divisors). *Given $f \in \mathbb{K}(\mathcal{E})$, $f \neq 0$, the principal divisor $\mathrm{Div}(f)$ of f is defined as the formal (finite) sum:*

$$\mathrm{Div}(f) = \sum_{P \in \mathcal{E}}(\mathrm{ord}_P(f))(P).$$

Note that $\mathrm{Div}(f)$ is well defined because a rational function has only finitely many zeros and poles. We write $\mathrm{Prin}(\mathcal{E})$ for the set of principal divisors.

The set $\mathrm{Prin}(\mathcal{E})$ forms a subgroup of $\mathrm{Div}(\mathcal{E})$, the set of formal sums over $\mathcal{E}$, a notion that we define now.

Definition 6. *A divisor on an elliptic curve $\mathcal{E}$ is a formal sum of points*

$$D = \sum_{P \in \mathcal{E}} n_P(P),$$

where $n_P \in \mathbb{Z}$, only finitely many nonzero. In other words, a divisor is any expression taken in the free abelian group generated over $\mathcal{E}(\mathbb{K})$. The domain of D is $\mathrm{dom}(D) = \{P \mid n_P \neq 0\}$, *and its degree is* $\deg(D) = \sum_{P \in E} n_P$. *For any point P, the coefficient of P in D is* $\mathrm{coeff}(P, D) = n_P$.

We write $\mathrm{Div}(\mathcal{E})$ *for the set of divisors on $\mathcal{E}$, and* $\mathrm{Div}^0(\mathcal{E})$ *its subgroup composed of divisors of degree 0.*

The set of divisors on $\mathcal{E}$ is an abelian group. The zero divisor is the unique divisor with all its coefficient set to 0, whereas the sum of two divisors is defined as the point-wise addition.

Based on the `quotient` libraries of SSReflect, we develop the theory of free abelian groups. Let T be a type. We first define the type of *pre-free group* as the collection of all sequences s of type `int * T` s.t. no pair of the form `(0, _)` can appear in s and for any `z : T`, a pair of the form `(_, z)` can appear at most once in s.

```
Definition reduced (D : seq (int * T)) :=
    (uniq [seq zx.2 | zx <- D])
 && (all [pred zx | zx.1 != 0] D).

Record prefreeg : Type := mkPrefreeg {
  seq_of_prefreeg : seq (int * T);
  _ : reduced seq_of_prefreeg
}.
```

The intent of `prefreeg` is to give a unique representation of a free-group expression, up to the order of the coefficients. For instance, if $D = k_1 x_1 + \cdots + k_n x_n$ (with all the x_i's pairwise distinct and all the k_i's in $\mathbb{Z}^*$), then the reduced sequence `s = [:: (k_1, x_1), ..., (k_n, x_n)]`, or any sequence equal up to a permutation to s, is a valid representation of D. The type `freeg` of free-groups is then obtained by quotienting `prefreeg` by the `perm_eq` equivalence relation.

From there, we equip the type `freeg` with a group structure (the operation is noted additively), and define all the usual notions related to free groups (domain, coefficient, degree, ...). For instance, assume `G : zmodType` (G is a $\mathbb{Z}$-module) and `f : T -> G`. Then, f defines a unique group homomorphism from `freeg` to G that can be defined as follows:

```
Definition prelift (D : seq (int * T)) : G :=
  \sum_(x <- D) (f x.2) * x.1.

Definition lift (s : prefreeg T) : G := prelift s.
Definition fglift (D : {freeg T}) := lift (repr D).
```

One can check that the `fglift` function defines the homomorphism

$$\sum_{(z,x) \in D} z f(x)$$

The coefficient `coeff` and degree `deg` functions can be then defined as:

```
Definition coeff (t : T) (D : {freeg T}) :=
  fglift (fun x => (x == t)) D.

Definition deg (D : {freeg K}) : int :=
  fglift (fun x => 1) D.
```

The Group of Principal Divisors. Returning to principal divisors, one can now check that $\mathrm{Prin}(\mathcal{E})$ is a subgroup of $\mathrm{Div}^0(\mathcal{E})$. Indeed, i) $\deg(div f) = 0$ by Lemma 3, and ii) since the order function is multiplicative $(\mathrm{ord}_p(f/g) = \mathrm{ord}_p(f) - \mathrm{ord}_p(g))$, we have $div(f/g) = div(f) - div(g)$.

Moreover, it is now clear that the coefficients associated in $\mathrm{Div}(f)$, to each point P, is the order of the function f at P, highlighting the fact that a divisor wraps up the zeros and poles of f.

Formally, we define principal divisors for polynomials on the curve with the function `ecdivp`:

```
Definition ecdivp (f : ecring) : {freeg (point)} :=
  \sum_(p <- ecroots f)
      << (order f (p.1, p.2)) * (p.1, p.2) >>
    + << order f EC_Inf * EC_Inf >>.
```

where `<< z * P >>` stands for the divisor $z(P)$ and the function `ecroots` takes a polynomial of $\mathbb{K}[\mathcal{E}]$ and returns the list of its finite zeros:

```
Definition ecroots f : seq (K * K) :=
  let forx := fun x =>
    let sqrts := roots ('X^2 - ('X^3 + A *: 'X + B).[x]) in
      [seq (x, y) | y <- sqrts & f.[x, y] == 0]
  in
    undup (flatten ([seq forx x | x <- roots (norm f)])).
```

The function `ecroots` relies on $\mathrm{norm}(f)$, a polynomial in $\mathbb{K}[x]$ associated to f that has the following property: (x, y) is a zero of f if and only if x is a zero of $\mathrm{norm}(f)$ and $y^2 = x^3 + ax + b$.

Next, we lift the definition of principal divisors to $\mathbb{K}(\mathcal{E})$, prove its correctness and recast the key Lemma 3 (`deg_ecdiv_eq0`):

```
Notation "\n_f" := (numerator f).
Notation "\d_f" := (denominator f).

Definition ecdiv_r (f : {ratio ecring}) :=
  if \n_f == 0 then 0 else (ecdivp \n_f) - (ecdivp \d_f).

Definition ecdiv := lift_fun1 {fraction ecring} ecdiv_r.
```

```
Lemma ecdiv_coeffE (f : {fraction ecring}) p:
  coeff p (ecdiv f) = order f p.
```

```
Lemma deg_ecdiv_eq0 (f : {fraction ecring}): deg (ecdiv f) = 0.
```

3.4   Divisor of a Line

Before moving to the definition of the Picard group, we characterize the divisors of some specific rational functions. These divisors will later help formalize the construction of the Picard group:

Definition 7. *A line $l \in \mathbb{K}(\mathcal{E})$ is any rational function of the form $l(x, y) = ax + by + c$ with $a, b, c \in \mathbb{K}$ not all zero.*

For instance, if (PQ) is the line intersecting the curve at P and Q, then we know that (PQ) intersects $\mathcal{E}$ at exactly three points (counting multiplicities): P, Q and $P \boxdot Q$. Assuming that P, Q and $P \boxdot Q$ are all finite, these three points are the unique zeros of the rational function l associated to (PQ) and $\mathrm{Div}(l) = (P)+(Q)+(P\boxdot Q)-3(\mathcal{O})$. This relation still holds when one or several of these three points are equal to $\mathcal{O}$. For instance, $\mathrm{Div}(x - x_P) = (P)+(-P)-2(\mathcal{O})$, where $x - x_P$ is the line intersecting $\mathcal{E}$ at P, $-P$ and $\mathcal{O}$.

3.5   The Picard Group

Definition 8. *The Picard group $\mathrm{Pic}(\mathcal{E})$ is the group quotient $\mathrm{Div}^0(\mathcal{E})/\mathrm{Prin}(\mathcal{E})$. Note that the degree is well defined on the divisor class group since if $D_1 = D_2 + \mathrm{Div}(f)$ then $\deg D_1 = \deg D_2 + \deg(\mathrm{Div}(f)) = \deg D_2 + 0 = \deg D_2$.*

In other words, $\mathrm{Pic}(\mathcal{E})$ is defined as the quotient of $\mathrm{Div}^0(\mathcal{E})$ by the following equivalence relation $\sim$:

$$D_1 \sim D_2 \text{ if and only if } \exists f \in \mathbb{K}(\mathcal{E}) \text{ such that } div(f) = D_1 - D_2.$$

a notion that we formalize as follows:

```
Definition ecdeqv D1 D2 :=
  (exists f : {fraction ecring}, ecdiv f = D1 - D2).
Notation "D1 :~: D2" := (ecdeqv D1 D2).
```

We do not give a direct construction of $\mathrm{Pic}(\mathcal{E})$ but instead prove that any class of $\mathrm{Pic}(\mathcal{E})$ can be represented by a divisor of the form $(P) - (\mathcal{O})$.

The construction of this representative is based on a procedure called *Linear Reduction*. Assume that P and Q are two finite points of $\mathcal{E}(\mathbb{K})$. We know that the divisor of the line l intersecting $\mathcal{E}$ at P and Q is $\mathrm{Div}(l) = (P) + (Q) + (P \boxdot Q) - 3(\mathcal{O})$. Likewise, the divisor of the line l' intersecting $\mathcal{E}$ at $P \boxdot Q$ and $-(P \boxdot Q)$ $(= P + Q)$ is $\mathrm{Div}(l') = (P + Q) + (P \boxdot Q) - 2(\mathcal{O})$. Hence,

$$\mathrm{Div}(l/l') = \mathrm{Div}(l) - \mathrm{Div}(l')$$
$$= (P) + (Q) - (P+Q) - (\mathcal{O})$$

and, $(P) + (Q) \sim (P+Q) + (\mathcal{O})$.

Iterating this procedure, we can reduce any divisor of the form:

$$(P_1) + \cdots + (P_n) - (Q_1) - \cdots - (Q_k) + r(\mathcal{O})$$

to an equivalent one $(P) - (Q) + r'(\mathcal{O})$, with $r' \in \mathbb{Z}$. Using one more time the same construction, one can show that $(P) - (Q) + n'(\mathcal{O})$ is equivalent to $(P-Q) + n''(\mathcal{O})$ where $n', n'' \in \mathbb{Z}$.

The lr function formally defines the linear reduction procedure:

```
Definition fgpos (D : {freeg K}) :=
  \sum_(p <- dom D | coeff p D > 0) coeff p D.
Definition fgneg (D : {freeg K}) :=
  \sum_(p <- dom D | coeff p D < 0) -(coeff p D).

Definition lr_r (D : {freeg point}) :=
  let iter p n := iterop _ n + p EC_Inf in
    \sum_(p <- dom D | p != EC_Inf) (iter p '|coeff p D|).

Definition lr (D : {freeg point}) : point :=
  let: (Dp, Dn) := (fgpos D, fgneg D) in
    lr_r Dp - lr_r Dn.

Lemma ecdeqv_lr D: all oncurve (dom D) ->
  D :~: << lr D >> + << deg D - 1 *g EC_Inf >>.
```

where (Dp, Dn) := (fgpos D, fgneg D) is the decomposition of D into its negative and positive parts.

The lemma ecdeqv_lr states that any divisor is equivalent to a divisor of the form $(P) + (\deg D - 1)(\mathcal{O})$. In the context of $\mathrm{Pic}(\mathcal{E})$, this means that any class contains a divisor of the form $(P) - (\mathcal{O})$ (recall that $\mathrm{Pic}(\mathcal{E})$ is a group quotient of $\mathrm{Div}^0(\mathcal{E})$ — the divisors of degree 0). In the next section, we end the construction of the Picard group by proving that at most one such representative can be found in each class of $\mathrm{Pic}(\mathcal{E})$.

4   Linking $\mathrm{Pic}(\mathcal{E})$ to $\mathcal{E}(\mathbb{K})$

In this section, we finish our formal construction of the Picard group and prove the existence of an isomorphism between $\mathrm{Pic}(\mathcal{E})$ and $\mathcal{E}(\mathbb{K})$. We start by defining a canonical representative for the classes of $\mathrm{Pic}(\mathcal{E})$:

Lemma 4. *For every class of* $\mathrm{Pic}(\mathcal{E})$*, there exists a unique representative of the form* $(P) - (\mathcal{O})$ *with* $P \in \mathcal{E}(\mathbb{K})$.

From Section 3.5, we already know that each class of $\text{Pic}(\mathcal{E})$ contains one such representative. Assume now that $(P) - (\mathcal{O})$ and $(Q) - (\mathcal{O})$ are two representatives of a class of $\text{Pic}(\mathcal{E})$ with $P \neq Q$. Such an assumption allows us to find a rational function $h \in \mathbb{K}(\mathcal{E})$ s.t. every rational function $f \in \mathbb{K}(\mathcal{E})$ can be expressed as a polynomial fraction of h. This implies that $K(\mathcal{E})$ and $K(x)$ are isomorphic. However, since $K(\mathcal{E})$ is a field extension of degree 2 of $K(x)$, such an h cannot exist. Hence, $P = Q$:

Lemma lr_uniq: << p >> :~: << q >> -> p = q.

The Picard group can now be formally defined as the set of divisors of the form $(P) - (\mathcal{O})$. It remains to prove the existence of a bijection between $\text{Pic}(\mathcal{E})$ and $\mathcal{E}(\mathbb{K})$. Namely, the function

$$\phi : \mathcal{E}(\mathbb{K}) \to \text{Pic}(\mathcal{E})$$
$$P \mapsto [(P) - (\mathcal{O})]$$

is our isomorphism. Indeed, ϕ is clearly bijective and from the results of Section 3:

$$\phi(P_1) - \phi(P_2) = [(P_1) - (\mathcal{O})] - [(P_2) - (\mathcal{O})] = [(P_1) - (P_2)]$$
$$= [(P_1 - P_2) - (\mathcal{O})] = \phi(P_1 - P_2).$$

In our formalization, we directly use the linear reduction function lr in place of ϕ^{-1}. For instance, we prove that lr commutes with the curve operations and maps $(P) - (\mathcal{O})$ to $P \in \mathbb{K}(\mathcal{E})$:

Lemma lrB: forall (D1 D2: {freeg point},
    deg D1 = 0 -> all oncurve (dom D1) ->
    deg D2 = 0 -> all oncurve (dom D2) ->
    lr (D1 - D2) = lr D1 - lr D2.

Lemma lrpi: forall p : point,
    oncurve p -> lr (<<p>> - <<EC_Inf>>) = p.

This allows us to transport the structure from $\text{Pic}(\mathcal{E})$ to $\mathcal{E}(\mathbb{K})$, proving that $\mathcal{E}(\mathbb{K})$ is a group.

5   Related Work

Hurd et al. [13] formalize elliptic curves in higher order logic using the HOL-4 proof assistant. Their goal is to create a "gold-standard" set of elliptic curve operations mechanized in HOL-4, which can be used afterwards to verify ec-algorithms for scalar multiplication. They define datatypes to represent elliptic curves on arbitrary fields (in both projective and affine representation), rational points and the elliptic curve group operation, although they do not provide a proof that the operation indeed satisfies the group properties. In the end, they state the theorem that expresses the functional correctness of the ElGamal encryption scheme for elliptic curves.

Smith et al. [18] use the Verifun proof assistant to prove that two representations of an elliptic curve in different coordinate systems are isomorphic. Their theory applies to elliptic curves on prime fields. They define data structures for affine and projective points and the functions that compute the elliptic curve operations in affine and Jacobian coordinates. In their formalization there is no datatype for elliptic curves, an elliptic curve is a set of points that satisfy a set of conditions. They define the transformation functions between the two systems of coordinate and prove that for elliptic curve points the transformation functions commute with the operations and that both representations of elliptic curves in affine or Jacobian coordinates are isomorphic.

Théry [20] present a formal proof that an elliptic curve is a group using the Coq proof assistant. The proof that the operation is associative relies heavily on case analysis and requires handling of elementary but subtle geometric transformations and therefore uses computer-algebra systems to deal with non-trivial computation. In our development, we give a different proof of the associativity of the elliptic curve group law: we define an algebraic structure (the Picard group of divisors) and proceed to prove that the elliptic curve is isomorphic to this structure. Our formalization is more structural than [20] in the sense that it involves less computation and the definition of new algebraic structures.

As in [13] and [18] we wish to develop libraries that will enable the formal analysis of elliptic curve algorithms and our proofs follow textbook mathematics. As in [20], we give a formal proof of the group law for elliptic curves. Nevertheless, the content of our development is quite different from the related work. To the extent of our knowledge this is the first formalization of divisors and rational functions of a curve, which are objects of study of algebraic geometry. Such libraries may allow the formalization of non-trivial algorithms that involve divisors (such as the Miller algorithm for pairings [16]), isogenies (such as [3], [8]) or endomorphisms on elliptic curves (such as the GLV algorithm for scalar multiplication [10]).

6   Future Work

This paper presents a formalization of the elementary elliptic curve theory in the SSReflect extension of Coq. Our central result is the formal proof of the Picard theorem which is a structure theorem for elliptic curves. A direct implication of this theorem is the associativity of the elliptic curve group operation. Our development includes generic libraries formalizing divisors and rational functions that are designed to enable the formal verification of elliptic curve cryptographic algorithms. Our formalization required 10k lines of code out of which 6.5k lines were required for the proof of the Picard theorem. The SSReflect features and methodology for the formalization of algebraic structures have been very helpful to our development.

The proof layout follows the ones that can be found in most graduate text books about smooth algebraic curves, but instantiated to the case of elliptic curves. Generalizing our development should not deeply change the general structure, but will certainly require the development of a lot of background theory:

affine spaces, multinomials, rational maps, ring quotients, valuation rings, formal differentials, ... to name some of them.

To further validate our development, we are working on the formal proof of correctness of an implementation of the GLV algorithm [10]. The GLV algorithm is a non-generic scalar multiplication algorithm that uses endomorphisms on elliptic curves to accelerate computation. It is composed of three independent algorithms: parallel exponentiation, decomposition of the scalar, computation of endomorphisms on elliptic curves. The third algorithm involves background from algebraic geometry that can be provided by the rational functions' libraries presented in this paper. We aim to generate a certified implementation of the GLV algorithm based on the methodology described in [7]. Another algorithm that would be interesting to formalize is the Miller algorithm for bilinear pairings [16], which would rely on our divisors' library to compute pairings by evaluating linear functions on divisors.

In the development presented in this paper, we chose to represent elliptic curves in an affine coordinate system. However, projective, Jacobian and other coordinate systems are widely used in practice mainly for reasons of efficiency. Indeed, nowadays the search of optimal coordinate-systems is an active domain of research in elliptic curve cryptography. Future work is to extend our libraries to isomorphic coordinate representations in order to allow formal analysis of algorithms in different coordinate systems. Furthermore, in our development we treat elliptic curves on fields with characteristic different from 2 and 3 although in cryptography binary fields are used often. Generalizing our development to more general curves is a natural extension.

Acknowledgements. We thank Philippe Guillot and Benjamin Smith for their discussions on the theory of elliptic curves. We also thank Cyril Cohen, Assia Mahboubi and all the SSReflect experts for spending time explaining the internals of the SSReflect libraries. Finally, we thank Karthikeyan Bhargavan, Cătălin Hrițcu, Robert Kunnemann, Alfredo Pironti, Ben Smyth for their feedback and discussions on this paper.

References

1. The mathematical components project,
   http://www.msr-inria.fr/projects/mathematical-components/
2. Avalle, M., Pironti, A., Sisto, R.: Formal verification of security protocol implementations: a survey. In: Formal Aspects of Computing, pp. 1–25 (2012)
3. Brier, E., Joye, M.: Fast point multiplication on elliptic curves through isogenies. In: Fossorier, M.P.C., Høholdt, T., Poli, A. (eds.) AAECC 2003. LNCS, vol. 2643, pp. 43–50. Springer, Heidelberg (2003)
4. Brumley, B.B., Barbosa, M., Page, D., Vercauteren, F.: Practical realisation and elimination of an ecc-related software bug attack. Cryptology ePrint Archive, Report 2011/633 (2011), http://eprint.iacr.org/
5. Charlap, L.S., Robbins, D.P.: Crd expository report 31 an elementary introduction to elliptic curves (1988)

6. Cohen, C.: Pragmatic quotient types in COQ. In: Blazy, S., Paulin-Mohring, C., Pichardie, D. (eds.) ITP 2013. LNCS, vol. 7998, pp. 213–228. Springer, Heidelberg (2013)
7. Dénès, M., Mörtberg, A., Siles, V.: A refinement-based approach to computational algebra in COQ. In: Beringer, L., Felty, A. (eds.) ITP 2012. LNCS, vol. 7406, pp. 83–98. Springer, Heidelberg (2012)
8. Doche, C., Icart, T., Kohel, D.R.: Efficient scalar multiplication by isogeny decompositions. In: Yung, M., Dodis, Y., Kiayias, A., Malkin, T. (eds.) PKC 2006. LNCS, vol. 3958, pp. 191–206. Springer, Heidelberg (2006)
9. Fulton, W.: Algebraic curves - an introduction to algebraic geometry (reprint vrom 1969). Advanced book classics. Addison-Wesley (1989)
10. Gallant, R.P., Lambert, R.J., Vanstone, S.A.: Faster point multiplication on elliptic curves with efficient endomorphisms. In: Kilian, J. (ed.) CRYPTO 2001. LNCS, vol. 2139, pp. 190–200. Springer, Heidelberg (2001)
11. Gonthier, G., Mahboubi, A., Tassi, E.: A Small Scale Reflection Extension for the Coq system. Research Report RR-6455, INRIA (2008)
12. Guillot, P.: Courbes Elliptiques, une présentation élémentaire pour la cryptographie. Lavoisier (2010)
13. Hurd, J., Gordon, M., Fox, A.: Formalized elliptic curve cryptography. High Confidence Software and Systems (2006)
14. Koblitz, N.: Elliptic curve cryptosystems. Mathematics of Computation 48(177), 203–209 (1987)
15. Miller, V.S.: Use of elliptic curves in cryptography. In: Williams, H.C. (ed.) CRYPTO 1985. LNCS, vol. 218, pp. 417–426. Springer, Heidelberg (1986)
16. Miller, V.S.: Short programs for functions on curves. IBM Thomas J. Watson Research Center (1986)
17. Shamir, A.: Identity-based cryptosystems and signature schemes. In: Blakely, G.R., Chaum, D. (eds.) CRYPTO 1984. LNCS, vol. 196, pp. 47–53. Springer, Heidelberg (1985)
18. Smith, E.W., Dill, D.L.: Automatic formal verification of block cipher implementations. In: FMCAD, pp. 1–7 (2008)
19. The Coq development team. The Coq Proof Assistant Reference Manual Version 8.4 (2013), http://coq.inria.fr
20. Théry, L.: Proving the group law for elliptic curves formally. Technical Report RT-0330, INRIA (2007)

Truly Modular (Co)datatypes for Isabelle/HOL

Jasmin Christian Blanchette[1], Johannes Hölzl[1], Andreas Lochbihler[2],
Lorenz Panny[1], Andrei Popescu[1,3], and Dmitriy Traytel[1]

[1] Fakultät für Informatik, Technische Universität München, Germany
[2] Institute of Information Security, ETH Zurich, Switzerland
[3] Institute of Mathematics Simion Stoilow of the Romanian Academy, Bucharest, Romania

Abstract. We extended Isabelle/HOL with a pair of definitional commands for datatypes and codatatypes. They support mutual and nested (co)recursion through well-behaved type constructors, including mixed recursion–corecursion, and are complemented by syntaxes for introducing primitively (co)recursive functions and by a general proof method for reasoning coinductively. As a case study, we ported Isabelle's Coinductive library to use the new commands, eliminating the need for tedious ad hoc constructions.

1 Introduction

Coinductive methods are becoming widespread in computer science. In proof assistants such as Agda, Coq, and Matita, codatatypes and coinduction are intrinsic to the logical calculus [2]. Formalizations involving programming language semantics, such as the CompCert verified C compiler [17], use codatatypes to represent potentially infinite execution traces. The literature also abounds with "coinductive pearls," which demonstrate how coinductive methods can lead to nicer solutions than traditional approaches.

Thus far, provers based on higher-order logic (HOL) have mostly stood on the sidelines of these developments. Isabelle/HOL [24, Part 1; 25] provides a few manually derived codatatypes (e.g., lazy lists) in the Coinductive entry of the *Archive of Formal Proofs* [18]. This library forms the basis of JinjaThreads [19], a verified compiler for a Java-like language, and of the formalization of the Java memory model [21]. The manual constructions are heavy, requiring hundreds of lines for each codatatype.

Even in the realm of datatypes, there is room for improvement. Isabelle's datatype package was developed by Berghofer and Wenzel [4], who could draw on the work of Melham [23], Gunter [11, 12], Paulson [28], and Harrison [14]. The package supports positive recursion through functions and reduces nested recursion through datatypes to mutual recursion, but otherwise allows no nesting. It must reject definitions such as

datatype $\alpha\ tree_{FS} = \mathsf{Tree_{FS}}\ \alpha\ (\alpha\ tree_{FS}\ fset)$

where *fset* designates finite sets (a non-datatype). Moreover, the reduction of nested to mutual recursion makes it difficult to specify recursive functions truly modularly.

We introduce a definitional package for datatypes and codatatypes that addresses the issues noted above. The key notion is that of a *bounded natural functor* (BNF), a type constructor equipped with map and set functions and a cardinality bound (Section 2). BNFs are closed under composition and least and greatest fixpoints and are expressible in HOL. Users can register well-behaved type constructors such as *fset* as BNFs.

G. Klein and R. Gamboa (Eds.): ITP 2014, LNAI 8558, pp. 93–110, 2014.

The BNF-based **datatype** and **codatatype** commands provide many conveniences such as automatically generated discriminators, selectors, map and set functions, and relators (Sections 3 and 4). Thus, the command

codatatype (lset: α) *llist* (**map**: lmap **rel**: lrel) =
lnull: LNil | LCons (lhd: α) (ltl: α *llist*)

defines the type α *llist* of lazy lists over α, with constructors LNil :: α *llist* and LCons :: $\alpha \Rightarrow \alpha$ *llist* $\Rightarrow \alpha$ *llist*, a discriminator lnull :: α *llist* $\Rightarrow$ *bool*, selectors lhd :: α *llist* $\Rightarrow \alpha$ and ltl :: α *llist* $\Rightarrow \alpha$ *llist*, a set function lset :: α *llist* $\Rightarrow \alpha$ *set*, a map function lmap :: $(\alpha \Rightarrow \beta) \Rightarrow \alpha$ *llist* $\Rightarrow \beta$ *llist*, and a relator lrel :: $(\alpha \Rightarrow \beta \Rightarrow bool) \Rightarrow \alpha$ *llist* $\Rightarrow \beta$ *llist* $\Rightarrow$ *bool*. Intuitively, the **codatatype** keyword indicates that the constructors can be applied repeatedly to produce infinite values—e.g., LCons 0 (LCons 1 (LCons 2 ...)).

Nesting makes it possible to mix recursion and corecursion arbitrarily. The next commands introduce the types of Rose trees with finite or possibly infinite branching (*list* vs. *llist*) and with finite or possibly infinite paths (**datatype** vs. **codatatype**):

datatype α *tree* = Tree (lab: α) (sub: α *tree list*)
datatype α *tree*$_\omega$ = Tree$_\omega$ (lab$_\omega$: α) (sub$_\omega$: α *tree*$_\omega$ *llist*)
codatatype α *ltree* = LTree (llab: α) (lsub: α *ltree list*)
codatatype α *ltree*$_\omega$ = LTree$_\omega$ (llab$_\omega$: α) (lsub$_\omega$: α *ltree*$_\omega$ *llist*)

Primitively (co)recursive functions can be specified using **primrec** and **primcorec** (Sections 5 and 6). The function below constructs a possibly infinite tree by repeatedly applying $f :: \alpha \Rightarrow \alpha$ *llist* to x. It relies on lmap to construct the nested *llist* modularly:

primcorec iterate_ltree$_\omega$ :: $(\alpha \Rightarrow \alpha$ *llist*$) \Rightarrow \alpha \Rightarrow \alpha$ *ltree*$_\omega$ **where**
iterate_ltree$_\omega$ $f x$ = LTree$_\omega$ x (lmap (iterate_ltree$_\omega$ f) $(f x)$)

An analogous definition is possible for α *ltree*, using *list*'s map instead of lmap.

For datatypes that recurse through other datatypes, and similarly for codatatypes, old-style mutual definitions are also allowed. For the above example, this would mean defining iterate_ltree$_\omega$ by mutual corecursion with iterate_ltrees$_\omega$ :: $(\alpha \Rightarrow \alpha$ *llist*$) \Rightarrow \alpha$ *llist* $\Rightarrow \alpha$ *ltree*$_\omega$ *llist*. Despite its lack of modularity, the approach is useful both for compatibility and for expressing specifications in a more flexible style. The package generates suitable (co)induction rules to facilitate reasoning about the definition.

Reasoning coinductively is needlessly tedious in Isabelle, because the *coinduct* method requires the user to provide a witness relation. Our new *coinduction* method eliminates this boilerplate; it is now possible to have one-line proofs **by** *coinduction auto* (Section 7). To show the package in action, we present a theory of stream processors, which combine a least and a greatest fixpoint (Section 8). In addition, we describe our experience porting the Coinductive library to use the new package (Section 9). A formal development accompanies this paper [6].

The package has been part of Isabelle starting with version 2013. The implementation is a significant piece of engineering, at over 18 000 lines of Standard ML code and 1 000 lines of Isabelle formalization. The features described here are implemented in the development repository and are expected to be part of version 2014. (In the current implementation, the BNF-based **datatype** command is suffixed with **_new** to avoid a clash with the old package.) The input syntax and the generated constants and theorems are documented in the user's manual [7].

2 Low-Level Constructions

At the lowest level, each (co)datatype has a single unary constructor. Multiple curried constructors are modeled by disjoint sums ($+$) of products ($\times$). A (co)datatype definition corresponds to a fixpoint equation. For example, the equation $\beta = \mathit{unit} + \alpha \times \beta$ specifies either (finite) lists or lazy lists, depending on which fixpoint is chosen.

Bounded natural functors (BNFs) are a semantic criterion for where (co)recursion may appear on the right-hand side of an equation. The theory of BNFs is described in a previous paper [33] and in Traytel's M.Sc. thesis [32]. We refer to either of these for a discussion of related work. Here, we focus on implementational aspects.

There is a large gap between the low-level view and the end products presented to the user. The necessary infrastructure—including support for multiple curried constructors, generation of high-level characteristic theorems, and commands for specifying functions—constitutes a new contribution and is described in Sections 3 to 6.

Bounded Natural Functors. An n-ary BNF is a type constructor equipped with a map function (or functorial action), n set functions (or natural transformations), and a cardinal bound that satisfy certain properties. For example, llist is a unary BNF. Its relator lrel extends binary predicates over elements to binary predicates over lazy lists:

$$\mathsf{lrel}\, R\, xs\, ys = (\exists zs.\ \mathsf{lset}\, zs \subseteq \{(x, y) \mid R\, x\, y\} \wedge \mathsf{lmap}\, \mathsf{fst}\, zs = xs \wedge \mathsf{lmap}\, \mathsf{snd}\, zs = ys)$$

Additionally, lbd bounds the number of elements returned by the set function lset; it may not depend on α's cardinality. To prove that llist is a BNF, the greatest fixpoint operation discharges the following proof obligations:[1]

$$
\begin{array}{lll}
\mathsf{lmap}\ \mathsf{id} = \mathsf{id} & \mathsf{lmap}\ (f \circ g) = \mathsf{lmap}\ f \circ \mathsf{lmap}\ g & \bigwedge x.\ x \in \mathsf{lset}\ xs \Longrightarrow f\, x = g\, x \\
|\mathsf{lset}\ xs| \leq_o \mathsf{lbd} & \mathsf{lset} \circ \mathsf{lmap}\ f = \mathsf{image}\ f \circ \mathsf{lset} & \overline{\qquad \mathsf{lmap}\ f\ xs = \mathsf{lmap}\ g\ xs \qquad} \\
\aleph_0 \leq_o \mathsf{lbd} & \mathsf{lrel}\ R \circ\!\circ \mathsf{lrel}\ S \sqsubseteq \mathsf{lrel}\ (R \circ\!\circ S) &
\end{array}
$$

(The operator $\leq_o$ is a well-order on ordinals [8], $\sqsubseteq$ denotes implication lifted to binary predicates, and $\circ\!\circ$ denotes the relational composition of binary predicates.) Internally, the package stores BNFs as an ML structure that combines the functions, the basic properties, and derived facts such as $\mathsf{lrel}\ R \circ\!\circ \mathsf{lrel}\ S = \mathsf{lrel}\ (R \circ\!\circ S)$, $\mathsf{lrel}\ (\mathrm{op} =) = (\mathrm{op} =)$, and $R \sqsubseteq S \Longrightarrow \mathsf{lrel}\ R \sqsubseteq \mathsf{lrel}\ S$.

Given an n-ary BNF, the n type variables associated with set functions, and on which the map function acts, are *live*; any other variables are *dead*. The notation $\sigma\,\langle \overline{\alpha} \mid \Delta \rangle$ stands for a BNF of type σ depending on the (ordered) list of live variables $\overline{\alpha}$ and the set of dead variables Δ. Nested (co)recursion can only take place through live variables.

A two-step procedure introduces (co)datatypes as solutions to fixpoint equations:

1. Construct the BNFs for the right-hand sides of the equations by composition.
2. Perform the least or greatest fixpoint operation on the BNFs.

Whereas codatatypes are necessarily nonempty, some datatype definitions must be rejected in HOL. For example, **codatatype** $\alpha\ \mathit{stream} = \mathsf{SCons}\ (\mathsf{shd}\colon \alpha)\ (\mathsf{stl}\colon \alpha\ \mathit{stream})$,

---

[1] The list of proof obligations has evolved since our previous work [33]. The redundant cardinality condition $|\{xs \mid \mathsf{lset}\ xs \subseteq A\}| \leq_o (|A| + 2)^{\mathsf{lbd}}$ has been removed, and the preservation of weak pullbacks has been reformulated as a simpler property of the relator.

the type of infinite streams, can be defined only as a codatatype. In the general BNF setting, each functor must keep track of its nonemptiness witnesses [9].

The Fixpoint Operations. The LFP operation constructs a least fixpoint solution $\tau_1, \ldots, \tau_n$ to n mutual fixpoint equations $\beta_j = \sigma_j$. Its input consists of n BNFs sharing the same live variables [32, 33]:

LFP : n $(m+n)$-ary BNFs $\sigma_j \langle \overline{\alpha}, \overline{\beta} \,|\, \Delta_j \rangle \rightarrow$
- n m-ary BNFs $\tau_j \langle \overline{\alpha} \,|\, \Delta_1 \cup \cdots \cup \Delta_n \rangle$ for newly defined types τ_j
- n constructors ctor_τ_j :: $\sigma_j[\overline{\beta} \mapsto \overline{\tau}] \Rightarrow \tau_j$
- n iterators iter_τ_j :: $(\sigma_1 \Rightarrow \beta_1) \Rightarrow \cdots \Rightarrow (\sigma_n \Rightarrow \beta_n) \Rightarrow \tau_j \Rightarrow \beta_j$
- characteristic theorems including an induction rule

(The fixpoint variables β_j are harmlessly reused as result types of the iterators.) The contract for GFP, the greatest fixpoint, is identical except that coiterators and coinduction replace iterators and induction. The coiterator coiter_τ_j has type $(\beta_1 \Rightarrow \sigma_1) \Rightarrow \cdots \Rightarrow (\beta_n \Rightarrow \sigma_n) \Rightarrow \beta_j \Rightarrow \tau_j$. An iterator *consumes* a datatype, peeling off one constructor at a time; a coiterator *produces* a codatatype, delivering one constructor at a time.

LFP defines algebras and morphisms based on the equation system. The fixpoint, or initial algebra, is defined abstractly by well-founded recursion on a sufficiently large cardinal. The operation is defined only if the fixpoint is nonempty. In contrast, GFP builds a concrete tree structure [33]. This asymmetry is an accident of history—an abstract approach is also possible for GFP [30].

Nesting BNFs scales much better than the old package's reduction to mutual recursion [32, Appendix B]. On the other hand, LFP and GFP scale poorly in the number of mutual types; a 12-ary LFP takes about 10 minutes of CPU time on modern hardware. Reducing mutual recursion to nested recursion would circumvent the problem.

The ML functions that implement BNF operations all adhere to the same pattern: They introduce constants, state their properties, and discharge the proof obligations using dedicated tactics. About one fifth of the code base is devoted to tactics. They rely almost exclusively on resolution and unfolding, which makes them fast and reliable.

Methodologically, we developed the package in stages, starting with the formalization of a fixed abstract example $\beta = (\alpha, \beta, \gamma) F_0$ and $\gamma = (\alpha, \beta, \gamma) G_0$ specifying αF and αG. We axiomatized the BNF structure and verified the closure under LFP and GFP using structured Isar proofs. We then expanded the proofs to detailed **apply** scripts [6]. Finally, we translated the scripts into tactics and generalized them for arbitrary m and n.

The Composition Pipeline. Composing functors together is widely perceived as being trivial, and accordingly it has received little attention in the literature, including our previous paper [33]. Nevertheless, an implementation must perform a carefully orchestrated sequence of steps to construct BNFs (and discharge the accompanying proof obligations) for the types occurring on the right-hand sides of fixpoint equations. This is achieved by four operations:

COMPOSE : m-ary BNF $\sigma \langle \overline{\alpha} \,|\, \Delta \rangle$ and m n-ary BNFs $\tau_i \langle \overline{\beta} \,|\, \Theta_i \rangle \rightarrow$
n-ary BNF $\sigma[\overline{\alpha} \mapsto \overline{\tau}] \langle \overline{\beta} \,|\, \Delta \cup \Theta_1 \cup \cdots \cup \Theta_m \rangle$

KILL : m-ary BNF $\sigma \langle \overline{\alpha} \,|\, \Delta \rangle$ and $k \leq m \rightarrow$
$(m-k)$-ary BNF $\sigma \langle \alpha_{k+1}, \ldots, \alpha_m \,|\, \Delta \cup \{\alpha_1, \ldots, \alpha_k\} \rangle$

LIFT : m-ary BNF $\sigma \langle \overline{\alpha} \,|\, \Delta \rangle$ and n fresh type variables $\overline{\beta}$ →
          $(m+n)$-ary BNF $\sigma \langle \overline{\beta}, \overline{\alpha} \,|\, \Delta \rangle$

PERMUTE : m-ary BNF $\sigma \langle \overline{\alpha} \,|\, \Delta \rangle$ and permutation π of $\{1, \ldots, m\}$ →
             m-ary BNF $\sigma \langle \alpha_{\pi(1)}, \ldots, \alpha_{\pi(m)} \,|\, \Delta \rangle$

COMPOSE operates on BNFs normalized to share the same live variables; the other operations perform this normalization. Traytel's M.Sc. thesis [32] describes all of them in more detail. Complex types are proved to be BNFs by applying normalization followed by COMPOSE recursively. The base cases are manually registered as BNFs. These include constant $\alpha \langle \,|\, \alpha \rangle$, identity $\alpha \langle \alpha \,|\, \rangle$, sum $\alpha + \beta \langle \alpha, \beta \,|\, \rangle$, product $\alpha \times \beta \langle \alpha, \beta \,|\, \rangle$, and restricted function space $\alpha \Rightarrow \beta \langle \beta \,|\, \alpha \rangle$. Users can register further types, such as those introduced by the new package for non-free datatypes [31].

As an example, consider the type $(\alpha \Rightarrow \beta) + \gamma \times \alpha$. The recursive calls on the arguments to $+$ return the BNFs $\alpha \Rightarrow \beta \langle \beta \,|\, \alpha \rangle$ and $\gamma \times \alpha \langle \gamma, \alpha \,|\, \rangle$. Since α is dead in $\alpha \Rightarrow \beta$, it must be killed in $\gamma \times \alpha$ as well. This is achieved by permuting α to be the first variable and killing it, yielding $\gamma \times \alpha \langle \gamma \,|\, \alpha \rangle$. Next, both BNFs are lifted to have the same set of live variables: $\alpha \Rightarrow \beta \langle \gamma, \beta \,|\, \alpha \rangle$ and $\gamma \times \alpha \langle \beta, \gamma \,|\, \alpha \rangle$. Another permutation ensures that the live variables appear in the same order: $\alpha \Rightarrow \beta \langle \beta, \gamma \,|\, \alpha \rangle$. At this point, the BNF for $+$ can be composed with the normalized BNFs to produce $(\alpha \Rightarrow \beta) + \gamma \times \alpha \langle \beta, \gamma \,|\, \alpha \rangle$.

The compositional approach to BNF construction keeps the tactics simple at the expense of performance. By inlining intermediate definitions and deriving auxiliary BNF facts lazily, we were able to address the main bottlenecks. Nevertheless, Brian Huffman has demonstrated in a private prototype that a noncompositional, monolithic approach is also feasible and less heavy, although it requires more sophisticated tactics.

Nested-to-Mutual Reduction. The old **datatype** command reduces nested recursion to mutual recursion, as proposed by Gunter [11]. Given a nested datatype specification such as $\alpha\ tree$ = Tree α ($\alpha\ tree\ list$), the old command first unfolds the definition of *list*, resulting in the mutual specification of trees and "lists of trees," as if the user had entered

datatype $\alpha\ tree$ = Tree α ($\alpha\ treelist$) **and** $\alpha\ treelist$ = Nil | Cons ($\alpha\ tree$) ($\alpha\ treelist$)

In a second step, the package translates all occurrences of $\alpha\ treelist$ into the more palatable $\alpha\ tree\ list$ via an isomorphism. As a result, the induction principle and the input syntax to **primrec** have an unmistakable mutual flavor.

For compatibility, and for the benefit of users who prefer the mutual approach, the new package implements a nested-to-mutual reduction operation, N2M, that constructs old-style induction principles and iterators from those produced by LFP:

N2M : n $(m+n)$-ary BNFs $\sigma_j \langle \overline{\alpha}, \overline{\beta} \,|\, \Delta_j \rangle$ and n (new-style) datatypes τ_j →
- n iterators n2m_iter_τ_j :: $(\sigma_1 \Rightarrow \beta_1) \Rightarrow \cdots \Rightarrow (\sigma_n \Rightarrow \beta_n) \Rightarrow \tau_j \Rightarrow \beta_j$
- characteristic theorems including an induction rule

Like LFP and GFP, the N2M operation takes a system of equations $\beta_j = \sigma_j$ given as normalized BNFs. In addition, it expects a list of datatypes τ_j produced by LFP that solve the equations and that may nest each other (e.g., $\alpha\ tree$ and $\alpha\ tree\ list$). The operation is dual for codatatypes; its implementation is a single ML function that reverses some of the function and implication arrows when operating on codatatypes.

The **primrec** and **primcorec** commands invoke N2M when they detect nested (co)datatypes used as if they were mutual. In addition, **datatype_compat** relies on N2M to register new-style nested datatypes as old-style datatypes, which is useful for interfacing with existing unported infrastructure. In contrast to Gunter's approach, N2M does not introduce any new types. Instead, it efficiently composes artifacts of the fixpoint operations: (co)iterators and (co)induction rules. By giving N2M a similar interface to LFP and GFP, we can use it uniformly in the rest of the (co)datatype code. The description below is admittedly rather technical, because there is (to our knowledge) no prior account of such an operation in the literature.

As an abstract example that captures most of the complexity of N2M, let $(\alpha, \beta) F_0$ and $(\alpha, \beta) G_0$ be arbitrary BNFs with live type variables α and β. Let αF be the LFP of $\beta = (\alpha, \beta) F_0$ and αG be the LFP of $\beta = (\alpha, \beta F) G_0$ (assuming they exist).[2] These two definitions reflect the modular, nested view: First αF is defined as an LFP, becoming a BNF in its own right; then αG is defined as an LFP using an equation that nests F. The resulting iterator for αG, iter_G, has type $((\alpha, \gamma F) G_0 \Rightarrow \gamma) \Rightarrow \alpha G \Rightarrow \gamma$, and its characteristic equation recurses through the F components of G using map_F.

If we instead define αG_M ($\simeq \alpha G$) and αGF_M ($\simeq \alpha G F$) together in the old-style, mutually recursive fashion as the LFP of $\beta = (\alpha, \gamma) G_0$ and $\gamma = (\beta, \gamma) F_0$, we obtain two iterators with the following types:

$$\text{iter\_G}_M \; :: ((\alpha, \gamma) G_0 \Rightarrow \beta) \Rightarrow ((\beta, \gamma) F_0 \Rightarrow \gamma) \Rightarrow \alpha G_M \; \Rightarrow \beta$$
$$\text{iter\_GF}_M :: ((\alpha, \gamma) G_0 \Rightarrow \beta) \Rightarrow ((\beta, \gamma) F_0 \Rightarrow \gamma) \Rightarrow \alpha GF_M \Rightarrow \gamma$$

These are more flexible: iter_GF$_M$ offers the choice of indicating recursive behavior other than a map for the αGF_M components of αG_M. The gap is filled by N2M, which defines mutual iterators by combining the standard iterators for F and G. It does not introduce any new types αG_M and αGF_M but works with the existing ones:

$$\text{n2m\_iter\_G} \quad :: ((\alpha, \gamma) G_0 \Rightarrow \beta) \Rightarrow ((\beta, \gamma) F_0 \Rightarrow \gamma) \Rightarrow \alpha G \quad \Rightarrow \beta$$
$$\text{n2m\_iter\_G\_F} :: ((\alpha, \gamma) G_0 \Rightarrow \beta) \Rightarrow ((\beta, \gamma) F_0 \Rightarrow \gamma) \Rightarrow \alpha G F \Rightarrow \gamma$$

$$\text{n2m\_iter\_G} \quad g \, f = \text{iter\_G} \, (g \circ \text{map\_G}_0 \, \text{id} \, (\text{iter\_F} \, f))$$
$$\text{n2m\_iter\_G\_F} \, g \, f = \text{iter\_F} \, (f \circ \text{map\_F}_0 \, (\text{n2m\_iter\_G} \, g \, f) \, \text{id})$$

N2M also outputs a corresponding mutual induction rule. For each input BNF, the operation first derives the low-level relator induction rule—a higher-order version of parallel induction on two values of the same shape (e.g., lists of the same length):

$$\frac{\bigwedge x \, x'. \; \text{rel\_G}_0 \, P \, (\text{rel\_F} \, R) \, x \, x' \Longrightarrow R \, (\text{ctor\_G} \, x) \, (\text{ctor\_G} \, x')}{\text{rel\_G} \, P \sqsubseteq R}$$

$$\frac{\bigwedge y \, y'. \; \text{rel\_F}_0 \, R \, S \, y \, y' \Longrightarrow S \, (\text{ctor\_F} \, y) \, (\text{ctor\_F} \, y')}{\text{rel\_F} \, R \sqsubseteq S}$$

The binary predicates R and S are the properties we want to prove on αG and $\alpha G F$. The left-hand sides of the of lifted implications $\sqsubseteq$ ensure that the two values related by R or S have the same shape. The binary predicate P relates α elements. The antecedents of the rules are the induction steps. The left-hand sides of the implications $\Longrightarrow$ are the

---

[2] It may help to think of these types more concretely by taking

$$F := list \quad G := tree \quad (\alpha, \beta) F_0 := unit + \alpha \times \beta \quad (\alpha, \beta) G_0 := \alpha \times \beta$$

induction hypotheses; they ensure that R and S hold for all parallel, direct subterms with types $\alpha\,G$ and $\alpha\,G\,F$ of the values for which we need to prove the step.

The relators are compositional, enabling a modular proof of the mutual relator induction rule from the above rules and relator monotonicity of G_0 and F_0:

$$\frac{\bigwedge x\,x'.\ \mathsf{rel\_G_0}\ P\ S\ x\ x' \Longrightarrow R\ (\mathsf{ctor\_G}\ x)\ (\mathsf{ctor\_G}\ x') \quad \bigwedge y\,y'.\ \mathsf{rel\_F_0}\ R\ S\ y\ y' \Longrightarrow S\ (\mathsf{ctor\_F}\ y)\ (\mathsf{ctor\_F}\ y')}{\mathsf{rel\_G}\ P \sqsubseteq R \wedge \mathsf{rel\_F}\ (\mathsf{rel\_G}\ P) \sqsubseteq S}$$

The standard induction rule is derived by instantiating $P :: \alpha \Rightarrow \alpha' \Rightarrow bool$ with equality, followed by some massaging. Coinduction is dual, with $\Longrightarrow$ and $\sqsubseteq$ reversed.

3 Types with Free Constructors

Datatypes and codatatypes are instances of types equipped with free constructors. Such types are useful in their own right, regardless of whether they support induction or coinduction; for example, pattern matching requires only distinctness and injectivity.

We have extended Isabelle with a database of freely constructed types. Users can enter the **free_constructors** command to register custom types, by listing the constructors and proving exhaustiveness, distinctness, and injectivity. In exchange, Isabelle generates constants for case expressions, discriminators, and selectors—collectively called *destructors*—as well as a wealth of theorems about constructors and destructors. Our new **datatype** and **codatatype** commands use this functionality internally.

The case constant is defined via the definite description operator ($\imath$)—for example, $\mathsf{case\_list}\ n\ c\ xs = (\imath z.\ xs = \mathsf{Nil} \wedge z = n \vee (\exists y\,ys.\ xs = \mathsf{Cons}\ y\,ys \wedge z = c\,y\,ys))$. Syntax translations render $\mathsf{case\_list}\ n\ c\ xs$ as an ML-style case expression.

Given a type τ constructed by $C_1, \dots, C_m$, its *discriminators* are constants $\mathsf{is\_C_1}, \dots,$ $\mathsf{is\_C_m} :: \tau \Rightarrow bool$ such that $\mathsf{is\_C_i}\ (C_j\ \bar{x})$ if and only if $i = j$. No discriminators are needed if $m = 1$. For the $m = 2$ case, Isabelle generates a single discriminator and uses its negation for the second constructor by default. For nullary constructors C_i, Isabelle can be told to use $\lambda x.\ x = C_i$ as the discriminator in the theorems it generates.

In addition, for each n-ary constructor $C_i :: \tau_1 \Rightarrow \cdots \Rightarrow \tau_n \Rightarrow \tau$, n *selectors* $\mathsf{un\_C_{ij}} ::$ $\tau \Rightarrow \tau_j$ extract its arguments. Users can reuse selector names across constructors. They can also specify a default value for constructors on which a selector would otherwise be unspecified. The example below defines four selectors and assigns reasonable default values. The mid selector returns the third argument of $\mathsf{Node2}\ x\,l\,r$ as a default:

> **datatype** $\alpha\ tree_{23} =$
> Leaf (**defaults** left: Leaf mid: Leaf right: Leaf)
> | Node2 (val: α) (left: $\alpha\ tree_{23}$) (right: $\alpha\ tree_{23}$) (**defaults** mid: $\lambda x\,l\,r.\,r$)
> | Node3 (val: α) (left: $\alpha\ tree_{23}$) (mid: $\alpha\ tree_{23}$) (right: $\alpha\ tree_{23}$)

4 (Co)datatypes

The **datatype** and **codatatype** commands share the same input syntax, consisting of a list of mutually (co)recursive types to define, their desired constructors, and optional information such as custom names for destructors. They perform the following steps:

1. Formulate and solve the fixpoint equations using LFP or GFP.
2. Define the constructor constants.
3. Generate the destructors and the free constructor theorems.
4. Derive the high-level map, set, and relator theorems.
5. Define the high-level (co)recursor constants.
6. Derive the high-level (co)recursor theorems and (co)induction rules.

Step 1 relies on the fixpoint and composition operations described in Section 2 to produce the desired types and low-level constants and theorems. Step 2 defines high-level constructors that untangle sums of products—for example, $\mathsf{Nil} = \mathsf{ctor\_list}\,(\mathsf{Inl}\,())$ and $\mathsf{Cons}\,x\,xs = \mathsf{ctor\_list}\,(\mathsf{Inr}\,(x, xs))$. Step 3 amounts to an invocation of the **free_ constructors** command described in Section 3. Step 4 reformulates the low-level map, set, and relator theorems in terms of constructors; a selection is shown for α *list* below:

list.map: $\mathsf{map}\,f\,\mathsf{Nil} = \mathsf{Nil}$ $\mathsf{map}\,f\,(\mathsf{Cons}\,x\,xs) = \mathsf{Cons}\,(f\,x)\,(\mathsf{map}\,f\,xs)$

list.set: $\mathsf{set}\,\mathsf{Nil} = \{\}$ $\mathsf{set}\,(\mathsf{Cons}\,x\,xs) = \{x\} \cup \mathsf{set}\,xs$

list.rel_inject: $\mathsf{rel}\,R\,\mathsf{Nil}\,\mathsf{Nil}$ $\mathsf{rel}\,R\,(\mathsf{Cons}\,x\,xs)\,(\mathsf{Cons}\,y\,ys) \longleftrightarrow R\,x\,y \wedge \mathsf{rel}\,R\,xs\,ys$

Datatypes and codatatypes differ at step 5. For an m-constructor datatype, the high-level iterator takes m curried functions as arguments (whereas the low-level version takes one function with a sum-of-product domain). For convenience, a *recursor* is defined in terms of the iterator to provide each recursive constructor argument's value both before and after the recursion. The list recursor has type $\beta \Rightarrow (\alpha \Rightarrow \alpha\,list \Rightarrow \beta \Rightarrow \beta) \Rightarrow \alpha\,list \Rightarrow \beta$. The corresponding induction rule has one hypothesis per constructor:

list.rec: $\mathsf{rec\_list}\,n\,c\,\mathsf{Nil} = n$ $\mathsf{rec\_list}\,n\,c\,(\mathsf{Cons}\,x\,xs) = c\,x\,xs\,(\mathsf{rec\_list}\,n\,c\,xs)$

list.induct: $$\frac{P\,\mathsf{Nil} \quad \bigwedge x\,xs.\;P\,xs \Longrightarrow P\,(\mathsf{Cons}\,x\,xs)}{P\,t}$$

For nested recursion beyond sums of products, the map and set functions of the type constructors through which recursion takes place appear in the high-level theorems:

tree$_\omega$.rec: $\mathsf{rec\_tree}_\omega\,f\,(\mathsf{Tree}_\omega\,x\,ts) = f\,x\,(\mathsf{lmap}\,(\lambda t.\,(t, \mathsf{rec\_tree}_\omega\,f\,t))\,ts)$

tree$_\omega$.induct: $$\frac{\bigwedge x\,ts.\;(\bigwedge t.\,t \in \mathsf{lset}\,ts \Longrightarrow P\,t) \Longrightarrow P\,(\mathsf{Tree}_\omega\,x\,ts)}{P\,t}$$

As for corecursion, given an m-constructor codatatype, $m-1$ predicates sequentially determine which constructor to produce. Moreover, for each constructor argument, a function specifies how to construct it from an abstract value of type α. For corecursive arguments, the function has type $\alpha \Rightarrow \tau + \alpha$ and returns either a value that stops the corecursion or a tuple of arguments to a corecursive call. The high-level corecursor presents such functions as three arguments $stop :: \alpha \Rightarrow bool$, $end :: \alpha \Rightarrow \tau$, and $continue :: \alpha \Rightarrow \alpha$, abbreviated to s, e, c below. Thus, the high-level corecursor for lazy lists has the type $(\alpha \Rightarrow bool) \Rightarrow (\alpha \Rightarrow \beta) \Rightarrow (\alpha \Rightarrow bool) \Rightarrow (\alpha \Rightarrow \beta\,llist) \Rightarrow (\alpha \Rightarrow \alpha) \Rightarrow \alpha \Rightarrow \beta\,llist$:

llist.corec: $n\,a \Longrightarrow \mathsf{corec\_llist}\,n\,h\,s\,e\,c\,a = \mathsf{LNil}$

$\neg\,n\,a \Longrightarrow \mathsf{corec\_llist}\,n\,h\,s\,e\,c\,a =$

$\mathsf{LCons}\,(h\,a)\,(\text{if } s\,a \text{ then } e\,a \text{ else } \mathsf{corec\_llist}\,n\,h\,s\,e\,c\,(c\,a))$

Nested corecursion is expressed using the map functions of the nesting type constructors. The coinduction rule uses the relators to lift a coinduction witness R. For example:

$ltree.corec$: $\text{corec\_ltree } l\ s\ a =$
$\qquad \text{LTree } (l\ a)\ (\text{map } (\text{case\_sum } (\lambda t.\ t)\ (\text{corec\_ltree } l\ s))\ (s\ a))$

$ltree.coinduct$: $$\frac{R\ t\ u \quad \bigwedge t\ u.\ R\ t\ u \Longrightarrow \text{llab } t = \text{llab } u \wedge \text{rel } R\ (\text{lsub } t)\ (\text{lsub } u)}{t = u}$$

5 Recursive Functions

Primitively recursive functions can be defined by providing suitable arguments to the
recursors. The **primrec** command automates this process: From recursive equations
specified by the user, it synthesizes a recursor-based definition.

The main improvement of the new implementation of **primrec** over the old one is its
support for nested recursion through map functions [27]. For example:

primrec $\text{height\_tree}_{\text{FS}} :: \alpha\ tree_{\text{FS}} \Rightarrow nat$ **where**
$\quad \text{height\_tree}_{\text{FS}}\ (\text{Tree}_{\text{FS}}\ \_\ T) = 1 + \bigsqcup \text{fset } (\text{fimage height\_tree}_{\text{FS}}\ T)$

In the above, $\alpha\ tree_{\text{FS}}$ is the datatype constructed by $\text{Tree}_{\text{FS}} :: \alpha \Rightarrow \alpha\ tree_{\text{FS}}\ fset \Rightarrow \alpha\ tree_{\text{FS}}$
(Section 1), $\bigsqcup N$ stands for the maximum of N, fset injects $\alpha\ fset$ into $\alpha\ set$, and the
map function fimage gives the image of a finite set under a function. From the specified
equation, the command synthesizes the definition

$\quad \text{height\_tree}_{\text{FS}} = \text{rec\_tree}_{\text{FS}}\ (\lambda\_\ TN.\ 1 + \bigsqcup \text{fset } (\text{fimage snd } TN))$

From this definition and the $tree_{\text{FS}}.rec$ theorems, it derives the original specification as a
theorem. Notice how the argument $T :: \alpha\ tree_{\text{FS}}\ fset$ becomes $TN :: (\alpha\ tree_{\text{FS}} \times nat)\ fset$,
where the second pair components store the result of the corresponding recursive call.

Briefly, constructor arguments x are transformed as follows. Nonrecursive arguments
appear unchanged in the recursor and can be used directly. Directly or mutually recursive arguments appear as two values: the original value x and the value y after the
recursive call to f. Calls f x are replaced by y. Nested recursive arguments appear as a
single argument but with pairs inside the nesting type constructors. The syntactic transformation must follow the map functions and eventually apply fst or snd, depending on
whether a recursive call takes place. Naked occurrences of x without map are replaced
by a suitable "map fst" term; for example, if the constant 1 were changed to fcard T in
the specification above, the definition would have fcard (fimage fst TN) in its place.

The implemented procedure is somewhat more complicated. The recursor generally
defines functions of type $\alpha\ tree_{\text{FS}} \Rightarrow \beta$, but **primrec** needs to process n-ary functions that
recurse on their jth argument. This is handled internally by moving the jth argument to
the front and by instantiating β with an $(n-1)$-ary function type.

For recursion through functions, the map function is function composition ($\circ$). Instead of $f \circ g$, **primrec** also allows the convenient (and backward compatible) syntax
$\lambda x.\ f\ (g\ x)$. More generally, $\lambda x_1 \ldots x_n.\ f\ (g\ x_1 \ldots x_n)$ expands to $(\text{op} \circ (\ldots (\text{op} \circ f) \ldots))\ g$.

Thanks to the N2M operation described in Section 2, users can also define mutually
recursive functions on nested datatypes, as they would have done with the old package:

primrec height_tree :: α *tree* $\Rightarrow$ *nat* **and** height_trees :: α *tree list* $\Rightarrow$ *nat* **where**
  height_tree (Tree _ *ts*)   = 1 + height_trees *ts*
  | height_trees Nil   = 0
  | height_trees (Cons *t ts*) = height_tree *t* $\sqcup$ height_trees *ts*

Internally, the following steps are performed:

1. Formulate and solve the fixpoint equations using N2M.
2. Define the high-level (co)recursor constants.
3. Derive the high-level (co)recursor theorems and (co)induction rules.

Step 1 produces low-level constants and theorems. Steps 2 and 3 are performed by the same machinery as when declaring mutually recursive datatypes (Section 4).

6   Corecursive Functions

The **primcorec** command is the main mechanism to introduce functions that produce potentially infinite codatatype values; alternatives based on domain theory and topology are described separately [22]. The command supports three competing syntaxes, or *views*: *destructor*, *constructor*, and *code*. Irrespective of the view chosen for input, the command generates the characteristic theorems for all three views [27].

The Destructor View. The coinduction literature tends to favor the destructor view, perhaps because it best reflects the duality between datatypes and codatatypes [1, 16]. The append function on lazy lists will serve as an illustration:

primcorec lapp :: α *llist* $\Rightarrow$ α *llist* $\Rightarrow$ α *llist* **where**
  lnull *xs* $\Longrightarrow$ lnull *ys* $\Longrightarrow$ lnull (lapp *xs ys*)
  | lhd (lapp *xs ys*) = lhd (if lnull *xs* then *ys* else *xs*)
  | ltl (lapp *xs ys*) = (if lnull *xs* then ltl *ys* else lapp (ltl *xs*) *ys*)

The first formula, called the *discriminator formula*, gives the condition on which LNil should be produced. For an m-constructor datatype, up to m discriminator formulas can be given. If exactly $m - 1$ formulas are stated (as in the example above), the last one is implicitly understood, with the complement of the other conditions as its condition.

The last two formulas, the *selector equations*, describe the behavior of the function when an LCons is produced. They are implicitly conditional on $\neg$ lnull *xs* $\vee$ $\neg$ lnull *ys*. The right-hand sides consist of 'let', 'if', or 'case' expressions whose leaves are either corecursive calls or arbitrary non-corecursive terms. This restriction ensures that the definition qualifies as primitively corecursive. The selector patterns on the left ensure that the function is productive and hence admissible [16].

With nesting, the corecursive calls appear under a map function, in much the same way as for **primrec**. Intuitive λ syntaxes for corecursion via functions are supported. The nested-to-mutual reduction is available for corecursion through codatatypes.

Proof obligations are emitted to ensure that the conditions are mutually exclusive. These are normally given to *auto* but can also be proved manually. Alternatively, users can specify the **sequential** option to have the conditions apply in sequence.

The conditions need not be exhaustive, in which case the function's behavior is left underspecified. If the conditions are syntactically detected to be exhaustive, or if the user enables the **exhaustive** option and discharges its proof obligation, the package generates stronger theorems—notably, discriminator formulas with $\longleftrightarrow$ instead of $\Longrightarrow$.

The Constructor View. The constructor view can be thought of as an abbreviation for the destructor view. It involves a single conditional equation per constructor:

> **primcorec** lapp :: α *llist* $\Rightarrow$ α *llist* $\Rightarrow$ α *llist* **where**
> lnull xs $\Longrightarrow$ lnull ys $\Longrightarrow$ lapp xs ys = LNil
> | _ $\Longrightarrow$ lapp xs ys = LCons (lhd (if lnull xs then ys else xs))
>   (if lnull xs then ltl ys else lapp (ltl xs) ys)

The wildcard _ stands for the complement of the previous conditions.

This view is convenient as input and sometimes for reasoning, but the equations are generally not suitable as simplification rules since they can loop. Compare this with the discriminator formulas and the selector equations of the destructor view, which can be safely registered as simplification rules.

The Code View. The code view is a variant of the constructor view in which the conditions are expressed using 'if' and 'case' expressions. Its primary purpose is for interfacing with Isabelle's code generator, which cannot cope with conditional equations.

The code view that **primcorec** generates from a destructor or constructor view is simply an equation that tests the conditions sequentially using 'if':

> lapp xs ys = (if lnull xs $\wedge$ lnull ys then LNil
> else LCons (lhd (if lnull xs then ys else xs)) (if lnull xs then ltl ys else lapp (ltl xs) ys))

If the cases are not known to be exhaustive, an additional 'if' branch ensures that the generated code throws an exception when none of the conditions are met.

The code view has a further purpose besides code generation: It provides a more flexible input format, with nested 'let', 'if', and 'case' expressions outside the constructors, multiple occurrences of the same constructors, and non-corecursive branches without constructor guards. This makes the code view the natural choice for append:

> **primcorec** lapp :: α *llist* $\Rightarrow$ α *llist* $\Rightarrow$ α *llist* **where**
> lapp xs ys = (case xs of LNil $\Rightarrow$ ys | LCons x xs $\Rightarrow$ LCons x (lapp xs ys))

The package reduces this specification to the following constructor view:

> lnull xs $\Longrightarrow$ lnull ys $\Longrightarrow$ lapp xs ys = LNil
> _ $\Longrightarrow$ lapp xs ys = LCons (case xs of LNil $\Rightarrow$ lhd ys | LCons x _ $\Rightarrow$ x)
>   (case xs of LNil $\Rightarrow$ ltl ys | LCons _ xs $\Rightarrow$ lapp xs ys)

In general, the reduction proceeds as follows:

1. Expand branches t of the code equation that are not guarded by a constructor to the term (case t of C_1 $\bar{x}_1$ $\Rightarrow$ C_1 $\bar{x}_1$ | $\cdots$ | C_m $\bar{x}_m$ $\Rightarrow$ C_m $\bar{x}_m$), yielding an equation χ.
2. Gather the conditions Φ_i associated with the branches guarded by C_i by traversing χ, with 'case' expressions recast as 'if's.
3. Generate the constructor equations $\bigvee \Phi_i$ $\bar{x}$ $\Longrightarrow$ f $\bar{x}$ = C_i (un_C_{i1} χ) $\ldots$ (un_C_{ij} χ), taking care of moving the un_C_{ij}'s under the conditionals and of simplifying them.

For the append example, step 1 expands the ys in the first 'case' branch to the term (case ys of LNil $\Rightarrow$ LNil | LCons y ys $\Rightarrow$ LCons y ys).

Finally, although **primcorec** does not allow pattern matching on the left-hand side, the **simps_of_case** command developed by Gerwin Klein and Lars Noschinski can be used to generate the pattern-matching equations from the code view—in our example, lapp LNil $ys = ys$ and lapp (LCons x xs) $ys =$ LCons x (lapp xs ys).

7   Coinduction Proof Method

The previous sections focused on the infrastructure for defining coinductive objects. Also important are the user-level proof methods, the building blocks of reasoning. The new method *coinduction* provides more automation over the existing *coinduct*, following a suggestion formulated in Lochbihler's Ph.D. thesis [20, Section 7.2]. The method handles arbitrary predicates equipped with suitable coinduction theorems. In particular, it can be used to prove equality of codatatypes by exhibiting a bisimulation.

A coinduction rule for a codatatype contains a free bisimulation relation variable R in its premises, which does not occur in the conclusion. The *coinduct* method crudely leaves R uninstantiated; the user is expected to provide the instantiation. However, the choice of the bisimulation is often canonical, as illustrated by the following proof:

> **lemma**
>   **assumes** infinite (lset xs)
>   **shows** lapp xs $ys = xs$
> **proof** (*coinduct xs*)
>   **def** [*simp*]: $R \equiv \lambda l\, r.\ \exists xs.\ l =$ lapp xs $ys \wedge r = xs \wedge$ infinite (lset xs)
>   **with** *assms* **show** R (lapp xs ys) xs **by** *auto*
>
>   **fix** $l\, r$ **assume** $R\, l\, r$
>   **then obtain** xs **where** $l =$ lapp xs $ys \wedge r = xs \wedge$ infinite (lset xs) **by** *auto*
>   **thus** lnull $l =$ lnull $r \wedge (\neg$ lnull $l \longrightarrow \neg$ lnull $r \longrightarrow$ lhd $l =$ lhd $r \wedge R$ (ltl l) (ltl r))
>   **by** *auto*
> **qed**

The new method performs the steps highlighted in gray automatically, making a one-line proof possible: **by** (*coinduction arbitrary: xs*) *auto*.

In general, given a goal $P \Longrightarrow$ q $t_1 \ldots t_n$, the method selects the rule q.*coinduct* and takes $\lambda z_1 \ldots z_n.\ \exists x_1 \ldots x_m.\ z_1 = t_1 \wedge \cdots \wedge z_n = t_n \wedge P$ as the coinduction witness R. The variables x_i are those specified as being *arbitrary* and may freely appear in $P, t_1, \ldots, t_n$. After applying the instantiated rule, the method discharges the premise $R\, t_1 \ldots t_n$ by reflexivity and using the assumption P. Then it unpacks the existential quantifiers from R.

8   Example: Stream Processors

Stream processors were introduced by Hancock et al. [13] and have rapidly become the standard example for demonstrating mixed fixpoints [1, 3, 10, etc.]. Thanks to the new (co)datatype package, Isabelle finally joins this good company.

A stream processor represents a continuous transformation on streams—that is, a function of type $\alpha\ stream \Rightarrow \beta\ stream$ that consumes at most a finite prefix of the input stream before producing an element of output. The datatype sp_1 captures a single iteration of this process. The codatatype sp_ω nests sp_1 to produce an entire stream:

datatype $(\alpha, \beta, \delta)\, sp_1 = \text{Get}\ (\alpha \Rightarrow (\alpha, \beta, \delta)\, sp_1) \mid \text{Put}\,\beta\,\delta$

codatatype $(\alpha, \beta)\, sp_\omega = \text{SP}\ (\text{unSP}\colon (\alpha, \beta, (\alpha, \beta)\, sp_\omega)\, sp_1)$

Values of type sp_1 are finite-depth trees with inner nodes Get and leaf nodes Put. Each inner node has $|\alpha|$ children, one for each possible input α. The Put constructor carries the output element of type β and a continuation of type δ. The definition of sp_ω instantiates the continuation type to a stream processor $(\alpha, \beta)\, sp_\omega$.

The semantics of sp_ω is given by two functions: run_1 recurses on sp_1 (i.e., consumes an sp_1), and run_ω, corecurses on *stream* (i.e., produces a *stream*, defined in Section 2):

primrec $\text{run}_1 \colon\colon (\alpha, \beta, \delta)\, sp_1 \Rightarrow \alpha\ stream \Rightarrow (\beta \times \delta) \times \alpha\ stream$ **where**
    $\text{run}_1\ (\text{Get } f)\quad s = \text{run}_1\ (f\ (\text{shd } s))\ (\text{stl } s)$
    $\text{run}_1\ (\text{Put } x\ q)\ s = ((x, q), s)$

primcorec $\text{run}_\omega \colon\colon (\alpha, \beta)\, sp_\omega \Rightarrow \alpha\ stream \Rightarrow \beta\ stream$ **where**
    $\text{run}_\omega\ q\ s = (\text{let } ((x, q'), s') = \text{run}_1\ (\text{unSP } q)\ s \text{ in } \text{SCons } x\ (\text{run}_\omega\ q'\ s'))$

These definitions illustrate some of the conveniences of **primrec** and **primcorec**. For run_1, the modular way to nest the recursive call of run_1 through functions would rely on composition—i.e., $(\text{run}_1 \circ f)\ (\text{shd } s)\ (\text{stl } s)$. The **primrec** command allows us not only to expand the term $\text{run}_1 \circ f$ to $\lambda x.\ \text{run}_1\ (f\ x)$ but also to β-reduce it. For run_ω, the constructor view makes it possible to call run_1 only once, assign the result in a 'let', and use this result to specify both arguments of the produced constructor.

The stream processor copy outputs the input stream:

primcorec copy $\colon\colon (\alpha, \alpha)\, sp_\omega$ **where** copy $= \text{SP}\ (\text{Get}\ (\lambda a.\ \text{Put } a\ \text{copy}))$

The nested sp_1 value is built directly with corecursion under constructors as an alternative to the modular approach: copy $= \text{SP}\ (\text{map\_sp}_1\ \text{id}\ (\lambda\_.\ \text{copy})\ (\text{Get}\ (\lambda a.\ \text{Put } a\ ())))$. The lemma run_ω copy $s = s$ is easy to prove using *coinduction* and *auto*.

Since stream processors represent functions, it makes sense to compose them:

function $\circ_1 \colon\colon (\beta, \gamma, \delta)\, sp_1 \Rightarrow (\alpha, \beta, (\alpha, \beta)\, sp_\omega)\, sp_1 \Rightarrow (\alpha, \gamma, \delta \times (\alpha, \beta)\, sp_\omega)\, sp_1$ **where**
    $\text{Put } b\ q \circ_1 p\qquad\quad = \text{Put } b\ (q, \text{SP } p)$
    $\text{Get } f\quad \circ_1 \text{Put } b\ q = f\ b \circ_1 \text{unSP } q$
    $\text{Get } f\quad \circ_1 \text{Get } g\ = \text{Get}\ (\lambda a.\ \text{Get } f \circ_1 g\ a)$
by *pat_completeness auto*
termination by (*relation* lex_prod sub sub) *auto*

primcorec $\circ_\omega \colon\colon (\beta, \gamma)\, sp_\omega \Rightarrow (\alpha, \beta)\, sp_\omega \Rightarrow (\alpha, \gamma)\, sp_\omega$ **where**
    $\text{unSP}\ (q \circ_\omega q') = \text{map\_sp}_1\ (\lambda b.\ b)\ (\lambda(q, q').\ q \circ_\omega q')\ (\text{unSP } q \circ_1 \text{unSP } q')$

The corecursion applies $\circ_\omega$ nested through the map function map_sp$_1$ to the result of finite preprocessing by the recursion $\circ_1$. The $\circ_1$ operator is defined using **function**, which emits proof obligations concerning pattern matching and termination.

Stream processors are an interesting example to compare Isabelle with other proof assistants. Agda does not support nesting, but it supports the simultaneous mutual definition of sp_1 and sp_ω with annotations on constructor arguments indicating whether they are to be understood coinductively [10]. Least fixpoints are always taken before greatest fixpoints, which is appropriate for this example but is less flexible than nesting in general. PVS [26] and Coq [5] support nesting of datatypes through datatypes and codatatypes through codatatypes, but no nontrivial mixtures.[3]

9   Case Study: Porting the Coinductive Library

To evaluate the merits of the new definitional package, and to benefit from them, we have ported existing coinductive developments to the new approach. The Coinductive library [18] defines four codatatypes and related functions and comprises a large collection of lemmas. Originally, the codatatypes were manually defined as follows:

- extended naturals *enat* as **datatype** *enat* = enat *nat* | ∞;
- lazy lists α *llist* using Paulson's construction [29];
- terminated lazy lists (α, β) *tllist* as the quotient of α *llist* $\times \beta$ over the equivalence relation that ignores the second component if and only if the first one is infinite;
- streams α *stream* as the subtype of infinite lazy lists.

Table 1 presents the types and the evaluation's statistics. The third column gives the lines of code for the definitions, lemmas, and proofs that were needed to define the type, the constructors, the corecursors, and the case constants, and to prove the free constructor theorems and the coinduction rule for equality. For *enat*, we kept the old definition because the datatype view is useful. Hence, we still derive the corecursor and the coinduction rules manually, but we generate the free constructor theorems with the **free_constructors** command (Section 3), saving 6 lines. In contrast, the other three types are now defined with **codatatype** in 33 lines instead of 774, among which 28 are for *tllist* because the default value for TNil's selector applied to TCons requires unbounded recursion. However, we lost the connection between *llist*, *tllist*, and *stream*, on which the function definitions and proofs relied. Therefore, we manually set up the Lifting and Transfer tools [15]; the line counts are shown behind plus signs (+).

The type definitions are just a small fraction of the library; most of the work went into proving properties of the functions. The fourth column shows the number of lemmas that we have proved for the functions on each type. There are 36% more than before, which might be surprising at first, since the old figures include the manual type constructions. Three reasons explain the increase. First, changes in the views increase the counts. Coinductive originally favored the code and constructor views, following Paulson [29], whereas the new package expresses coinduction and other properties in terms of the destructors (Sections 4 and 6). We proved additional lemmas for our functions that reflect the destructor view. Second, the manual setup for Lifting and Transfer

---

[3] In addition, Coq allows modifications of the type arguments under the constructors (e.g., for powerlists α *plist* = $\alpha + (\alpha \times \alpha)$ *plist*), which Isabelle/HOL cannot support due to its more restrictive type system.

Table 1. Statistics on porting Coinductive to the new package (before → after)

Codatatype	Constructors	Lines of code for definition	Number of lemmas	Lines of code per lemma
enat	0 | eSuc *enat*	200 → 194	31 → 57	8.42 → 5.79
α *llist*	LNil | LCons α (α *llist*)	503 → 3	527 → 597	9.86 → 6.44
(α, β) *tllist*	TNil β | TCons α ((α, β) *tllist*)	169 → 28 + 120	121 → 200	6.05 → 4.95
α *stream*	SCons α (α *stream*)	102 → 2 + 96	64 → 159	3.11 → 3.47
Total		974 → 227 + 216	743 → 1013	8.60 → 5.65

accounts for 36 new lemmas. Third, the porting has been distributed over six months such that we continuously incorporated our insights into the package's implementation. During this period, the *stream* part of the library grew significantly and *tllist* a little. (Furthermore, **primcorec** was not yet in place at the time of the porting. It is now used, but the gains it led to are not captured by the statistics.)

Therefore, the absolute numbers should not be taken too seriously. It is more instructive to examine how the proofs have changed. The last column of Table 1 gives the average length of a lemma, including its statement and its proof; shorter proofs indicate better automation. Usually, the statement takes between one and four lines, where two is the most common case. The port drastically reduced the length of the proofs: We now prove 36% more lemmas in 11% fewer lines.

Two improvements led to these savings. First, the *coinduction* method massages the proof obligation to fit the coinduction rule. Second, automation for coinduction proofs works best with the destructor view, as the destructors trigger rewriting. With the code and constructor style, we formerly had to manually unfold the equations, and pattern-matching equations obtained by **simps_of_case** needed manual case distinctions.

The destructor view also has some drawbacks. The proofs rely more on Isabelle's classical reasoner to solve subgoals that the simplifier can discharge in the other styles, and the reasoner often needs more guidance. We have not yet run into scalability issues, but we must supply a lengthy list of lemmas to the reasoner. The destructor style falls behind when we leave the coinductive world. For example, the recursive function lnth :: $nat \Rightarrow \alpha$ *llist* $\Rightarrow \alpha$ returns the element at a given index in a lazy list; clearly, there are no destructors on lnth's result type to trigger unfolding. Since induction proofs introduce constructors in the arguments, rewriting with pattern-matching equations obtained from the code view yields better automation. In summary, all three views are useful.

10   Conclusion

Codatatypes and corecursion have long been missing features in proof assistants based on higher-order logic. Isabelle's new (co)datatype definitional package finally addresses this deficiency, while generalizing and modularizing the support for datatypes. Within the limitations of the type system, the package is the most flexible one available in any proof assistant. The package is already highly usable and is used not only for the Coinductive library but also in ongoing developments by the authors. Although Isabelle is our vehicle, the approach is equally applicable to the other provers in the HOL family.

For future work, our priority is to integrate the package better with other Isabelle subsystems, including the function package (for well-founded recursive definitions), the Lifting and Transfer tools, and the counterexample generators Nitpick and Quickcheck. Another straightforward development would be to have the package produce even more theorems, notably for parametricity. There is also work in progress on supporting more general forms of corecursion and mixed recursion–corecursion. Finally, we expect that BNFs can be generalized to define non-free datatypes, including nominal types, but this remains to be investigated.

Acknowledgment. Tobias Nipkow and Makarius Wenzel encouraged us to implement the new package. Florian Haftmann and Christian Urban provided general advice on Isabelle and package writing. Brian Huffman suggested simplifications to the internal constructions, many of which have yet to be implemented. Stefan Milius and Lutz Schröder found an elegant proof to eliminate one of the BNF assumptions. René Thiemann contributed precious benchmarks from his IsaFoR formalization, helping us optimize the BNF operations. Sascha Böhme, Lars Hupel, Tobias Nipkow, Mark Summerfield, and the anonymous reviewers suggested many textual improvements.

Blanchette is supported by the Deutsche Forschungsgemeinschaft (DFG) project Hardening the Hammer (grant Ni 491/14-1). Hölzl is supported by the DFG project Verification of Probabilistic Models (grant Ni 491/15-1). Popescu is supported by the DFG project Security Type Systems and Deduction (grant Ni 491/13-2) as part of the program Reliably Secure Software Systems (RS^3, priority program 1496). Traytel is supported by the DFG program Program and Model Analysis (PUMA, doctorate program 1480). The authors are listed alphabetically.

References

1. Abel, A., Pientka, B.: Wellfounded recursion with copatterns: A unified approach to termination and productivity. In: Morrisett, G., Uustalu, T. (eds.) ICFP 2013, pp. 185–196. ACM (2013)
2. Asperti, A., Ricciotti, W., Coen, C.S., Tassi, E.: A compact kernel for the calculus of inductive constructions. Sādhanā 34, 71–144 (2009)
3. Atkey, R., McBride, C.: Productive coprogramming with guarded recursion. In: Morrisett, G., Uustalu, T. (eds.) ICFP 2013, pp. 197–208. ACM (2013)
4. Berghofer, S., Wenzel, M.: Inductive datatypes in HOL—Lessons learned in formal-logic engineering. In: Bertot, Y., Dowek, G., Hirschowitz, A., Paulin, C., Théry, L. (eds.) TPHOLs 1999. LNCS, vol. 1690, pp. 19–36. Springer, Heidelberg (1999)
5. Bertot, Y., Castéran, P.: Interactive Theorem Proving and Program Development: Coq'Art: The Calculus of Inductive Constructions. Texts in Theoretical Computer Science. Springer (2004)
6. Blanchette, J.C., Hölzl, J., Lochbihler, A., Panny, L., Popescu, A., Traytel, D.: Formalization accompanying this paper (2014),
http://www21.in.tum.de/~blanchet/codata_impl.tar.gz
7. Blanchette, J.C., Panny, L., Popescu, A., Traytel, D.: Defining (co)datatypes in Isabelle/HOL (2014),http://isabelle.in.tum.de/dist/Isabelle/doc/datatypes.pdf
8. Blanchette, J.C., Popescu, A., Traytel, D.: Cardinals in Isabelle/HOL. In: Klein, G., Gamboa, R. (eds.) ITP 2014. LNCS (LNAI), vol. 8558, pp. 111–127. Springer, Heidelberg (2014)

9. Blanchette, J.C., Popescu, A., Traytel, D.: Witnessing (co)datatypes (2014), http://www21.in.tum.de/~blanchet/wit.pdf

10. Danielsson, N.A., Altenkirch, T.: Subtyping, declaratively. In: Bolduc, C., Desharnais, J., Ktari, B. (eds.) MPC 2010. LNCS, vol. 6120, pp. 100–118. Springer, Heidelberg (2010)

11. Gunter, E.L.: Why we can't have SML-style datatype declarations in HOL. In: Claesen, L.J.M., Gordon, M.J.C. (eds.) TPHOLs 1992. IFIP Transactions, vol. A-20, pp. 561–568. North-Holland (1993)

12. Gunter, E.L.: A broader class of trees for recursive type definitions for HOL. In: Joyce, J.J., Seger, C.-J.H. (eds.) HUG 1993. LNCS, vol. 780, pp. 141–154. Springer, Heidelberg (1994)

13. Hancock, P., Ghani, N., Pattinson, D.: Representations of stream processors using nested fixed points. Log. Meth. Comput. Sci. 5(3) (2009)

14. Harrison, J.: Inductive definitions: Automation and application. In: Schubert, E.T., Windley, P.J., Alves-Foss, J. (eds.) TPHOLs 1995. LNCS, vol. 971, pp. 200–213. Springer, Heidelberg (1995)

15. Huffman, B., Kunčar, O.: Lifting and Transfer: A modular design for quotients in Isabelle/HOL. In: Gonthier, G., Norrish, M. (eds.) CPP 2013. LNCS, vol. 8307, pp. 131–146. Springer, Heidelberg (2013)

16. Jacobs, B., Rutten, J.: A tutorial on (co)algebras and (co)induction. Bull. EATCS 62, 222–259 (1997)

17. Leroy, X.: A formally verified compiler back-end. J. Autom. Reas. 43(4), 363–446 (2009)

18. Lochbihler, A.: Coinductive. In: Klein, G., Nipkow, T., Paulson, L. (eds.) Archive of Formal Proofs (2010), http://afp.sf.net/entries/Coinductive.shtml

19. Lochbihler, A.: Verifying a compiler for Java threads. In: Gordon, A.D. (ed.) ESOP 2010. LNCS, vol. 6012, pp. 427–447. Springer, Heidelberg (2010)

20. Lochbihler, A.: A Machine-Checked, Type-Safe Model of Java Concurrency: Language, Virtual Machine, Memory Model, and Verified Compiler. Ph.D. thesis, Karlsruher Institut für Technologie (2012)

21. Lochbihler, A.: Making the Java memory model safe. ACM Trans. Program. Lang. Syst. 35(4), 12:1–12:65 (2014)

22. Lochbihler, A., Hölzl, J.: Recursive functions on lazy lists via domains and topologies. In: Klein, G., Gamboa, R. (eds.) ITP 2014. LNCS (LNAI), vol. 8558, pp. 341–357. Springer, Heidelberg (2014)

23. Melham, T.F.: Automating recursive type definitions in higher order logic. In: Birtwistle, G., Subrahmanyam, P.A. (eds.) Current Trends in Hardware Verification and Automated Theorem Proving, pp. 341–386. Springer (1989)

24. Nipkow, T., Klein, G.: Concrete Semantics: A Proof Assistant Approach. Springer (to appear), http://www.in.tum.de/~nipkow/Concrete-Semantics

25. Nipkow, T., Paulson, L.C., Wenzel, M.: Isabelle/HOL. LNCS, vol. 2283. Springer, Heidelberg (2002)

26. Owre, S., Shankar, N.: A brief overview of PVS. In: Mohamed, O.A., Muñoz, C., Tahar, S. (eds.) TPHOLs 2008. LNCS, vol. 5170, pp. 22–27. Springer, Heidelberg (2008)

27. Panny, L.: Primitively (Co)recursive Function Definitions for Isabelle/HOL. B.Sc. thesis draft, Technische Universität München (2014)

28. Paulson, L.C.: A fixedpoint approach to implementing (co)inductive definitions. In: Bundy, A. (ed.) CADE 1994. LNCS, vol. 814, pp. 148–161. Springer, Heidelberg (1994)

29. Paulson, L.C.: Mechanizing coinduction and corecursion in higher-order logic. J. Log. Comput. 7(2), 175–204 (1997)

30. Rutten, J.J.M.M.: Universal coalgebra: A theory of systems. Theor. Comput. Sci. 249, 3–80 (2000)
31. Schropp, A., Popescu, A.: Nonfree datatypes in Isabelle/HOL: Animating a many-sorted metatheory. In: Gonthier, G., Norrish, M. (eds.) CPP 2013. LNCS, vol. 8307, pp. 114–130. Springer, Heidelberg (2013)
32. Traytel, D.: A Category Theory Based (Co)datatype Package for Isabelle/HOL. M.Sc. thesis, Technische Universität München (2012)
33. Traytel, D., Popescu, A., Blanchette, J.C.: Foundational, compositional (co)datatypes for higher-order logic: Category theory applied to theorem proving. In: LICS 2012, pp. 596–605. IEEE (2012)

Cardinals in Isabelle/HOL

Jasmin Christian Blanchette[1], Andrei Popescu[1,2], and Dmitriy Traytel[1]

[1] Fakultät für Informatik, Technische Universität München, Germany
[2] Institute of Mathematics Simion Stoilow of the Romanian Academy, Bucharest, Romania

Abstract. We report on a formalization of ordinals and cardinals in Isabelle/HOL. A main challenge we faced is the inability of higher-order logic to represent ordinals canonically, as transitive sets (as done in set theory). We resolved this into a "decentralized" representation that identifies ordinals with wellorders, with all concepts and results proved to be invariant under order isomorphism. We also discuss two applications of this general theory in formal developments.

1 Introduction

Set theory is the traditional framework for ordinals and cardinals. Axiomatizations such as Zermelo–Fraenkel (ZF) and von Neumann–Bernays–Gödel (NBG) permit the definition of ordinals as transitive sets well ordered by membership as the strict relation and by inclusion as the nonstrict counterpart. Ordinals form a class Ord which is itself well ordered by membership. Basic constructions and results in the theory of ordinals and cardinals make heavy use of Ord, employing definitions and proofs by transfinite recursion and induction. In short, Ord conveniently captures the notion of wellorder.

The situation is quite different in higher-order logic (HOL, Section 2). There is no support for infinite transitive sets, since the type system permits only finite iterations of the powerset. Consequently, membership cannot be used to implement ordinals and cardinals. Another difficulty is that there is no single type that can host a complete collection of canonical representatives for wellorders.

A natural question to ask is: Can we still develop in HOL a theory of cardinals? The answer depends on the precise goals. Our criterion for the affirmative answer is the possibility to prove general-purpose theorems on cardinality for the working mathematician, such as: Given any two types, one can be embedded into the other; given any infinite type, the type of lists over it has the same cardinality; and so on.

We present a formalization in Isabelle/HOL [14] that provides such general-purpose theorems, as well as some more specialized results and applications. We follow a "decentralized" approach, identifying ordinals with arbitrary wellorders and developing all the concepts up to (order-preserving) isomorphism (Section 3). Cardinals are defined, again up to isomorphism, as the minimum ordinals on given underlying sets (Section 4).

The concepts are more abstract than in set theory: Ordinal equality is replaced by a polymorphic relation $=_o$ stating the existence of an order isomorphism, and membership is replaced by a polymorphic operator $<_o$ stating the existence of a strict order embedding (with a nonstrict counterpart $\le_o$). This abstract view takes more effort to maintain than the concrete implementation from set theory (Section 5), since all the defined operations need to be shown compatible with the new equality and most of them

G. Klein and R. Gamboa (Eds.): ITP 2014, LNAI 8558, pp. 111–127, 2014.
© Springer International Publishing Switzerland 2014

need to be shown monotonic with respect to the new ordering. For example, $|A|$, the cardinal of A, is defined as *some* cardinal order on A and then proved to be isomorphic to *any* cardinal order on A; similarly, $r_1 +_c r_2$, the sum of cardinals r_1 and r_2, is defined as the cardinal of the sum of r_1's and r_2's fields, before it is proved compatible with $=_o$ and $\leq_o$. Moreover, since the collection of all ordinals does not fit in one type, we must predict the size of the constructed objects and choose suitably large support types.

Our development validates the following thesis:

> *The basics of cardinals can be developed independently of membership-based implementation details and the existence of large classes from set theory.*

This was not clear to us when we started formalizing, since we could not find any textbook or formal development that follows this abstract approach. Most introductions to cardinals rely quite heavily on set theory, diving at will into the homogeneous ether provided by the class of all ordinals.[1]

The initial infrastructure and general-purpose theorems were incorporated in the *Archive of Formal Proofs* [19] in 2009, together with thorough documentation, but was not otherwise published. Since then, the formalization has evolved to help specific applications: Cofinalities and regular cardinals were added for a formalization of syntax with bindings [20] (Section 6), and cardinal arithmetic was developed to support Isabelle's new (co)datatype package [24] (Section 7). Moreover, Breitner employed our cardinal infrastructure to formalize free group theory [2].

Most of the theory of cardinals is already included in the 2012 edition of Isabelle, but some features will be available starting with the forthcoming 2014 release.

Related Work. Ordinals have been developed in HOL before. Harrison [7] formalized ordinals in HOL88 and proved theorems such as Zermelo, Zorn, and transfinite induction. Huffman [11] formalized countable ordinals in Isabelle/HOL, including arithmetic and the Veblen hierarchies; the countability assumption made it possible to fix a type of ordinals. Recently, Norrish and Huffman [15] independently redeveloped in HOL4 much of our theory of ordinals and beyond, covering advanced ordinal arithmetic including Cantor normal form. In contrast to them, we see the ordinals mostly as a stepping stone toward the cardinals and concentrate on these.

Whereas the HOL-based systems have extensive support for finite cardinal reasoning, general cardinals have received little attention. The only account we are aware of is part of the HOL Light library [9], but it employs cardinals only as "virtual objects" [4], not defining a notion of cardinal but directly relating sets via injections and bijections.

Beyond HOL, Paulson and Grabczewski [17] have formalized some ordinal and cardinal theory in Isabelle/ZF following the usual set-theoretic path, via the class of ordinals with membership. Their main objective was to formalize several alternative statements of the axiom of choice, and hence they preferred constructive arguments for most of the cardinal theory. In our development, Hilbert's choice operator (effectively enforcing a bounded version of the axiom of choice) is pervasive.

---

[1] Notable exceptions are Taylor's category-theory-oriented foundation for ordinals [23] and Forster's implementation-independent analysis of ordinals and cardinals [4]. The latter was brought to our attention by an anonymous reviewer.

2 Higher-Order Logic

Isabelle/HOL implements classical higher-order logic with Hilbert choice, the axiom of infinity, and rank-1 polymorphism. HOL is based on Church's simple type theory [3]. It is the logic of Gordon's system of the same name [5] and of its many successors. HOL is roughly equivalent to ZF without support for classes and with the axiom of comprehension taking the place of the axiom of replacement. We refer to Nipkow and Klein [13, Part 1] for a modern introduction.

Types in HOL are either atomic types (e.g., unit, nat, and bool), type variables α, β, or fully applied type constructors (e.g., nat list and nat set). The binary type constructors $\alpha \to \beta$, $\alpha + \beta$, and $\alpha \times \beta$ for function space, disjoint sum, and product are written in infix notation. All types are nonempty. New types can be introduced by carving out nonempty subsets of existing types. A constant c of type τ is indicated as $c : \tau$.

The following types and constants from the Isabelle library are heavily used in our formalization. UNIV : α set is the universe set, the set of all elements of type α. 0 and Suc are the constructors of the type nat. Elements of the sum type are constructed by the two embeddings Inl : $\alpha \to \alpha + \beta$ and Inr : $\beta \to \alpha + \beta$.

The function id : $\alpha \to \alpha$ is the identity. $f \cdot A$ is the image of $A : \alpha$ set through $f : \alpha \to \beta$, i.e., the set $\{f a. \ a \in A\}$. The predicates inj_on $f A$ and bij_betw $f A B$ state that $f : \alpha \to \beta$ is an injection on $A : \alpha$ set and that $f : \alpha \to \beta$ is a bijection between $A : \alpha$ set and $B : \beta$ set. The type $(\alpha \times \alpha)$ set of binary relations on α is abbreviated to α rel. Id : α rel is the identity relation. Given $r : \alpha$ rel, Field $r : \alpha$ set is its field (underlying set), i.e., the union between its domain and its codomain: $\{a. \ \exists b. \ (a,b) \in r\} \cup \{b. \ \exists a. \ (a,b) \in r\}$. The following predicates operate on relations, where $A : \alpha$ set and $r : \alpha$ rel:

REFLEXIVE	refl_on $A r \equiv r \subseteq A \times A \wedge \forall x \in A. \ (x,x) \in r$
TRANSITIVE	trans $r \equiv \forall abc. \ (a,b) \in r \wedge (b,c) \in r \to (a,c) \in r$
ANTISYMMETRIC	antisym $r \equiv \forall ab. \ (a,b) \in r \wedge (b,a) \in r \to a = b$
TOTAL	total_on $A r \equiv \forall (a \in A) \ (b \in A). \ a \neq b \to (a,b) \in r \vee (b,a) \in r$
WELLFOUNDED	wf $r \equiv \forall P. \ (\forall a. \ (\forall b. \ (b,a) \in r \to P b) \to P a) \to (\forall a. \ P a)$
PARTIAL ORDER	partial_order_on $A r \equiv$ refl_on $A r \wedge$ trans $r \wedge$ antisym r
LINEAR ORDER	linear_order_on $A r \equiv$ partial_order_on $A r \wedge$ total_on $A r$
WELLORDER	well_order_on $A r \equiv$ linear_order_on $A r \wedge$ wf $(r - $ Id$)$

If r is a partial order, then $r - $ Id is its associated strict partial order. Some of the above definitions are slightly nonstandard, but they can be proved equivalent to standard ones. For example, well-foundedness is given here a higher-order definition useful in proofs as an induction principle, while it is usually equivalently defined as the nonexistence of infinite chains $a : $ nat $\to \alpha$ with $(a (\text{Suc } i), a i) \in r$ for all i.

Note that refl_on $A r$ (and hence well_order_on $A r$) implies Field $r = A$. We abbreviate well_order_on (Field r) r to Wellorder r and well_order_on UNIV r to wellorder r.

3 Ordinals

This section presents some highlights of our formalization of ordinals. In a break with tradition, we work with abstract ordinals—i.e., with wellorders—making no assumption about their underlying implementation.

3.1 Infrastructure

We represent a wellorder as a relation $r : \tau$ rel, where τ is some type. Although some of the lemmas below hold for arbitrary relations, we generally assume that r, s, and t range over wellorders. The following operators are pervasive in our constructions: under $r\ a$ is the set of all elements less than or equal to a, or "under" a, with respect to r; underS $r\ a$ gives the elements strictly under a. We call these *under-* and *strict-under*-intervals:

$$\text{under} : \alpha \text{ rel} \to \alpha \to \alpha \text{ set} \qquad\qquad \text{underS} : \alpha \text{ rel} \to \alpha \to \alpha \text{ set}$$
$$\text{under } r\ a \equiv \{b \mid (b,a) \in r\} \qquad\qquad \text{underS } r\ a \equiv \{b \mid (b,a) \in r \wedge b \neq a\}$$

A wellorder is a linear order relation r such that its strict version, $r - \text{Id}$, is a well-founded relation. Well-founded induction and recursion are well supported by Isabelle's library. We define slight variations of these notions tailored for wellorders.

Lemma 1. *If* $\forall a \in \text{Field } r.\ (\forall a' \in \text{underS } r\ a.\ P\ a') \to P\ a$, *then* $\forall a \in \text{Field } r.\ P\ a$.

When proving a property P for all elements of r's field, wellorder induction allows us to show P for fixed $a \in \text{Field } r$, assuming P holds for elements strictly r-smaller than a.

Wellorder recursion is similar, except that it allows us to define a function f on Field r instead of to prove a property. For each $a \in \text{Field } r$, we assume f already defined on underS $r\ a$ and specify $f\ a$. This is technically achieved by a "wellorder recursor" operator wo_rec$_r$: $((\alpha \to \beta) \to \alpha \to \beta) \to \alpha \to \beta$ and an admissibility predicate adm_wo$_r$: $((\alpha \to \beta) \to \alpha \to \beta) \to$ bool defined by

$$\text{adm\_wo}_r\ H \equiv \forall f\ g\ a.\ (\forall a' \in \text{underS } r\ a.\ f\ a' = g\ a') \to H\ f\ a = H\ g\ a$$

A recursive definition is represented by a function $H : (\alpha \to \beta) \to \alpha \to \beta$, where $H\ f$ maps a to a value based on the values of f on underS $r\ a$. A more precise type for H would be $\prod_{a \in \text{Field } r}(\text{underS } r\ a \to \beta) \to \beta$, but this is not possible in HOL. Instead, H is required to be admissible, i.e., not dependent on the values of f outside underS $r\ a$. The defined function wo_rec H is then a fixpoint of H on Field r.

Lemma 2. *If* adm_wo$_r$ H, *then* $\forall a \in \text{Field } r.$ wo_rec$_r$ $H\ a = H\ (\text{wo\_rec}_r\ H)\ a$.

An (*order*) *filter* on r, also called an *initial segment* of r if r is a wellorder, is a subset A of r's field such that whenever A contains a, it also contains all elements under a:

$$\text{ofilter} : \alpha \text{ rel} \to \alpha \text{ set} \to \text{bool}$$
$$\text{ofilter } r\ A \equiv A \subseteq \text{Field } r \wedge (\forall a \in A.\ \text{under } r\ a \subseteq A)$$

Both the under- and the strict-under-intervals are filters of r. Moreover, every filter of r is either its whole field or a strict-under-interval.

Lemma 3. (1) ofilter r (under $r\ a$) $\wedge$ ofilter r (underS $r\ a$);
(2) ofilter $r\ A \leftrightarrow A = \text{Field } r \vee (\exists a \in \text{Field } r.\ A = \text{underS } r\ a)$.

3.2 Embedding and Isomorphism

Wellorder embeddings, strict embeddings, and isomorphisms are defined as follows:

$$\text{embed, embedS, iso} : \alpha \text{ rel} \to \beta \text{ rel} \to (\alpha \to \beta) \to \text{bool}$$
$$\text{embed}\quad r\ s\ f \equiv \forall a \in \text{Field } r.\ \text{bij\_betw } f\ (\text{under } r\ a)\ (\text{under } s\ (f\ a))$$

embedS $r\ s\ f\ \equiv$ embed $r\ s\ f\ \wedge \neg$ bij_betw f (Field r) (Field s)

iso $r\ s\ f\ \equiv$ embed $r\ s\ f\ \wedge$ bij_betw f (Field r) (Field s)

We read embed $r\ s\ f$ as "f embeds r into s." It is defined by stating that for all $a \in$ Field r, f establishes a bijection between the under-intervals of a in r and those of $f\ a$ in s. The more conventional definition (stating that f is injective, order preserving, and maps Field r into a filter of s) is derived as a lemma:

Lemma 4. embed $r\ s\ f \longleftrightarrow$ compat $r\ s\ f\ \wedge$ inj_on f (Field r) $\wedge$ ofilter s ($f\ \cdot$ Field r), *where* compat $r\ s\ f$ *expresses order preservation of* f ($\forall a\ b.\ (a,b) \in r \rightarrow (f\ a,\ f\ b) \in s$).

Every embedding is either an (order) isomorphism or a strict embedding (i.e., iso $r\ s\ f\ \vee$ embedS $r\ s\ f$), depending on whether f is a bijection. These notions yield the following relations between wellorders:

$$\leq_o, <_o, =_o : (\alpha\ \text{rel} \times \beta\ \text{rel})\ \text{set}$$
$$\leq_o \equiv \{(r,s).\ \text{Wellorder}\ r \wedge \text{Wellorder}\ s \wedge (\exists f.\ \text{embed}\ r\ s\ f)\}$$
$$<_o \equiv \{(r,s).\ \text{Wellorder}\ r \wedge \text{Wellorder}\ s \wedge (\exists f.\ \text{embedS}\ r\ s\ f)\}$$
$$=_o \equiv \{(r,s).\ \text{Wellorder}\ r \wedge \text{Wellorder}\ s \wedge (\exists f.\ \text{iso}\ r\ s\ f)\}$$

We abbreviate $(r,s) \in \leq_o$ to $r \leq_o s$, and similarly for $<_o$ and $=_o$. These notations are fairly intuitive; for example, $r \leq_o s$ means that r is smaller than or equal to s, in that it can be embedded in s. The relations are also well behaved.

Theorem 5. *The following properties hold:*

(1) $r =_o r$

(2) $r =_o s \rightarrow s =_o r$

(3) $r =_o s \wedge s =_o t \rightarrow r =_o t$

(4) $r \leq_o r$

(5) $r \leq_o s \wedge s \leq_o t \rightarrow r \leq_o t$

(6) $\neg\ r <_o r$

(7) $r <_o s \wedge s <_o t \rightarrow r <_o t$

(8) $r \leq_o s \longleftrightarrow r <_o s \vee r =_o s$

(9) $r =_o s \longleftrightarrow r \leq_o s \wedge s \leq_o r$

If we restrict the types of these relations from $(\alpha\ \text{rel} \times \beta\ \text{rel})$ set to $(\alpha\ \text{rel})$ rel (by taking $\beta = \alpha$), we obtain that $=_o$ is an equivalence (1–3) and $\leq_o$ is a preorder (4–5). Moreover, $<_o$ is the strict version of $\leq_o$ with respect to $=_o$ (6–8). If we think of $=_o$ as the equality, $\leq_o$ becomes a partial order (9) and $<_o$ a strict partial order.

The above relations establish an order between the wellorders similar to the standard one on the class of ordinals but distributed across types and, as a consequence, only up to isomorphism. What is still missing is a result corresponding to the class of ordinals being itself well ordered. To this end, we first show that $\leq_o$ is total.

Theorem 6. $r \leq_o s \vee s \leq_o r$.

Proof idea. In textbooks, totality of $\leq_o$ follows from the fact that every wellorder is isomorphic to an ordinal and that the class of ordinals Ord is totally ordered. To show the former, one starts with a wellorder r and provides an embedding of r into Ord.

In our distributed setting, we must start with two wellorders $r : \alpha$ rel and $s : \beta$ rel, without a priori knowing which one is larger, hence which should embed which. Our proof proceeds by defining a function by transfinite recursion on r that embeds r into s if $r \leq_o s$ and that is the inverse of an embedding of s into r otherwise.   □

This total order is a wellorder. Equivalently, its strict counterpart $<_o$ is well founded.

Theorem 7. wf $(<_o : (\alpha\ \text{rel})\ \text{rel})$.

Theorems 5, 6, and 7 yield that for any fixed type, its wellorders are themselves well ordered up to isomorphism. This paves the way for introducing cardinals.

3.3 Ordinal Arithmetic

Most textbooks define operations on ordinals (sum, product, exponentiation) by transfinite recursion. Yet these operations admit direct, nonrecursive definitions, which are particularly suited to arbitrary wellorders. In Holz et al. [10], these direct definitions are presented as "visual" descriptions.

We define the ordinal sum $+_o$ by concatenating the two argument wellorders r and s such that elements of Field r come below those of Field s:

$$+_o : \alpha\ \text{rel} \to \beta\ \text{rel} \to (\alpha + \beta)\ \text{rel}$$
$$r +_o s \equiv (\text{Inl} \otimes \text{Inl}) \bullet r \cup (\text{Inr} \otimes \text{Inr}) \bullet s \cup \{(\text{Inl}\ x, \text{Inr}\ y).\ x \in \text{Field}\ r \wedge y \in \text{Field}\ s\}$$

In the above, the operator $\otimes : (\alpha_1 \to \beta_1) \to (\alpha_2 \to \beta_2) \to (\alpha_1 \times \alpha_2 \to \beta_1 \times \beta_2)$ is the map function for products: $(f_1 \otimes f_2)\ (a_1, a_2) = (f_1\ a_1, f_2\ a_2)$.

Similarly, ordinal multiplication $\times_o$ is defined as the anti-lexicographic ordering on the product type:

$$\times_o : \alpha\ \text{rel} \to \beta\ \text{rel} \to (\alpha \times \beta)\ \text{rel}$$
$$r \times_o s \equiv \{((x_1, y_1), (x_2, y_2)).\ x_1, x_2 \in \text{Field}\ r \wedge y_1, y_2 \in \text{Field}\ s \wedge$$
$$(y_1 \neq y_2 \wedge (y_1, y_2) \in s \vee y_1 = y_2 \wedge (x_1, x_2) \in r)\}$$

For ordinal exponentiation $r \wedge_o s$, the underlying set consists of the functions of finite support from Field s to Field r. Assuming $f \neq g$, the finite support ensures that there exists a maximum $z \in$ Field s (with respect to s) such that $f\ z \neq g\ z$. We make f smaller than g if $(f\ z, g\ z) \in r$:

$$\wedge_o : \alpha\ \text{rel} \to \beta\ \text{rel} \to (\beta \to \alpha)\ \text{rel}$$
$$r \wedge_o s \equiv \{(f, g).\ f, g \in \text{FinFunc}\ (\text{Field}\ s)\ (\text{Field}\ r) \wedge$$
$$(f = g \vee (\text{let}\ z = \max_s\{x \in \text{Field}\ s.\ f\ x \neq g\ x\}\ \text{in}\ (f\ z, g\ z) \in r))\}$$

The definition rests on the auxiliary notion of a function of finite support from $B : \alpha$ set to $A : \beta$ set. FinFunc $B\ A$ carves out a suitable subspace of the total function space $\beta \to \alpha$ by requiring that functions are equal to a particular unspecified value $\bot$ outside their intended domains. In addition, finite support means that only finitely many elements of B are mapped to elements other than the minimal element 0_r of the wellorder r:

$$\text{Func, FinFunc} : \beta\ \text{set} \to \alpha\ \text{set} \to (\beta \to \alpha)\ \text{set}$$
$$\text{Func}\ B\ A \quad \equiv \{f.\ f \bullet B \subseteq A \wedge (\forall x \notin B.\ f\ x = \bot)\}$$
$$\text{FinFunc}\ B\ A \equiv \text{Func}\ B\ A \cap \{f.\ \text{finite}\ \{x \in B.\ f\ x \neq 0_r\}\}$$

All three constructions yield wellorders. Moreover, they satisfy various arithmetic properties, including those listed below.

Theorem 8. (1) Wellorder $(r +_o s)$; (2) Wellorder $(r \times_o s)$; (3) Wellorder $(r \wedge_o s)$.

Lemma 9 (Lemma 1.4.3 in Holz et al. [10]). *Let 0 be the empty wellorder and 1 be the singleton wellorder. The following properties hold:*

(1) $0 +_o r =_o r =_o r +_o 0$

(2) $s \leq_o r +_o s$

(3) $s <_o t \to r +_o s <_o r +_o t$

(4) $(r +_o s) +_o t =_o r +_o (s +_o t)$

(5) $r \leq_o s \to r +_o t \leq_o s +_o t$

(6) $0 \times_o r =_o 0 =_o r \times_o 0$

(7) $(r \times_o s) \times_o t =_o r \times_o (s \times_o t)$

(8) $r \leq_o s \to r \times_o t \leq_o s \times_o t$

(9) $1 \times_o r =_o r =_o r \times_o 1$

(10) $r \times_o (s +_o t) =_o r \times_o s +_o r \times_o t$

(11) $0 <_o r \land s <_o t \to r \times_o s <_o r \times_o t$

(12) $0 <_o r \to 0 \wedge_o r =_o 0$

(13) $(r \wedge_o s) \wedge_o t =_o r \wedge_o (s \times_o t)$

(14) $r \leq_o s \to r \wedge_o t \leq_o s \wedge_o t$

(15) $1 <_o r \to s \leq_o r \wedge_o s$

(16) $1 \wedge_o r =_o 1$

(17) $r \wedge_o s +_o t =_o r \wedge_o s \times_o r \wedge_o t$

(18) $1 <_o r \land s <_o t \to r \wedge_o s <_o r \wedge_o t$

An advantage of the standard definitions of these operations by transitive recursion is that the above arithmetic facts can then be nicely proved by corresponding transfinite induction. With direct definitions, we aim as much as possible at direct proofs via the explicit indication of suitable isomorphisms or embeddings, as in the definitions of $=_o$, $\leq_o$, and $<_o$. This approach works well for the equations ($=_o$-identities) and for right-monotonicity properties of the operators (where one assumes equality on the left arguments and ordering of the right arguments). For example, to prove $0 <_o r \land s <_o t \to r \times_o s <_o r \times_o t$, we use the definition of $<_o$ to obtain from $s <_o t$ a strict embedding f of s into t. The desired strict embedding of $r \times_o s$ into $r \times_o t$ is then $\mathrm{id} \otimes f$.

In contrast, left-monotonicity properties such as $r \leq_o s \to r \times_o t \leq_o s \times_o t$ no longer follow smoothly, because it is not clear how to produce an embedding of $r \times_o t$ into $s \times_o t$ from one of r into s. An alternative characterization of $\leq_o$ is called for:

Lemma 10. $r \leq_o s \leftrightarrow$ Wellorder $r \land$ Wellorder $s \land (\exists f. \forall a \in$ Field $r.\ f\,a \in$ Field $s \land f \cdot$ underS $r\,a \subseteq$ underS $s\,(f\,a))$.

Thus, to show $r \leq_o s$, it suffices to provide an order embedding, which need not be a wellorder embedding (an embedding of Field r as a filter of s). This dramatically simplifies the proof. To show the left-monotonicity property $r \times_o t \leq_o s \times_o t$ assuming an embedding f of r into s, the obvious order embedding $f \otimes \mathrm{id}$ meets the requirements. Surprisingly, this technique is not mentioned in the textbooks.

Right-monotonicity holds for both $<_o$ and $\leq_o$, whereas left-monotonicity holds only for $\leq_o$. This is fortunate in a sense, because Lemma 10 is not adaptable to $<_o$.

4 Cardinals

With the ordinals in place, we can develop a theory of cardinals, which endows HOL with many conveniences of cardinality reasoning, including basic cardinal arithmetic.

4.1 Bootstrapping

We define cardinal orders on a set (or cardinals) as those wellorders that are minimal with respect to $=_o$. This is our HOL counterpart of the standard definition of cardinals as "ordinals that cannot be mapped one-to-one onto smaller ordinals" [10, p. 42]:

card_order_on $A\,r \equiv$ well_order_on $A\,r \land (\forall s.$ well_order_on $A\,s \to r \leq_o s)$

We abbreviate card_order_on (Field r) r to Card_order r and card_order_on UNIV r to card_order r. By definition, card_order_on A r implies $A =$ Field r, allowing us to write Card_order r when we want to omit A.

Cardinals are useful to measure sets. There exists a cardinal on every set, and it is unique up to isomorphism.

Theorem 11. (1) $\exists r$. card_order_on A r;
(2) card_order_on A $r \wedge$ card_order_on A $s \rightarrow r =_o s$.

We define the cardinality of a set $|\_| : \alpha$ set $\rightarrow \alpha$ rel using Hilbert's choice operator to pick an arbitrary cardinal order on A: $|A| \equiv \varepsilon r$. card_order_on A r. The order exists and is irrelevant by Theorem 11. We can prove that the cardinality operator behaves as expected; in particular, it is monotonic. We can also connect it to the more elementary comparisons in terms of functions.

Lemma 12. *The following properties hold:*

(1) card_order_on A $|A|$
(2) Field $|A| = A$

(3) $A \subseteq B \rightarrow |A| \leq_o |B|$
(4) $r \leq_o s \rightarrow |$Field $r| \leq_o |$Field $s|$

Lemma 13. *The following equivalences hold:*

(1) $|A| =_o |B| \leftrightarrow (\exists f.$ bij_betw f A $B)$
(2) $|A| \leq_o |B| \leftrightarrow (\exists f.$ inj_on f $A \wedge f \cdot A \subseteq B)$
(3) $A \neq \emptyset \rightarrow (|A| \leq_o |B| \leftrightarrow (\exists g. A \subseteq g \cdot B))$

Lemma 13, in conjunction with Theorem 6, allows us to prove the following interesting order-free fact for the working mathematician, mentioned in Section 1.

Theorem 14. *Given any two types σ and τ, one is embeddable in the other: There exists an injection either from σ to τ or from τ to σ.*

4.2   Cardinality of Set and Type Constructors

We analyze the cardinalities of several standard type constructors: $\alpha + \beta$ (disjoint sum), $\alpha \times \beta$ (binary product), α set (powertype), and α list (lists). In the interest of generality, we consider the homonymous set-based versions of these constructors, which take the form of polymorphic constants:

$$+ : \alpha \text{ set} \rightarrow \beta \text{ set} \rightarrow (\alpha + \beta) \text{ set} \qquad \times : \alpha \text{ set} \rightarrow \beta \text{ set} \rightarrow (\alpha \times \beta) \text{ set}$$
$$A + B \equiv \{\text{Inl } a \mid a \in A\} \cup \{\text{Inr } b \mid b \in B\} \quad A \times B \equiv \{(a,b) \mid a \in A \wedge b \in B\}$$

$$\text{Pow} : \alpha \text{ set} \rightarrow (\alpha \text{ set}) \text{ set} \qquad \text{lists} : \alpha \text{ set} \rightarrow (\alpha \text{ list}) \text{ set}$$
$$\text{Pow } A \equiv \{X \mid X \subseteq A\} \qquad \text{lists } A \equiv \{as \mid \text{set } as \subseteq A\}$$

(Such operators can be generated automatically from the specification of a (co)datatype, as we will see in Section 7.) The cardinalities of these operators are compatible with isomorphism and embedding.

Lemma 15. *Let K be any of $+$, $\times$,* Pow, *and* lists, *let $n \in \{1,2\}$ be its arity, and let θ be either $=_o$ or $\leq_o$. If $\forall i \in \{1,\dots,n\}$. $|A_i| \, \theta \, |B_i|$, then $|K \, A_1 \, \dots \, A_n| \, \theta \, |K \, B_1 \, \dots \, B_n|$.*

Lemma 16. *The following orderings between cardinalities hold:*

(1) $|A| \leq_o |A+B|$
(2) $|A+B| \leq_o |A \times B|$ *if both A and B have at least two elements*
(3) $|A| <_o |\mathsf{Pow}\, A|$
(4) $|A| \leq_o |\mathsf{lists}\, A|$

If one of the involved sets is infinite, some embeddings collapse to isomorphisms.

Lemma 17. *Assuming* infinite A, *the following equalities between cardinals hold:*

(1) $|A \times A| =_o |A|$
(2) $|A| =_o |\mathsf{lists}\, A|$
(3) $|A+B| =_o$ (if $A \leq_o B$ then $|A|$ else $|B|$)
(4) $B \neq \emptyset \to |A \times B| =_o$ (if $A \leq_o B$ then $|A|$ else $|B|$)

The formalization of property (1) required a significant effort. Its proof relies on the so-called bounded product construction, which is extensively discussed by Paulson and Grabczewski [17] in the context of Isabelle/ZF.

In Isabelle/HOL, the Cartesian product is a special case of the indexed sum (or disjoint union) operator:

$$\Sigma : \alpha \text{ set} \to (\alpha \to \beta \text{ set}) \to (\alpha \times \beta) \text{ set}$$
$$\Sigma A f \equiv \bigcup_{a \in A} \bigcup_{b \in f a} \{(a,b)\}$$

We write $\Sigma_{a \in A} f a$ for $\Sigma A f$. The properties of $\times$ given above carry over to Σ. In addition, Σ can be used to prove cardinality bounds of indexed unions:

Lemma 18. (1) $|\bigcup_{a \in A} f a| \leq_o |\Sigma_{a \in A} f a|$;
(2) infinite $B \wedge |A| \leq_o |B| \wedge (\forall a \in A. \, |f a| \leq_o |B|) \to |\bigcup_{a \in A} f a| \leq_o |B|$.

4.3 $\aleph_0$ and the Finite Cardinals

Our $\aleph_0$ is the standard order $\leq$: nat rel on natural numbers, which we denote by natLeq. It behaves as expected of $\aleph_0$; in particular, it is $\leq_o$-minimal among infinite cardinals. Proper filters of natLeq are precisely the finite sets of the first consecutive numbers.

Lemma 19. (1) infinite $A \leftrightarrow$ natLeq $\leq_o |A|$;   (2) Card_order natLeq;
(3) Card_order $r \wedge$ infinite (Field r) $\to$ natLeq $\leq_o r$;
(4) ofilter natLeq $A \leftrightarrow A = (\mathsf{UNIV} : \text{nat set}) \vee (\exists n. \, A = \{0,\dots,n\})$.

The finite cardinals are obtained as restrictions of natLeq: natLeq_on $n \equiv$ natLeq $\cap$ $\{0,\dots,n\} \times \{0,\dots,n\}$. These behave like the finite cardinals (up to isomorphism):

Lemma 20. (1) card_order (natLeq_on n);   (2) finite $A \wedge |A| =_o |B| \to$ finite B;
(3) finite $A \leftrightarrow (\exists n. \, |A| =_o$ natLeq_on n).

For finite cardinalities, we prove backward compatibility with the preexisting cardinality operator card : α set $\to$ nat, which maps infinite sets to 0:

Lemma 21. *Assuming* finite $A \wedge$ finite B:

(1) $|A| =_o |B| \leftrightarrow \text{card } A = \text{card } B$ (2) $|A| \leq_o |B| \leftrightarrow \text{card } A \leq \text{card } B$

The card operator has extensive library support in Isabelle. It is still the preferred cardinality operator for finite sets, since it refers to numbers with order and equality rather than the more bureaucratic order embeddings and isomorphisms.

cardSuc preserves finiteness and behaves as expected for finite cardinals:

Lemma 22. (1) $\text{Card\_order } r \to (\text{finite } (\text{cardSuc } r) \leftrightarrow \text{finite } (\text{Field } r))$;
(2) $\text{cardSuc } (\text{natLeq\_on } n) =_o \text{natLeq\_on } (\text{Suc } n)$.

4.4   Cardinal Arithmetic

To define cardSuc r, the successor of a cardinal $r : \alpha$ rel, we first choose a type that is large enough to contain a cardinal greater than r, namely, α set. The successor cardinal is then defined as a cardinal that is greater than r and that is $\leq_o$-minimal among all cardinals on the chosen type α set:

$$\text{isCardSuc} : \alpha \text{ rel} \to (\alpha \text{ set}) \text{ rel} \to \text{bool}$$
$$\text{isCardSuc } r\, s \equiv \text{Card\_order } s \wedge r <_o s \wedge$$
$$(\forall t : (\alpha \text{ set}) \text{ rel. Card\_order } t \wedge r <_o t \to s \leq_o t)$$

The choice of the second argument's type, together with Theorem 7, ensures that such a cardinal exists:

Lemma 23. $\exists s.\ \text{isCardSuc } r\, s$.

This allows us to define the function cardSuc : α rel $\to$ (α set) rel that yields an arbitrary successor cardinal of its argument r: $\text{cardSuc } r \equiv \varepsilon s.\ \text{isCardSuc } r\, s$. The chosen cardinal is really a successor cardinal:

Lemma 24. $\text{isCardSuc } r\, (\text{cardSuc } r)$

To obtain the desired characteristic properties of successor cardinals in full generality, we must prove that cardSuc r is minimal not only among the cardinals on α set but among all cardinals. This is achieved by a tedious process of making isomorphic copies.

Theorem 25. *Assuming* $\text{Card\_order } (r : \alpha \text{ rel})$ *and* $\text{Card\_order } (s : \beta \text{ rel})$:

(1) $r <_o \text{cardSuc } r$ (2) $r <_o s \to \text{cardSuc } r \leq_o s$

Finally, we prove that cardSuc is compatible with isomorphism and is monotonic.

Theorem 26. *Assuming* $\text{Card\_order } r$ *and* $\text{Card\_order } s$:

(1) $\mathsf{cardSuc}\ r =_o \mathsf{cardSuc}\ s \leftrightarrow r =_o s$ (2) $\mathsf{cardSuc}\ r <_o \mathsf{cardSuc}\ s \leftrightarrow r <_o s$

In summary, we first introduced the successor in a type-specific manner, asserting minimality within a chosen type, since HOL would not allow us to proceed more generally. Then we proved the characteristic property in full generality, and finally we showed that the notion is compatible with $=_o$ and $\leq_o$.

This approach is certainly more bureaucratic than the traditional set theoretic constructions, but it achieves the desired effect. The same technique is used to introduce all the standard cardinal operations (e.g, $+_c : \alpha\ \mathsf{rel} \to \beta\ \mathsf{rel} \to (\alpha + \beta)\ \mathsf{rel}$), for which we prove the basic arithmetic properties.

Lemma 27 (Lemma 1.5.10 in Holz et al. [10]). *The following properties hold:*

(1) $(r +_c s) +_c t =_o r +_c (s +_c t)$ (2) $r +_c s =_o s +_c r$

(3) $(r \times_c s) \times_c t =_o r \times_c (s \times_c t)$ (6) $r \times_c 1 =_o r$
(4) $r \times_c s =_o s \times_c r$ (7) $r \times_c (s +_c t) =_o r \times_c s +_c r \times_c t$
(5) $r \times_c 0 =_o 0$

(8) $r \,^{\wedge}c\, (s +_c t) =_o r \,^{\wedge}c\, s \times_c r \,^{\wedge}c\, t$ (12) $r \,^{\wedge}c\, 1 =_o r$
(9) $(r \,^{\wedge}c\, s) \,^{\wedge}c\, t =_o r \,^{\wedge}c\, (s \times_c t)$ (13) $1 \,^{\wedge}c\, r =_o 1$
(10) $(r \times_c s) \,^{\wedge}c\, t =_o r \,^{\wedge}c\, t \times_c s \,^{\wedge}c\, t$ (14) $r \,^{\wedge}c\, 2 =_o r \times_c r$
(11) $\neg\, r =_o 0 \to r \,^{\wedge}c\, 0 =_o 1 \wedge 0 \,^{\wedge}c\, r =_o 0$

(15) $r \leq_o s \wedge t \leq_o u \to r +_c t \leq_o s +_c u$ (17) $r \leq_o s \wedge t \leq_o u \to r \times_c t \leq_o s \times_c u$
(16) $r \leq_o s \wedge t \leq_o u \wedge \neg\, t =_o 0 \to r \,^{\wedge}c\, t \leq_o s \,^{\wedge}c\, u$

Another useful cardinal operation is the maximum of two cardinals, $\mathsf{cmax}\ r\ s$, which is well defined by the totality of $\leq_o$. Thanks to Lemma 17(1), it behaves like both sum and product for infinite cardinals:

Lemma 28. (infinite (Field r) $\wedge$ Field $s \neq \emptyset$) $\vee$ (infinite (Field s) $\wedge$ Field $r \neq \emptyset$) $\to$ $\mathsf{cmax}\ r\ s =_o r +_c s =_o r \times_c s$.

4.5 Regular Cardinals

A set $A : \alpha\ \mathsf{set}$ is *cofinal* for $r : \alpha\ \mathsf{rel}$, written $\mathsf{cofinal}\ A\ r$, if $\forall a \in \mathsf{Field}\ r.\ \exists b \in A.\ a \neq b\ \wedge$ $(a, b) \in r$; and r is *regular*, written regular r, if $\forall A.\ A \subseteq \mathsf{Field}\ r \wedge \mathsf{cofinal}\ A\ r \to |A| =_o r$.

Regularity is a generalization of the property of natLeq of not being "coverable" by smaller cardinals—indeed, no finite set A of numbers fulfills $\forall m.\ \exists n \in A.\ m < n$. The infinite successor cardinals are further examples of regular cardinals.

Lemma 29. (1) regular natLeq;
(2) Card_order $r \wedge$ infinite (Field r) $\to$ regular ($\mathsf{cardSuc}\ r$).

A property of regular cardinals useful in applications is the following: Inclusion of a set of smaller cardinality in a union of a chain indexed by the cardinal behaves similarly to membership, in that it amounts to inclusion in one of the sets in the chain.

Lemma 30. *Assume* Card_order r, *regular* r, $\forall i\, j.\ (i, j) \in r \to A\, i \subseteq A\, j$, $|B| <_o r$, *and* $B \subseteq \bigcup_{i \in \text{Field } r} A\, i$. *Then* $\exists i \in \text{Field } r.\ B \subseteq A\, i$.

Finally, regular cardinals are stable under unions. They cannot be covered by a union of sets of smaller cardinality indexed by a set of smaller cardinality.

Lemma 31. *Assuming* Card_order r, *regular* r, $|I| <_o r$, *and* $\forall i \in I.\ |A\, i| <_o r$, *we have* $|\bigcup_{i \in I} A\, i| <_o r$.

We also proved the converse: The above property is not only necessary but also sufficient for regularity.

5   Discussion of the Formalization

Figure 1 shows the main theory structure of our development. The overall development amounts to about 14 000 lines of scripts, excluding the applications. We also formalized many basic facts about wellorders and (order-)isomorphic transfer across bijections. When we started, Isabelle's library had extensive support for orders based on type classes [6]. However, working with the wellorder type class was not an option, since we need several wellorders for the same type—for example, the cardinal of a type is defined as the minimum of all its wellorders.

Reasoning about the modified version of equality and order ($=_o$, $\leq_o$, and $<_o$) was probably the most tedious aspect of the formalization effort. The standard Isabelle proof methods (*auto*, *blast*, etc.) are optimized for reasoning about actual equality and order. Some of the convenience could be recovered via an appropriate setup; for example, declaring these relations as transitive enables calculational reasoning in Isar [1].

For the initial version of the formalization, developed in 2009, Isabelle's Sledgehammer tool for deploying external automatic theorem provers [16] did not help much. The proofs required a careful combination of facts on orders and isomorphic transfer, and Sledgehammer was not as powerful as it is today. In contrast, cardinal arithmetic was developed later and largely automated in this way.

Throughout the paper, we have illustrated our effort to adapt the theory of cardinals to the HOL types, doing without a canonical class of ordinals ordered by membership. Another limitation of HOL is its inability to quantify over types except universally and at the statements' top level. A notorious example originates from the formalizations of the FOL completeness theorem (e.g., Harrison [8]): A sentence is provable if and only if it is true in all models. The 'if' direction is not expressible in HOL, because the right-hand side quantifies over all carrier types of all models, which amounts to an existential type quantification at the top of the formula. But one can express and prove a stronger statement: Based on the language cardinality, one fixes a witness type so that satisfaction in all models on that type already ensures provability. Our formalization abounds in such apparently inexpressible statements. One example is the definition of the successor cardinal from Section 4.4. Another is the claimed converse of Lemma 31. Each time, we needed to select a suitable witness type in an ad hoc fashion.

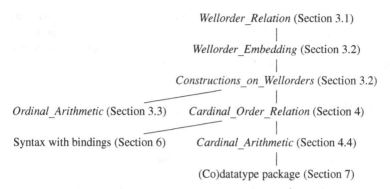

Fig. 1. Theory structure

6   Application: Syntax with Bindings

Popescu has formalized a general theory of syntax with bindings, parameterized over a binding signature with possibly infinitary operation symbols [20–22]. Cardinals were crucially needed for supporting infinitary syntax.

We illustrate the problem and solution on an example. Let index and var be types representing indices and variables, and consider the freely generated type of terms

> **datatype** term = Var var | Lam var term | Sum (index → term)

Thus, a term is either (an injection of) a variable, a λ-abstraction, or an indexed sum of a family of terms. The standard operators of free variables fvars : term → var set and capture-avoiding substitution $\_[\_/\_]$: term → term → var → term are defined below:

$$
\begin{aligned}
\text{fvars (Var } x) &= \{x\} & \text{(Var } x)[s/y] &= \text{(if } x = y \text{ then } s \text{ else Var } x)\\
\text{fvars (Lam } x\, t) &= \text{fvars } t - \{x\} & \text{(Lam } x\, t)[s/y] &= \text{(let } x' = \text{pickFresh } [\text{Var } y, s]\\
\text{fvars (Sum } f) &= \bigcup_{i \in I} \text{fvars } (f\, i) & & \quad \text{in Lam } x'\ (t[x'/x][s/y]))\\
& & \text{(Sum } f)[s/y] &= \text{Sum } (\lambda i.\ (f\, i)[s/y])
\end{aligned}
$$

To avoid capture, the Lam case of substitution renames the variable x to x'. The new name is chosen by the pickFresh operator, which takes a list of terms ts as argument and returns some variable not occurring freely in ts. But how can we be sure that such a choice exists? The standard solution of making the type var infinite does not suffice here: The Sum constructor introduces possibly infinite branching on index, and therefore fvars t may return an infinite set of variables, potentially even UNIV.

Fortunately, the essence of the standard solution can be generalized to the infinitary situation. Finitely branching syntax relies on the observation that no n-ary constructor violates the finiteness of the set of free variables, since a finite union of finite sets is finite. Lemma 31 generalizes this notion to regular cardinals. Hence, we simply need to define var so that it has a regular cardinal greater than index: var = cardSuc |index|.

Lemma 32. regular $|\text{var}| \wedge |\text{index}| <_o |\text{var}| \rightarrow (\forall t.\ |\text{fvars } t| <_o |\text{var}|)$.

Proof idea. By structural induction on t, using Lemma 31.   □

After passing this milestone, a theory of substitution and free variables proceeds similarly to the finitary case [20]. Most current frameworks for syntax with bindings, including nominal logic [12, 18], assume finiteness of the syntactic objects. Regular cardinals provide a foundation for an infinitary generalization.

7   Application: Bounded Functors and the (Co)datatype Package

Isabelle's new (co)datatype package draws on both category theory and cardinal theory. It maintains a class of functors with additional structure, called *bounded natural functors* (BNFs), for which it constructs initial algebras (datatypes) and final coalgebras (codatatypes). The category theory underlying the package is described in Traytel et al. [24]; here, we focus on the cardinality aspects.

BNFs are type constructors equipped with functorial actions, n natural transformations, and a cardinality bound. A unary BNF consists of a type constructor α F, a constant Fmap : $(\alpha \to \beta) \to \alpha$ F $\to \beta$ F, a constant Fset : α F $\to \alpha$ set that is natural with respect to F, and a cardinal Fbd such that $\forall x.\ |\text{Fset } x| \leq_o \text{Fbd}$. We define Fin : α set $\to (\alpha\,F)$ set, the set-based version of F, by Fin $A = \{x \mid \text{Fset } x \subseteq A\}$—this is a common generalization of the specific set-based operators from Section 4.2.

An algebra for F is a triple $\mathscr{A} = (T, A : T \text{ set}, s : T \text{ F} \to T)$ (where T is a type) such that $\forall x \in \text{Fin } A.\ s\ x \in A$. The condition ensures that s is a function from Fin A to A, and we allow ourselves to write $s : \text{Fin } A \to A$. The set A is the *carrier* of $\mathscr{A}$, and s is the *structural map* of A. The structural map models the operations of the algebra. For example, if α F = unit $+ \alpha \times \alpha$, an algebra $\mathscr{A}$ consists of a set $A : T$ set with a constant and a binary operation on it, encoded as $s : \text{unit} + \alpha \times \alpha \to \alpha$.

This notion accommodates standard algebraic constructions. One forms the *product* $\prod_{i \in I} \mathscr{A}_i$ of a family of algebras (of type T) by taking the product of the carrier sets and defining the structural map $s : \text{Fin } (\prod_{i \in I} A_i) \to \prod_{i \in I} A_i$ as $s\ x = (s_i\ (\text{Fmap proj}_i\ x))_{i \in I}$. A *stable part* of $\mathscr{A}$ is any set $A' \subseteq A$ such that $\forall x \in \text{Fin } A'.\ s\ x \in A'$. Since the intersection of stable parts is a stable part, we can define an algebra $\text{Min}(\mathscr{A})$, the *minimal algebra* of $\mathscr{A}$, by taking its carrier to be the intersection of all stable parts and its structural map to be (the restriction of) s. This corresponds to the notion of subalgebra generated by $\emptyset$. A *morphism* between two algebras $\mathscr{A}$ and $\mathscr{A}'$ is a function $h : A \to A'$ that commutes with the structural maps, in that $\forall x \in \text{Fin } A.\ h\ (s\ x) = s'\ (\text{Fmap } h\ x)$.

Building the initial algebra of F (an algebra such that for any algebra $\mathscr{A}$, there exists precisely one morphism between it and $\mathscr{A}$) can be naively attempted as follows: First we take $\mathscr{R} = \prod \{\mathscr{A} \mid \mathscr{A} \text{ algebra}\}$, the product of all algebras. Given an algebra $\mathscr{A}$, there must exist a morphism h from $\mathscr{R}$ to $\mathscr{A}$—the corresponding projection. The restriction of h to $\text{Min}(\mathscr{R})$ is then the desired unique morphism from $\text{Min}(\mathscr{R})$ to $\mathscr{A}$, and $\text{Min}(\mathscr{R})$ is our desired initial algebra.

This naive approach fails since we cannot construct the product of all algebras in HOL—and even if we could, it would not be an algebra itself due to its size. Fortunately, it suffices to define the morphism h from $\mathscr{R}$ not to $\mathscr{A}$ but to $\text{Min}(\mathscr{A})$. Hence, we can take $\mathscr{R}$ as the product of all *minimal* algebras and consider only a complete collection of representatives (up to isomorphism). This is where the bound on F comes into play. If we know that all minimal algebras of all algebras had cardinality smaller than a given bound r_0, we can choose a type T_0 of cardinality r_0 and define $\mathcal{R}$ as the product of all algebras on T_0: $\mathcal{R} = \prod \{\mathcal{A} \mid \mathcal{A} = (T_0, A : T_0 \text{ set}, s : T_0 \text{ F} \to T_0) \text{ algebra}\}$. A suitable cardinal bound is $r_0 = 2 \wedge_c k$, where $k = \text{cardSuc} (\text{Fbd} +_c |\text{Fin} (\text{Field Fbd})|)$. To prove this, we first establish the following consequence of the BNF boundedness property:[2]

Lemma 33. $|A| \geq_o 2 \to |\text{Fin } A| \leq_o |A| \wedge_c k$.

Theorem 34. *For all algebras $\mathcal{A}$, let M be the carrier of $\text{Min}(\mathcal{A})$. Then $|M| \leq_o 2 \wedge_c k$.*

Proof idea. The definition of $\text{Min}(\mathcal{A})$ performs a construction of M "from above," as an intersection, yielding no cardinality information. We must produce an alternative construction "from below," exploiting the internal structure of F. Let $N = \bigcup_{i \in \text{Field } k} N_i$, where each N_i is defined by wellorder recursion as follows: $N_i = \bigcup_{j \in \text{underS} k i} s \bullet \text{Fin } N_j$. We first prove that N is a stable part of $\mathcal{A}$, and hence $M \subseteq N$. Let $x \in \text{Fin } N$. Then $\text{Fset } x \subseteq N = \bigcup_{i \in \text{Field } k} N_i$, and since k is regular by Lemma 29(2), we use Lemma 30 to obtain $i \in \text{Field } k$ such that $\text{Fset } x \subseteq N_i$ (i.e., $x \in \text{Fin } N_i$). Hence, $s\, x \in N_{\text{succ } k\, i} \subseteq N$, as desired. Conversely, $N \subseteq M$ follows by wellorder induction. Thus, we have $M = N$. The inequality $|N| \leq_o 2 \wedge_c k$ follows by wellorder induction, using Lemma 33 and cardinal arithmetic to keep the passage from N_i to $\text{Fin } N_i$ bounded. Knowing $|N_i| \leq_o 2 \wedge_c k$, we obtain $|\text{Fin } N_i| \leq_o |N_i| \wedge_c k \leq_o (2 \wedge_c k) \wedge_c k =_o 2 \wedge_c (k \times_c k) =_o 2 \wedge_c k$. ☐

Cardinal arithmetic is also used throughout the package for showing that the various constructions on BNFs (composition, initial algebra, and final coalgebra) yield BNFs.

8 Conclusion

We have formalized in Isabelle/HOL a theory of cardinals, proceeding locally and abstractly, up to wellorder isomorphism. The theory has been applied to reason about infinitary objects arising in syntax with bindings and (co)datatypes.

We hope our experiment will be repeated by the other HOL provers, where a theory of cardinals seems as useful as in any other general-purpose framework for mathematics. Indeed, the theory provides working mathematicians with the needed injections and bijections (e.g., between lists over an infinite type, or the square of an infinite type, and the type itself) without requiring them to perform awkward encodings.

An open question is whether the quotient construction performed by Norrish and Huffman (briefly discussed in the introduction) would have helped the cardinal formalization. With their approach, we would still need to change the underlying type of cardinals to accommodate for increasingly large sizes. HOL offers no way to reason about arbitrary cardinals up to equality, so isomorphism appears to be the right compromise.

Acknowledgment. Tobias Nipkow made this work possible. Stefan Milius and Lutz Schröder contributed an elegant proof of Lemma 33. The anonymous reviewers suggested many improvements to the paper. Blanchette is supported by the DFG (Deutsche

---

[2] Initially, we had maintained a slight variation of this property as an additional BNF requirement [24, Section 4], not realizing that it is redundant. Removing it has simplified the package code substantially.

Forschungsgemeinschaft) project Hardening the Hammer (grant Ni 491/14-1). Popescu is supported by the DFG project Security Type Systems and Deduction (grant Ni 491/ 13-2) as part of the program Reliably Secure Software Systems (RS³, priority program 1496). Traytel is supported by the DFG program Program and Model Analysis (PUMA, doctorate program 1480). The authors are listed alphabetically regardless of individual contributions or seniority.

References

1. Bauer, G., Wenzel, M.: Calculational reasoning revisited (An Isabelle/Isar experience). In: Boulton, R.J., Jackson, P.B. (eds.) TPHOLs 2001. LNCS, vol. 2152, pp. 75–90. Springer, Heidelberg (2001)
2. Breitner, J.: Free groups. In: Klein, G., Nipkow, T., Paulson, L. (eds.) Archive of Formal Proofs (2011), http://afp.sf.net/entries/Free-Groups.shtml
3. Church, A.: A formulation of the simple theory of types. J. Symb. Logic 5(2), 56–68 (1940)
4. Forster, T.E.: Reasoning about Theoretical Entities. World Scientific (2003)
5. Gordon, M.J.C., Melham, T.F. (eds.): Introduction to HOL: A Theorem Proving Environment for Higher Order Logic. Cambridge University Press (1993)
6. Haftmann, F., Wenzel, M.: Constructive type classes in Isabelle. In: Altenkirch, T., McBride, C. (eds.) TYPES 2006. LNCS, vol. 4502, pp. 160–174. Springer, Heidelberg (2007)
7. Harrison, J.: The HOL wellorder library (1992),
   http://www.cl.cam.ac.uk/~jrh13/papers/wellorder-library.html
8. Harrison, J.: Formalizing basic first order model theory. In: Grundy, J., Newey, M. (eds.) TPHOLs 1998. LNCS, vol. 1479, pp. 153–170. Springer, Heidelberg (1998)
9. The HOL Light theorem prover (2014),
   http://www.cl.cam.ac.uk/~jrh13/hol-light/index.html
10. Holz, M., Steffens, K., Weitz, E.: Introduction to Cardinal Arithmetic. Birkhäuser Advanced Texts. Birkhäuser (1999)
11. Huffman, B.: Countable ordinals. In: Klein, G., Nipkow, T., Paulson, L. (eds.) Archive of Formal Proofs (2005), http://afp.sf.net/entries/Ordinal.shtml
12. Huffman, B., Urban, C.: Proof pearl: A new foundation for Nominal Isabelle. In: Kaufmann, M., Paulson, L.C. (eds.) ITP 2010. LNCS, vol. 6172, pp. 35–50. Springer, Heidelberg (2010)
13. Nipkow, T., Klein, G.: Concrete Semantics: A Proof Assistant Approach. Springer (to appear), http://www.in.tum.de/~nipkow/Concrete-Semantics
14. Nipkow, T., Paulson, L.C., Wenzel, M.: Isabelle/HOL. LNCS, vol. 2283. Springer, Heidelberg (2002)
15. Norrish, M., Huffman, B.: Ordinals in HOL: Transfinite arithmetic up to (and beyond) ω_1. In: Blazy, S., Paulin-Mohring, C., Pichardie, D. (eds.) ITP 2013. LNCS, vol. 7998, pp. 133–146. Springer, Heidelberg (2013)
16. Paulson, L.C., Blanchette, J.C.: Three years of experience with Sledgehammer, a practical link between automatic and interactive theorem provers. In: Sutcliffe, G., Schulz, S., Ternovska, E. (eds.) IWIL 2010. EPiC Series, vol. 2, pp. 1–11. EasyChair (2012)
17. Paulson, L.C., Grabczewski, K.: Mechanizing set theory. J. Autom. Reasoning 17(3), 291–323 (1996)
18. Pitts, A.M.: Nominal logic, a first order theory of names and binding. Inf. Comput. 186(2), 165–193 (2003)
19. Popescu, A.: Ordinals and cardinals in HOL. In: Klein, G., Nipkow, T., Paulson, L. (eds.) Archive of Formal Proofs (2009),
   http://afp.sf.net/entries/Ordinals_and_Cardinals.shtml

20. Popescu, A.: Contributions to the theory of syntax with bindings and to process algebra. Ph.D. thesis, University of Illinois at Urbana-Champaign (2010)
21. Popescu, A., Gunter, E.L.: Recursion principles for syntax with bindings and substitution. In: Chakravarty, M.M.T., Hu, Z., Danvy, O. (eds.) ICFP 2011, pp. 346–358. ACM (2011)
22. Popescu, A., Gunter, E.L., Osborn, C.J.: Strong normalization of System F by HOAS on top of FOAS. In: LICS 2010, pp. 31–40. IEEE (2010)
23. Taylor, P.: Intuitionistic sets and ordinals. J. Symb. Log. 61(3), 705–744 (1996)
24. Traytel, D., Popescu, A., Blanchette, J.C.: Foundational, compositional (co)datatypes for higher-order logic: Category theory applied to theorem proving. In: LICS 2012, pp. 596–605. IEEE (2012)

Verified Abstract Interpretation Techniques for Disassembling Low-level Self-modifying Code*

Sandrine Blazy[1], Vincent Laporte[1], and David Pichardie[2]

[1] Université Rennes 1, IRISA, Inria
[2] ENS Rennes, IRISA, Inria

Abstract Static analysis of binary code is challenging for several reasons. In particular, standard static analysis techniques operate over control flow graphs, which are not available when dealing with self-modifying programs which can modify their own code at runtime. We formalize in the Coq proof assistant some key abstract interpretation techniques that automatically extract memory safety properties from binary code. Our analyzer is formally proved correct and has been run on several self-modifying challenges, provided by Cai et al. in their PLDI 2007 paper.

1 Introduction

Abstract interpretation [9] provides advanced static analysis techniques with strong semantic foundations. It has been applied on a large variety of programming languages. Still, specific care is required when adapting these techniques to low-level code, specially when the program to be analyzed comes in the form of a sequence of bits and must first be disassembled. Disassembling is the process of translating a program from a machine friendly binary format to a textual representation of its instructions. It requires to *decode* the instructions (i.e., understand which instruction is represented by each particular bit pattern) but also to precisely locate the instructions in memory. Indeed instructions may be interleaved with data or arbitrary padding. Moreover once encoded, instructions may have various byte sizes and may not be well aligned in memory, so that a single byte may belong to several instructions.

To thwart the problem of locating the instructions in a program, one must follow its control flow. However, this task is not easy because of the indirect jumps, whose targets are unknown until runtime. A static analysis needs to know precisely enough the values that the expression denoting the jump target may evaluate to. In addition, instructions may be produced at runtime, as a result of the very execution of the program. Such programs are called *self-modifying* programs; they are commonly used in security as an obfuscation technique, as well as in just-in-time compilation. Analyzing a binary code is mandatory when this code is the only available part of a software. Most of standard reverse engineering tools (e.g., IDA Pro) cannot disassemble and analyze self-modifying programs. In order to disassemble

---

* This work was supported by Agence Nationale de la Recherche, grant number ANR-11-INSE-003 Verasco.

G. Klein and R. Gamboa (Eds.): ITP 2014, LNAI 8558, pp. 128–143, 2014.
© Springer International Publishing Switzerland 2014

and analyze such programs, one must very precisely understand which instructions are written and where. And for all programs, one must check every single memory write to decide whether it modifies the program code.

Self-modifying programs are also beyond the scope of the vast majority of formal semantics of programming languages. Indeed a prerequisite in such semantics is the isolation and the non-modification of code in memory. Turning to verified static analyses, they operate over toy languages [5, 16] or more recently over realistic C-like languages [18, 3], but they assume that the control-flow graph is extracted by a preliminary step, and thus they do not encompass techniques devoted to self-modifying code.

In this paper, we formalize with the Coq proof assistant, key static analysis techniques to predict the possible targets of the computed jumps and make precise which instructions alter the code and how, while ensuring that the other instructions do not modify the program. Our static analysis techniques rely on two main components classically used in abstract interpretation, abstract domains and fixpoint iterators, that we detail in this paper. The complete Coq development is available online [8].

Our formalization effort is divided in three parts. Firstly, we formalize a small binary language in which code is handled as regular mutable data. Secondly, we formalize and prove correct an abstract interpreter that takes as input an initial memory state, computes an over-approximation of the reachable states that may be generated during the program execution, and then checks that all reachable states maintain memory safety. Finally, we extract from our formalization an executable OCaml tool that we run on several self-modifying challenges, provided by Cai et al. [6].

The paper makes the following contributions.

– We push further the limit in terms of verified static analysis by tackling the specific challenge of binary self-modifying programs, such as fixpoint iteration without control-flow graph and simple trace partitioning [12].

– We provide a complementary approach to [6] by automatically inferring the required state invariants that enforce memory safety. Indeed, the axiomatic semantics of [6] requires programs to be manually annotated with invariants written in a specific program logic.

The remainder of this paper is organized as follows. First, Section 2 briefly introduces the static analysis techniques we formalized. Then, Section 3 details our formalization: it defines the semantics of our low-level language and details our abstract interpreter. Section 4 describes some improvements that we made to the abstract interpreter, as well as the experimental evaluation of our implementation. Related work is discussed in Section 5, followed by concluding remarks.

2   Disassembling by Abstract Interpretation

We now present the main principles of our analysis on the program shown in Figure 1. It is printed as a sequence of bytes (on the extreme left) as well as under a disassembled form (on the extreme right) for readability purposes. This program, as we will see, is self-modifying, so these bytes correspond to the initial content of the memory from addresses 0 to 11. The remaining of the memory (addresses in $[-2^{32}; -1] \cup [12; 2^{32} - 1]$), as well as the content of the registers, is unknown and can be regarded as the program input.

All our example programs target a machine operating over a low-level memory made of 2^{32} cells, eight registers (R0, ... R7), and flags — boolean registers that are set by comparison instructions. Each memory cell or register stores a 32 bits integer value, that may be used as an address in the memory. Programs are stored as regular data in the memory; their execution starts from address zero. Nevertheless, throughout this paper we write the programs using the following custom syntax. The instruction cst v → r loads register r with the given value v; cmp r, r' denotes the comparison of the contents of registers r and r'; gotoLE d is a conditional jump to d, it is taken if in the previous comparison the content of r' was less than or equal to the one of r; goto d is an unconditional jump to d. The instruction load *r → r' and store r' → *r denote accesses to memory at the address given in register r; and halt r halts the machine with as final value the content of register r.

The programming language we consider is inspired from x86 assembly; notably instructions have variable size (one or two bytes, e.g., the length of the instruction stored at line 1 is two bytes) and conditional jumps rely on flags. In this setting, a program is no more than an initial memory state, and a program point is simply the address of the next instruction to execute.

Initial program	Possible final program	Initial assembly listing
07000607	07000607	0: cmp R6, R7
03000000	03000000	1: gotoLE 5
00000005	**00000004**	2:
00000000	00000000	3: halt R0
00000100	00000100	4: halt R1
09000000	09000000	5: cst 4 → R0
00000004	00000004	6:
09000002	09000002	7: cst 2 → R2
00000002	00000002	8:
05000002	05000002	9: store R0 → *R2
04000000	04000000	10: goto 1
00000001	00000001	11:

Fig. 1. A self-modifying program: as a byte sequence (left); after some execution steps (middle); assembly source (right)

In order to understand the behavior of this program, one can follow its code as it is executed starting from the entry point (byte 0). The first instruction compares the (statically unknown) content of two registers. This comparison modifies only the states of the flags. Then, depending on the outcome of this comparison, the execution proceeds either on the following instruction (stored at byte 3), or from byte 5. Executing the block from byte 5 will modify the byte 2 belonging to the gotoLE instruction (highlighted in Figure 1); more precisely it will change the jump destination from 5 to 4: the store R0 → *R2 instruction writes the content of register R0 (namely 4) in memory at the address given in register R2 (namely 2). Notice that a program may directly read from or write to any memory cell: we assume that there is no protection mechanism as provided by usual operating systems. After the modification is performed, the execution jumps back to the modified instruction, jumps to byte 4 then halts, with final value the content of register R1.

This example highlights that the code of a program (or its control-flow graph) is not necessarily a static property of this program: it may vary as the program runs. To correctly analyze such a program, one must discover, during the fixpoint iteration, the two possible states of the instruction at locations 1 and 2 and its two possible targets. More specially, we need at least to know, for each program point (i.e., memory location), which instructions may be decoded from there when the execution reaches this point. This in turn requires to know what are the values that the program operates on. We therefore devise a value analysis that computes, for each reachable program point (i.e., in a *flow sensitive* way) an over-approximation of the content of the memory and the registers, and the state of the flags when the execution reaches that point.

The analysis relies on a numeric abstract domain $N^\sharp$ that provides a representation for sets of machine integers ($\gamma_N \in N^\sharp \to \mathcal{P}(\text{int})$) and abstract arithmetic operations. Relying on such a numeric domain, one can build abstract transformers that model the execution of each instruction over an abstract memory that maps locations (i.e., memory addresses[1] and registers) to abstract numeric values. An abstract state is then a mapping that attaches such an abstract memory to each program point of the program, and thus belongs to $\text{addr} \to ((\text{addr} + \text{reg}) \to N^\sharp)$.

To perform one abstract execution step, from a program point pp and an abstract memory state $m^\sharp$ that is attached to pp, we first enumerate all instructions that may be decoded from the set $\gamma_N(m^\sharp(\text{pp}))$. Then for each of such instructions, we apply the matching abstract transformer. This yields a new set of successor states whose program points are dynamically discovered during the fixpoint iteration.

The abstract interpretation of a whole program iteratively builds an approximation executing all reachable instructions until nothing new is learned. This iterative process may not terminate, since there might be infinite increasing chains in the abstract search space. As usual in abstract interpretation, we accelerate the iteration using widening operations [9]. Once a stable approximation is finally reached, an approximation of the program listing or control-flow graph can be produced.

To illustrate this process, Figure 2 shows how the analysis of the program from Figure 1 proceeds. We do not expose a whole abstract memory but only the underlying control-flow graph it represents. On this specific example, three

---

[1] Type addr is a synonym of int, the type of machine integers.

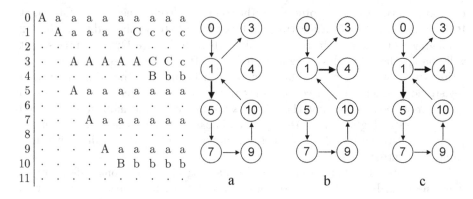

Fig. 2. Iterative fixpoint computation

different graphs are encountered during the analysis. For each program point pp, we represent a node with same name and link it with all the possible successor nodes according to the decoding of the set $\gamma_N(m^\sharp(pp))$. The array shows the construction of the fixpoint: each line represents a program point and the columns represent the iterations of the analysis. In each array cell lies the name of the control-flow graph representing the abstract memory for the given program point during the given iteration; a dot stands for an unreachable program point. The array cells whose content is in upper case highlight the program points that need to be analyzed: they are the worklist.

Initially, at iteration 0, only program point 0 is known to be reachable and the memory is known to exactly contain the program, denoted by the first control-flow graph. The only successor of point 0 is point 1 and it is updated at the next iteration. After a few iterations, point 9 is reached and the abstract control-flow graph a is updated into a control-flow graph b that is propagated to point 10. After a few more iterations, the process converges.

In addition to a control-flow graph or an assembly listing, more properties can be deduced from the analysis result. We can prove safety properties about the analyzed program, like the fact that its execution is never stuck. Since the semantics only defines the good behaviors of programs, unsafe programs reach states that are not final and from which no further execution step is possible (e.g., the byte sequence at current program point is not the valid encoding of an instruction).

The analysis produces an over-approximation of the set of reachable states. In particular, a superset of the reachable program points is computed, and for each of these program points, an over-approximation of the memory state when the execution reaches this program point is available. Thus we can check that for every program point that may be reached, the next execution step from this point cannot be stuck. This verification procedure is formally verified, as described in the following section.

3   Formalization

The static analyzer is specified, programmed and proved correct using the Coq proof assistant. This involves several steps that are described in this section: first, define the semantics of a binary language, then design abstract domains and abstract transformers, as well as write a fixpoint iterator, and lastly state and prove soundness properties about the results of the static analysis.

3.1   Concrete Syntax and Semantics

The programming language in which are written the programs to analyze is formalized using the syntax shown on Figure 3. So as to model a binary language, we introduce a decoding function dec (mem: (addr → int)) (pp: int) : option (instruction * nat) that given a memory mem (i.e., a function from addresses to values) and an address pp yields the instruction stored from this address along with its byte size. Since not all integer sequences are valid encodings, this decoding may fail (hence the option type[2]). In order to be able to conveniently write programs, there is also a matching encoding function. However the development does not depend on it at all.

```
Inductive reg  := R0 | R1 | R2 | R3 | R4 | R5 | R6 | R7.
Inductive flag := FLE | FLT | FEQ.
Inductive instruction :=
(* arithmetic *)
| ICst (v:int) (dst:reg) | ICmp (src dst: reg)
| IBinop (op: int_binary_operation) (src dst: reg)
(* memory *)
| ILoad  (src dst: reg) | IStore (src dst: reg)
(* control *)
| IGoto (tgt: addr) | IGotoInd (r: reg) | IGotoCond (f: flag) (tgt: addr)
| ISkip | IHalt (r: reg).
```

Fig. 3. Language syntax

The language semantics is given as a small-step transition relation between machine states. A machine state may be ⟨pp, f, r, m⟩ where pp is the current program point (address of the next instruction to be executed), f is the current flag state, r is the current register state, and m is the current memory. Such a tuple is called a machine *configuration* (type machine_config). Otherwise, a machine state is ⌈v⌉, meaning that the program stopped returning the value v.

The semantics is defined as a set of rules of the following shape:

$$\frac{\text{dec m pp} = \lfloor(\text{i, z})\rfloor}{\langle\text{pp, f, r, m}\rangle \leadsto \langle\text{pp', f', r', m'}\rangle} .$$

---

[2] Values of type option A are either None or ⌊a⌋ with a a value of type A.

The premise states that decoding the bytes in memory m from address pp yields the instruction i whose size in memory is z. Then each rule describes how to execute a particular instruction at program point pp in memory m with flag state f and register state r. In each case, most of the state is kept unchanged. Instructions that are not branching proceed their execution at program point pp+z (since z is the size of this instruction once encoded). The whole set of rules can be found in the Coq development [8]. We describe only some of them. Instruction ICmp rs rd updates the flag state according to the comparison of the values held by the two involved registers. Conditional jump instruction IGotoCond c v jumps to address v or falls through to pp+z depending on the current state of flag c. Indirect jump instruction IGotoInd rd proceeds at the program point found in register rd.

Finally, we define the semantics $[\![P]\!]$ of a program P as the set of states that are reachable from an initial state, with current program point zero and memory P (where $\rightsquigarrow^*$ denotes the reflexive-transitive closure of the small-step relation).

$$[\![P]\!] = \{s \mid \exists f\ r,\ \langle 0,\ f,\ r,\ P \rangle \rightsquigarrow^* s\}$$

3.2   Abstract Interpreter

In order to analyze programs, we build an abstract interpreter, i.e., an executable semantics that operates over abstract elements, each of them representing many concrete machine configurations. Such an abstract domain provides operators that model basic concrete operations: read a value from a register, store some value at some address in memory, and so on. The static analyzer then computes a fixpoint within the abstract domain, that over-approximates all reachable states of the analyzed program.

We first describe our abstract domain before we head to the abstract semantics and fixpoint computation. An abstract memory domain is a carrier type along with some primitive operators whose signatures are given in Figure 4. The ab_num type refers to a numeric abstract domain, as described in [3]: we only require that this type is equipped with a concretization to sets of machine integers and abstract transformers corresponding to arithmetic operations.

The carrier type ab_mc is equipped with a lattice structure. An object of this type represents a set of concrete machine states, as described by the primitive gamma. It can be queried for the values stored in some register (var) or at some known memory address (load_single); these operators return an abstract numeric value. Other operators enable us to alter an abstract state, like assign that sets the contents of a register to a given abstract numeric value, and store_single that similarly updates the memory at a given address.

All these operators obey some specifications. As an example, the load_sound property states that given a concrete state m in the concretization of an abstract state ab, the concrete value stored at any address a in m is over-approximated by the abstract value returned by the matching abstract load. The γ symbol is overloaded through the use of type classes: its first occurrence refers to the concretization from the abstract memory domain (the gamma field of record mem_dom) and its second occurrence is the concretization from the numeric domain ab_num.

```
Record mem_dom (ab_num ab_mc: Type) :=
{ as_wl: weak_lattice ab_mc
; var: ab_mc → reg → ab_num
; load_single: ab_mc → addr → ab_num
; store_single: ab_mc → addr → ab_num → ab_mc
; assign: ab_mc → reg → ab_num → ab_mc
(* more abstract operators omitted *)
; gamma: gamma_op ab_mc machine_config
; as_adom : adom ab_mc machine_config as_wl gamma
; load_sound: ∀ ab:ab_mc, ∀ m: machine_config,
        m ∈ γ(ab) → ∀ a:addr, m(a) ∈ γ(load_single ab a)
(* more properties omitted *)  }.
```

Fig. 4. Signature of abstract memory domains (excerpt)

Such an abstract memory domain is implemented using two maps: from registers to abstract numeric values to represent the register state and from values to abstract numeric values to represent the memory.

```
Record ab_machine_config :=
    { ab_reg: Map [ reg, ab_num ] ; ab_mem: Map [ addr, ab_num ] }.
```

To prevent the domain of the ab_mem map from infinitely growing, we bound it by a finite set computed before the analysis: the analysis will try to compute some information only for the memory addresses found in this set [1]. The content of this set does not alter its soundness: the values stored at addresses not in it are unknown and the analyzer makes no assumptions about them. On the other hand, the success of the analysis and its precision depend on it. In particular, the analyzed set must cover the whole code segment.

As a second layer, we build abstract transformers over any such abstract domain. Consider for instance the abstract load presented in Figure 5; it is used to analyze any ILoad instruction (T denotes a record of type mem_dom ab_num ab_mc). The source address may not be exactly known, but only represented by an abstract numeric value a. Since any address in $\gamma(a)$ may be read, we have to query all of them and take the least upper bound of all values that may be stored at any of these addresses: $\bigsqcup \{T.(\text{load\_single}) \ m \ x \mid x \in \gamma(a)\}$. However the set of concrete addresses may be huge and care must be taken: if the size of this set exceeds some threshold, the analysis gives up on this load and yields top, representing all possible values.

We build enough such abstract transformers to be able to analyze any instruction (function ab_post_single, shown in Figure 6). This function returns a list of possible next states, each of which being either Hlt v (the program halts returning a value approximated by v) or Run pp m (the execution proceeds at program point pp in a configuration approximated by m) or GiveUp (the analysis is too imprecise to compute anything meaningful). The computed jump (IGotoInd) also has a dedicated abstract transformer (inlined in Figure 6): in order to know from where to continue the analysis, we have to enumerate all possible targets.

```
Inductive botlift (A:Type) : Type := Bot | NotBot (x:A).
Definition load_many (m: ab_mc) (a: ab_num) : botlift ab_num :=
  match concretize_with_care a with
  | Just addr_set ⇒ IntSet.fold
        (λ acc addr, acc ⊔ NotBot (T.(load_single) m addr)) addr_set Bot
  | All ⇒ NotBot top end.
```

Fig. 5. Example of abstract transformer

```
Inductive ab_post_res := Hlt(v:ab_num) | Run(pp:addr)(m:ab_mc) | GiveUp.
Definition ab_post_single (m:ab_mc) (pp:addr) (instr:instruction * nat)
  : list ab_post_res := match instr with
  | (IHalt rs, z) ⇒ Hlt (T.(var) m rs) :: nil
  | (ISkip, z) ⇒ Run (pp + z) m :: nil
  | (IGoto v, z) ⇒ Run v m :: nil
  | (IGotoInd rs, z) ⇒ match concretize_with_care (T.(var) m rs) with
      | Just tgt ⇒ IntSet.fold (λ acc addr, Run addr m :: acc) tgt nil
      | All ⇒ GiveUp :: nil end
  | (IStore rs rd, z) ⇒
      Run (pp + z) (store_many m (T.(var) m rd) (T.(var) m rs)) :: nil
  | (ILoad rs rd, z) ⇒ match load_many m (T.(var) m rs) with
      | NotBot v ⇒ Run (pp + z) (T.(assign) m rd v) :: nil
      | Bot ⇒ nil end
  | (ICmp rs rd, z) ⇒ Run (pp + z) (T.(compare) m rs rd ) :: nil
  | (ICst v rd, z) ⇒ Run (pp + z) (T.(assign) m rd v) :: nil
  (* ... *) end.
Definition ab_post_many (pp: addr) (m:ab_mc) : list ab_post_res :=
  match abstract_decode_at pp m with
  | Just instr ⇒ flat_map (ab_post_single m pp) instr
  | All ⇒ GiveUp :: nil end.
```

Fig. 6. Abstract small-step semantics (excerpt)

Then, function `ab_post_many` performs one execution step in the abstract. To do so, we first need to identify what is the next instruction, i.e., to decode in the abstract memory from the current program point. This may require to enumerate all concrete values that may be stored at this address. Therefore this abstract decoding either returns a set of possible next instructions or gives up. In such a case, the whole analysis will abort since the analyzed program is unknown.

Finally, the abstract semantics is iteratively applied until a fixpoint is reached following a worklist algorithm as the one found in [1, § 3.4]. However there may be infinite ascending chains, so to ensure termination we need to apply widening operators instead of regular joins frequently enough during the search. In our setting, with no control-flow graph available, the widening is applied on every back edge, but the implementation makes it easy to try different widening strategies. So as to convince Coq that the analysis indeed terminates, we rely on a counter (known as fuel) that obviously decreases at each iteration; when it reaches zero, the analyzer must give up.

To enhance the precision, we have introduced three more techniques: a dedicated domain to abstract the flag state, a partitioning of the state space, and a use of abstract instructions. They will be described in the next section.

3.3   Soundness of the Abstract Interpreter

We now describe the formal verification of our analyzer. The soundness property we ensure is that the result of the analysis of a program P over-approximates its semantics $[\![P]\!]$. This involves on one hand a proof that the analysis result is indeed a fixpoint of the abstract semantics and on the other hand a proof that the abstract semantics is correct with respect to the concrete one.

The soundness of the abstract semantics is expressed by the following lemma, which reads: given an abstract state ab and a concrete one m in the concretization of ab, for each concrete small-step m ⤳ m', there exists a result ab' in the list ab_post_single m.(pc) ab that over-approximates m'. Our use of Coq type classes enables us to extensively overload the γ notation and write this statement in a concise way as follows.

Lemma ab_post_many_correct :
  ∀ (m:machine_config) (m':machine_state) (ab:ab_mc),
    m ∈ γ(ab) → m ⤳ m' → m' ∈ γ(ab_post_single m.(pc) ab).

The proof of this lemma follows from the soundness of the various abstract domains (as load_sound in Figure 4), transformers and decoder.

Lemma abstract_decode_at_sound : ∀ (m:machine_config)(ab:ab_mc)(pp:addr),
    m ∈ γ(ab) → dec m.(mc_mem) pp ∈ γ(abstract_decode_at pp ab).

The proof that the analyzer produces a fixpoint is not done directly. Instead, we rely on *a posteriori* verification: we do not trust the fixpoint computation and instead program and prove a checker called validate_fixpoint. Its specification, proved thanks to the previous lemma, reads as follows.

Lemma validate_correct : ∀ (P: memory) (dom: list addr) (E: AbEnv),
    validate_fixpoint P dom E = true → $[\![P]\!]$ ⊆ γ(E).

Going through this additional programming effort has various benefits: a direct proof of the fixpoint iterator would be very hard; we can adapt the iteration strategy, optimize the algorithm and so on with no additional proof effort.

This validation checks two properties of the result E: that it over-approximates the initial state; and that it is a post-fixpoint of the abstract semantics, i.e., for each abstract state in the result, performing one abstract step leads to abstract states that are already included in the result. These properties, combined to the soundness of the abstract semantics, ensure the conclusion of this lemma.

Finally we pack together the iterator and the checker with another operation performed on sound results that checks for its safety. The resulting analysis enjoys the following property: if, given a program P, it outputs some result, then that program is safe.

Theorem analysis_sound : ∀ (P: memory) (dom: list addr) (fuel: nat)
(ab_num: num_dom_index), analysis ab_num P dom fuel ≠ None → safe P.

The arguments of the analysis program are the program to analyze, the list of addresses in memory to track, the counter that enforces termination and the name of the numeric domain to use. We provide two numeric domains: intervals with congruence information and finite sets.

4   Case Studies and Analysis Extensions

The extraction mechanism of Coq enables us to generate an OCaml program from our development and to link it with a front-end. Hence we can automatically analyze programs and prove them safe. This section shows the behavior of our analyzer on chosen examples, most of them taken from [6] (they have been rewritten to fit our custom syntax). All examples are written in an assembly-like syntax with some syntactic sugar: labels refer to byte offsets in the encoded program, the enc(I) notation denotes the encoding of the instruction I. The study of some examples highlights the limits of the basic technique presented before and suggests to refine the analyzer as we describe below. The source code of all the examples that are mentioned thereafter is available on the companion web site [8].

4.1   Basic Example

The multilevel runtime code generation program of Figure 7 is a program that, when executed, writes some code to line gen on and runs it; this generated program, in turn, writes some more code at line ggen and runs it. Finally execution starts again from the beginning. Moreover, at each iteration, register R6 is incremented.

The analysis of such a program follows its concrete execution and exactly computes the content of each register at each program point. It thus correctly tracks what values are written and where, so as to be able to analyze the program as it is generated.

However, when the execution reaches program point loop again, both states that may lead to that program point are merged. And the analysis of the loop body starts again. After the first iteration, the program text is exactly known, but each iteration yields more information about the dynamic content of register R6. Therefore we apply widening steps to ensure the termination of the analysis. Finally, the set of reachable program points is exactly computed and for each of them, we know what instruction will be executed from there.

Many self-modifying programs are successfully analyzed in a similar way: opcode modification, code obfuscation, and code checking [8].

4.2   A First Extension: Dealing with Flags

The example program in Figure 8 illustrates the abstract domain for the flags. This program stores the content of R0 in an array (stored in memory from address −128 to address −96) at the offset given in register R1. Before that store,

```
        cst 0 → R6
        cst 1 → R5
loop:   add R5 → R6
        cst gen → R0
        cst enc(store R1 → *R2) → R1
        store R1 → *R0
        cst enc(goto R2) → R1
        cst gen + 1 → R0
        store R1 → *R0
        cst ggen → R2
        cst loop → R0
        cst enc(goto R0) → R1
        goto gen
gen:    skip
        skip
ggen:   skip
```

```
cst -128 → R6
add R6 → R1
cmp R6, R1
gotoLT ko
cst -96 → R7
cmp R1, R7
gotoLE ko
store R0 → *R1
ko:halt R0
```

Fig. 8. Array bounds check

Fig. 7. Multilevel Runtime Code Generation

checks are performed to ensure that the provided offset lies inside the bounds of the array. The destination address is compared against the lowest and highest addresses of the array; if any of the comparisons fails, then the store is bypassed.

To properly analyze this program, we need to understand that the store does not alter the code. When analyzing a conditional branch instruction, the abstract state is refined differently at its two targets. However, the only information we have is about one flag, whereas the comparison that sets this flag operated on the content of registers. We therefore need to keep the link between the flags and the registers.

To this end, we extend our ab_machine_config record with a field containing an optional pair of registers ab_reg: option (reg * reg). It enables the analyzer to remember which registers were involved in the last comparison (the None value is used when this information is unknown). With such information available, even though the conditional jump is not directly linked to the comparison operation, we can gain some precision in the various branches.

Indeed, when we assume that the first conditional branch is not taken, the flag state is abstracted by the pair $\lfloor$(R6,R1)$\rfloor$, so we refine our knowledge about register R1: its content is not less than the -128. Similarly, when we assume that the second conditional branch is not taken, the abstract flag state is $\lfloor$(R1,R7)$\rfloor$, so we can finally infer that the content of register R1 is in the bounds.

This extension of the abstract domain increases a lot the precision of the analyzer on some programs, yet has little impact on the formalization: we need to explain its lattice structure (top element, order and least upper bound) and define its concretization. Then it enables us to program more precise primitives (namely compare and assume) that we must prove correct. No other part of the development is modified.

4.3   A Second Extension: Trace Partitioning

Some self-modifying programs store in the same memory space various pieces of their code. Successfully analyzing such programs requires not to merge these different code fragments, i.e., we need to distinguish in the execution which code is being run: flow sensitivity is not enough. To this end we use a specific form of *trace partitioning* [12] that makes an analysis sensitive to the value of a particular memory location.

Consider as an example the *polymorphic* program [8] that we briefly describe below. Polymorphism here refers to a technique used by for instance viruses that change their code while preserving their behavior, so as to hide their presence. The main loop of this program repeatedly adds forty-two to register R3. However, it is obfuscated in two ways. First, the source code initially contains a jump to some random address. But this instruction will be overwritten before it is executed. Second, this bad instruction is written back, but at a different address. So when the execution reaches the beginning of the loop, the program stored in memory is one of two different versions, both featuring the unsafe jump.

When analyzing this program, the abstract state computed at the beginning of the loop must over-approximate the two program versions. Unfortunately it is not possible to analyze the mere superposition of both versions, in which the unsafe jump may occur. The two versions can be distinguished through, for instance, the value at address 12. We therefore prevent the merging of any two states that disagree on the value stored at this address. Two different abstract states are then computed at each program point in the loop, as if the loop were unrolled once.

More generally, the analysis is parametrized by a partitioning criterion δ: ab_mc$\rightarrow$int that maps abstract states to values (the criterion used in this example maps an abstract state m to the value stored at address 12 in all concrete states represented by m; or to an arbitrary constant if there may be many values at this address). No abstract states that differ according to this criterion are merged. Taking a constant criterion amounts to disabling this partitioning. The abstract interpreter now computes for each program point, a map from criterion values to abstract states (rather than only one abstract state). Given such an environment E, a program point pp, and a value v, if there is an abstract state m such that $E(\text{pp})(v) = \lfloor m \rfloor$, then $\delta(m) = v$. Such an environment E represents the following set of machine configurations:

$$\gamma(E) = \{c \in \text{machine\_config} \mid \exists v, \ c \in \gamma\left(E(c.\text{pc})(v)\right)\}$$

To implement this technique, we do not need to modify the abstract domain, but only the iterator and fixpoint checker. The worklist holds pairs (program point, criterion value) rather than simple program points, and the iterator and fixpoint checker (along with its proof) straightforwardly adapted. The safety checker does not need to be updated since we can forget the partitioning before applying the original safety check.

Thanks to this technique, we can selectively enhance the precision of the analysis and correctly handle challenging self-modifying programs: control-flow modification, mutual modification, and code encryption [8]. However, the analyst must manually pick a suitable criterion for each program to analyze.

4.4   A Third Extension: Abstract Decoding

The program in Figure 9 computes the nth Fibonacci number in register R2, where n is an input value read from address -1 and held in register R0. There is a for-loop in which register R1 goes from 1 to n and some constant value is added to register R2. The trick is that the actual constant (which is encoded as part of an instruction and is stored at the address held in R6) is overwritten at each iteration by the previous value of R2.

When analyzing this program, we cannot infer much information about the content of the patched cell. Therefore, we cannot enumerate all instructions that may be stored at the patched point. So we introduce abstract instructions: instructions that are not exactly known, but of which some part is abstracted by a suitable abstract domain. Here we only need to abstract values using a numeric domain. With such a tool, we can decode *in the abstract*: the analyzer does not recover the exact instructions of the program, but only the information that some (unknown) value is loaded into register R4, which is harmless (no stores and no jumps depend on it).

This self-modifying code pattern, in which only part of an instruction is overwritten occurs also in the vector dot product example [8] where specialized multiplication instructions are emitted depending on an input vector.

The techniques presented here enable us to automatically prove the safety of various self-modifying programs including almost all the examples of [6]. Out of twelve, only two cannot be dealt with. The self-replicating example is a program that fills the memory with copies of itself: the code, being infinite, cannot be represented with our abstract domain. The bootloader example does not fit in the considered machine model, as it calls BIOS interrupts and reads files. Our Coq development [8] features all the extensions along with their correctness proofs.

```
        cst -1 → R7          gotoLE last          add R4 → R2
        load *R7 → R0        cst 1 → R7           store R3 → *R6
        cst key+1 → R6       add R7 → R1           goto loop
        cst 1 → R1           cst 0 → R3      last: halt R2
        cst 1 → R2           add R2 → R3
loop:   cmp R1, R0      key: cst 0 → R4
```

Fig. 9. Fibonacci

5   Related Work

Most of the previous works on mechanized verification of static analyzes focused on standard data-flow frameworks [13] or abstract interpretation for small imperative structured languages [5, 16]. In a previous work [3], we formally verified a value analysis for an intermediate language of the Compcert C compiler toolchain. The current work shares the same notion of abstract numerical domain but develops its own notion of memory abstraction, dynamic control-flow graph reconstruction and trace partitioning.

The current work formalizes more advanced abstract interpretation techniques, targeting self-modifying low-level code, and is based on several recent non-verified static analyses. A large amount of work was done by Balakrishnan et al. in this area [1]. Control-flow graph reconstruction was specially studied by Kinder et al. [12] and Bardin et al. [2]. Still, these works are unsound with respect to self-modifying code. Bonfante et al. provide a paper-and-pencil operational semantics for self-modifying programs [4].

Our current work tackles a core subset of a self-modifying low-level programming language. More realistic formalizations of x86 semantics were proposed [15, 14, 11] but none of them handles the problem of disassembling self-modifying programs. Our work complements other verification efforts of low-level programs [7, 6, 10] based on program logics. While we provide automatic inference of loop invariants, they are able to handle more expressive correctness properties.

6   Conclusion and Perspectives

This work provides the first verified static analyis for self-modifying programs. In order to tackle this challenge, we formalized original techniques such as control-flow graph reconstruction and partitioning. We formalized these techniques on a small core language but we managed to verify ten out of twelve of the challenges proposed in [6].

An important further work is to scale these technique on more realistic Coq language models [14, 11]. Developing directly an analyzer on these representations may be a huge development task because of the number of instructions to handle. One strategy could be to relate on a good intermediate representation such as the one proposed by Rocksalt [14]. Our current work does not consider the specific challenge of call stack reconstruction [1] that may require some form of verified alias analysis [17]. This is an important place for further work.

References

[1] Balakrishnan, G., Reps, T.W.: WYSINWYX: What you see is not what you eXecute. ACM Trans. Program. Lang. Syst. 32(6) (2010)
[2] Bardin, S., Herrmann, P., Védrine, F.: Refinement-Based CFG Reconstruction from Unstructured Programs. In: Jhala, R., Schmidt, D. (eds.) VMCAI 2011. LNCS, vol. 6538, pp. 54–69. Springer, Heidelberg (2011)

[3] Blazy, S., Laporte, V., Maroneze, A., Pichardie, D.: Formal Verification of a C Value Analysis Based on Abstract Interpretation. In: Logozzo, F., Fähndrich, M. (eds.) SAS 2013. LNCS, vol. 7935, pp. 324–344. Springer, Heidelberg (2013)

[4] Bonfante, G., Marion, J.Y., Reynaud-Plantey, D.: A Computability Perspective on Self-Modifying Programs. In: SEFM, pp. 231–239 (2009)

[5] Cachera, D., Pichardie, D.: A Certified Denotational Abstract Interpreter. In: Kaufmann, M., Paulson, L.C. (eds.) ITP 2010. LNCS, vol. 6172, pp. 9–24. Springer, Heidelberg (2010)

[6] Cai, H., Shao, Z., Vaynberg, A.: Certified Self-Modifying Code. In: PLDI, pp. 66–77. ACM (2007)

[7] Chlipala, A.: Mostly-automated verification of low-level programs in computational separation logic. In: PLDI. ACM (2011)

[8] Companion website, http://www.irisa.fr/celtique/ext/smc

[9] Cousot, P., Cousot, R.: Abstract interpretation: a unified lattice model for static analysis of programs by construction or approximation of fixpoints. In: POPL, pp. 238–252. ACM (1977)

[10] Jensen, J., Benton, N., Kennedy, A.: High-Level Separation Logic for Low-Level Code. In: POPL. ACM (2013)

[11] Kennedy, A., et al.: Coq: The world's best macro assembler? In: PPDP, pp. 13–24. ACM (2013)

[12] Kinder, J.: Towards static analysis of virtualization-obfuscated binaries. In: WCRE, pp. 61–70 (2012)

[13] Klein, G., Nipkow, T.: A Machine-Checked Model for a Java-Like Language, Virtual Machine and Compiler. ACM TOPLAS 28(4), 619–695 (2006)

[14] Morrisett, G., et al.: RockSalt: better, faster, stronger SFI for the x86. In: PLDI, pp. 395–404 (2012)

[15] Myreen, M.O.: Verified just-in-time compiler on x86. In: POPL, pp. 107–118. ACM (2010)

[16] Nipkow, T.: Abstract Interpretation of Annotated Commands. In: Beringer, L., Felty, A. (eds.) ITP 2012. LNCS, vol. 7406, pp. 116–132. Springer, Heidelberg (2012)

[17] Robert, V., Leroy, X.: A Formally-Verified Alias Analysis. In: Hawblitzel, C., Miller, D. (eds.) CPP 2012. LNCS, vol. 7679, pp. 11–26. Springer, Heidelberg (2012)

[18] Stewart, G., Beringer, L., Appel, A.W.: Verified heap theorem prover by paramodulation. In: ICFP, pp. 3–14. ACM (2012)

Showing Invariance Compositionally for a Process Algebra for Network Protocols

Timothy Bourke[1,2], Robert J. van Glabbeek[3,4], and Peter Höfner[3,4]

[1] Inria Paris-Rocquencourt, France
[2] Ecole normale supérieure, Paris, France
[3] NICTA, Sydney, Australia
[4] Computer Science and Engineering, UNSW, Sydney, Australia

Abstract. This paper presents the mechanization of a process algebra for Mobile Ad hoc Networks and Wireless Mesh Networks, and the development of a compositional framework for proving invariant properties. Mechanizing the core process algebra in Isabelle/HOL is relatively standard, but its layered structure necessitates special treatment. The control states of reactive processes, such as nodes in a network, are modelled by terms of the process algebra. We propose a technique based on these terms to streamline proofs of inductive invariance. This is not sufficient, however, to state and prove invariants that relate states across multiple processes (entire networks). To this end, we propose a novel compositional technique for lifting global invariants stated at the level of individual nodes to networks of nodes.

1 Introduction and Related Work

The Algebra for Wireless Networks (AWN) is a process algebra developed for modelling and analysing protocols for Mobile Ad hoc Networks (MANETs) and Wireless Mesh Networks (WMNs) [6, §4]. This paper reports on both its mechanization in Isabelle/HOL [15] and the development of a compositional framework for showing invariant properties of models.[1] The techniques we describe are a response to problems encountered during the mechanization of a model and proof—presented elsewhere [4]—of an RFC-standard for routing protocols. Despite the existence of extensive research on related problems [18] and several mechanized frameworks for reactive systems [5,10,14], we are not aware of other solutions that allow the compositional statement and proof of properties relating the states of different nodes in a message-passing model—at least not within the strictures imposed by an Interactive Theorem Prover (ITP).

But is there really any need for yet another process algebra and associated framework? AWN provides a unique mix of communication primitives and a treatment of data structures that are essential for studying MANET and WMN protocols with dynamic topologies and sophisticated routing logic [6, §1]. It supports communication primitives for one-to-one (unicast), one-to-many (group-cast), and one-to-all (broadcast) message passing. AWN comprises distinct layers

---

[1] The Isabelle/HOL source files can be found in the Archive of Formal Proofs [3].

G. Klein and R. Gamboa (Eds.): ITP 2014, LNAI 8558, pp. 144–159, 2014.

for expressing the structure of nodes and networks. We exploit this structure, but we also expect the techniques proposed in Sections 3 and 4 to apply to similar layered modelling languages. Besides this, our work differs from other mechanizations for verifying reactive systems, like UNITY [10], TLA$^+$ [5], or I/O Automata [14] (from which we drew the most inspiration), in its explicit treatment of control states, in the form of process algebra terms, as distinct from data states. In this respect, our approach is close to that of Isabelle/Circus [7], but it differs in (1) the treatment of operators for composing nodes, which we model directly as functions on automata, (2) the treatment of recursive invocations, which we do not permit, and (3) our inclusion of a framework for compositional proofs. Other work in ITPs focuses on showing traditional properties of process algebras, like, for instance, the treatment of binders [1], that bisimulation equivalence is a congruence [9,11], or properties of fix-point induction [20], while we focus on what has been termed 'proof methodology' [8], and develop a compositional method for showing correctness properties of protocols specified in a process algebra. Alternatively, Paulson's inductive approach [16] can be applied to show properties of protocols specified with less generic infrastructure. But we think it to be better suited to systems specified in a 'declarative' style as opposed to the strongly operational models we consider.

Structure and contributions. Section 2 describes the mechanization of AWN. The basic definitions are routine but the layered structure of the language and the treatment of operators on networks as functions on automata are relatively novel and essential to understanding later sections. Section 3 describes our mechanization of the theory of inductive invariants, closely following [13]. We exploit the structure of AWN to generate verification conditions corresponding to those of pen-and-paper proofs [6, §7]. Section 4 presents a compositional technique for stating and proving invariants that relate states across multiple nodes. Basically, we substitute 'open' Structural Operational Semantics (SOS) rules over the global state for the standard rules over local states (Section 4.1), show the property over a single sequential process (Section 4.2), 'lift' it successively over layers that model message queueing and network communication (Section 4.3), and, ultimately, 'transfer' it to the original model (Section 4.4).

2   The Process Algebra AWN

AWN comprises five layers [6, §4]. We treat each layer as an automaton with states of a specific form and a given set of transition rules. We describe the layers from the bottom up over the following sections.

2.1   Sequential Processes

Sequential processes are used to encode protocol logic. Each is modelled by a *(recursive) specification* Γ of type 'p $\Rightarrow$ ('s, 'p, 'l) seqp, which maps process names of type 'p to terms of type ('s, 'p, 'l) seqp, also parameterized by 's, data states, and 'l, labels. States of sequential processes have the form (ξ, p) where ξ is a data state of type 's and p is a control term of type ('s, 'p, 'l) seqp.

$\{|\}[u]$ p $'l \Rightarrow ('s \Rightarrow 's) \Rightarrow ('s, 'p, 'l)$ seqp $\Rightarrow ('s, 'p, 'l)$ seqp

$\{|\}\langle g \rangle$ p $'l \Rightarrow ('s \Rightarrow 's$ set$) \Rightarrow ('s, 'p, 'l)$ seqp $\Rightarrow ('s, 'p, 'l)$ seqp

$\{|\}$unicast(s_{ip}, s_{msg}) . p $\rhd$ q $'l \Rightarrow ('s \Rightarrow ip) \Rightarrow ('s \Rightarrow msg) \Rightarrow ('s, 'p, 'l)$ seqp $\Rightarrow$
                                                             $('s, 'p, 'l)$ seqp $\Rightarrow ('s, 'p, 'l)$ seqp

$\{|\}$broadcast(s_{msg}) . p $'l \Rightarrow ('s \Rightarrow msg) \Rightarrow ('s, 'p, 'l)$ seqp $\Rightarrow ('s, 'p, 'l)$ seqp

$\{|\}$groupcast(s_{ips}, s_{msg}) . p $'l \Rightarrow ('s \Rightarrow ip$ set$) \Rightarrow ('s \Rightarrow msg) \Rightarrow ('s, 'p, 'l)$ seqp $\Rightarrow$
                                                             $('s, 'p, 'l)$ seqp

$\{|\}$send(s_{msg}) . p $'l \Rightarrow ('s \Rightarrow msg) \Rightarrow ('s, 'p, 'l)$ seqp $\Rightarrow ('s, 'p, 'l)$ seqp

$\{|\}$receive(u_{msg}) . p $'l \Rightarrow (msg \Rightarrow 's \Rightarrow 's) \Rightarrow ('s, 'p, 'l)$ seqp $\Rightarrow ('s, 'p, 'l)$ seqp

$\{|\}$deliver(s_{data}) . p $'l \Rightarrow ('s \Rightarrow data) \Rightarrow ('s, 'p, 'l)$ seqp $\Rightarrow ('s, 'p, 'l)$ seqp

$p_1 \oplus p_2$ $('s, 'p, 'l)$ seqp $\Rightarrow ('s, 'p, 'l)$ seqp $\Rightarrow ('s, 'p, 'l)$ seqp

call(pn) $'p \Rightarrow ('s, 'p, 'l)$ seqp

(a) Term constructors for $('s, 'p, 'l)$ seqp.

$$\frac{\xi' = u \, \xi}{((\xi, \{|\}[u] \text{ p}), \tau, (\xi', \text{p})) \in \text{seqp-sos } \Gamma} \qquad \frac{((\xi, \text{p}), a, (\xi', \text{p}')) \in \text{seqp-sos } \Gamma}{((\xi, \text{p} \oplus \text{q}), a, (\xi', \text{p}')) \in \text{seqp-sos } \Gamma}$$

$$\frac{((\xi, \Gamma \text{ pn}), a, (\xi', \text{p}')) \in \text{seqp-sos } \Gamma}{((\xi, \text{call(pn)}), a, (\xi', \text{p}')) \in \text{seqp-sos } \Gamma} \qquad \frac{((\xi, \text{q}), a, (\xi', \text{q}')) \in \text{seqp-sos } \Gamma}{((\xi, \text{p} \oplus \text{q}), a, (\xi', \text{q}')) \in \text{seqp-sos } \Gamma}$$

$$((\xi, \{|\}\text{unicast}(s_{ip}, s_{msg}) \text{ . p} \rhd \text{q}), \text{unicast } (s_{ip} \, \xi) \, (s_{msg} \, \xi), (\xi, \text{p})) \in \text{seqp-sos } \Gamma$$

$$((\xi, \{|\}\text{unicast}(s_{ip}, s_{msg}) \text{ . p} \rhd \text{q}), \neg\text{unicast } (s_{ip} \, \xi), (\xi, \text{q})) \in \text{seqp-sos } \Gamma$$

(b) SOS rules for sequential processes: examples from seqp-sos.

Fig. 1. Sequential processes: terms and semantics

Process terms are built from the constructors that are shown with their types[2] in Figure 1a. The inductive set seqp-sos, shown partially in Figure 1b, contains one or two SOS rules for each constructor. It is parameterized by a specification Γ and relates triples of source states, actions, and destination states.

The 'prefix' constructors are each labelled with an $\{|\}$. Labels are used to strengthen invariants when a property is only true in or between certain states; they have no influence on control flow (unlike in [13]). The prefix constructors are *assignment*, $\{|\}[u]$ p, which transforms the data state deterministically according to the function u and performs a τ action, as shown in Figure 1b; *guard/bind*, $\{|\}\langle g \rangle$ p, with which we encode both guards, $\langle \lambda \xi.$ if g ξ then $\{\xi\}$ else $\emptyset \rangle$ p, and variable bindings, as in $\langle \lambda \xi. \{\xi(no := n) \mid n < 5\}$ p;[3] *network synchronizations*, receive/unicast/broadcast/groupcast, of which the rules for unicast are characteristic and shown in Figure 1b—the environment decides between a successful unicast i m and an unsuccessful ¬unicast i; and, *internal communications*, send/receive/deliver.

The other constructors are unlabelled and serve to 'glue' processes together: *choice*, $p_1 \oplus p_2$, takes the union of two transition sets; and, *call*, call(pn), affixes a term from the specification (Γ pn). The rules for both are shown in Figure 1b.

---

[2] Leading abstractions are omitted, for example, λl fa p. $\{|\}[u]$ p is written $\{|\}[u]$ p.

[3] Although it strictly subsumes assignment we prefer to keep both.

We introduce the specification of a simple 'toy' protocol as a running example:

Γ_{Toy} PToy = labelled PToy (receive(λmsg' ξ. ξ (| msg := msg' |)).   {PToy-:0}
$\quad$ [[$\lambda\xi$. ξ (|nhip := ip ξ|)]]   {PToy-:1}
$\quad$ (⟨is-newpkt⟩   {PToy-:2}
$\qquad$ [[$\lambda\xi$. ξ (|no := max (no ξ) (num ξ)|)]]   {PToy-:3}
$\qquad$ broadcast($\lambda\xi$. pkt(no ξ, ip ξ)). Toy()   {PToy-:4,5}
$\quad \oplus$ ⟨is-pkt⟩   {PToy-:2}
$\qquad$ (⟨$\lambda\xi$. if num $\xi \geq$ no ξ then $\{\xi\}$ else $\{\}$⟩   {PToy-:6}
$\qquad\quad$ [[$\lambda\xi$. ξ (|no := num ξ|)]]   {PToy-:7}
$\qquad\quad$ [[$\lambda\xi$. ξ (|nhip := sip ξ|)]]   {PToy-:8}
$\qquad\quad$ broadcast($\lambda\xi$. pkt(no ξ, ip ξ)). Toy()   {PToy-:9,10}
$\qquad \oplus$ ⟨$\lambda\xi$. if num $\xi <$ no ξ then $\{\xi\}$ else $\{\}$⟩   {PToy-:6}
$\qquad\quad$ Toy()))) ,   {PToy-:11}

where PToy is the process name, is-newpkt and is-pkt are guards that unpack the contents of msg, and Toy() is an abbreviation that clears some variables before a call(PToy). The function labelled associates its argument PToy paired with a number to every prefix constructor. There are two types of messages: newpkt (data, dst), from which is-newpkt copies data to the variable num, and pkt (data, src), from which is-pkt copies data into num and src into sip.

The corresponding sequential model is an automaton—a record[4] of two fields: a set of initial states and a set of transitions—parameterized by an address i:

$$\text{ptoy i} = (|\text{init} = \{(\text{toy-init i}, \Gamma_{\mathsf{Toy}} \text{ PToy})\}, \text{trans} = \text{seqp-sos } \Gamma_{\mathsf{Toy}}|) ,$$

where toy-init i yields the initial data state (|ip = i, no = 0, nhip = i, msg = SOME x. True, num = SOME x. True, sip = SOME x. True|). The last three variables are initialized to arbitrary values, as they are considered local—they are explicitly reinitialized before each call(PToy). This is the biggest departure from the original definition of AWN; it simplifies the treatment of call, as we show in Section 3.1, and facilitates working with automata where variable locality makes little sense.

2.2   Local Parallel Composition

Message sending protocols must nearly always be input-enabled, that is, nodes should always be in a state where they can receive messages. To achieve this, and to model asynchronous message transmission, the protocol process is combined with a queue model, qmsg, that continually appends received messages onto an

$$\frac{(s, a, s') \in S \quad \bigwedge m.\ a \neq \text{receive } m}{((s, t), a, (s', t)) \in \text{parp-sos } S\ T} \qquad \frac{(t, a, t') \in T \quad \bigwedge m.\ a \neq \text{send } m}{((s, t), a, (s, t')) \in \text{parp-sos } S\ T}$$

$$\frac{(s, \text{receive } m, s') \in S \quad (t, \text{send } m, t') \in T}{((s, t), \tau, (s', t')) \in \text{parp-sos } S\ T}$$

Fig. 2. SOS rules for parallel processes: parp-sos

---

[4] The generic record has type ('s, 'a) automaton, where the type 's is the domain of states, here pairs of data records and control terms, and 'a is the domain of actions.

$$\frac{(s, \text{groupcast D m, s'}) \in S}{(s_R^i, (R \cap D):\text{*cast}(m), s'_R) \in \text{node-sos } S}$$

$$\frac{(s, \text{receive m, s'}) \in S}{(s_R^i, \{i\}\neg\emptyset:\text{arrive}(m), s'_R) \in \text{node-sos } S}$$

$$(s_R^i, \emptyset\neg\{i\}:\text{arrive}(m), s_R^i) \in \text{node-sos } S$$

$$(s_R^i, \text{connect}(i, i'), s_{R \cup \{i'\}}^i) \in \text{node-sos } S$$

Fig. 3. SOS rules for nodes: examples from node-sos

internal list and offers to send the head message to the protocol process: ptoy i ⟨⟨ qmsg. The *local parallel* operator is a function over automata:

$$s \langle\langle t = (\text{init} = \text{init } s \times \text{init } t, \text{trans} = \text{parp-sos (trans s) (trans t)}).$$

The rules for parp-sos are shown in Figure 2.

2.3 Nodes

At the node level, a local process np is wrapped in a layer that records its address i and tracks the set of neighbouring node addresses, initially R_i:

$$\langle i : np : R_i \rangle = (\text{init} = \{s_{R_i}^i \mid s \in \text{init } np\}, \text{trans} = \text{node-sos (trans np)}).$$

Node states are denoted s_R^i. Figure 3 presents rules typical of node-sos. Output network synchronizations, like groupcast, are filtered by the list of neighbours to become *cast actions. The H¬K:arrive(m) action—in Figure 3 instantiated as $\emptyset\neg\{i\}$:arrive(m), and $\{i\}\neg\emptyset$:arrive(m)—is used to model a message m received by nodes in H and not by those in K. The connect(i, i') adds node i' to the set of neighbours of node i; disconnect(i, i') works similarly.

2.4 Partial Networks

Partial networks are specified as values of type net-tree, that is, as a node $\langle i; R_i \rangle$ with address i and a set of initial neighbours R_i, or a composition of two net-trees $p_1 \parallel p_2$. The function pnet maps such a value, together with the process np i to execute at each node i, here parameterized by an address, to an automaton:

pnet np $\langle i; R_i \rangle$ = $\langle i : np\ i : R_i \rangle$
pnet np $(p_1 \parallel p_2)$ = $(\text{init} = \{s_1 \| s_2 \mid s_1 \in \text{init (pnet np } p_1) \wedge s_2 \in \text{init (pnet np } p_2)\},$
                         $\text{trans} = \text{pnet-sos (trans (pnet np } p_1)) (\text{trans (pnet np } p_2))),$

The states of such automata mirror the tree structure of the network term; we denote composed states $s_1 \| s_2$. This structure, and the node addresses, remain constant during an execution. These definitions suffice to model an example three node network of toy processes:

pnet $(\lambda i.\ \text{ptoy i} \langle\langle \text{qmsg}) (\langle A; \{B\}\rangle \parallel \langle B; \{A, C\}\rangle \parallel \langle C; \{B\}\rangle).$

Figure 4 presents rules typical of pnet-sos. There are rules where only one node acts, like the one shown for τ, and rules where all nodes act, like those for *cast and arrive. The latter ensure—since qmsg is always ready to receive m—that a partial network can always perform an H¬K:arrive(m) for any combination of H and K consistent with its node addresses, but that pairing with an R:*cast(m) restricts the possibilities to the one consistent with the destinations in R.

$$\frac{(s, \ R{:}{*}cast(m), \ s') \in S \quad (t, \ H\neg K{:}arrive(m), \ t') \in T \quad H \subseteq R \quad K \cap R = \emptyset}{(s \parallel t, \ R{:}{*}cast(m), \ s' \parallel t') \in \text{pnet-sos } S \ T}$$

$$\frac{(s, \ H\neg K{:}arrive(m), \ s') \in S \quad (t, \ H'\neg K'{:}arrive(m), \ t') \in T}{(s \parallel t, \ (H \cup H')\neg (K \cup K'){:}arrive(m), \ s' \parallel t') \in \text{pnet-sos } S \ T} \qquad \frac{(s, \ \tau, \ s') \in S}{(s \parallel t, \ \tau, \ s' \parallel t) \in \text{pnet-sos } S \ T}$$

Fig. 4. SOS rules for partial networks: examples from pnet-sos

2.5   Complete Networks

The last layer closes a network to further interactions with an environment; the
*cast action becomes a τ and H¬K:arrive(m) is forbidden:

$$\text{closed } A = A(\!|\text{trans} := \text{cnet-sos (trans } A)|\!).$$

The rules for cnet-sos are straight-forward and not presented here.

3   Basic Invariance

This paper only considers proofs of invariance, that is, properties of reachable
states. The basic definitions are classic [14, Part III].

Definition 1 (reachability). *Given an automaton A and an assumption I over
actions, reachable A I is the smallest set defined by the rules:*

$$\frac{s \in init \ A}{s \in reachable \ A \ I} \qquad \frac{s \in reachable \ A \ I \quad (s, a, s') \in trans \ A \quad I \ a}{s' \in reachable \ A \ I}$$

Definition 2 (invariance). *Given an automaton A and an assumption I, a
predicate P is* invariant, *denoted* $A \models (I \rightarrow) P$, *iff* $\forall s \in reachable \ A \ I. \ P \ s$.

We state reachability relative to an assumption on (input) actions I. When I is
$\lambda{-}.$ True, we write simply $A \models P$.

Definition 3 (step invariance). *Given an automaton A and an assumption I,
a predicate P is* step invariant, *denoted* $A \models (I \rightarrow) P$, *iff*
$$\forall a. \ I \ a \longrightarrow (\forall s \in reachable \ A \ I. \ \forall s'. \ (s, a, s') \in trans \ A \longrightarrow P \ (s, a, s')).$$

Our invariance proofs follow the compositional strategy recommended in [18,
§1.6.2]. That is, we show properties of sequential process automata using the
induction principle of Definition 1, and then apply generic proof rules to successively lift such properties over each of the other layers. The inductive assertion
method, as stated in rule INV-B of [13], requires a finite set of transition schemas,
which, together with the obligation on initial states yields a set of sufficient verification conditions. We develop this set in Section 3.1 and use it to derive the
main proof rule presented in Section 3.2 together with some examples.

3.1 Control Terms

Given a specification Γ over finitely many process names, we can generate a finite set of verification conditions because transitions from ('s, 'p, 'l) seqp terms always yield subterms of terms in Γ. But, rather than simply consider the set of all subterms, we prefer to define a subset of 'control terms' that reduces the number of verification conditions, avoids tedious duplication in proofs, and corresponds with the obligations considered in pen-and-paper proofs. The main idea is that the $\oplus$ and call operators serve only to combine process terms: they are, in a sense, executed recursively by seqp-sos to determine the actions that a term offers to its environment. This is made precise by defining a relation between sequential process terms.

Definition 4 ($\leadsto_\Gamma$). *For a (recursive) specification Γ, let $\leadsto_\Gamma$ be the smallest relation such that $(p_1 \oplus p_2) \leadsto_\Gamma p_1$, $(p_1 \oplus p_2) \leadsto_\Gamma p_2$, and $(call(pn)) \leadsto_\Gamma \Gamma\, pn$.*

We write $\leadsto_\Gamma^*$ for its reflexive transitive closure. We consider a specification to be *well formed*, when the inverse of this relation is well founded:

$$\text{wellformed } \Gamma = \text{wf } \{(\mathsf{q}, \mathsf{p}) \mid \mathsf{p} \leadsto_\Gamma \mathsf{q}\}.$$

Most of our lemmas only apply to well formed specifications, since otherwise functions over the terms they contain cannot be guaranteed to terminate. Neither of these two specifications is well formed: $\Gamma_a(1) = \mathsf{p} \oplus call(1)$; $\Gamma_b(n) = call(n + 1)$.

We will also need a set of 'start terms'—the subterms that can act directly.

Definition 5 (sterms). *Given a wellformed Γ and a sequential process term p, sterms $\Gamma\, p$ is the set of maximal elements related to p by the reflexive transitive closure of the $\leadsto_\Gamma$ relation[5]:*

sterms $\Gamma\ (p_1 \oplus p_2)$ = sterms $\Gamma\ p_1 \cup$ sterms $\Gamma\ p_2$,
sterms $\Gamma\ (call(pn))$ = sterms $\Gamma\ (\Gamma\ pn)$, and,
sterms $\Gamma\ p$ = $\{p\}$ otherwise.

We also define 'local start terms' by stermsl $(p_1 \oplus p_2)$ = stermsl $p_1 \cup$ stermsl p_2 and otherwise stermsl $p = \{p\}$ to permit the sufficient syntactic condition that a specification Γ is well formed if call(pn') $\notin$ stermsl $(\Gamma\ pn)$.

Similarly to the way that start terms act as direct sources of transitions, we define 'derivative terms' giving possible active destinations of transitions.

Definition 6 (dterms). *Given a wellformed Γ and a sequential process term p, dterms p is defined by:*

dterms $\Gamma\ (p_1 \oplus p_2)$ = dterms $\Gamma\ p_1 \cup$ dterms $\Gamma\ p_2$,
dterms $\Gamma\ (call(pn))$ = dterms $\Gamma\ (\Gamma\ pn)$,
dterms $\Gamma\ (\{l\}[u]\ p)$ = sterms $\Gamma\ p$,
dterms $\Gamma\ (\{l\}$unicast$(s_{ip}, s_{msg}) \cdot p \rhd q)$ = sterms $\Gamma\ p \cup$ sterms $\Gamma\ q$, and so on.

---

[5] This characterization is equivalent to $\{q \mid p \leadsto_\Gamma^* q \land (\nexists q'.\ q \leadsto_\Gamma q')\}$. Termination follows from wellformed Γ, that is, wellformed $\Gamma \implies$ sterms-dom (Γ, p) for all p.

These derivative terms overapproximate the set of reachable sterms, since they do not consider the truth of guards nor the willingness of communication partners.

These auxiliary definitions lead to a succinct definition of the set of control terms of a specification.

Definition 7 (cterms). *For a specification* Γ*, cterms is the smallest set where:*

$$\frac{p \in sterms\ \Gamma\ (\Gamma\ pn)}{p \in cterms\ \Gamma} \qquad \frac{pp \in cterms\ \Gamma \qquad p \in dterms\ \Gamma\ pp}{p \in cterms\ \Gamma}$$

It is also useful to define a local version independent of any specification.

Definition 8 (ctermsl). *Let ctermsl be the smallest set defined by:*

$ctermsl\ (p_1 \oplus p_2) = ctermsl\ p_1 \cup ctermsl\ p_2,$
$ctermsl\ (call(pn)) = \{call(pn)\},$
$ctermsl\ (\{l\}[u]\ p) = \{\{l\}[u]\ p\} \cup ctermsl\ p, \ and\ so\ on.$

Including call terms ensures that $q \in stermsl\ p$ implies $q \in ctermsl\ p$, which facilitates proofs. For wellformed Γ, ctermsl allows an alternative definition of cterms,

$$cterms\ \Gamma = \{p \mid \exists\, pn.\ p \in ctermsl\ (\Gamma\ pn) \wedge not\text{-}call\ p\}. \tag{1}$$

While the original definition is convenient for developing the meta-theory, due to the accompanying induction principle, this one is more useful for systematically generating the set of control terms of a specification, and thus, we will see, sets of verification conditions. And, for wellformed Γ, we have as a corollary

$$cterms\ \Gamma = \{p \mid \exists\, pn.\ p \in subterms\ (\Gamma\ pn) \wedge not\text{-}call\ p \wedge not\text{-}choice\ p\}, \tag{2}$$

where subterms, not-call, and not-choice are defined in the obvious way.

We show that cterms over-approximates the set of reachable control states.

Lemma 1. *For wellformed* Γ *and automaton A where control-within* Γ *(init A) and trans A = seqp-sos* Γ*, if* $(\xi, p) \in$ *reachable A I and* $q \in sterms\ \Gamma\ p$ *then* $q \in cterms\ \Gamma$.

The predicate control-within $\Gamma\ \sigma = \forall\, (\xi, p) \in \sigma.\ \exists\, pn.\ p \in subterms\ (\Gamma\ pn)$ serves to state that the initial control state is within the specification.

3.2 Basic Proof Rule and Invariants

Using the definition of invariance (Definition 2), we can state a basic property of an instance of the toy process:

$$ptoy\ i \models onl\ \Gamma_{Toy}\ (\lambda(\xi, l).\ l \in \{PToy\text{-}:2..PToy\text{-}:8\} \longrightarrow nhip\ \xi = ip\ \xi), \tag{3}$$

This invariant states that between the lines labelled PToy-:2 and PToy-:8, that is, after the assignment of PToy-:1 until before the assignment of PToy-:8, the values of nhip and ip are equal; onl Γ P, defined as $\lambda(\xi, p).\ \forall\, l \in labels\ \Gamma\ p.\ P\ (\xi, l)$, extracts labels from control states.[6] Invariants like these are solved using a procedure whose soundness is justified as a theorem. The proof exploits (1) and Lemma 1.

---

[6] Using labels in this way is standard, see, for instance, [13, Chap. 1], or the 'assertion networks' of [18, §2.5.1]. Isabelle rapidly dispatches all the uninteresting cases.

Theorem 1. *To prove* $A \models (I \rightarrow)$ *onl* Γ P, *where wellformed* Γ, *simple-labels* Γ, *control-within* Γ *(init A), and trans* $A = $ *seqp-sos* Γ, *it suffices*

(init)   *for arbitrary* $(\xi, p) \in$ *init A and* $I \in$ *labels* Γ *p, to show* P (ξ, I), *and,*

(step)   *for arbitrary* $p \in$ *ctermsl* $(\Gamma$ *pn), but not-call p, and* $I \in$ *labels* Γ *p, given that* $p \in$ *sterms* Γ *pp for some* $(\xi, pp) \in$ *reachable A I, to assume* P (ξ, I) *and* I *a, and then for any* (ξ', q) *such that* $((\xi, p), a, (\xi', q)) \in$ *seqp-sos* Γ *and* $I' \in$ *labels* Γ *q, to show* P (ξ', I').

Here, simple-labels $\Gamma = \forall$ pn. $\forall$ p $\in$ subterms $(\Gamma$ pn). $\exists! I.$ labels Γ p $= \{I\}$: each control term must have exactly one label, that is, $\oplus$ terms must be labelled consistently.

We incorporate this theorem into a tactic that (1) applies the introduction rule, (2) replaces p $\in$ ctermsl $(\Gamma$ pn) by a disjunction over the values of pn, (3) applies Definition 8 and repeated simplifications of Γs and eliminations on disjunctions to generate one subgoal (verification condition) for each control term, (4) replaces control term derivatives, the subterms in Definition 6, by fresh variables, and, finally, (5) tries to solve each subgoal by simplification. Step 4 replaces potentially large control terms by their (labelled) heads, which is important for readability and prover performance. The tactic takes as arguments a list of existing invariants to include after having applied the introduction rule and a list of lemmas for trying to solve any subgoals that survive the final simplification. There are no schematic variables in the subgoals and we benefit greatly from Isabelle's PARALLEL_GOALS tactical [22].

In practice, one states an invariant, applies the tactic, and examines the resulting goals. One may need new lemmas for functions over the data state or explicit proofs for difficult goals. That said, the tactic generally dispatches the uninteresting goals, and the remaining ones typically correspond with the cases treated explicitly in manual proofs [4].

For step invariants, we show a counterpart to Theorem 1, and declare it to the tactic. Then we can show, for our example, that the value of no never decreases:

$$\text{ptoy i} \models (\lambda((\xi, \text{-}), \text{-}, (\xi', \text{-})). \text{ no } \xi \leq \text{no } \xi').$$

4   Open Invariance

The analysis of network protocols often requires 'inter-node' invariants, like

wf-net-tree n $\Longrightarrow$ closed (pnet (λi. ptoy i $\langle\langle$ qmsg) n) $\models$

$$\text{netglobal } (\lambda\sigma. \forall i. \text{ no } (\sigma \ i) \leq \text{no } (\sigma \ (\text{nhip } (\sigma \ i)))), \quad (4)$$

which states that, for any net-tree with disjoint node addresses (wf-net-tree n), the value of no at a node is never greater than its value at the 'next hop'—the address in nhip. This is a property of a global state σ mapping addresses to corresponding data states. Such a global state is readily constructed with:

netglobal P $=$ λs. P (default toy-init (netlift fst s)),

default df f $=$ (λi. case f i of None $\Rightarrow$ df i | Some s $\Rightarrow$ s), and

netlift sr (s_R^i) $= [i \mapsto$ fst (sr s)]

netlift sr $(s_{||}t) = $ netlift sr s $++$ netlift sr t.

The applications of fst elide the state of qmsg and the protocol's control state.[7]

While we can readily state inter-node invariants of a complete model, showing them compositionally is another issue. Sections 4.1 and 4.2 present a way to state and prove such invariants at the level of sequential processes—that is, with only ptoy i left of the turnstile. Sections 4.3 and 4.4 present, respectively, rules for lifting such results to network models and for recovering invariants like (4).

4.1   The Open Model

Rather than instantiate the 's of ('s, 'p, 'l) seqp with elements ξ of type state, our solution introduces a global state σ of type ip $\Rightarrow$ state. This necessitates a stack of new SOS rules that we call the *open model*; Figure 5 shows some representatives.

The rules of oseqp-sos are parameterized by an address i and constrain only that entry of the global state, either to say how it changes (σ' i = u (σ i)) or that it does not (σ i = σ i). The rules for oparp-sos only allow the first sub-process to constrain σ. This choice is disputable: it precludes comparing the states of qmsgs (and any other local filters) across a network, but is also simplifies the mechanics and use of this layer of the framework.[8] The sets onode-sos and opnet-sos need not be parameterized since they are generated inductively from lower layers. Together they constrain subsets of elements of σ. This occurs naturally for rules like those for arrive and *cast, where the synchronous communication serves as a conjunction of constraints on sub-ranges of σ. But for others that normally only constrain a single element, like those for τ, assumptions ($\forall$j $\neq$ i. σ' j = σ j) are introduced here and later dispatched (Section 4.4). The rules for ocnet-sos, not shown, are similar—elements not addressed within a model may not change.

The stack of operators and model layers described in Section 2 is refashioned to use the new transition rules and to distinguish the global state, which is preserved as the fst element across layers, from the local state elements which are combined in the snd element as before.

For instance, a sequential instance of the toy protocol is defined as

$$\text{optoy i} = (\!|\text{init} = \{(\text{toy-init}, \Gamma_{\text{Toy}} \text{ PToy})\}, \text{trans} = \text{oseqp-sos } \Gamma_{\text{Toy}} \text{ i}|\!)\,,$$

combined with the standard qmsg process using the operator

$$\text{s} \langle\!\langle_i \text{ t} = (\!|\text{init} = \{(\sigma, (\text{s}_l, \text{t}_l)) \mid (\sigma, \text{s}_l) \in \text{init s} \wedge \text{t}_l \in \text{init t}\},$$
$$\text{trans} = \text{oparp-sos i (trans s) (trans t)}|\!)\,,$$

and lifted to the node level via the open node constructor

$$\langle\text{i} : \text{onp} : R_i\rangle_o = (\!|\text{init} = \{(\sigma, \text{s}^i_{R_i}) \mid (\sigma, \text{s}) \in \text{init onp}\}, \text{trans} = \text{onode-sos (trans onp)}|\!)\,.$$

Similarly, to map a net-tree term to an open model we define:

$$\text{opnet onp } \langle\text{i}; R_i\rangle \quad = \langle\text{i} : \text{onp i} : R_i\rangle_o$$
$$\text{opnet onp } (\text{p}_1 \parallel \text{p}_2) = (\!|\text{init} = \{(\sigma, \text{s}_1 {}_{\text{II}}\text{s}_2) \mid (\sigma, \text{s}_1) \in \text{init (opnet onp p}_1)$$
$$\wedge (\sigma, \text{s}_2) \in \text{init (opnet onp p}_2)$$
$$\wedge \text{net-ips s}_1 \cap \text{net-ips s}_2 = \emptyset\},$$
$$\text{trans} = \text{opnet-sos (trans(opnet onp p}_1)) (\text{trans(opnet onp p}_2))|\!)\,.$$

---

[7] The formulation here is a technical detail: sr corresponds to netlift as np does to pnet.

[8] The treatment of the other layers is completely independent of this choice.

$$\frac{\sigma' \; i = u \, (\sigma \; i)}{((\sigma, \; \{l\}[\![u]\!] \; p), \; \tau, \; (\sigma', \; p)) \in \mathsf{oseqp\text{-}sos} \; \varGamma \; i} \qquad \frac{((\sigma, \; p), \; a, \; (\sigma', \; p')) \in \mathsf{oseqp\text{-}sos} \; \varGamma \; i}{((\sigma, \; p \oplus q), \; a, \; (\sigma', \; p')) \in \mathsf{oseqp\text{-}sos} \; \varGamma \; i}$$

$$\frac{\sigma' \; i = \sigma \; i}{((\sigma, \; \{l\}\mathsf{unicast}(s_{ip}, s_{msg}) \; . \; p \triangleright q), \; \mathsf{unicast} \; (s_{ip} \, (\sigma \; i)) \; (s_{msg} \, (\sigma \; i)), \; (\sigma', p)) \in \mathsf{oseqp\text{-}sos} \; \varGamma \; i}$$

(a) Sequential processes: examples from oseqp-sos.

$$\frac{((\sigma, \; s), \; \mathsf{receive} \; m, \; (\sigma', \; s')) \in S \qquad (t, \; \mathsf{send} \; m, \; t') \in T}{((\sigma, \; (s, \; t)), \; \tau, \; (\sigma', \; (s', \; t'))) \in \mathsf{oparp\text{-}sos} \; i \; S \; T}$$

(b) Parallel processes: example from oparp-sos.

$$\frac{((\sigma, s), \; \mathsf{receive} \; m, \; (\sigma', s')) \in S}{((\sigma, s_R^i), \; \{i\}\neg\emptyset{:}\mathsf{arrive}(m), \; (\sigma', s_R'^i)) \in \mathsf{onode\text{-}sos} \; S} \qquad \frac{((\sigma, s), \tau, (\sigma', s')) \in S \quad \forall j \neq i. \; \sigma' j = \sigma j}{((\sigma, s_R^i), \tau, (\sigma', s_R'^i)) \in \mathsf{onode\text{-}sos} \; S}$$

(c) Nodes: examples from onode-sos.

$$\frac{((\sigma, s), \; H \neg K{:}\mathsf{arrive}(m), \; (\sigma', \; s')) \in S \qquad ((\sigma, t), \; H' \neg K'{:}\mathsf{arrive}(m), \; (\sigma', \; t')) \in T}{((\sigma, \; s \|t), \; (H \cup H')\neg(K \cup K'){:}\mathsf{arrive}(m), \; (\sigma', \; s'\|t')) \in \mathsf{opnet\text{-}sos} \; S \; T}$$

(d) Partial networks: example from opnet-sos.

Fig. 5. SOS rules for the open model (cf. Figures 1, 2, 3, and 4)

   This definition is non-empty only for well-formed net-trees (net-ips gives the set of node addresses in the state of a partial network). Including such a constraint within the open model, rather than as a separate assumption like the wf-net-tree n in (4), eliminates an annoying technicality from the inductions described in Section 4.3. As with the extra premises in the open SOS rules, we can freely adjust the open model to facilitate proofs but each 'encoded assumption' becomes an obligation to be discharged in the transfer lemma of Section 4.4.

   An operator for adding the last layer is also readily defined by

$$\mathsf{oclosed} \; A = A(\!|\mathsf{trans} := \mathsf{ocnet\text{-}sos} \; (\mathsf{trans} \; A)|\!),$$

giving all the definitions necessary to turn a standard model into an open one.

4.2   Open Invariants

The basic definitions of reachability and invariance, Definitions 1–3, apply to open models, but constructing a compositional proof requires considering the effects of both synchronized and interleaved actions of possible environments.

Definition 9 (open reachability). *Given an automaton A and assumptions S and U over, respectively, synchronized and interleaved actions, oreachable A S U is the smallest set defined by the rules:*

$$\frac{(\sigma, \; p) \in \mathsf{init} \; A}{(\sigma, \; p) \in \mathsf{oreachable} \; A \; S \; U} \qquad \frac{(\sigma, \; p) \in \mathsf{oreachable} \; A \; S \; U \qquad U \, \sigma \, \sigma'}{(\sigma', \; p) \in \mathsf{oreachable} \; A \; S \; U}$$

$$\frac{(\sigma, \; p) \in \mathsf{oreachable} \; A \; S \; U \qquad ((\sigma, \; p), \; a, \; (\sigma', \; p')) \in \mathsf{trans} \; A \qquad S \, \sigma \, \sigma' \, a}{(\sigma', \; p') \in \mathsf{oreachable} \; A \; S \; U}$$

In practice, we use restricted forms of the assumptions S and U, respectively,

$$\text{otherwith E N I } \sigma \sigma' \text{ a} = (\forall i.\ i \notin \text{N} \longrightarrow \text{E } (\sigma\ i)\ (\sigma'\ i)) \wedge \text{I } \sigma \text{ a}, \tag{5}$$

$$\text{other F N } \sigma \sigma' = \forall i.\ \text{if } i \in \text{N then } \sigma'\ i = \sigma\ i \text{ else F } (\sigma\ i)\ (\sigma'\ i). \tag{6}$$

The former permits the restriction of possible environments (E) and also the extraction of information from shared actions (I). The latter restricts (F) the effects of interleaved actions, which may only change non-local state elements.

Definition 10 (open invariance). *Given an automaton A and assumptions S and U over, respectively, synchronized and interleaved actions, a predicate P is an* open invariant, *denoted A ⊨ (S, U →) P, iff* ∀ s ∈ oreachable A S U. P s.

It follows easily that existing invariants can be made open: most invariants can be shown in the basic context but still exploited in the more complicated one.

Lemma 2. *Given an invariant A ⊨ (I →) P where* trans A = seqp-sos Γ, *and any F, there is an open invariant A' ⊨ (λ- -. I, other F {i} →) (λ(σ, p). P (σ i, p)) where* trans A' = oseqp-sos Γ i, *provided that* init A = {(σ i, p) | (σ, p) ∈ init A'}.

Open step invariance and a similar transfer lemma are defined similarly. The meta theory for basic invariants is also readily adapted, in particular,

Theorem 2. *To show A ⊨ (S, U →) onl Γ P, in addition to the conditions and the obligations **(init)** and **(step)** of Theorem 1, suitably adjusted, it suffices,*

(env)   *for arbitrary (σ, p) ∈ oreachable A S U and l ∈ labels Γ p, to assume both P (σ, l) and U σ σ', and then to show P (σ', l).*

This theorem is declared to the tactic described in Section 3.2 and proofs proceed as before, but with the new obligation to show invariance over interleaved steps.

We finally have sufficient machinery to state (and prove) Invariant (4) at the level of a sequential process:

$$\text{optoy i } \models \text{ (otherwith nos-inc } \{i\} \text{ (orecvmsg msg-ok), other nos-inc } \{i\} \to)$$
$$(\lambda(\sigma, \text{-}).\ \text{no } (\sigma\ i) \le \text{no } (\sigma\ (\text{nhip } (\sigma\ i)))) , \tag{7}$$

where nos-inc $\xi \xi'$ = no $\xi \le$ no ξ', orecvmsg applies its given predicate to receive actions and is otherwise true, msg-ok σ (pkt (data, src)) = (data $\le$ no (σ src)), and msg-ok σ (newpkt (data, dst)) = True. So, given that the variables no in the environment never decrease and that incoming pkts reflect the state of the sender, there is a relation between the local node and the next hop. Similar invariants occur in proofs of realistic protocols [4].

4.3   Lifting Open Invariants

The next step is to lift Invariant (7) over each composition operator of the open model. We mostly present the lemmas over oreachable, rather than those for open invariants and step invariants, which follow more or less directly.

The first lifting rule treats composition with the qmsg process. It mixes ore-achable and reachable predicates: the former for the automaton being lifted, the latter for properties of qmsg. The properties of qmsg—only received messages are added to the queue and sent messages come from the queue—are shown using the techniques of Section 3.

Lemma 3 (qmsg lifting). *Given* $(\sigma, (s, (q, t))) \in$ *oreachable* $(A \langle\langle_i$ qmsg$)$ S U, *where predicates* S = *otherwith* E {i} (*orecvmsg* R) *and* U = *other* F {i}, *and provided (1)* A $\models$ (S, U →) $(\lambda((\sigma, -), -, (\sigma', -))$. F $(\sigma\ i)$ $(\sigma'\ i))$, *(2) for all* ξ, ξ', E ξ ξ' *implies* F ξ ξ', *(3) for all* $\sigma, \sigma', m, \forall j$. F $(\sigma\ j)$ $(\sigma'\ j)$ *and* R σ m *imply* R σ' m, *and, (4)* F *is reflexive, then* $(\sigma, s) \in$ *oreachable* A S U *and* $(q, t) \in$ *reachable* qmsg (*recvmsg* (R σ)), *and furthermore* $\forall m \in$ set q. R σ m.

The key intuition is that every message m received, queued, and sent by qmsg satisfies R σ m. The proof is by induction over oreachable. The R's are preserved when the external environment acts independently (3, 4), when it acts synchronously (2), and when the local process acts (1, 3).

The rule for lifting to the node level adapts assumptions on receive actions (orecvmsg) to arrive actions (oarrivemsg).

Lemma 4 (onode lifting). *If, for all* ξ *and* ξ', E ξ ξ' *implies* F ξ ξ', *then given* $(\sigma, s_R^i) \in$ *oreachable* $(\langle i : A : R_i \rangle_o)$ (*otherwith* E {i} (*oarrivemsg* I)) (*other* F {i}) *it follows that* $(\sigma, s) \in$ *oreachable* A (*otherwith* E {i} (*orecvmsg* I)) (*other* F {i}).

The sole condition is needed because certain node-level actions—namely connect, disconnect, and $\emptyset\neg\{i\}$:arrive(m)—synchronize with the environment (giving E ξ ξ') but appear to 'stutter' (requiring F ξ ξ') relative to the underlying process.

The lifting rule for partial networks is the most demanding. The function net-tree-ips, giving the set of addresses in a net-tree, plays a key role.

Lemma 5 (opnet lifting). *Given* $(\sigma, s_{\,\|\,}t) \in$ *oreachable* (opnet onp $(p_1 \| p_2)$) S U, *where* S = *otherwith* E (net-tree-ips $(p_1 \| p_2)$) (*oarrivemsg* I), U = *other* F (net-tree-ips $(p_1 \| p_2)$), *and* E *and* F *are reflexive, for arbitrary* p i *of the form* $\langle i : $ onp $ i : R \rangle_o$, p i $\models$ $(\lambda\sigma$ -. oarrivemsg I σ, other F {i} →) $(\lambda((\sigma, -), a, (\sigma', -))$. castmsg (I σ) a), *and similar step invariants for* E $(\sigma\ i)$ $(\sigma'\ i)$ *and* F $(\sigma\ i)$ $(\sigma'\ i)$, *then it follows that both* $(\sigma, s) \in$ *oreachable* (opnet onp p_1) S_1 U_1 *and* $(\sigma, t) \in$ *oreachable* (opnet onp p_2) S_2 U_2, *where* S_1 *and* U_1 *are over* p_1, *and* S_2 *and* U_2 *are over* p_2.

The proof is by induction over oreachable. The initial and interleaved cases are trivial. For the local case, given open reachability of (σ, s) and (σ, t) for p_1 and p_2, respectively, and $((\sigma, s_{\,\|\,}t), a, (\sigma', s'_{\,\|\,}t')) \in$ trans (opnet onp $(p_1 \| p_2)$), we must show open reachability of (σ', s') and (σ', t'). The proof proceeds by cases of a. The key step is to have stated the lemma without introducing cyclic dependencies between (synchronizing) assumptions and (step invariant) guarantees. For a synchronizing action like arrive, Definition 9 requires satisfaction of S_1 to advance in p_1 and of S_2 to advance in p_2, but the assumption S only holds for addresses j $\notin$ net-tree-ips $(p_1 \| p_2)$. This is why the step invariants required of nodes only assume oarrivemsg I σ of the environment, rather than an S over node address {i}.

This is not unduly restrictive since the step invariants provide guarantees for individual local state elements and not between network nodes. The assumption oarrivemsg I σ is never cyclic: it is either assumed of the environment for paired arrives, or trivially satisfied for the side that *casts. The step invariants are lifted from nodes to partial networks by induction over net-trees. For non-synchronizing actions, we exploit the extra guarantees built into the open SOS rules.

The rule for closed networks is similar to the others. Its important function is to eliminate the synchronizing assumption (S in the lemmas above), since messages no longer arrive from the environment. The conclusion of this rule has the form required by the transfer lemma of the next section.

4.4 Transferring Open Invariants

The rules in the last section extend invariants over sequential processes, like that of (7), to arbitrary, open network models. All that remains is to transfer the extended invariants to the standard model. We do so using a locale [12] openproc np onp sr where np has type ip $\Rightarrow$ ('s, 'm seq-action) automaton, onp has type ip $\Rightarrow$ ((ip $\Rightarrow$ 'g) $\times$ 'l, 'm seq-action) automaton, and sr has type 's $\Rightarrow$ 'g $\times$ 'l. The automata use the actions of Section 2.1 with arbitrary messages ('m seq-action).

The openproc locale relates an automaton np to a corresponding 'open' automaton onp, where sr splits the states of the former into global and local components. Besides two technical conditions on initial states, this relation requires assuming σ i = fst (sr s), σ' i = fst (sr s') and (s, a, s') $\in$ trans (np i), and then showing ((σ, snd (sr s)), a, (σ', snd (sr s'))) $\in$ trans (onp i)—that is, that onp simulates np. For our running example, we show openproc ptoy optoy id, and then lift it to the composition with qmsg, using a generic relation on openproc locales.

Lemma 6 (transfer). *Given np, onp, and sr such that openproc np onp sr, then for any wf-net-tree n and s $\in$ reachable (closed (pnet np n)) (λ-. True), it follows that (default (someinit np sr) (netlift sr s), netliftl sr s)*

$$\in \text{oreachable (oclosed (opnet onp n)) } (\lambda\text{- - -. True) U.}$$

This lemma uses two openproc constants: someinit np sr i chooses an arbitrary initial state from np (SOME x. x $\in$ (fst o sr) ' init (np i)), and

$$\text{netliftl sr } (s_R^i) \ = (\text{snd (sr s)})_R^i$$
$$\text{netliftl sr } (s_{II} t) = (\text{netliftl sr s})_{II} (\text{netliftl sr t}).$$

The proof of the lemma 'discharges' the assumptions incorporated into the open SOS rules. An implication from an open invariant on an open model to an invariant on the corresponding standard model follows as a corollary.

Summary. The technicalities of the lemmas in this and the preceding section are essential for the underlying proofs to succeed. The key idea is that through an open version of AWN where automaton states are segregated into global and local components, one can reason locally about global properties, but still, using the so called transfer and lifting results, obtain a result over the original model.

5   Concluding Remarks

We present a mechanization of a modelling language for MANET and WMN protocols, including a streamlined adaptation of standard theory for showing invariants of individual reactive processes, and a novel and compositional framework for lifting such results to network models. The framework allows the statement and proof of inter-node properties. We think that many elements of our approach would apply to similarly structured models in other formalisms.

It is reasonable to ask whether the basic model presented in Section 2 could not simply be abandoned in favour of the open model of Section 4.1. But we believe that the basic model is the most natural way of describing what AWN means, proving semantic properties of the language, showing 'node-only' invariants, and, potentially, for showing refinement relations. Having such a reference model allows us to freely incorporate assumptions into the open SOS rules, knowing that their soundness must later be justified.

The Ad hoc On-demand Distance Vector (AODV) case study. The framework we present in this paper was successfully applied in the mechanization of a proof of loop freedom [6, §7] of the AODV protocol [17], a widely-used routing protocol designed for MANETs, and one of the four protocols currently standardized by the IETF MANET working group. The model has about 100 control locations across 6 different processes, and uses about 40 functions to manipulate the data state. The main property (loop freedom) roughly states that 'a data packet is never sent round in circles without being delivered'. To establish this property, we proved around 400 lemmas. Due to the complexity of the protocol logic and the length of the proof, we present the details elsewhere [4]. The case study shows that the presented framework can be applied to verification tasks of industrial relevance.

Acknowledgments. We thank G. Klein and M. Pouzet for support and complaisance, and M. Daum for participation in discussions. Isabelle/jEdit [21], Sledgehammer [2], parallel processing [22], and the TPTP project [19] were invaluable.

NICTA is funded by the Australian Government through the Department of Communications and the Australian Research Council through the ICT Centre of Excellence Program.

References

1. Bengtson, J., Parrow, J.: Psi-calculi in Isabelle. In: Berghofer, S., Nipkow, T., Urban, C., Wenzel, M. (eds.) TPHOLs 2009. LNCS, vol. 5674, pp. 99–114. Springer, Heidelberg (2009)
2. Blanchette, J.C., Böhme, S., Paulson, L.C.: Extending Sledgehammer with SMT solvers. In: Bjørner, N., Sofronie-Stokkermans, V. (eds.) CADE 2011. LNCS (LNAI), vol. 6803, pp. 116–130. Springer, Heidelberg (2011)
3. Bourke, T.: Mechanization of the Algebra for Wireless Networks (AWN). In: Archive of Formal Proofs (2014), http://afp.sf.net/entries/AWN.shtml

4. Bourke, T., van Glabbeek, R.J., Höfner, P.: A mechanized proof of loop freedom of the (untimed) AODV routing protocol. See authors' webpages (2014)
5. Chaudhuri, K., Doligez, D., Lamport, L., Merz, S.: Verifying safety properties with the TLA$^+$ proof system. In: Giesl, J., Hähnle, R. (eds.) IJCAR 2010. LNCS, vol. 6173, pp. 142–148. Springer, Heidelberg (2010)
6. Fehnker, A., van Glabbeek, R.J., Höfner, P., McIver, A., Portmann, M., Tan, W.L.: A process algebra for wireless mesh networks used for modelling, verifying and analysing AODV. Technical Report 5513, NICTA (2013), http://arxiv.org/abs/1312.7645
7. Feliachi, A., Gaudel, M.-C., Wolff, B.: Isabelle/Circus: A process specification and verification environment. In: Joshi, R., Müller, P., Podelski, A. (eds.) VSTTE 2012. LNCS, vol. 7152, pp. 243–260. Springer, Heidelberg (2012)
8. Fokkink, W., Groote, J.F., Reniers, M.: Process algebra needs proof methodology. EATCS Bulletin 82, 109–125 (2004)
9. Göthel, T., Glesner, S.: An approach for machine-assisted verification of Timed CSP specifications. Innovations in Systems and Software Engineering 6(3), 181–193 (2010)
10. Heyd, B., Crégut, P.: A modular coding of UNITY in COQ. In: Goos, G., Hartmanis, J., van Leeuwen, J., von Wright, J., Grundy, J., Harrison, J. (eds.) TPHOLs 1996. LNCS, vol. 1125, pp. 251–266. Springer, Heidelberg (1996)
11. Hirschkoff, D.: A full formalisation of π-calculus theory in the Calculus of Constructions. In: Gunter, E.L., Felty, A.P. (eds.) TPHOLs 1997. LNCS, vol. 1275, pp. 153–169. Springer, Heidelberg (1997)
12. Kammüller, F., Wenzel, M., Paulson, L.C.: Locales - A sectioning concept for Isabelle. In: Bertot, Y., Dowek, G., Hirschowitz, A., Paulin, C., Théry, L. (eds.) TPHOLs 1999. LNCS, vol. 1690, pp. 149–165. Springer, Heidelberg (1999)
13. Manna, Z., Pnueli, A.: Temporal Verification of Reactive Systems: Safety. Springer (1995)
14. Müller, O.: A Verification Environment for I/O Automata Based on Formalized Meta-Theory. PhD thesis, TU München (1998)
15. Nipkow, T., Paulson, L.C., Wenzel, M.: Isabelle/HOL. LNCS, vol. 2283. Springer, Heidelberg (2002)
16. Paulson, L.C.: The inductive approach to verifying cryptographic protocols. J. Computer Security 6(1-2), 85–128 (1998)
17. Perkins, C.E., Belding-Royer, E.M., Das, S.R.: Ad hoc on-demand distance vector (AODV) routing. RFC 3561 (Experimental), Network Working Group (2003)
18. de Roever, W.-P., de Boer, F., Hannemann, U., Hooman, J., Lakhnech, Y., Poel, M., Zwiers, J.: Concurrency Verification: Introduction to Compositional and Non-compositional Methods. Cambridge Tracts in Theor. Comp. Sci., vol. 54. CUP (2001)
19. Sutcliffe, G.: The TPTP problem library and associated infrastructure: The FOF and CNF parts, v3.5.0. J. Automated Reasoning 43(4), 337–362 (2009)
20. Tej, H., Wolff, B.: A corrected failure divergence model for CSP in Isabelle/HOL. In: Fitzgerald, J.S., Jones, C.B., Lucas, P. (eds.) FME 1997. LNCS, vol. 1313, pp. 318–337. Springer, Heidelberg (1997)
21. Wenzel, M.: Isabelle/jEdit – A prover IDE within the PIDE framework. In: Jeuring, J., Campbell, J.A., Carette, J., Dos Reis, G., Sojka, P., Wenzel, M., Sorge, V. (eds.) CICM 2012. LNCS, vol. 7362, pp. 468–471. Springer, Heidelberg (2012)
22. Wenzel, M.: Shared-memory multiprocessing for interactive theorem proving. In: Blazy, S., Paulin-Mohring, C., Pichardie, D. (eds.) ITP 2013. LNCS, vol. 7998, pp. 418–434. Springer, Heidelberg (2013)

A Computer-Algebra-Based Formal Proof
of the Irrationality of $\zeta(3)$

Frédéric Chyzak[1], Assia Mahboubi[1], Thomas Sibut-Pinote[2], and Enrico Tassi[1]

[1] Inria, France
[2] ENS de Lyon, France

Abstract. This paper describes the formal verification of an irrationality proof of $\zeta(3)$, the evaluation of the Riemann zeta function, using the Coq proof assistant. This result was first proved by Apéry in 1978, and the proof we have formalized follows the path of his original presentation. The crux of this proof is to establish that some sequences satisfy a common recurrence. We formally prove this result by an a posteriori verification of calculations performed by computer algebra algorithms in a Maple session. The rest of the proof combines arithmetical ingredients and some asymptotic analysis that we conduct by extending the Mathematical Components libraries. The formalization of this proof is complete up to a weak corollary of the Prime Number Theorem.

1 Introduction

The irrationality status of the evaluations of the Riemann ζ-function at positive odd integers is a long-standing challenge of number theory. To date, $\zeta(3)$ is the only one known to be irrational, although recent advances obtained by Rivoal [19] and Zudilin [24] showed that one at least of the numbers $\zeta(5), \ldots, \zeta(11)$ must be irrational. The number $\zeta(3)$ is sometimes referred to as the *Apéry constant*, after Roger Apéry who first proved that it is irrational [3]. As reported by van der Poorten [21], Apéry announced this astonishing result by giving a rather obscure lecture that raised more skepticism than enthusiasm among the audience. His exposition indeed involved a number of suspicious assertions, proclaimed without a proof, among which was a mysterious common recurrence for two given sequences (see Lemma 2). After two months of work, however, Cohen, Lenstra, and van der Poorten completed, with the help of Zagier, a verification of Apéry's proof.

Theorem 1 (Apéry, 1978). *The constant $\zeta(3)$ is irrational.*

Almost at the same time, symbolic computation was emerging as a scientific area of its own, getting fame with the Risch algorithm [18] for indefinite integration. It gradually provided efficient computer implementations and got attention in experimental mathematics. Beside commutative algebra, differential and recurrence equations remained a central research topic of computer algebra over the years. In particular, the sequences used by Apéry in his proof belong to a class

G. Klein and R. Gamboa (Eds.): ITP 2014, LNAI 8558, pp. 160–176, 2014.

of objects well known to combinatorialists and computer-algebraists. Following seminal work of Zeilberger's [22], algorithms have been designed and implemented in computer-algebra systems, which are able to obtain linear recurrences for these sequences. For instance the Maple packages gfun and Mgfun (both distributed as part of the Algolib [2] library) implement these algorithms, among other. Basing on this implementation, Salvy wrote a Maple worksheet [20] that follows Apéry's original method but interlaces Maple calculations with human-written parts, illustrating how parts of this proof, including the discovery of Apéry's mysterious recurrence, can be performed by computations.

In the present paper, we describe a formal proof of Theorem 1, based on a Maple session, in the Coq proof assistant. The computer-algebra system is used in a skeptical way [15], to produce conjectures that are a posteriori proved formally. Alternative proofs are known for Theorem 1, as for instance the elegant one proposed by Beukers [5] shortly after Apéry. Our motivation however was to devise a protocol to obtain formal proofs of computer-algebra-generated recurrences in a systematic way. Interestingly, this work challenges the common belief in the computer-algebra community that such an a posteriori checking can be automatized. In addition to the formal verification of these computer-algebra-produced assertions, we have also machine-checked the rest of the proof of irrationality, which involves both elementary number theory and some asymptotic analysis. The latter part of the proof essentially consists in a formal study of the asymptotic behaviors of some sums and of tails of sums. Our formal proof is complete, up to a weak corollary of the repartition of prime numbers that we use as an assumption.

In Section 2, we outline a proof of Theorem 1. Section 3 presents the algorithms which are run in the Maple session we base on. Section 4 describes the formalization of the formal proof we obtain from the data produced by the computer-algebra system. Section 5 provides some concluding remarks and some perspectives for future work.

Our Maple and Coq scripts will be found at http://specfun.inria.fr/ zeta-of-3/.

2   From Apéry's Recurrence to the Irrationality of $\zeta(3)$

In this section, we outline the path we have followed in our formalization, highlighting the places where we resorted to more elementary variants than Salvy or van der Poorten. In particular, Section 2.3 describes a simple argument we devised to simplify the proof of asymptotic considerations.

2.1   Overview

In all what follows, a Cauchy real (number) x is a sequence of rational numbers $(x_n)_{n\in\mathbb{N}}$ for which there exists a function $m_x : \mathbb{Q} \to \mathbb{N}$, such that for any $\epsilon > 0$ and any indices i and j, having $i \geq m_x(\epsilon)$ and $j \geq m_x(\epsilon)$ implies $|x_i - x_j| \leq \epsilon$.

Proposition 1. *The sequence* $z_n = \sum_{m=1}^{n} \frac{1}{m^3}$ *is a Cauchy real.*

The Cauchy real of Proposition 1 is our definition for $\zeta(3)$. Consider the two sequences a and b of rational numbers defined as:

$$a_n = \sum_{k=0}^{n} \binom{n}{k}^2 \binom{n+k}{k}^2, \qquad b_n = a_n z_n + \sum_{k=1}^{n} \sum_{m=1}^{k} \frac{(-1)^{m+1}\binom{n}{k}^2\binom{n+k}{k}^2}{2m^3\binom{n}{m}\binom{n+m}{m}}. \tag{1}$$

Introducing the auxiliary sequences of *real* numbers:

$$\delta_n = a_n\zeta(3) - b_n, \quad \sigma_n = 2\ell_n^3\delta_n, \quad \text{for } \ell_n \text{ the lcm of the integers } 1,\ldots,n, \tag{2}$$

the proof goes by showing that the sequence $(\sigma_n)_{n\in\mathbb{N}}$ has positive values and tends to zero. Now if $\zeta(3)$ was a rational number, then for n large enough, every σ_n would be a (positive) *integer*, preventing σ from tending to zero.

2.2 Arithmetics, Number Theory

We extend the usual definition of binomial coefficients $\binom{n}{k}$ for $n, k \in \mathbb{N}$ to $n, k \in \mathbb{Z}$ by enforcing the Pascal triangle recurrence $\binom{n+1}{k+1} = \binom{n}{k+1} + \binom{n}{k}$ for all $n, k \in \mathbb{Z}$. Although this extension is not required by the present proof, it spares us some spurious considerations about subtraction over $\mathbb{N}$. Binomial coefficients being integers, a_n is also an integer for any nonnegative $n \in \mathbb{N}$.

An important property of the sequence $(b_n)_{n\in\mathbb{N}}$ is that for any $n \in \mathbb{N}$, the product $2\ell_n^3 b_n$ is an integer. Therefore if $\zeta(3)$ were a rational number, then $\ell_n\zeta(3)$, and hence $\sigma_n = 2\ell_n^3(a_n\zeta(3) - b_n)$, would be an integer for n larger than the denominator of $\zeta(3)$. We follow the argument described by Salvy in [20], and show that each summand in the double sum defining b_n has a denominator that divides $2\ell_n^3$: after a suitable re-organization in the expression of the summand, which uses standard properties of binomial coefficients, this follows easily from the following slightly less standard property of theirs:

Lemma 1. *For any integers i, j, n such that $1 \leq j \leq i \leq n$, $j\binom{i}{j}$ divides ℓ_n.*

Lemma 1 is considered as folklore in number theory. Its proof consists in showing that for any prime p, the p-valuation of $j\binom{i}{j}$ is smaller than the one of ℓ_n.

Standard presentations of Apéry's proof make use of the asymptotic bound $\ell_n = e^{n(1+o(1))}$, which is a corollary of the distribution of the prime numbers. A bound 3^n is however tight enough for our purpose and has been proved by several independent and elementary proofs, for instance by Hanson [13] and Feng [11]. However, we have not yet formalized any proof of this ingredient, which is completely independent from the rest of the irrationality proof. More precisely, our formal proof is parametrized by the following assumption:

Proposition 2. *There exists two positive rationals K and r, with $r^3 < 33$, such that for any large enough integer n, $\ell_n < Kr^n$.*

2.3 Consequences of Apéry's Recurrence

The Cauchy sequence $(b_n/a_n)_{n\in\mathbb{N}}$ tends to $\zeta(3)$, thus δ_n tends to zero. In this section, we prove that it does so fast enough to compensate for ℓ_n^3, while being positive. The starting point is Apéry's recurrence, (3) below:

Lemma 2. *For $n \geq 0$, the sequences $(a_n)_{n\in\mathbb{N}}$ and $(b_n)_{n\in\mathbb{N}}$ satisfy the* same *second-order recurrence:*

$$(n+2)^3 y_{n+2} - (17n^2 + 51n + 39)(2n+3)y_{n+1} + (n+1)^3 y_n = 0. \qquad (3)$$

Salvy's worksheet [20] demonstrates in particular how to obtain this common recurrence by Maple calculations, performed by the Algolib library [2]. Following van der Poorten [21], we next use Lemma 2 (and initial conditions) to obtain a closed form of the Casoratian $w_n = b_{n+1}a_n - b_n a_{n+1}$. Indeed, we prove $w_n = \frac{6}{(n+1)^3}$ for $n \geq 2$. From this, we prove that δ_n, and hence σ_n, is positive for any $n \geq 2$. We also use the closed form to estimate the growth of δ in terms of a. The result is that there exists a positive rational number K such that $\delta_n \leq \frac{K}{a_n}$ for large enough n. Finally, the zero limit of σ_n follows from Proposition 2, the behaviour of δ, and Lemma 3 below, which quantifies that a grows fast enough.

Lemma 3. $33^n \in O(a_n)$.

Proof. Introduce the sequence $\rho_n = a_{n+1}/a_n$ and observe that $\rho_{51} > 33$. We now show that ρ is increasing. Define rational functions α and β so that the conclusion of Lemma 2 for a_n rewrites to $a_{n+2} - \alpha(n)a_{n+1} + \beta(n)a_n = 0$ for $n \geq 0$. Now, for any $n \in \mathbb{N}$, introduce the homography $h_n(x) = \alpha(n) - \frac{\beta(n)}{x}$, so that $\rho_{n+1} = h_n(\rho_n)$. Let x_n be the largest root of $x^2 - \alpha(n)x + \beta(n)$. The result follows by induction on n from the fact that $h([1, x_n]) \subset [1, x_n]$ and from the observation that $\rho_2 \in [1, x_2]$. $\qquad\square$

3 Algorithms on Sequences in Computer Algebra

Lemma 2 is the bottleneck in Apéry's proof. Both sums a_n and b_n in there are instances of *parametrised summation*: they follow the pattern $F_n = \sum_{k=\alpha(n)}^{\beta(n)} f_{n,k}$ in which the summand $f_{n,k}$, potentially the bounds, and thus the sum, depend on a parameter n. This makes it appealing to resort to the algorithmic paradigm of *creative telescoping*, which was developed for this situation in computer algebra.

In order to operate on sequences, computer algebra substitutes implicit representations for explicit representations in terms of named sequences (factorial, binomial, etc). This is the topic of Section 3.1. A typical example of parametrised summation by this approach is provided by the identity $\sum_{k=0}^{n} \binom{n}{k} = 2^n$: from an encoding of the summand $\binom{n}{k}$ by the recurrences

$$\binom{n+1}{k} = \frac{n+1}{n+1-k}\binom{n}{k}, \qquad \binom{n}{k+1} = \frac{n-k}{k+1}\binom{n}{k}, \qquad (4)$$

deriving the relation, with finite difference with respect to k in right-hand side,

$$\binom{n+1}{k} - 2\binom{n}{k} = \left(\binom{n+1}{k+1} - \binom{n}{k+1}\right) - \left(\binom{n+1}{k} - \binom{n}{k}\right) \qquad (5)$$

is sufficient to derive the explicit form 2^n, as will be explained below.

3.1   Recurrences as a Data Structure for Sequences

The implicit representation fruitfully introduced by computer algebra to deal with sequences are systems of linear recurrences. In this spirit, ∂-*finite sequences* are algebraic objects that model mathematical sequences and enjoy nice algorithmic properties. Notably, the finiteness property of their definition makes algorithmic most operations under which the class of ∂-finite sequences is stable.

A ∂-finite sequence (see [6] for a complete exposition of the subject) is an element of a module over the non-commutative ring $\mathcal{A}$ of skew polynomials in the indeterminates S_n and S_k, with coefficients in the rational-function field $\mathbb{Q}(n,k)$, and commutation rule $S_n^i S_k^j c(n,k) = c(n+i, k+j) S_n^i S_k^j$. A skew polynomial $P = \sum_{(i,j) \in I} p_{i,j}(n,k) S_n^i S_k^j \in \mathcal{A}$ acts on a "sequence" f by $(P \cdot f)_{n,k} = \sum_{(i,j) \in I} p_{i,j}(n,k) f_{n+i, k+j}$, where subscripts denote evaluation. For example for $f_{n,k} = \binom{n}{k}$, the recurrences (4) once rewritten as equalities to zero can be represented as $P \cdot f = 0$ for $P = S_n - \frac{n+1}{n+1-k}$ and $P = S_k - \frac{n-k}{k+1}$, respectively.

To any ∂-finite sequence f, one associates the set of skew polynomials that annihilate it. This set, $\{P \in \mathcal{A} : P \cdot f = 0\}$ is a left ideal of $\mathcal{A}$, named the *annihilating ideal* of f, and denoted $\operatorname{ann} f$. A non-commutative extension of the usual Gröbner-basis theory is available, together with algorithmic analogues. In this setting, a good representation of a ∂-finite sequence is obtained as a Gröbner basis of $\operatorname{ann} f$ for a suitable ordering on the monomials in S_n and S_k. For the example of $f_{n,k} = \binom{n}{k}$, a Gröbner basis consists of both already-mentioned skew polynomials encoding (4). In general, a Gröbner basis provides us with a (vectorial) basis of the quotient module $\mathcal{A}/\operatorname{ann} f$. This basis can be explicitly written in the form $B = \{f_{n+i, k+j}\}_{(i,j) \in \mathcal{U}}$, where the finite set $\mathcal{U}$ of indices is given as the part under the classical stair shape of the Gröbner-basis theory. Given a Gröbner basis GB for $\operatorname{ann} f$, the normal form $\operatorname{NF}(p, \text{GB})$ is unique for any $p \in \mathcal{A}$. Again in the binomial example, the finite set is $\mathcal{U} = \{(0,0)\}$, and normal forms are rational functions.

This is the basis of algorithms for a number of operations under which the ∂-finite class is stable, which all process by looking for enough dependencies between normal forms: application of an operator, addition, product. The case of summing a sequence $(f_{n,k})$ into a parametrised sum $F_n = \sum_{k=0}^{n} f_{n,k}$ is more involved: it performs according to the *method of creative telescoping* [23], in two stages. First, an *algorithmic* step determines pairs (P, Q) satisfying

$$P \cdot f = (S_k - 1) Q \cdot f \tag{6}$$

with $P \in \mathcal{A}'$ and $Q \in \mathcal{A}$, where $\mathcal{A}'$ is the subalgebra $\mathbb{Q}(n)\langle S_n \rangle$ of $\mathcal{A}$. To continue with our example $f_{n,k} = \binom{n}{k}$, Eq. (5) can be recast into this framework by choosing $P = S_n - 2$ and $Q = S_n - 1$. Second, a *systematic* but not fully algorithmic step follows: summing (6) for k between 0 and $n + \deg_{S_n} P$ yields

$$(P \cdot F)_n = (Q \cdot f)_{k=n+\deg_{S_n} P+1} - (Q \cdot f)_{k=0}. \tag{7}$$

Continuing with our binomial example, summing (5) (or its equivalent form (6)) for k from 0 to $n + 1$ (and taking special values into account) yields

Table 1. Construction of a_n and b_n: At each step, the Gröbner basis named in column GB, which annihilates the sequence given in explicit form, is obtained by the corresponding operation *on ideals*, with input(s) given on the last column

step	explicit form	GB	operation	input(s)
1	$c_{n,k} = \binom{n}{k}^2 \binom{n+k}{k}^2$	C	direct	
2	$a_n = \sum_{k=1}^{n} c_{n,k}$	A	creative telescoping	C
3	$d_{n,m} = \frac{(-1)^{m+1}}{2m^3 \binom{n}{m}\binom{n+m}{m}}$	D	direct	
4	$s_{n,k} = \sum_{m=1}^{k} d_{n,m}$	S	creative telescoping	D
5	$z_n = \sum_{m=1}^{n} \frac{1}{m^3}$	Z	direct	
6	$u_{n,k} = z_n + s_{n,k}$	U	addition	Z and S
7	$v_{n,k} = c_{n,k} u_{n,k}$	V	product	C and U
8	$b_n = \sum_{k=1}^{n} v_{n,k}$	B	creative telescoping	V

$\sum_{k=0}^{n+1} \binom{n+1}{k} - 2\sum_{k=0}^{n} \binom{n}{k} = 0$, a special form of (7) with right-hand side canceling to zero. The formula (7) in fact assumes several hypotheses that hold not so often in practice; this will be formalized by Eq. (8) below.

3.2   Apéry's Sequences are ∂-finite Constructions

The sequences a and b in (1) are ∂-finite: they have been announced to be solutions of (3). But more precisely, they can be viewed as constructed from "atomic" sequences by operations under which the class of ∂-finite sequences is stable. This is summarised in Table 1.

Both systems C and D are first-order systems obtained directly as easy consequences of (4); they consist respectively of expressions for $c_{n+1,k}$ and $c_{n,k+1}$ in terms of $c_{n,k}$ and of expressions for $d_{n+1,k}$ and $d_{n,k+1}$ in terms of $d_{n,k}$. The case of Z is almost the same: it is not a parametrised summation but an indefinite summation. A (univariate, second-order) recurrence is easily obtained for it, without referring to any creative telescoping.

For each of C, D, and V, which undergo a summation operation, we obtain creative-telescoping pairs (P, Q): one for C for a set $\mathcal{U} = \{(0,0)\}$; one for V for a set $\mathcal{U} = \{(0,0),(1,0),(0,1)\}$; four for D for the same set. In all cases, we have had our computer-algebra program (informally) ensure that the corresponding P cancels the sum. For C and V, the single P thus obtained is (trivially) a Gröbner basis of the annihilating ideal of A or B, respectively. But for D, the four P have to be recombined, leading to three operators.

It should be observed that Gröbner bases is the only data used by computer-algebra algorithms in the program above, including when simplifying the right-hand side in (7) and in its generalization to come, Eq. (8) below. For instance, computer algebra computes the system V from the systems C and U alone, without resorting to any other knowledge of particular values of c and u in Table 1, and so does our formal proof to verify that the pointwise product xy is annihilated by V whenever C and U respectively annihilate sequences x and y.

In other words, although a ∂-finite sequence is fully determined by a system of recurrences and sufficiently many initial conditions (that is, values of $u_{i,j}$ for small values of i and j), we do not maintain those values along our proofs.

Our formal proof as well models each sequence by a system of recurrences obtained solely from the operators of the Gröbner basis. We hence bet that the computer-algebra implementation of the algorithmic operations of addition, product, and summation, as well as the parts of our Maple script relying on less algorithmic operations do not take decisions based on private knowledge they could have on their input, viewed as a specific solution to the recurrence system used to encode it. Would the implementation do so without letting us know, then our a posteriori verification would have required guessing an appropriate description of this additional knowledge, like operators for specializations of the sequences. Fortunately, the Mgfun package we used has the wanted property.

3.3   Provisos and Sound Creative Telescoping

Observe the denominators in (4): they prevent the rules to be used, respectively when $k = n + 1$ and $k = -1$. For example, one can "almost prove" Pascal's triangle rule by

$$\binom{n+1}{k+1} - \binom{n}{k+1} - \binom{n}{k} = \left(\frac{n+1}{n-k}\frac{n-k}{k+1} - \frac{n-k}{k+1} - 1\right)\binom{n}{k} = 0 \times \binom{n}{k} = 0,$$

but this requires $k \neq -1$ and $k \neq n$. Therefore, this does not prove Pascal's rule for all n and k. The phenomenon is general: computer algebra is unable to take denominators into account. This incomplete modelling of sequences by algebraic objects may cast doubt on these computer-algebra proofs, in particular when it comes to the output of creative-telescoping algorithms.

By contrast, in our formal proofs, we augmented the recurrences with provisos that restrict their applicability. In this setting, we validate a candidate identity like the Pascal triangle rule by a normalization modulo the elements of a Gröbner basis plus a verification that this normalization only involves legal instances of the recurrences. In the case of creative telescoping, Eq. (6) takes the form:

$$(n, k) \notin \Delta \Rightarrow (P \cdot f_{-,k})_n = (Q \cdot f)_{n,k+1} - (Q \cdot f)_{n,k}, \tag{8}$$

where $\Delta \subset \mathbb{Z}^2$ guards the relation and where $f_{-,j}$ denotes the univariate sequence obtained by specializing the second argument of f to j. Thus our formal analogue of Eq. (7) takes this restriction into account and has the shape

$$(P \cdot F)_n = \left((Q \cdot f)_{n,n+\beta+1} - (Q \cdot f)_{n,\alpha}\right) + \sum_{i=1}^{r}\sum_{j=1}^{i} p_i(n)\, f_{n+i,n+\beta+j}$$
$$+ \sum_{\alpha \leq k \leq n+\beta \,\wedge\, (n,k)\in\Delta} (P \cdot f_{-,k})_n - (Q \cdot f)_{n,k+1} + (Q \cdot f)_{n,k}, \tag{9}$$

for F the sequence with general term $F_n = \sum_{k=\alpha}^{n+\beta} f_{n,k}$. The proof of identity (9) is a straightforward reordering of the terms of the left-hand side, $(P \cdot F)_n =$

$\sum_{i=0}^{r} p(n) F_{n+i}$, after unfolding the definition of F and applying relation (8) everywhere allowed in the interval $\alpha \leq k \leq n + \beta$. The first part of the right-hand side is the usual difference of border terms, already present in Eq. (7). The middle part is a collection of terms that arise from the fact that the upper bound of the sum defining F_n depends linearly on n and that we do not assume any nullity of the summand outside the summation domain. The last part, which we will call the singular part, witnesses the possible partial domain of validity of relation (8). The operator P is a valid recurrence for the sequence F if the right-hand side of Eq. (9) normalizes to zero, at least outside of an algebraic locus that will guard the recurrence.

4 Formal Proof of the Common Recurrence

This section describes the computer-algebra-aided formal proof of Lemma 2, based on a Maple session implementing the program described in Table 1.

4.1 Generated Operators, Hand-Written Provisos, and Formal Proofs

For each step in Table 1, we make use of the data computed by the Maple session in a systematic way. Figure 1 illustrates this pattern on the example of step 7. As mentioned in Section 3.3, we annotate each operator produced by the computer-algebra program with provisos (see below) and turn it this way into a conditional recurrence predicate on sequences. To each sequence in the program corresponds a file defining the corresponding conditional recurrences, for instance annotated_recs_c, annotated_recs_u, and annotated_recs_v for c, u, and v, respectively. More precisely these files contain *all* the operators obtained by the Maple script for a given sequence, not only the Gröbner basis. We use rounded boxes to depict the files that store the *definitions* of these predicates. These are generated by the Maple script which pretty-prints its output in Coq syntax, with the exception of the definition of provisos. Throughout this section, a maple leaf tags the files that are generated by our Maple script. Yet automating these annotations is currently out of reach.

In our formal proof, each step in Table 1 consists in proving that some conditional recurrences on a composed sequence can be proved from some conditional recurrences known for the arguments of the operation. We use square boxes to depict the files that store these *formal proofs*. The statement of the theorems proved in these files are composed from the predicates defined in the round boxes: a dashed line points to (predicates used to state) conclusions and a labelled solid line points to (predicates used to state) hypotheses.

4.2 Definitions of Conditional Recurrence Predicates

All files defining the conditional recurrence predicates obtained from the operators annihilating sequences of the program share the same structure. An excerpt

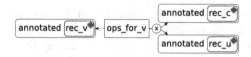

Fig. 1. Proving that V is $C \times U$

of the generated part of the file `annotated_recs_c` is displayed on Listing 1.1. The constants `Sn`, `Sk`, and `CT_premise` are recurrences predicates, defined in terms of a bound variable c. Constants `Sn` and `Sk` are elements of the Gröbner basis. The definition of these recurrences is named to reflect the term it rewrites, e.g., the left-hand sides in (4): these names are the result of pretty-printing the (skew) monomial that encodes these left-hand sides, the prefix S standing for "shift". For example `Sn` is the name of a recurrence defining $c_{n+1,k}$, while `SnSk` would be for $c_{n+1,k+1}$. Rewriting a given term with such an equation makes the term decrease for the order associated with the Gröbner basis. Another part of the file defines the recurrences obtained from a creative-telescoping pair (P, Q) generated for the purpose of the summation defining the sequence a.

```
(* Coefficients of every recurrence, P, and Q. *)
Definition Sn₀₀ n k := (n + 1 + k)² / (-n - 1 + k)².
Definition Sk₀₀ n k := (-n + k)² * (n + 1 + k)² / (k + 1)⁴.
Definition P₀ n := (n + 1)³.
...
(* Conditional recurrences. *)
Definition Sn c := ∀ n k, precond.Sn n k → c (n + 1) k = Sn₀₀ n k * c n k
Definition Sk c := ∀ n k, precond.Sk n k → c n (k + 1) = Sk₀₀ n k * c n k

(* Operators P and Q. *)
Definition P c n := P₀ n * c n + P₁ n * c (n + 1) + P₂ n * c (n + 2).
...
(* Statement P = Δₖ Q. *)
Definition CT_premise c := ∀ n k, precond.CT_premise n k →
  P (c ⊔ k) n = Q c n (k + 1) - Q c n k.
```

Listing 1.1. Generated part of annotated_rec_c

Observe that these generated definitions feature named provisos that are in fact placeholders. In the preamble of the file, displayed on Listing 1.2, we provide by a manual annotation a concrete definition for the proviso of each recurrence defined in the generated part. Observe however that part of these definitions can be inferred from the coefficients of the recurrences. For example the $k \neq n + 1$ condition in `precond.Sn`, the proviso of recurrence `Sn`, is due to the denominator $(-n - 1 + k)^2$ of the coefficient (Sn_{00}n k).

```
Module precond.
Definition Sn n k := (k ≠ n + 1) ∧ (n ≠ -1).
Definition Sk n k := (k + 1 ≠ 0) ∧ (n ≠ 0).
Definition CT_premise n k := (n ≥ 0) ∧ (k ≥ 0) ∧ (k < n).
End precond.
```

Listing 1.2. Hand-written provisos in annotated_rec_c

In the last part of the file, see Listing 1.3, a record collects the elements of the Gröbner basis C. Maple indeed often produces a larger set of annihilators for a given sequence, for instance CT_premise in Listing 1.1 is related to a creative telescoping pair but not to the Gröbner basis. Also, the Gröbner basis can be obtained by refining a first set of annihilators, which happens at step 4 of Table 1.

```
(* Choice of recurrences forming a Groebner basis. *)
Record Annihilators c := { Sn : Sn c; Sk : Sk c }.
```

Listing 1.3. Selection of a Gröbner basis

4.3   Formal Proofs of a Conditional Recurrence

We take as a running example the file ops_for_a, which models step 2 in Table 1. This file proves theorems about an arbitrary sequence c satisfying the recurrences in the Gröbner basis displayed on Listing 1.3, and about the sequence a by definite summation over c.

```
Require Import annotated_recs_c.
Variables (c : int → int → rat) (ann_c : Annihilators c).

Theorem P_eq_Delta_k_Q : CT_premise c. Proof. ... Qed.

Let a n := \sum_(0 ≤ k < n + 1) c n k.
```

The formal proof of lemma P_eq_Delta_k_Q is an instance of Eq. 8. Using this property, we prove that the sequence a verifies a conditional recurrence associated to the operator P. As suggested in Section 3.1, this proof consists in applying the lemma punk.sound_telescoping, which formalizes a sound creative telescoping and in normalizing to zero the resulting right-hand side of Eq. 9. Listing 1.4 displays the first lines of the corresponding proof script, which select and name the three components of the right-hand side of Eq. 9, with self-explanatory names. The resulting proof context is displayed on Listing 1.5.

```
Theorem recApery_a n (nge2 : n ≥ 2) : P a n = 0.
Proof.
rewrite (punk.sound_telescoping P_eq_Delta_k_Q).
set boundary_part := (X in X + _ + _).
set singular_part := (X in _ + X + _).
set overhead_part := (X in _ + _ + X).
```

Listing 1.4. Begining of a proof of sound creative telescoping

```
boundary_part := Q c n (n + 1) - Q c n 0
singular_part := \sum_(0 ≤ i < n + 1 | precond.CT_premise n i)
        P (c ⊔ i) n - (Q c n (i + 1) - Q c n i)
overhead_part := \sum_(0 ≤ i < degree P)
        \sum_(0 ≤ j < i) Pᵢ n * c (n + i) (n + j + 1)
==============================
 boundary_part + singular_part + overhead_part = 0
```

Listing 1.5. Corresponding goal

In Listing 1.5, (c ⊔ i) denotes the expression (fun x => c x i), P_i denotes the i-th coefficient of the polynomial P, and degree P the degree of P (two in this specific case). Note that we have access to degree P because in addition to the definition displayed on Listing 1.1, we also have at our disposal a list representation [:: P₀; P₁] of the same operator.

The proof of the goal of Listing 1.5, proceeds in three steps. The first step is to inspect the terms in singular_part and to chase ill-formed denominators, like $n - n$. These can arise from the specialisations, like $k = n$, induced when unrolling the definition of (the negation of) precond.CT_premise. In our formalization, a division by zero is represented by a conventional value: we check that these terms vanish by natural compensations, independently of the convention, and we keep only the terms in singular_part that represent genuine rational numbers. The second step consists in using the annihilator ann_c of the summand to reduce the resulting expression under the stairs of the Gröbner basis. In fact, this latter expression features several collections of terms, that will be reduced to as many independent copies of the stairs. In the present example, we observe two such collections: *(i)* terms that are around the lower bound $(n, 0)$ of the sum, of the form $c_{n,0}, \ldots c_{n,s}$; *(ii)* terms that are around the upper bound (n, n) of the summation, of the form $c_{n,n}, \ldots, c_{n,n+s}$ for a constant s. The border terms induce two such collections but there might be more, depending in particular on the shape of the precond.CT_premise proviso. For example, the sum $\sum_{k=0}^{n}(-1)^k \binom{n}{k}\binom{3k}{n} = (-3)^n$ leads to a proviso involving $n = 3k + 1$ and similar terms: an additional category of terms around $(n, n/3)$ drifts away from both $(n, 0)$ and (n, n) when n grows.

```
==============================
P₂ n * c (n + 2) n + P₁ n * c (n + 1) n +
P₀ n * c n n + Q₀₀ n n * c n n + P₁ n * c (n + 1) (n + 1) +
P₂ n * c (n + 2) (n + 1) + P₂ n * c (n + 2) (n + 2) = 0
```

Listing 1.6. Terms around the upper bound

The collection of terms around the upper bound in our running example is displayed on Listing 1.6. The script of Listing 1.7 reduces this collection under the stairs of ann_c, producing the expression displayed on Listing 1.8. The premise of each rule in this basis being an integer linear arithmetic expression, we check its satisfiability using our front-end intlia to the lia proof command [4], which automates the formal proof of first-order formulae of linear arithmetics.

```
rewrite (ann_c.Sk (n + 2) (n + 1)); last by intlia.
rewrite (ann_c.Sk (n + 2) n); last by intlia.
rewrite (ann_c.Sk (n + 1) n); last by intlia.
rewrite (ann_c.Sn (n + 1) n); last by intlia.
rewrite (ann_c.Sn n n); last by intlia.
set cnn := c n n.
Fail set no_more_c := c _ _.
```

Listing 1.7. Reduction modulo the Gröbner basis of c

```
==============================
P2 n * Sn00 (n + 1) n * Sn00 n n * cnn + P1 n * Sn00 n n * cnn +
P0 n * cnn + Q00 n n * cnn + P1 n * Sk00 (n + 1) n * Sn00 n n * cnn +
P2 n * Sk00 (n + 2) n * Sn00 (n + 1) n * Sn00 n n * cnn +
P2 n * Sk00 (n + 2) (n + 1) * Sk00 (n + 2) n *
  Sn00 (n + 1) n * Sn00 n n * cnn = 0
```

Listing 1.8. Rational function with folded coefficients

The third and last step consists in checking that the rational-function coefficient of every remaining evaluation of c is zero. For this purpose, we start by unfolding the definitions of the coefficients P_2, Sn_0. Previous steps kept them carefully folded as these values play no role in the previous normalizations but can lead to pretty large expressions if expanded, significantly slowing down any other proof command. The resulting rational function is proved to be zero by a combination of the `field` [10] and `lia` [4] proof commands. The former reduces the rational equation into a polynomial one between the cross product of two rational functions. This equation is then solved by the `ring` proof command [17]. The algebraic manipulations performed by `field` produce a set of non-nullity conditions for the denominators. These are solved by the `lia` proof command. To this end, our Maple script generates rational fractions with factored denominators, that happen to feature only linear factors in these examples.

4.4   Composing Closures and Reducing the Order of B

Figure 2 describes the global dependencies of the files proving all the steps in Table 1. In order to complete the formal proof of Lemma 2, we verify formally in file `algo_closures` that each sequence involved in the construction of a_n and b_n is a solution of the corresponding Gröbner system of annotated recurrence, starting from c_n, d_n, and z_n and applying the lemmas proved in the `ops_for_*` files all the way to the the final conclusions of `ops_for_a` and `ops_for_b`. This proves that a_n is a solution of the recurrence (3) but provides only a recurrence of order four for b_n. In file `reduce_order`, we prove that b as well satisfies the recurrence (3) using four evaluations b_0, b_1, b_2, b_3 that we compute in file `initial_conds`.

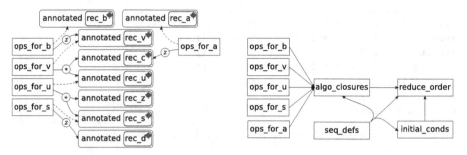

Fig 2. Formal proofs of Table 1   **Fig 3.** Formal proof of Lemma 2

5   Conclusion

5.1   Formally Proving the Consequences of Apéry's Recurrence

The present paper focuses on the computer-algebra-aided part of our formalization. Another significant part of it addresses the asymptotic study of some sequences of rational numbers. Formalizing this part of the proof requires introducing a type with more inhabitants than just rational numbers, for it deals with the properties of limits of sequences, notably the constant $\zeta(3)$ itself: we hence need to include at least computable real numbers. The construction of Cauchy reals proposed by Cohen in his PhD [8] turned out to perfectly suit our needs, although we had to enrich the existing library with a few basic properties. Indeed, we use the type of Cauchy real numbers mostly to state our theorems, like Theorem 1, but the proofs only involve reasoning with *rational* values of the sequences. We also benefited from the proof commands provided by Cohen [8] to postpone the definition of "large enough" bounds or moduli of convergence.

The proof of Lemma 3 however involves non-rational real numbers that are not defined as converging sequences but as roots of polynomials. We resort in this case to Cohen's construction of real algebraic numbers [7] and benefit constructively from the decidability of comparison relations. Navigating between the different natures of numbers this irrationality proof features, integer, rational numbers, real algebraic numbers, and Cauchy reals was made possible by the genericity of the core Mathematical Component libraries [1].

The proof of the same Lemma 3 involves two computational steps, namely the observations that $\rho_{51} > 33$ and that $\rho_2 \in [1, x_2]$. These numbers are rather large: for instance the normalized fraction ρ_{51} features integers with more than 70 digits. The issue here is to obtain a formal proof of these facts that both fits with the rest of the proof of Lemma 3 and does not take ages to compute. We used the CoqEAL [9] library and the framework it provides for writing parametric code which can be executed either symbolically, in abstract parts of a proof, or with appropriate data structures for efficient computations. The programs that implement the computation of ρ_n and x_n are very short and boil down to evaluating a (rather small) rational fraction defined recursively. At this place and when computing evaluations of the sequence (see Section 4.4), computations use (interpreted) binary integer arithmetics and are almost instantaneous.

5.2 Asymptotic Behavior of lcm$(1, \ldots, n)$

We plan to complete our irrationality proof by providing a formal proof of Proposition 2 following the proof of Hanson [13]. This involves considerations of elementary number theory as well as asymptotic-analysis studies that are rather similar to the ones we dealt with in the present work. The Prime Number Theorem has been machine-checked by Harrison [14] in HOL-Light. We considered importing this result into Coq by using the automated translation tool developed by Keller and Werner [16]. Unfortunately, according to Keller, the import of such an involved proof is currently beyond reach because of its size.

5.3 Formal Proofs on Objects of Size Typical of Computer Algebra

The size of the mathematical objects in our formalization makes the interactive writing of proof scripts quite unusual and challenging. For example, the recurrence $P \cdot y = 0$ of order four satisfied by the sequence b spans over 8,000 lines when pretty-printed. Both proofs of the fact $P \cdot b = 0$, named recApery_b in our formalization, and of its premise $P \cdot v = (S_k - 1)Q \cdot v$, named P_eq_Delta_k_Q, end by normalizing rational functions to 0. Figure 4 reports on the size of the polynomials that occur in these proofs, together with the amount of time needed by Coq's field proof command to check that the rational functions are zero.

| lemma | lines | terms | # of digits | | time |
			avg	max	(seconds)
P_eq_Delta_k_Q	1006	1714	6	13	247
recApery_b		7811	18602	2	13	179

Fig 4. Statistics about the polynomials in ops_for_b and their normalization

These objects can be easily manipulated by a computer-algebra system: Maple normalizes $P \cdot b$ to zero in less than 2 seconds. Coq, in its latest stable version (8.4pl3), takes roughly 4 minutes, but of course it produces and checks a formal proof of such normalization. To achieve this reasonable timing, we contributed to version 8.4pl3 a patch to substantially improve the performances of rational-function normalization: in Coq 8.4pl2 the very same recurrence requires a bit less than one hour to be normalized to zero.

Navigating expressions of size typical of computer algebra. Under the circumstances described above, where the goal to be proved cannot even be displayed, the need for uncommon proof commands arises. The pattern matching facilities offered by the set command [12] can be used to *probe* the goal for a subterm of the shape u (n + _) (k + _), or to identify all the occurrences of a closed terms.

Unfortunately not all uncommon needs are easily satisfiable with standard proof commands. A typical one is to invoke a computer-algebra system to rearrange an expression in order decide how to proceed with the proof. This was performed by hand, massaging text files to accommodate the small syntactic differences between the syntax of Coq and Maple. This task has been automated in the past in the Coq-Maple bridge developed by Delahaye and Mayero [10], for the purpose of enhancing proof automation in Coq. We foresee to generalise this work by allowing arbitrary Maple commands to let one explore algebraic expressions, and not necessarily prove the correctness of the results.

Proof-language implementation. The size of the mathematical expressions poses additional efficiency challenges when combined with the ability of the Coq logic to identify terms up to computation. For example a pattern like (3 * _) is matched by searching for the head symbol * verbatim, but its non-wildcard argument, 3 here, is compared taking computation into account. Given the size of the expressions, the fast filtering performed on the head symbol is not sufficient to discard enough false positives. Comparing two terms up to computation often triggers a full normalization before failing and this is an expensive operation.

The definitions of integer and rational numbers that are part of the Mathematical Components library are designed for the purpose of a comprehensive library of theorems, and not for fast computations: this makes the aforementioned pattern-matching operation often intractable. The commonly accepted idea of having two representations of the same concept, one good for reasoning and one good for computing, linked by a morphism does not work for expressions of this size, as switching from one representation to the other one relies again on pattern matching to push the morphism through the huge expression. For instance, we had to craft boilerplate code, which carefully controls computations and use locked casts for large constant numbers, in order to make the `field` and `lia` tactic work on these data-structures.

One possible work-around would be to have available, in the proof language, two distinct pattern-matching facilities: one that eases reasoning steps by identifying terms up to computation, and another one that performs a dumb syntactic comparison and is efficient even when applied to large, homogeneous expressions.

5.4   Theoretical and Practical Limitations of Data Structures

Algebraic rewriting with provisos An obstruction to turning our approach into a complete protocol is that we do not know how to determine the provisos other than by trial and error. This is connected to the fact that recurrences that have well understood rewriting properties when forming a Gröbner basis (a priori) lose all this nice structure when decorated with provisos. We do not understand what the new critical pairs have to be and how to ensure we give ourselves complete information with our selection of rules-with-provisos that simply lift a Gröbner basis. To the best of our knowledge, there is no literature on the topic yet.

We have been lucky with the few cases in the present formalization, in that we could just guide our reduction by the usual strategy of division by a Gröbner basis: all implied integer premises could be solved, without us understanding why they would.

Polynomial data structures. Manipulating polynomials in computer algebra is made natural by the ease to keep them in collected form, whether automatically or by easy-to-call procedures. On the other hand, our treatment of recurrences amounts mostly to collecting polynomials and skew polynomials, through the encoding suggested in Section 3.1. A (skew) polynomial data structure, with computational product and division, and the corresponding theory, would help keeping goals tidy and more manageable, especially during the rewriting steps. In particular, the current strategy of reserving the call to `field` to the end is certainly a cause of blow-up of the goals and inefficiency of the computations.

In addition, making it generic with respect to the nature of skew-polynomial commutation would provide similar functionality for both recurrence and differential equations, paving the way to the formalization of integral identities by a known differential counterpart to creative telescoping.

5.5   Building on Top of State-of-the-art Libraries

It was a help to have at our disposal the broad scope of the Mathematical Components library. It was a cause for trouble, too. This research started as an interdisciplinary activity: half of the authors were newcomers to formal proofs. As a matter of fact understanding the sophisticated design of the library turned out to be a very arduous task for them.

We can identify in the combined use of implicit coercions, notations, and structure inference the major source of complexity. Their formalization activity was additionally hindered by the fact that various features of the Coq system do not interact well with the design patterns employed in the Mathematical Components library. The most notable example being search, that is almost ineffective in the context of the algebraic hierarchy.

We believe such problems could be alleviated by: introductory material specific to the Mathematical Components library and written with newcomers in mind; a "teaching/debug" mode in which type inference explicates steps at the desired level of detail; finally a proper searching facility.

References

[1] Mathematical Components Libraries. Version 1.4. For Coq 8.4pl3 (2013), http://www.msr-inria.fr/projects/mathematical-components

[2] Algolib. Version 17.0. For Maple 17 (2013), http://algo.inria.fr/libraries/

[3] Apéry, R.: Irrationalité de $\zeta(2)$ et $\zeta(3)$. In: Astérisque, vol. 61. Société Mathématique de France (1996)

[4] Besson, F.: Fast reflexive arithmetic tactics the linear case and beyond. In: Altenkirch, T., McBride, C. (eds.) TYPES 2006. LNCS, vol. 4502, pp. 48–62. Springer, Heidelberg (2007)

[5] Beukers, F.: A note on the irrationality of $\zeta(2)$ and $\zeta(3)$. Bull. London Math. Soc. 11(3), 268–272 (1979)

[6] Chyzak, F., Salvy, B.: Non-commutative elimination in Ore algebras proves multivariate identities. J. Symbolic Comput. 26(2), 187–227 (1998)

[7] Cohen, C.: Construction of real algebraic numbers in CoQ. In: Beringer, L., Felty, A. (eds.) ITP 2012. LNCS, vol. 7406, pp. 67–82. Springer, Heidelberg (2012)

[8] Cohen, C.: Formalized algebraic numbers: construction and first-order theory. PhD thesis, École polytechnique (November 2012)

[9] Cohen, C., Dénès, M., Mörtberg, A.: Refinements for free! In: Gonthier, G., Norrish, M. (eds.) CPP 2013. LNCS, vol. 8307, pp. 147–162. Springer, Heidelberg (2013)

[10] Delahaye, D., Mayero, M.: Dealing with algebraic expressions over a field in Coq using Maple. J. Symb. Comput. 39(5), 569–592 (2005)

[11] Feng, B.-Y.: A simple elementary proof for the inequality $d_n < 3^n$. Acta Math. Appl. Sin. Engl. Ser. 21(3), 455–458 (2005)

[12] Gonthier, G., Mahboubi, A., Tassi, E.: A small scale reflection extension for the Coq system, RR-6455. Version 12 (2013)

[13] Hanson, D.: On the product of the primes. Canad. Math. Bull. 15, 33–37 (1972)

[14] Harrison, J.: Formalizing an analytic proof of the Prime Number Theorem. Journal of Automated Reasoning 43, 243–261 (2009)

[15] Harrison, J., Théry, L.: A skeptic's approach to combining HOL and Maple. J. Automat. Reason. 21(3), 279–294 (1998)

[16] Keller, C., Werner, B.: Importing HOL light into Coq. In: Kaufmann, M., Paulson, L.C. (eds.) ITP 2010. LNCS, vol. 6172, pp. 307–322. Springer, Heidelberg (2010)

[17] Grégoire, B., Mahboubi, A.: Proving equalities in a commutative ring done right in Coq. In: Hurd, J., Melham, T. (eds.) TPHOLs 2005. LNCS, vol. 3603, pp. 98–113. Springer, Heidelberg (2005)

[18] Risch, R.H.: The solution of the problem of integration in finite terms. Bull. Amer. Math. Soc. 76, 605–608 (1970)

[19] Rivoal, T.: Propriétés diophantiennes de la fonction zéta de Riemann aux entiers impairs. PhD thesis, Université de Caen (2001)

[20] Salvy, B.: An Algolib-aided version of Apéry's proof of the irrationality of $\zeta(3)$ (2003), http://algo.inria.fr/libraries/autocomb/Apery2-html/apery.html

[21] van der Poorten, A.: A proof that Euler missed: Apéry's proof of the irrationality of $\zeta(3)$. Math. Intelligencer 1(4), 195–203 (1979), An informal report

[22] Zeilberger, D.: A holonomic systems approach to special functions identities. J. Comput. Appl. Math. 32(3), 321–368 (1990)

[23] Zeilberger, D.: The method of creative telescoping. J. Symbolic Comput. 11(3), 195–204 (1991)

[24] Zudilin, V.V.: One of the numbers $\zeta(5)$, $\zeta(7)$, $\zeta(9)$, $\zeta(11)$ is irrational. Uspekhi Mat. Nauk 56(4(340)), 149–150 (2001)

From Operational Models to Information Theory; Side Channels in pGCL with Isabelle

David Cock

NICTA and University of New South Wales, Australia
David.Cock@nicta.com.au

Abstract. In this paper, we formally derive the probabilistic security predicate (expectation) for a guessing attack against a system with side-channel leakage, modelled in pGCL. Our principal theoretical contribution is to link the process-oriented view, where attacker and system execute particular model programs, and the information-theoretic view, where the attacker solves an optimal-decoding problem, viewing the system as a noisy channel. Our practical contribution is to illustrate the selection of probabilistic loop invariants to verify such security properties, and the demonstration of a mechanical proof linking traditionally distinct domains.

1 Introduction

This paper presents a formally-verified proof of the optimality of a Bayesian attack against an interactive system with side-channel leakage. This is an unsurprising result from a security and machine-learning perspective, and would typically just be assumed. Formalising the proof, however, requires effort. The point, of course, is that the proof is straightforward exactly where it should be, in the mechanical plugging-together of pieces, but interesting (or difficult), only where it must be, namely in choosing a model that faithfully represents our intuitive understanding of the problem, and in the 'creative moment' of choosing an appropriate loop invariant. The result itself will be unsurprising to anybody familiar with information theory or machine learning, but is the critical step in linking this typically mathematical field with the more pragmatic world of concrete operational semantics. As we will briefly detail in the final section of this paper, and as we have presented elsewhere(Cock, 2014), we can then apply the tools of one domain to the other, to tease out deeper theoretical results.

The contribution of this paper is a worked mechanical proof that conducting an optimal guessing attack, in the presence of side-channel leakage, reduces to an optimal decision problem on a corresponding noisy channel, expressed in purely probabilistic terms, with all artefacts of the operational model eliminated. This model, in turn, is carefully selected to make the connection to the intuitive security property as clear as possible, but also flexible enough to incorporate realistic-scale systems in the manner demonstrated in our previous work(Cock, 2013). To reiterate, a straightforward proof of such a cross-cutting result, requiring deep

G. Klein and R. Gamboa (Eds.): ITP 2014, LNAI 8558, pp. 177–192, 2014.

thought only at those points where it really should, demonstrates that the infrastructure of mechanisation is mature enough to allow us to transfer a security result from a detailed, fully-implemented system, through a probabilistic operational semantics(Cock, 2012), to a fully probabilistic or information-theoretic setting.

1.1   Side Channels and Guessing Attacks

The context of this work is our ongoing project(Cock, 2014) to demonstrate that we can transfer probabilistic information-flow properties from an abstract information-theoretic model, down to a concrete verified implementation, leveraging an existing correctness proof (for example that of the seL4 microkernel(Klein et al., 2009)). Previously(Cock, 2013), we demonstrated that the top-level result of that refinement proof could be embedded in the probabilistic language pGCL(McIver and Morgan, 2004), and lemmas about the abstract specification lifted into a probabilistic context. We now demonstrate that we can take one further step, and lift a probabilistic security property away from implementation details entirely, and reason solely at an abstract level.

The reason that this approach works, and that we are able to claim that it is applied to a real system is precisely the embedding result just mentioned. We were able to take the top-level specification of seL4, and embed it into pGCL in such a way that annotations of the original specification are preserved, and the refinement orders correspond. Therefore, any refinement-sound result on the probabilistic specification (such as that shown in this paper), is preserved by the full seL4 refinement stack and therefore applies to the low-level kernel implementation.

This security property is derived from our *threat model*: A guessing attack, against a system with a side channel. We define a side channel to be any information provided by the system to a client that a) depends on a *secret*, and b) is not allowed by the system's specification.

As a motivating example, consider an authentication system in which the client supplies some shared secret (for example a password) to the system in order to prove its identity. The system must compare the supplied secret to its stored version, and then either grant or deny access. Such a system has some degree of unavoidable leakage: a malicious client can guess secrets one-by-one and (assuming the set is finite), must eventually get the right one. In practice, we hope that the space of secrets is sufficiently large that such an attack is impractical. More interesting is whether the system leaks anything *beyond* this unavoidable amount, and to what extent any such additional *side-channel* leakage increases the vulnerability of the system (or the attacker's chance of success).

This leakage might be in the form of variable runtime, power consumption, radiation, or any other property not covered by the system's specification. We assume that continuous channel values are measured with some finite precision by the attacker, and are thus effectively quantised into some finite set of *outputs*. A common property of such side-channel outputs is that they are *stochastic*: there is no functional relation between the secret and the observed side-channel value. There may, however, exist some *probabilistic* relationship.

If there were a functional relation between the secret, $S \in \mathbb{S}$, and observation $o \in \mathbb{O}$, i.e. we are able to find some $F :: \mathbb{S} \Rightarrow \mathbb{O} \Rightarrow \textit{bool}$, such that[1]

$$\forall S.\ \exists! o.\ F\ S\ o$$

then we could proceed to reason about whether any information about S is leaked by revealing o. In particular, if F is invariant under changing S, or $\forall S\,S'\,o.\ F\ S\ o = F\ S'\ o$, then we can say with confidence that o gives no information about S.

In the probabilistic case, in general, no such relation exists and we must be satisfied with a probabilistic relationship, the *conditional probability* $P :: \mathbb{S} \Rightarrow \mathbb{O} \Rightarrow [0,1]$, where:

$$\forall S.\ \sum_o P\ S\ o = 1$$

Here, we interpret $P\ S\ o$ as the probability of producing observation o, assuming that the secret is S, for which we write $P[o|S]$, or 'the probability of o given S'.

Note that exactly the same invariance reasoning applies: if $\forall S\,S'\,o.\ P\ S\ o = P\ S'\ o$ i.e. the distribution of outcomes does not depend on S, then observing o tells the attacker nothing about S. Note also that if we define the boolean embedding operator $\langle\!\langle \cdot \rangle\!\rangle :: (\alpha \Rightarrow \textit{bool}) \Rightarrow \alpha \Rightarrow [0,1]$ by:

$$\langle\!\langle Q \rangle\!\rangle S = \text{if } Q\ S \text{ then } 1 \text{ else } 0$$

then we can view a functional relation $(\lambda S\,o.\ F\ S\ o)$ as the special case of a probabilistic relation $(\lambda S\,o.\ \langle\!\langle F\ S \rangle\!\rangle\ o)$ which, for each S, assigns 100% probability to exactly one o.

Given a non-injective relation, (or overlapping distributions), the attacker will generally not be able to determine the secret with certainty: As long as at least two secrets can produce the observed output with non-zero probability, then both are possible. In the probabilistic case however (and in contrast to the functional case), the attacker may be able to distinguish possible secrets by *likelihood*. All other things being equal, if secret S produces the observed output o with *greater conditional probability* than S' i.e. $P[o|S] > P[o|S']$, then S is a more likely candidate than S'.

Intuitively, if the attacker wants to *use* the secret that it has guessed, say to attempt to authenticate, then it should use the one that it considers to be *most likely to be correct*, in order to maximise its probability of success. The attacker thus needs to order possible secrets by $P[S|o]$, or the probability that the secret is S, having observed o. In order to calculate this, however, the attacker needs to know $P[S]$, or the prior probability of a given secret: even if $P[o|S] > P[o|S']$, and so S is more likely that S' on the basis of the evidence, if the secret is overwhelmingly more likely to be S' than S in the first case (perhaps S' is the

---

[1] The Isabelle/HOL syntax $\exists! x.\ P\ x$ reads 'there exists a *unique* x satisfying P'.

password '12345'), then even given the evidence it may still be more likely than S, albeit less so than initially.

The calculation of this inverse conditional probability $P[S|o]$, from the forward probability $P[o|S]$, is made via Bayes' rule:

$$P[S|o] = \frac{P[o, S]}{P[o]}$$

This states that the probability of the hypothesis (that the secret is S) given the evidence (that o was observed), is the joint probability of the hypothesis and the evidence ($P[o, S] = P[o|S]P[S]$), divided by the probability of the evidence ($P[o]$, or the likelihood of observing o across all secrets).

Thus, as $P[o]$ does not depend on S, the most likely secret (given that we have observed o), is that which maximises $P[o, S]$. Selecting S so as to maximise $P[o, S]$ in this way is known as the *maximum a posteriori* (MAP) rule: selecting the hypothesis with the highest a posteriori (after the fact) probability. In information theory, this is the optimal solution to the *decoding problem*: If the conditional probability $P[o|S]$ summarises the *channel matrix* of some noisy channel, i.e. $P[o|S]$ is the probability that symbol o is received, when S is transmitted, the decoding problem is that faced by the receiver: to choose the most likely transmitted symbol with which to decode each received symbol. This is exactly analogous to the challenge facing an attacker attempting to guess the secret given a side-channel observation, and both are solved optimally if the MAP rule is observed.

There is one way, however, in which the attacker is not like the receiver on a noisy channel: where the receiver is *passively* listening to the output, and making a best-effort reconstruction of the input, the attacker is *actively* probing the system, and has a second source of information in knowing whether it manages to authenticate, or is refused. The additional leakage implies that the most-likely secret (on the current evidence) is not necessarily the right one to guess. In particular, the attacker should never guess the same thing twice: if it's been refused once, it'll be refused again.

The criterion for success is also slightly different. In the noisy channel example, the decoder is successful if it chooses the correct S, and fails otherwise: there are no second chances. In the authentication example, however, if the attacker fails on its first guess (as it will most likely do), it can simply try again. Given a finite space of secrets therefore, an attacker that does not repeat guesses is guaranteed to eventually guess correctly. The trick is, of course, that in a large enough space, the chance of doing so in any practical timeframe is essentially zero. The security measure is therefore not 'the probability that the attacker guesses correctly' (which is 100%, eventually), but 'the probability that the attacker guesses correctly *in no more than n tries*'. The parameter n can then be set to an appropriate value for the system in question. For example, a login service may allow only n attempts before blacklisting a particular client, in which case we are asking for the probability that the system is compromised before the countermeasure (blacklisting) kicks in.

1.2   What the Proof Shows

The proof in Section 2 shows that what we expect, based on the informal reasoning above is, in fact, exactly what we get. As previously described, the bulk of the proof is relatively mechanical but, thanks to mechanisation, rather short. The interesting points, on which we will focus, are, again as already mentioned: in establishing the faithfulness of the formal model to the above intuitive description, and in the choice of loop invariant, given which the rest of the result follows straightforwardly.

The last thing we must mention before diving in is our two major assumptions, which are *implicit* in the above discussion, and in our choice of model, but need to be stated explicitly:

- We assume that multiple observations are independent, given the secret. The underlying assumption is that there is no lasting effect between two invocations: the result of one probe has no influence on the result of the next:
$$P(o, o'|S) = P(o|S)P(o'|S) \tag{1}$$
  This assumption could be lifted, for example by moving to an n-Markov model, which should be expressible in the same framework, if this turns out to be necessary in modelling a real system. For simplicity however, we start with the '0-Markov', or independent case.
- We assume that the side-channel output (o) depends only on the secret (S), and not on the attacker-supplied input.
  This assumption is more important, and more limiting, than the first. It is easy to construct systems where side-channel output depends on attacker-supplied input, and in making this assumption, we are excluding them from consideration. Such fully-dynamic attacks are the subject of current research, and there is as yet no elegant information-theoretic interpretation. We must thus be content for now with our partially-dynamic attack, where the attacker's actions may be dynamically determined, but the conditional probabilities are not.

1.3   An Overview of pGCL

The probabilistic imperative language pGCL is a descendent of GCL(Dijkstra, 1975), that incorporates both classical nondeterminism and probability. Common primitives are provided, including:

- Sequential composition: $a\,;;b$, or "do a then b".
- Variable update: $x:=(\lambda s.\ e\ s)$, or "update variable x, with the value of expression e in the current state, s".
- Looping: $do\ G \longrightarrow body\ od$, or "execute *body* for as long as G (the guard) holds".

We may also make both classically nondeterministic and probabilistic choices, which we present here in the form of choices over values:

- Unconstrained (demonic) choice: *any x*, or "assign any value to variable *x*".
- Constrained (probabilistic) choice: *any x at (λs v. P s v)*, or "choose value *v*, with probability *P* (which may depend on the state), and assign it to variable *x*".
- Constrained name binding (let expressions): *bind n at (λs v. P s v) in a n*. This is almost equivalent to the previous operation, but instead of modifying the state, simply binds the name *n*, which parameterises the program *a*. As indicated, this is simply a probabilistic let expression.

Expectations. In pGCL, we are no longer constrained to reasoning about boolean predicates on the state (pre- and post-conditions, for example), but can now consider any bounded, non-negative real-valued function (an *expectation*). We can annotate program fragments just as in GCL:

$$Q \Vdash \text{wp } a \ R$$

Traditionally, we would read this predicate-entailment relation as a Hoare triple: if *Q* holds initially, then *R* will hold after executing *a*. The literal meaning being that *Q* implies the *weakest precondition* of *R*. In pGCL, both *Q* and *R* are real-valued, and implication is replaced by pointwise comparison. The weakest precondition is replaced by the weakest pre-*expectation*, such that the equation now reads: *Q* is everywhere lower than the weakest preexpectation of *R*.

The action of programs is defined such that the weakest preexpectation of *R* is the expected value of *R* after executing *a*, or the sum over all possible states *s* of *R s*, weighted by the probability of ending in state *s*. If *R* is the probability function associated with some predicate i.e. *R* is 1 whenever the predicate holds, and 0 otherwise, then the weakest preexpectation is now the *probability*, calculated in the initial state, that the predicate holds in the final state (the expected value of a probability is itself a probability). In this case, the above annotation states that the probability that the predicate holds finally, if we begin in state *s* is *at least Q s*. The embedding, in the above fashion, of a predicate *S* as an expectation is written «*S*».

Note that if *Q s* is 1, this reduces to the classical case: If *Qs* is true (has probability 1), then *Rs* will hold (with probability at least[2] 1.).

Invariants. A loop *invariant* in pGCL is an expectation that is preserved by every iteration of the loop:

$$I * «G» \Vdash \text{wp } body \ I$$

In the standard (non-probabilistic) setting, this states that if the invariant *I* and guard *G* hold before the body executes, the invariant still holds afterwards. In our probabilistic setting, we instead have that (as *G* is {0, 1}-valued), the probability

---

[2] The healthiness conditions for the underlying semantics ensures that the probability is in fact exactly one.

of establishing I from any state where the loop body executes (satisfying G), is at least I, or that I is a *fixed point* of the guarded loop body. This seemingly weaker condition in fact suffices to provide a full annotation of the loop (see Lemma 1).

Further Information. This summary of pGCL is incomplete, and covers only those elements essential to the following exposition. For full details of the language and its formal semantics, the interested reader is directed to McIver and Morgan (2004), or for details of the mechanisation in Isabelle/HOL to Cock (2012, 2013).

2 The Proof

To model the system, we fix two distributions: the prior distribution on secrets, $P[S]$, and the conditional distribution of observation-given-secret, $P[o'|S]$. We assume throughout that these are valid distributions:

$$\sum S \in UNIV . \ P[S] = 1 \qquad\qquad 0 \le P[s]$$
$$\sum o' \in UNIV . \ P[o'|S] = 1 \qquad\qquad 0 \le P[S|o']$$

We capture these, and the assumption that both the space of secrets and of observations is finite, in an Isabelle local theory (locale).

2.1 Modelling the Guessing Attack

The state of the attack after some number of guesses (perhaps 0) is simply the list of observations that the attacker has made so far, or $\mathcal{O}$. In order to capture the system's (probabilistic) choice of secret, and the attacker's choice of strategy, however, we add two extra state components: $\mathcal{S}$ — the secret, and τ — the strategy. The system state is thus represented by the following record[3]:

record $('s, 'o)$ *attack-state* =
$\quad \mathcal{S} :: 's$
$\quad \mathcal{O} :: 'o \ list$
$\quad \tau :: 'o \ list \Rightarrow 's$

The use of a record as the state type allows us to take advantage of the field-update syntax introduced in our previous work(Cock, 2013).

We model the attack as a loop, with the attacker making one guess and one observation per iteration. This specification of a single iteration of the loop body demonstrates the syntax for probabilistically binding a name (without updating the state), and updating a single state component as if it were a variable:

---

[3] A tuple with named fields.

$body \equiv$
  $bind\ obs\ at\ (\lambda s.\ P[obs|\mathcal{S}\ s])\ in$
  $\mathcal{O} := (\lambda s.\ obs \cdot (\mathcal{O}\ s))$

The full attack has three phases: first the attacker decides on a strategy, knowing both the distribution on secrets ($P[S]$), and the conditional probability of observations, given secrets ($P[o'|S]$). This is expressed as an unconstrained nondeterministic choice (any), in a context where the names P-s and P-os are already bound (these are the underlying constants corresponding to the syntax $P[S]$ and $P[o'|S]$).

Next, the system generates a secret at random, according to the distribution $P[S]$, expressed using a *probabilistically-constrained* choice[4]. It is critical that this step comes *after* the first, as otherwise the attacker could simply choose a strategy where $\tau\ s\ [] = \mathcal{S}\ s$. Semantically, we avoid this as the value $\mathcal{S}\ s$ is only bound *after* the choice of τ.

Finally, starting with an empty observation list, the attacker probes the system, collecting observations, until it terminates by guessing correctly. Note that there is no guarantee that this loop terminates, for example if the attacker repeats an incorrect guess again and again. We will detail shortly how this affects the formalisation, as it interacts with our definition of vulnerability.

The full attack is expressed in pGCL as follows:

$attack =$
  $any\ \tau\ ;;$
  $any\ \mathcal{S}\ at\ (\lambda s\ S.\ P[S])\ ;;$
  $\mathcal{O}:= []\ ;;$
  $do\ G \longrightarrow body\ od$

where

$G\ s = (\tau\ s\ (\mathcal{O}\ s) \neq \mathcal{S}\ s)$

We now need a security predicate: a judgement on states, declaring them either secure or insecure. While a predicate in pGCL can take any (non-negative) real value, we begin by cleanly distinguishing secure states. Our security predicate is thus the embedding of a boolean predicate, and therefore takes only the values 0 (for insecure) and 1 (for secure). Under the action of the above program (interpreted as an *expectation transformer*), this 0,1-valued predicate (in the postcondition) is transformed to its *expected value*: the value of the weakest pre-expectation is (a lower bound on) the *probability* that we end in a secure state. The complement is therefore the *probability of compromise*, if the system begins in that state.

---

[4] The alternatives in a probabilistic choice need not sum to 1: the remainder of the probability is assigned to a nondeterministic choice among all possibilities. Thus, any branch is taken with *at least* the specified probability, and perhaps more. In this way, nondeterministic choice is a special case of (sub-)probabilistic choice, where all branches have probability 0.

We consider the system to be secure in its final state if the attacker has guessed correctly: $\tau\ s\ (\mathcal{O}\ s) = \mathcal{S}\ s$, but has taken at least n incorrect guesses to do so: $n < |\mathcal{O}\ s| \wedge (\forall i \leq n.\ \tau\ s\ (tail\ i\ (\mathcal{O}\ s)) \neq \mathcal{S}\ s)$.

The first of these conditions may appear odd, in that we are asserting that the system is only secure if the attacker knows the secret! The technical reason is that this term is the negation of the loop guard, G. The trick, and the reason that we can safely include this conjunct, is that this predicate is only applied to *final* states i.e those where the loop has terminated, which, by definition, only occurs once the guard is invalidated. This term thus only ever appears in a conjuction with the guard, leaving us with a term of the form $\langle\!\langle\ G\ \rangle\!\rangle\ s * \langle\!\langle\ G\ \rangle\!\rangle\ s * x$, which simply collapses to $\langle\!\langle\ G\ \rangle\!\rangle\ s * x$, the security predicate.

If the attacker has not yet guessed correctly, then our definition of security depends not only on this predicate, but also on the interpretation on non-terminating programs, which we will address shortly.

$secure\ n\ s =$
$\langle\!\langle\ \lambda s.\ \tau\ s\ (\mathcal{O}\ s) = \mathcal{S}\ s\ \rangle\!\rangle\ s *$
$\langle\!\langle\ \lambda s.\ n < |\mathcal{O}\ s| \wedge (\forall i \leq n.\ \tau\ s\ (tail\ i\ (\mathcal{O}\ s))) \neq \mathcal{S}\ s)\ \rangle\!\rangle\ s$

Finally, we define vulnerability as the complement of security: the prior vulnerability is the probability that the system will end in an insecure state, given at most (n) guesses. This is, of course, the complement of the probability that the system ends in a *secure* state, thus the definition:

$V_n\ n = 1 - wlp\ attack\ (secure\ n)\ (SOME\ x.\ True)$

The term $SOME\ x.\ True$ is the Hilbert choice on the universal set. This simply expresses our intention that the vulnerability does not depend on the initial state.

Our definition of vulnerability is made in this slightly roundabout fashion ('the complement of the likelihood of success', rather than simply 'the likelihood of failure'), in order to ensure that it is preserved under *refinement*.

It is a well-recognised problem(Morgan, 2006) that many security properties (for example noninterference) are not preserved under refinement: a nondeterministic choice is refined by any of its branches, in particular, a nondeterministic choice of guess, is refined by the program that simply guesses the correct secret on the first try. The same problem does not occur with probabilistic choice in pGCL: it is already maximal in the refinement order. This is clearer on inspecting the destruction rule for refinement:

$$\frac{prog \sqsubseteq prog' \qquad nneg\ P}{wp\ prog\ P\ s \leq wp\ prog'\ P\ s}$$

Here we see that for any non-negative expectation P, and any initial state s, a refinement assigns at least as high a value (or probability, in the case of a boolean post-expectation) to the pre-expectation as the original program did. Thus, by phrasing our postcondition as 'the system is secure', any refinement will only increase the probability of remaining secure, and thus decrease vulnerability. If we instead chose 'the system is compromised', then refinement would work against us: a refinement could have greater vulnerability than its specification.

The final wrinkle is that, while refinement now acts in the 'right' direction, nontermination doesn't. Under the strict semantics (wp), the weakest preexpectation of *any* postcondition in an initial state from which the program diverges is 0—the probability of terminating *and* establishing the postcondition is zero. The solution to this dilemma is to switch to the liberal semantics (wlp). Here we ask for the probability of establishing the postcondition *if* terminating, rather then establishing it *and* terminating, as in the strict case. The only difference between the two is in the treatment of non-terminating programs. In particular the refinement order on terminating programs is unaffected, and thus our property is still preserved by refinement. This choice also matches our intuitive expectation: if the attacker never guesses the secret (either because it repeats failed guesses, or because either party fails or enters an infinite loop and stops responding), the system is secure.

Having established that the model meets our expectations, we turn to the second of the two instances in which cleverness is required: annotating the loop.

2.2   Annotating the Attack Loop

Annotating a probabilistic loop is very similar to annotating a classical loop: Any invariant, combined with the guard, gives an annotation, via the loop rule:

Lemma 1. *If the expectation I is an* invariant *of the loop do $G \longrightarrow$ body od:*

$$\ll G \gg s * I\ s \leq wlp\ body\ I\ s$$

then the following annotation holds for the loop itself:

$$I\ s \leq wlp\ do\ G \longrightarrow body\ od\ (\lambda s.\ \ll \mathcal{N}\ G \gg s * I\ s)\ s \qquad (2)$$

Proof. See McIver and Morgan (2004)

The verification challenge is also the same: to find an invariant that is simultaneously strong enough to imply the desired postcondition, and weak enough to be established by the given precondition. Here, we take a well-known tactic from conventional invariant selection, and modify it to suit a probabilistic setting.

The trick is to split the invariant into a conjunction of two parts: a predicate that represents the intended postcondition, as it would look in the current state, and a separate predicate that 'prunes' the set of future traces. The first half is chosen so that it evaluates to the postcondition (in our case the security predicate) in any terminating state, and the second to only allow traces that preserve the first half. Given that we are manipulating probabilities, and not truth values, we use a product rather than a conjunction. The result is equivalent, as a glance at the truth table will demonstrate[5]. The first 'conjunct' (the portion

---

[5] Note that multiplication is not the only choice that preserves boolean conjunction: the 'probabilistic conjunction' $p\ .\&\ q \equiv max\ 0\ (p + q - 1)$ would also work, and

on the left-hand side of the product) is thus logically equivalent to our security predicate.

The right conjunct is rather different, however. Rather than choosing a predicate that prunes the undesired traces, we instead take the weighted sum over all traces (lists of observations), of the *probability* that the given trace preserves the first conjunct. The weight assigned to a trace is in turn the probability of it occurring which, by the assumption of independence (Equation 1), is simply the product of the conditional probability of each observation, given the secret. We thus construct our invariant as *the sum over all possible futures, weighted by the probability of establishing the postcondition.*

Note that the syntax $\sum l[..n]$. f l refers to the sum over all lists of length n, and R b to the embedding of a boolean value as either 0 or 1:

$$I\ n\ s =$$
$$(\prod i{=}0..min\ n\ |\mathcal{O}\ s|.\ R\ (\tau\ s\ (tail\ i\ (\mathcal{O}\ s)) \neq \mathcal{S}\ s))$$
$$*\ (\sum ol[..n - |\mathcal{O}\ s|\ ].$$
$$\prod i = |\mathcal{O}\ s| + 1\ ..\ n.$$
$$P[((ol\ @\ (\mathcal{O}\ s))\ !\ (n{-}i))|(\mathcal{S}\ s)]$$
$$*\ R\ (\tau\ s\ (tail\ i\ (ol\ @\ (\mathcal{O}\ s)))) \neq \mathcal{S}\ s))$$

The proof that this is indeed an invariant is straightforward (with full details of this and all other proofs available in the accompanying theory source).

Lemma 2. *The expectation I is an invariant of the loop do G $\longrightarrow$ body od i.e.:*

$$\ll G \gg s * I\ n\ s \leq wlp\ body\ (I\ n)\ s$$

Proof. By unfolding. □

Moreover, the combination of the invariant and the guard is precisely the security predicate:

$$\ll \mathcal{N}\ G \gg s * I\ n\ s =$$
$$\ll \lambda s.\ \tau\ s\ (\mathcal{O}\ s) = \mathcal{S}\ s \gg s *$$
$$\ll \lambda s.\ n < |\mathcal{O}\ s| \wedge (\forall i{\leq}n.\ \tau\ s\ (tail\ i\ (\mathcal{O}\ s)) \neq \mathcal{S}\ s) \gg s$$

Thus, applying Lemma 1, we have:

$$I\ n\ s \leq wlp\ do\ G \longrightarrow body\ od\ (secure\ n)\ s \tag{3}$$

---

is used in other parts of the pGCL formalisation. The reason for instead choosing multiplication is that we will later manipulate it algebraically to create a product of probabilities.

2.3 Annotating the Initialisation

The loop initialisation (assigning the empty list of observations) is annotated just as in a classical setting, by instantiating the pre-expectation, and thus:

Lemma 3 (Initialisation).

$\sum ol[..n]. (\prod i = 1..n. \ P[ol_{[n-i]}|S\ s]) *$
$\qquad (\prod i = 0..n. \ R \ (\tau\ s\ (tail\ i\ ol) \neq S\ s))$
$\leq wlp\ (\mathcal{O} := [] \ ;;$
$\qquad\qquad do\ G \longrightarrow body\ od)$
$\qquad (secure\ n)\ s$

Proof. By instantiating Equation 3, we have:

$R\ (\tau\ s\ [] \neq S\ s) *$
$(\sum ol[..n]. \prod i = 1..n. \ P[ol_{[n-i]}|S\ s] * R\ (\tau\ s\ (tail\ i\ ol) \neq S\ s))$
$\leq wlp\ (\mathcal{O} := [] \ ;;$
$\qquad\qquad do\ G \longrightarrow body\ od)$
$\qquad (secure\ n)\ s$

whence we rearrange the pre-expectation by splitting the product and distributing over the summation:

$R\ (\tau\ s\ [] \neq S\ s) *$
$(\sum ol[..n]. \prod i = 1..n. \ P[ol_{[n-i]}|S\ s] * R\ (\tau\ s\ (tail\ i\ ol) \neq S\ s)) =$
$(\sum ol[..n]. (\prod i = 1..n. \ P[ol_{[n-i]}|S\ s]) *$
$\qquad\qquad (\prod i = 0..n. \ R\ (\tau\ s\ (tail\ i\ ol) \neq S\ s)))$

at which point the result follows.   □

Making the choice over secrets, this becomes the weighted sum:

$\sum S. \ P[S] *$
$\qquad (\sum ol[..n]. (\prod i = 1..n. \ P[ol_{[n-i]}|S]) *$
$\qquad\qquad (\prod i = 0..n. \ R\ (\tau\ s\ (tail\ i\ ol) \neq S)))$

We manipulate this expression to produce a sum of probabilities, by first defining the *guess set* (Γ) of a strategy—the set of all guesses produced by terminal sublists of the given list of observations:

$\sigma_{tr}\ \sigma\ [] = [([], \sigma\ [])] \ |$
$\sigma_{tr}\ \sigma\ (o' \cdot os) = (o' \cdot os, \sigma\ (o' \cdot os)) \cdot \sigma_{tr}\ \sigma\ os$
$\Gamma\ \sigma\ ol = snd\ `\ set\ (\sigma_{tr}\ \sigma\ ol)$

We can now annotate the entire attack, including the nondeterministic choice of strategy:

Lemma 4 (The Attack). *We have:*

$(\bigcap x.\ 1 - (\sum ol[..n].\ \sum S{\in}\Gamma\ x\ ol.\ P[ol,S])) \leq wlp\ attack\ (secure\ n)\ s$

Proof. We begin by noting that the pre-expectation of the choice over secrets:

$\sum S.\ P[S]\ *$
$\quad(\sum ol[..n].\ (\prod i = 1..n.\ P[ol_{[n - i]}|S])\ *$
$\qquad\qquad(\prod i = 0..n.\ R\ (\sigma\ (tail\ i\ ol) \neq S)))$

can be rewritten (by changing the order of summation, and distributing multiplication) to:

$\sum ol[..n].\ \sum S.\ P[S]\ *\ (\prod i = 1..n.\ P[ol_{[n - i]}|S])\ *$
$\qquad(\prod i = 0..n.\ R\ (\sigma\ (tail\ i\ ol) \neq S))$

We then note that the innermost product is simply the joint probability of the secret and the list of observations, and thus we have:

$\sum ol[..n].\ \sum S.\ P[ol,S]\ *\ (\prod i = 0..n.\ R\ (\sigma\ (tail\ i\ ol) \neq S))$

Finally, we note that by the definition of Γ, if $|ol| = n$ then

$((\prod i = 0..n.\ R\ (\sigma\ (tail\ i\ ol) \neq S)) = 0) = (S \in \Gamma\ \sigma\ ol)$

and thus we have

$1 - (\sum ol[..n].\ \sum S{\in}\Gamma\ \sigma\ ol.\ P[ol,S])$

as the reworked precondition for the choice over secrets. The result then follows from the definition of nondeterministic choice as the infimum *over the branches.*

□

2.4   The Top-Level Theorem

Our ultimate result then follows:

Theorem 1. *The vulnerability of the system is bounded above by a sum over the joint probability of the full list of observations, and the set of the attacker's guesses for all initial sublists of observations (intermediate states), maximised over possible choices of strategy.*

$V_n\ n \leq (\bigsqcup x.\ \sum ol[..n].\ \sum S{\in}\Gamma\ x\ ol.\ P[ol,S])$

Proof. Follows from Lemma 4, the definition of V_n and the algebraic properties of the infimum.

□

From this theorem, a number of facts are immediately apparent. First, as every term in the sum is non-negative, vulnerability is (weakly) monotonic in the size of the guess set. Thus (assuming that there are enough distinct secrets for it to be possible), a strategy that repeats can be extended into a *non-repeating* strategy (by replacing repeated guesses with fresh secrets), with at least as high a vulnerability. Thus we need only consider non-repeating strategies in calculating the supremum.

Secondly, and more importantly, for any strategy, vulnerability is a sum of the joint probability of a list of observations and a guess. It is thus obvious that a strategy that maximises this, also maximises vulnerability. This is, of course, the MAP strategy that we introduced in Section 1.2. In fact, the above prohibition on repeating strategies can be rephrased in terms of distributions: the optimal strategy maximises the joint probability over the distribution where the probability of any secret that has already been guessed is set to zero (and the rest scaled appropriately).

We have now come full circle, eliminating our operational model entirely, and finishing with a formulation solely in terms of the statistical properties of the system that, as we see, matches our expectations. Importantly, we didn't *assume* anything about the optimality of the non-repeating MAP strategy: it fell out as a *consequence* of the model.

3   Using the Result

We conclude with a sketch to illustrate how this result fits into the overall theory, and provides the connection between the operational and information-theoretic models.

As indicated at the end of Section 2, the pen-and-paper proof picks up at Theorem 1, establishing the optimality of the MAP strategy. Given this, we are able to establish a number of results extant in the literature. For example V_1, or the chance of compromise in a single guess, is simply the probability of the most likely secret:

$$V_1 = \max_S P[[\,], S]$$
$$= \max_S P[S]$$

This is closely related to the *min-entropy* measure (H_∞), which has recently begun to supplant Shannon entropy in the formal analysis of vulnerability to information leakage(Smith, 2009). Specifically:

$$H_\infty(P) = -\log_2(\max_S P[S]) = -\log_2 V_1$$

From this, we synthesise a *leakage measure*, or a bound on how vulnerability changes over time, by analogy with the min-leakage: $L_\infty = H_\infty(P_1) - H_\infty(P_2)$. We estimate the rate of leakage by the ratio of the vulnerability given 2 guesses, to that given only 1. This multiplicative measure becomes additive, once we rephrase it in terms of entropy:

$$L = \frac{V_2}{V_1}$$

$$\log_2 L = \log_2 \frac{V_2}{V_1}$$

$$= \log_2 V_2 - \log_2 V_1$$

$$= H_\infty(P_1) - H_\infty(P_2)$$

Ultimately, we extend the model to allow nondeterministic choice not only over strategies, but also over the distributions $P[S]$ and $P[o|S]$ themselves, taken from two sets, Q_S and Q_{oS} respectively. Thanks to our definition of vulnerability, this is equivalent to asking for the greatest (strictly, supremum) vulnerability over all distributions in Q_S and Q_{oS}.

A particularly interesting case occurs when Q_S is the set of all distributions of *Shannon* entropy H_1, and Q_{oS} the set of all conditional distributions with *channel capacity* C. We thus express the worst case vulnerability, given that we only know the Shannon entropy of the distribution (this is not generally straightforward to calculate, see Smith (2009)). Moreover, on average (assuming that the space of secrets is large enough that the effect of the individual yes/no answers is small), the entropy remaining after a single guess is just $H - C$. We thus iterate the model, recalculating V_n, this time setting Q_S to the set of all distributions of entropy $H - C$.

Finally, consider what happens to the vulnerability as we vary Q_S. As the size of the set increases, the supremum is taken over more possibilities, and thus we should expect the vulnerability to increase. Likewise, smaller sets should give lower vulnerability. This is indeed precisely what we see as, from the semantics of nondeterminism in pGCL, the choice over some $S \subseteq T$ is a refinement of the choice over T. Recall that a refinement assigns a *higher* value to every state than the original program and thus, as it appears negated in our definition of vulnerability, the choice over a smaller set gives a *lower* vulnerability, and vice versa.

We thus see that the order on the vulnerability bounds is determined by the subset order on the set of distributions. In particular, this gives us a complete lattice of bounds, since $V_1 \emptyset = 0$ (by definition) and $V_1 \top = 1$ (this includes every distribution that assigns 100% probability to a single secret, which the attacker will then guess with certainty). We thus link the refinement order on our operational models, with a (dual) order on this complete lattice of bounds. For the full derivation see Cock (2014).

4   Related Work

This work draws on our own previous work on the mechanisation of probabilistic reasoning, together with results in both the programming-language semantics and the security literature.

We build on our own previous work in mechanising(Cock, 2012) the pGCL semantics of McIver and Morgan (2004), and in demonstrating the feasibility of carrying probabilistic results down to real systems by incorporating large existing proof results(Cock, 2013; Klein et al., 2009). We extend this by demonstrating that we can take the approach one step further: into the domain of information theory.

The concept of entropy, and the solution to the optimal decoding problem both date to the earliest years of information theory(Shannon, 1948), while the shift to *min*-entropy in the security domain largely follows the insight of Smith (2009) that the single-guess vulnerability of a distribution is only loosely connected to its Shannon entropy. The further generalisation of this idea is a subject of active research(Espinoza and Smith, 2013; Alvim et al., 2012; McIver et al., 2010). Our own previous work include the rigorous evaluation of the *guessing attack* as a threat model(Cock, 2014).

Acknowledgements. NICTA is funded by the Australian Government through the Department of Communications and the Australian Research Council through the ICT Centre of Excellence Program.

References

Alvim, M.S., Chatzikokolakis, K., Palamidessi, C., Smith, G.: Measuring information leakage using generalized gain functions. In: 25th CSF, pp. 265–279. IEEE (2012), doi:10.1109/CSF.2012.26

Cock, D.: Verifying probabilistic correctness in Isabelle with pGCL. In: 7th SSV, Sydney, Australia, pp. 1–10 (November 2012), doi:10.4204/EPTCS.102.15

Cock, D.: Practical probability: Applying pGCL to lattice scheduling. In: Blazy, S., Paulin-Mohring, C., Pichardie, D. (eds.) ITP 2013. LNCS, vol. 7998, pp. 311–327. Springer, Heidelberg (2013)

Cock, D.: Leakage in Trustworthy Systems. PhD thesis, School Comp. Sci. & Engin., Sydney, Australia (2014)

Dijkstra, E.W.: Guarded commands, nondeterminacy and formal derivation of programs. CACM 18(8), 453–457 (1975), doi:10.1145/360933.360975, ISSN 0001-0782

Espinoza, B., Smith, G.: Min-entropy as a resource. Inform. & Comput. 226, 57–75 (2013), doi:10.1016/j.ic.2013.03.005, ISSN 0890-5401.

Klein, G., Elphinstone, K., Heiser, G., Andronick, J., Cock, D., Derrin, P., Elkaduwe, D., Engelhardt, K., Kolanski, R., Norrish, M., Sewell, T., Tuch, H., Winwood, S.: seL4: Formal verification of an OS kernel. In: SOSP, Big Sky, MT, USA, pp. 207–220. ACM (October 2009), doi:10.1145/1629575.1629596

McIver, A., Morgan, C.: Abstraction, Refinement and Proof for Probabilistic Systems. Springer (2004), doi:10.1007/b138392, ISBN 978-0-387-40115-7

McIver, A., Meinicke, L., Morgan, C.: Compositional closure for bayes risk in probabilistic noninterference. In: Abramsky, S., Gavoille, C., Kirchner, C., Meyer auf der Heide, F., Spirakis, P.G. (eds.) ICALP 2010. LNCS, vol. 6199, pp. 223–235. Springer, Heidelberg (2010)

Morgan, C.: The shadow knows: Refinement of ignorance in sequential programs. In: Uustalu, T. (ed.) MPC 2006. LNCS, vol. 4014, pp. 359–378. Springer, Heidelberg (2006)

Shannon, C.E.: A mathematical theory of communication. In: The Bell Syst. Techn. J. (1948), doi:10.1145/584091.584093, Reprinted in SIGMOBILE Mobile Computing and Communications Review 5(1), 3–55 (2001)

Smith, G.: On the foundations of quantitative information flow. In: de Alfaro, L. (ed.) FOSSACS 2009. LNCS, vol. 5504, pp. 288–302. Springer, Heidelberg (2009)

A Coq Formalization
of Finitely Presented Modules

Cyril Cohen and Anders Mörtberg

Department of Computer Science and Engineering
Chalmers University of Technology and University of Gothenburg, Sweden
{cyril.cohen,anders.mortberg}@cse.gu.se

Abstract. This paper presents a formalization of constructive module theory in the intuitionistic type theory of Coq. We build an abstraction layer on top of matrix encodings, in order to represent finitely presented modules, and obtain clean definitions with short proofs justifying that it forms an abelian category. The goal is to use it as a first step to get certified programs for computing topological invariants, like homology groups and Betti numbers.

Keywords: Formalization of mathematics, Homological algebra, Constructive algebra, Coq, SSReflect.

1 Introduction

Homological algebra is the study of linear algebra over rings instead of fields, this means that one considers modules instead of vector spaces. Homological techniques are ubiquitous in many branches of mathematics like algebraic topology, algebraic geometry and number theory. Homology was originally introduced by Henri Poincaré in order to compute topological invariants of spaces [23], which provides means for testing whether two spaces cannot be continuously deformed into one another. This paper presents a formalization[1] of constructive module theory in type theory, using the Coq proof assistant [7] together with the Small Scale Reflection (SSREFLECT) extension [12], which provides a potential core of a library of certified homological algebra.

A previous work, that one of the authors was involved in, studied ways to compute homology groups of vector spaces [18,17] in Coq. When generalizing this to commutative rings the universal coefficient theorem of homology [15] states that most of the homological information of an R-module over a ring R can be computed by only doing computations with elements in $\mathbb{Z}$. This means that if we were only interested in computing homology it would not really be necessary to develop the theory of R-modules in general, but instead do it for $\mathbb{Z}$-modules which are well behaved because any matrix can be put in Smith normal form. However, by developing the theory for general rings it should be possible

---

[1] The formal development is at: http://perso.crans.org/cohen/work/fpmods/

G. Klein and R. Gamboa (Eds.): ITP 2014, LNAI 8558, pp. 193–208, 2014.
© Springer International Publishing Switzerland 2014

to implement and reason about other functors like cohomology, Ext and Tor as in the HOMALG computer algebra package [3].

In [13], Georges Gonthier shows that the theory of finite dimensional vector spaces can be elegantly implemented in CoQ by using matrices to represent subspaces and morphisms, as opposed to an axiomatic approach. The reason why abstract finite dimensional linear algebra can be concretely represented by matrices is because any vector space has a basis (a finite set of generators with no relations among the generators) and any morphism can be represented by a matrix in this canonical basis. However, for modules over rings this is no longer true: consider the ideal (X, Y) of $k[X, Y]$, it is a module generated by X and Y which is not free because $XY = YX$. This means that the matrix-based approach cannot be directly applied when formalizing module theory. This is why we restrict our attention to finitely generated modules that are *finitely presented*, that is, modules with a finite number of generators and a finite number of relations among these. In constructive module theory one usually restricts attention to this class of modules and all algorithms can be described by manipulating the presentation matrices [10,14,20,22]. This paper can hence be seen as a generalization of the formalization of Gonthier to modules over rings instead over fields.

At the heart of the formalization of Gonthier is an implementation of Gaussian elimination which is used in all subspace constructions. Using it we can compute the kernel which characterizes the space of solutions of a system of linear equations. However when doing module theory over arbitrary rings, there is no general algorithm for solving systems of linear equations. Because of this we restrict our attention to modules over *coherent* and *strongly discrete* rings, as is customary in constructive algebra [20,22], which means that we can solve systems of equations.

The main contributions of this paper are the representation of finitely presented modules over coherent strongly discrete rings (Sect. 2), basic operations on these modules (Sect. 3) and the formalization that the collection of these modules and morphisms forms an abelian category (Sect. 4), which means that it is a suitable setting for developing homological algebra. We have also proved that, over elementary divisor rings (*i.e.* rings with an algorithm to compute the Smith normal form of matrices), it is possible to test if two finitely presented modules represent isomorphic modules (Sect. 5). (Examples of such rings include principal ideal domains, in particular $\mathbb{Z}$ and $k[X]$ where k is a field).

2   Finitely Presented Modules

As mentioned in the introduction, a module is finitely presented if it can be given by a finite set of generators and relations. This is traditionally expressed as:

Definition 1. *An R-module $\mathcal{M}$ is **finitely presented** if there is an exact sequence:*

$$R^{m_1} \xrightarrow{M} R^{m_0} \xrightarrow{\pi} \mathcal{M} \longrightarrow 0$$

Recall that R^m is the type of m-tuples of elements in R. More precisely, π is a surjection and M a matrix representing the m_1 relations among the m_0 generators of the module $\mathcal{M}$. This means that $\mathcal{M}$ is the cokernel of M:

$$\mathcal{M} \simeq \mathrm{coker}(M) = R^{m_0}/im(M)$$

Hence a module has a finite presentation if it can be expressed as the cokernel of a matrix. As all information about a finitely presented module is contained in its presentation matrix we will omit the surjection π when giving presentations of modules.

Example 1. The $\mathbb{Z}$-module $\mathbb{Z} \oplus \mathbb{Z}/2\mathbb{Z}$ is given by the presentation:

$$\mathbb{Z} \xrightarrow{\begin{pmatrix} 0 & 2 \end{pmatrix}} \mathbb{Z}^2 \longrightarrow \mathbb{Z} \oplus \mathbb{Z}/2\mathbb{Z} \longrightarrow 0$$

because if $\mathbb{Z} \oplus \mathbb{Z}/2\mathbb{Z}$ is generated by (e_1, e_2) there is one relation, namely $0e_1 + 2e_2 = 2e_2 = 0$.

Operations on finitely presented modules can now be implemented by manipulating the presentation matrices, for instance if $\mathcal{M}$ and $\mathcal{N}$ are finitely presented R-modules given by presentations:

$$R^{m_1} \xrightarrow{M} R^{m_0} \longrightarrow \mathcal{M} \longrightarrow 0 \qquad R^{n_1} \xrightarrow{N} R^{n_0} \longrightarrow \mathcal{N} \longrightarrow 0$$

the presentation of $\mathcal{M} \oplus \mathcal{N}$ is:

$$R^{m_1+n_1} \xrightarrow{\begin{pmatrix} M & 0 \\ 0 & N \end{pmatrix}} R^{m_0+n_0} \longrightarrow \mathcal{M} \oplus \mathcal{N} \longrightarrow 0$$

We have represented finitely presented modules in Coq using the datastructure of matrices from the SSREFLECT library which is defined as:

```
Inductive matrix R m n := Matrix of {ffun 'I_m * 'I_n -> R}.
```

```
(* With notations: *)
(* 'M[R]_(m,n) = matrix R m n *)
(* 'rV[R]_m = 'M[R]_(1,m) *)
(* 'cV[R]_m = 'M[R]_(m,1) *)
```

where 'I_m is the type ordinal m which represents all natural numbers smaller than m. This type has exactly m inhabitants and can be coerced to the type of natural numbers, nat. Matrices are then represented as finite functions over the finite set of indices, which means that dependent types are used to express well-formedness. Finitely presented modules are now conveniently represented using a record containing a matrix and its dimension:

```
Record fpmodule := FPModule {
  nbrel : nat;
  nbgen : nat;
  pres : 'M[R]_(nbrel, nbgen)
}.
```

The direct sum of two finitely presented modules is now straightforward to implement:

```
Definition dsum (M N : fpmodule R) :=
  FPModule (block_mx (pres M) 0 0 (pres N)).
```

Here `block_mx` forms the block matrix consisting of the four submatrices. We now turn our attention to morphisms of finitely presented modules.

2.1   Morphisms

As for vector spaces we represent morphisms of finitely presented modules using matrices. The following lemma states how this can be done:

Lemma 1. *If $\mathcal{M}$ and $\mathcal{N}$ are finitely presented R-modules given by presentations:*

$$R^{m_1} \xrightarrow{M} R^{m_0} \longrightarrow \mathcal{M} \longrightarrow 0 \qquad R^{n_1} \xrightarrow{N} R^{n_0} \longrightarrow \mathcal{N} \longrightarrow 0$$

and $\varphi : \mathcal{M} \to \mathcal{N}$ a module morphism then there is a $m_0 \times n_0$ matrix φ_G and a $m_1 \times n_1$ matrix φ_R such that the following diagram commutes:

$$
\begin{array}{ccccccc}
R^{m_1} & \xrightarrow{\ M\ } & R^{m_0} & \longrightarrow & \mathcal{M} & \longrightarrow & 0 \\
\downarrow{\scriptstyle \varphi_R} & & \downarrow{\scriptstyle \varphi_G} & & \downarrow{\scriptstyle \varphi} & & \\
R^{n_1} & \xrightarrow{\ N\ } & R^{n_0} & \longrightarrow & \mathcal{N} & \longrightarrow & 0
\end{array}
$$

For a proof of this see Lemma 2.1.25 in [14]. This means that morphisms between finitely presented modules can be represented by pairs of matrices. The intuition why two matrices are needed is that the morphism affects both the generators and relations of the modules, hence the names φ_G and φ_R.

In order to be able to compute for example the kernel of a morphism of finitely presented modules we need to add some constraints on the ring R since, in general, there is no algorithm for solving systems of equations over arbitrary rings. The class of rings we consider are *coherent* and *strongly discrete* which means that it is possible to solve systems of equations. In HOMALG these are called *computable rings* [2] and form the basis of the system.

2.2   Coherent and Strongly Discrete Rings

Given a ring R (in our setting commutative but it is possible to consider non-commutative rings as well [2]) we want to study the problem of solving linear systems over R. If R is a field we have a nice description of the space of solutions by a basis of solutions. Over an arbitrary ring R there is in general no basis. For instance over the ring $R = k[X, Y, Z]$ where k is a field, the equation $pX + qY + rZ = 0$ has no basis of solutions. It can be shown that a generating system of solutions is given by $(-Y, X, 0)$, $(Z, 0, -X)$, $(0, -Z, Y)$. An important weaker property than having a basis is that there is a finite number of solutions which generate all solutions.

Definition 2. *A ring is **(left) coherent** if for any matrix M it is possible to compute a matrix L such that:*

$$XM = 0 \leftrightarrow \exists Y. X = YL$$

This means that L generates the module of solutions of $XM = 0$, hence L is the kernel of M. For this it is enough to consider the case where M has only one column [20]. Note that the notion of coherent rings is not stressed in classical presentations of algebra since Noetherian rings are automatically coherent, but in a computationally meaningless way. It is however a fundamental notion, both conceptually [20,22] and computationally [3].

Coherent rings have previously been represented in Coq [8], the only difference is that in the previous presentation, composition was read from right to left, whereas here we adopt the SSREFLECT convention that composition is read in diagrammatic order (*i.e.* left to right).

In the development, coherent rings have been implemented using the design pattern of [11], using packed classes and the canonical structure mechanism to help Coq automatically infer structures. As matrices are represented using dependent types denoting their size this needs to be known when defining coherent rings. In general the size of L cannot be predicted, so we include an extra function to compute this:

```
Record mixin_of (R : ringType) : Type := Mixin {
  dim_ker : forall m n, 'M[R]_(m,n) -> nat;
  ker : forall m n (M : 'M_(m,n)), 'M_(dim_ker M,m);
  _ : forall m n (M : 'M_(m,n)) (X : 'rV_m),
    reflect (exists Y, X = Y *m ker M) (X *m M == 0)
}.
```

Here *m denotes matrix multiplication and == is the boolean equality of matrices, so the specification says that this equality is equivalent to the existence statement. An alternative to having a function computing the size would be to output a dependent pair but this has the undesirable behavior that the pair has to be destructed when stating lemmas about it, which in turn would make these lemmas cumbersome to use as it would not be possible to rewrite with them directly.

An algorithm that can be implemented using `ker` is the kernel modulo a set of relations, that is, computing $\ker(R^m \xrightarrow{M} \operatorname{coker}(N))$. This is equivalent to computing an X such that $\exists Y, XM + YN = 0$, which is the same as solving $(X\ Y)(M\ N)^T = 0$ and returning the part of the solution that corresponds to XM. In the paper this is written as $\ker_N(M)$ and in the formalization as N.-ker(M). Note that this is a more fundamental operation than taking the kernel of a matrix as $XM = 0$ is also equivalent to $\exists Y, X = Y \ker_0(M)$

In order to conveniently represent morphisms we also need to be able to solve systems of the kind $XM = B$ where B is not zero. In order to do this we need to introduce another class of rings that is important in constructive algebra:

Definition 3. *A ring R is **strongly discrete** if membership in finitely generated ideals is decidable and if $x \in (a_1, \ldots, a_n)$ there is an algorithm computing $w_1, \ldots, w_n$ such that $x = \sum_i a_i w_i$.*

Examples of such rings are multivariate polynomial rings over fields with decidable equality (via Gröbner bases) and Bézout domains (for instance $\mathbb{Z}$ and $k[X]$ with k a field).

If a ring is both coherent and strongly discrete it is not only possible to solve homogeneous systems $XM = 0$ but also any system $XM = B$ where B is an arbitrary matrix with the same number of columns as M. This operation can be seen as division of matrices as:

```
Lemma dvdmxP m n k (M : 'M[R]_(n,k)) (B : 'M[R]_(m,k)) :
  reflect (exists X, X *m M = B) (M %| B).
```

Here %| is notation for the function computing a particular solution to $XM = B$, returning None in the case no solution exists. We have developed a library of divisibility of matrices with lemmas like

```
Lemma dvdmxD m n k (M : 'M[R]_(m,n)) (N K : 'M[R]_(k,n)) :
  M %| N -> M %| K -> M %| N + K.
```

which follow directly from dvdmxP. This can now be used to represent morphisms of finitely presented modules and the division theory of matrices gives short and elegant proofs about operations on morphisms.

2.3 Finitely Presented Modules Over Coherent Strongly Discrete Rings

Morphisms between finitely presented R-modules $\mathcal{M}$ and $\mathcal{N}$ can be represented by a pair of matrices. However when R is coherent and strongly discrete it suffices to only consider the φ_G matrix as φ_R can be computed by solving $XN = M\varphi_G$, which is the same as testing $N \mid M\varphi_G$. In CoQ this means that morphisms between two finitely presented modules can be implemented as:

```
Record morphism_of (M N : fpmodule R) := Morphism {
  matrix_of_morphism : 'M[R]_(nbgen M,nbgen N);
  _ : pres N %| pres M *m matrix_of_morphism
}.

(* With notation: *)
(* 'Mor(M,N) := morphism_of M N *)
```

Using this representation we can define the identity morphism (`idm`) and composition of morphisms (`phi ** psi`) and show that these form a category. We also define the zero morphism (`0`) between two finitely presented modules, the sum (`phi + psi`) of two morphisms and the negation (`- phi`) of a morphism, respectively given by the zero matrix, the sum and the negation of the underlying matrices. It is straightforward to prove using the divisibility theory of matrices that this is a pre-additive category (i.e. the hom-sets form abelian groups).

However, morphisms are not uniquely represented by an element of type `'Mor(M,N)`, but it is possible to test if two morphisms φ $\psi : M \to N$ are equal by checking if $\varphi - \psi$ is zero modulo the relations of N.

```
Definition eqmor (M N : fpmodule R) (phi psi : 'Mor(M,N)) :=
  pres N %| phi%:m - psi%:m.
```

```
(* With notation: *)
(* phi %= psi = eqmor phi psi *)
```

As this is an equivalence relation it would be natural to either use the Coq setoid mechanism [4,24] or quotients [6] in order to avoid applying symmetry, transitivity and compatibility with operators (e.g. addition and multiplication) by hand where it would be more natural to use rewriting. We have begun to rewrite the library with quotients as we would get a set of morphisms (instead of a setoid), which is closer to the standard category theoretic notion.

3   Monomorphisms, Epimorphisms and Operations on Morphisms

A monomorphism is a morphism $\varphi : B \to C$ such that whenever there are ψ_1, $\psi_2 : A \to B$ with $\psi_1\varphi = \psi_2\varphi$ then $\psi_1 = \psi_2$. When working in pre-additive categories the condition can be simplified to, whenever $\psi\varphi = 0$ then $\psi = 0$.

```
Definition is_mono (M N : fpmodule R) (phi : 'Mor(M,N)) :=
  forall (P : fpmodule R) (psi : 'Mor(P, M)),
    psi ** phi %= 0 -> psi %= 0.
```

It is convenient to think of monomorphisms $B \to C$ as defining B as a subobject of C, so a monomorphism $\varphi : M \to N$ can be thought of as a representation of a submodule M of N. However, submodules are not uniquely represented by monomorphisms even up to equality of morphisms (`%=`). Indeed, multiple monomorphisms with different sources can represent the same submodule. Although "representing the same submodule" is decidable in our theory, we chose not to introduce the notion of submodule, because it is not necessary to develop the theory.

Intuitively monomorphisms correspond to injective morphisms (*i.e.* with zero kernel). The dual notion to monomorphism is epimorphism, which intuitively corresponds to surjective morphism (*i.e.* with zero cokernel). For finitely presented modules, mono- (resp. epi-) morphisms coincide with injective (resp. surjective) morphisms, but this is not clear *a priori*. The goal of this section is to clarify this by defining when a finitely presented module is zero, showing how to define kernels and cokernels, and explicit the correspondence between injective (resp. surjective) morphisms and mono- (resp. epi-) morphisms.

3.1   Testing if Finitely Presented Modules Are Zero

As a finitely presented module is the cokernel of a presentation matrix we have that if the presentation matrix of a module is the identity matrix of dimension $n \times n$ the module is isomorphic to n copies of the zero module. Now consider the following diagram:

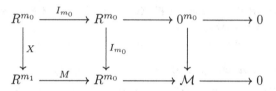

which commutes if $\exists X, XM = I_{m_0}$, *i.e.* when $M \mid I_{m_0}$. Hence this gives a condition that can be tested in order to see if a module is zero or not.

3.2   Defining the Kernel of a Morphism

In order to compute the kernel of a morphism the key observation is that there is a commutative diagram:

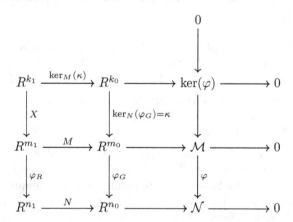

It is easy to see that κ is a monomorphism, which means that the kernel is a submodule of $\mathcal{M}$ as expected. In CoQ this is easy to define:

```
Definition kernel (M N : fpmodule R) (phi : 'Mor(M,N)) :=
  mor_of_mx ((pres N).-ker phi).
```

Where `mor_of_mx` takes a matrix K with as many columns as N and builds a morphism from $ker_N(K)$ to M. Using this it is possible to test if a morphism is injective:

```
Definition injm (M N : fpmodule R) (phi : 'Mor(M,N)) :=
  kernel phi %= 0.
```

We have proved that a morphism is injective if and only if it is a monomorphism:

```
Lemma monoP (M N : fpmodule R) (phi : 'Mor(M,N)) :
  reflect (is_mono phi) (injm phi).
```

Hence we can define monomorphisms as:

```
Record monomorphism_of (M N : fpmodule R) := Monomorphism {
  morphism_of_mono :> 'Mor(M, N);
  _ : injm morphism_of_mono
}.
```

```
(* With notation: *)
(* 'Mono(M,N) = monomorphism_of M N *)
```

The reason why we use injm instead of is_mono is that injm is a boolean predicate, which makes monomorphisms a subtype of morphisms, thanks to Hedberg's theorem [16].

3.3   Defining the Cokernel of a Morphism

The presentation of the cokernel of a morphism can also be found by looking at a commutative diagram:

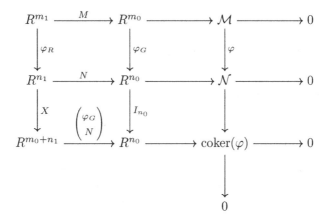

Note that the canonical surjection onto the cokernel is given by the identity matrix. The fact that this is a morphism is clear as X may be $\begin{pmatrix} 0 & I_{n_1} \end{pmatrix}$. However, before defining this we can define the more general operation of quotienting a module by the image of a morphism by stacking matrices:

```
Definition quot_by (M N : fpmodule R) (phi : 'Mor(M, N)) :=
  FPModule (col_mx (pres N) phi)
```

Now the cokernel is the canonical surjection from N onto `quot_by` phi. Since it maps each generator to itself, the underlying matrix is the identity matrix.

```
Definition coker : 'Mor(N, quot_by) :=
  Morphism1 (dvd_quot_mx (dvdmx_refl _)).
```

We can now test if a morphism is surjective by comparing the cokernel of phi with the zero morphism, which coincides with epimorphisms:

```
Definition surjm (M N : fpmodule R) (phi : 'Mor(M,N)) :=
  coker phi %= 0.
```

```
Lemma epiP (M N : fpmodule R) (phi : 'Mor(M,N)) :
  reflect (is_epi phi) (surjm phi).
```

Now we have algorithms deciding both if a morphism is injective and surjective we can easily test if it is an isomorphism:

```
Definition isom (M N : fpmodule R) (phi : 'Mor(M,N)) :=
  injm phi && surjm phi.
```

A natural question to ask is if we get an inverse from this notion of isomorphism. In order to show this we have introduced the notion of `isomorphisms` that take two morphisms and express that they are mutual inverse of each other, in the sense that given $\varphi : M \to N$ and $\psi : N \to M$ then $\varphi\psi = 1_M$ modulo the relations in M. Using this we have proved:

```
Lemma isoP (M N : fpmodule R) (phi : 'Mor(M,N)) :
  reflect (exists psi, isomorphisms phi psi) (isom phi).
```

Hence isomorphisms are precisely the morphisms that are both mono and epi. Note that this does not mean that we can decide if two modules are isomorphic, what we can do is testing if a given morphism is an isomorphism or not.

3.4   Defining Homology

The homology at $\mathcal{N}$ is defined as the quotient $\ker(\psi)/\text{im}(\varphi)$, in

$$\mathcal{M} \xrightarrow{\ \varphi\ } \mathcal{N} \xrightarrow{\ \psi\ } \mathcal{K} \qquad \text{where } \varphi\psi = 0.$$

As $\varphi\psi = 0$, we have that $\text{im}(\varphi) \subset \ker(\psi)$ so the quotient makes sense and we have an injective map $\iota : \text{im}(\varphi) \to \ker(\psi)$. The homology at $\mathcal{N}$ is the cokernel of this map. We can hence write:

```
Hypothesis mul_phi_psi (M N K : fpmodule R) (phi : 'Mor(M,N))
  (psi : 'Mor(N,K)) : phi ** psi %= 0.
```

```
Definition homology (M N K : fpmodule R) (phi : 'Mor(M,N))
  (psi : 'Mor(N,K)) := kernel psi %/ phi.
```

Where %/ is a notation for taking the quotient of a monomorphism by a morphism with the same target.

In the next section, we show that these operations satisfy the axioms of abelian categories.

4 Abelian Categories

As mentioned in the end of Sect. 2 the collection of morphisms between two finitely presented modules forms an abelian group. This means that the category of finitely presented modules and their morphisms is a **pre-additive category**. It is easy to show that the dsum construction provides both a product and coproduct. This means that the category is also **additive**.

In order to show that we have a **pre-abelian** category we need to show that morphisms have both a kernel and cokernel in the sense of category theory. A morphism $\varphi : A \to B$ has a kernel $\kappa : K \to A$ if $\kappa\varphi = 0$ and for all $\psi : Z \to A$ with $\psi\varphi = 0$ the following diagram commutes:

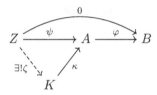

This means that any morphism with $\psi\varphi = 0$ factors uniquely through the kernel κ. The dual statement for cokernels state that any morphism ψ with $\varphi\psi = 0$ factors uniquely through the cokernel of φ. The specification of the kernel can be written.

```
Definition is_kernel (M N K : fpmodule R) (phi : 'Mor(M,N))
  (k : 'Mor(K,M)) :=
  (k ** phi %= 0) *
  forall L (psi : 'Mor(L,M)),
    reflect (exists Y, Y ** k %= psi) (psi ** phi %= 0).
```

We have proved that our definition of kernel satisfies this specification:

```
Lemma kernelP (M N : fpmodule R) (phi : 'Mor(M,N)) :
  is_kernel phi (kernel phi).
```

We have also proved the dual statement for cokernels. The only properties left in order to have an **abelian category** is that every mono- (resp. epi-) morphism is normal which means that it is the kernel (resp. cokernel) of some morphism. We have shown that if φ is a monomorphism then its cokernel satisfies the specification of kernels:

```
Lemma mono_ker (M N : fpmodule R) (phi : 'Mono(M,N)) :
  is_kernel (coker phi) phi.
```

This means that φ is a kernel of $coker(\varphi)$ if φ is a monomorphism, hence are all monomorphisms normal. We have also proved the dual statement for epimorphisms which means that we indeed have an abelian category.

It is interesting to note that many presentations of abelian categories say that phi is `kernel(coker phi)`, but this is not even well-typed as:

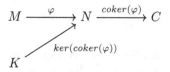

One cannot just subtract φ and ker(coker(φ)) as they have different sources. This abuse of language is motivated by the fact that kernels are limits which are unique up to unique isomorphism which is why many authors speak of *the* kernel of a morphism. However, in order to express this formally we need to exhibit the isomorphism between M and K explicitly and insert it in the equation.

Note that if we introduced a notion of submodule, we could have defined the kernel as a unique submodule of N. The reason is that the type of submodules of N would be the quotient of monomorphisms into N by the equivalence relation which identifies them up to isomorphism.

5 Smith Normal Form

As mentioned before, it is in general not possible to decide if two presentations represent isomorphic modules, even when working over coherent strongly discrete rings. When the underlying ring is a field it is possible to represent a finite dimensional vector space in a canonical way as they are determined up to isomorphism by their dimension (*i.e.* the rank of the underlying matrix) which can be computed by Gaussian elimination [13]. A generalization of this is the class of rings, called *elementary divisor rings* by Kaplansky [19], where any matrix is equivalent to a matrix in Smith normal form. Recall that a matrix M is *equivalent* to a matrix D if there exist invertible matrices P and Q such that $PMQ = D$.

Definition 4. *A matrix is in* Smith normal form *if it is a diagonal matrix of the form:*

$$\begin{pmatrix} d_1 & & 0 & \cdots & \cdots & 0 \\ & \ddots & & & & \vdots \\ 0 & & d_k & 0 & \cdots & 0 \\ \vdots & & 0 & 0 & & \vdots \\ \vdots & & \vdots & & \ddots & \vdots \\ 0 & \cdots & 0 & \cdots & \cdots & 0 \end{pmatrix}$$

where $d_i \mid d_{i+1}$ *for* $1 \leqslant i < k$.

The connection between elementary divisor rings and finitely presented modules is that the existence of a Smith normal form for the presentation matrix gives us:

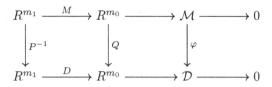

Now φ is an isomorphism as P and Q are invertible. In order to represent this in COQ we need to represent diagonal matrices. For this we use the function `diag_mx_seq`. It is a function that takes two numbers m and n, a list s and returns a matrix of type `'M[R]_(m,n)` where the elements of the diagonal are the elements of s. It is defined as follows:

```
Definition diag_mx_seq m n (s : seq R) :=
  \matrix_(i < m, j < n) (s'_i *+ (i == j :> nat)).
```

This means that the ith diagonal element of the matrix is the ith element of the list and the rest are zero. Now if M is a matrix, our algorithm for computing the Smith normal form should return a list s and two matrices P and Q such that:

1. s is sorted by division and its length is less than m and n,
2. P *m M *m Q = diag_mx_seq m n s and
3. P and Q are invertible.

Any elementary divisor ring is coherent as the existence of an algorithm computing Smith normal form implies that we can compute kernels. Hence we only need to assume that R is a strongly discrete elementary divisor ring to be able to consider finitely presented modules over R. As P is invertible it is obvious that Q defines a morphism from M to `diag_mx_seq m n s`. Also P^-1 defines a morphism in the other direction that is inverse to P which means that M and `diag_mx_seq m n s` are isomorphic.

Bézout Domains

We now assume that all rings have explicit divisibility, that is, we can decide if $a \mid b$ and moreover produce x such that $b = xa$. Two elements a and b are *associate* if $a \mid b$ and $b \mid a$. Since we are working over integral domains, a and b are associate if and only if there exists a unit $u \in R$ such that $a = bu$.

Definition 5. *An integral domain R is a **Bézout domain** if every finitely generated ideal is principal (generated by a single element).*

This is equivalent to requiring that R has a GCD operation and a function computing the elements of the Bézout identity. This means that given a and b one can compute x and y such that $xa + by$ and $gcd(a, b)$ are associate.

We have formalized a proof that Bézout domains of Krull dimension less than or equal to 1 (in particular principal ideal domains like $\mathbb{Z}$ and $k[X]$ with k a field) are elementary divisor rings, however as this paper is concerned with finitely presented modules we do not go into the details of this proof here. The reason why we restrict our attention to rings of Krull dimension less than or equal to 1 is that it is still an open problem whether all Bézout domains are elementary divisor rings or not [21].

Combining this with finitely presented modules we get a constructive generalization to the classification theorem of finitely generated modules over principal ideal domains. This theorem states that any finitely presented R-module $\mathcal{M}$ over a principal ideal domain R can be decomposed into a direct sum of a free module and cyclic modules, that is, there exists $n \in \mathbb{N}$ and elements $d_1, \ldots, d_k \in R$ such that:

$$\mathcal{M} \simeq R^n \oplus R/(d_1) \oplus \cdots \oplus R/(d_k)$$

with the additional property that $d_i \mid d_{i+1}$ for $1 \leqslant i < k$.

In [5], it is formally proved in COQ that the Smith normal form is unique up to multiplication by units for rings with a GCD operation. This means that for any matrix M equivalent to a diagonal matrix D in Smith normal form, each of the diagonal elements of the Smith normal form of M will be associate to the corresponding diagonal element in D. This implies that the decomposition of finitely presented modules over elementary divisor rings is unique up to multiplication by units. This also gives a way for deciding if two finitely presented modules are isomorphic.

6   Conclusions and Future Work

In this paper we have presented a formalization of the category of finitely presented modules over coherent strongly discrete rings and shown that it is an abelian category. The fact that we can represent everything using matrices makes is possible for us to reuse basic results on these when building the abstraction layer of modules on top. The division theory of matrices over coherent strongly discrete rings makes it straightforward for us to do reasoning modulo a set of relations.

It is not only interesting that we have an abelian category because it provides us with a setting to do homological algebra, but also because it is proved in [9] that in order to show that abelian groups (and hence the category of R-modules) form an abelian category in COQ one needs the principle of unique choice. As our formalization is based on the Mathematical Components hierarchy [11] of algebraic structures, we inherit a form of axiom of choice in the structure of discrete rings. However, we speculate that this axiom is in fact not necessary for our proof that the category of **finitely presented** modules over **coherent strongly discrete** rings is abelian.

In Homotopy Type Theory [25] there is a distinction between pre-categories and univalent categories (just called categories in [1]). A *pre-category* is a category where the collection of morphisms forms a set in the sense of homotopy type theory, that is, they satisfy the uniqueness of identity proofs principle. Our category of finitely presented modules satisfy the uniqueness of morphism equivalence (phi %= psi) proofs (by Hedberg's theorem [16]), but morphisms form a setoid instead of a set. If we quotiented morphisms by the equivalence relation on morphisms we would get a set, and thus our category of finitely presented modules would become a pre-category.

A *univalent category* on the other hand is a pre-category where the equality of objects coincides with isomorphism. As we have shown that for elementary divisor rings there is a way to decide isomorphism, we speculate that we would also get a univalent category by quotienting modules by isomorphisms. It would be interesting to develop these ideas further and define the notion of univalent abelian category and study its properties. Note that in Homotopy Type Theory, it may be no longer necessary to have the decidability of the equivalence relation to form the quotient, so we would not need to be in an elementary divisor ring to get a univalent category.

Since we have shown that we have an abelian category it would now be very interesting to formally study more complex constructions from homological algebra. It would for instance be straightforward to define resolutions of modules. We can then define define the *Hom* and tensor functors in order to get derived functors like *Tor* and *Ext*. It would also be interesting to define graded objects like chain complexes and graded finitely presented modules, and prove that they also form abelian categories.

Acknowledgments. The authors are grateful to Bassel Mannaa for his comments on early versions of the paper, and to the anonymous reviewers for their helpful comments.

References

1. Ahrens, B., Kapulkin, C., Shulman, M.: Univalent categories and the Rezk completion (2013), http://arxiv.org/abs/1303.0584 (preprint)
2. Barakat, M., Lange-Hegermann, M.: An axiomatic setup for algorithmic homological algebra and an alternative approach to localization. J. Algebra Appl. 10(2), 269–293 (2011)
3. Barakat, M., Robertz, D.: HOMALG – A Meta-Package for Homological Algebra. J. Algebra Appl. 7(3), 299–317 (2008)
4. Barthe, G., Capretta, V., Pons, O.: Setoids in type theory. Journal of Functional Programming 13(2), 261–293 (2003)
5. Cano, G., Dénès, M.: Matrices à blocs et en forme canonique. JFLA - Journées Francophones des Langages Applicatifs (2013)
6. Cohen, C.: Pragmatic Quotient Types in Coq. In: Blazy, S., Paulin-Mohring, C., Pichardie, D. (eds.) ITP 2013. LNCS, vol. 7998, pp. 213–228. Springer, Heidelberg (2013)
7. Coq development team: The Coq Proof Assistant Reference Manual, version 8.4. Tech. rep., Inria (2012)
8. Coquand, T., Mörtberg, A., Siles, V.: Coherent and strongly discrete rings in type theory. In: Hawblitzel, C., Miller, D. (eds.) CPP 2012. LNCS, vol. 7679, pp. 273–288. Springer, Heidelberg (2012)

9. Coquand, T., Spiwack, A.: Towards constructive homological algebra in type theory. In: Kauers, M., Kerber, M., Miner, R., Windsteiger, W. (eds.) MKM/CALCULEMUS 2007. LNCS (LNAI), vol. 4573, pp. 40–54. Springer, Heidelberg (2007)
10. Decker, W., Lossen, C.: Computing in Algebraic Geometry: A Quick Start using SINGULAR. Springer (2006)
11. Garillot, F., Gonthier, G., Mahboubi, A., Rideau, L.: Packaging mathematical structures. In: Berghofer, S., Nipkow, T., Urban, C., Wenzel, M. (eds.) TPHOLs 2009. LNCS, vol. 5674, pp. 327–342. Springer, Heidelberg (2009)
12. Gonthier, G., Mahboubi, A.: A Small Scale Reflection Extension for the Coq system. Tech. rep., Microsoft Research INRIA (2009)
13. Gonthier, G.: Point-Free, Set-Free Concrete Linear Algebra. In: van Eekelen, M., Geuvers, H., Schmaltz, J., Wiedijk, F. (eds.) ITP 2011. LNCS, vol. 6898, pp. 103–118. Springer, Heidelberg (2011)
14. Greuel, G.M., Pfister, G.: A Singular Introduction to Commutative Algebra, 2nd edn (2007)
15. Hatcher, A.: Algebraic Topology, 1st edn. Cambridge University Press (2001), http://www.math.cornell.edu/~hatcher/AT/AT.pdf
16. Hedberg, M.: A Coherence Theorem for Martin-Löf's Type Theory. Journal of Functional Programming 8(4), 413–436 (1998)
17. Heras, J., Dénès, M., Mata, G., Mörtberg, A., Poza, M., Siles, V.: Towards a certified computation of homology groups for digital images. In: Ferri, M., Frosini, P., Landi, C., Cerri, A., Di Fabio, B. (eds.) CTIC 2012. LNCS, vol. 7309, pp. 49–57. Springer, Heidelberg (2012)
18. Heras, J., Coquand, T., Mörtberg, A., Siles, V.: Computing persistent homology within Coq/SSReflect. ACM Transactions on Computational Logic 14(4), 26 (2013)
19. Kaplansky, I.: Elementary divisors and modules. Transactions of the American Mathematical Society 66, 464–491 (1949)
20. Lombardi, H., Quitté, C.: Algèbre commutative, Méthodes constructives: Modules projectifs de type fini. Calvage et Mounet (2011)
21. Lorenzini, D.: Elementary divisor domains and bézout domains. Journal of Algebra 371(0), 609–619 (2012)
22. Mines, R., Richman, F., Ruitenburg, W.: A Course in Constructive Algebra. Springer (1988)
23. Poincaré, H.: Analysis situs. Journal de l' École Polytechnique 1, 1–123 (1895)
24. Sozeau, M.: A new look at generalized rewriting in type theory. Journal of Formalized Reasoning 2(1), 41–62 (2009)
25. The Univalent Foundations Program: Homotopy Type Theory: Univalent Foundations of Mathematics. Institute for Advanced Study (2013), http://homotopytypetheory.org/book/

Formalized, Effective Domain Theory in Coq

Robert Dockins

Portland State University, Portland, Oregon, USA
rdockins@pdx.edu

Abstract. I present highlights from a formalized development of domain theory in the theorem prover Coq. This is the first development of domain theory that is *effective, formalized* and that supports all the usual constructions on domains. In particular, I develop constructive models of both the unpointed profinite and the pointed profinite domains. Standard constructions (e.g., products, sums, the function space, and powerdomains) are all developed. In addition, I build the machinery necessary to compute solutions to recursive domain equations.

1 Introduction and Related Work

The term "domain theory" refers to a class of mathematical techniques that are used to develop computational models suitable for reasoning about the semantics of general purpose programming languages. Proofs done in domain theory often have an elegant, compositional nature, and the much of the modern thinking about programming languages (especially functional languages) can be traced to domain-theoretic roots. Domain theory has a long lineage, starting from the classic work of Dana Scott on continuous lattices [22]. Domain theory was intensively studied in the 1970's and 80's, producing far more research than I have space here to review; see Abramsky and Jung's account for a good overview and further sources [2].

Unfortunately, using domain theory for language semantics requires specialized mathematical knowledge. Many proofs that are purely "domain theoretic" in nature must be done before domain theory can be used on a language problem of interest. Furthermore, the wealth of academic literature on domain theory is actually a liability in some ways: the sheer volume of knowledge on the subject and the myriad minor variations can be overwhelming to the novice.

It would thus be desirable to package up the difficult mathematics underlying domain theory into a library of facts and constructions that can be pulled off the shelf and used by those interested in programming languages, but not in domain theory *per se.* Such a library should choose, from among the many varieties and presentations in the literature, a selection of domain theoretic techniques that are suitable for practical use. The development of domain theory described in this paper is my attempt to develop such a library for the Coq proof assistant.

My own effort is far from the first attempt to make domain theory more accessible. One of the earliest mechanical proof systems, Milner's Logic for Computable Functions (LCF) [15,16], is a syntactic presentation of a logic for

G. Klein and R. Gamboa (Eds.): ITP 2014, LNAI 8558, pp. 209–225, 2014.

reasoning about functions defined via least-fixed-point semantics in Scott domains.[1]

In terms of the mathematics formalized, the most closely related work is probably the formalization of bifinite domains in Isabelle/HOL by Brian Huffman [13], building on the HOLCF system by Müller *et al.* [17]. One major difference between this work and that of Huffman, besides the choice of theorem prover, is that my development of domain theory is *effective*. There are two different senses in which this statement is true. First, the objects and constructions in the theory are effective; I formalize categories of effective algebraic domains (whose bases are countable and have decidable ordering), similar to those considered by Egli and Constable [9]. Second, the *metalogic* in which these constructions and proofs are carried out is the purely constructive metalogic of Coq with no additional axioms. As a result, I can forgo the explicit development of recursive function theory, relying instead on the innate "effectiveness" of functions definable in type theory. In contrast, HOL is a classical logic with strong choice principles. There are also some differences in the proof strategy — Huffman uses definitions of bifinite domains based on finitary deflations, whereas I use properties of Plotkin orders. I found it difficult to obtain appropriately constructive definitions of deflations, whereas Plotkin orders proved more amenable to constructive treatment. In addition, the uniform presentation of pointed and unpointed domains I used (discussed below) is novel.

In the Coq community, the most closely-related work appears to be the development of constructive complete partial orders (CPOs) by Benton *et al.* [7], which is based on earlier work by Paulin-Mohring [18]. The CPOs built by Benton *et al.* are constructive; however, they lack bases and thus are not actually *domains* according to the usage of the term used by some authors [2]. This line of work develops a constructive version of the lift construction by using coinductive ε-streams. In contrast, the main tool I use for achieving effective constructions is the enumerable set. Proofs using enumerable sets are, in my opinion, both easier to understand and easier to develop than those based on coinductive structures (`cofix`, with its syntactic guardedness condition, is especially troublesome).

My development of domain theory is able to handle some constructions that Benton *et al.* cannot handle; for example, they report difficulty defining the smash product, which presents no problem for me. Powerdomains can also be constructed in my system, whereas they cannot be on basic CPOs.

The main contribution of this paper is a formalization of the pointed profinite and unpointed profinite domains, together with common operations on them like products, sums, function space, and powerdomains. I also build the machinery necessary to solve recursive domain equations. The constructions and proofs largely follow the development of the profinite domains given by Carl Gunter in his thesis [12]; however, all the constructions are modified to be "effectively given," similar to [9].

The primary novelty of my approach is related to proof-engineering advances that both clarify and simplify portions of the proof. The main instance of this is

---

[1] Milner [16] credits the design of LCF to an unpublished note of Dana Scott.

a unification of the constructions for the pointed profinite domains (those with a least element) and the unpointed profinite domains (those *not necessarily* containing a least element). These two categories of domains are very similar; in my proof development, this similarity is aggressively exploited by constructing both categories of domains at the same time as two instantiations of a single parametrized construction. Common operations on domains (sums, products, function space, powerdomains) are also constructed uniformly for both categories. This saves a great deal of otherwise redundant work and highlights the tight connection between these categories of domains. This unification is made possible by a minor deviation from the usual definitions dealing with algebraic domains and compact elements, as explained later in this paper.

A second proof-engineering advance is the fact that I do not formalize domains in the usual way (as a subcategory of complete partial orders with Scott-continuous functions), but instead develop the category formed by the *bases* of profinite domains with approximable relations as arrows. This category is equivalent to the category of profinite domains via ideal completion. I can save some effort because rather than building constructions on CPOs as well as on their bases, I need only perform constructions on bases. With the notable exception of the function space, constructions on the bases of algebraic domains are simpler than corresponding constructions on CPOs (especially in the pointed case). This formalization of profinite domains via their bases (the Plotkin orders) mirrors the development of Scott domains via the Scott information systems [25]. I am not aware of any mechanized development of Scott information systems.

The goal of this work is ultimately to provide a solid foundation of domain theoretic constructions and theorems that can be used for practical program semantics in Coq. My focus on constructive mathematics, however, stems from a personal philosophy that values constructive methods *per se*; this perspective is especially well-justified for domain theory, which aims to be a mathematical model of computation. Nonetheless, I derive a practical benefit, which is that my development of domain theory asserts *no additional axioms at all* over the base metatheory of Coq — this makes my development compatible with any axiomatic extension of Coq, even unusual ones (say, anticlassical extensions).

In this paper, I will explain the high-level ideas of the proof and present the main definitions. The interested reader is encouraged to consult the formal proof development for details; it may be found at the author's website.[2] Knowledge of Coq is not required to understand the main ideas of this paper; however, a novice grasp of category theory concepts will be required for some sections.

2   Basic Definitions

We start with a key difference between my proof development and textbook references on domain theory. Usually, domain theory is concerned with certain classes of *partial orders*: types equipped with an order relation that is transitive, reflexive and antisymmetric. I will be working instead with the *preorders*: types

---

[2] http://rwd.rdockins.name/domains/

equipped with an order relation that is merely reflexive and transitive. The antisymmetry condition is dropped. This leads us to working with *setoids*, sets equipped with an equivalence relation, and setoid homomorphisms, functions that respect setoid equivalence. Using setoids is a common technique in Coq because quotients are not readily available [5].

Definition 1 (Setoid). *Suppose A is a type, and $\approx$ is a binary relation on A. We say $\langle A, \approx \rangle$ is a setoid provided $\approx$ is reflexive, transitive and symmetric.*

Throughout this paper (and the vast majority of the formal proof), the only notion of equality we will be interested in is the $\approx$ relation of setoids. In the formal proof, many of the definitions require axioms stating that various operations preserve $\approx$ equality. In this paper, I will elide such axioms and concerns about the preservation of equality.

Definition 2 (Preorder). *Suppose A is a type and $\sqsubseteq$ is a binary relation on A. We say $\langle A, \sqsubseteq \rangle$ is a preorder provided $\sqsubseteq$ is reflexive and transitive. Furthermore, we automatically assign to A a setoid where $x \approx y$ iff $x \sqsubseteq y \wedge y \sqsubseteq x$.*

Every preorder automatically induces a setoid; because we work up to $\approx$, we obtain "antisymmetry on preorders" by convention.

Definition 3 (Finite set). *Suppose $\langle A, \sqsubseteq \rangle$ is a preorder.[3] Then we can consider the type* list A *to be a preorder of finite sets of A. We say that an element x is in the finite set l, and write $x \in l$ provided $\exists x'.$* In x' $l \wedge x \approx x'$. *Here* In *refers to the Coq standard library predicate for list membership. We can then equip finite sets with the inclusion order where $l_1 \subseteq l_2$ iff $\forall x.\ x \in l_1 \rightarrow x \in l_2$.*

Note that finite set membership is defined so that it respects $\approx$ equality.

The difference between unpointed directed-complete partial orders (which might not have a least element) and pointed DCPOs (those having a least element) can be expressed entirely in terms of whether certain finite sets are allowed to be empty or not. Likewise, the Scott-continuous functions are distinguished from the *strict* Scott-continuous functions; and the profinite domains from the pointed profinite domains. Therefore, we make the following technical definition that allows us to uniformly state definitions that are valid in both settings.

Definition 4 (Conditionally inhabited). *Suppose A is a preorder, l is a finite set of A and h is a boolean value. Then we say l is* conditionally inhabited *and write* inh$_h$ l *provided either $h = false$ or $h = true \wedge \exists x.\ x \in l$.*

When h is false, l need not be inhabited; however, when h is true, inh$_h$ l requires l to be inhabited. This strange little definition turns out to be the key to achieving a uniform presentation of pointed and unpointed domains.

---

[3] It suffices to suppose A is a setoid, but the objects of interest are always preorders, so this is mildly simpler. Likewise for the enumerable sets below.

Definition 5 (Enumerable set). *Suppose A is a preorder. Then the functions $\mathbb{N} \to$ option A are enumerable sets of A. If X is an enumerable set and x is an element of A, we write $x \in X$ iff $\exists x'\ n.\ X(n) =$ Some $x' \wedge x \approx x'$. As with finite sets, we write $X \subseteq Y$ iff $\forall x.\ x \in X \to x \in Y$.*

Here $\mathbb{N}$ is the Coq standard library type of binary natural numbers and option A is an inductive type containing either Some x for some x in A or the distinguished element None. If we interpret Coq's metalogic as a constructive logic of computable functions then the enumerable sets defined here are a good analogue for the recursively enumerable sets (recall a set is recursively enumerable iff it is the range of a partial recursive function). I avoid using the terms *recursive* or *recursively enumerable* so as not to invoke the machinery of recursive function theory, which I have not explicitly developed.

Throughout this development of domain theory, the enumerable sets will fill in where, in more classical presentations, "ordinary" sets would be. The use of enumerable sets is one of the primary ways I achieve an effective presentation of domain theory. In the rest of this paper, I will freely use set-comprehension notation when defining sets; the diligent reader may wish to verify for himself that sets so defined are actually represented by some concrete enumeration function.

Definition 6 (Directed set). *Suppose A is a preorder, h is a boolean value, and X is an enumerable set of A. We say that X is directed (or h-directed), and write $\mathrm{directed}_h\ X$, if every finite subset $l \subseteq X$ where $\mathrm{inh}_h\ l$ has an upper bound in X.*

This definition differs from the standard one. My definition agrees when $h = false$; but when $h = true$, only the *inhabited* finite subsets must have upper bounds in X. As a consequence, when $h = false$, $\mathrm{directed}_h\ X$ implies X is nonempty (because the empty set must have an upper bound in X); but when $h = true$, $\mathrm{directed}_h\ X$ holds even when X is empty (in which case the condition holds vacuously because there are no finite inhabited subsets of the empty set).

Definition 7 (Directed-complete partial order). *Suppose A is a preorder and let h be a boolean value. Let $\bigsqcup X$ be an operation that, for every h-directed enumerable set X of A, calculates the least upper bound of X. Then we say $\langle A, \bigsqcup \rangle$ is a directed-complete partial order (with respect to h). The category of directed-complete partial orders is named $\mathbf{DCPO}_h$ when paired with the Scott-continuous functions.*

Definition 8 (Scott-continuous function). *Suppose A and B are in $\mathbf{DCPO}_h$ and $f : A \to B$ is a function from A to B. Then we say f is Scott-continuous if f is monotone and satisfies the following for all h-directed sets X:*

$$f(\bigsqcup X) \sqsubseteq \bigsqcup(\text{image } f\ X)$$

Pause once again to reflect on the role of the parameter h. When $h = false$ the empty set is not considered directed, and thus may not have a supremum;

this gives unpointed DCPOs. On the other hand when $h = true$, the empty set must have a supremum, which is *a fortiori* the least element of the DCPO. Likewise for Scott-continuous functions, $h = false$ gives the standard Scott-continuous functions on unpointed DCPOs; whereas $h = true$ gives the *strict* Scott-continuous functions on pointed DCPOs.

Next we move on to definitions relating to algebraic domains. The main technical definition here is the "way below" relation, which gives rise to the compact elements.

Definition 9 (Way-below, compact element). *Suppose A is in $\mathbf{DCPO}_h$ and let x and y be elements of A. Then we say x is way-below y (and write $x \ll y$) provided that, for every h-directed set X where $y \sqsubseteq \bigsqcup X$, there exists x' such that $x \sqsubseteq x'$ and $x' \in X$. An element x of A is compact if $x \ll x$.*

This is the standard statement of the way-below relation (also called the "order of approximation") and compact elements, except that we have specified the h-directed sets instead of the directed sets. With $h = false$, this is just the standard definition of way-below and compact elements for unpointed DCPOs. However, when $h = true$ it means that the least element of a pointed DCPO is *not* compact (because X may be the empty set). Although readers already familiar with domain theory may find this puzzling, it actually is quite good, as it simplifies constructions on pointed domains (see §4).

Definition 10 (Effective preorder). *Suppose A is a preorder. Then we say A is an effective preorder provided it is equipped with a decision procedure for $\sqsubseteq_A$ and an enumerable set containing every element of A.*

In order to ensure that all the constructions we want to perform can be done effectively, we need to limit the scope of our ambitions to the effective preorders and effective domains.

Definition 11 (Effective algebraic domain). *Suppose A is in $\mathbf{DCPO}_h$. A is an algebraic domain provided that for all x in A, the set $\{b \mid b \ll x \land b \ll b\}$ is enumerable, h-directed and has supremum x. A is furthermore called effective if the set of compact elements is enumerable, and the $\ll$ relation is decidable on compact elements.*

Said another way, a DCPO is an algebraic domain if every element arises as the supremum of the set of compact elements way-below it. The set of compact elements of an algebraic domain is also called the basis.[4] Note that an effective domain is *not* necessarily an effective preorder; merely its basis is effective.

Definition 12 (Basis). *Suppose A is an effective algebraic domain. Then let basis(A) be the preorder consisting of the compact elements of A, with $\ll$ as the ordering relation. Note that $\ll$ is always transitive; furthermore, $\ll$ is reflexive on the compact elements of A by definition. basis(A) is an effective preorder because A is an effective domain.*

---

[4] The more general class of *continuous* domains may have a basis that is distinct from the set of compact elements; I do not consider continuous domains.

The compact elements of an effective algebraic domain form an effective pre-order. Furthermore, the basis preorder uniquely (up to isomorphism) determines an algebraic domain.

Definition 13 (Ideal completion). *Suppose A is an effective preorder. We say an enumerable subset X of A is an* ideal *provided it is h-directed and down-ward closed with respect to $\sqsubseteq_A$. Let* ideal(A) *be the preorder of ideals of A ordered by subset inclusion. Then* ideal(A) *is a DCPO (with enumerable set union as the supremum operation) and an effective algebraic domain.*

The proof that ideal(A) is an algebraic domain is standard (see [2, §2.2.6]), as is the following theorem.

Theorem 1. *The effective algebraic domains are in one-to-one correspondence with their bases. In particular A is isomorphic to* ideal$($basis$(A))$ *for all algebraic domains A and B is isomorphic (as a preorder) to* basis$($ideal$(B))$ *for all effective preorders B.*

This justifies our position, in the rest of this paper, of considering *just* the bases of domains rather than algebraic domains *per se*. Although the formal proofs we develop will involve only bases and certain kinds of *approximable relations* between bases, the reader may freely draw intuition from the more well-known category of algebraic domains and Scott-continuous functions.

3   Profinite Domains and Plotkin Orders

The profinite domains are fairly well-known. They are very closely related to Plotkin's category SFP (Sequences of Finite inductive Posets) [20]. In fact, when limited to effective bases, the category of pointed profinite domains has the same objects as SFP; SFP, however, is equipped with the *nonstrict* continuous functions, whereas the pointed profinite domains are equipped with the strict continuous functions. Unpointed profinite domains are the largest cartesian closed full subcategory of algebraic domains with countable bases [14], which justifies our interest in them. I suspect that the pointed profinite domains are likewise the largest *monoidal* closed subcategory of countably-based pointed algebraic domains when considered with strict continuous functions.[5]

My development of the profinite domains (more properly, their bases: the Plotkin orders) roughly follows the strategy found in chapter 2 of Gunter's thesis [12], incorporating some modifications due to Abramsky [1]. In addition, I have modified some definitions to make them more constructive and to incorporate the h parameter for selecting pointed or unpointed domains.

The central concept in this strategy is the *Plotkin order*. Plotkin orders are preorders with certain completeness properties based on *minimal* upper bounds.

---

[5] A nearby result holds: the pointed profinite domains are the largest cartesian closed category of countably-based algebraic domains when considered with the nonstrict continuous functions.

Definition 14 (Minimal upper bound). *Suppose A is a preorder, and let X be a set of elements of A. We say m is a* minimal upper bound *of X provided that m is an upper bound for X; and for every other upper bound of X, m′ where m′ ⊑ m, we have m ≈ m′.*

Note that minimal upper bounds are subtly different than *least* upper bounds. There may be several distinct minimal upper bounds (MUBs) of a set; in contrast, *least* upper bounds (AKA suprema) must be unique (up to ≈ as usual).

Definition 15 (MUB complete preorder). *Suppose A is a preorder. Then we say A is* MUB complete *if for every finite subset l of elements of A where* inh_h *l and where z is an upper bound of l, there exists a minimal upper bound m of l with m ⊑ z.*

In a MUB-complete preorder every finite set bounded-above by some z has a minimal upper bound below z. Here we merely assert the desired MUB to exist, but it can actually be computed when A is an effective Plotkin order.

Definition 16 (MUB closed set). *Suppose A is a preorder and X is a subset of A. We say X is* MUB closed *if for every finite subset l ⊆ X with* inh_h *l, every minimal upper bound of l is in X.*

Definition 17 (Plotkin order). *Suppose A is a preorder, and let Θ be an operation (called the MUB closure) from finite sets of A to finite sets of A. Then we say ⟨A, Θ⟩ is a* Plotkin order *(with respect to h) provided the following hold for all finite sets l of A:*

1. *A is MUB complete;*
2. *l ⊆ Θ(l);*
3. *Θ(l) is MUB closed; and*
4. *Θ(l) is the smallest set satisfying the above.*

Note that for a given preorder A, Θ is uniquely determined when it exists. As such, we will rarely give Θ explicitly, preferring instead simply to assert that a suitable Θ can be constructed.

Some of the constructions we wish to do require additional computational content from domain bases; we therefore require them to be *effective preorders* in addition to being Plotkin. Thus, the effective Plotkin preorders will be the domain structures we are interested in. However, it still remains to define the arrows between domains. Unlike the usual case with concrete categories, the arrows between domains are not functions, but certain kinds of relations. These *approximable relations*, however, induce functions on ideal completions in a unique way (see [12, §2.1]), so they nonetheless behave very much like functions.

Definition 18 (Approximable relation). *Let A and B be effective Plotkin orders, and let R ⊆ A × B be an enumerable relation (i.e., enumerable set of pairs). Then we say R is an* approximable relation *provided the following hold:*

- $x \sqsubseteq x' \wedge y' \sqsubseteq y \wedge (x,y) \in R$ implies $(x',y') \in R$; and
- the set $\{y \mid (x,y) \in R\}$ is h-directed for all x.

These requirements seem mysterious at first, but they are precisely what is required to ensure that approximable relations give rise to Scott-continuous functions via the ideal completion construction.

Theorem 2. *Suppose A is an effective Plotkin order. Then $\{(x,y) \mid y \sqsubseteq_A x\}$ is an approximable relation; call it id_A.*

Suppose A, B, and C are effective Plotkin orders; let R be an approximable relation from A to B and S be an approximable relation from B to C. Then the composed relation $S \circ R \equiv \{(x,z) \mid \exists y.\ (x,y) \in R \wedge (y,z) \in S\}$ is an approximable relation.

Definition 19 (PLT$_h$). *Let h be a boolean value. Then $\mathbf{PLT}_h$ is the category whose objects are the effective Plotkin orders (with parameter h) and whose arrows are the approximable relations (again with h). The previous theorem gives the construction for the identity and composition arrows; the associativity and unit axioms are easily proved.*

$\mathbf{PLT}_{false}$ is equivalent to the category of unpointed profinite domains with Scott-continuous functions and $\mathbf{PLT}_{true}$ is equivalent to the category of pointed profinite domains with strict Scott-continuous functions via the ideal completion construction discussed above.

Definition 20 (Undefined relation). *Let A and B be objects of $\mathbf{PLT}_{true}$. Let $\perp \equiv \emptyset$ be the empty approximable relation between A and B. $\perp$ corresponds to the undefined function that sends every argument to the least element of B.*

Note that, while $\perp \equiv \emptyset$ is a perfectly valid approximable relation in $\mathbf{PLT}_{true}$, the empty relation is *not* an approximable relation in $\mathbf{PLT}_{false}$ (unless A is empty). This is because of the second property of approximable relations, which requires $\{y \mid (x,y) \in R\}$ to be h-directed for each x. The empty set is h-directed for $h = true$, but not for $h = false$. Further, note that in $\mathbf{PLT}_{true}$, $f \circ \perp \approx \perp \approx \perp \circ g$, so composition is strict on both sides.

4   Constructions on Plotkin Orders

Now that we have defined the category of effective Plotkin orders, we can begin to build some of the standard constructions: products, sums and the function space. Our strategy of unifying pointed and unpointed domains will now begin to pay dividends, as these constructions need be performed only once.

Definition 21 (Unit and empty orders). *Let 0 represent the empty preorder. It is easy to see that 0 is effective and Plotkin. Let 1 represent the unit preorder, having a single element; 1 is also effective and Plotkin.*

Definition 22 (Products). *Suppose A, B and C are effective Plotkin orders with parameter h. Then $A \times B$ (product preorder) is an effective Plotkin preorder with parameter h. Set $\pi_1 \equiv \{((x, y), x') \mid x' \sqsubseteq x\}$ and $\pi_2 \equiv \{((x, y), y') \mid y' \sqsubseteq y\}$. Suppose $f : C \to A$ and $f : C \to B$ are approximable relations. Then set $\langle f, g \rangle \equiv \{(z, (x, y)) \mid (z, x) \in f \land (z, y) \in g\}$. π_1 and π_2 represent pair projections and $\langle f, g \rangle$ is the paring operation.*

Theorem 3. *The product construction is the categorical product in $\mathbf{PLT}_{false}$. In addition, 1 is the terminal object; making $\mathbf{PLT}_{false}$ a cartesian category. In $\mathbf{PLT}_{false}$, the product is denoted $A \times B$.*

Theorem 4. *The product construction is not the categorical product in $\mathbf{PLT}_{true}$. It is instead the "smash" product, or the strict pair; we denote the smash product as $A \otimes B$. Although not the categorical product, $\otimes$ gives $\mathbf{PLT}_{true}$ the structure of a symmetric monoidal category, with 1 as the unit object.*

In $\mathbf{PLT}_{true}$, π_1, π_2 and $\langle f, g \rangle$ are all still useful operations; they are just not the projection and pairing arrows for the categorical product. They are instead "strict" versions that satisfy laws like $\langle x, \bot \rangle \approx \bot$ and $\pi_1 \circ \langle x, y \rangle \sqsubseteq x$.

Definition 23 (Sums). *Suppose A, B and C are effective Plotkin preorders with parameter h. Then $A + B$ (disjoint sum) is an effective Plotkin preorder with parameter h. Set $\iota_1 \equiv \{(x, \text{inl } x') \mid x' \sqsubseteq x\}$ and $\iota_2 \equiv \{(y, \text{inr } y') \mid y' \sqsubseteq y\}$. Suppose $f : A \to C$ and $f : B \to C$ are approximable relations. Then set $[f, g] \equiv \{(\text{inl } x, z) \mid (x, z) \in f\} \cup \{(\text{inr } y, z) \mid (y, z) \in g\}$. ι_1 and ι_2 are the sum injections and $[f, g]$ is the case analysis operation.*

Theorem 5. *In $\mathbf{PLT}_{false}$, the above sum construction is the categorical coproduct, which we denote $A + B$. In addition, 0 is the initial object. Thus $\mathbf{PLT}_{false}$ is a cocartesian category.*

Theorem 6. *In $\mathbf{PLT}_{true}$, the sum construction above gives the "coalesced sum," which identifies the bottom elements of the two objects; it is denoted $A \oplus B$. Like $+$, $\oplus$ is the the categorical coproduct in $\mathbf{PLT}_{true}$. Furthermore 0 serves as the initial object. Thus $\mathbf{PLT}_{true}$ is also a cocartesian category.*

Theorem 7. *In $\mathbf{PLT}_{true}$, the empty preorder 0 is also the terminal object.*

This series of results reveals some deep connections between the structure $\mathbf{PLT}_{false}$ and $\mathbf{PLT}_{true}$. Not only are $\times$ and $\otimes$ intuitively closely related, they are literally the same construction. Likewise for $+$ and $\oplus$. This coincidence goes even further; the "function space" construction in both categories is likewise the same. This construction is based on the concept of "joinable" relations. My definition is a minor modification of the one given by Abramsky [1].

Definition 24 (Joinable relation). *Suppose A and B are objects of $\mathbf{PLT}_h$. Let R be a finite set of pairs in $A \times B$. We say R is a joinable relation if the following hold:*

- inh$_h$ R; and
- for all finite sets G with $G \subseteq R$ and inh$_h$ G, and for all x where x is a minimal upper bound of image π_1 G, there exists y where y is an upper bound of image π_2 G and $(x, y) \in R$.

Unfortunately, this definition is highly technical and difficult to motivate, except by the fact that it yields the expected exponential object. The rough idea is that R is supposed to be a finite fragment of an approximable relation. The complicated second requirement ensures that joinable relations are "complete enough" that the union of a directed collection of joinable relations makes an approximable relation and that we can compute finite MUB closures.

Definition 25 (Function space). *Suppose A and B are objects of* **PLT**$_h$. *The joinable relations from A to B form a preorder where we set $G \sqsubseteq H$ iff* $\forall x\ y.\ (x, y) \in G \to \exists x'\ y'.\ (x', y') \in H \wedge x' \sqsubseteq x \wedge y \sqsubseteq y'$. *Moreover, this preorder is effective and Plotkin; we denote it by $A \Rightarrow B$.*

Now, suppose $f : C \times A \to B$ is an approximable relation.[6] Then curry f : $C \to (A \Rightarrow B)$ *and* app $: (A \Rightarrow B) \times A \to B$ *are approximable relations as defined below:*

$$\text{curry } f \equiv \{(c, R) \mid \forall x\ y.\ (x, y) \in R \to ((c, x), y) \in f\}$$
$$\text{app} \equiv \{((R, x), y) \mid \exists x'\ y'.\ (x', y',) \in R \wedge x' \sqsubseteq x \wedge y \sqsubseteq y'\}$$

The proof that the function space construction forms a Plotkin order is one of the most involved proofs in this entire development. My proof takes elements from those of Gunter [12] and Abramsky [1]. The reader may consult the formal proof development for details.

Theorem 8. $\Rightarrow$ *is the exponential object in* **PLT**$_{false}$ *and makes* **PLT**$_{false}$ *into a cartesian closed category.*

Theorem 9. *In* **PLT**$_{true}$, $\Rightarrow$ *constructs the exponential object with respect to $\otimes$ and makes* **PLT**$_{true}$ *a monoidal closed category. In* **PLT**$_{true}$, *we use the symbol* $\multimap$ *instead of* $\Rightarrow$.

Here is a huge payoff for our strategy of giving uniform constructions for **PLT**$_h$; the proofs and constructions leading to $\Rightarrow$ and the MUB closure properties are technical and lengthy. Furthermore, the pointed and unpointed cases differ only in a few localized places. With this proof strategy, these difficult proofs need only be done once.

5   Lifting and Adjunction

One standard construction we have not yet seen is "lifting," which adds a new bottom element to a domain. In our setting, lifting is actually split into two

---

[6] Note, here $\times$ refers generically to the product construction in **PLT**$_h$, not just the categorical product of **PLT**$_{false}$.

pieces: a forgetful functor from pointed to unpointed domains, and a lifting functor from unpointed to pointed domains. These functors are adjoint, which provides a tight and useful connection between these two categories of domains.

However, working with bases instead of algebraic domains *per se* causes an interesting inversion to occur. When working with $\mathbf{PLT}_h$ instead of profinite domains as such, the functor that is the forgetful functor and the one that actually does lifting exchange places.

First let us consider the functor that passes from pointed domains ($\mathbf{PLT}_{true}$) to unpointed domains ($\mathbf{PLT}_{false}$). This is the one usually known as the forgetful functor; it forgets the fact that the domains are pointed and that the functions are strict.

Definition 26 ("Forgetful" functor). *Suppose A is an object of $\mathbf{PLT}_{true}$. Then let option A be the preorder that adjoins to A a new bottom element, None. Then option A is an element of $\mathbf{PLT}_{false}$. Furthermore, suppose $f : A \rightarrow B$ is an approximable relation in $\mathbf{PLT}_{true}$. Then g : option $A \rightarrow$ option B is an approximable relation in $\mathbf{PLT}_{false}$ defined by:*

$$g \equiv \{(x, \mathsf{None})\} \cup \{(\mathsf{Some}\ x, \mathsf{Some}\ y) \mid (x, y) \in f\}.$$

These operations produce a functor $U : \mathbf{PLT}_{true} \rightarrow \mathbf{PLT}_{false}$.

Why is adding a new element the right thing to do? Recall from earlier that our definition of the way-below relation and compact elements *excludes* the bottom element of pointed domains, in contrast to the usual definitions. When we consider the bases of pointed domains, the bottom element is implicit; is is the *empty set* of basis elements (which is an ideal when $h = true$) that represents bottom. It is because the bottom element is implicit that makes all the constructions from the previous section work uniformly in both categories.

When passing to unpointed domains, the empty set is no longer directed and the implicit bottom element must become *explicit*. This is why the forgetful functor actually adds a new basis element. In contrast, the "lifting" functor is more like a forgetful functor in that there is nothing really to do. Passing from unpointed to pointed domains automatically adds the new implicit bottom element, so the basis does not change.

Definition 27 ("Lifting" functor). *Suppose A is an object of $\mathbf{PLT}_{false}$; then A is also an object of $\mathbf{PLT}_{true}$. Furthermore, if $f : A \rightarrow B$ is an approximable relation in $\mathbf{PLT}_{false}$ then f is also an approximable relation in $\mathbf{PLT}_{true}$. These observations define a functor $L : \mathbf{PLT}_{false} \rightarrow \mathbf{PLT}_{true}$.*

Theorem 10. *The lifting functor L is left adjoint to the forgetful functor U.*

Said another way, the adjunction between L and U means that there is a one-to-one correspondence between the strict $\mathbf{PLT}_{true}$ arrows $L(X) \rightarrow Y$ and the nonstrict $\mathbf{PLT}_{false}$ arrows $X \rightarrow U(Y)$. This adjunction induces a structure on $\mathbf{PLT}_h$ that is a model of dual intuitionistic linear logic (DILL) [6,4], which can be used to combine the features of strict and nonstrict computation into a nice, unified theory.

6 Powerdomains

Powerdomains provide operators on domains that are analogues to the standard powerset operation on sets [20]; powerdomains can be used to give semantics to nondeterminism and to set-structured data (like relational tables). Each of the three standard powerdomains operations (upper, lower and convex) [2] can be constructed in both $\mathbf{PLT}_{false}$ and in $\mathbf{PLT}_{true}$, for a total of six powerdomain operators. Again, our uniform presentation provides a significant savings in work.

Definition 28 (Powerdomains). *Suppose X is an element of $\mathbf{PLT}_h$. Let the finite h-inhabited sets of X be the elements of the powerdomain. The preorder on domain elements is one of the following: $\sqsubseteq^\flat$ for the lower powerdomain, $\sqsubseteq^\sharp$ for the upper powerdomain, and $\sqsubseteq^\natural$ for the convex powerdomain.*

$$a \sqsubseteq^\flat b \equiv \forall x \in a.\ \exists y \in b.\ x \sqsubseteq_X y$$
$$a \sqsubseteq^\sharp b \equiv \forall y \in b.\ \exists x \in a.\ x \sqsubseteq_X y$$
$$a \sqsubseteq^\natural b \equiv a \sqsubseteq^\flat b \wedge a \sqsubseteq^\sharp b$$

In each case, the resulting preorder is effective and Plotkin, making it again an object of $\mathbf{PLT}_h$.

Of these, the most mathematically natural (that is, most like the powerset operation) is probably the convex powerdomain in unpointed domains ($\mathbf{PLT}_{false}$). Unlike the convex powerdomain in pointed domains, the unpointed version has a representation for the empty set.

7 Solving Recursive Domain Equations

To get a usable domain for semantics, we frequently want to be able to solve recursive domain equations. Indeed, much of the impetus for domain theory was originally motivated by the desire to build a semantic models for the lambda calculus [22]. The classic example is the simple recursive equation $D \cong (D \Rightarrow D)$.

My approach to this problem is the standard one, which is to take bilimits of continuous functors expressed in categories of embedding-projection pairs [21]. The main advantage of this technique is that it turns mixed-variance functors on $\mathbf{PLT}_h$ (like the function space) into covariant functors on the category of EP-pairs. Such covariant functors can then be handled via standard fixpoint theorems. Furthermore, isomorphisms constructed in categories of EP-pairs yield isomorphisms in the base category.

Thus, we need to construct the category of EP-pairs over $\mathbf{PLT}_h$. However, it is significantly more convenient to work in a different, but equivalent category: the category of basis embeddings. To the best of my knowledge, this category has not been previously defined.

Definition 29 (Basis embedding). *Suppose A and B are objects of $\mathbf{PLT}_h$ and let f be a function (N.B., not a relation) from A to B. Then we say f is a basis embedding if the following hold:*

- $a \sqsubseteq_A b$ iff $f(a) \sqsubseteq_B f(b)$ (f is monotone and reflective);
- for all y, the set $\{x \mid f(x) \sqsubseteq_B y\}$ is h-directed.

Let $\mathbf{BE}_h$ represent the category of effective Plotkin orders with basis embeddings as arrows.

Every basis embedding gives rise to an embedding-projection pair in $\mathbf{PLT}_h$, and likewise every EP-pair gives rise to a basis embedding[7]; these are furthermore in one-to-one correspondence. This is sufficient to show that $\mathbf{BE}_h$ and the category of EP-pairs on $\mathbf{PLT}_h$ are equivalent categories.

Now we can use a standard fixpoint theorem (essentially, the categorical analogue of Kleene's fixpoint theorem) to build least fixpoints of *continuous* functors. Continuous functors are those that preserve directed colimits [2].

Theorem 11. *Suppose $\mathcal{C}$ is a category with initial object 0 that has directed colimits for all directed systems, and let $F : \mathcal{C} \to \mathcal{C}$ be a continuous functor. Then F has an initial algebra D and $D \cong F(D)$.*

When combined with the next theorem, this allows us to construct fixpoints of functors in $\mathbf{BE}_{true}$, the category basis embeddings over pointed domains.

Theorem 12. $\mathbf{BE}_h$ *has directed colimits for all directed systems.*

In fact, $\mathbf{PLT}_{true}$ is in the class of CPO-algebraically ω-compact categories [10, §7], which are especially well-behaved for building recursive types.[8]

We cannot apply the same strategy to $\mathbf{BE}_{false}$, because the category of embeddings over unpointed domains fails to have an initial object. However, this is not a problem, because we can import constructions from $\mathbf{BE}_{true}$ into $\mathbf{BE}_{false}$ by passing through the adjoint functors L and U. Passing through L may add "extra" bottom elements, but it does so in places that are consistent with standard practice (e.g., the canonical model for lazy λ-calculus [3]). Indeed, this setup partially explains *why* those extra bottoms appear.

The following theorems allow us to find fixpoint solutions to recursive domain equations for many interesting functors by building up continuous functors from elementary pieces.

Theorem 13. *The identity and constant functors are continuous.*

Theorem 14. *The composition of two continuous functors is continuous.*

Theorem 15. $\times$, $+$ *and* $\Rightarrow$ *extend to continuous functors on* $\mathbf{BE}_{false}$.

Theorem 16. $\otimes$, $\oplus$ *and* $\multimap$ *extend to continuous functors on* $\mathbf{BE}_{true}$.

Theorem 17. *The forgetful functor* $U : \mathbf{BE}_{true} \to \mathbf{BE}_{false}$ *is continuous.*

---

[7] Surprisingly, (to me, at least) this can be done in an entirely constructive way using an indefinite description principle that can be proved for enumerable sets.

[8] This is not yet formalized; initial attempts have triggered universe consistency issues for which I have yet to find a solution.

Theorem 18. *The lifting functor* $L : \mathbf{BE}_{false} \to \mathbf{BE}_{true}$ *is continuous.*

Theorem 19. *The lower, upper and convex powerdomains are all continuous.*

Now we can construct a wide variety of interesting recursive semantic domains for both pointed and unpointed domains. For example, we can construct the domain $D \cong (D \multimap D)$, representing eager λ-terms; and we can also construct $E \cong L(U(E) \Rightarrow U(E))$, the canonical model of lazy λ-terms. Furthermore, algebraic data types in the style of ML or Haskell can be constructed using combinations of sums and products. Sophisticated combinations of strict and nonstrict features can be built using the adjunction functors U and L, and nondeterminism may be modeled using the powerdomains.

8   Implementation

The entire proof development consists of a bit over 30KLOC, not including the examples. This includes a development of some elementary category theory and the fragments of set theory we need to work with finite and enumerable sets. The proof development has been built using Coq version 8.4; it may be found at the author's personal website.[9] The development includes examples demonstrating soundness and adequacy for four different systems (the simply-typed SKI combinator calculus with and without fixpoints, and the simply-typed λ-calculus with and without fixpoints). Additional examples are in progress.

9   Conclusion and Future Work

I have presented a high-level overview of a formal development of domain theory in the proof assistant Coq. The basic trajectory of the proof follows lines well-established by prior work. My presentation is fully constructive and mechanized — the first such presentation of domain theory based on profinite domains. I show how some minor massaging of the usual definitions leads to a nice unification for most constructions in pointed and unpointed domains. The system is sufficient to build denotational models for a variety of programming language semantics.

Currently I have no explicit support for parametric polymorphism. Polymorphism requires fairly complex machinery, and I have not yet undertaken the task. There seem to be two possible paths forward: one approach is explained by Paul Taylor in his thesis [23]; the other is the $PILL_Y$ models of Birkedal *et al.* [8]. This latter method has the significant advantage that it validates the parametricity principle, which underlies interesting results like Wadler's free theorems [24]. I hope to build one or both of these systems for polymorphism in the future.

The Coq formalization of Benton *et al.* [7] uses a more modern method for building recursive domains based on locally-continuous bifunctors in compact categories [11]. Their approach has some advantages; Pitts's invariant relations [19], especially, provide useful reasoning principles for recursively defined domains. I hope to incorporate these additional techniques in future work.

---

[9] http://rwd.rdockins.name/domains/

References

1. Abramsky, S.: Domain theory in logical form. Annals of Pure and Applied Logic 51, 1–77 (1991)
2. Abramsky, S., Jung, A.: Domain Theory. In: Handbook of Logic in Computer Science, vol. 3, pp. 1–168. Clarendon Press (1994)
3. Abramsky, S., Ong, C.-H.L.: Full abstraction in the lazy lambda calculus. Information and Computation 105, 159–267 (1993)
4. Barber, A.: Linear Type Theories, Semantics and Action Calculi. Ph.D. thesis, Edinburgh University (1997)
5. Barthe, G., Capretta, V.: Setoids in type theory. Journal of Functional Programming 13(2), 261–293 (2003)
6. Benton, N.: A mixed linear and non-linear logic: Proofs, terms and models. In: Pacholski, L., Tiuryn, J. (eds.) CSL 1994. LNCS, vol. 933, pp. 121–135. Springer, Heidelberg (1995)
7. Benton, N., Kennedy, A., Varming, C.: Some domain theory and denotational semantics in Coq. In: Berghofer, S., Nipkow, T., Urban, C., Wenzel, M. (eds.) TPHOLs 2009. LNCS, vol. 5674, pp. 115–130. Springer, Heidelberg (2009)
8. Birkedal, L., Møgelberg, R., Petersen, R.: Domain-theoretical models of parametric polymorphism. Theoretical Computer Science 288, 152–172 (2007)
9. Egli, H., Constable, R.L.: Computability concepts for programming language semantics. Theoretical Computer Science 2, 133–145 (1976)
10. Fiore, M.P.: Axiomatic Domain Theory in Categories of Partial Maps. Ph.D. thesis, University of Edinburgh (1994)
11. Freyd, P.: Remarks on algebraically compact categories. In: Applications of Categories in Computers Science. London Mathematical Society Lecture Note Series, vol. 177, pp. 95–106. Cambridge University Press (1991)
12. Gunter, C.: Profinite Solutions for Recursive Domain Equations. Ph.D. thesis, Carnegie-Mellon University (1985)
13. Huffman, B.: A purely definitional universal domain. In: Berghofer, S., Nipkow, T., Urban, C., Wenzel, M. (eds.) TPHOLs 2009. LNCS, vol. 5674, pp. 260–275. Springer, Heidelberg (2009)
14. Jung, A.: Cartesian Closed Categories of Domains. Ph.D. thesis, Centrum voor Wiskunde en Informatica, Amsterdam (1988)
15. Milner, R.: Logic for computable functions: Description of a machine implementation. Tech. Rep. STAN-CS-72-288, Stanford University (May 1972)
16. Milner, R.: Models of LCF. Tech. Rep. STAN-CS-73-332, Stanford (1973)
17. Müller, O., Nipkiw, T., von Oheimb, D., Slotosch, O.: HOLCF = HOL + LCF. Journal of Functional Programming 9 (1999)
18. Paulin-Mohring, C.: A constructive denotational semantics for Kahn networks. In: From Semantics to Computer Sciences. Essays in Honour of G. Kahn. Cambridge University Press (2009)
19. Pitts, A.M.: Relational properties of domains. Information and Computation 127 (1996)
20. Plotkin, G.D.: A powerdomain construction. SIAM J. of Computing 5, 452–487 (1976)
21. Plotkin, G., Smyth, M.: The category theoretic solution of recursive domain equations. Tech. rep., Edinburgh University (1978)

22. Scott, D.: Outline of a mathematical theory of computation. Tech. Rep. PRG02, OUCL (November 1970)
23. Taylor, P.: Recursive Domains, Indexed Category Theory and Polymorphism. Ph.D. thesis, University of Cambridge (1986)
24. Wadler, P.: Theorems for free! In: Intl. Conf. on Functional Programming and Computer Architecture (1989)
25. Winskell, G.: The Formal Semantics of Programming Languages: An Introduction. MIT Press (1993)

Completeness and Decidability Results for CTL in Coq

Christian Doczkal and Gert Smolka

Saarland University, Saarbrücken, Germany
{doczkal,smolka}@ps.uni-saarland.de

Abstract. We prove completeness and decidability results for the temporal logic CTL in Coq/Ssreflect. Our basic result is a constructive proof that for every formula one can obtain either a finite model satisfying the formula or a proof in a Hilbert system certifying the unsatisfiability of the formula. The proof is based on a history-augmented tableau system obtained as the dual of Brünnler and Lange's cut-free sequent calculus for CTL. We prove the completeness of the tableau system and give a translation of tableau refutations into Hilbert refutations. Decidability of CTL and completeness of the Hilbert system follow as corollaries.

1 Introduction

We are interested in a formal and constructive metatheory of the temporal logic CTL [6]. We start with the definitions of formulas, models, and a satisfiability relation relating models and formulas. The models are restricted such that the satisfiability relation is classical. We then formalize a Hilbert proof system and prove it sound for our models. Up to this point everything is straightforward. Our basic result is a constructive proof that for every formula one can obtain either a finite model satisfying the formula or a derivation in the Hilbert system certifying the unsatisfiability of the formula. As corollaries of this result we obtain the completeness of the Hilbert system, the finite model property of CTL, and the decidability of CTL.

Informal and classical proofs of our corollaries can be found in Emerson and Halpern's work on CTL [7,5]. Their proofs are of considerable complexity as it comes to the construction of models and Hilbert derivations. As is, their completeness proof for the Hilbert system is not constructive and it is not clear how to make it constructive.

Brünnler and Lange [3] present a cut-free sequent system for CTL satisfying a finite subformula property. Due to the subformula property, the sequent system constitutes a decision method for formulas that yields finite counter-models for non-valid formulas. The sequent system is non-standard in that formulas are annotated with histories, which are finite sets of formulas. Histories are needed to handle eventualities (e.g., until formulas) with local rules.

We base the proof of our main result on a tableau system that we obtain by dualizing Brünnler and Lange's sequent system. This is the first tableau system

G. Klein and R. Gamboa (Eds.): ITP 2014, LNAI 8558, pp. 226–241, 2014.
© Springer International Publishing Switzerland 2014

for CTL employing only local rules. Existing tableau methods for CTL [7,5] combine local rules with global model checking of eventualities. Given a formula, the tableau system either constructs a finite model satisfying the formula or a tableau refutation. We give a translation from tableau refutations to Hilbert refutations, thereby showing the completeness of the Hilbert system and the soundness of the tableau system. The translation is compositional in that it is defined by structural recursion on tableau refutations. For the translation it is essential that the tableau system has only local rules.

With our results it should not be difficult to obtain formal and constructive proofs of the soundness and completeness of Brünnler and Lange's original system.

The standard definition [5] of the satisfiability relation of CTL employs infinite paths, which are difficult to handle in a constructive setting. We avoid infinite paths by capturing the semantics of eventualities with induction and the semantics of co-eventualities with coinduction.

Our formal development consists of about 3500 lines of Coq/Ssreflect. There are three subtasks of considerable complexity. One complex subtask is the construction of finite models from intermediate structures we call demos. Our demos play the role of the pseudo-Hintikka structures in Emerson [5] and are designed such that they go well with the tableau system. Another complex subtask is the construction of a demo from the tableau-consistent clauses in a subformula universe. Finally, the translation of tableau refutations to Hilbert refutations is of considerable complexity, in particular as it comes to the application of the induction axioms of the Hilbert system.

Given the practical importance of CTL and the complex proofs of the metatheoretic results for CTL, we think that the metatheory of CTL is an interesting and rewarding candidate for formalization. No such formalization exists in the literature. In previous work [4] we have prepared this work by proving related results for a weaker modal logic. As it comes to eventualities, which are responsible for the expressiveness and the complexity of the logic, our previous work only captured the simplest eventuality saying that a state satisfying a given formula is reachable.

Our development is carried out in Coq [13] with the Ssreflect [9] extension. We build a library for finite sets on top of Ssreflect's countable types and use it to capture the subformula property. We also include a fixpoint theorem for finite sets and use it to show decidability of tableau derivability.

In each section of the paper, we first explain the mathematical ideas behind the proofs and then comment briefly on the difficulties we faced in the formalization. For additional detail, we refer the reader to Coq development.[1]

2   CTL in Coq

We define the syntax and semantics of CTL as we use it in our formalization. We fix a countable alphabet $\mathcal{AP}$ of atomic propositions p and define formulas as follows:

---

[1] http://www.ps.uni-saarland.de/extras/itp14.

$$s,t := p \mid \bot \mid s \to t \mid \mathsf{AX}\,s \mid \mathsf{A}(s\,\mathsf{U}\,t) \mid \mathsf{A}(s\,\mathsf{R}\,t)$$

We define the remaining propositional connectives using $\to$ and $\bot$. We also use the following defined modal operators: $\mathsf{EX}\,s \equiv \neg\,\mathsf{AX}\,\neg s$, $\mathsf{A}^+(s\,\mathsf{U}\,t) \equiv \mathsf{AX}\,\mathsf{A}(s\,\mathsf{U}\,t)$, $\mathsf{E}(s\,\mathsf{U}\,t) \equiv \neg\,\mathsf{A}(\neg s\,\mathsf{R}\,\neg t)$, $\mathsf{E}^+(s\,\mathsf{U}\,t) \equiv \mathsf{EX}\,\mathsf{E}(s\,\mathsf{U}\,t)$, $\mathsf{E}(s\,\mathsf{R}\,t) \equiv \neg\,\mathsf{A}(\neg s\,\mathsf{U}\,\neg t)$, and $\mathsf{EG}\,t \equiv \mathsf{E}(\bot\,\mathsf{R}\,t)$.

The formulas of CTL are interpreted over transition systems where the states are labeled with proposition symbols. Unlike most of the literature on CTL [5,7,1], where the semantics of CTL formulas is defined in terms of infinite paths, we define the semantics of CTL using induction and coinduction. Our semantics is classically equivalent to the standard infinite path semantics but better suited for a constructive formalization.

Let W be a type, $R : W \to W \to Prop$ a relation, and $P, Q : W \to Prop$ predicates. We require that R is serial, i.e., that every $w : W$ has some R-successor. We define the eventuality AU ("always until") inductively as:

$$\frac{Q\,w}{\mathsf{AU}\,R\,P\,Q\,w} \qquad \frac{P\,w \quad \forall v.\,R\,w\,v \implies \mathsf{AU}\,R\,P\,Q\,v}{\mathsf{AU}\,R\,P\,Q\,w}$$

Further, we define AR ("always release") coinductively.

$$\frac{Q\,w \quad P\,w}{\mathsf{AR}\,R\,P\,Q\,w} \qquad \frac{Q\,w \quad R\,w\,v \quad \mathsf{AR}\,R\,P\,Q\,v}{\mathsf{AR}\,R\,P\,Q\,w}$$

Now let $L : \mathcal{AP} \to W \to Prop$ be a labeling function. We evaluate CTL formulas to predicates on W:

$$eval\ p = L\,p \qquad eval\ (s \to t) = \lambda w.\,eval\ s\ w \implies eval\ t\ w$$

$$eval\ \bot = \lambda\_.\,False \qquad eval\ (\mathsf{AX}\,s) = \lambda w.\,\forall v.\,R\,w\,v \implies eval\ t\ v$$

$$eval\ (\mathsf{A}(s\,\mathsf{U}\,t)) = \mathsf{AU}\,R\,(eval\ s)\,(eval\ t)$$

$$eval\ (\mathsf{A}(s\,\mathsf{R}\,t)) = \mathsf{AR}\,R\,(eval\ s)\,(eval\ t)$$

We say w *satisfies* a formula s, written $w \models s$, if we have $eval\ s\ w$. Similar to [4], we consider as models only those serial transition systems (W, R, L) for which

$$\forall s \forall w \in W.\,w \models s \vee w \not\models s \tag{1}$$

is provable. When $\mathcal{M}$ is a model, we write $\to_{\mathcal{M}}$ for the transition relation of $\mathcal{M}$ and $w \in \mathcal{M}$ if w is a state of $\mathcal{M}$.

Note that having to prove (1) severely restricts our ability to construct infinite models. However, since CTL has the small model property it suffices to construct finite models for our completeness results. For these models (1) is easy to prove. Formalizing models this way allows us to reason about the classical object logic CTL without assuming any classical axioms.

The Hilbert axiomatization we use in our formalization is a variant of the Hilbert system given by Emerson and Halpern [7]. The rules and axioms of the Hilbert axiomatization are given in Figure 1. We write $\vdash s$ if s is provable from the axioms and call a proof of $\neg s$ a *Hilbert refutation* of s.

$$
\begin{array}{ll}
\text{K} & s \to t \to s \\
\text{S} & ((u \to s \to t) \to (u \to s) \to u \to t) \\
\text{DN} & ((s \to \bot) \to \bot) \to s \\
\text{N} & \text{AX}(s \to t) \to \text{AX}\,s \to \text{AX}\,t \\
\text{U1} & t \to \text{A}(s\,\text{U}\,t) \\
\text{U2} & s \to \text{AX}\,\text{A}(s\,\text{U}\,t) \to \text{A}(s\,\text{U}\,t) \\
\text{R1} & \text{A}(s\,\text{R}\,t) \to t \\
\text{R2} & \text{A}(s\,\text{R}\,t) \to (s \to \bot) \to \text{AX}\,\text{A}(s\,\text{R}\,t) \\
\text{AX} & \text{AX}\,\bot \to \bot
\end{array}
$$

$$
\dfrac{s \qquad s \to t}{t}\ \text{MP} \qquad
\dfrac{s}{\text{AX}\,s}\ \text{Nec} \qquad
\dfrac{t \to u \qquad s \to \text{AX}\,u \to u}{\text{A}(s\,\text{U}\,t) \to u}\ \text{AU}_{\text{ind}}
$$

$$
\dfrac{u \to t \qquad u \to (s \to \bot) \to \text{AX}\,u}{u \to \text{A}(s\,\text{R}\,t)}\ \text{AR}_{\text{ind}}
$$

Fig. 1. Hilbert Axiomatization of CTL

Theorem 2.1. *If $\vdash s$ then $w \models s$ for all models $\mathcal{M}$ and states $w \in \mathcal{M}$.*

Proof. Induction on the derivation of $\vdash s$, using (1) for the cases corresponding to DN and AR_{ind}. □

We are now ready to state our basic theorem.

Theorem 2.2 (Certifying Decision Method). *For every formula we can construct either a finite model or a Hilbert refutation.*

3 A History-Based Tableau System for CTL

The tableau system we use as the basis for our certifying decision method employs signed formulas [11]. A *signed formula* is either s^+ or s^- where s is a formula. Signs bind weaker than formula constructors, so $s \to t^+$ is to be read as $(s \to t)^+$. We write σ for arbitrary signs and $\bar{\sigma}$ for the sign opposite to σ. A state satisfies a signed formula s^σ if it satisfies $\lfloor s^\sigma \rfloor$ where $\lfloor s^+ \rfloor = s$ and $\lfloor s^- \rfloor = \neg s$.

We refer to positive until formulas and negative release formulas as *eventualities*. For the eventuality $\text{A}(s\,\text{R}\,t)^-$ to be satisfied at a state, there must be a path from this state to a state satisfying $\neg t$ that satisfies $\neg s$ on every state along the way.

A *clause* is a finite set of signed formulas and a *history* is a finite set of clauses. The letters C and D range over clauses and the letter H ranges over histories. For the rest of this paper, sets are always assumed to be finite. An *annotated eventuality* is a formula of the form

$$
\text{A}(s\,\text{U}_H t)^+ \mid \text{A}^+(s\,\text{U}_H t)^+ \mid \text{A}(s\,\text{R}_H t)^- \mid \text{A}^+(s\,\text{R}_H t)^-
$$

An *annotation* is either an annotated eventuality or the empty annotation ".".
The letter a ranges over annotations. An *annotated clause* is a pair $C|a$ of a
clause C and an annotation a.

We give the semantics of annotated clauses by interpreting clauses, histories,
and annotations as formulas. If an object with an associated formula appears
in the place of a formula, it is to be interpreted as its associatend formula. The
associated formula of a clause C is $\bigwedge_{s^\sigma \in C} \lfloor s^\sigma \rfloor$. The *associated formula* of a
history H is the formula $\bigwedge_{C \in H} \neg C$. The *associated formula* of an annotation is
defined as follows:

$$\mathsf{af}(\cdot) = \top$$
$$\mathsf{af}(\mathsf{A}(s \cup_H t)^+) = \mathsf{A}((s \wedge H) \cup (t \wedge H))$$
$$\mathsf{af}(\mathsf{A}^+(s \cup_H t)^+) = \mathsf{A}^+((s \wedge H) \cup (t \wedge H))$$
$$\mathsf{af}(\mathsf{A}(s \, \mathsf{R}_H \, t)^-) = \mathsf{E}((\neg s \wedge H) \cup (\neg t \wedge H))$$
$$\mathsf{af}(\mathsf{A}^+(s \, \mathsf{R}_H \, t)^-) = \mathsf{E}^+((\neg s \wedge H) \cup (\neg t \wedge H))$$

The meaning of an annotated eventuality can be understood as follows: a state
satisfies $\mathsf{A}(s \cup_H t)^+$ if it satisfies $\mathsf{A}(s \cup t)$ without satisfying any clause from H
along the way. For $\mathsf{A}(s \, \mathsf{R}_H \, t)^-$ we push the negation introduced by the sign down
to s and t before adding the history. A state satisfies the annotated clause $C|a$,
if it satisfies the formula $C \wedge a$.

The *request* of a clause is the set $\mathcal{R}C := \{ s^+ \mid \mathsf{AX}\, s^+ \in C \}$. The request of
annotations is defined such that $r\,(\mathsf{A}^+(s \cup_H t)) = \mathsf{A}(s \cup_H t)$ and $r\,a = \cdot$ for all
other annotations. The intuition behind requests is that if a state satisfies $C|a$,
then every successor state must satisfy $\mathcal{R}C|r\,a$.

Our tableau calculus derives unsatisfiable clauses. The rules of the calculus can
be found in Figure 2. The notation C, s^σ is to be read as $C \cup \{s^\sigma\}$. If C, s^σ appears
in the conclusion of a rule, we refer to C as the *context* and to s^σ as the *active
formula*. The tableau system is essentially dual to the sequent calculus CT [3].
While CT derives valid disjunctions, our tableau calculus derives unsatisfiable
conjunctions. Aside from syntactic changes, the main difference between CT and
the tableau calculus is that in CT all the rules carry the proviso that the active
formula in the conclusion does not appear in the context. We impose no such
restriction. The reason for this is simply convenience. Our completeness proof
does not rely on this added flexibility.

The history mechanism (last two rows in Figure 2) works by recording all
contexts encountered while trying to fulfill one eventuality. If a context reappears
further up in the derivation, we can close this branch since every eventuality that
can be fulfilled, can be fulfilled without going through cycles. If all branches lead
to cycles, the eventuality cannot be fulfilled and the clause is unsatisfiable.

In our formalization, we do not argue soundness of the tableau system directly
using models. Instead, we show the following translation theorem:

Theorem 3.1. *If $C|a$ is tableau derivable, then $\vdash \neg(C \wedge a)$.*

Corollary 3.2. *If $C|a$ is tableau derivable, then $C|a$ is unsatisfiable.*

$$\frac{}{C, p^+, p^-|a} \qquad \frac{}{C, \bot^+|a} \qquad \frac{C, s^-|a \quad C, t^+|a}{C, s \to t^+|a} \to^+ \qquad \frac{C, s^+, t^-|a}{C, s \to t^-|a} \to^-$$

$$\frac{\mathcal{R}\,C|r\,a}{C|a}\, \mathsf{X} \qquad \frac{\mathcal{R}C, u^-|r\,a}{C, \mathsf{AX}\,u^-|a}\, \mathsf{AX}^- \qquad \frac{\mathcal{R}C|\,\mathsf{A}(s\,\mathsf{U}_H t)^-}{C|\mathsf{A}^+(s\,\mathsf{U}_H t)^-}\, \mathsf{R}_\mathsf{H}^+$$

$$\frac{C, t^+|a \quad C, s^+, \mathsf{A}^+(s\,\mathsf{U}\,t)^+|a}{C, \mathsf{A}(s\,\mathsf{U}\,t)^+|a}\, \mathsf{U}^+ \qquad \frac{C, t^-, s^-|a \quad C, t^-, \mathsf{A}^+(s\,\mathsf{U}\,t)^-|a}{C, \mathsf{A}(s\,\mathsf{U}\,t)^-|a}\, \mathsf{U}^-$$

$$\frac{C, s^+, t^+|a \quad C, t^+, \mathsf{A}^+(s\,\mathsf{R}\,t)^+|a}{C, \mathsf{A}(s\,\mathsf{R}\,t)^+|a}\, \mathsf{R}^+ \qquad \frac{C, t^-|a \quad C, s^-, \mathsf{A}^+(s\,\mathsf{R}\,t)^-|a}{C, \mathsf{A}(s\,\mathsf{R}\,t)^-|a}\, \mathsf{R}^-$$

$$\frac{C|\mathsf{A}(s\,\mathsf{U}_\emptyset t)^+}{C, \mathsf{A}(s\,\mathsf{U}\,t)^+|\cdot}\, \mathsf{A}_\emptyset \qquad \frac{C, t^+|\cdot \quad C, s^+|\mathsf{A}^+(s\,\mathsf{U}_{H,C}t)^+}{C|\mathsf{A}(s\,\mathsf{U}_H t)^+}\, \mathsf{A}_\mathsf{H} \qquad \frac{}{C|\mathsf{A}(s\,\mathsf{U}_{H,C}t)^+}\, \overline{\mathsf{A}}$$

$$\frac{C|\mathsf{A}(s\,\mathsf{R}_\emptyset\,t)^-}{C, \mathsf{A}(s\,\mathsf{R}\,t)^-|\cdot}\, \mathsf{R}_\emptyset \qquad \frac{C, t^-|\cdot \quad C, s^-|\mathsf{A}^+(s\,\mathsf{R}_{H,C}\,t)^-}{C|\mathsf{A}(s\,\mathsf{R}_H\,t)^-}\, \mathsf{R}_\mathsf{H} \qquad \frac{}{C|\mathsf{A}(s\,\mathsf{R}_{H,C}\,t)}\, \overline{\mathsf{R}}$$

Fig. 2. Tableau System for CTL

We defer the proof of Theorem 3.1 to Section 6.

Even though it is not part of our formal development, we still argue soundness of the tableau system informally (and classically) to give some intuition how the history mechanism works. Soundness of all the rules except A_H and R_H is easy to see. The case for A_H is argued (in the dual form) by Brünnler and Lange [3]. So we argue soundness of R_H here. Assume that $C|\mathsf{A}(s\,\mathsf{R}_H\,t)^-$ is satisfiable and $C, t^-|\cdot$ is unsatisfiable. Then the situation looks as follows:

There exists some state satisfying $C \wedge \mathsf{E}(\neg s \wedge H \,\mathsf{U}\, \neg t \wedge H)$. Hence, there exists a path satisfying $\neg s \wedge H$ at every state until it reaches a state satisfying $\neg t \wedge H$. Since $C, t^-|\cdot$ is unsatisfiable, this state must also satisfy $\neg C$. Therefore, the path consists of at least 2 states. The last state on the path that satisfies C (left circle) also satisfies $\neg s$ and $\mathsf{E}^+((\neg s \wedge H \wedge \neg C) \,\mathsf{U}\, (\neg t \wedge H \wedge \neg C))$ and therefore $C, s^-|\mathsf{A}^+(s\,\mathsf{R}_{H,C}\,t)^-$.

Note that, although the R_H rule looks similar to the local rule R^-, the soundness argument is non-local; if there is state satisfying the conclusion of the rule, the state satisfying one of the premises may be arbitrarily far away in the model.

As noted by Brünnler and Lange [3], the calculus is sound for all annotated clauses but only complete for clauses with the empty annotation. Consider the clause $\emptyset|\, A(p \, U_{\{\{p^+\}\}}p)^+$. The clause is underivable, but its associated formula is equivalent to the unsatisfiable formula $A((p \wedge \neg p) \, U \, (p \wedge \neg p))$. To obtain a certifying decision method, completeness for history-free clauses is sufficient.

3.1 Decidability of Tableau Derivability

For our certifying decision method, we need to show that tableau derivability is decidable. The proof relies on the subformula property, i.e., the fact that backward application of the rules stays within a finite syntactic universe. We call a set of signed formulas *subformula closed*, if it satisfies the following conditions:

S1. If $(s \to t)^\sigma \in \mathcal{F}$, then $\{s^{\overline{\sigma}}, t^\sigma\} \subseteq \mathcal{F}$.
S2. If $AX\, s^\sigma \in \mathcal{F}$, then $s^\sigma \in \mathcal{F}$.
S3. If $A(s \, U \, t)^\sigma \in \mathcal{F}$, then $\{s^\sigma, t^\sigma, A^+(s \, U \, t)^\sigma\} \subseteq \mathcal{F}$.
S4. If $A(s \, R \, t)^\sigma \in \mathcal{F}$, then $\{s^\sigma, t^\sigma, A^+(s \, R \, t)^\sigma\} \subseteq \mathcal{F}$.

It is easy to define a recursive function *ssub* that computes for a signed formula s^σ a finite subformula closed set containing s^σ. The *subformula closure* of a clause C is defined as $sfc \, C := \bigcup_{s \in C} ssub\, s$ and is always a subformula closed extension of C. Now let $\mathcal{F}$ be a subformula closed set. The *annotations for* $\mathcal{F}$, written $\mathcal{A}(\mathcal{F})$, consist of $\cdot$ and eventualities from $\mathcal{F}$ annotated with histories $H \subseteq \mathcal{P}(\mathcal{F})$, where $\mathcal{P}(\mathcal{F})$ is the powerset of $\mathcal{F}$. We define the *universe for* $\mathcal{F}$ as $\mathcal{U}(\mathcal{F}) := \mathcal{P}(\mathcal{F}) \times \mathcal{A}(\mathcal{F})$.

Lemma 3.3. *1. If $\mathcal{F}$ is subformula closed, the set $\mathcal{U}(\mathcal{F})$ is closed under backward application of the tableau rules.*

2. For every annotated clause $C|a$ there exists a subformula closed set $\mathcal{F}$, such that $C|a \in \mathcal{U}(\mathcal{F})$.

3. Derivability of annotated clauses is decidable.

Proof. Claim (1) follows by inspection of the individual rules. For (2) we reason as follows: If $a = \cdot$, we take $\mathcal{F}$ to be $sfc\, C$. If $a = A(s \, U_{\, H} t)$, one can show that $C|\, A(s \, U_{\, H} t) \in \mathcal{U}(sfc\, (C, A(s \, U \, t) \cup \bigcup_{D \in H} D))$. All other cases are similar.

For (3) consider the annotated clause $C|a$. By (2) we know that $C|a \in \mathcal{U}(\mathcal{F})$ for some $\mathcal{F}$. We now compute the least fixpoint of one-step tableau derivability inside $\mathcal{U}(\mathcal{F})$. By (1) the annotated clause $C|a$ is derivable iff it is contained in the fixpoint. $\square$

3.2 Finite Sets in Coq

To formalize the tableau calculus and the decidability proof, we need to formalize clauses and histories. The Ssreflect libraries [8] contain a library for finite sets. However, the type of sets defined there requires that the type over which the sets are formed is a finite type, i.e., a type with finitely many elements. This is clearly not the case for the type of signed formulas.

We want a library providing extensional finite sets over countable types (e.g., signed formulas) providing all the usual operations including separation ($\{\, x \in$

$A \mid p\,x\,\})$, replacement ($\{\,f\,x \mid x \in A\,\}$), and powerset. We could not find a library satisfying all our needs, so we developed our own.

Our set type is a constructive quotient over lists. We use the choice operator provided by Ssreflect to define a normalization function that picks some canonical duplicate-free list to represent a given set. This normalization function is the main primitive for constructing sets. On top of this we build a library providing all the required operations. Our lemmas and notations are inspired by Ssreflect's finite sets and we port most of the lemmas that apply to the setting with infinite base types. We instantiate Ssreflect's big operator library [2], which provides us with indexed unions.

Our library also contains a least fixpoint construction. For every bounded monotone function from sets to sets we construct its least fixpoint and show the associated induction principle. This is used in the formalization of Lemma 3.3 to compute the set of derivable clauses over a given subformula universe.

4   Demos

We now define demos. In the completeness proof of the tableau calculus, demos serve as the interface between the model construction and the tableau system. Our demos are a variant of the pseudo-Hintikka structures used by Emerson [5]. Instead of Hintikka clauses, we use literal clauses and the notion of support [10].

A signed formula is a *literal*, if it is of the form p^σ, $\perp^\sigma$, or $\mathsf{AX}\,s^\sigma$. A *literal clause* is a clause containing only literals. A literal clause is *locally consistent* if it contains neither $\perp^+$ nor both p^+ and p^- for any p. A clause *supports* a signed formula, written $C \triangleright s^\sigma$, if

$$C \triangleright l \iff l \in C \qquad\qquad \text{if } l \text{ is a literal}$$
$$C \triangleright (s \to t)^+ \iff C \triangleright s^- \lor C \triangleright t^+$$
$$C \triangleright (s \to t)^- \iff C \triangleright s^+ \land C \triangleright t^-$$
$$C \triangleright \mathsf{A}(s\,\mathsf{U}\,t)^+ \iff C \triangleright t^+ \lor (C \triangleright s^+ \land C \triangleright \mathsf{A}^+(s\,\mathsf{U}\,t)^+)$$
$$C \triangleright \mathsf{A}(s\,\mathsf{U}\,t)^- \iff C \triangleright t^- \land (C \triangleright s^- \lor C \triangleright \mathsf{A}^+(s\,\mathsf{U}\,t)^-)$$
$$C \triangleright \mathsf{A}(s\,\mathsf{R}\,t)^+ \iff C \triangleright t^+ \land (C \triangleright s^+ \lor C \triangleright \mathsf{A}^+(s\,\mathsf{R}\,t)^+)$$
$$C \triangleright \mathsf{A}(s\,\mathsf{R}\,t)^- \iff C \triangleright t^- \lor (C \triangleright s^- \land C \triangleright \mathsf{A}^+(s\,\mathsf{R}\,t)^-)$$

We define $C \triangleright D := \forall s^\sigma \in D.\ C \triangleright s^\sigma$.

A *fragment* is a finite, rooted, and acyclic directed graph labeled with literal clauses. If G is a fragment, we write $x \in G$ to say that x is a node of G and $x \to_G y$ if there is a G-edge from x to y. A node $x \in G$ is *internal* if it has some successor and a *leaf* otherwise. If $x \in G$, we write Λ_x for the literal clause labeling x. We also write x_{root} for the root of a graph if the graph can be inferred from the context. A fragment is *nontrival* if its root is not a leaf.

We fix some subformula closed set $\mathcal{F}$ for the rest of this section. and write $\mathcal{L}$ for the set of locally consistent literal clauses over $\mathcal{F}$. We also fix some set $\mathcal{D} \subseteq \mathcal{L}$. Let $L \in \mathcal{D}$ be a clause. A fragment G is a $\mathcal{D}$-*fragment for* L if:

F1. If $x \in G$ is a leaf, then $\Lambda_x \in \mathcal{D}$ and $\Lambda_x \in \mathcal{L}$ otherwise.

F2. The root of G is labeled with L.

F3. If $x \to_G y$, then $\Lambda_y \triangleright \mathcal{R}(\Lambda_x)$.

F4. If $x \in G$ is internal and $\mathsf{AX}\, s^- \in \Lambda_x$, then $x \to_G y$ and $\Lambda_y \triangleright \mathcal{R}(\Lambda_x), s^-$ for some $y \in G$.

A $\mathcal{D}$-fragment G for L is a $\mathcal{D}$-*fragment for L and u* if whenever $L \triangleright u$ then:

E1. If $u = \mathsf{A}(s \cup t)^+$, then $L \triangleright t^+$ or $\Lambda_x \triangleright s^+$ for every internal $x \in G$ and $\Lambda_y \triangleright t^+$ for all leaves $y \in G$.

E2. If $u = \mathsf{A}(s \,\mathsf{R}\, t)^-$, then $L \triangleright t^-$ or $\Lambda_x \triangleright s^-$ every internal $x \in G$ and $\Lambda_y \triangleright t^-$ for some $y \in G$.

Note that if u is an eventuality and $L \triangleright u$, then u is fulfilled in every $\mathcal{D}$-fragment for L and u. The conditions $L \triangleright t^+$ in (E1) and $L \triangleright t^-$ in (E2) are required to handle the case of an eventuality that is fulfilled in L and allow for the construction of nontrivial fragments in this case. A *demo* for $\mathcal{D}$ is an indexed collection of nontrivial fragments $(G(u, L))_{u \in \mathcal{F}, L \in \mathcal{D}}$ where each $G(u, L)$ is a $\mathcal{D}$-fragment for L and u.

4.1   Demos to Finite Models

Assume that we are given some demo $(G(u, L))_{u \in \mathcal{F}, L \in \mathcal{D}}$. We construct a model $\mathcal{M}$ satisfying all labels occurring in the demo. If $\mathcal{F}$ is empty, there is nothing to show, so we can assume that $\mathcal{F}$ is nonempty.

The states of $\mathcal{M}$ are the nodes of all the fragments in the demo, i.e., every state of $\mathcal{M}$ is a dependent triple (u, L, x) with $u \in \mathcal{F}$, $L \in \mathcal{D}$, and $x \in G(u, L)$. A state (u, L, x) is labeled with atomic proposition p iff $p^+ \in \Lambda_x$.

To define the transitions of $\mathcal{M}$, we fix an ordering $u_0, \ldots, u_n$ of the signed formulas in $\mathcal{F}$. We write u_{i+1} for the successor of u_i in this ordering. The successor of u_n is taken to be u_0. The transitions of $\mathcal{M}$ are of two types. First, we lift all the internal edges of the various fragments to transitions in $\mathcal{M}$. Second, if x is a leaf in $G(u_i, L_j)$ that is labeled with L, we add transitions from (u_i, L_j, x) to all successors of the root of $G(u_{i+1}, L)$. Thus, the fragments in the demo can be thought of as arranged in a matrix as shown in Figure 3 where the L_i are the clauses in $\mathcal{D}$. Note that every root has at least one successor, since demos contain only nontrivial fragments. Thus, the resulting transition system is serial and hence a model. We then show that every state of $\mathcal{M}$ satisfies all signed formulas it supports.

Lemma 4.1. *If $(u, L, x) \in \mathcal{M}$ and $\Lambda_x \triangleright s^\sigma$, then $(u, L, x) \models \lfloor s^\sigma \rfloor$.*

Proof. The proof goes by induction on s. We sketch the case for $\mathsf{A}(s \cup t)^+$. The case for $\mathsf{A}(s \,\mathsf{R}\, t)^-$ is similar and all other cases are straightforward.

Let $w = (u_i, L_j, x) \in \mathcal{M}$ and assume $\Lambda_x \triangleright \mathsf{A}(s \cup t)^+$. By induction hypothesis it suffices to show $\mathsf{AU}_{\mathcal{M}}\, s\, t\, w$ where

$$\mathsf{AU}_{\mathcal{M}}\, s\, t\, w := \mathsf{AU}\, (\to_{\mathcal{M}})\, (\lambda(\_,\_, y).\Lambda_y \triangleright s^+)\, (\lambda(\_,\_, y).\Lambda_y \triangleright t^+)\, w$$

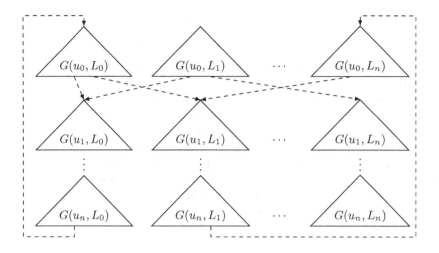

Fig. 3. Matrix of Fragments

To show $AU_{\mathcal{M}} s\, t\, w$ it suffices to show $AU_{\mathcal{M}} s\, t\, (u_{i+1}, L, x_{root})$ for all L satisfying $L \rhd A(s \cup t)^+$ since by (F3) the property of supporting $A(s \cup t)^+$ gets propagated down to the leaves of $G(u_i, L_j)$ on all paths that do not support t^+ along the way.

Without loss of generality, we can assume $A(s \cup t)^+ \in \mathcal{F}$. Thus, we can prove $AU_{\mathcal{M}} s\, t\, (u_{i+1}, L, x_{root})$ by induction on the distance from u_{i+1} to $A(s \cup t)^+$ according to the ordering of $\mathcal{F}$. If $u_{i+1} = A(s \cup t)^+$, we have $AU_{\mathcal{M}} s\, t\, (u_{i+1}, L, x_{root})$ by (E1). Otherwise, the claim follows by induction, deferring to the next row of the matrix as we did above. □

Theorem 4.2. *If $(G_{u,L})_{u \in F, L \in \mathcal{D}}$ is a demo for $\mathcal{D}$, there exists a finite model satisfying every label occurring in $(G_{u,L})_{u \in F, L \in \mathcal{D}}$.*

4.2 Formalizing the Model Construction

Our representation of fragments is based on finite types. We represent finite labeled graphs as relations over some finite type together with a labeling function. We then represent fragments using clause labeled graphs with a distinguished root element.

We turn the finite set $\mathcal{F} \times \mathcal{D}$ into a finite type I. Except for the transitions connecting the leaves of one row to the next row, the model is then just the disjoint union of a collection of graphs indexed by I. Let $G : I \to \textit{graph}$ be such a collection. We lift the internal edges of G by defining a predicate

$$\textit{liftEdge} \; : \; (\Sigma i{:}I.\, G\, i) \to (\Sigma i{:}I.\, G\, i) \to \textit{bool}$$

on the dependent pairs of an index and a node of the respective graph satisfying

$$liftEdge\,(i,x)\,(i,y) \iff x \to_{Gi} y$$
$$i \neq j \implies \neg liftEdge\,(i,x)\,(j,y)$$

The definition of *liftEdge* uses dependent types in a form that is well supported by Ssreflect.

Our model construction differs slightly from the construction used by Emerson and Halpern [5]. In Emerson's handbook article, every leaf of a fragment is replaced by the root with the same label on the next level. Thus, only the internal nodes of every fragment become states of the model. This would amount to using a Σ-type on the vertex type of every dag. In our model construction, we connect the leaves of one row to the successors of the equally labeled root of the next row, thus, avoiding a Σ-type construction. This makes use of the fact that CTL formulas cannot distinguish different states that have the same labels and the same set of successors.

5 Tableaux to Demos

An annotated clause is *consistent* if it is not derivable, and a clause C is consistent if $C|\cdot$ is consistent. Let $\mathcal{F}$ be a subformula closed set. We now construct a demo for the consistent literal clauses over $\mathcal{F}$. We define

$$\mathcal{D} := \{\, L \subseteq \mathcal{F} \mid L \text{ consistent}, L \text{ literal} \,\}$$

We now have to construct for every pair $(u,L) \in \mathcal{F} \times \mathcal{D}$ a nontrivial $\mathcal{D}$-fragment for L and u. We will construct a demo, where all the fragments are trees. To bridge the gap between the tableau, which works over arbitrary clauses, and $\mathcal{D}$-fragments, which are labeled with literal clauses only, we need the following lemma:

Lemma 5.1. *If $C|a \in \mathcal{U}(\mathcal{F})$ is consistent, we can construct a literal clause $L \subseteq \mathcal{F}$ such that $L \rhd C$ and $L|a$ is consistent.*

Proof. The proof proceeds by induction on the total size of the non-literal formulas in C. If this total size is 0, then C is a literal clause and there is nothing to show. Otherwise there exists some non-literal formula $u^\sigma \in C$. Thus $C|a = C \setminus \{u^\sigma\}, u^\sigma|a$ and we can apply the local rule for u^σ. Consider the case where $u^\sigma = s \to t^+$. By rule $\to^+$ we know that $C \setminus \{s \to t^+\}, s^-|a$ or $C \setminus \{s \to t^+\}, t^+$ is consistent. In either case we obtain a literal clause L supporting C by induction hypothesis. The other cases are similar.

Before we construct the fragments, we need one more auxiliary definition

$$\mathcal{R}^- C := \mathcal{R}C, \{\, \mathcal{R}C, s^- \mid \mathsf{AX}\, s^- \in C \,\}$$

The set of clauses $\mathcal{R}^- C$ serves the dual purpose of the request $\mathcal{R}C$. It contains all the clauses that must be supported at the successors of C to satisfy (F4).

The demo for $\mathcal{D}$ consists of three kinds of fragments. The easiest fragments are those for a pair (u,L) where $L \not\rhd u$ or u is not an eventuality. In this case, a $\mathcal{D}$-fragment for L is also a $\mathcal{D}$-fragment for L and u.

Lemma 5.2. *If $L \in \mathcal{D}$, we can construct a nontrivial $\mathcal{D}$-fragment for L.*

Proof. By assumption $L|\cdot$ is consistent. According to rules X and AX^-, $C|\cdot$ is consistent for every clause $C \in R^- L$. Note that there is at least one such clause. By Lemma 5.1, we can obtain for every $C \in \mathcal{R}^- L$ some clause $L_C \in \mathcal{D}$. The $\mathcal{D}$-fragment for L consists of a single root labeled with L and one successor labeled with L_C for every $C \in \mathcal{R}^- L$. □

Next, we deal with the case of a pair $(\mathsf{A}(s \cup t)^+, L)$ where $L \rhd \mathsf{A}(s \cup t)^+$. This is the place where we make use of the history annotations.

Lemma 5.3. *If $C|A(s \cup_H t)^+ \in \mathcal{U}(\mathcal{F})$ is consistent, we can construct a $\mathcal{D}$-fragment G for L such that $\Lambda_x \rhd s^+$ for every internal node $x \in G$ and $\Lambda_y \rhd t^+$ for all leaves $y \in G$ where L is some clause supporting $C, \mathsf{A}(s \cup t)^+$.*

Proof. Induction on the slack of H, i.e., the number of clauses from $\mathcal{P}(\mathcal{F})$ that are not in H. Since $C|A(s \cup_H t)^+$ is consistent, we know $C \notin H$. According to rule A_H, there are two cases to consider:

- $C, t^+|\cdot \in \mathcal{U}(\mathcal{F})$ is consistent: By Lemma 5.1 we obtain a literal clause L such that $L \rhd C, t^+$ and $L|\cdot$ is consistent. The trivial fragment with a single node labeled with L satisfies all required properties.
- $C, s^+|A^+(s \cup_{H,C} t)^+ \in \mathcal{U}(\mathcal{F})$ is consistent: By Lemma 5.1 we obtain a literal clause L such that $L \rhd C, s^+$ and $L|A^+(s \cup_{H,C} t)^+$ is consistent. In particular, $L, \mathsf{A}^+(s \cup t)^+$ is locally consistent and supports C as well as $\mathsf{A}(s \cup t)^+$. By induction hypothesis, we obtain a $\mathcal{D}$-fragment for every clause in $\mathcal{R}^- L$. Putting everything together, we obtain a $\mathcal{D}$-fragment for $L, \mathsf{A}^+(s \cup t)^+$ satisfying all required properties □

Lemma 5.4. *If $L \in \mathcal{D}$, we can construct a nontrivial $\mathcal{D}$-fragment for L and $\mathsf{A}(s \cup t)^+$.*

Proof. Without loss of generality we can assume that $L \rhd s^+$ and $\mathsf{A}^+(s \cup t)^+ \in L$. All other cases are covered by Lemma 5.2. Using rules X, AX^-, and $\mathsf{A}_\emptyset$, we show for every $C \in \mathcal{R}^- L$ that $C|\mathsf{A}(s \cup_\emptyset t)^+$ is a consistent clause in $\mathcal{U}(\mathcal{F})$. By Lemma 5.3 we obtain a $\mathcal{D}$-fragment for every such clause. Putting a root labeled with L on top as in the proof of Lemma 5.2, we obtain a nontrivial $\mathcal{D}$-fragment for L and $\mathsf{A}(s \cup t)^+$ as required. □

Lemma 5.5. *If $L \in \mathcal{D}$, we can construct a nontrivial $\mathcal{D}$-fragment for L and $\mathsf{A}(s \mathsf{R} t)^-$.*

The proof of Lemma 5.5 is similar to the proof of Lemma 5.4 and uses a similar auxiliary lemma.

Theorem 5.6. *1.   There exists a $\mathcal{D}$-demo.*
2.   If $C|\cdot$ is consistent, then $w \models C$ for some finite model $\mathcal{M}$ with $w \in \mathcal{M}$.

Note that by Theorem 4.2 all the locally consistent labels of the internal nodes of the constructed fragments are satisfiable and hence must be consistent. However, at the point in the proof of Lemma 5.3 where we need to show local consistency of $L, \mathsf{A}^+(s \cup t)$ from consistency of $L|\mathsf{A}^+(s \cup_H t)$ showing local consistency is all we can do.

All fragments constructed in this section are trees. In the formalization, we state the lemmas from this section using an inductively defined tree type, leaving the sets of nodes and edges implicit. Thus, trees can be composed without updating an edge relation or changing the type of vertices. Even using this tailored representation, the formalization of Lemma 5.3 is one of the most complex parts of our development.

For Theorem 5.6, we convert the constructed trees to rooted dags. To convert a tree T to a dag, we turn the list of subtrees of T into a finite type and use this as the type of vertices. We then add edges from every tree to its immediate subtrees. This construction preserves all fragment properties even though identical subtrees of T are collapsed in into a single vertex.

6   Tableau Refutations to Hilbert Refutations

We now return to the proof of Theorem 3.1. For this proof, we will translate the rules of the tableau calculus to lemmas in the Hilbert calculus. For this we need a number of basic CTL lemmas. The lemmas to which we will refer explicitly can be found in Figure 4. In formulas, we let $s \cup_H t$ abbreviate $(s \wedge H) \cup (t \wedge H)$.

We present the translation lemmas for the rules $\mathsf{A_H}$ and $\mathsf{R_H}$. Given the non-local soundness argument sketched in Section 3, it should not come as a surprise that the translation of both rules requires the use of the corresponding induction rule from the Hilbert axiomatization. For both lemmas we use the respective induction rule in dualized form as shown in Figure 4.

Lemma 6.1. *If* $\vdash t \to \neg C$ *and* $\vdash \mathsf{E}^+(s \cup_{H,C} t) \to s \to \neg C$, *then* $\vdash \neg(C \wedge \mathsf{E}(s \cup_H t))$.

Proof. Assume we have (a) $\vdash t \to \neg C$ and (b) $\vdash \mathsf{E}^+(s \cup_{H,C} t) \to s \to \neg C$. By propositional reasoning, it suffices to show

$$\vdash \mathsf{E}(s \cup_H t) \to \neg C \wedge \mathsf{E}(s \cup_{H,C} t)$$

Applying the $\mathsf{EU_{ind}}$ rule leaves us with two things to prove. The first one is $\vdash t \wedge H \to \neg C \wedge E(s \cup_{H,C} t)$ and can be shown using (a) and $\mathsf{E1}$. The other is

$$\vdash s \wedge H \to \mathsf{EX}(\neg C \wedge \mathsf{E}(s \cup_{H,C} t)) \to \neg C \wedge \mathsf{E}(s \cup_{H,C} t)$$

The second assumption can be weakened to $\mathsf{E}^+(s \cup_{H,C} t)$. Thus, we also have $\neg C$ by assumption (b). Finally, we obtain $\mathsf{E}(s \cup_{H,C} t)$ using Lemma $\mathsf{E2}$.   □

Lemma 6.2. *If* $\vdash C \to \neg t$ *and* $\vdash C \to s \to \neg A^+(s \cup_{H,C} t)$, *then* $\vdash \neg(C \wedge \mathsf{A}(s \cup_H t))$.

A1 $\vdash A(s\,\mathsf{U}\,t) \leftrightarrow t \vee s \wedge A^+(s\,\mathsf{U}\,t)$
A2 $\vdash \mathsf{EG}\,\neg t \to \neg A(s\,\mathsf{U}\,t)$
A3 $\vdash A((s \wedge u)\,\mathsf{U}\,(t \wedge u)) \to u$
E1 $\vdash t \to \mathsf{E}(s\,\mathsf{U}\,t)$
E2 $\vdash s \to \mathsf{E}^+(s\,\mathsf{U}\,t) \to \mathsf{E}(s\,\mathsf{U}\,t)$
AE $\vdash \mathsf{AX}\,s \to \mathsf{EX}\,t \to \mathsf{EX}(s \wedge t)$
$\mathsf{EU}_{\mathsf{ind}}$ If $\vdash t \to u$ and $\vdash s \to \mathsf{EX}\,u \to u$, then $\vdash \mathsf{E}(s\,\mathsf{U}\,t) \to u$
$\mathsf{EG}_{\mathsf{ind}}$ If $\vdash u \to s$ and $\vdash u \to \mathsf{EX}\,u$, then $\vdash u \to \mathsf{EG}\,s$

Fig. 4. Basic CTL Lemmas

Proof. Assume we have (a) $\vdash C \to \neg t$ and (b) $\vdash C \to s \to \neg A^+(s\,\mathsf{U}_{H,C}t)$. We set $u := \neg t \wedge A^+(s\,\mathsf{U}_H t) \wedge \neg A^+(s\,\mathsf{U}_{H,C}t)$. We first argue that it suffices to show (1) $\vdash u \to \mathsf{EG}\,\neg t$. Assume we have C and $A(s\,\mathsf{U}_H t)$. By (a) we also know $\neg t$ and thus we have $s \wedge H$ and $A^+(s\,\mathsf{U}_H t)$ by A1. Using (b) and (1), we obtain $\mathsf{EG}\,\neg t$ which contradicts $A(s\,\mathsf{U}_H t)$ according to A2.

We show (1) using the $\mathsf{EG}_{\mathsf{ind}}$ rule. Showing $\vdash u \to \neg t$ is trivial so it remains to show $\vdash u \to \mathsf{EX}\,u$. Assume u. By Lemma AE we have $\mathsf{EX}(A(s\,\mathsf{U}_H t) \wedge \neg A(s\,\mathsf{U}_{H,C}t))$. It remains to show

$$\vdash A(s\,\mathsf{U}_H t) \wedge \neg A(s\,\mathsf{U}_{H,C}t) \to u$$

We reason as follows:

1. $A(s\,\mathsf{U}_H t)$	assumption
2. $\neg A(s\,\mathsf{U}_{H,C}t)$	assumption
3. $\neg t \vee \neg H \vee C$	2, A1
4. $\neg s \vee \neg H \vee C \vee \neg A^+(s\,\mathsf{U}_{H,C}t)$	2, A1
5. H	1,A3
6. $\underline{\neg t}$	3, 5, (a)
7. $s \wedge A^+(s\,\mathsf{U}_H t)$	1,6,A1
8. $\underline{\neg A^+(s\,\mathsf{U}_{H,C}t)}$	4, 5, 7, (b)

This finishes the proof.   □

Proof (of Theorem 3.1). Let $C|a$ be derivable. We prove the claim by induction on the derivation of $C|a$. All cases except those for the rules $\mathsf{R_H}$ and $\mathsf{A_H}$ are straightforward. The former follows with Lemma 6.1 the latter with Lemma 6.2.   □

To formalize this kind of translation argument, we need some infrastructure for assembling Hilbert proofs as finding proofs in the bare Hilbert system can be a difficult task. We extend the infrastructure we used in our previous work [4] to CTL. We use conjunctions over lists of formulas to simulate context. We also use Coq's generalized (setoid) rewriting [12] with the preorder $\{(s,t) \mid \vdash s \to t\}$.

Putting our results together we obtain a certifying decision method for CTL.

Proof (of Theorem 2.2). By Lemma 3.3, derivability of the clause $s^+|\cdot$ is decidable. If $s^+|\cdot$ is derivable we obtain a proof of $\neg s$ with Theorem 3.1. Otherwise, we obtain a finite model satisfying s with Theorem 5.6 and Theorem 4.2. By Theorem 2.1, the two results are mutually exclusive.   □

Corollary 6.3 (Decidability). *Satisfiability of formulas is decidable.*

Corollary 6.4 (Completeness). *If $\forall \mathcal{M}.\forall w \in \mathcal{M}.w \models s$, then $\vdash s$.*

7 Conclusion

Our completeness proof for the tableau calculus differs considerably from the corresponding completeness proof for the sequent system given by Brünnler and Lange [3]. Their proof works by proving the completeness of another more restrictive sequent calculus which is ad-hoc in the sense that it features a rule whose applicability is only defined for backward proof search. We simplify the proof by working directly with the tableau rules and by using the model construction of Emerson [5].

The proof of Theorem 3.1 relies on the ability to express the semantics of history annotations in terms of formulas. This allows us to show the soundness of the tableau calculus by translating the tableau derivations in a compositional way. While this works well for CTL, this is not necessarily the case for other modal logics. As observed previously [4], the tableau system for CTL can be adapted to modal logic with transitive closure (K^+). However, K^+ cannot express the "until" operator used in the semantics of annotated eventualities. It therefore appears unlikely that the individual rules of a history-augmented tableau system for K^+ can be translated one by one to the Hilbert axiomatization. The tableau system we used to obtain a certifying decision method for K^+ [4] uses a complex compound rule instead of the more fine-grained history annotations employed here. Thus, even though the logic CTL is more expressive than K^+, the results presented here do not subsume our previous results.

An alternative to our construction of a certifying decision method could be to replace the tableau calculus with a pruning-based decision procedure like the one described by Emerson and Halpern [7]. In fact Emerson and Halpern's completeness proof for their Hilbert axiomatization of CTL is based on this algorithm. While their proof is non-constructive, we believe that it can be transformed into a constructive proof. In any case, we believe that the formal analysis of our history-augmented tableau calculus is interesting in its own right.

The proofs we present involve a fair amount of detail, most of which is omitted in the paper for reasons of space. Having a formalization thus not only ensures that the proofs are indeed correct, but also gives the reader the possibility to look at the omitted details.

For our formal development, we profit much from Ssreflect's handling of countable and finite types. Countable types form the basis for our set library and finite types are used heavily when we assemble the fragments of a demo into a finite model. Altogether our formalization consists of roughly 3500 lines. The included set library consists of about 700 lines, the remaining lines are split almost evenly over the proofs of Theorems 3.1, 5.6, and 4.2 and the rest of the development.

References

1. Baier, C., Katoen, J.P.: Principles of Model Checking. MIT Press (2008)
2. Bertot, Y., Gonthier, G., Biha, S.O., Pasca, I.: Canonical big operators. In: Mohamed, O.A., Muñoz, C., Tahar, S. (eds.) TPHOLs 2008. LNCS, vol. 5170, pp. 86–101. Springer, Heidelberg (2008)
3. Brünnler, K., Lange, M.: Cut-free sequent systems for temporal logic. J. Log. Algebr. Program. 76(2), 216–225 (2008)
4. Doczkal, C., Smolka, G.: Constructive completeness for modal logic with transitive closure. In: Hawblitzel, C., Miller, D. (eds.) CPP 2012. LNCS, vol. 7679, pp. 224–239. Springer, Heidelberg (2012)
5. Emerson, E.A.: Temporal and modal logic. In: van Leeuwen, J. (ed.) Handbook of Theoretical Computer Science. Formal Models and Sematics (B), vol. B, pp. 995–1072. Elsevier (1990)
6. Emerson, E.A., Clarke, E.M.: Using branching time temporal logic to synthesize synchronization skeletons. Sci. Comput. Programming 2(3), 241–266 (1982)
7. Emerson, E.A., Halpern, J.Y.: Decision procedures and expressiveness in the temporal logic of branching time. J. Comput. System Sci. 30(1), 1–24 (1985)
8. Gonthier, G., Mahboubi, A., Rideau, L., Tassi, E., Théry, L.: A modular formalisation of finite group theory. In: Schneider, K., Brandt, J. (eds.) TPHOLs 2007. LNCS, vol. 4732, pp. 86–101. Springer, Heidelberg (2007)
9. Gonthier, G., Mahboubi, A., Tassi, E.: A small scale reflection extension for the Coq system. Research Report RR-6455, INRIA (2008),
   http://hal.inria.fr/inria-00258384/en/
10. Kaminski, M., Smolka, G.: Terminating tableaux for hybrid logic with eventualities. In: Giesl, J., Hähnle, R. (eds.) IJCAR 2010. LNCS, vol. 6173, pp. 240–254. Springer, Heidelberg (2010)
11. Smullyan, R.M.: First-Order Logic. Springer (1968)
12. Sozeau, M.: A new look at generalized rewriting in type theory. Journal of Formalized Reasoning 2(1) (2009)
13. The Coq Development Team, http://coq.inria.fr

Hypermap Specification and Certified Linked Implementation Using Orbits

Jean-François Dufourd*

ICUBE, Université de Strasbourg et CNRS,
Pôle API, Boulevard S. Brant, BP 10413, 67412 Illkirch, France
jfd@unistra.fr

Abstract. We propose a revised constructive specification and a certified hierarchized linked implementation of combinatorial hypermaps using a general notion of orbit. Combinatorial hypermaps help to prove theorems in algebraic topology and to develop algorithms in computational geometry. Orbits unify the presentation at conceptual and concrete levels and reduce the proof effort. All the development is formalized and verified in the Coq proof assistant. The implementation is easily proved observationally equivalent to the specification and translated in C language. Our method is transferable to a great class of algebraic specifications implemented into complex data structures with hierarchized linear, circular or symmetric linked lists, and pointer arrays.

1 Introduction

We propose a revised constructive specification and a certified hierarchized linked implementation of 2-dimensional combinatorial hypermaps using a general notion of orbit. *Combinatorial hypermaps* [8] are used to algebraically describe meshed topologies at any dimension. They have been formalized to interactively prove great mathematical results, e.g. the famous Four-Colour theorem [20] or the discrete Jordan curve theorem [10], with the help of a proof assistant. They are also at the root of the certification of functional algorithms in computational geometry, e.g. Delaunay triangulation [14]. Once implemented, hypermaps (or derivatives) are a basic data structure in geometric libraries, e.g. Topofil [2] or CGAL [27]. *Orbits*, whose formal study is presented in [12], allow us to deal with trajectories, at conceptual and concrete levels. A precise correspondence between hypermaps and their linked implementation remained a challenge which we have decided to address. The novalties of this work are:

- An entire development formalized and verified in the *Coq proof assistant* [1]. Nothing is added to its higher-order calculus, except the *axiom of extensionality* and another one for *address generation* of new allocated memory cells. The first says that two functions are equal if they are equal at any point, and the second that an address generated for a block allocation is necessarily fresh and non-null;

---

* This work was supported in part by the French ANR project GALAPAGOS.

G. Klein and R. Gamboa (Eds.): ITP 2014, LNAI 8558, pp. 242–257, 2014.
© Springer International Publishing Switzerland 2014

- An extensive use of *orbits*, which nicely unifies and simplifies the presentation of specifications and linked implementations and reduces the proof effort;
- An intricate pointer implementation in a general simple memory model, and concrete operations described in a functional form which is easy to translate in a "true" programming language, e.g. in C;
- A proof of observational equivalence between specification and pointer implementation thank to morphisms.

The underlying method is transferable to a great class of complex data structures with hierarchized linear, circular or symmetric lists, and pointer arrays. That is greatly due to the unification provided by orbits. In Sect. 2, we recall the orbit formalization. In Sect. 3 and 4, we present and specify the hypermaps. In Sect. 5, we formalize the general memory model, then we describe in Sect. 6 the linked implementation. In Sect. 7, we prove the observational equivalence between specification and implementation. We discuss related work in Sect. 8 and conclude in Sect. 9. The whole formalization process is described but the proof scripts are out of the scope of this article. The Coq development, including ref. [12] and orbit library files, may be downloaded [11].

2   Orbits for Functions in Finite Domains

General Definitions and Properties. In Coq, all objects are strongly typed, `Prop` is the type of propositions, and `Type` can be viewed as the "type of types" [1]. We work in a *context* composed of: `X:Type`, any type whose built-in equality `=` is equipped with `eqd X`, a *decision function*, `exc X:X`, an element chosen as *exception*, `f:X -> X`, any *total function* on X, and D, a *finite* (sub)domain of X never containing `exc X`. For technical reasons, D is described as a finite list of type `list X` with *no repetitive element*. We write `In z l` when z occurs in the list l, `nil` for the empty list, and `~` for not. For any `z:X` and `k >= 0`, we consider `zk := Iter f k z`, the k-*th iterate* of z by f (with `z0 := z`). Since D is finite, during the iteration process calculating `zk`, a time necessarily comes where `zk` goes outside D or encounters an iterate already met [12].

Definition 1. *(Orbital sequence, length, orbit, limit, top)*
(i) The orbital sequence *of z by* f *at the order* k `>= 0`*, denoted by* `orbs k z`*, is the list containing* `z(k-1)`*, ..., * `z1`*, * `z0`*, written from first to last elements.*
(ii) The length *of z's orbit by* f *w.r.t.* D*, denoted by* `lorb z`*, is the least integer p such that* `~ In zp D` *or* `In zp (orbs p z)`*.*
(iii) The orbit *of z by* f *w.r.t.* D *is* `orbs (lorb z) z`*, in short* `orb z`*.*
(iv) The limit *of z by* f *w.r.t.* D*, is* `zp` − *or* `z(lorb z)` −*, in short* `lim z`*.*
(v) When `In z D`*, the* top *of z by* f *w.r.t.* D *is* `z(lorb z - 1)`*, in short* `top z`*.*

So, an orbit is a bounded list *without repetition*, possibly empty, which can be viewed as a finite set when it is more convenient. Necessarily the *shape* of z's orbit (Fig. 1(Left)) is: *(i) empty when* `~ In z D`; *(ii) a line when* `~ In (lim z) D`,

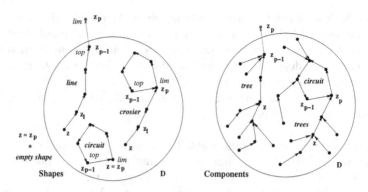

Fig. 1. Orbit Shapes (for 4 positions of z) / Components (for 2 positions of z)

what is denoted by the predicate `inv_line z`; *(iii)* a *crosier* when `In (lim z)` `(orb z)` what is denoted by `inv_crosier z`; *(iv)* a *circuit* when `lim z = z`, what is denoted by `inv_circ z`. An empty orbit is a line and a circuit as well, and a circuit is a crosier, since `lim z = z` entails `In (lim z) (orb z)`. The existence of an *orbital path* from `z` to `t`, i.e. the fact that `t` is in `z`'s orbit, is written `expo z t`, where `expo` is a binary *reflexive transitive* relation, which is *symmetric* only if `z`'s shape is a circuit. A lot of lemmas express the variation of `lorb`, `orb` and `lim` along `z`'s orbit, depending on its shape [11]. In fact, `(D, f)` forms a *functional graph* whose *connected components* are *trees* or *circuits with grafted trees* in D. Fig. 1(Right) shows components for two positions of `z` in D.

When the orbit of `z` is a non-empty circuit or `z` has exactly one `f`-predecessor in D, the definition of an *inverse* for `f`, denoted by `f_1 z`, is immediate (see [11]). That is the case for any `z` having an `f`-predecessor in D when `f` is *partially injective* w.r.t. D in the following sense (Here, `->` denotes the implication):

```
forall z t, In z D -> In t D -> f z <> exc X -> f z = f t -> z = t
```

That is the usual injection characterization for `f`, but only with `f z <> exc X` to fully capture the features of linked lists in memories. In this case, the orbit shapes are only lines and circuits, and the connected components only (linear) *branches* and *circuits*. The branch of `z` is obtained by prolongation of `z`'s orbit with the `f`-ancestors of `z` in D (Fig. 2(a)). When `f` is a partial injection, its inverse `f_1` enjoys expected properties, e.g. `f_1 (f z) = f (f_1 z) = z` and orbit shape similarities with `f` [11]. Fig. 2(b) shows the *inversion* of a branch connected component (Fig. 2(a)) of `z`. The previous notions are given in a *context*, in fact a *Coq section*, where X, `f` and D are *variables*. Outside the section, each notion is parameterized by X, `f` and D. For instance, `lorb z` becomes `lorb X f D z` and is usable *for any type, function and domain*. This is necessary when `f` or D are changing as in what follows.

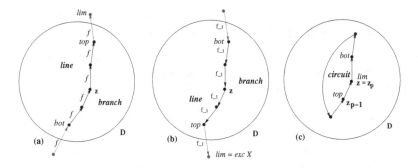

Fig. 2. (a) Branch containing z / (b) Inversion / (c) Closure

So, when f is *partially injective*, if we want to work only with circuits, we can *close* all the "open orbits", in fact the branches. This operation (Fig. 2(c)), which modifies f but not D, uses top (Sect. 2) and bot (Fig. 2(a,b)):

```
Definition bot f D z := top (f_1 X f D) D z.
Definition Cl f D z := if In_dec X z D then if In_dec X (f z) D then f z else bot f D z else z.
```

In Cl f D, all the orbits are circuits and the reachability expo is symmetrically obtained from this for f. Now, we outline orbit updating operations.

• **Addition/deletion.** An *addition* inserts a new element a in D, while the (total) function f:X -> X remains the same. We require that ~ In a D, we pose Da := a :: D and a1 := f a. Regarding the orbit variation, for any z:X, two cases arise: *(i)* When lim D f z <> a, the orbit of z is entirely preserved; *(ii)* Otherwise, great changes can occur (See [12]). But, if we suppose that ~ In a1 D (Fig. 3(Left)), which corresponds to most practical cases, z's new orbit is the previous one completed by a only. The "inverse" operation is the *deletion* of an element a from D. Regarding the orbit variation for any z:X, two cases arise: *(i)*

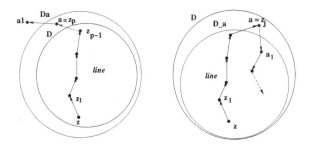

Fig. 3. Addition (Left)/Deletion (Right) effect on an orbit

When In a (orb D f z), z's orbit is cut into a *line* with limit a (Fig. 3(Right), where D_a:= remove D a); *(ii)* Otherwise it is entirely preserved.

• **Mutation.** A *mutation* modifies the *image* by f of an element u, i.e. f u, into an element, named u1, while all the other elements, and D, are unchanged. The new function, named Mu f u u1, is defined by:

```
Definition Mu(f:X->X)(u u1:X)(z:X):X := if eqd X u z then u1 else f z.
```

If u is not in D nothing important occurs. If u is in D, two cases arise to determine the new orbits of u1 and u (Fig. 4): *(i)* When ˜ In u (orb f D u1), the orbit of u1 does not change and the new orbit of u is this of u1 plus u itself (Fig. 4(Case A)); *(ii)* Otherwise, the mutation closes a *new circuit* which, say, starts from u1, goes to u by the old path, then goes back to u1 in one step (Fig. 4(Case B)). Then, the new orbit of any z:X can be obtained similarly. Different cases arise depending on the respective positions of z, u1 and u (See [12]).

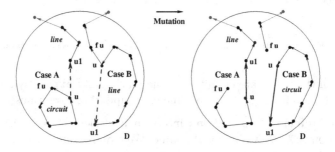

Fig. 4. Mutation: Cases A and B

• **Transposition.** A *transposition* consists in *exchanging* the images by f of two elements which must belong to circuits. In fact, just one element, u, and its *new successor*, u1, are provided in the transposition, whose definition is:

```
Definition Tu(f:X->X)(D: list X)(u u1:X)(z:X):X :=
    if eqd X u z then u1 else if eqd X (f_1 X f D u1) z then f u else f z.
```

The definition also says that the new successor of f_1 X f D u1, i.e. u1's predecessor, is f u, i.e. the old u's successor, and that nothing is changed for the other elements (Fig. 5). In the case where u1 = f u, the transposition has no effect. This operation is usable if u and u1 are in the same circuit or not (Fig. 5, Left and Right). Intuitively, in the first case, the unique circuit is *split* into two circuits, and, in the second case, the two circuits are *merged* into a unique one. That is the intuition, but the formal proof of these facts is far from being easy. Of course, for any z, it is proved that z's orbit (and connected components) are not modified by Tu if it contains neither u nor f_1 X f D u1.

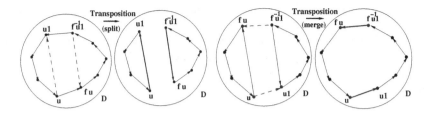

Fig. 5. Split (Left) and merge (Right)

3 Combinatorial Hypermaps

Combinatorial hypermaps [8] describe space subdivision topologies at any dimension. We focus to the dimension 2, and surface subdivisions, like in [20,10].

Definition 2. *(Hypermap)*
A (2-dimensional combinatorial) hypermap, $M = (D, \alpha_0, \alpha_1)$, is an algebraic structure where D is a finite set, the elements of which are called dart, *and α_0, α_1 are permutations on D, being indexed by a dimension, 0 or 1.*

Fig. 6(Left) shows a hypermap with $D = \{1, ..., 16\}$ and α_0, α_1 given by a table. It is *embedded* in the plane, its darts being represented by half-Jordan arcs (labeled by the darts) which are oriented from a bullet to a small transverse stroke. It z is a dart, $\alpha_0 z$ shares z's stroke, $\alpha_1 z$ shares z's bullet. A hypermap describes surface subdivision topologies, the cells of which can be defined by orbits, where $X = dart$, the type of the darts, and $D \subset dart$ in Def. 1.

Definition 3. *(Hypermap orbits)*
Let $M = (D, \alpha_0, \alpha_1)$ be a hypermap and $z \in D$ any dart in M. The edge *(resp.* vertex, face*) of z is the orbit of z by α_0 (resp. α_1, $\phi = \alpha_1^{-1} \circ \alpha_0^{-1}$) w.r.t. D.*

D	1	2	3	4	5	6	7	8	9	10	11	12	13	14	15	16
α_0	1	5	3	2	4	7	6	8	10	9	12	11	14	13	16	15
α_1	2	3	1	7	6	5	8	4	16	11	10	13	12	15	14	9

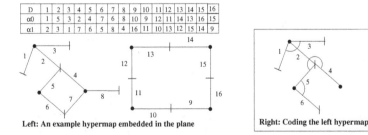

Left: An example hypermap embedded in the plane   Right: Coding the left hypermap

Fig. 6. 2-combinatorial hypermap and partial coding in Coq

Since α_0, α_1 and ϕ are bijections on D, then, for any $z \in D$, the orbits of z by them w.r.t. D are *circuits*. It is the same for their inverses – α_0^{-1}, α_1^{-1} and ϕ^{-1} – which are also defined over all D. So, in this cyclic case, if t is in z's orbit by

f, t's orbit and z's orbit may be identified into a *unique circuit*, defined *modulo* a cyclic permutation. So, the common orbit only counts for 1 in the number of orbits by f. In fact, these cyclic orbits modulo permutation correspond exactly to the *connected components* generated by f in D. Cells and connected components (w.r.t. $\{\alpha_0, \alpha_1\}$) can be counted to classify hypermaps according to their *Euler characteristic*, *genus* and *planarity* [9,12].

4   Hypermap Coq Formalization

Preliminaries. For simplicity, `dart` is the Coq library type `nat` of the natural numbers. The `eq_dart_dec` decision function of `dart` equality is `eq_nat_dec`, the decision function of `nat` equality, and the `dart` exception is `nild := 0`. Then, the *dimensions* are the two constants `zero` and `one` of the `dim` *enumerated* type, a simple case of *inductive* type.

Free Maps. As in [9,10,14], the 2-dimensional combinatorial hypermap specification begins with the inductive definition of a type, called `fmap`, of *free maps*, i.e. without any constraint:

```
Inductive fmap:Type :=
    V : fmap | I : fmap->dart->fmap | L : fmap->dim->dart->dart->fmap.
```

It has three *constructors*: V, for the *empty* − or *void* − free map; I m x, for the *insertion* in the free map m of an *isolated* dart x; and L m k x y, for the *linking* at dimension k of the dart x to the dart y: y becomes the k-successor of x, and x the k-predecessor of y. Any hypermap can be built by using these constructors, i.e. viewed as a *term* combining V, I and L. For instance, the six-darts part m2 (Fig. 6(Right), with k-links by L being symbolized by small circle arcs) of the hypermap of Fig. 6(Left) is built by:

```
m1 := I ( I (I (I ( I (I V 1) 2) 3) 4) 5) 6.
m2 := L (L (L (L (L m1 zero 4 2) zero 2 5) one 1 2) one 2 3) one 6 5.
```

Observers. Some *observers* (or *selectors*) can be easily defined on `fmap`. So, the *existence* in a free map m of a dart z is a predicate defined by structural induction on `fmap` by (True (resp. False) is the predicate always (resp. never) satisfied):

```
Fixpoint exd(m:fmap)(z:dart){struct m}: Prop :=
    match m with V => False | I m0 x => x = z \/ exd m0 z | L m0 _ _ _ => exd m0 z end.
```

The k-*successor*, pA m k z, of any dart z in m is similarly defined. It is nild if m is empty or contains no k-link from z. A similar definition is written for pA_1 m k z, the k-*predecessor*. To avoid returning nild in case of exception, A and A_1, *closures* of pA and pA_1, are defined in a mutual recursive way ([11]). They exactly simulate the expected behavior of the α_k permutations (Def. 2).

Hypermaps. To build *(well-formed) hypermaps*, I and L must be used only when the following preconditions are satisfied. Then, the *invariant* `inv_hmap m`, inductively defined on m, completely characterizes the *hypermaps* [11]:

```
Definition prec_I (m:fmap)(x:dart) := x <> nild /\ ~ exd m x .
Definition prec_L (m:fmap)(k:dim)(x y:dart) :=
  exd m x /\ exd m y /\ pA m k x = nild /\ pA_1 m k y = nild /\ A m k x <> y.
Fixpoint inv_hmap(m:fmap):Prop:= match m with
   V => True | I m0 x => inv_hmap m0 /\ prec_I m0 x
  | L m0 k0 x y => inv_hmap m0 /\ prec_L m0 k0 x y end.
```

When m is a well-formed hypermap, the last constraint of prec_L, A m k x <> y, entails that, for any z, the orbit for pA m k and pA_1 m k are *never closed*, i.e. are never circuits, and always remain *lines* (Sect. 2). For instance, our m2 example respects these preconditions and satisfies inv_hmap m2, so the circle arcs in Fig. 6(Right) do not form full circles. This choice is motivated by *normal form* considerations allowing inductive proofs of topological results [9,10].

Hypermap properties. When inv_hmap m is satisfied, it is proved that A m k and A_1 m k are *inverse bijective* operations. Then, considering the function m2s m which sends the free map m into its support "set" − or instead finite list, it is proved that pA m k and A m k are *partial injections* on m2s m. So, for any dart z, the orbit of z for pA m k w.r.t. m2s m always is a *line* (ended by nild). It is proved that A m k is really the *closure* of pA m k in the orbit meaning (Sect. 2), so the orbits for A m k and A_1 m k are circular:

```
Theorem inv_line_pA: forall m k z, inv_hmap m -> inv_line dart (pA m k) (m2s m) z.
Theorem A_eq_Cl: forall m k, inv_hmap m -> A m k = Cl dart (pA m k) (m2s m).
Theorem inv_circ_A: forall m k z, inv hmap m -> inv_circ dart (A m k) (m2s m) z.
Lemma A_1_eq_f_1: forall m k z,
  inv_hmap m -> exd m z -> A_1 m k z = f_1 dart (A m k) (m2s m) z.
```

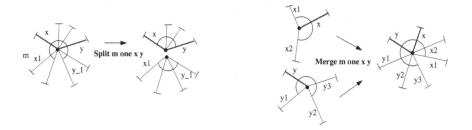

Fig. 7. Split (Left) and Merge (Right) at dimension one

Then, orbit results on *connectivity* [11] for A m k, A_1 m k and their compositions apply, allowing recursive definitions of the numbers of edges, vertices, faces. The number of connected components, w.r.t. {A m zero, A m one}, is defined similarly. All this leads to an incremental definition of the *Euler characteristic*, to the *Genus theorem* and to *full constructive planarity criteria* [9,11].

High-level Operations. Operations to *break* a k-link, *delete* an (isolated) dart, and *shift* the hole of a k-orbit, are specified inductively. They preserve the hypermap invariant. Finally, `Split m k x y` *splits* an orbit for `A m k` into two parts and `Merge m k x y` *merges* two distinct orbits for `A m k` (Fig. 7 for `k:= one`). For both, the goal is to insert a new k-link from `x` to `y`, while restoring the consistency of the resulting hypermap. The two operations correspond exactly with an orbit *transposition* (Sect. 2):

```
Lemma A_Split_eq_Tu: forall m k x y,
  inv_hmap m -> prec_Split m k x y -> A (Split m k x y) k = Tu dart (A m k) (m2s m) x y.
Lemma A_Merge_eq_Tu: forall m k x y,
  inv_hmap m -> prec_Merge m k x y -> A (Merge m k x y) k = Tu dart (A m k) (m2s m) x y.
```

5   General Memory Model

Goal We want to derive imperative programs using "high-level" C type facilities, i.e. `typedef`. So, in our memory C manager simulation, we avoid "low-level" features – bits, bytes, words, registers, blocks, offsets, etc – as in projects about compilers or distributed systems [21,6]. In C, an allocation is realized by `malloc(n)` where `n` is the byte-size of the booked area. A C *macro*, we call `MALLOC(T)`, hides the size and allows us to allocate an area for any object of type `T` and to return a pointer (of type `(T *)`) on this object:

```
#define MALLOC(T) ((T *) malloc(sizeof(T)))
```

Our (simple) *memory model* simulates it, with natural numbers as potential *addresses* and the possibility to always obtain a *fresh* address to store any object.

Coq Formalization. In Coq, the address type `Addr` is `nat`, with the exception `null` equal to 0. In a Coq section, we first declare *variables*: `T:Type`, for any type, and `undef`, exception of type `T`. Then we define a *memory generic type* `Mem T`, for data of type `T`, inductively with two constructors: `initm`, for the empty memory; and `insm M a c`, for the insertion in the memory `M` at the address `a` of an object `c:T`. It is easy to recursively define `valid M a`, the predicate which expresses that `a` is a *valid* address (i.e. pointing to a stored object) in `M`, and `dom M`, the *validity domain* of `M`, a finite address set – or instead, list. A precondition derives for `insm` and an invariant `inv_Mem M` for each memory `M`. Finally, the parameter function `adgen` returns a fresh address from a memory, since it satisfies the *axiom* `adgen_axiom`, which is the second and last one (after *extensionality*) of our full development. This mechanism looks like the *axiom of choice* introduction which does not affect Coq's consistency:

```
Variables (T:Type) (undef:T).
Inductive Mem: Type:= initm : Mem | insm : Mem -> Addr -> T -> Mem.
Fixpoint valid(M:Mem)(z:Addr): Prop :=
    match M with initm => False | insm M0 a c => a = z \/ valid M0 z end.
Definition prec_insm M a := ~valid M a /\ a <> null.
Parameter adgen: Mem -> Addr.
Axiom adgen_axiom: forall(M:Mem), let a := adgen M in ~valid M a /\ a <> null.
```

Memory Operations. The *allocation* of a block for an object of type T (recall that T is an implicit parameter) is defined by (%type forces Coq to consider * as the Cartesian type product):

```
Definition alloc(M:Mem):(Mem * Addr)%type:= let a := adgen M in (insm M a undef, a).
```

It returns a pair composed of a new memory and a fresh address for an allocated block containing undef. It formalizes the behavior of the above MALLOC. The other operations, which are "standard", are proved *conservative* w.r.t. to inv_Mem: load M z returns the data at the address z if it is valid, undef otherwise; free M z releases, if necessary, the address z and its corresponding block in M; and mut M z t changes in M the value at z into t.

6   Hypermap Linked Implementation

Linked cell memory As usual in geometric libraries, we orient the implementation toward linked cells. First, at each *dart* is associated a structure of type cell, defined thanks to a Coq Record scheme. So, a dart is itself represented by an address. Then, the type Mem, which was parameterized by T, is instantiated into Memc to contain cell data. In the following, all the generic memory operations are instantiated like Memc, with names suffixed by "c":

```
Record cell:Type:= mkcell { s : dim -> Addr; p : dim -> Addr; next : Addr }.
Definition Memc := Mem cell.
```

So, s and p are viewed as *functions*, in fact *arrays* indexed by dim, to represent *pointers* to the successors and predecessors by A, and next is a *pointer*, intended for a *next* cell, used for hypermap full traversals (e.g., see Fig. 8 or 9).

Main linked list A 2-dimensional hypermap is represented by a pair (M, h), where M is such a memory and h the *head pointer* of a *singly-linked linear list* of cells representing exactly the darts of the hypermap. The corresponding type is called Rhmap. Then, a lot of functions are defined for this representation. Their names start with "R", for "Representation", and their meaning is immediate:

```
Definition Rhmap := (Memc * Addr)%type.
Definition Rnext M z := next (loadc M z).
Definition Rorb Rm := let (M, h) := Rm in orb Addr (Rnext M) (domc M) h.
Definition Rlim Rm := let (M, h) := Rm in lim Addr (Rnext M) (domc M) h.
Definition Rexd Rm z := In z (Rorb Rm).
Definition RA M k z := s (loadc M z) k.
Definition RA_1 M k z := p (loadc M z) k.
```

So, Rnext M z gives the address of z's *successor* in the list, Rorb Rm and Rlim Rm are z's *orbit* and *limit* for Rnext M w.r.t. domc M when Rm = (M, h). Operations Rexd, RA and RA_1 are intended to be the representations of exd, A and A_1 of the hypermap specification. However, this will have to be proved.

Invariants and Consequences. To manage well-defined pointers and lists, our hypermap representation must be restricted. We have grouped constraints in a *representation invariant*, called inv_Rhmap, which is the conjunction of inv_Rhmap1 and inv_Rhmap2 dealing with orbits:

```
Definition inv_Rhmap1 (Rm:Rhmap) := let (M, h) := Rm in
  inv_Memc M /\ (h = null \/ In h (domc M)) /\ lim Addr (Rnext M) (domc M) h = null.
Definition inv_Rhmap2 (Rm:Rhmap) := let (M, h) := Rm in
  forall k z, Rexd Rm z ->
    inv_circ Addr (RA M k) (Rorb Rm) z /\ RA_1 M k z = f_1 Addr (RA M k) (Rorb Rm) z.
Definition inv_Rhmap (Rm:Rhmap) := inv_Rhmap1 Rm /\ inv_Rhmap2 Rm.
```

For a hypermap representation $Rm = (M,h)$: *(i)* inv_Rhmap1 Rm prescribes that M is a well-formed cell memory, h is null or in M, and the limit of h by Rnext M w.r.t. the domain of M is null. Therefore, the corresponding orbit, called Rorb Rm, is a *line*; *(ii)* inv_Rhmap2 Rm stipulates that, for all dimension k, and address z in Rorb Rm, the orbit of z by RA M k w.r.t. Rorb Rm is a *circuit*, and RA_1 M k z is the *inverse image* of z for RA M k.

So, for any address z in Rorb Rm, i.e. of a cell in the main list, we immediately have that RA and RA_1 are *inverse operations* and that the orbit of z by RA_1 M k w.r.t. Rorb Rm is also a *circuit*. Consequently, for k = zero or one, the fields (s k) and (p k) are inverse pointers which determine *doubly-linked circular lists*, each corresponding to a hypermap edge or vertex, which can be traversed in forward and backward directions. Moreover, these lists are *never empty*, and, for each k and direction, they determine a *partition* of the main simply-linked list.

User Operations. Now, a *complete kernel* of user concrete operations may be defined, exactly as in geometric libraries, e.g. Topofil [2], preserving inv_Rhmap and hiding dangerous pointer manipulations. We "program" it by using Coq functional forms whose translation in C is immediate.

• Firstly, RV returns an *empty* hypermap representation in any well-formed cell memory M. So, no address z points in the representation:

```
Definition RV(M:Memc): Rhmap := (M, null).
Lemma Rexd_RV: forall M z, inv_Memc M -> ~ Rexd (RV M) z.
```

• Secondly, RI Rm *inserts* at the head of Rm = (M, h) a new cell whose generated address is x, initializes it with convenient pointers (ficell x initializes s's and p's pointers with x), and returns the new memory, M2, and head pointer, x. Fig. 8 illustrates RI when Rm is not empty:

```
Definition RI(Rm:Rhmap):Rhmap :=
  let (M, h) := Rm in
  let (M1, x) := allocc M in
  let M2 := mutc M1 x (modnext (ficell x) h) in (M2, x).
```

Fig. 8. Operation RI: Inserting a new dart in Rm (Left) giving RI Rm (Right)

Numerous properties are proved by following the *state changes* simulated by the variables assignments. Of course, RI does what it should do: adding a new cell pointed by x, and creating *new fixpoints* at x for RA and RA_1 at any dimension. The corresponding orbits of x are *loops*, while the other orbits are unchanged (eq_Addr_dec is the address comparison):

```
Lemma Rorb_RI: Rorb (RI Rm) = Rorb Rm ++ (x :: nil).
Lemma Rexd_RI: forall z, Rexd (RI Rm) z <-> Rexd Rm z \/ z = adgenc M.
Lemma RA_RI: forall k z, RA (fst (RI Rm)) k z = if eq_Addr_dec x z then x else RA M k z.
Lemma RA_1_RI: forall k z, RA_1 (fst (RI Rm)) k z = if eq_Addr_dec x z then x else RA_1 M k z.
```

Proofs use in a crucial way properties of orbit operations (Sect. 2): Rnext, RA and RA_1 can be viewed through additions and mutations in orbits, e.g.:

```
Lemma Rnext_M2_Mu: Rnext M2 = Mu Addr (Rnext M) z_1 (Rnext M z).
Lemma RA_M2_Mu: forall k, RA M2 k = Mu Addr (RA M k) x x.
```

• Thirdly, RL m k x y performs a *transposition* involving the orbits of x and y at dimension k. It replaces, for function RA Rm k, the successor of x by y, and the successor of RA_1 m k y by RA m k x, and also modifies the predecessors consistently. This operation, which is the composition of four memory mutations, is illustrated in Fig. 9 for k:= one. It is proved that the expected missions of RL are satisfied, for Rexd, RA m k and RA_1 m k. So, the orbits for RA m k and RA_1 m k remains *circuits*, and RL realizes either a *merge* or a *split*. The proofs extensively use relations established with the orbit operations, particularly with the generic orbit Tu transposition (Sect. 2), e.g., M6 being the final memory state:

```
Lemma RA_M6_eq_Tu: RA M6 k = Tu Addr (RA M k) (Rorb (M,h)) x y.
```

• Fourthly, RD Rm z H achieves a *deletion* of the pointer x (which must be a fixpoint for RA and RA_1) and a *deallocation* of the corresponding cell, if any. The additional formal parameter H is a proof of inv_Rhmap1 Rm, which is used to guarantee the *termination* of x's searching in Rm. Indeed, in the general case, one has to find x_1, the *predecessor* address of x in the linked list, thanks to a *sequential search*. It is proved that, after RD, x is always invalid, and Rnext, RA and RA_1 are updated conveniently. Once again, the proofs [11] rely on properties of orbit deletion and mutation (Sect. 2).

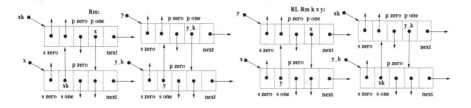

Fig. 9. Operation RL: Transposing two darts of Rm (Left) giving Rm k x y (Right)

7 Equivalence Hypermap Specification / Implementation

Abstraction Function. We want to go from a hypermap representation, Rm, to its meaning − its *semantics* − in terms of hypermap specification by an *abstraction function*. From a user's viewpoint, the construction of Rm is realized exclusively from a well-formed memory M throughout RV, RI, RL and RD calls, satisfying the preconditions. That is expressed by a predicate, called CRhmap Rm, which is inductively defined (for technical reasons of recursion principle generation) in Type [11]. If Rm satisfies CRhmap Rm, then it is proved to be a consistent hypermap representation. Then, the *abstraction function*, called Abs, is recursively defined by a matching on a proof CRm of this predicate, m0 := Abs Rm0 H0 being the result from the previous hypermap, Rm0, in the recursion, if any:

```
Fixpoint Abs (Rm: Rhmap) (CRm : CRhmap Rm) {struct CRm}: fmap :=
  match CRm with
    CRV M H0 => V
  | CRI Rm0 H0 => let m0 := Abs Rm0 H0 in I m0 (adgenc (fst Rm0))
  | CRL Rm0 k x y H0 H1 => let m0 := Abs Rm0 H0 in
       if expo_dec Addr (RA (fst Rm0) k) (Rorb Rm0) x y
       then if eqd Addr (A m0 k x) y then m0 else Split m0 k x y
       else Merge m0 k x y
  | CRD Rm0 x Inv H0 H1 => let m0 := Abs Rm0 H0 in D m0 x.
  end.
```

So, we prove that Abs leads to a real abstract hypermap, with an exact correspondence between exd, A and A_1 of Abs Rm CRm and Rexd, RA and RA_1 of Rm.

Representation Function. Conversely, we go from a hypermap to its representation in a given memory M. However, the mapping is possible only if the darts correspond to addresses which would be generated by successive cell allocations from M. So, the *representation function* Rep returns from a hypermap a pair (Rm, Pm), where Rm:Rhmap is a hypermap representation and Pm:Prop is a proposition saying if the preceding property is satisfied or not:

```
Fixpoint Rep (M:Memc)(m:fmap): (Rhmap * Prop)%type :=
  match m with
    V => (RV M, True)
  | I m0 x t p => let (Rm0, P0) := Rep M m0 in (RI Rm0 t p, P0 /\ x = adgenc (fst Rm0))
  | L m0 k x y => let (Rm0, P0) := Rep M m0 in (RL Rm0 k x y, P0)
  end.
```

Consequently, the representation into Rm succeeds *if and only if* Pm is satisfied. In this case, Rm satisfies inv_Rhmap, with an exact correspondence between exd, A and A_1 of m and Rexd, RA and RA_1 of Rm. These results make Abs and Rep *morphisms*. So, we have an *observational equivalence* specification-representation.

8 Related Work and Discussion

Static Proofs of Programs. They are mostly rooted in Floyd-Hoare logic, which evaluates predicates over a variable *stack* and a memory *heap* throughout program running paths. To overcome difficulties of conjoint local and global reasoning, Reynolds's *separation logic* [26] considers heap *regions* − combined by *, a *conjunction operator* −, to link predicates to regions. In fact, predicates often refer to underlying *abstract* data types. Then, the reasoning concerns at the same time low- and high-level concepts, which is quite difficult to control. These approaches are rather used to prove isolated already written programs.

Algebraic Specification. To specify and correctly implement entire *libraries*, we prefer starting with a formal specification, then deriving an implementation, and proving they correspond well. So, like [3,24,23,17,7], we find it better to specify with *inductive datatype* definitions in the spirit of *algebraic specifications* [28]. They must be constrained by *preconditions* inducing datatype *invariants*. Of course, definitions and proofs by *structural induction* must be supplemented by *Nœtherian* definitions and reasoning, to deal with termination problems. **Memory and Programming Models.** To implement, we focus on

C, in which our geometric programs are mostly written [2]. Here, fine storage considerations [21,6] are not necessary: A simple "type-based" memory model is good enough. Coq helps to simulate C memory management, expressions and commands at high level, following the "good rules" of structured (sequential) programming. Moreover, we try to converge to *tail-recursive definitions*, which are directly translatable into C loops. Of course, additional *variant* parameters to deal with termination, e.g. in RD, must be erased, as in the Coq *extraction* mechanism [1].

Separation and Collision. Burstall [4] defines *list systems* to verify linked list algorithms. Bornat [3] deals with address sequences like in our orbits. Mehta and Nipkow propose relational abstractions of linear or circular linked lists with some separation properties in Hoare logic embedded in Isabelle/HOL [24]. These questions are systematized in *separation logic* [26,25] which is well suited to distributed or embedded systems [18] and to composite data structures [17]. Enea et al. propose a separation logic extension for program manipulating "overlaid" and nested linked lists [16]. They introduce, $*_w$, a field separating conjunction operator to compose structures sharing some objects. Their examples turn around nested *singly*-linked lists and list arrays. In fact, our orbit notion seems sufficient for all singly- or doubly-linked linear or cyclic data structures, at several levels, and predicates can be expressed by using Coq logic. Separation is expressed by *disjunction* of orbits, whereas collision (and aliasing) is described by their *coalescence*, orbits often playing the role of *heaps* in separation logic.

Hypermap Specification. To specify hypermaps [8] in Coq/SSReflect, Gonthier et al. [20] adopt an *observational* viewpoint with a definition similar to Def. 2 (Sect. 3). So, they quickly have many properties on permutations. Their final result, i.e. the proof of the Four-Colour theorem, is resounding [20]. Our approach is *constructive*, starting from a free inductive type, which is constrained into a hypermap type. It complicates the true beginning but favors structural induction, algorithms and verifications in discrete topology [9,10,14].

Dedicated Proof Systems. The Ynot project is entirely based on Coq [15,22]: Higher-order imperative programs (with pointers) are constructed and verified in a Coq specific environment based on *Hoare Type Theory*, separation logic, monads, tactics and Coq to OCaml extraction. Thanks to special annotations, proof obligations are largely automated. Our approach is more pragmatic and uses Coq, and orbits, without new logic features. The proofs are interactive, but specification and implementation are distinct, their links being morphisms, and the simulated imperative programs are close to C programs. Many verification platforms, e.g. Why3 [19] or Frama-C [5], are based on Hoare logic, and federate solvers, mainly SAT and SMT, often in first-order logic. The Bedrock system by Chlipala [6] is adapted to implementation, specification and verification of low-level programs. It uses Hoare logic, separation logic and Coq to reason about pointers, mostly automatically. However, it cannot tolerate any abstraction beyond that of assembly languages (words, registers, branching, etc).

9   Conclusion

We presented a study in Coq for the entire certified development of a hypermap library. We follow a data refinement by algebraic specifications and reuse at each stage orbit features [12]. To certify pointer implementations, we use only Coq on a simple memory model, instead of Hoare logic and separation logic. The Coq development for this hypermap application − including the memory model, contains about 9,000 lines (60 definitions, 630 lemmas and theorems) [11]. It imports a Coq generic orbit library [12] reusable to certify implementations with singly- or doubly-linked lists, alone, intermixed or nested.

In the future, we will generalize orbit results, in extension of [13], to deal with general trees or graphs. Then, we will try to establish an accurate relation with *separation logic*, where orbits could provide the basis of new predicates on complex structures. The *fragment of C* which we cover by Coq must be precised to characterize reachable programs and develop a translator to C. The use of automatic provers, often jointed in platforms is questionnable. It would be interesting to carry orbit notions in their specification languages, e.g. ACSL in Frama-C [5]. At high level, we will investigate 3-dimensional hypermaps to deal with 3D functional or imperative computational geometry programs, like the 3D Delaunay triangulation which remains a formidable challenge.

References

1. Bertot, Y., Casteran, P.: Interactive Theorem Proving and Program Development - Coq'Art: The Calculus of Inductive Constructions. Springer (2004)
2. Bertrand, Y., Dufourd, J.-F., Françon, J., Lienhardt, P.: Algebraic Specification and Development in Geometric Modeling. In: Gaudel, M.-C., Jouannaud, J.-P. (eds.) CAAP 1993, FASE 1993, and TAPSOFT 1993. LNCS, vol. 668, pp. 75–89. Springer, Heidelberg (1993)
3. Bornat, R.: Proving Pointer Programs in Hoare Logic. In: Backhouse, R., Oliveira, J.N. (eds.) MPC 2000. LNCS, vol. 1837, pp. 102–126. Springer, Heidelberg (2000)
4. Burstall, R.M.: Some Techniques for Proving Correctness of Programs which Alter Data Structures. Machine Intelligence 7, 23–50 (1972)
5. CEA-LIST and INRIA-Saclay-Proval Team. Frama-C Project (2013), http://frama-c.com/about.html
6. Chlipala, A.: Mostly-Automated Verification of Low-level Programs in Computational Separation Logic. In: Int. ACM Conf. PLDI 2011, pp. 234–245 (2011)

7. Conway, C.L., Barrett, C.: Verifying Low-Level Implementations of High-Level Datatypes. In: Touili, T., Cook, B., Jackson, P. (eds.) CAV 2010. LNCS, vol. 6174, pp. 306–320. Springer, Heidelberg (2010)
8. Cori, R.: Un code pour les graphes planaires et ses applications. Soc. Math. de France, Astérisque 27 (1970)
9. Dufourd, J.-F.: Polyhedra genus theorem and Euler formula: A hypermap-formalized intuitionistic proof. Theor. Comp. Science 403(2-3), 133–159 (2008)
10. Dufourd, J.-F.: An Intuitionistic Proof of a Discrete Form of the Jordan Curve Theorem Formalized in Coq with Combinatorial Hypermaps. J. of Automated Reasoning 43(1), 19–51 (2009)
11. Dufourd, J.-F.: Hmap Specification and Implementation - On-line Coq Development (2013), http://dpt-info.u-strasbg.fr/~jfd/Hmap.tar.gz
12. Dufourd, J.-F.: Formal Study of Functional Orbits in Finite Domains. Submitted to TCS, 40 pages (2013)
13. Dufourd, J.-F.: Dérivation de l'Algorithme de Schorr-Waite par une Méthode Algébrique. In: JFLA 2012, INRIA, hal-00665909, Carnac, 15 p. (February 2012)
14. Dufourd, J.-F., Bertot, Y.: Formal Study of Plane Delaunay Triangulation. In: Kaufmann, M., Paulson, L.C. (eds.) ITP 2010. LNCS, vol. 6172, pp. 211–226. Springer, Heidelberg (2010)
15. Chlipala, A., et al.: Effective Interactive Proofs for Higher-Order Imperative programs. In: ICFP 2009, pp. 79–90 (2009)
16. Enea, C., Saveluc, V., Sighireanu, M.: Compositional Invariant Checking for Overlaid and Nested Linked Lists. In: Felleisen, M., Gardner, P. (eds.) ESOP 2013. LNCS, vol. 7792, pp. 129–148. Springer, Heidelberg (2013)
17. Berdine, J., Calcagno, C., Cook, B., Distefano, D., O'Hearn, P.W., Wies, T., Yang, H.: Shape Analysis for Composite Data Structures. In: Damm, W., Hermanns, H. (eds.) CAV 2007. LNCS, vol. 4590, pp. 178–192. Springer, Heidelberg (2007)
18. Marti, N., Affeldt, R.: Formal Verification of the Heap Manager of an Operating System Using Separation Logic. In: Liu, Z., Kleinberg, R.D. (eds.) ICFEM 2006. LNCS, vol. 4260, pp. 400–419. Springer, Heidelberg (2006)
19. Filliâtre, J.-C.: Verifying Two Lines of C with Why3: An exercise in program verification. In: Joshi, R., Müller, P., Podelski, A. (eds.) VSTTE 2012. LNCS, vol. 7152, pp. 83–97. Springer, Heidelberg (2012)
20. Gonthier, G.: Formal Proof - the Four Color Theorem. Notices of the AMS 55(11), 1382–1393 (2008)
21. Leroy, X., Blazy, S.: Formal Verification of a C-like Memory Model and Its Uses for Verifying Program Transformations. J. of Autom. Reas. 41(1), 1–31 (2008)
22. Malecha, G., Morrisett, G.: Mechanized Verification with sharing. In: Cavalcanti, A., Deharbe, D., Gaudel, M.-C., Woodcock, J. (eds.) ICTAC 2010. LNCS, vol. 6255, pp. 245–259. Springer, Heidelberg (2010)
23. Marché, C.: Towards Modular Algebraic Specifications for Pointer Programs: A Case Study. In: Comon-Lundh, H., Kirchner, C., Kirchner, H. (eds.) Jouannaud Festschrift. LNCS, vol. 4600, pp. 235–258. Springer, Heidelberg (2007)
24. Mehta, F., Nipkow, T.: Proving pointer programs in higher-order logic. Information and Computation 199(1-2), 200–227 (2005)
25. O'Hearn, P.W., Yang, H., Reynolds, J.C.: Separation and information hiding. ACM TOPLAS 31(3) (2009)
26. Reynolds, J.C.: Separation Logic: A Logic for Shared Mutable Data Structures. In: LICS 2002, pp. 55–74 (2002)
27. CGAL Team. Computational Geometry Algorithms Library Project, Chapter 27: Combinatorial Maps (2013), http://www.cgal.org
28. Wirsing, M.: Algebraic Specification. In: Handbook of TCS, vol. B. Elsevier/MIT Press (1990)

A Verified Generate-Test-Aggregate Coq Library for Parallel Programs Extraction

Kento Emoto[1], Frédéric Loulergue[2], and Julien Tesson[3]

[1] Kyushu Institute of Technology, Japan
emoto@ai.kyutech.ac.jp
[2] Univ. Orléans, INSA Centre Val de Loire, LIFO EA 4022, France
Frederic.Loulergue@univ-orleans.fr
[3] Université Paris Est, LACL, UPEC, France
Julien.Tesson@univ-paris-est.fr

Abstract. The integration of the generate-and-test paradigm and semi-rings for the aggregation of results provides a parallel programming framework for large scale data-intensive applications. The so-called GTA framework allows a user to define an inefficient specification of his/her problem as a composition of a generator of all the candidate solutions, a tester of valid solutions, and an aggregator to combine the solutions. Through two calculation theorems a GTA specification is transformed into a divide-and-conquer efficient program that can be implemented in parallel. In this paper we present a verified implementation of this framework in the Coq proof assistant: efficient bulk synchronous parallel functional programs can be extracted from naive GTA specifications. We show how to apply this framework on an example, including performance experiments on parallel machines.

Keywords: List homomorphism, functional programming, automatic program calculation, semi-ring computation, bulk synchronous parallelism, Coq.

1 Introduction

Nowadays parallel architectures are everywhere. However parallel programming is still reserved to experienced programmers. There is an urgent need of programming abstractions, programming methodologies as well as support for the verification of parallel applications, in particular for distributed memory models. Our goal is to provide a framework to *ease* the *systematic* development of *correct* parallel programs. We are particularly interested in large scale data-intensive applications.

Such a framework should provide programming building blocks whose semantics is easy to understand for users. These building blocks should come with an equational theory that allows to transform programs towards more efficient versions. These more efficient versions should be parallelisable in a transparent way for the user.

G. Klein and R. Gamboa (Eds.): ITP 2014, LNAI 8558, pp. 258–274, 2014.

GTA (generate-test-aggregate) [4,3] provides such a framework, integrating the generate-and-test paradigm and semi-rings for the aggregation of results. Through two calculation theorems a GTA specification is transformed into an efficient program. It can then be implemented in parallel as a composition of algorithmic skeletons [1] which can be seen as higher-order functions implemented in parallel.

In this paper we present a verified implementation of the framework in the Coq proof assistant. The contributions of this paper are:

- the formalisation of the GTA paradigm in the Coq proof assistant,
- the proof of correctness of the program transformations,
- an application of the framework to produce a parallel program for the knapsack problem and experiments with its execution on parallel machines.

We first present the generate-test-aggregate paradigm (Section 2), and its formalisation in Coq including the proof of the calculation theorems (Section 3). We then explain how to obtain a parallel code from the result of a program calculation using Coq extraction mechanism (Section 4). In Section 5 we describe experiments performed on parallel machines. Comparison with related work (Section 6) and conclusion (Section 7) follow.

2   GTA: Generate, Test and Aggregate

We briefly review the Generate-Test-and-Aggregate (GTA) paradigm [4,3]. It has been proposed as an algorithmic way to synthesise efficient programs from naive specifications (executable naive programs) in the following GTA form.

$$GTAspec = aggregate \circ test \circ generate$$

In this section we first give a simple example of how to specify naive algorithms in the GTA form. We give a clear but inefficient specification of the knapsack problem following this structure.

The knapsack problem is to fill a knapsack with items, each of certain non-negative value and weight, such that the total value of packed items is maximal while adhering to a weight restriction of the knapsack. For example, if the maximum total weight of our knapsack is 3kg and there are three items ($10, 1kg), ($20, 2kg), and ($30, 2kg) then the best we can do is to pick the selection ($10, 1kg), ($30, 2kg) with total value $40 and weight 3kg because all selections with larger value exceed the weight restriction.

The function *knapsack*, which takes as input a list of value-weight pairs (both positive integers) and computes the maximum total value of a selection of items not heavier than a total weight W, can be written in the GTA form:

$$knapsack\ W = maxvalue \circ validWeight\ W \circ subs$$

The function *subs* is the generator. From the given list of pairs it computes all possible selections of items, that is, all 2^n sublists if the input list has length n.

The function $validWeight\ W = \mathsf{filter_B}((<= W) \circ weight)$ is the tester. It discards all generated sublists whose total weight exceeds W and keeps the rest. The function $maxvalue$ is the aggregator. From the remaining sublists adhering to the weight restriction it computes the maximum of all total values.

The function $subs$ can be defined as follows:

$$subs = \mathsf{fold\_right}\ (\lambda\ x\ s.(\wp[]\!\int \uplus \wp[x]\!\int) \times_{+\!\!+} s)\ (\wp[]\!\int)$$

The result of the generator $subs$ is a bag of lists which we denote using $\wp$ and $\int$. The symbol $\uplus$ denotes bag union, e.g., $\wp[],[x]\int = \wp[]\int \uplus \wp[x]\int$, and $\times_{+\!\!+}$ the lifting of list concatenation to bags, concatenating every list in one bag with every list in the other. Here is an example application of $subs$: $subs\ [1,3,3] = \wp[],[1]\int \times_{+\!\!+} \wp[],[3]\int \times_{+\!\!+} \wp[],[3]\int = \wp[],[1],[1,3],[1,3],[1,3,3],[3],[3],[3,3]\int$. We took the liberty to reorder bag elements lexicographically because bags are unordered collections. Note that elements may occur more than once as witnessed by $[1,3]$.

The function $validWeight$ is filter operation $\mathsf{filter_B}$ with a predicate to check the weight restriction in which we use $weight = \mathsf{fold\_right}\ (\lambda(v,w)s.w + s)\ 0$ to find the total weight of a given item list:

$$validWeight = \mathsf{filter_B}\ ((<= w) \circ weight)$$

Here, it is easily seen that $weight$ satisfies the follwoing three equations:

$$
\begin{aligned}
weight\ [] &= 0 \\
weight\ [(v,w)] &= w \\
weight\ (xs +\!\!+ ys) &= weight\ xs + weight\ ys
\end{aligned}
$$

It simply replaces list constructors $[]$ and $+\!\!+$ with 0 and $+$. A function satisfying three equations of this form is called monoid homomorphism. Note that a monoid is a mathematical structure made of an associative binary operator with its identity element: The constructors and operators in the equations describe monoids.

Finally, the aggregator $maxvalue$ computes the maximum of summing up the values of each list in a bag using the maximum operator $\uparrow$, which is defined as a recursive function (natural fold operation) on bags:

$$
\begin{aligned}
maxvalue\ \wp\int &= -\infty \\
maxvalue\ \wp l\int &= \mathsf{fold\_right}\ (\lambda(v,w)s.v + s)\ 0\ l \\
maxvalue\ (b \uplus b') &= maxvalue\ b \uparrow maxvalue\ b'
\end{aligned}
$$

It is easily seen that $maxvalue$ satisfies the following five equations; it simply replaces constructors $\uplus$, $\times_{+\!\!+}$, $\wp[]\int$ and $\wp\int$ with $\uparrow$, $+$, 0 and $-\infty$, respectively. A function satisfying five equations of this form is called semiring homomorphism. Note that a semiring is a mathematical structure of a monoid; a commutative monoid and a distributivity law over the operators of those monoids. The constructors and operators in the following equations form a semiring.

$$
\begin{aligned}
maxvalue\ \langle\!\langle\ \rangle\!\rangle &= -\infty \\
maxvalue\ \langle\!\langle [] \rangle\!\rangle &= 0 \\
maxvalue\ \langle\!\langle [(v, w)] \rangle\!\rangle &= v \\
maxvalue\ (b \uplus b') &= maxvalue\ b \uparrow maxvalue\ b' \\
maxvalue\ (b \times_{++} b') &= maxvalue\ b + maxvalue\ b'
\end{aligned}
$$

Now, we have defined a naive program, i.e., a GTA specification of the problem.

Next, let us consider an efficient algorithm $knapsack'$ to solve the problem. The GTA provides two theorems to derive mechanically such an efficient program from a GTA specification, but we start with the derived efficient program to understand what the theorems do. We will see the theorems later.

We may use linear dynamic programming of the following form that repeatedly updates a map denoted by using $\{$ and $\}$, in which each entry $w \mapsto v$ means that there is a selection of items with the best total value v with the total weight w:

$$
knapsack'\ W = \pi \circ \mathsf{fold\_right}\ (\lambda(v, w)m.(\{0 \mapsto 0\} \oplus_W \{w \mapsto v\}) \otimes_W m)\ \{0 \mapsto 0\}
$$

Here, $\oplus_W$ merges two maps by taking the maximum value of two entries of the same weight, while $\otimes_W$ makes every possible combination of entries of two maps and merges the results by $\oplus_W$. The post-process function π extracts the final result from the final map.

For example, we have $knapsack'\ 3\ [] = \pi(\{0 \mapsto 0\}) = 0$. Here, the map $\{0 \mapsto 0\}$ means that we have one selection with the best total value 0 and the total weight 0 since we have no item, and π extracts the only value 0 as the final answer. Similarly, we have $knapsack'\ 3\ [(30, 2)] = \pi(\{0 \mapsto 0, 2 \mapsto 30\}) = 30$. The map $\{0 \mapsto 0, 2 \mapsto 30\}$ represents the selection possibilities about the item $(30, 2)$: either it is selected (entry $2 \mapsto 30$ that results in the best value 30 with total weight 2), or not $(0 \mapsto 0)$. The work of $\otimes_3$ can be seen clearly in the following sub-computation in $knapsack'\ 3\ [(10, 1), (20, 2), (30, 2)] = \pi(\{0 \mapsto 0, 1 \mapsto 10, 2 \mapsto 30, 3 \mapsto 40, \_ \mapsto 60\}) = 40$. Here, $\_$ represents entries with "more than 3" to be ignored by π, and thus a map has at most 5 entries.

$$
\begin{aligned}
&\{0 \mapsto 0, 1 \mapsto 10\} \otimes_3 (\{0 \mapsto 0, 2 \mapsto 20\} \otimes_3 \{0 \mapsto 0, 2 \mapsto 30\}) \\
&= \{0 \mapsto 0, 1 \mapsto 10\} \otimes_3 (\{(0 + 0) \mapsto (0 + 0)\} \oplus_3 \{(2 + 0) \mapsto (20 + 0)\} \\
&\qquad\qquad\qquad\qquad \oplus_3 \{(0 + 2) \mapsto (0 + 30)\} \oplus_3 \{(2 + 2) \mapsto (20 + 30)\}) \\
&= \{0 \mapsto 0, 1 \mapsto 10\} \otimes_3 (\{0 \mapsto 0\} \oplus_3 \{2 \mapsto 20\} \oplus_3 \{2 \mapsto 30\} \oplus_3 \{\_ \mapsto 50\}) \\
&= \{0 \mapsto 0, 1 \mapsto 10\} \otimes_3 \{0 \mapsto 0, 2 \mapsto 30, \_ \mapsto 50\} \\
&= \{(0{+}0) \mapsto (0{+}0)\} \oplus_3 \{(1{+}0) \mapsto (10{+}0)\} \oplus_3 \{(0{+}2) \mapsto (0{+}30)\} \oplus_3 \cdots \\
&= \{0 \mapsto 0, 1 \mapsto 10, 2 \mapsto 30, 3 \mapsto 40, \_ \mapsto 60\}
\end{aligned}
$$

It should be noted that $\mathsf{fold\_right}$ used in $knapsack'$ is parallelisable because of the associativity of the operator $\otimes_W$.

We have got two programs $knapsack$ and $knapsack'$ to solve the knapsack problem. What is the relationship between these two? Comparing $subs$ and $knapsack'$ we can find that both $subs$ and $knapsack'$ can be written with a new function $poly\_subs\ f\ (\oplus)\ (\otimes)\ \imath_\oplus\ \imath_\otimes = \mathsf{fold\_right}\ (\lambda xm.(\imath_\otimes \oplus f\ x) \otimes m)\ \imath_\otimes$:

$$subs \qquad = poly\_subs \ (\lambda x.\wr[x]\wr) \ (\uplus) \ (\times_{+\!\!\!+}) \ (\wr\wr) \ (\wr\wr\wr)$$
$$knapsack' \ W = \pi \circ poly\_subs \ (\lambda(v,w).\{w \mapsto v\}) \ (\oplus_W) \ (\otimes_W) \ (\{\}) \ (\{0 \mapsto 0\})$$

Here, *poly_subs* is polymorphic over the result type of semiring operators $\oplus$ and $\otimes$, and *subs* is an instantiation of this polymorphic function with the constructors $\uplus$ and $\times_{+\!\!\!+}$. Such a generator is called a polymorphic semiring generator.

Now, we are ready to see the fusion theorems for mechanical derivation of *knapsack'* from *knapsack*. The GTA provides two theorems, namely, semiring-fusion and filter-embedding. The filter-embedding gives a way to derive the operators into $\oplus_W$ and $\otimes_W$ on maps, while the semiring-fusion gives a way to substitute these operators into the generator.

Theorem 1 (Filter Embedding [4]). *Given a monoid homomorphism mhom, a semiring homomorphism agg, and a function ok, there exist a function π and a semiring homomorphism agg' and the following equation holds:*

$$agg \circ \mathsf{filter_B} \ (ok \circ mhom) = \pi \circ agg'$$

Theorem 2 (Semiring Fusion [4]). *Given a semiring homomorphism agg, which replaces the constructors with f, $\oplus$, $\otimes$, $\imath_\oplus$, and $\imath_\otimes$, and a polymorphic semiring generator gen, the following equation holds:*

$$agg \circ gen \ (\lambda x \to \wr[x]\wr) \ (\uplus) \ (\times_{+\!\!\!+}) \ (\wr\wr) \ (\wr\wr\wr) = gen \ f \ (\oplus) \ (\otimes) \ \imath_\oplus \ \imath_\otimes$$

These two theorems derive *knapsack'* from *knapsack* as follows. Here, *maxvalue'* is a semiring homomorphism that replaces the constructors with $\lambda(v,w).\{w \mapsto v\}$, $\oplus_W$, $\otimes_W$, $\{\}$ and $\{0 \mapsto 0\}$, and is mechanically derived from *maxvalue* and *validWeight W* by the filter-embedding.

$$
\begin{aligned}
knapsack \ W &= maxvalue \circ validWeight \ W \circ subs \\
&= \{ \quad \text{Filter-embedding} \quad \} \\
&\quad \pi \circ maxvalue' \circ subs \\
&= \{ \quad \text{Semiring-fusion} \quad \} \\
&\quad \pi \circ poly\_subs \ (\lambda(v,w).\{w \mapsto v\}) \ (\oplus_W) \ (\otimes_W) \ (\{\}) \ (\{0 \mapsto 0\}) \\
&= knapsack' \ W
\end{aligned}
$$

It should be noted that by using GTA one can easily develop an efficient parallel programs to solve many variants of problems such as a statistics probelm to find a most likely sequence of hidden events [7], a combinatorial problem to find the best period (contiguous subsequence) in a time series [2], and so on [3], by simply defining testers to specify variant problems with additional conditions.

3   Verified GTA Library

In this section we introduce our verified GTA library with an automatic fusion mechanism that allows a user to get an efficient Coq code freely from his/her naive Coq code in the GTA form. The library mainly consists of three parts: user interface, proof of the fusion theorems, and the automatic fusion mechanism.

3.1   User Interface: Writing Your Naive Code

This part defines a variety of items used to define a GTA specification as a Coq program, which includes axiomatization of the bag (i.e., multi-set) data structure and mathematical properties of components in a GTA specification.

Bag Axiomatisation. Since in a GTA specification a generator produces a bag of lists, we need a bag data structure in Coq to define a GTA specification. We axiomatised the bag data structure as a module type, and implemented a module of this type by using the list data structure as its underlying structure. A bag module has three constructors: `empty` for an empty bag, `singleton` to make a singleton bag of the given element, and `union` (⊎ in the mathematical notation) to merge two bags. It is also equipped with a decidable equivalence relation, under which the usual semantics of bags (multi-sets) holds, and the natural fold operation `homB` respecting the equivalence relation. Interested readers may refer to the source code [15]. We also defined a module that, using the constructors and fold operation, implements computations on bags, such as map, filter, operator $\times_{+\!\!+}$ (`mapB`, `filterB`, and `cross` in Coq code), and function `singleBag` to make a bag of a singleton list of the given element. In addition, we showed properties of these operators, such as the semiring properties of the operators `union` and `cross` with their identities `empty` and `nilBag` (a bag of nil).

Generators. The first component of a GTA specification is a generator that produces a bag of lists and has to be an instance with ⊎ and $\times_{+\!\!+}$ of a polymorphic function over the result type of semiring operators $\oplus$ and $\otimes$. Figure 1 shows Coq code related to generators. The polymorphism condition for each generator is captured by an instance of typeclass `isSemiringPolymorphicGenerator`, which connects a generator `gen` and its polymorphic function `pgen`. The typeclass also has a field to show fusable property based on the polymorphism. We omit the details here, but the concept is that a polymorphic function determines computation structure independent of given arguments. This could be shown by the free theorem [17] about Coq, but Coq cannot prove such a property about himself. Thus, the library asks a user to show the property by hand. We may provide tactics to support this part because such a proof can be systematic. It is planed in our future work.

For example, the generator `subs` in Section 2 for the knapsack problem can be defined as an instance of the polymorphic function `poly_subs` as follows. We can show its fusable condition by a simple induction.

```
Fixpoint poly_subs
        (V:Type) (f:T→V) (oplus otimes:V→V→V) (ep et :V) (l:list T) :=
    match l with
    | nil  ⇒ et
    | a::l' ⇒ otimes (oplus (f a) et) (poly_subs f oplus otimes ep et l')
    end.
Definition subs := poly_subs singleBag union cross empty nilBag.
Global Program Instance subs_is_polymorphic_generator
    : isSemiringPolymorphicGenerator subs (@poly_subs).
```

```
Definition semiringPolymorphicType (B : Type) : Type
  := ∀{V:Type} (f : T →V) (oplus : V→V→V) (otimes : V→V→V) (ep et : V), list B →V.

Class isSemiringPolymorphicGenerator '{A : Type}
  (gen : list A →bag (list T)) (pgen : semiringPolymorphicType A) := {
  SemiringGenEquiv : ∀l, gen l = pgen (bag (list T)) singleBag union cross empty nilBag l;
  isSemiringPolymorphicGenerator_Polymorphism : (* snip *)
}.
```

Fig. 1. Formalization of polymorphic semiring generators

```
Next Obligation.
  apply Build_isSemiringPolymorphicFunction.
  induction l.
  - unfold poly_subs. simpl. reflexivity.
  - simpl. intros. rewrite IHl. reflexivity.
Defined.
```

Testers. The second component is a tester to discard invalid lists in the bag produced by a generator. Figure 2 shows Coq code related to testers. To be successfully fused by the theorems, a tester has to be a filter and its predicate is a composition (denoted by :o: in the code) of a simple decidable predicate and a monoid homomorphism. These conditions are straightforwardly captured by typeclasses isHomomorphicFilter and isMonoidHomomorphism.

For example, the tester validWeight in Section 2 for the knapsack problem can be defined by using the filter filterB on bags as follows. Here, decidability of its predicate (named weightOk) is also defined to be used by filterB.

```
Inductive Item := item : nat →nat →Item.
Definition getVal i := match i with item v _ ⇒ v end.
Definition getWeight i := match i with item _ w ⇒ w end.
Definition atmost (w:nat) := fun a ⇒ a <= w.
Definition weight (l:list Item):= fold_right (fun a w⇒ getWeight a+w) 0 l.
Definition weightOk (w:nat) := atmost w :o: weight.
Lemma atmost_dec (w:nat): ∀(a:nat), {atmost w a}+{~atmost w a}.(* snip *)
Lemma weightOk_dec (w:nat):
                 ∀(l:list Item), {weightOk w l}+{~weightOk w l}.(* snip *)
Definition validWeight (w:nat) := filterB (weightOk w) (weightOk_dec w).
```

For successful fusion we also need an instance of isMonoidHomomorphism as well as the Proper instance of atmost, while we do not need an instance of isHomomorphicFilter because validWeight is directly defined by fitlerB:

```
Program Instance proper_atmost: Proper (eq⟹ eq⟹ iff) atmost.(* snip *)
Program Instance weightOk_monoidHom:
  isMonoidHomomorphism (T:=Item) weight getWeight plus 0. (* snip *)
```

Note that for the performance of the final fused program it is better to finitise the domain of the monoid homomorphism (e.g., to use the finite set $\{n : \mathbb{N} \mid n \leq w\}$ as the domain) before applying fusions. This can be done by hand or

```
Context '{eqDecT : @EqDec T eqT equivT} '{eqDecM : @EqDec M eqM equivM}.
(* Monoid is a class to hold a monoid operator with its identity *)
Class isMonoidHomomorphism (h : list T →M) (f : T →M) (oplus : M →M →M) (e : M) := {
  isMH_Monoid : Monoid oplus e;
  isMH_CondAppend : ∀x y, h (app x y) === oplus (h x) (h y);
  isMH_CondSingle: ∀a, h (a::nil) === f a;
  isMH_CondNil: h nil === e;
  isMH_proper_oplus: Proper (eqM ⟹ eqM ⟹ eqM) oplus;
  isMH_proper_f: Proper (eqT ⟹ eqM) f
}.
(* "dec_comp f g dec_f" derives decidability of (f :o: g) from that of f, namely, dec_f . *)
Class isHomomorphicFilter (tes : bag (list T) →bag (list T)) (mhom : list T →M) (h : T →M)
  (odot : M →M →M) (e : M) (ok : M →Prop) (dec : ∀m : M, {ok m} + {~ ok m}) := {
  isHF_isMonoidHomomorphism: isMonoidHomomorphism (eqT:=eqT) mhom h odot e;
  isHF_spec: ∀b:bag(list T),tes b === filterB(ok:o:mhom)(dec_comp mhom ok dec)b;
  isHF_proper_ok: Proper (eqM ⟹ iff) ok
}.
```

Fig. 2. Formalization of monoid homomorphisms and homomorphic filters

even by an automatic mechanism in some cases. We omit this here for the space limitation, but the experiment was conducted on the finitised version. Interested readers may refer to papers [4,3] and the source code [15].

Aggregators. The final component is an aggregator to make a summary using semiring operators from lists passing testers. Figure 3 shows Coq code related to aggregators. To be fused by the semiring-fusion, an aggregator has to be a semiring homomorphism. This is captured by typeclass isSemiringHomomorphism saying that the given function agg is a semiring homomorphism made of the function f, the operators oplus and otimes with ep as zero and et as one. The library provides a simple way to make a semiring homomorphism: Given semiring operators, nested fold operation semiringHom is the semiring homomorphism.

For example, the aggregator maxvalue in Section 2 to find the maximum sum value can be defined by using the fold operation semiringHom with predefined semiring semiring_max'_plus' of the max and plus operators with the minus infinity (i.e., the zero of the semiring):

```
Definition maxvalue
        := semiringHom (fun a:Item ⟹ Num(getVal a)) semiring_max'_plus'.
```

Now, we have got a GTA specification as a naive Coq program, and we can check its correctness by running it with a small example:

```
Definition knapsack (w : nat) := maxvalue :o: validWeight w :o: subs.
Definition items := [item 10 1; item 20 2; item 30 2 ].
Eval compute in (knapsack 3 items).
(* = Num 40 : nat_minf *)
```

This is basically all that a user has to do in the GTA paradigm. The rest is to call an interface function (a field of a specific class) to trigger automatic fusion, which is shown later.

```
Context '{S : Type} '{eqDecS : @EqDec S eqS equivS} '{eqDecT : @EqDec T eqT equivT}.
(* eqBag is the equivalence relation given in a Bag module, taking an equivalence relation on elements.*)
(* Semiring is a typeclass to hold semiring operators with identities. *)
Class isSemiringHomomorphism (agg:bag(list T)→S)(f:T→S)(oplus otimes:S→S→S)(ep et:S):={
  isSH_Semiring : Semiring oplus otimes ep et;
  isSH_CondUnion : ∀x y, agg (union x y) === oplus (agg x) (agg y);
  isSH_CondCross : ∀x y, agg (cross x y) === otimes (agg x) (agg y);
  isSH_CondSingle : ∀a, agg (singleBag a) === f a;
  isSH_CondEmpty : agg (empty) === ep;
  isSH_CondNil : agg (nilBag) === et;
  isSH_proper_f : Proper (eqT ⟹ eqS) f;
  isSH_proper_oplus : Proper (eqS ⟹ eqS ⟹ eqS) oplus;
  isSH_proper_otimes : Proper (eqS ⟹ eqS ⟹ eqS) otimes;
  isSH_proper : Proper (eqBag (list T) ⟹ eqS) agg
}.

(* short−hand to make a semiring homomorphism, i.e., nested fold operations *)
Definition semiringHom (f : T →S) '(semiring : Semiring oplus otimes ep et)
  := homB oplus (fold_right (fun a r ⟹ otimes (f a) r) et) ep.

Global Program Instance semiringHom_is_semiringHomomorphism
  '{semiring : Semiring oplus otimes ep et}
  '{proper_oplus : Proper (eqS ⟹ eqS ⟹ eqS) oplus }
  '{proper_otimes : Proper (eqS ⟹ eqS ⟹ eqS) otimes }
  '{proper_f : Proper (eqT ⟹ eqS) f}
  : isSemiringHomomorphism (semiringHom f oplus otimes ep et semiring) f oplus otimes ep et.
```

Fig. 3. Formalization of semiring homomorphisms

3.2   The Core: Proof of the GTA Fusion Theorems

The core of the library is proof of two fusion theorems, namely, the semiring-fusion and filter-embedding (Theorems 2 and 1). These fusion theorems give mechanical rules to transform a GTA specification (naive program) into an efficient program, whose automatic mechanism will be shown later.

The filter-embedding fusion transforms a composition of a tester and an aggregator into a new aggregator followed by a simple projection function. This eliminates the tester from a GTA specification. Since the new aggregator uses a semiring (so-called *monoid semiring* [12]) on finite maps, we need formalization of finite maps to formalise this theorem. Coq's standard library has formalization of maps, but it requires a module for each key type to show its decidability, which prevents us from an automatic optimisation mechanism that may change the key type. Therefore, we reformalised maps as a module type, in which the decidability of keys is given as an instance of a specific typeclass while functions and properties are the same as the standard library except for additional induction principles. The library also provides a list-based implementation of the module type. Interested readers may refer to the source code [15].

On top of the map formalisation we proved properties of the semiring on maps, which is the most crucial and difficult part of proving the filter-embedding. We also introduced disjoint-sum version of maps whose semiring properties are easily shown, and used an equivalence correspondence between the original and disjoint maps to prove the difficult part clearly. Interested readers may find a formalisation of the disjoint-sum maps in paper [12].

```
Context '{S : Type} '{eqDecS : @EqDec S eqS equivS}.

Theorem filterEmbeddingFusion
  '(shomAgg : isSemiringHomomorphism agg f oplus otimes ep et)
  '(homFilter : isHomomorphicFilter tes mhom h odot e ok dec)
  :∀ l,(agg:o:tes) l === (postproc ok dec ep oplus :o:
                          semiringHom(embed oplus h f)(monoid_semiring_of shomAgg homFilter))l.
Theorem semiringFusion
  '(polyGen : isSemiringPolymorphicGenerator A gen pgen)
  '(shomAgg : isSemiringHomomorphism agg f oplus otimes ep et)
  : ∀l, (agg :o: gen) l === (pgen S f oplus otimes ep et) l.
```

Fig. 4. Two fusion theorems of the GTA

Once the properties of the semiring are shown, the filter-embedding theorem can be shown straightforwardly as the previous papers [4,3] did. The proof of semiring fusion is also straightforward once we are given an instance of isSemiringPolymorphicGenerator. Figure 4 shows the theorems proved in Coq, in which postproc is a simple projection function to extract the final result from a map and monoid_semiring_of is the semiring on maps built from the semiring in the aggregator agg and the monoid in the tester tes.

3.3 Automatic Fusion Mechanism

To allow a user to get an efficient Coq code freely from his/her naive GTA specification, the library implements an automatic fusion mechanism based on the typeclass mechanism. He/she can get efficient code by calling the function fuse on his/her specification as follows.

Definition knapsack' (w : nat) := fused (f := knapsack w).

The automatic fusion mechanism is implemented by instances of two classes Fusion and Fuser shown in Fig. 5. Instances of Fusion form knowledge database of fusion, i.e., a set of function triples (*consumer, producer, fused*) such that *consumer* ∘ *producer* is equivalent to *fused*, while the instance fuser of Fuser triggers a search to find an instance of Fusion that gives the result *fused* of fusing the given function composition $f = consumer \circ producer$. Figure 5 shows some of these instances, including fusion knowledge of the theorems in Fig. 4.

For example, the above call of fused on knapsack eventually finds the instance comp_1_fuser. It first calls fused on maxvalue :o: validWeight to get their fused result, and then calls another fused on the composition of the result and subs. The first call finds the instance filterEmbeddingFusionInstance to fuse maxvalue and validWeight, and returns a composition of postproc with the new aggregator, namely, semiringHom with the semiring operators on maps. The latter call finds the instance semiringFusionInstanceWithPP to get the final fused result equivalent to the efficient program shown in Section 2. It is worth noting that the mechanism works for multiple testers in a GTA specification, although it may take longer time to finish the fusion.

```
Class Fusion '{eqDec : EqDec D} '(producer : B →C) '(consumer : C →D) (fused : B →D)
    := { fusion_spec : ∀b, consumer (producer b) === fused b }.

Class Fuser '{eqDec : EqDec D} '(f : B →D) := { fused : B →D; fuser_spec : ∀
b, f b === fused b }.

(* An instance of Fuser to trigger a search for Fusion instances *)
Global Program Instance fuser '{fusion : Fusion producer consumer _fused}
    : Fuser (consumer :o: producer) := { fused := _fused }.

(* A fuser for multiple compositions: fused ( (f . g) . h ) = fused ( fused (f . g ) . h) *)
Global Program Instance comp_1_fuser
    '{fusion_gf : @Fusion D R equiv0 eqDec B C g f fg}
    '{fusion_hgf : @Fusion D R equiv0 eqDec A B h fg fgh}
    : Fusion h (f :o: g) fgh. (* snip proof. *)

Global Program Instance filterEmbeddingFusionInstance (* Knowledge of the filter−embedding *)
    '{shomAgg : isSemiringHomomorphism agg f oplus otimes ep et}
    '{homFilter : isHomomorphicFilter tes mhom h odot e ok dec}
    : Fusion (bag (list T)) (bag (list T)) tes agg
    ((postproc ok dec ep oplus)
        :o: (semiringHom (embed oplus h f) (monoid_semiring_of shomAgg homFilter))).

Global Program Instance semiringFusionInstanceWithPP (* A variant of the semiring−fusion *)
    '{polyGen : isSemiringPolymorphicGenerator A gen pgen}
    '{shomAgg : isSemiringHomomorphism agg f oplus otimes ep et}
    '{pp : S →X} (* projection function after an aggregator *)
    : Fusion gen (pp :o: agg) (pp :o: pgen S f oplus otimes ep et).
```

Fig. 5. Instances for automatic fusion mechanism

4   Extraction of Certified Efficient Parallel Code

The result of the fusion mechanism described in the previous section, is an instance of `poly_subs`. This function could actually be proven equivalent to a composition of a `map` and a `reduce` on lists. These two combinators could themselves be proven to correspond (in a sense we will describe later) to parallel implementations of `map` and `reduce` on *distributed* lists. In this way we obtain a parallel version of `poly_subs` and therefore of the `knapsack'` function. In order to do so, we need to be able to write (data) parallel programs in Coq.

Bulk Synchronous Parallel ML or BSML is a purely functional programming language [8]. It is currently implemented as a library for OCaml, on top of any C MPI [13] implementation. It thus allows execution on a wide variety of parallel architectures and is especially well suited as an extraction target for Coq development.

In BSML, the underlying architecture is supposed to be a Bulk Synchronous Parallel (BSP) [16] computer. It is an abstract architecture as, with the help of a software layer, any general purpose parallel computer can be seen as a BSP computer. Such a computer is a *distributed* memory architecture: a set of p processor-memory pairs, connected through a network that allows point-to-point communications, together with a global synchronisation unit. A BSP program is executed as a sequence of *super-steps*, each one divided into (at most) three successive and logically disjoint phases: (a) Each processor uses its local data (only) to perform sequential computations and to request data transfers to/from other nodes; (b) the network delivers the requested data transfers; (c) a global synchronisation barrier occurs, making the transferred data available for the next super-step.

BSML primitives are shallowly embedded in Coq. Their specifications are given in a module type PRIMITIVES. In this module, a partial description of a BSP architecture is as follows:

Section Processors.
  Parameter bsp_p : nat. (∗ *the number p of processors* ∗)
  Axiom bsp_pLtZero: 0 < bsp_p. (∗ *we have at least one processor* ∗)
  Definition processor : Type := { pid: nat | pid < bsp_p }.
End Processors.

BSML is based on a distributed data structure, named *parallel vector*. In OCaml its abstract type is 'a par, and in Coq we have par: Type →Type. We write informally $\langle v_0, \ldots, v_{p-1} \rangle$ a value of this type. There is *one* value per processor of the BSP computer, and nesting is not allowed: These values should be "sequential", i.e. they cannot be or contain parallel vectors. BSML offers a global view of a parallel program: It looks like a sequential program but manipulates parallel vectors. The implementation however is a parallel composition of sequential communicating programs.

There are four primitives to deal with parallel vectors. Their signatures and informal semantics follows:

mkpar:(processor→A)→par A   apply:par(A→B)→ par A→par
proj:par A→processor→A   put:par(processor→A)→par(processor→A)

$$
\begin{aligned}
\text{mkpar } f &= \langle f\ 0, \ldots, f\ (p-1) \rangle \\
\text{apply } \langle f_0, \ldots, f_{p-1} \rangle \langle v_0, \ldots, v_{p-1} \rangle &= \langle f_0\ v_0, \ldots, f_{p-1}\ v_{p-1} \rangle \\
\text{proj } \langle v_0, \ldots, v_{p-1} \rangle &= \lambda i \rightarrow v_i \\
\text{put } \langle f_0, \ldots, f_{p-1} \rangle &= \langle \lambda j \rightarrow f_j\ 0, \ldots, \lambda j \rightarrow f_j\ (p-1) \rangle
\end{aligned}
$$

mkpar is used to create parallel vectors whose values are given by a function f. The function f should be a sequential function, i.e. it should not call any of the BSML primitives. proj is its inverse. apply denotes the pointwise application of a parallel vector of functions to a parallel vector of values. put is used to exchange data between processors. In the input parallel vector of functions, each function encodes the messages to be sent to other processors. $(f_i\ j)$ is the message to be sent by processor i to processor j. In BSML OCaml implementation the first constant constructor of a type is considered as the "empty" message and incurs no communication cost. The functions in the output vector encode received messages. If the input and output vectors are thought as matrices (each of the p functions can produce p values when applied to the p processor names), put is matrix transposition. Note that at a given processor there is no way to directly access the value held by another processor: put is needed. Both proj and put require a full BSP super-step to be evaluated. mkpar and apply are evaluated during the computation phase of a super-step.

```
Module List.
  Include Coq.Lists.List.
  Definition reduce '(op:A →A →A) '{m: Monoid A op e} := fun l ⇒ fold_left op l e.
End List.
Module MR (Import Bsml : PRIMITIVES).
  Program Definition map '(f:A→B)'(v:par(list A)) : par(list B) :=
    apply (mkpar (fun _ ⇒ List.map f)) v.
  Program Definition reduce '(op:A →A →A)'{m: Monoid A op e}(v:par(list A)) : A :=
    List.reduce op (List.map (proj (apply (mkpar (fun _ ⇒ List.reduce op)) v)) processors).
End MR.
```

Fig. 6. Parallel map and reduce

In order to formalise this semantics in Coq, we need an "observer" function that is able to get the values held in a parallel vector, for which extensional equality implies equality:

Parameter get:∀ A: Type, par A→processor→A.
Axiom par_eq:∀(A:Type)(v w:par A),(∀(i:processor),get v i=get w i)→v=w.

It is then straightforward to formalise the semantics. For example:

Parameter put: ∀vf:par(processor→A),
{ X: par(processor→A) | ∀i j: processor, get X i j = get vf j i }.

Verified algorithmic skeletons. If we think of a parallel vector of lists par(list A) as a *distributed* list, a parallel BSML implementation of map and reduce is shown in Figure 6, where Monoid is a type class for monoids, and processors is the list of all the processors. We proved that these parallel implementations of map and reduce are correct with respect to their sequential counter-parts, i.e:

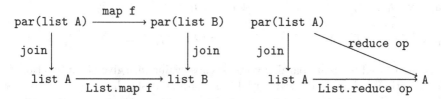

where join transforms a distributed list to the sequential list it represents:

Definition join '(v:par(list A)):list A:= List.flat_map (get v) processors.

These correspondences are stored in instances of type classes, so that when the user requires the parallelisation of a composition of sequential functions for which there exist corresponding parallel implementations, it is done automatically. We use this feature of our framework to parallelise the outcome of the fusion mechanism.

Code extraction. Using the extraction mechanism of Coq, for example on the module MR of Figure 6, we obtain an OCaml functor that need to be applied to an implementation of the (extraction of the) module type PRIMITIVES. Actually,

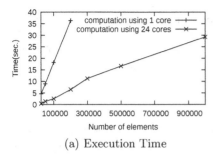
(a) Execution Time

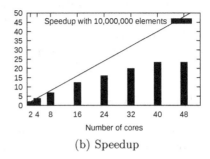
(b) Speedup

Fig. 7. Execution Time and Speedup for the parallel execution of the extracted program

the BSML library for OCaml provides several implementations of a module type named BSML, including a MPI-based parallel one.

This module type almost contains the extracted module type PRIMITIVES. The only difference is that in the extracted module, the type processor is nat but in BSML it is int. We thus provide a wrapper module BsmlNat that performs conversions between processor representations. This conversion is correct as far as the processor name remains below $2^{30} - 1$ (1073741823) or $2^{62} - 1$ (4611686018427387903), depending on the architecture.

BSML implementations in OCaml and C are not formally verified yet. However we implemented a sequential version of the module type PRIMITIVES in Coq (using Coq vectors for implementing par). After extraction this provides a verified reference sequential implementation. It could be used to test the unverified implementations of BSML in OCaml.

5   Experiment Results

The experiments were conducted on a shared memory machine containing 4 processors with 12 computer cores each (thus a total of 48 cores) and 64GB of memory. On each processors, there is two NUMA nodes, each node connecting 6 cores. On this particular architecture, we noticed in other experiments that there is a performance loss when there is active communication at the same time on the two NUMA nodes of one processor. The operating system is Ubuntu, the used languages and libraries are: Open MPI 1.5, BSML 0.5, and OCaml 3.12.1.

We extracted the parallel version of the knapsack' program in Section 3.3. We then measured the scalability of this programs run in parallel. Figure 7(a) shows timings for computations on different number of elements for a knapsack with a capacity of 30. The list contains elements of random weight and value (always lesser than 10). The computation time grows linearly with the number of elements. In the sequential case (poly_subs implementation), for lists over 200,000 the program fails due to stack overflow, as poly_subs is non tail recursive. For the parallel version however, the map we use is a tail recursive one. It is therefore possible to consider much bigger input lists. Figure 7(b) shows the mean (over 30 measures) *relative* (i.e. the reference implementation is the same program but executed with only one core) speedup for a computation over a list of 10,000,000 elements.

VerifiedGTA is one of the component of the SyDPaCC system. The full development (version "ITP2014") is available at the web [15].

6   Related Work

To our knowledge SyDPaCC [6], on which VerifiedGTA is based, is the only framework, that makes possible the extraction of compilable and scalable parallel programs from a development in a proof assistant.

Among the work on formal semantics for BSP computations, the only one used to generate actual programs is LOGS [18]. In this approach, parallel programs are built by composing sequential programs in parallel whereas we adopt a global view. The GTA specifications required from the user are much simpler than LOGS specifications, but less general. To our knowledge no part of LOGS is formally verified. BSP-Why [5] is an extension of Why2 for the verification of imperative BSP programs but it does not support program derivation.

The style of programming we follow is polymorphic. Mu et al. provides a framework for polytypic programming and program transformation in Coq [11]. Expressiveness is very interesting but it would be a challenge to provide correspondence between polytypic sequential functions and parallel ones.

Lupinski et al. [9] formalised the semantics of a skeletal parallel programming language. This work is based on a deep embedding: one formalisation provides the high-level semantics of the skeletons, another one provides a model of the implementation in Join-calculus. From this model an implementation of the skeletons in JoCaml is designed. The semantics of BSML being purely functional, it is possible to have a shallow embedding in Coq, and then to write BSML programs in Coq and finally extract OCaml/BSML programs. Neither BSML, nor JoCaml implementations have been proved correct with respect to their calculi.

Swierstra [14] formalised mutable arrays in Agda, and added explicit distributions to these arrays. He can then write and reason on algorithms on these distributed arrays. The two main examples are a distributed map, and a distributed sum. In BSML the distribution of parallel vectors is fixed. On the other hand, it is possible to defined a higher-level data structure on top of parallel vector and consider various distributions of the data structure in parallel vectors. BSML in Coq remains purely functional. It would be possible to consider parallel vectors of mutable arrays, and even extract such BSML imperative programs to OCaml as it was done for imperative programs by Malecha et al. [10].

7   Conclusion and Future Work

The verified GTA library in Coq allows to derive and extract an efficient Bulk Synchronous Parallel ML program from a naive program definition (a specification) as a composition of a generator, a tester and an aggregator. We experimented an extracted application on two parallel architectures.

Future work includes a specialisation of the framework for the case where the monoid of map keys is finite. This allows to replace the finite maps by tuples (or arrays) for which we have direct access to elements. We are also interested in extending the framework to handle GTA programs on other structures than lists, such as trees and graphs.

Acknowledgements. This work was partially supported by JSPS (KAKENHI Grant Number 24700025), ANR (ANR-2010-INTB-0205-02) and JST (10102704).

References

1. Cole, M.: Algorithmic Skeletons: Structured Management of Parallel Computation. MIT Press (1989), http://homepages.inf.ed.ac.uk/mic/Pubs
2. Corra, R.C., Farias, P.M., de Souza, C.P.: Insertion and sorting in a sequence of numbers minimizing the maximum sum of a contiguous subsequence. Journal of Discrete Algorithms 21, 1–10 (2013)
3. Emoto, K., Fischer, S., Hu, Z.: Filter-embedding semiring fusion for programming with MapReduce. Formal Aspects of Computing 24(4-6), 623–645 (2012)
4. Emoto, K., Fischer, S., Hu, Z.: Generate, Test, and Aggregate – A Calculation-based Framework for Systematic Parallel Programming with MapReduce. In: Seidl, H. (ed.) ESOP 2012. LNCS, vol. 7211, pp. 254–273. Springer, Heidelberg (2012)
5. Gava, F., Fortin, J., Guedj, M.: Deductive Verification of State-Space Algorithms. In: Johnsen, E.B., Petre, L. (eds.) IFM 2013. LNCS, vol. 7940, pp. 124–138. Springer, Heidelberg (2013)
6. Gesbert, L., Hu, Z., Loulergue, F., Matsuzaki, K., Tesson, J.: Systematic Development of Correct Bulk Synchronous Parallel Programs. In: International Conference on Parallel and Distributed Computing, Applications and Technologies (PDCAT), pp. 334–340. IEEE (2010)
7. Ho, T.J., Chen, B.S.: Novel extended viterbi-based multiple-model algorithms for state estimation of discrete-time systems with markov jump parameters. IEEE Transactions on Signal Processing 54(2), 393–404 (2006)
8. Loulergue, F., Hains, G., Foisy, C.: A Calculus of Functional BSP Programs. Science of Computer Programming 37(1-3), 253–277 (2000)
9. Lupinski, N., Falcou, J., Paulin-Mohring, C.: Sémantique d'une langage de squelettes (2012), http://www.lri.fr/~paulin/Skel/article.pdf
10. Malecha, G., Morrisett, G., Wisnesky, R.: Trace-based verification of imperative programs with i/o. J. Symb. Comput. 46(2), 95–118 (2011)
11. Mu, S.-C., Ko, H.-S., Jansson, P.: Algebra of programming using dependent types. In: Audebaud, P., Paulin-Mohring, C. (eds.) MPC 2008. LNCS, vol. 5133, pp. 268–283. Springer, Heidelberg (2008)
12. Otto, F., Sokratova, O.: Reduction relations for monoid semirings. Journal of Symbolic Computation 37(3), 343–376 (2004)
13. Snir, M., Gropp, W.: MPI the Complete Reference. MIT Press (1998)
14. Swierstra, W.: More dependent types for distributed arrays. Higher-Order and Symbolic Computation 23(4), 489–506 (2010)

15. SyDPaCC Home Page, http://traclifo.univ-orleans.fr/SyDPaCC
16. Valiant, L.G.: A bridging model for parallel computation. Commun. ACM 33(8), 103 (1990)
17. Wadler, P.: Theorems for free! In: Proceedings of the Fourth International Conference on Functional Programming Languages and Computer Architecture, pp. 347–359. ACM (1989)
18. Zhou, J., Chen, Y.: Generating C code from LOGS specifications. In: Van Hung, D., Wirsing, M. (eds.) ICTAC 2005. LNCS, vol. 3722, pp. 195–210. Springer, Heidelberg (2005)

Experience Implementing
a Performant Category-Theory Library in Coq

Jason Gross, Adam Chlipala, and David I. Spivak

Massachusetts Institute of Technology, Cambridge, MA, USA
jgross@mit.edu, adamc@csail.mit.edu, dspivak@math.mit.edu

Abstract. We describe our experience implementing a broad category-theory library in Coq. Category theory and computational performance are not usually mentioned in the same breath, but we have needed substantial engineering effort to teach Coq to cope with large categorical constructions without slowing proof script processing unacceptably. In this paper, we share the lessons we have learned about how to represent very abstract mathematical objects and arguments in Coq and how future proof assistants might be designed to better support such reasoning. One particular encoding trick to which we draw attention allows category-theoretic arguments involving *duality* to be internalized in Coq's logic with definitional equality. Ours may be the largest Coq development to date that uses the relatively new Coq version developed by homotopy type theorists, and we reflect on which new features were especially helpful.

Keywords: Coq, category theory, homotopy type theory, duality, performance.

1 Introduction

Category theory [14] is a popular all-encompassing mathematical formalism that casts familiar mathematical ideas from many domains in terms of a few unifying concepts. A *category* can be described as a directed graph plus algebraic laws stating equivalences between paths through the graph. Because of this spartan philosophical grounding, category theory is sometimes referred to in good humor as "formal abstract nonsense." Certainly the popular perception of category theory is quite far from pragmatic issues of implementation. This paper is an experience report on an implementation of category theory that has run squarely into issues of design and efficient implementation of type theories, proof assistants, and developments within them.

It would be reasonable to ask, what would it even mean to implement "formal abstract nonsense," and what could the answer have to do with optimized execution engines for functional programming languages? We mean to cover the whole scope of category theory, which includes many concepts that are not manifestly computational, so it does not suffice merely to employ the well-known folklore semantic connection between categories and typed functional programming [19].

G. Klein and R. Gamboa (Eds.): ITP 2014, LNAI 8558, pp. 275–291, 2014.

Instead, a more appropriate setting is a computer proof assistant. We chose to build a library for Coq [24], a popular system based on constructive type theory.

One might presume that it is a routine exercise to transliterate categorical concepts from the whiteboard to Coq. Most category theorists would probably be surprised to learn that standard constructions "run too slowly," but in our experience that is exactly the result of experimenting with naïve first Coq implementations of categorical constructs. It is important to tune the library design to minimize the cost of manipulating terms and proving interesting theorems.

This design experience is also useful for what it reveals about the consequences of design decisions for type theories themselves. Though type theories are generally simpler than widely used general-purpose programming languages, there is still surprising subtlety behind the few choices that must be made. Homotopy type theory [25] is a popular subject of study today, where there is intense interest in designing a type theory that makes proofs about topology particularly natural, via altered treatment of equality. In this setting and others, there remain many open questions about the consequences of type theoretical features for different sorts of formalization. Category theory, said to be "notoriously hard to formalize" [9], provides a good stress test of any proof assistant, highlighting problems in usability and efficiency.

Formalizing the connection between universal morphisms and adjunctions provides a typical example of our experience with performance. A *universal morphism* is a construct in category theory generalizing extrema from calculus. An *adjunction* is a weakened notion of equivalence. In the process of rewriting our library to be compatible with homotopy type theory, we discovered that cleaning up this construction conceptually resulted in a significant slow-down, because our first attempted rewrite resulted in a leaky abstraction barrier and, most importantly, large goals (Section 4.2). Plugging the holes there reduced goal sizes by two orders of magnitude[1], which led to a factor of ten speedup in that file (from 39s to 3s), but incurred a factor of three slow-down in the file where we defined the abstraction barriers (from 7s to 21s). Working around slow projections of Σ types (Section 2.5) and being more careful about code reuse each gave us back half of that lost time.

For reasons that we present in the course of the paper, we were unsatisfied with the feature set of released Coq version 8.4. We wound up adopting the Coq version under development by homotopy type theorists [22], making critical use of its stronger universe polymorphism (Section 2.2) and higher inductive types (Section 2.4). We hope that our account here provides useful data points for proof assistant designers about which features can have serious impact on proving convenience or performance in very abstract developments. The two features we mentioned earlier in the paragraph can simplify the Coq user experience dramatically, while a number of other features, at various stages of conception or implementation by Coq team members, can make proving much easier or improve proof script performance by orders of magnitude, generally by reducing term size (Section 4.2): primitive record projections (Section 2.5), internalized

---

[1] The word count of the larger of the two relevant goals went from 163,811 to 1,390.

proof irrelevance for equalities (Section 4.2), and η rules for records (Section 3.1) and equality proofs (Section 3.2).

Although pre-existing formalizations of category theory in proof assistants abound [1, 6, 15, 17, 18, 20, 23], we chose to implement our library [10] from scratch. Beginning from scratch allowed the first author to familiarize himself with both category theory and Coq, without simultaneously having to familiarize himself with a large pre-existing code base. Additionally, starting from scratch forced us to confront all of the decisions involved in designing such a library, and gave us the confidence to change the definitions of basic concepts multiple times to try out various designs, including fully rewriting the library at least three times. Although this paper is much more about the design of category theory libraries in general than our library in particular, we include a comparison of our library [10] with selected extant category theory libraries in Section 5. At present, our library subsumes many of the constructions in most other such Coq libraries, and is not lacking any constructions in other libraries that are of a complexity requiring significant type checking time, other than monoidal categories.

We begin our discussion in Section 2 considering a mundane aspect of type definitions that has large consequences for usability and performance. With the expressive power of Coq's logic Gallina, we often face a choice of making *parameters* of a type family explicit arguments to it, which looks like universal quantification; or of including them within values of the type, which looks like existential quantification. As a general principle, we found that the universal or *outside* style improves the user experience modulo performance, while the existential or *inside* style speeds up type checking. The rule that we settled on was: *inside* definitions for pieces that are usually treated as black boxes by further constructions, and *outside* definitions for pieces whose internal structure is more important later on.

Section 3 presents one of our favorite design patterns for categorical constructions: a way of coaxing Coq's definitional equality into implementing *proof by duality*, one of the most widely known ideas in category theory. In Section 4, we describe a few other design choices that had large impacts on usability and performance, often of a few orders of magnitude. Section 5 wraps up with a grid comparison of our library with others.

2   Issues in Defining the Type of Categories

We have chosen to use the outside style when we care more about the definition of a construct than about carrying it around as an opaque blob to fit into other concepts. The first example of this choice comes up in deciding how to define categories.

2.1   Dependently Typed Morphisms

In standard mathematical practice, a category $\mathcal{C}$ can be defined [2] to consist of:

- a class $Ob_\mathcal{C}$ of *objects*
- for all objects $a, b \in Ob_\mathcal{C}$, a class $Hom_\mathcal{C}(a, b)$ of *morphisms from a to b*
- for each object $x \in Ob_\mathcal{C}$, an *identity morphism* $1_x \in Hom_\mathcal{C}(x, x)$
- for each triple of objects $a, b, c \in Ob_\mathcal{C}$, a *composition function* $\circ : Hom_\mathcal{C}(b, c) \times Hom_\mathcal{C}(a, b) \to Hom_\mathcal{C}(a, c)$

satisfying the following axioms:

- associativity: for composable morphisms f, g, h, we have $f \circ (g \circ h) = (f \circ g) \circ h$.
- identity: for any morphism $f \in Hom_\mathcal{C}(a, b)$, we have $1_b \circ f = f = f \circ 1_a$

Following [25], we additionally require our morphisms to be 0-truncated (to have unique identity proofs). Without this requirement, we have a standard pre–homotopy type theory definition of a category.

We might[2] formalize the definition in Coq (if Coq had mixfix notation) as:

```
Record Category :=
  { Ob : Type;
    Hom : Ob → Ob → Type;
    _o_ : ∀ {a b c}, Hom b c → Hom a b → Hom a c;
    1 : ∀ {x}, Hom x x;
    Assoc : ∀ a b c d (f : Hom c d) (g : Hom b c) (h : Hom a b),
      f o (g o h) = (f o g) o h;
    LeftId : ∀ a b (f : Hom a b), 1 o f = f;
    RightId : ∀ a b (f : Hom a b), f o 1 = f;
    Truncated : ∀ a b (f g : Hom a b) (p q : f = g), p = q }.
```

We could just as well have replaced the classes $Hom_\mathcal{C}(a, b)$ with a single class of morphisms $Hom_\mathcal{C}$, together with functions defining the source and target of each morphism. Then it would be natural to define morphism composition to take a further argument, a proof of equal domain and codomain between the morphisms. Users of dependent types are aware that explicit manipulation of equality proofs can complicate code substantially, often to the point of obscuring what would be the heart of an argument on paper. For instance, the algebraic laws associated with categories must be stated with explicit computation of equality proofs, and further constructions only become more involved. Additionally, such proofs will quickly bloat the types of goals, resulting in slower type checking. For these reasons, we decided to stick with the definition of `Category` above, getting more lightweight help from the type checker in place of explicit proofs.

2.2   Complications from Categories of Categories

Some complications arise in applying the last subsection's definition of categories to the full range of common constructs in category theory. One particularly prominent example formalizes the structure of a collection of categories, showing that this collection itself may be considered as a category.

---

[2] The definition we actually use has some additional fields; see, e.g., Section 3.1.

The morphisms in such a category are *functors*, maps between categories consisting of a function on objects, a function on hom-types, and proofs that these functions respect composition and identity [2, 14, 25].

The naïve concept of a "category of all categories," which includes even itself, leads into mathematical inconsistencies, which manifest as universe inconsistency errors in Coq. The standard resolution is to introduce a hierarchy of categories, where, for instance, most intuitive constructions are considered *small* categories, and then we also have *large* categories, one of which is the category of small categories. Both definitions wind up with literally the same text in Coq, giving:

```
Definition SmallCat : LargeCategory :=
  {| Ob := SmallCategory;
     Hom C D := SmallFunctor C D; ... |}.
```

It seems a shame to copy-and-paste this definition (and those of `Category`, `Functor`, etc.) n times to define an n-level hierarchy. Coq 8.4 and some earlier versions support a flavor of *universe polymorphism* that allows the universe of a definition to vary as a function of the universes of its arguments. Unfortunately, it is not natural to parametrize `Cat` by anything but a universe level, which does not have first-class status in Coq anyway. We found the connection between universe polymorphism and arguments to definitions to be rather inconvenient, and it forced us to modify the definition of `Category` so that the record field `Ob` changes into a parameter of the type family. Then we were able to use the following slightly awkward construction:

```
Definition Cat_helper I ObOf (CatOf : ∀ i : I, Category (ObOf i))
  : Category I
  := {| Hom C D := Functor (CatOf C) (CatOf D); ... |}.
Notation Cat := (Cat_helper {T : Type & Category T} projT1 projT2).
```

Now the definition is genuinely reusable for an infinite hierarchy of sorts of category, because the `Notation` gives us a fresh universe each time we invoke it, but we have paid the price of adding extra parameters to both `Category` and `Cat_helper`, and this seemingly innocent change produces substantial blow-up in the sizes of proof goals arising during interesting constructions. So, in summary, we decided that the basic type theoretical design of Coq 8.4 did not provide very good support for pleasing definitions that can be reasoned about efficiently.

This realization (and a few more that will come up shortly) pushed us to become early adopters of the modified version of Coq developed by homotopy type theorists [22]. Here, an established kind of more general universe polymorphism [8], previously implemented only in NuPRL, is available, and the definitions we wanted from the start work as desired.

2.3   Arguments vs. Fields

Unlike most of our other choices, there is a range of possibilities in defining categories, with regards to arguments (on the outside) and fields (on the inside). At one extreme, everything can be made a field, with a type `Category` whose inhabitants are categories. At the other extreme, everything can be made an argument to a dummy function. Some authors [23] have chosen the intermediate option of making all of the computationally relevant parts (objects, morphisms, composition, and the identity morphism) arguments and the irrelevant proofs (associativity and left and right identity) fields. We discussed in Section 2.2 the option of parametrizing on just the type of objects. We now consider pros and cons of other common options; we found no benefits to the "outside" extreme.

Everything on the Inside. Once we moved to using the homotopy type theorists' Coq with its broader universe polymorphism, we decided to use fields for all of the components of a category. Switching from the version where the types of objects and morphisms were parameters brought a factor of three speed-up in compilation time over our whole development. The reason is that, at least in Coq, the performance of proof tree manipulations depends critically on their size (Section 4.2). By contrast, the size of the normal form of the term does not seem to matter much in most constructions; see Section 3 for an explanation and the one exception that we might have found. By using fields rather than parameters for the types of objects and morphisms, the type of functors goes from

$$\text{Functor} : \forall\ (ob_C : \text{Type})\ (ob_D : \text{Type})$$
$$(hom_C : ob_C \to ob_C \to \text{Type})\ (hom_D : ob_D \to ob_D \to \text{Type}),$$
$$\text{Category } ob_C\ hom_C \to \text{Category } ob_D\ hom_D \to \text{Type}$$

to

$$\text{Functor} : \text{Category} \to \text{Category} \to \text{Type}$$

The corresponding reduction for the type of natural transformations is even more remarkable, and with a construction that uses natural transformations multiple times, the term size blows up very quickly, even with only two parameters. If we had more parameters (for composition and identity), the term size would blow up even more quickly.

Usually, we do not care what objects and morphisms a particular category has; most of our constructions take as input arbitrary categories. Thus, there is a significant performance benefit to having all of the fields on the inside and so hidden from most theorem statements.

Relevant Things on the Outside. One of the main benefits to making all of the relevant components arguments, and requiring all of the fields to satisfy proof irrelevance, is that it allows the use of type-class resolution without having to worry about overlapping instances. Practically, this choice means that it is easier to get Coq to infer automatically the proofs that given types and operations assemble into a category, at least in simple cases. Although others [23] have found this approach useful, we have not found ourselves wishing we had type-class resolution when formalizing constructions, and there is a significant computational cost of exposing so many parameters in types. The "packed classes" of Ssreflect [7] alleviate this problem by combining this approach with the previous one, at the slight cost of more verbosity in initial definitions.

2.4 Equality

Equality, which has recently become a very hot topic in type theory [25] and higher category theory [13], provides another example of a design decision where most usage is independent of the exact implementation details. Although the question of what it means for objects or morphisms to be equal does not come up much in classical 1-category theory, it is more important when formalizing category theory in a proof assistant, for reasons seemingly unrelated to its importance in higher category theory. We consider some possible notions of equality.

Setoids. A setoid [5] is a carrier type equipped with an equivalence relation; a map of setoids is a function between the carrier types and a proof that the function respects the equivalence relations of its domain and codomain. Many authors [11, 12, 15, 18] choose to use a setoid of morphisms, which allows for the definition of the category of set(oid)s, as well as the category of (small) categories, without assuming functional extensionality, and allows for the definition of categories where the objects are quotient types. However, there is significant overhead associated with using setoids everywhere, which can lead to slower compile times. Every type that we talk about needs to come with a relation and a proof that this relation is an equivalence relation. Every function that we use needs to come with a proof that it sends equivalent elements to equivalent elements. Even worse, if we need an equivalence relation on the universe of "types with equivalence relations," we need to provide a transport function between equivalent types that respects the equivalence relations of those types.

Propositional Equality. An alternative to setoids is propositional equality, which carries none of the overhead of setoids, but does not allow an easy formulation of quotient types, and requires assuming functional extensionality to construct the category of sets.

Intensional type theories like Coq's have a built-in notion of equality, often called definitional equality or judgmental equality, and denoted as $x \equiv y$. This notion of equality, which is generally internal to an intensional type theory and therefore cannot be explicitly reasoned about inside of that type theory, is the equality that holds between $\beta\delta\iota\zeta\eta$-convertible terms.

Coq's standard library defines what is called *propositional equality* on top of judgmental equality, denoted $x = y$. One is allowed to conclude that propositional equality holds between any judgmentally equal terms.

Using propositional equality rather than setoids is convenient because there is already significant machinery made for reasoning about propositional equalities, and there is much less overhead. However, we ran into significant trouble when attempting to prove that the category of sets has all colimits, which amounts to proving that it is closed under disjoint unions and quotienting; quotient types cannot be encoded without assuming a number of other axioms.

Higher Inductive Types. The recent emergence of higher inductive types allows the best of both worlds. The idea of higher inductive types [25] is to allow inductive types to be equipped with extra proofs of equality between constructors. They originated as a way to allow homotopy type theorists to construct types with non-trivial higher paths. A very simple example is the interval type, from which functional extensionality can be proven [21].[3] The interval type consists of two inhabitants `zero : Interval` and `one : Interval`, and a proof `seg : zero = one`. In a hypothetical type theory with higher inductive types, the type checker does the work of carrying around an equivalence relation on each type for us, and forbids users from constructing functions that do not respect the equivalence relation of any input type. For example, we can, hypothetically, prove functional extensionality as follows:

```
Definition f_equal {A B x y} (f : A → B) : x = y → f x = f y.
Definition functional_extensionality {A B} (f g : A → B)
   : (∀ x, f x = g x) → f = g
 := λ (H : ∀ x, f x = g x)
      ⇒ f_equal (λ (i : Interval) (x : A)
            ⇒ match i with
               | zero ⇒ f x
               | one  ⇒ g x
               | seg  ⇒ H x
            end)
         seg.
```

Had we neglected to include the branch for `seg`, the type checker should complain about an incomplete match; the function λ i : Interval ⇒ match i with zero ⇒ true | one ⇒ false end of type Interval → bool should not typecheck for this reason.

The key insight is that most types do not need any special equivalence relation, and, moreover, if we are not explicitly dealing with a type with a special equivalence relation, then it is impossible (by parametricity) to fail to respect the equivalence relation. Said another way, the only way to construct a function that might fail to respect the equivalence relation would be by some eliminator like pattern matching, so all we have to do is guarantee that direct invocations of the eliminator result in functions that respect the equivalence relation.

As with the choice involved in defining categories, using propositional equality with higher inductive types rather than setoids derives many of its benefits from not having to deal with all of the overhead of custom equivalence relations in constructions that do not need them. In this case, we avoid the overhead by making the type checker or the metatheory deal with the parts we usually do

---

[3] This assumes a computational interpretation of higher inductives, an open problem.

not care about. Most of our definitions do not need custom equivalence relations, so the overhead of using setoids would be very large for very little gain. We plan to use higher inductive types[4] to define quotients, which are necessary to show the existence of certain functors involving the category of sets. We also currently use higher inductive types to define propositional truncation [25], which we use to define what it means for a function to be surjective, and prove that in the category of sets, being an isomorphism (an invertible morphism) is equivalent to being injective and surjective.

2.5   Records vs. Nested Σ Types

In Coq, there are two ways to represent a data structure with one constructor and many fields: as a single inductive type with one constructor (records), or as a nested Σ type. For instance, consider a record type with two type fields A and B and a function f from A to B. A logically equivalent encoding would be $\Sigma A. \Sigma B. A \to B$. There are two important differences between these encodings in Coq.

The first is that while a theorem statement may abstract over all possible Σ types, it may not abstract over all record types, which somehow have a less first-class status. Such a limitation is inconvenient and leads to code duplication.

The far more pressing problem, overriding the previous point, is that nested Σ types have horrendous performance, and are sometimes a few orders of magnitude slower. The culprit is projections from nested Σ types, which, when unfolded (as they must be, to do computation), each take almost the entirety of the nested Σ type as an argument, and so grow in size very quickly. Matthieu Sozeau is currently working on primitive projections for records for Coq, which would eliminate this problem by eliminating the arguments to the projection functions.[5]

3   Internalizing Duality Arguments in Type Theory

In general, we have tried to design our library so that trivial proofs on paper remain trivial when formalized. One of Coq's main tools to make proofs trivial is the definitional equality, where some facts follow by computational reduction of terms. We came up with some small tweaks to core definitions that allow a common family of proofs by *duality* to follow by computation.

Proof by duality is a common idea in higher mathematics: sometimes, it is productive to flip the directions of all the arrows. For example, if some fact about least upper bounds is provable, chances are that the same kind of fact about greatest lower bounds will also be provable in roughly the same way, by replacing "greater than"s with "less than"s and vice versa.

Concretely, there is a dualizing operation on categories that inverts the directions of the morphisms:

---

[4] We fake these in Coq using Yves Bertot's Private Inductive Types extension [4].

[5] We eagerly await the day when we can take advantage of this feature in our library.

Notation "C ᵒᵖ" := ({| Ob := Ob C; Hom x y := Hom C y x; ... |}).

Dualization can be used, roughly, for example, to turn a proof that Cartesian product is an associative operation into a proof that disjoint union is an associative operation; products are dual to disjoint unions.

One of the simplest examples of duality in category theory is initial and terminal objects. In a category $\mathcal{C}$, an initial object 0 is one that has a unique morphism $0 \to x$ to every object x in $\mathcal{C}$; a terminal object 1 is one that has a unique morphism $x \to 1$ from every object x in $\mathcal{C}$. Initial objects in $\mathcal{C}$ are terminal objects in $\mathcal{C}^{\mathrm{op}}$. The initial object of any category is unique up to isomorphism; for any two initial objects 0 and $0'$, there is an isomorphism $0 \cong 0'$. By flipping all of the arrows around, we can prove, by duality, that the terminal object is unique up to isomorphism. More precisely, from a proof that an initial object of $\mathcal{C}^{\mathrm{op}}$ is unique up to isomorphism, we get that any two terminal objects $1'$ and 1 in $\mathcal{C}$, which are initial in $\mathcal{C}^{\mathrm{op}}$, are isomorphic in $\mathcal{C}^{\mathrm{op}}$. Since an isomorphism $x \cong y$ in $\mathcal{C}^{\mathrm{op}}$ is an isomorphism $y \cong x$ in $\mathcal{C}$, we get that 1 and $1'$ are isomorphic in $\mathcal{C}$.

It is generally straightforward to see that there is an isomorphism between a theorem and its dual, and the technique of dualization is well-known to category theorists, among others. We discovered that, by being careful about how we defined things, we could make theorems be judgmentally equal to their duals! That is, when we prove a theorem

```
initial_ob_unique : ∀ C (x y : Ob C),

                is_initial_ob x → is_initial_ob y → x ≅ y,
```

we can define another theorem

```
terminal_ob_unique : ∀ C (x y : Ob C),

                is_terminal_ob x → is_terminal_ob y → x ≅ y
```

as

```
terminal_ob_unique C x y H H' := initial_ob_unique Cᵒᵖ y x H' H.
```

Interestingly, we found that in proofs with sufficiently complicated types, it can take a few seconds or more for Coq to accept such a definition; we are not sure whether this is due to peculiarities of the reduction strategy of our version of Coq, or speed dependency on the size of the normal form of the type (rather than on the size of the unnormalized type), or something else entirely.

In contrast to the simplicity of witnessing the isomorphism, it takes a significant amount of care in defining concepts, often to get around deficiencies of Coq, to achieve *judgmental* duality. Even now, we were unable to achieve this ideal for some theorems. For example, category theorists typically identify the functor category $\mathcal{C}^{\mathrm{op}} \to \mathcal{D}^{\mathrm{op}}$ (whose objects are functors $\mathcal{C}^{\mathrm{op}} \to \mathcal{D}^{\mathrm{op}}$ and whose morphisms are natural transformations) with $(\mathcal{C} \to \mathcal{D})^{\mathrm{op}}$ (whose objects are functors $\mathcal{C} \to \mathcal{D}$ and whose morphisms are flipped natural transformations). These categories are canonically isomorphic (by the dualizing natural transformations), and, with the univalence axiom [25], they are equal as categories! But we have not found a way to make them definitionally equal, much to our disappointment.

3.1 Duality Design Patterns

One of the simplest theorems about duality is that it is involutive; we have that $(\mathcal{C}^{\mathrm{op}})^{\mathrm{op}} = \mathcal{C}$. In order to internalize proof by duality via judgmental equality, we sometimes need this equality to be judgmental. Although it is impossible in general in Coq 8.4 (see dodging judgmental η on records below), we want at least to have it be true for any explicit category (that is, any category specified by giving its objects, morphisms, etc., rather than referred to via a local variable).

Removing Symmetry. Taking the dual of a category, one constructs a proof that $f \circ (g \circ h) = (f \circ g) \circ h$ from a proof that $(f \circ g) \circ h = f \circ (g \circ h)$. The standard approach is to apply symmetry. However, because applying symmetry twice results in a judgmentally different proof, we decided instead to extend the definition of `Category` to require both a proof of $f \circ (g \circ h) = (f \circ g) \circ h$ and a proof of $(f \circ g) \circ h = f \circ (g \circ h)$. Then our dualizing operation simply swaps the proofs. We added a convenience constructor for categories that asks only for one of the proofs, and applies symmetry to get the other one. Because we formalized 0-truncated category theory, where the type of morphisms is required to have unique identity proofs, asking for this other proof does not result in any coherence issues.

Dualizing the Terminal Category. To make everything work out nicely, we needed the terminal category, which is the category with one object and only the identity morphism, to be the dual of itself. We originally had the terminal category as a special case of the discrete category on n objects. Given a type T with uniqueness of identity proofs, the discrete category on T has as objects inhabitants of T, and has as morphisms from x to y proofs that $x = y$. These categories are not judgmentally equal to their duals, because the type $x = y$ is not judgmentally the same as the type $y = x$. As a result, we instead used the indiscrete category, which has `unit` as its type of morphisms.

Which Side Does the Identity Go On? The last tricky obstacle we encountered was that when defining a functor out of the terminal category, it is necessary to pick whether to use the right identity law or the left identity law to prove that the functor preserves composition; both will prove that the identity composed with itself is the identity. The problem is that dualizing the functor leads to a road block where either concrete choice turns out to be "wrong," because the dual of the functor out of the terminal category will not be judgmentally equal to another instance of itself. To fix this problem, we further extended the definition of category to require a proof that the identity composed with itself is the identity.

Dodging Judgmental η on Records. The last problem we ran into was the fact that sometimes, we really, really wanted judgmental η on records. The η rule for records says any application of the record constructor to all the projections of an object yields exactly that object; e.g. for pairs, $x \equiv (x_1, x_2)$ (where x_1

and x_2 are the first and second projections, respectively). For categories, the η rule says that given a category C, for a "new" category defined by saying that its objects are the objects of C, its morphisms are the morphisms of C, ..., the "new" category is judgmentally equal to C.

In particular, we wanted to show that any functor out of the terminal category is the opposite of some other functor; namely, any $F : 1 \to C$ should be equal to $(F^{\mathrm{op}})^{\mathrm{op}} : 1 \to (C^{\mathrm{op}})^{\mathrm{op}}$. However, without the judgmental η rule for records, a local variable C cannot be judgmentally equal to $(C^{\mathrm{op}})^{\mathrm{op}}$, which reduces to an application of the constructor for a category. To get around the problem, we made two variants of dual functors: given $F : C \to D$, we have $F^{\mathrm{op}} : C^{\mathrm{op}} \to D^{\mathrm{op}}$, and given $F : C^{\mathrm{op}} \to D^{\mathrm{op}}$, we have $F^{\mathrm{op}'} : C \to D$. There are two other flavors of dual functors, corresponding to the other two pairings of ^{op} with domain and codomain, but we have been glad to avoid defining them so far. As it was, we ended up having four variants of dual natural transformation, and are very glad that we did not need sixteen. We look forward to Coq 8.5, when we will hopefully only need one.

3.2   Moving Forward: Computation Rules for Pattern Matching

While we were able to work around most of the issues that we had in internalizing proof by duality, things would have been far nicer if we had more η rules. The η rule for records is explained above. The η rule for equality says that the identity function is judgmentally equal to the function $f : \forall x\, y, x = y \to x = y$ defined by pattern matching on the first proof of equality; this rule is necessary to have any hope that applying symmetry twice is judgmentally the identity transformation. Matthieu Sozeau is currently working on giving Coq judgmental η for records with one or more fields, though not for equality.

Section 4.1 will give more examples of the pain of manipulating pattern matching on equality. Homotopy type theory provides a framework that systematizes reasoning about proofs of equality, turning a seemingly impossible task into a manageable one. However, there is still a significant burden associated with reasoning about equalities, because so few of the rules are judgmental.

We are currently attempting to divine the appropriate computation rules for pattern matching constructs, in the hopes of making reasoning with proofs of equality more pleasant.[6]

4   Other Design Choices

A few other pervasive strategies made non-trivial differences for proof performance or simplicity.

---

[6] See https://coq.inria.fr/bugs/show_bug.cgi?id=3179 and https://coq.inria.fr/bugs/show_bug.cgi?id=3119.

4.1   Identities vs. Equalities; Associators

There are a number of constructions that are provably equal, but which we found more convenient to construct transformations between instead, despite the increased verbosity of such definitions. This is especially true of constructions that strayed towards higher category theory. For example, when constructing the Grothendieck construction of a functor to the category of categories, we found it easier to first generalize the construction from functors to pseudofunctors. The definition of a pseudofunctor results from replacing various equalities in the definition of a functor with isomorphisms (analogous to bijections between sets or types), together with proofs that the isomorphisms obey various coherence properties. This replacement helped because there are fewer operations on isomorphisms (namely, just composition and inverting), and more operations on proofs of equality (pattern matching, or anything definable via induction); when we were forced to perform all of the operations in the same way, syntactically, it was easier to pick out the operations and reason about them.

Another example was defining the (co)unit of adjunction composition, where instead of a proof that $F \circ (G \circ H) = (F \circ G) \circ H$, we used a natural transformation, a coherent mapping between the actions of functors. Where equality-based constructions led to computational reduction getting stuck at casts, the constructions with natural transformations reduce in all of the expected contexts.

4.2   Opacity; Linear Dependence of Speed on Term Size

Coq is slow at dealing with large terms. For goals around 175,000 words long[7], we have found that simple tactics like `apply f_equal` take around 1 second to execute, which makes interactive theorem proving very frustrating.[8] Even more frustrating is the fact that the largest contribution to this size is often arguments to irrelevant functions, i.e., functions that are provably equal to all other functions of the same type. (These are proofs related to algebraic laws like associativity, carried inside many constructions.)

Opacification helps by preventing the type checker from unfolding some definitions, but it is not enough: the type checker still has to deal with all of the large arguments to the opaque function. Hash-consing might fix the problem completely.

Alternatively, it would be nice if, given a proof that all of the inhabitants of a type were equal, we could forget about terms of that type, so that their sizes would not impose any penalties on term manipulation. One solution might be irrelevant fields, like those of Agda, or implemented via the Implicit CiC [3, 16].

4.3   Abstraction Barriers

In many projects, choosing the right abstraction barriers is essential to reducing mistakes, improving maintainability and readability of code, and cutting down

---

[7] When we had objects as arguments rather than fields (see Section 2.3), we encountered goals of about 219,633 words when constructing pointwise Kan extensions.

[8] See also `https://coq.inria.fr/bugs/show_bug.cgi?id=3280`.

on time wasted by programmers trying to hold too many things in their heads at once. This project was no exception; we developed an allergic reaction to constructions with more than four or so arguments, after making one too many mistakes in defining limits and colimits. Limits are a generalization, to arbitrary categories, of subsets of Cartesian products. Colimits are a generalization, to arbitrary categories, of disjoint unions modulo equivalence relations.

Our original flattened definition of limits involved a single definition with 14 nested binders for types and algebraic properties. After a particularly frustrating experience hunting down a mistake in one of these components, we decided to factor the definition into a larger number of simpler definitions, including familiar categorical constructs like terminal objects and comma categories. This refactoring paid off even further when some months later we discovered the universal morphism definition of adjoint functors. With a little more abstraction, we were able to reuse the same decomposition to prove the equivalence between universal morphisms and adjoint functors, with minimal effort.

Perhaps less typical of programming experience, we found that picking the right abstraction barriers could drastically reduce compile time by keeping details out of sight in large goal formulas. In the instance discussed in the introduction, we got a factor of ten speed-up by plugging holes in a leaky abstraction barrier![9]

5   Comparison of Category-Theory Libraries

We present here a table comparing the features of various category-theory libraries. Our library is the first column. Gray dashed check-marks ($\checkmark$) indicate features in progress. Library [18] is in Agda; the rest are in Coq. A check-mark with n stars (*) indicates a construction taking $20n$ seconds to compile on a 64-bit server with a 2.40 GHz CPU and 16 GB of RAM.

Construction	[10]	[15]	[20]	[18]	[1]
Mostly automated (with custom Ltac)	$\checkmark$				
Uses HoTT	$\checkmark$				$\checkmark$
Uses type classes		$\checkmark$			
Setoid of morphisms		$\checkmark$	$\checkmark$	$\checkmark$	
Uses higher inductive types	$\checkmark$				
Assumes UIP or equivalent				$\checkmark$	
Category of sets	$\checkmark$		$\checkmark$	$\checkmark$	$\checkmark$
Initial/Terminal objects	$\checkmark$	$\checkmark$	$\checkmark$	$\checkmark$	$\checkmark$
(co)limits	$\checkmark$		$\checkmark$	$\checkmark$	$\checkmark$
(co)limit functor	$\checkmark$				
(co)limit adjoint to Δ	$\checkmark$				
Fully faithful functors	$\checkmark$			$\checkmark$	$\checkmark$
Essentially surjective functors	$\checkmark$	$\checkmark$		$\checkmark$	$\checkmark$

---

[9] See https://github.com/HoTT/HoTT/commit/eb0099005171 for the exact change.

Construction	[10]	[15]	[20]	[18]	[1]
Unit-Counit Adjunctions	✓	✓	✓	✓	✓
Hom Adjunctions	✓		✓		
Universal morphism adjunctions	✓		✓		
Adjoint composition laws	✓**			✓* 10	
Monoidal categories	✓	✓**********	✓*******		
Enriched categories	✓	✓***:	✓		
2-categories				✓*	
Category of (strict) categories	✓		✓	✓	
Hom functor	✓	✓	✓	✓	✓
Profunctors	✓			✓	
Pseudofunctors	✓*:				
Kan extensions	✓			✓	✓
Pointwise Kan extensions	✓				
$\mathcal{C}^{\mathcal{D}^{\mathcal{E}}} \cong \mathcal{C}^{\mathcal{D}\times\mathcal{E}}$; $(\mathcal{C}\times\mathcal{D})^{\mathcal{E}} \cong \mathcal{C}^{\mathcal{E}}\times\mathcal{D}^{\mathcal{E}}$	✓**				
Adjoint Functor Theorem			✓		
Yoneda	✓		✓	✓***:	✓
dep. product (oplax lim $F : \mathcal{C} \to \mathrm{Cat}$)	✓				
dep. sum (oplax colim $F : \mathcal{C} \to \mathrm{Cat}$)	✓*:			✓*	
$(\_/\_)$ functor $(\mathcal{C}^{\mathcal{A}})^{\mathrm{op}} \times \mathcal{C}^{\mathcal{B}} \to \mathrm{Cat}_{/\mathcal{A}\times\mathcal{B}}$	✓******				
Rezk completion					✓
Mean lines per file	78	126	133	98	407
Total compilation time	490s	517s	21s	717s	62s[11]
Total time w/o monoidal	490s	43s	21s	579s	62s
Median file compilation time	0.3s	0.4s	0.1s	1.5s	0.9s
Total number of files	147	36	105	143	13
Total number of definitions	578	214	995	396	367

In summary, our library includes many constructions from past formalizations, plus a few rather complex new ones. We test the limits of Coq by applying mostly automated Ltac proofs for these constructions, taking advantage of ideas from homotopy type theory and extensions built to support such constructions. In most cases, we found that term size had the biggest impact on speed. We have summarized our observations on using new features from that extension and on other hypothetical features that could make an especially big difference in our development, and we hope these observations can help guide the conversation on the design of future versions of Coq and other proof assistants.

Acknowledgments. This work was supported in part by the MIT big-data@CSAIL initiative, NSF grant CCF-1253229, ONR grant N000141310260,

---

[10] The use of proof-irrelevant fields speeds up this construction significantly in Agda.

[11] Nearly 75% of the time in this library is spent on properties of functor composition. Nearly 50% of this time is spent closing sections, for an as-yet unknown reason.

and AFOSR grant FA9550-14-1-0031. We also thank Benedikt Ahrens, Daniel R. Grayson, Robert Harper, Bas Spitters, and Edward Z. Yang for feedback on this paper.

References

1. Ahrens, B., Kapulkin, C., Shulman, M.: Univalent categories and the Rezk completion. In: ArXiv e-prints (March 2013)
2. Awodey, S.: Category theory, 2nd edn. Oxford University Press
3. Barras, B., Bernardo, B.: The implicit calculus of constructions as a programming language with dependent types. In: Amadio, R.M. (ed.) FOSSACS 2008. LNCS, vol. 4962, pp. 365–379. Springer, Heidelberg (2008)
4. Bertot, Y.: Private Inductive Types: Proposing a language extension (April 2013), http://coq.inria.fr/files/coq5_submission_3.pdf
5. Bishop, E.: Foundations of Constructive Analysis. McGraw-Hill series in higher mathematics. McGraw-Hill (1967)
6. Formalizations of category theory in proof assistants. MathOverflow, http://mathoverflow.net/questions/152497/formalizations-of-category-theory-in-proof-assistants
7. Garillot, F., Gonthier, G., Mahboubi, A., Rideau, L.: Packaging Mathematical Structures. In: Berghofer, S., Nipkow, T., Urban, C., Wenzel, M. (eds.) TPHOLs 2009. LNCS, vol. 5674, pp. 327–342. Springer, Heidelberg (2009)
8. Harper, R., Pollack, R.: Type checking with universes. Theoretical Computer Science 89(1), 107–136 (1991), ISSN: 0304-3975
9. Harrison, J.: Formalized mathematics. TUCS technical report. Turku Centre for Computer Science (1996) ISBN: 9789516508132
10. HoTT/HoTT Categories, https://github.com/HoTT/HoTT/tree/master/theories/categories
11. Huet, G., Saïbi, A.: Constructive category theory. In: Proof, Language, and Interaction, pp. 239–275. MIT Press (2000)
12. Krebbers, R., Spitters, B., van der Weegen, E.: Math Classes
13. Leinster, T.: Higher Operads, Higher Categories. Cambridge Univ. Press
14. Mac Lane, S.: Categories for the working mathematician
15. Megacz, A.: Category Theory Library for Coq. Coq, http://www.cs.berkeley.edu/~megacz/coq-categories/
16. Miquel, A.: The Implicit Calculus of Constructions. In: Abramsky, S. (ed.) TLCA 2001. LNCS, vol. 2044, pp. 344–359. Springer, Heidelberg (2001)
17. O'Keefe, G.: Towards a readable formalisation of category theory. Electronic Notes in Theoretical Computer Science 91, 212–228 (2004)
18. Peebles, D., Vezzosi, A., Cook, J.: copumpkin/categories, https://github.com/copumpkin/categories
19. Pierce, B.: A taste of category theory for computer scientists. Tech. rep.
20. Saïbi, A.: Constructive Category Theory, http://coq.inria.fr/pylons/pylons/contribs/view/ConCaT/v8.4
21. Shulman, M.: An Interval Type Implies Function Extensionality, http://homotopytypetheory.org/2011/04/04

22. Sozeau, M., et al.: HoTT/coq, `https://github.com/HoTT/coq`
23. Spitters, B., van der Weegen, E.: Developing the algebraic hierarchy with type classes in Coq. In: Kaufmann, M., Paulson, L.C. (eds.) ITP 2010. LNCS, vol. 6172, pp. 490–493. Springer, Heidelberg (2010)
24. The Coq Proof Assistant. INRIA, `http://coq.inria.fr`
25. The Univalent Foundations Program. Homotopy Type Theory: Univalent Foundations of Mathematics (2013)

A New and Formalized Proof
of Abstract Completion[*]

Nao Hirokawa[1], Aart Middeldorp[2], and Christian Sternagel[2]

[1] JAIST, Japan
hirokawa@jaist.ac.jp
[2] University of Innsbruck, Austria
{aart.middeldorp,christian.sternagel}@uibk.ac.at

Abstract. Completion is one of the most studied techniques in term rewriting. We present a new proof of the correctness of abstract completion that is based on peak decreasingness, a special case of decreasing diagrams. Peak decreasingness replaces Newman's Lemma and allows us to avoid proof orders in the correctness proof of completion. As a result, our proof is simpler than the one presented in textbooks, which is confirmed by our Isabelle/HOL formalization. Furthermore, we show that critical pair criteria are easily incorporated in our setting.

1 Introduction

Knuth and Bendix' completion procedure [11] is a landmark result in term rewriting. Given an equational system $\mathcal{E}$ and a reduction order, the completion procedure aims to construct a complete (terminating and confluent) term rewrite system that is equivalent to $\mathcal{E}$, thereby providing a general solution to the validity problem. Completion has had significant impact on various areas of computer science, in particular automated theorem proving.

The completion process is non-trivial and showing its correctness is a challenge [8]. Bachmair, Dershowitz, and Hsiang [4] introduced abstract inference rules that capture the essence of completion and introduced a new proof technique based on proof orders and persistent sets. This became the de facto standard and has been adopted in textbooks on term rewriting [1,17]. Very recently, this proof was further simplified for finite runs and formalized in Isabelle/HOL by Sternagel and Thiemann [16]. Still, we do not hesitate to point to the intricacy of these proofs, especially when practical critical pair criteria [2,3] are incorporated in completion.

Contribution. In this paper we present a new and formalized correctness proof of abstract completion for finite runs. We introduce a new confluence criterion for abstract rewriting, which we name peak decreasingness, allowing us to abstract from proof orders in order to obtain a simple and elementary proof of the

[*] Supported by JSPS KAKENHI Grant Number 25730004 and the Austrian Science Fund (FWF) projects I963 and J3202.

G. Klein and R. Gamboa (Eds.): ITP 2014, LNAI 8558, pp. 292–307, 2014.

correctness of completion. Moreover the proof incorporates a critical pair criterion: it suffices to consider prime critical pairs. Our formalization was conducted using Isabelle/HOL [12].

Formalization. Our formalization is available as part of IsaFoR (an Isabelle/HOL formalization of rewriting) version 2.14.[1] To have a look at the actual formalization visit IsaFoR's website and follow the link *Mercurial repository* under *Downloads*. Alternatively you can download the provided *.tgz file. Either way, all the relevant theory files are to be found in the subdirectory IsaFoR/. The content of this paper comprises the following theory files: Renaming formalizes permutations and permutation types, and proves useful facts about them; Renaming_Interpretations gives permutation type instances for terms, rules, substitutions, TRSs, etc., i.e., allows us to apply permutations to them; Peak_Decreasingness defines labeled conversions and peak decreasingness, and contains the proof that the latter implies confluence; CP defines overlaps, critical peaks, and critical pairs, and proves the critical pair lemma as well as the fact that for finite TRSs only finitely many representatives of critical pairs have to be considered; Prime_Critical_Pairs defines prime critical pairs and proves an important result about peaks that allows us to restrict to prime critical pairs for fairness; Abstract_Completion defines the inference rules of abstract completion, and proves their soundness; finally Completion_Fairness proves soundness of abstract completion when restricting fairness to prime critical pairs. For the benefit of a general audience we present all the proofs in the following on a high level and using standard mathematical notation. Nevertheless the proofs are exactly along the lines of our formalization and from time to time we sprinkle the text with comments directed at Isabelle initiates. But those are not essential for understanding.

Organization. The remainder of the paper is organized as follows. In the next section we recall some rewriting preliminaries. In Section 3 we discuss our formalization of variable renamings. Peak decreasingness is introduced in Section 4. The critical pair lemma, or rather: a formalized critical peak lemma, is the subject of Section 5. Our new correctness proof for abstract completion is presented in full detail in Section 6. Related work is discussed in Section 7 before we conclude in Section 8.

2   Preliminaries

We assume familiarity with term rewriting and all that (e.g., [1]) and only shortly recall notions that are used in the following. An *abstract rewrite system* (ARS for short) $\mathcal{A}$ is a set A, also called the carrier, equipped with a binary relation $\rightarrow$. Sometimes we partition the binary relation into parts according to a set I of indices (or labels). Then we write $\mathcal{A} = \langle A, \{\rightarrow_\alpha\}_{\alpha \in I}\rangle$ where we denote the part of the relation with label α by $\rightarrow_\alpha$, i.e., $\rightarrow = \bigcup_{\alpha \in I} \rightarrow_\alpha$. In our formalization

---

[1] http://cl-informatik.uibk.ac.at/software/ceta

ARSs are just relations which are represented by sets of pairs in Isabelle/HOL, i.e., of type $(\alpha \times \alpha)$ *set*, and their carrier is given implicitly by the type α.

Terms are defined inductively as follows: a *term* is either a variable x from the set $\mathcal{V}$ or is constructed by applying a function symbol $f \in \mathcal{F}$ to a list of argument terms $f(t_1, \ldots, t_n)$. Here $\mathcal{F}$ is called the *signature*. The set of all terms built over $\mathcal{F}$ and $\mathcal{V}$ is denoted by $\mathcal{T}(\mathcal{F}, \mathcal{V})$. In our formalization terms are represented by the datatype

> **datatype** (α, β) *term* $=$ *Var* β | *Fun* α $((\alpha, \beta)$ *term list*$)$

that is, the signature as well as the set of variables is given implicitly by the type parameters α and β, respectively. The *set of variables* of a term t is denoted by $\mathcal{V}ar(t)$ (this is easily extended to rules, term rewrite systems, etc.). *Positions* are finite lists of positive natural numbers where the empty position (or *root position*) is denoted by ϵ. The set of positions of a term t is denoted by $\mathcal{P}os(t)$ and partitioned into *function positions* $\mathcal{P}os_{\mathcal{F}}$ and *variable positions* $\mathcal{P}os_{\mathcal{V}}$. Positions are partially ordered by the prefix order, i.e., $p \leqslant q$ if p is a prefix of q (we also say that p is above q). Two positions p and q for which neither $p \leqslant q$ nor $q \leqslant p$ are called *parallel*, denoted by $p \parallel q$. Whenever $p \leqslant q$, by $q \setminus p$ we denote position q without its prefix p. The subterm of t at position p is denoted by $t|_p$ and replacing this term by s is denoted by $t[s]_p$.

A *substitution* is a mapping σ from variables to terms such that its domain $\{x \in \mathcal{V} \mid \sigma(x) \neq x\}$ is finite. Applying a substitution to a term is written $t\sigma$.

A pair of terms (s, t) is sometimes considered an *equation*, then we write $s \approx t$, and sometimes a *(rewrite) rule*, then we write $s \to t$. In the latter case we assume that the left-hand side is not a variable and that the variables of the right-hand side t are all contained in the left-hand side s. This assumption we call the *variable condition*.

A set $\mathcal{E}$ of equations is called an *equational system* (ES for short) and a set $\mathcal{R}$ of rules a *term rewrite system* (TRS for short). In the following we assume the variable condition for all rules of a TRS. Sets of pairs of terms induce a *rewrite relation* by closing their components under contexts and substitutions. More precisely the rewrite relation of $\mathcal{R}$, denoted by $\to_{\mathcal{R}}$, is defined inductively by $s \to_{\mathcal{R}} t$ whenever there are a rule $\ell \to r \in \mathcal{R}$, a position p, and a substitution σ such that $s|_p = \ell\sigma$ and $t = s[r\sigma]_p$. In the following we sometimes drop $\mathcal{R}$ in $\to_{\mathcal{R}}$ if it is clear from the context. Moreover, individual steps are sometimes annotated with additional information (the employed rule, the corresponding position, etc.).

A binary relation $\to$ is well-founded (or *terminating*) if it does not admit any infinite descending sequence $a_1 \to a_2 \to a_3 \to \cdots$. A well-founded order that is closed under contexts and substitutions is called a *reduction order*. Thus a reduction order that strictly orients all rules of a TRS establishes termination of the induced rewrite relation.

3   Renaming Variables

This section is not essential for understanding the remainder of the paper. However, it addresses a typical problem that arises when formalizing proofs involving variable renaming. Thus it is mostly interesting for users of proof assistants.

One thing that is often neglected or treated implicitly in paper proofs is renaming of variables, which is often necessary to make sure that two given objects do not contain any common variables. While on the one hand, this is clearly not good enough for a formalization using a proof assistant; on the other hand, a thorough treatment quickly leads to tedious reasoning (which is typically left as an exercise in textbooks; see for example the proof of the statement that two most general unifiers only differ by a renaming in [17, Chapter 2] and [1, Chapter 4]).

We aim for a setup that allows us to argue along the lines of a paper proof also in its formalization. The advantage of doing so is that not only the result is certified to be correct, but also the proof itself. Moreover, simulating the implicit reasoning used in a paper proof should be as painless as possible.

To this end it turns out that a slight modification of a previous Isabelle/HOL formalization of permutations and permutation types by Urban *et al.* [9,18] is very useful. Here a permutation (also called renaming) is just a bijective function f such that $\{x \mid f(x) \neq x\}$ is finite, and a *permutation type* is a type whose elements support applying a permutation to them. The mentioned modification consists in parameterizing permutation types over the type of atoms (which we call variables in the following) in addition to the type of elements (which may be terms, substitutions, rules, TRSs, etc.) To make this possible we have to switch from Isabelle's type classes to locales and therefore to reformalize a theory of permutations (thankfully, most proofs are not much different from the type class version). Moreover many useful results only hold under the assumption that we have an infinite set of variables, e.g., that we can always rename the variables of two finite objects apart. This we express in Isabelle by demanding that the type of variables is in the type class *infinite* (whose only assumption says that the universe of all values of the corresponding type be infinite). Permutation types are expressed as follows in Isabelle (see theory **Renaming**):

> **locale** *pt* =
> **fixes**   ·   ::   *(α :: infinite) perm* $\Rightarrow$ β $\Rightarrow$ β
> **assumes**   *id · x = x*   **and** $(\pi_1 \circ \pi_2) \cdot x = \pi_1 \cdot (\pi_2 \cdot x)$

where *id* is the empty permutation, ∘ denotes function composition, and $\pi \cdot x$ denotes applying the renaming π to the element x. Also note that $\alpha :: infinite$ requires the type α to contain infinitely many elements. Associated with each permutation type is the notion of *support*. Since we will not give any more details about permutation types, suffice it to say that for permutation types whose elements have finite support, this notion corresponds to the set of free variables.

The most important result about finitely supported permutation types for our purposes is that we can always find a permutation π which makes the support of x disjoint from a given finite set of variables. Intuitively, this means that we can always rename variables apart.

After a general theory of renamings we have to create concrete instances for terms, rules, substitutions, TRSs, etc. (see theory `Renaming_Interpretations`). Using this machinery we have formalized the following results (the interested reader is referred to the formalization for proofs).

Lemma 1. *Let σ and τ be substitutions.*[2]

1. *If σ' and τ' are substitutions with $\sigma\sigma' = \tau$ and $\tau\tau' = \sigma$ then there is a renaming π such that $\pi \cdot \sigma = \tau$.*
2. *If σ and τ are two most general unifiers of a set of equations, then they are variants of each other.*

In the remainder, whenever we say that a term t is a *variant* of a term u, what we formally mean is that there is a permutation π such that $\pi \cdot t = u$. Of course, this also works for rules, substitutions, TRSs, etc.

4   Peak Decreasingness

In this section we present the abstract result that replaces Newman's Lemma in the proof of the correctness of abstract completion.

Definition 2. *An ARS $\mathcal{A} = \langle A, \{\rightarrow_\alpha\}_{\alpha\in I}\rangle$ is peak decreasing if there exists a well-founded order $>$ on I such that for all $\alpha, \beta \in I$ the inclusion*

$$_\alpha\!\leftarrow \cdot \rightarrow_\beta \ \subseteq \ \xleftrightarrow[\vee\alpha\beta]{*}$$

holds. Here $\vee\alpha\beta$ denotes the set $\{\gamma \in I \mid \alpha > \gamma \text{ or } \beta > \gamma\}$ and if $J \subseteq I$ then $\xrightarrow[J]{}$ denotes the union of all $\xrightarrow[\gamma]{}$ with $\gamma \in J$ and $\xleftrightarrow[J]{*}$ denotes a conversion consisting of $\xrightarrow[J]{}$ steps.

Peak decreasingness is a special case of decreasing diagrams [13], which is known as a powerful confluence criterion. Correctness of decreasing diagrams has been formally verified in Isabelle/HOL by Zankl [20] and it should in principle be possible to obtain our results on peak decreasingness as a special case. However, for the sake of simplicity we present its easy direct proof (which we also formalized in order to verify its correctness). We denote by $\mathcal{M}(J)$ the set of all multisets over a set J.

Lemma 3. *Every peak decreasing ARS is confluent.*

Proof. Let $>$ be a well-founded order on I which shows that the ARS $\mathcal{A} = \langle A, \{\rightarrow_\alpha\}_{\alpha\in I}\rangle$ is peak decreasing. With every conversion C in $\mathcal{A}$ we associate the multiset M_C consisting of the labels of its steps. These multisets are compared by the multiset extension $>_{\mathsf{mul}}$ of $>$, which is a well-founded order on $\mathcal{M}(I)$.

---

[2] Finiteness of substitution domains is used (only) for this lemma (in this paper).

We prove $\leftrightarrow^* \subseteq \downarrow$ by well-founded induction on $>_{\mathsf{mul}}$. Consider a conversion C between a and b. We either have $a \downarrow b$ or $a \leftrightarrow^* \cdot \leftarrow \cdot \rightarrow \cdot \leftrightarrow^* b$. In the former case we are done. In the latter case there exist labels $\alpha, \beta \in I$ and multisets $\Gamma_1, \Gamma_2 \in \mathcal{M}(I)$ such that $M_C = \Gamma_1 \uplus \{\alpha, \beta\} \uplus \Gamma_2$. By the peak decreasingness assumption there exists a conversion C' between a and b such that $M_{C'} = \Gamma_1 \uplus \Gamma \uplus \Gamma_2$ with $\Gamma \in \mathcal{M}(\vee\alpha\beta)$. We obviously have $\{\alpha, \beta\} >_{\mathsf{mul}} \Gamma$ and hence $M_C >_{\mathsf{mul}} M_{C'}$. We obtain $a \downarrow b$ from the induction hypothesis. $\qquad\square$

What we informally state as *with every conversion we associate the multiset of the labels of its steps* in the proof above is formalized as an inductive predicate $\leftrightarrow\hspace{-3pt}\rightarrow$ defined by the rules

$$\frac{}{a \leftrightarrow\hspace{-3pt}\rightarrow_{\{\}} a} \qquad\qquad \frac{\alpha \in I \qquad a \overset{\alpha}{\leftrightarrow} b \qquad b \leftrightarrow\hspace{-3pt}\rightarrow_M c}{a \leftrightarrow\hspace{-3pt}\rightarrow_{M \uplus \{\alpha\}} c}$$

together with the fact that for all a and b, we have $a \leftrightarrow^* b$ if and only if there is a multiset M such that $a \leftrightarrow\hspace{-3pt}\rightarrow_M b$. (This predicate is called *conv* in theory `Peak_Decreasingness`.)

5   Critical Pair Lemma

Completion is based on critical pair analysis. In this section we present a version of the critical pair lemma that incorporates primality. The correctness proof is based on peak decreasingness.

Definition 4. *An overlap of a TRS $\mathcal{R}$ is a triple $\langle \ell_1 \to r_1, p, \ell_2 \to r_2 \rangle$ satisfying the following properties:*

- *there are renamings π_1 and π_2 such that $\pi_1 \cdot (\ell_1 \to r_1), \pi_2 \cdot (\ell_2 \to r_2) \in \mathcal{R}$,*
- $\mathcal{V}ar(\ell_1 \to r_1) \cap \mathcal{V}ar(\ell_2 \to r_2) = \varnothing$,
- $p \in \mathcal{P}os_{\mathcal{F}}(\ell_2)$,
- *ℓ_1 and $\ell_2|_p$ are unifiable,*
- *if $p = \epsilon$ then $\ell_1 \to r_1$ and $\ell_2 \to r_2$ are not variants.*

In general this definition may lead to an infinite set of overlaps, since there are infinitely many possibilities of taking variable disjoint variants of rules. Fortunately it can be shown (and has been formalized) that overlaps that originate from the same two rules are variants of each other (see theory `CP`). Overlaps give rise to critical peaks and pairs.

Definition 5. *Suppose $\langle \ell_1 \to r_1, p, \ell_2 \to r_2 \rangle$ is an overlap of a TRS $\mathcal{R}$. Let σ be a most general unifier of ℓ_1 and $\ell_2|_p$. The term $\ell_2\sigma[\ell_1\sigma]_p = \ell_2\sigma$ can be reduced in two different ways:*

$$\ell_2\sigma[r_1\sigma]_p \xleftarrow[\quad p \quad]{\ell_1 \to r_1} \ell_2\sigma[\ell_1\sigma]_p = \ell_2\sigma \xrightarrow[\quad \epsilon \quad]{\ell_2 \to r_2} r_2\sigma$$

We call the quadruple $(\ell_2\sigma[r_1\sigma]_p, p, \ell_2\sigma, r_2\sigma)$ a critical peak *and the equation $\ell_2\sigma[r_1\sigma]_p \approx r_2\sigma$ a* critical pair *of $\mathcal{R}$, obtained from the overlap. The set of all critical pairs of $\mathcal{R}$ is denoted by* CP($\mathcal{R}$).

In our formalization we do not use an arbitrary most general unifier in the above definition. Instead we use *the* most general unifier that is computed by the formalized unification algorithm that is part of IsaFoR (thereby removing one degree of freedom and making it easier to show that only finitely many critical pairs have to be considered for finite TRSs).

A critical peak (t, p, s, u) is usually denoted by $t \overset{p}{\leftarrow} s \overset{\epsilon}{\rightarrow} u$. It can be shown (and has been formalized) that different critical peaks and pairs obtained from two variants of the same overlap are variants of each other. Since rewriting is equivariant under permutations, it is enough to consult finitely many critical pairs or peaks for finite TRSs (one for each pair of rules and each appropriate position) in order to conclude rewriting related properties (like joinability or fairness, see below) for all of them.

We present a variation of the well-known critical pair lemma for critical peaks and its formalized proof. The slightly cumbersome statement is essential to avoid gaps in the proof of Lemma 9 below.

Lemma 6. *Let $\mathcal{R}$ be a TRS. If $t \underset{\mathcal{R}}{\overset{p_1}{\leftarrow}} s \overset{p_2}{\rightarrow}_\mathcal{R} u$ then one of the following holds:*

(a) $t \downarrow_\mathcal{R} u$,

(b) $p_2 \leqslant p_1$ and $t|_{p_2} \overset{p_1 \backslash p_2}{\leftarrow} s|_{p_2} \overset{\epsilon}{\rightarrow} u|_{p_2}$ is an instance of a critical peak,

(c) $p_1 \leqslant p_2$ and $u|_{p_1} \overset{p_2 \backslash p_1}{\leftarrow} s|_{p_1} \overset{\epsilon}{\rightarrow} t|_{p_1}$ is an instance of a critical peak.

Proof. Consider an arbitrary peak $t \underset{p_1, \ell_1' \rightarrow r_1', \sigma_1'}{\leftarrow} s \rightarrow_{p_2, \ell_2 \rightarrow r_2, \sigma_2} u$. If $p_1 \parallel p_2$ then $t \rightarrow_{p_2, \ell_2 \rightarrow r_2, \sigma_2} t[r_2\sigma_2]_{p_2} = u[r_1'\sigma_1']_{p_1} \underset{p_1, \ell_1' \rightarrow r_1', \sigma_1'}{\leftarrow} u$. If the positions of the contracted redexes are not parallel then one of them is above the other. Without loss of generality we assume that $p_1 \geqslant p_2$. Let $p = p_1 \backslash p_2$. Moreover, let π be a permutation such that $\ell_1 \rightarrow r_1 = \pi \cdot (\ell_1' \rightarrow r_1')$ and $\ell_2 \rightarrow r_2$ have no variables in common. Such a permutation exists since we only have to avoid the finitely many variables of $\ell_2 \rightarrow r_2$ and assume an infinite set of variables. Furthermore, let $\sigma_1 = \pi^{-1} \cdot \sigma_1'$. We have $t = s[r_1\sigma_1]_{p_1} = s[\ell_2\sigma_2[r_1\sigma_1]_p]_{p_2}$ and $u = s[r_2\sigma_2]_{p_2}$. We consider two cases depending on whether $p \in \mathcal{P}os_{\mathcal{F}}(\ell_2)$ in conjunction with the fact that whenever $p = \epsilon$ then $\ell_1 \rightarrow r_1$ and $\ell_2 \rightarrow r_2$ are not variants, is true or not.

- Suppose $p \in \mathcal{P}os_{\mathcal{F}}(\ell_2)$ and $p = \epsilon$ implies that $\ell_1 \rightarrow r_1$ and $\ell_2 \rightarrow r_2$ are not variants. Let $\sigma'(x) = \sigma_1(x)$ for $x \in \mathcal{V}ar(\ell_1 \rightarrow r_1)$ and $\sigma'(x) = \sigma_2(x)$, otherwise. The substitution σ' is a unifier of $\ell_2|_p$ and ℓ_1: $(\ell_2|_p)\sigma' = (\ell_2\sigma_2)|_p = \ell_1\sigma_1 = \ell_1\sigma'$. Then $\langle \ell_1 \rightarrow r_1, p, \ell_2 \rightarrow r_2 \rangle$ is an overlap. Let σ be a most general unifier of $\ell_2|_p$ and ℓ_1. Hence $\ell_2\sigma[r_1\sigma]_p \overset{p}{\leftarrow} \ell_2\sigma \overset{\epsilon}{\rightarrow} r_2\sigma$ is a critical peak and there exists a substitution τ such that $\sigma' = \sigma\tau$. Therefore

$$\ell_2\sigma_2[r_1\sigma_1]_p = (\ell_2\sigma[r_1\sigma]_p)\tau \overset{p}{\leftarrow} (\ell_2\sigma)\tau \overset{\epsilon}{\rightarrow} (r_2\sigma)\tau = r_2\sigma_2$$

  and thus (b) is obtained.

- Otherwise, either $p = \epsilon$ and $\ell_1 \to r_1$, $\ell_2 \to r_2$ are variants, or $p \notin \mathcal{P}os_{\mathcal{F}}(\ell_2)$. In the former case it is easy to show that $r_1\sigma_1 = r_2\sigma_2$ and hence $t = u$. In the latter case, there exist positions q_1, q_2 such that $p = q_1 q_2$ and $q_1 \in \mathcal{P}os_{\mathcal{V}}(\ell_2)$. Let $\ell_2|_{q_1}$ be the variable x. We have $\sigma_2(x)|_{q_2} = \ell_1\sigma_1$. Define the substitution σ_2' as follows:

$$\sigma_2'(y) = \begin{cases} \sigma_2(y)[r_1\sigma_1]_{q_2} & \text{if } y = x \\ \sigma_2(y) & \text{if } y \neq x \end{cases}$$

Clearly $\sigma_2(x) \to_{\mathcal{R}} \sigma_2'(x)$, and thus $r_2\sigma_2 \to^* r_2\sigma_2'$. We also have

$$\ell_2\sigma_2[r_1\sigma_1]_p = \ell_2\sigma_2[\sigma_2'(x)]_{q_1} \to^* \ell_2\sigma_2' \to r_2\sigma_2'$$

Consequently, $t \to^* s[r_2\sigma_2']_{p_2} {}^* \!\leftarrow u$. Hence, (c) is concluded. $\qquad\square$

An easy consequence of the above lemma is that for every peak $t \; {}_{\mathcal{R}}\!\!\leftarrow s \to_{\mathcal{R}} u$ we have $t \downarrow_{\mathcal{R}} u$ or $t \leftrightarrow_{\mathsf{CP}(\mathcal{R})} u$. It might be interesting to note that in our formalization of the above proof we do actually not need the fact that left-hand sides of rules are not variables.

Definition 7. *A critical peak $t \overset{p}{\leftarrow} s \overset{\epsilon}{\to} u$ is* prime *if all proper subterms of $s|_p$ are in normal form. A critical pair is called* prime *if it is derived from a prime critical peak. We write $\mathsf{PCP}(\mathcal{R})$ to denote the set of all prime critical pairs of a TRS $\mathcal{R}$.*

Below we prove that non-prime critical pairs need not be computed. In the proof we use a new ternary relation on terms. It expresses the condition under which a conversion between two terms is considered harmless (when it comes to proving confluence of terminating TRSs). This relation is also used in the new correctness proof of abstract completion that we present in the next section.

Definition 8. *Given a TRS $\mathcal{R}$ and terms s, t, and u, we write $t \nabla_s u$ if $s \to_{\mathcal{R}}^+ t$, $s \to_{\mathcal{R}}^+ u$, and $t \downarrow_{\mathcal{R}} u$ or $t \leftrightarrow_{\mathsf{PCP}(\mathcal{R})} u$.*

Lemma 9. *Let $\mathcal{R}$ be a TRS. If $t \overset{p}{\leftarrow} s \overset{\epsilon}{\to} u$ is a critical peak then $t \nabla_s^2 u$.*

Proof. First suppose that all proper subterms of $s|_p$ are in normal form. Then $t \approx u \in \mathsf{PCP}(\mathcal{R})$ and thus $t \nabla_s u$. Since also $u \nabla_s u$, we obtain the desired $t \nabla_s^2 u$. This leaves us with the case that there is a proper subterm of $s|_p$ that is not in normal form. By considering an innermost redex in $s|_p$ we obtain a position $q > p$ and a term v such that $s \overset{q}{\to} v$ and all proper subterms of $s|_q$ are in normal form. Now, if $v \overset{q}{\leftarrow} s \overset{\epsilon}{\to} u$ is an instance of a critical peak then $v \to_{\mathsf{PCP}(\mathcal{R})} u$. Otherwise, $v \downarrow_{\mathcal{R}} u$ by Lemma 6, since $q \not\leqslant \epsilon$. In both cases we obtain $v \nabla_s u$. Finally, we analyze the peak $t \overset{p}{\leftarrow} s \overset{q}{\to} v$ by another application of Lemma 6.

1. If $t \downarrow_{\mathcal{R}} v$, we obtain $t \nabla_s v$ and thus $t \nabla_s^2 u$, since also $v \nabla_s u$.
2. Since $p < q$, only the case that $v|_p \overset{q\backslash p}{\leftarrow} s|_p \overset{\epsilon}{\to} t|_p$ is an instance of a critical peak remains. Moreover, all proper subterms of $s|_q$ are in normal form and thus we have an instance of a prime critical peak. Hence $t \leftrightarrow_{\mathsf{PCP}(\mathcal{R})} v$ and together with $v \nabla_s u$ we conclude $t \nabla_s^2 u$. $\qquad\square$

Corollary 10. *Let $\mathcal{R}$ be a TRS. If $t \,_\mathcal{R}{\leftarrow} s \rightarrow_\mathcal{R} u$ then $t \nabla_s^2 u$.*

Proof. From Lemma 6, either $t \downarrow_\mathcal{R} u$ and we are done, or $t \,_\mathcal{R}{\leftarrow} s \rightarrow_\mathcal{R} u$ contains a (possibly reversed) instance of a critical peak. By Lemma 9 we conclude the proof, since rewriting is closed under substitutions and contexts. □

The following result is due to Kapur *et al.* [10, Corollary 4].

Corollary 11. *A terminating TRS is confluent if and only if all its prime critical pairs are joinable.*

Proof. Let $\mathcal{R}$ be a terminating TRS such that $\mathsf{PCP}(\mathcal{R}) \subseteq \downarrow_\mathcal{R}$. We label rewrite steps by their starting term and we claim that $\mathcal{R}$ is peak decreasing. As well-founded order we take $> \,=\, \rightarrow_\mathcal{R}^+$. Consider an arbitrary peak $t \,_\mathcal{R}{\leftarrow} s \rightarrow_\mathcal{R} u$. Lemma 10 yields a term v such that $t \nabla_s v \nabla_s u$. From the assumption $\mathsf{PCP}(\mathcal{R}) \subseteq \downarrow_\mathcal{R}$ we obtain $t \downarrow_\mathcal{R} v \downarrow_\mathcal{R} u$. Since $s \rightarrow_\mathcal{R}^+ v$, all steps in the conversion $t \downarrow_\mathcal{R} v \downarrow_\mathcal{R} u$ are labeled with a term that is smaller than s. Since the two steps in the peak receive the same label s, peak decreasingness is established and hence we obtain the confluence of $\mathcal{R}$ from Lemma 3. The reverse direction is trivial. □

Note that unlike for ordinary critical pairs, joinability of prime critical pairs does not imply local confluence.

Example 12. Consider the following TRS $\mathcal{R}$:

$$\mathsf{f}(\mathsf{a}) \rightarrow \mathsf{b} \qquad\qquad \mathsf{f}(\mathsf{a}) \rightarrow \mathsf{c} \qquad\qquad \mathsf{a} \rightarrow \mathsf{a}$$

The set $\mathsf{PCP}(\mathcal{R})$ consists of the pairs $\mathsf{f}(\mathsf{a}) \approx \mathsf{b}$ and $\mathsf{f}(\mathsf{a}) \approx \mathsf{c}$, which are trivially joinable. But $\mathcal{R}$ is not locally confluent because the peak $\mathsf{b} \,_\mathcal{R}{\leftarrow} \mathsf{f}(\mathsf{a}) \rightarrow_\mathcal{R} \mathsf{c}$ is not joinable.

6   Abstract Completion

The abstract completion procedure for which we give a new and formalized correctness proof is presented in the following definition.

Definition 13. *The inference system KB operates on pairs consisting of an ES $\mathcal{E}$ and a TRS $\mathcal{R}$ over a common signature $\mathcal{F}$. It consists of the following six inference rules:*

$$\text{deduce } \frac{\mathcal{E},\mathcal{R}}{\mathcal{E} \cup \{s \approx t\},\mathcal{R}} \text{ if } s \,_\mathcal{R}{\leftarrow} \cdot \rightarrow_\mathcal{R} t \qquad \text{compose } \frac{\mathcal{E},\mathcal{R} \uplus \{s \rightarrow t\}}{\mathcal{E},\mathcal{R} \cup \{s \rightarrow u\}} \text{ if } t \rightarrow_\mathcal{R} u$$

$$\text{orient } \begin{array}{c} \dfrac{\mathcal{E} \uplus \{s \approx t\},\mathcal{R}}{\mathcal{E},\mathcal{R} \cup \{s \rightarrow t\}} \text{ if } s > t \\[1.2em] \dfrac{\mathcal{E} \uplus \{s \approx t\},\mathcal{R}}{\mathcal{E},\mathcal{R} \cup \{t \rightarrow s\}} \text{ if } t > s \end{array} \qquad \text{simplify } \begin{array}{c} \dfrac{\mathcal{E} \uplus \{s \approx t\},\mathcal{R}}{\mathcal{E} \cup \{u \approx t\},\mathcal{R}} \text{ if } s \rightarrow_\mathcal{R} u \\[1.2em] \dfrac{\mathcal{E} \uplus \{s \approx t\},\mathcal{R}}{\mathcal{E} \cup \{s \approx u\},\mathcal{R}} \text{ if } t \rightarrow_\mathcal{R} u \end{array}$$

$$\text{delete } \frac{\mathcal{E} \uplus \{s \approx s\},\mathcal{R}}{\mathcal{E},\mathcal{R}} \qquad\qquad \text{collapse } \frac{\mathcal{E},\mathcal{R} \uplus \{t \rightarrow s\}}{\mathcal{E} \cup \{u \approx s\},\mathcal{R}} \text{ if } t \rightarrow_\mathcal{R} u$$

Here $>$ is a fixed reduction order on $\mathcal{T}(\mathcal{F}, \mathcal{V})$.

Inference rules for completion were introduced by Bachmair, Dershowitz, and Hsiang in [4]. The version above differs from most of the inference systems in the literature (e.g. [2,3]) in that we do not impose any encompassment restriction in collapse. The reason is that only *finite* runs will be considered here (cf. [16]).

We write $(\mathcal{E}, \mathcal{R})$ for the pair $\mathcal{E}, \mathcal{R}$ when it increases readability. We write $(\mathcal{E}, \mathcal{R}) \vdash_{\mathsf{KB}} (\mathcal{E}', \mathcal{R}')$ if $(\mathcal{E}', \mathcal{R}')$ can be obtained from $(\mathcal{E}, \mathcal{R})$ by applying one of the inference rules of Definition 13.

According to the following lemma the equational theory induced by $\mathcal{E} \cup \mathcal{R}$ is not affected by application of the inference rules of KB. This is well-known, but our formulation is new and paves the way for a simple correctness proof.

Lemma 14. *Suppose $(\mathcal{E}, \mathcal{R}) \vdash_{\mathsf{KB}} (\mathcal{E}', \mathcal{R}')$.*

1. *If $s \xrightarrow[\mathcal{E} \cup \mathcal{R}]{} t$ then $s \xrightarrow[\mathcal{R}']{=} \cdot \xrightarrow[\mathcal{E}' \cup \mathcal{R}']{=} \cdot \xleftarrow[\mathcal{R}']{=} t$.*

2. *If $s \xrightarrow[\mathcal{E}' \cup \mathcal{R}']{} t$ then $s \xleftrightarrow[\mathcal{E} \cup \mathcal{R}]{*} t$.*

Proof. By inspecting the inference rules of KB we easily obtain the following inclusions:

deduce
$$\mathcal{E} \cup \mathcal{R} \subseteq \mathcal{E}' \cup \mathcal{R}' \qquad\qquad \mathcal{E}' \cup \mathcal{R}' \subseteq \mathcal{E} \cup \mathcal{R} \cup \xleftarrow[\mathcal{R}]{} \cdot \xrightarrow[\mathcal{R}]{}$$

orient
$$\mathcal{E} \cup \mathcal{R} \subseteq \mathcal{E}' \cup \mathcal{R}' \cup (\mathcal{R}')^{-1} \qquad\qquad \mathcal{E}' \cup \mathcal{R}' \subseteq \mathcal{E} \cup \mathcal{R} \cup \mathcal{E}^{-1}$$

delete
$$\mathcal{E} \cup \mathcal{R} \subseteq \mathcal{E}' \cup \mathcal{R}' \cup = \qquad\qquad \mathcal{E}' \cup \mathcal{R}' \subseteq \mathcal{E} \cup \mathcal{R}$$

compose
$$\mathcal{E} \cup \mathcal{R} \subseteq \mathcal{E}' \cup \mathcal{R}' \cup \xrightarrow[\mathcal{R}']{} \cdot \xleftarrow[\mathcal{R}']{} \qquad\qquad \mathcal{E}' \cup \mathcal{R}' \subseteq \mathcal{E} \cup \mathcal{R} \cup \xrightarrow[\mathcal{R}]{} \cdot \xrightarrow[\mathcal{R}]{}$$

simplify
$$\mathcal{E} \cup \mathcal{R} \subseteq \mathcal{E}' \cup \mathcal{R}' \cup \xrightarrow[\mathcal{R}']{} \cdot \xrightarrow[\mathcal{E}']{} \cup \xrightarrow[\mathcal{E}']{} \cdot \xleftarrow[\mathcal{R}']{} \qquad \mathcal{E}' \cup \mathcal{R}' \subseteq \mathcal{E} \cup \mathcal{R} \cup \xleftarrow[\mathcal{R}]{} \cdot \xrightarrow[\mathcal{E}]{} \cup \xrightarrow[\mathcal{E}]{} \cdot \xrightarrow[\mathcal{R}]{}$$

collapse
$$\mathcal{E} \cup \mathcal{R} \subseteq \mathcal{E}' \cup \mathcal{R}' \cup \xrightarrow[\mathcal{R}']{} \cdot \xrightarrow[\mathcal{E}']{} \qquad\qquad \mathcal{E}' \cup \mathcal{R}' \subseteq \mathcal{E} \cup \mathcal{R} \cup \xleftarrow[\mathcal{R}]{} \cdot \xrightarrow[\mathcal{R}]{}$$

Consider for instance the collapse rule and suppose that $s \approx t \in \mathcal{E} \cup \mathcal{R}$. If $s \approx t \in \mathcal{E}$ then $s \approx t \in \mathcal{E}'$ because $\mathcal{E} \subseteq \mathcal{E}'$. If $s \approx t \in \mathcal{R}$ then either $s \approx t \in \mathcal{R}'$ or $s \to_{\mathcal{R}} u$ with $u \approx t \in \mathcal{E}'$ and thus $s \to_{\mathcal{R}'} \cdot \to_{\mathcal{E}'} t$. This proves the inclusion on the left. For the inclusion on the right the reasoning is similar. Suppose that $s \approx t \in \mathcal{E}' \cup \mathcal{R}'$. If $s \approx t \in \mathcal{R}'$ then $s \approx t \in \mathcal{R}$ because $\mathcal{R}' \subseteq \mathcal{R}$. If $s \approx t \in \mathcal{E}'$ then either $s \approx t \in \mathcal{E}$ or there exists a rule $u \to t \in \mathcal{R}$ with $u \to_{\mathcal{R}} s$ and thus $s_{\mathcal{R}}\!\leftarrow \cdot \to_{\mathcal{R}} t$.

Since rewrite relations are closed under contexts and substitutions, the inclusions in the right column prove statement (2). Because each inclusion in the left column is a special case of

$$\mathcal{E} \cup \mathcal{R} \subseteq \xrightarrow[\mathcal{R}']{=} \cdot \xrightarrow[\mathcal{E}' \cup \mathcal{R}']{=} \cdot \xleftarrow[\mathcal{R}']{=}$$

also statement (1) follows from the closure under contexts and substitutions of rewrite relations. $\qquad\square$

Corollary 15. *If* $(\mathcal{E}, \mathcal{R}) \vdash^*_{\mathsf{KB}} (\mathcal{E}', \mathcal{R}')$ *then* $\xleftrightarrow[\mathcal{E} \cup \mathcal{R}]{*} = \xleftrightarrow[\mathcal{E}' \cup \mathcal{R}']{*}$.   □

The next lemma states that termination of $\mathcal{R}$ is preserved by applications of the inference rules of KB. It is the final result in this section whose proof refers to the inference rules.

Lemma 16. *If* $(\mathcal{E}, \mathcal{R}) \vdash^*_{\mathsf{KB}} (\mathcal{E}', \mathcal{R}')$ *and* $\mathcal{R} \subseteq >$ *then* $\mathcal{R}' \subseteq >$.

Proof. We consider a single step $(\mathcal{E}, \mathcal{R}) \vdash_{\mathsf{KB}} (\mathcal{E}', \mathcal{R}')$. The statement of the lemma follows by a straightforward induction proof. Observe that deduce, delete, and simplify do not change the set of rewrite rules and hence $\mathcal{R}' = \mathcal{R} \subseteq >$. For collapse we have $\mathcal{R}' \subsetneq \mathcal{R} \subseteq >$. In the case of orient we have $\mathcal{R}' = \mathcal{R} \cup \{s \to t\}$ with $s > t$ and hence $\mathcal{R}' \subseteq >$ follows from the assumption $\mathcal{R} \subseteq >$. Finally, consider an application of compose. So $\mathcal{R} = \mathcal{R}'' \uplus \{s \to t\}$ and $\mathcal{R}' = \mathcal{R}'' \cup \{s \to u\}$ with $t \to_{\mathcal{R}} u$. We obtain $s > t$ from the assumption $\mathcal{R} \subseteq >$. Since $>$ is a reduction order, $t > u$ follows from $t \to_{\mathcal{R}} u$. Transitivity of $>$ yields $s > u$ and hence $\mathcal{R}' \subseteq >$ as desired.   □

To guarantee that the result of a finite KB derivation is a complete TRS equivalent to the initial $\mathcal{E}$, KB derivations must satisfy the fairness condition defined below. Fairness requires that prime critical pairs of the final TRS $\mathcal{R}_n$ which were not considered during the run are joinable in $\mathcal{R}_n$.

Definition 17. *A run for a given ES $\mathcal{E}$ is a finite sequence*

$$\mathcal{E}_0, \mathcal{R}_0 \vdash_{\mathsf{KB}} \mathcal{E}_1, \mathcal{R}_1 \vdash_{\mathsf{KB}} \cdots \vdash_{\mathsf{KB}} \mathcal{E}_n, \mathcal{R}_n$$

such that $\mathcal{E}_0 = \mathcal{E}$ and $\mathcal{R}_0 = \varnothing$. The run fails *if $\mathcal{E}_n \neq \varnothing$. The run is* fair *if*

$$\mathsf{PCP}(\mathcal{R}_n) \subseteq \downarrow_{\mathcal{R}_n} \cup \bigcup_{i=0}^{n} \leftrightarrow_{\mathcal{E}_i}$$

The reason for writing $\leftrightarrow_{\mathcal{E}_i}$ instead of $\mathcal{E}_i$ in the definition of fairness is that critical pairs are ordered, so in a fair run a (prime) critical pair $s \approx t$ of $\mathcal{R}_n$ may be ignored by deduce if $t \approx s$ was generated, or more generally, if $s \leftrightarrow_{\mathcal{E}_i} t$ holds at some point in the run. Non-prime critical pairs can always be ignored.

According to the main result of this section (Theorem 20), a completion procedure that produces fair runs is correct. The challenge is the confluence proof of $\mathcal{R}_n$. We show that $\mathcal{R}_n$ is peak decreasing by labeling rewrite steps (not only in $\mathcal{R}_n$) with multisets of terms. As well-founded order on these multisets we take the multiset extension of $>$.

Definition 18. *Let $\to$ be a rewrite relation and M a finite multiset of terms. We write $s \xrightarrow{M} t$ if $s \to t$ and there exist terms $s', t' \in M$ such that $s' \geqslant s$ and $t' \geqslant t$. Here $\geqslant$ denotes the reflexive closure of the given reduction order $>$.*

Lemma 19. *Let* $(\mathcal{E}, \mathcal{R}) \vdash_{\mathsf{KB}} (\mathcal{E}', \mathcal{R}')$. *If* $s \xleftarrow{M}{}^*_{\mathcal{E} \cup \mathcal{R}} t$ *and* $\mathcal{R}' \subseteq >$ *then* $s \xleftarrow{M}{}^*_{\mathcal{E}' \cup \mathcal{R}'} t$.

Proof. We consider a single $(\mathcal{E} \cup \mathcal{R})$-step from s to t. The statement of the lemma follows then by induction on the length of the conversion between s and t. According to Lemma 14(1) there exist terms u and v such that

$$s \xrightarrow[\mathcal{R}']{=} u \xrightarrow[\mathcal{E}' \cup \mathcal{R}']{=} v \xleftarrow[\mathcal{R}']{=} t$$

We claim that the (non-empty) steps can be labeled by M. There exist terms $s', t' \in M$ with $s' \geqslant s$ and $t' \geqslant t$. Since $\mathcal{R}' \subseteq >$, $s \geqslant u$ and $t \geqslant v$ and thus also $s' \geqslant u$ and $t' \geqslant v$. Hence

$$s \xrightarrow[\mathcal{R}']{M} = u \xrightarrow[\mathcal{E}' \cup \mathcal{R}']{M} = v = \xleftarrow[\mathcal{R}']{M} t$$

and thus also $s \xleftrightarrow[\mathcal{E}' \cup \mathcal{R}']{M}{}^{*} t$. $\square$

After these preliminaries we are ready for the main result of this section. A TRS $\mathcal{R}$ is called a *representation* of an ES $\mathcal{E}$ if $\leftrightarrow^{*}_{\mathcal{R}}$ and $\leftrightarrow^{*}_{\mathcal{E}}$ coincide.

Theorem 20. *For every fair non-failing run γ*

$$\mathcal{E}_0, \mathcal{R}_0 \vdash_{\mathsf{KB}} \mathcal{E}_1, \mathcal{R}_1 \vdash_{\mathsf{KB}} \cdots \vdash_{\mathsf{KB}} \mathcal{E}_n, \mathcal{R}_n$$

the TRS $\mathcal{R}_n$ is a complete representation of $\mathcal{E}$.

Proof. We have $\mathcal{E}_n = \varnothing$. From Corollary 15 we know that $\leftrightarrow^{*}_{\mathcal{E}} = \leftrightarrow^{*}_{\mathcal{R}_n}$. Lemma 16 yields $\mathcal{R}_n \subseteq >$ and hence $\mathcal{R}_n$ is terminating. It remains to prove that $\mathcal{R}_n$ is confluent. Let

$$t \xleftarrow[\mathcal{R}_n]{M_1} s \xrightarrow[\mathcal{R}_n]{M_2} u$$

From Lemma 10 we obtain $t \nabla_s^2 u$. Let $v \nabla_s w$ appear in this sequence (so $t = v$ or $w = u$). We obtain

$$(v, w) \in {\downarrow}_{\mathcal{R}_n} \cup \bigcup_{i=0}^{n} \leftrightarrow_{\mathcal{E}_i}$$

from the definition of ∇_s and fairness of γ. We label all steps between v and w with the multiset $\{v, w\}$. Because $s > v$ and $s > w$ we have $M_1 >_{\mathsf{mul}} \{v, w\}$ and $M_2 >_{\mathsf{mul}} \{v, w\}$. Hence by repeated applications of Lemma 19 we obtain a conversion in $\mathcal{R}_n$ between v and w in which each step is labeled with a multiset that is smaller than both M_1 and M_2. It follows that $\mathcal{R}_n$ is peak decreasing. $\square$

A completion procedure is a program that generates KB runs. In order to ensure that the final outcome $\mathcal{R}_n$ is a complete representation of the initial ES, fair runs should be produced. Fairness requires that prime critical pairs of $\mathcal{R}_n$ are considered during the run. Of course, $\mathcal{R}_n$ is not known during the run, so to be on the safe side, prime critical pairs of any $\mathcal{R}$ that appears during the run should be generated by deduce. (If a critical pair is generated from a rewrite rule that disappears at a later stage, it can be safely deleted from the run.) In particular, there is no need to deduce equations that are not prime critical pairs. So we may strengthen the condition $s \mathrel{{}_{\mathcal{R}}\leftarrow} \cdot \rightarrow_{\mathcal{R}} t$ of deduce to $s \approx t \in \mathsf{PCP}(\mathcal{R})$ without affecting Theorem 20.

7   Related Work

Formalizations of the Critical Pair Lemma. There is previous work on formalizing the Critical Pair Lemma. The first such formalization that we are aware of is by Ruiz-Reina *et al.* in ACL2 [15]. Details of the formalization are not presented in the paper, however, the authors state the following:

> The main proof effort was done to handle noncritical (or variable) overlaps. It is interesting to point out that in most textbooks and surveys this case is proved pictorially. Nevertheless, in our mechanical proof [it] turns out to be the most difficult part and it even requires the design of an induction scheme not discovered by the heuristics of the prover.

In contrast our proof of Lemma 6 handles the variable overlap case rigorously but still without excessive complexity (also in the formalization).

Another formalization of the Critical Pair Lemma was conducted by Galdino and Ayala-Rincón in PVS [7]. Here renamings are handled as substitutions satisfying further restrictions. While this was also our first approach in our own formalization, it leads to rather cumbersome proof obligations where basically the same kind of proofs have to be done for every object that we want to permute, i.e., terms, rules, substitutions, TRSs, etc. Most of those obligations can be handled in the abstract setting of permutation types once and for all and thus freely carry over to any concrete instance. Moreover the obtained set of critical pairs is infinite but there is no formalized proof that it suffices to look at only finitely many representatives for finite TRSs.

The latest formalization of the Critical Pair Lemma we are aware of is by Sternagel and Thiemann in Isabelle/HOL [16]. It consists of a rather involved proof. Moreover, it is restricted to *strings* and relies on concrete renaming functions. Thus it is not so convenient to use in an abstract setting. A big advantage, however, is that this formalization is executable and the obtained set of critical pairs is finite (for finite TRSs) by construction. The good news is that it should be possible to prove the soundness of the same executable function also via our abstract formalization, which would combine the advantages of executability and an abstract theory.

Soundness of Completion. Bachmair *et al.* [4] consider an infinite fair run to characterize the output system as the pair $(\mathcal{E}_\infty, \mathcal{R}_\infty)$ of the *persistent sets*:

$$\mathcal{E}_\infty = \bigcup_{i \geqslant 0} \bigcap_{j \geqslant i} \mathcal{E}_i \qquad\qquad \mathcal{R}_\infty = \bigcup_{i \geqslant 0} \bigcap_{j \geqslant i} \mathcal{R}_i$$

When proving confluence of $\mathcal{R}_\infty$, conversions $s_1 \leftrightarrow \cdots \leftrightarrow s_n$ in $\mathcal{E}_\infty \cup \mathcal{R}_\infty$ are compared by comparing the corresponding multisets $\{\mathsf{cost}(s_i, s_{i+1}) \mid i < n\}$ using the *proof order* given by $((>_{\mathsf{mul}}, \rhd, >)_{\mathsf{lex}})_{\mathsf{mul}}$. Here the function cost is defined as

$$\mathsf{cost}(s,t) = \begin{cases} (\{s,t\}, -, -) & \text{if } s \leftrightarrow_{\mathcal{E}_\infty} t \\ (\{s\}, \ell, t) & \text{if } s \rightarrow_{\mathcal{R}_\infty} t \\ (\{t\}, \ell, s) & \text{if } t \rightarrow_{\mathcal{R}_\infty} s \end{cases}$$

Table 1. Comparison between existing Isabelle/HOL formalizations

	$\sim$ LoI new	$\sim$ LoI [16]
renaming (+ interpretations)	2000	–
peak decreasingness	400	–
critical peak/pair lemma	300	300
soundness of completion	600	1600
longest proof	80	900
soundness with PCPs	120	–

where – is an arbitrary fixed element. Whenever a conversion contains a local peak, one can find a conversion that is smaller in the proof order. In this way confluence is obtained.

Sternagel and Thiemann [16] observed that the encompassment restriction in the collapse inference rule is unnecessary for finite runs. Based on this observation they simplified the cost function (for runs of length n) to

$$\mathsf{cost}(s,t) = \begin{cases} (\{s,t\},-) & \text{if } s \leftrightarrow_{\mathcal{E}_\infty} t \\ (\{s\}, n - o(\ell \to r)) & \text{if } s \to_{\ell \to r} t \text{ and } \ell \to r \in \mathcal{R}_\infty \\ (\{t\}, n - o(\ell \to r)) & \text{if } t \to_{\ell \to r} s \text{ and } \ell \to r \in \mathcal{R}_\infty \end{cases}$$

where $o(\ell \to r)$ denotes the highest $i \leqslant n$ such that $\ell \to r \in \mathcal{R}_i$. The proof order is $((>_{\mathsf{mul}}, >_{\mathbb{N}})_{\mathsf{lex}})_{\mathsf{mul}}$. In our new proof the second ingredient of the cost is replaced by mathematical induction in Lemma 19, and the proof order is hidden behind the abstract notion of peak decreasingness.

For a more detailed comparison between our current formalization and the one of Sternagel and Thiemann consult Table 1, where we compare *Lines of Isabelle code* (LoI for short). A general theory of renamings (plus special instances for terms, rules, TRSs, etc.) is a big part of our formalization and not present in the previous formalization at all. However this theory should be useful in future proofs. Moreover, its absence restricts the previous work to strings as variables. Peak decreasingness is also exclusive to our formalization. Concerning the critical pair lemma, both formalizations are approximately the same size, but note that our formalization is concerned with critical peaks instead of critical pairs (which is more general and actually needed in later proofs). As for soundness of abstract completion, our new formalization is drastically shorter (both columns include all definitions and intermediate lemmas that are needed for the final soundness result). Another interesting observation might be that in our new formalization of soundness the longest proof (confluence of the final TRS via peak decreasingness) is a mere 80 LoI, whereas the longest proof in the previous formalization is more than 900 LoI long (and concerned with the fact that applying an inference rule strictly decreases the cost). Finally, on top of the previous result the soundness of completion via prime critical pairs is an easy extension.

In the literature (e.g. [2,3]) critical pair criteria (like primality) are formulated as fairness conditions for completion, and correctness proofs are a combination of proof orders and a confluence criterion known as *connected-below* due to Winkler and Buchberger [19]. Our new approach avoids this detour.

8 Conclusion

In this paper we presented a new and formalized correctness proof of abstract completion which is significantly simpler than the existing proofs in the literature. Unlike earlier formalizations of the critical pair lemma and abstract completion, our formalization follows the paper proof included in this paper. This was made possible by extending IsaFoR with an abstract framework for handling variable renamings inspired by and based on a previous formalization for Nominal Isabelle.

Furthermore, our formalization of completion is the first that incorporates critical pair criteria. The key to the simple proof is the notion of peak decreasingness, a very mild version of the decreasing diagrams technique for proving confluence in the absence of termination.

There are several important variations of completion. We anticipate that the presented approach can be adapted for them, in particular ordered completion [5].

Acknowledgments. We want to give special thanks to the team around Sledgehammer and Nitpick [6] two indispensable Isabelle tools, the former increasing productivity while proving by a factor of magnitude, and the latter often pointing out slightly wrong statements that could cost hours, if not days, of a formalization attempt.

References

1. Baader, F., Nipkow, T.: Term Rewriting and All That. Cambridge University Press (1998)
2. Bachmair, L.: Canonical Equational Proofs. Birkhäuser (1991)
3. Bachmair, L., Dershowitz, N.: Equational inference, canonical proofs, and proof orderings. Journal of the ACM 41(2), 236–276 (1994), doi:10.1145/174652.174655
4. Bachmair, L., Dershowitz, N., Hsiang, J.: Orderings for equational proofs. In: Proc. 1st IEEE Symposium on Logic in Computer Science, pp. 346–357 (1986)
5. Bachmair, L., Dershowitz, N., Plaisted, D.A.: Resolution of Equations in Algebraic Structures: Completion without Failure, vol. 2, pp. 1–30. Academic Press (1989)
6. Blanchette, J.C., Bulwahn, L., Nipkow, T.: Automatic proof and disproof in Isabelle/HOL. In: Tinelli, C., Sofronie-Stokkermans, V. (eds.) FroCoS 2011. LNCS, vol. 6989, pp. 12–27. Springer, Heidelberg (2011), doi:10.1007/978-3-642-24364-6_2
7. Galdino, A.L., Ayala-Rincón, M.: A formalization of the Knuth-Bendix(-Huet) critical pair theorem. Journal of Automated Reasoning 45(3), 301–325 (2010), doi:10.1007/s10817-010-9165-2
8. Huet, G.: A complete proof of correctness of the Knuth-Bendix completion algorithm. Journal of Computer and System Sciences 23(1), 11–21 (1981), doi:10.1016/0022-0000(81)90002-7

9. Huffman, B., Urban, C.: A new foundation for Nominal Isabelle. In: Kaufmann, M., Paulson, L.C. (eds.) ITP 2010. LNCS, vol. 6172, pp. 35–50. Springer, Heidelberg (2010), doi:10.1007/978-3-642-14052-5_5

10. Kapur, D., Musser, D.R., Narendran, P.: Only prime superpositions need be considered in the Knuth-Bendix completion procedure. Journal of Symbolic Computation 6(1), 19–36 (1988), doi:10.1016/S0747-7171(88)80019-1

11. Knuth, D.E., Bendix, P.: Simple word problems in universal algebras. In: Leech, J. (ed.) Computational Problems in Abstract Algebra, pp. 263–297 (1970)

12. Nipkow, T., Paulson, L.C., Wenzel, M. (eds.): Isabelle/HOL. LNCS, vol. 2283. Springer, Heidelberg (2002), doi:10.1007/3-540-45949-9

13. van Oostrom, V.: Confluence by decreasing diagrams. Theoretical Computer Science 126(2), 259–280 (1994), doi:10.1016/0304-3975(92)00023-K

14. van Raamsdonk, F. (ed.): Proc. 24th International Conference on Rewriting Techniques and Applications. Leibniz International Proceedings in Informatics, vol. 21. Schloss Dagstuhl – Leibniz-Zentrum für Informatik (2013)

15. Ruiz-Reina, J.-L., Alonso, J.-A., Hidalgo, M.-J., Martín-Mateos, F.-J.: Formal proofs about rewriting using ACL2. Annals of Mathematics and Artificial Intelligence 36(3), 239–262 (2002), doi:10.1023/A:1016003314081

16. Sternagel, C., Thiemann, R.: Formalizing Knuth-Bendix orders and Knuth-Bendix completion. In: van Raamsdonk [14], pp. 287–302, doi:10.4230/LIPIcs.RTA.2013.287

17. Terese: Term Rewriting Systems. Cambridge Tracts in Theoretical Computer Science, vol. 55. Cambridge University Press (2003)

18. Urban, C., Kaliszyk, C.: General bindings and alpha-equivalence in Nominal Isabelle. Logical Methods in Computer Science 8(2), 465–476 (2012), doi:10.2168/LMCS-8(2:14)2012

19. Winkler, F., Buchberger, B.: A criterion for eliminating unnecessary reductions in the Knuth-Bendix algorithm. In: Proc. Colloquium on Algebra, Combinatorics and Logic in Computer Science. Colloquia Mathematica Societatis J. Bolyai, vol. II, 42, pp. 849–869 (1986)

20. Zankl, H.: Decreasing diagrams – formalized. In: van Raamsdonk [14], pp. 352–367, doi:10.4230/LIPIcs.RTA.2013352

HOL with Definitions: Semantics, Soundness, and a Verified Implementation

Ramana Kumar[1], Rob Arthan[2], Magnus O. Myreen[1], and Scott Owens[3]

[1] Computer Laboratory, University of Cambridge, UK
[2] School of EECS, Queen Mary, University of London, UK
[3] School of Computing, University of Kent, UK

Abstract. We present a mechanised semantics and soundness proof for the HOL Light kernel including its definitional principles, extending Harrison's verification of the kernel without definitions. Soundness of the logic extends to soundness of a theorem prover, because we also show that a synthesised implementation of the kernel in CakeML refines the inference system. Our semantics is the first for Wiedijk's stateless HOL; our implementation, however, is stateful: we give semantics to the stateful inference system by translation to the stateless. We improve on Harrison's approach by making our model of HOL parametric on the universe of sets. Finally, we prove soundness for an improved principle of constant specification, in the hope of encouraging its adoption. This paper represents the logical kernel aspect of our work on verified HOL implementations; the production of a verified machine-code implementation of the whole system with the kernel as a module will appear separately.

1 Introduction

In this paper, we present a mechanised proof of the soundness of higher-order logic (HOL), including its principles for defining new types and new polymorphic constants, and describe production of a verified implementation of its inference rules. This work is part of a larger project, introduced in our rough diamond last year [11], to produce a verified machine-code implementation of a HOL prover. This paper represents the top half of the project: soundness of the logic, and a verified implementation of the logical kernel in CakeML [7].

What is the point of verifying a theorem prover and formalising the semantics of the logic it implements? One answer is that it raises our confidence in the correctness of the prover. A prover implementation usually sits at the centre of the trusted code base for verification work, so effort spent verifying the prover multiplies outwards. Secondly, it helps us understand our systems (logical and software), to the level of precision possible only via formalisation. Finally, a theorem prover is a non-trivial piece of software that admits a high-level specification and whose correctness is important: we see it as a catalyst for tools and methods aimed at developing complete verified systems, readying them for larger systems with less obvious specifications.

G. Klein and R. Gamboa (Eds.): ITP 2014, LNAI 8558, pp. 308–324, 2014.

The first soundness proof we present here is for Wiedijk's stateless HOL [16], in which terms carry their definitions; by formalising we hope to clarify the semantics of this system. We then show that traditional stateful HOL, where terms are understood in a context of definitions, is sound by a translation to the stateless inference system.

We build on Harrison's proof of the consistency of HOL without definitions [4], which shares our larger goal of verifying concrete HOL prover implementations, and advance this project by verifying an implementation of the HOL Light [5] kernel in CakeML, an ML designed to support fully verified applications. We discuss the merits of Harrison's model of set theory defined within HOL, and provide an alternative not requiring axiomatic extensions.

Our constant specification rule generalises the one found in the various HOL systems, adding support for implicit definitions with fewer constraints and no new primitives. We lack space here to justify its design in full detail, but refer to a proposal [2] by the second author. We hope our proof of its soundness will encourage its adoption.

The specific contributions of this paper are:

- a formal semantics for Wiedijk's stateless HOL (§4), against a new specification of set theory (§3),
- a proof of soundness (§4.2) for stateless HOL, including type definitions, a new rule for constant specification, and the three axioms used in HOL Light,
- a proof of soundness for stateful HOL by translation to stateless (§5), and
- a verified implementation of the HOL Light kernel in CakeML (§6) that should be a suitable basis for a verified implementation of the prover in machine-code.

All our definitions and proofs have been formalised in the HOL4 theorem prover [14] and are available from `https://cakeml.org`.[1]

2 Approach

At a high level, our semantics and verified implementation fit together as follows.

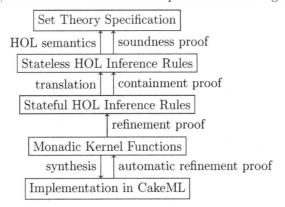

---

[1] Specifically, the `hol-light` directory of `https://github.com/xrchz/vml`.

The overall theorems we obtain are about evaluating the CakeML implementations of the HOL Light kernel functions in CakeML's operational semantics. For each kernel function, we prove that if the function is run in a good state on good arguments, it terminates in a good state and produces good results. Here "good" refers to our refinement invariants. In particular, a good value of type "thm" must refine a sequent in stateful HOL that translates to a sequent in stateless HOL that is valid according to the set-theoretic semantics.

We prove these results by composing the four proof layers in the diagram. Starting from the top, the HOL semantics interprets stateless HOL sequents in set theory, from which we obtain a definition of validity. The soundness proof says that each of the stateless HOL inference rules preserves validity of sequents.

In stateless HOL, defined types and terms carry syntactic tags describing their definitions, whereas in stateful HOL there is a context of definitions that is updated when a new definition is made. Our translation from stateful to stateless takes the definitions from the context and inlines them into the tags. Our containment proof then shows that whenever the stateful system proves a sequent, the stateless system proves the translation of the sequent.

As outlined in our rough diamond [11], we define shallowly-embedded HOL functions, using a state-exception monad, for each of the HOL Light kernel functions. These "monadic kernel functions" are written following the original OCaml code closely, then we prove that they implement the stateful inference rules. Specifically, if one of these functions is applied to good arguments, it terminates with a good result; any theorem result must refine a sequent that is provable in the stateful system.

Finally, using the method developed by Myreen and Owens [10] we synthesise CakeML implementations of the monadic kernel functions. This automatic translation from shallowly- to deeply-embedded code is proof-producing, and we use the certificate theorems to complete the refinement proof.

In the context of our larger project, the next steps include: a) proving, against CakeML's semantics, that our implementation of the kernel can be wrapped in a module to protect the key property, provability, of values of type "thm"; and b) using CakeML's verified compiler to generate a machine-code implementation of the kernel embedded in an interactive read-eval-print loop that is verified to never print a false theorem.

3   Set Theory

A rigorous but informal account of the semantics of HOL, due to Pitts, is given in the HOL4 documentation [12]. It assigns meanings to HOL constructs in a universe of sets satisfying Zermelo's axioms. We wish to do the same with a semantics developed using conservative extensions in HOL. Gödel's second incompleteness theorem implies that we cannot actually define a model of Zermelo set theory. However, we can define what properties such a model must have and for us this is sufficient. It is convenient to separate out the axioms of choice and infinity. A specification along these lines was developed previously by one of us [1] but without any formal proofs. We begin by defining a predicate

```
is_set_theory (mem :U -> U -> bool)
```

that says whether a membership relation defined on some universe U (represented by a type variable) satisfies the Zermelo axioms other than choice and infinity, namely the axioms of extensionality, separation (a.k.a. comprehension or specification), power set, union, and pairing. As we are working in HOL, we can use propositional functions in place of the metavariables required in a first-order presentation:

Definition 1 (Specification of Set Theory Axioms)

is_set_theory $mem \iff$
  extensional $mem \land (\exists sub.$ **is_separation** $mem\ sub) \land$
  $(\exists power.$ **is_power** $mem\ power) \land (\exists union.$ **is_union** $mem\ union) \land$
  $\exists upair.$ **is_upair** $mem\ upair$
extensional $mem \iff$
  $\forall x\ y.\ x = y \iff \forall a.\ mem\ a\ x \iff mem\ a\ y$
is_separation $mem\ sub \iff$
  $\forall x\ P\ a.\ mem\ a\ (sub\ x\ P) \iff mem\ a\ x \land P\ a$
is_power $mem\ power \iff$
  $\forall x\ a.\ mem\ a\ (power\ x) \iff \forall b.\ mem\ b\ a \Rightarrow mem\ b\ x$
is_union $mem\ union \iff$
  $\forall x\ a.\ mem\ a\ (union\ x) \iff \exists b.\ mem\ a\ b \land mem\ b\ x$
is_upair $mem\ upair \iff$
  $\forall x\ y\ a.\ mem\ a\ (upair\ x\ y) \iff a = x \lor a = y$

A relation mem satisfying the above axioms is sufficient to define the semantics of HOL without Hilbert choice or the axiom of infinity, that is, for the (polymorphic) simply typed λ-calculus with equality. For the remaining features of HOL, we need two more parameters: a choice function, and a distinguished infinite set for the individuals. We specify a complete model as follows.[2]

Definition 2 (Specification of a Model for HOL)

is_model $(mem, indset, ch) \iff$
  is_set_theory $mem \land$ **is_infinite** $mem\ indset \land$ **is_choice** $mem\ ch$
is_choice $mem\ ch \iff \forall x.\ (\exists a.\ mem\ a\ x) \Rightarrow mem\ (ch\ x)\ x$
is_infinite $mem\ s \iff$ **infinite** $\{a \mid mem\ a\ s\}$

3.1 Derived Concepts

In order to reuse Harrison's proofs [4] as much as possible, we define various constructions above our set theory model and prove the same theorems he did to characterise them. These theorems form the interface to set theory above which one can give a semantics to HOL. To save space, we do not list them all.

For function spaces, function application, and abstraction of a HOL function, we use the standard set-theoretic method of identifying functions with their

---

[2] `infinite` $(p : \alpha$ -> `bool`) abbreviates $\neg$`FINITE` p, with finiteness defined inductively for sets-as-predicates in HOL4's library.

graphs. For Booleans we define a distinguished set of two elements and name its members. We often use abbreviations to hide the *mem* argument to a function, for example `funspace` *s t* below actually abbreviates `funspace0` *mem s t*.

$\vdash$ `is_set_theory` *mem* $\Rightarrow$
    ($\forall f\ s\ t$.
      (($\forall x$. *mem* $x\ s$ $\Rightarrow$ *mem* ($f\ x$) t) $\Rightarrow$
        mem (`abstract` $s\ t\ f$) (`funspace` $s\ t$)) $\wedge$
      $\forall x$. *mem* $x\ s$ $\wedge$ *mem* ($f\ x$) t $\Rightarrow$ `apply` (`abstract` $s\ t\ f$) x = $f\ x$) $\wedge$
    $\forall x$. *mem* x `boolset` $\iff$ x = `true` $\vee$ x = `false`

3.2   Consistency

We wish to know that `is_model` is not an empty predicate, to protect against simple mistakes in the definition, and because the existence of a model will be an assumption on our soundness theorems. Since actually building a model in HOL would allow HOL to prove its own consistency, we will have to settle for something less. However, we wish to avoid axiomatic extensions if possible.

We tried using Harrison's construction [4] as witness, but unfortunately it uses what amounts to a type system to define a coherent notion of membership in terms of injections into a universe. (Harrison calls the types "levels".) For simplicity and familiarity our `is_set_theory` characterises an untyped set theory. In particular, we need extensionality to hold for all sets, while in Harrison's model empty sets of distinct types are distinct.

So instead we use a standard encoding of the hereditarily finite sets in HOL as natural numbers, which takes the universe to be the natural numbers, and *mem n m* to hold if the mth bit in the binary numeral for n is 1. With this model, we can introduce, by conservative extension, a universe type that satisfies `is_set_theory`, and, under the assumption that it contains an infinite set, that it satisfies `is_model` too. To be able to consistently assume the existence of infinite sets, the universe type has a free type variable.

Specifically, we define the universe as an arbitrary subset of α + `num` for which a suitable membership relation exists. We prove the existence of such a subset, namely all the numbers in the right of the sum, by using the standard encoding, which it is straightforward to show satisfies the set-theoretic axioms. Thus, we prove the following:

$\vdash$ $\exists$ (P :α + `num` -> `bool`) (*mem* :α + `num` -> α + `num` -> `bool`).
    `is_set_theory_pred` P *mem*

where `is_set_theory_pred` P is like `is_set_theory` but with all quantification relativised to P. We then feed this theorem into HOL4's constant and type specification machinery to derive a new type α V for our universe, and its membership relation `V_mem`. Our main lemma follows directly:

$\vdash$ `is_set_theory` `V_mem`

There is a natural choice function for non-empty sets in the standard encoding of finite sets, namely, the most significant bit in the binary numeral. But there are no infinite sets, so, as we would expect from Gödel's second incompleteness theorem at this point, no model for the set of individuals.

Now we use the facts that our specification of (`V_mem` :α `V -> ` α `V -> bool`) is loose—the only things that can be proved about it come from `is_set_theory` and not the specific construction of the model—and the type α `V`, being parametric, has no provable cardinality bound. Hence if τ is an unspecified type, it is consistent to assume that τ `V` includes infinite sets. We specify `V_indset` as an arbitrary infinite set under the assumption that one exists. We can then prove our desired theorem:

Theorem 1 (Example Model)

$\vdash$ ($\exists I$. `is_infinite V_mem` I) $\Rightarrow$
    `is_model (V_mem,V_indset,V_choice)`

An alternative to the general `is_model` characterisation of a suitable set-theoretic model is to define a particular universe of sets and then prove that it has all the desired properties. This is the approach taken by Harrison [4], who constructs by conservative extension a monomorphic type `V` equipped with a membership relation satisfying a typed analogue of our `is_set_theory`. `V` is countably infinite and it would be inconsistent to assert that it is a model of the axiom of infinity. Harrison observes that one could adapt his formalisation to give a model of the axiom of infinity using a non-conservative extension. Our approach allows us to work by conservative extension while remaining consistent with an assumption of the axiom of infinity.

4 Stateless HOL

Traditional implementations of HOL are stateful because they support the definition of new type and term constants by updating a context. Wiedijk [16] showed that this is not necessary if defined constants carry their definitions with them. Since there is no state, it was an appealing target for extension of Harrison's definitionless semantics [4].

4.1 Inference System

The distinguishing feature of stateless HOL syntax is the presence of tags, `const_tag` and `type_op`, on constants and types, containing information about how they were defined or whether they are primitive. Because of these tags, the datatypes for terms and types are mutually recursive.

```
term = Var of string * type
     | Const of string * type * const_tag
     | Comb of term * term
     | Abs of string * type * term
type = Tyvar of string | Tyapp of type_op * type list
type_op = Typrim of string * num | Tydefn of string * term
```

```
const_tag = Prim
          | Defn of num * (string × term) list * term
          | Tyabs of string * term
          | Tyrep of string * term
```

With the `Typrim` *name arity* and `Prim` tags we can build up HOL's primitive type operators and constants without baking them into the syntax, as the following abbreviations show.

`Bool`	for	`Tyapp (Typrim "bool" 0) []`
`Ind`	for	`Tyapp (Typrim "ind" 0) []`
`Fun` x y	for	`Tyapp (Typrim "fun" 2) [`x`; `y`]`
`Equal` ty	for	`Const "=" (Fun` ty `(Fun` ty `Bool)) Prim`
`Select` ty	for	`Const "@" (Fun (Fun` ty `Bool)` ty`) Prim`
s `===` t	for	`Comb (Comb (Equal (typeof` s`))` s`)` t

We will explain the tags for non-primitives when we describe the definitional rules, after introducing the inference system. We use similar notation to Harrison [4] wherever possible, for example we define well-typed terms and prove $\vdash$ `welltyped` tm $\iff$ tm `has_type (typeof` tm`)`, and we define the following concepts: `closed` tm, indicating that tm has no free variables; `tvars` tm and `tyvars` ty collecting the type variables appearing in a term or type, and `tyinst` $tyin$ ty instantiating type variables in a type.

We were able to reuse most of Harrison's stateful HOL rules for the stateless HOL inference system, defining provable sequents[3] inductively with a few systematic modifications. Changes were required to handle the fact that stateless syntax permits terms whose definitions are unsound because their tags do not meet the side-conditions required by the definitional principles. Therefore, we also define predicates picking out good types and terms, in mutual recursion with the provability relation.

For the most part, we define good terms implicitly as those appearing in provable sequents. We also need rules for the primitives, and for (de)constructing good terms and types. A few examples are shown:

$$\frac{hs \; \vdash \; c}{\text{member } t \; (c::hs)} \qquad \frac{\text{type\_ok } ty_1 \quad \text{type\_ok } ty_2}{\text{type\_ok (Fun } ty_1 \; ty_2)} \qquad \frac{\text{term\_ok (Comb } t_1 \; t_2)}{\text{term\_ok } t_1} \qquad \frac{\text{term\_ok } tm \quad tm \text{ has\_type } ty}{\text{type\_ok } ty}$$

$$\frac{}{\text{term\_ok } t}$$

We continue by specifying the inference rules as in [4], but restricted to good terms and types. For example, REFL, ASSUME, and INST_TYPE:

$$\frac{\text{term\_ok } t}{[] \; \vdash \; t === t} \qquad \frac{\text{term\_ok } p \quad p \text{ has\_type Bool}}{[p] \; \vdash \; p} \qquad \frac{hs \; \vdash \; c \quad \text{every type\_ok (map fst } tyin)}{\text{map (INST } tyin) \; hs \; \vdash \; \text{INST } tyin \; c}$$

To finish the inference system, we add the rules that extend Harrison's system – the principles of definition and the axioms.

---

[3] We write the relation we define as hs `|-` c. By contrast $\vdash$ p refers to theorems proved in HOL4.

Type Definitions. To define a new type in HOL, one chooses an existing type, called the representing type, and defines a subset via a predicate. HOL types are non-empty, so the principle of type definition requires a theorem as input that proves that the chosen subset of the representing type is non-empty.

In the stateless syntax for types, the tag `Tydefn` *name* p is found on a defined type operator. It contains the name of the new type and the predicate for which a theorem of the form `[]` `|-` `Comb` p w was proved as evidence that the representing type is inhabited.

The rule for defining new types also introduces two new constants representing injections between the new type and the representing type. In the syntax, these constants are tagged by `Tyabs` *name* p or `Tyrep` *name* p, with the name of the new type and the predicate as above defining the subset of the representing type. To show that these constants are injections and inverses, the rule produces two theorems. We show the complete provability clause for one of the theorems below.

```
closed p ∧ [] |- Comb p w ∧ rty = domain (typeof p) ∧
aty = Tyapp (Tydefn name p) (map Tyvar (sort (tvars p))) ⇒
  [] |-
  Comb (Const abs (Fun rty aty) (Tyabs name p))
    (Comb (Const rep (Fun aty rty) (Tyrep name p))
      (Var x aty)) === Var x aty
```

Because the new type and the two new constants appear in this theorem, there is no need to explicitly give rules showing that they are `type_ok` and `term_ok`.

Constant Specifications. Wiedijk [16] follows HOL Light in only admitting an equational definitional principle as a primitive, unlike other implementations of HOL which also provide a principle of constant specification that takes a theorem of the form $\exists x_1, \ldots, x_k \cdot P$ and introduces new constants $c_1, \ldots, c_k$ with $\vdash$ $P[c_1/x_1, \ldots, c_k/x_k]$ as their defining axiom. This is subject to certain restrictions on the types of the c_i. (The constant specification principle is supported in HOL Light, but as a derived rule, which, unfortunately, introduces an additional, less abstract form of the defining axiom.)

A disadvantage of this principle is that it presupposes a suitable definition of the existential quantifier, whereas we wish to give the semantics of the HOL language and use conservative extensions to define the logical operators. Our new constant specification mechanism overcomes this disadvantage and is less restrictive about the types of the new constants. See [2] for a fuller discussion of the new mechanism and the motivation for it. We describe it in mathematical notation rather than HOL4 syntax because the formalisation makes unwieldy but uninsightful use of list operations, since the rule may introduce multiple constants.

Given a theorem of the form $\{x_1 = t_1, \ldots, x_n = t_n\} \vdash p$, where the free variables of p are contained in $\{x_1, \ldots, x_n\}$, we obtain new constants $\{c_1, \ldots, c_n\}$ and a theorem $\vdash p[c_1/x_1, \ldots, c_n/x_n]$. The side-conditions are that the variables $x_1, \ldots, x_n$ are distinct and the type variables of each t_i are contained in its type.

In the stateless syntax, we use the tag Defn i *xts* p for the ith constant introduced by this rule when it is applied to the theorem with hypotheses map $(\lambda\,(x,t).$ Var x (typeof t) === t) *xts* and conclusion p,

Since the rule allows new constants to be introduced without appearing in any new theorems, we also add a clause for the new constants asserting term_ok (Const x ty (Defn i *xts* p)).

Axioms. We include the three mathematical axioms—ETA_AX (not shown), SELECT_AX, and INFINITY_AX—in our inference system directly:

```
p has_type (Fun ty Bool) ∧ h |- Comb p w ⇒
  h |- Comb p (Comb (Select ty) p)
```

```
[] |-
EXISTS "f" (Fun Ind Ind)
  (AND (ONE_ONE Ind Ind (Var "f" (Fun Ind Ind)))
     (NOT (ONTO Ind Ind (Var "f" (Fun Ind Ind)))))
```

Here EXISTS, AND, and so forth are abbreviations for defined constants in stateless HOL, and are built up following standard definitions of logical constants. For example,

```
NOT =
  Comb
    (Const1 "~" (Fun Bool Bool)
      (Abs "p" Bool (IMPLIES (Var "p" Bool) FALSE)))
```

where Const1 *name* *ty* *rhs* abbreviates

```
Const name ty (Defn 0 [(name,rhs)] (Var name (typeof rhs) === rhs))
```

and shows how the rule for new specification subsumes the traditional rule for defining a constant to be equal to an existing term.

4.2 Semantics

Just as we reused much of the inference system, we were able to reuse most of Harrison's proofs in establishing soundness of the stateless HOL inference system, again with systematic modifications. The main change, apart from our extensions, is that our semantics uses inductive relations rather than functions.

The purpose of the semantics is to interpret sequents. Subsidiary concepts include valuations interpreting variables and semantic relations interpreting types and terms. We differ from Harrison stylistically in making type and term valuations finite maps with explicit domains (he uses total functions). We briefly describe the proposition expressed by each piece of the semantics as follows:

`type_valuation` τ	τ maps type variables to non-empty sets
`typeset` τ ty mty	$(ty \;:\texttt{type})$ is interpreted by $(mty \;:\mathcal{U})$ in τ
`term_valuation` τ σ	σ maps each variable to an element of the interpretation of its type
`semantics` σ τ tm mtm	$(tm \;:\texttt{term})$ is interpreted by $(mtm \;:\mathcal{U})$ in τ and σ
`type_has_meaning` ty	ty has semantics in all closing valuations
`has_meaning` tm	tm has semantics in all closing valuations, and a pair of closing valuations exists
hs `|=` c	$c::hs$ are meaningful terms of type `Bool`, and, in all closing valuations where the semantics of each of the hs is `true`, so is the semantics of c

We prove that `semantics` σ τ and `typeset` τ are functional relations. Supporting definitions led us to prefer relations for two reasons. First, the semantics of defined constants in general requires type instantiation, and it is easier to state the condition it should satisfy than to calculate it explicitly. Second, defined types and constants are given semantics in terms of entities supplied by side-conditions on the definitional rules, so it is convenient to assume they hold. It made sense for Harrison to use total functions because without definitions, all terms (including ill-typed ones) can be handled uniformly.

Our inductive relations are mutually recursive: Harrison's had one-way dependency because the meaning of equality, for example, depends on the type. For definitions, we need the other way too because the meaning of a defined type depends on the term used to define it.

Another difference stems from our semantics being parametric on the choice of set theory model, $(mem, indset, ch)$. We always use the free variables mem, $indset$, and ch for the model, and we often leave these arguments implicit in our notation. So, for example, `typeset` τ ty mty above is actually an abbreviation for `typeset0` $(mem, indset, ch)$ τ ty mty.

A final addition we found helpful, especially for defined constants, is a treatment of type instantiation and variable substitution that is not complicated by the possibility of variable shadowing.

Now let us look at the new parts of the semantics in detail.

Semantics of Defined Types. A type operator is defined by a predicate on an existing type called the representing type. Its semantics is the subset of the representing type where the predicate holds, which must be non-empty.

We define a relation `inhab` τ p rty mty to express that the subset of the type rty carved out by the predicate p is non-empty and equal to mty. The semantics of rty and p are with respect to τ (and the empty term valuation). Then we formally define the semantics of an applied type operator as follows:

```
closed p ∧ p has_type (Fun rty Bool) ∧ length (tvars p) = length args ∧
pairwise (typeset τ) args ams ∧
(∀τ.
    type_valuation τ ∧ set (tvars p) ⊆ dom τ ⇒
      ∃mty. inhab τ p rty mty) ∧
inhab (sort (tvars p) ⇉ ams) p rty mty ⇒
    typeset τ (Tyapp (Tydefn op p) args) mty
```

The purpose of the type arguments is to provide interpretations for type variables appearing in the predicate, hence in the first argument to inhab we bind[4] sort (tvars p) to the interpretations of the arguments. The penultimate premise, requiring that p carve a non-empty subset of rty for any closing τ, is necessary to ensure that a badly defined type does not accidentally get semantics when applied to arguments that happen to produce a non-empty set.

Type definition also introduces two new constants, and they are given semantics as injections between the representing type and the subset carved out of it. We only show the rule for the function to the new type, which makes an arbitrary choice in case its argument is not already in the subset. (The other is the inclusion function.)

```
typeset τ (Tyapp (Tydefn op p) args) maty ∧ p has_type (Fun rty Bool) ∧
pairwise (typeset τ) args ams ∧ τi = sort (tvars p) ⇉ ams ∧
typeset τi rty mrty ∧ semantics ⊥ τi p mp ∧
tyin = sort (tvars p) ⇉ args ⇒
  semantics σ τ
    (Const s (Fun (tyinst tyin rty) (Tyapp (Tydefn op p) args))
      (Tyabs op p))
    (abstract mrty maty (λx. if Holds mp x then x else ch maty))
```

The type definition rule returns two theorems about the new constants, asserting that they form a bijection between the new type and the subset of the representing type defined by the predicate. It is straightforward to prove this rule sound since the semantics simply interprets the new type as the subset to which it must be in bijection.

Substitution and Instantiation. In Harrison's work, proving soundness for the two inference rules (INST_TYPE and INST) that use type instantiation and term substitution takes about 60% of the semantics.ml proof script by line count. These operations are complicated because they protect against unintended variable capture, e.g. instantiating α with bool in $\lambda(x : bool). (x : \alpha)$ triggers renaming of the bound variable. Since the semantics of defined constants uses type instantiation, we sought a simpler implementation.

The key observation is that there is always an α-equivalent term—with distinct variable names—for which instantiation is simple, and the semantics should be up to α-equivalence anyway. For any term tm and finite set of names s, we define fresh_term s tm as an arbitrary α-equivalent term with bound variable names that are all distinct and not in s.

---

[4] $ks \rightrightarrows vs$ is the finite map binding ks pairwise to vs.

We define unsafe but simple algorithms, `simple_inst` and `simple_subst`, which uniformly replace (type) variables in a term, ignoring capture. Then, under conditions that can be provided by `fresh_term`, namely, that bound variable names are distinct and not in the set of names appearing in the substitution, it is straightforward to show that `simple_subst` and `simple_inst` behave the same as the capture-avoiding algorithms, VSUBST and INST.

The inference rules use the capture-avoiding algorithms since they must cope with terms constructed by the user, but when we prove their soundness we first switch to a fresh term then use the simple algorithms. The theorems enabling this switch say that substitution (not shown) and instantiation respect α-equivalence:

$\vdash$ `welltyped` t_1 $\wedge$ `ACONV` t_1 t_2 $\Rightarrow$ `ACONV` (`INST` $tyin$ t_1) (`INST` $tyin$ t_2)

To prove these theorems, we appeal to a variable-free encoding of terms using de Bruijn indices. We define versions of VSUBST and INST that operate on de Bruijn terms, and prove that converting to de Bruijn then instantiating is the same as instantiating first then converting. The results then follow because α-equivalence amounts to equality of de Bruijn terms.

Semantics of Defined Constants. The semantics of the ith constant defined by application of our principle for new specification on $\{x_1 = t_1, \ldots, x_n = t_n\}$ $\vert$-p can be specified as the semantics of the term t_i. This choice might constrain the constant more than the predicate p does, but the inference system guarantees that all knowledge about the constant must be derivable from p. When t_i is polymorphic, we need to instantiate its type variables to match the type of the constant. The relevant clause of the semantics is as follows.

i < `length` eqs $\wedge$ `EL` i eqs = (s,t_i) $\wedge$ t = `fresh_term` $\varnothing$ t_i $\wedge$ `welltyped` t $\wedge$
`closed` t $\wedge$ `set` (`tvars` t) $\subseteq$ `set` (`tyvars` (`typeof` t)) $\wedge$
`tyinst` $tyin$ (`typeof` t) = ty $\wedge$ `semantics` $\perp$ τ (`simple_inst` $tyin$ t) mt $\Rightarrow$
  `semantics` σ τ (`Const` s ty (`Defn` i eqs p)) mt

To prove our new constant specification principle sound, we may assume $\{x_1 = t_1, \ldots, x_n = t_n\}$ $\vert$-p and need to show $\vert$-$p[c_1/x_1, \ldots, c_n/x_n]$. Given the semantics above and the interpretation of sequents, this reduces to proving the correctness of substitution, which we need to prove anyway for the INST rule.

Axioms, Soundness and Consistency. The axioms do not introduce new kinds of term or type, so do not affect the semantics. We just have to characterise the constants in the axiom of infinity using the semantics for defined constants. Since our interpretation of functions is natural, mapping functions to their graphs in the set theory, the soundness proofs for the axioms are straightforward.

We have described how we prove soundness for each of our additional inference rules (that is, for definitions and axioms). We prove soundness for all the other inference rules by adapting Harrison's proofs, with improvements where possible (*e.g.* for substitution and instantiation). Using the proofs for each rule, we obtain the main soundness theorem by induction on the inference system.

Theorem 2 (Soundness of Stateless HOL)

⊢ is_model $(mem, indset, ch)$ ⇒
 (∀ ty. type_ok ty ⇒ type_has_meaning ty) ∧
 (∀ tm. term_ok tm ⇒ has_meaning tm) ∧
 ∀ hs c. hs |- c ⇒ hs |= c

It is then straightforward to prove that there exist both provable and unprovable sequents (VARIANT here creates a distinct name by priming):

Theorem 3 (Consistency of Stateless HOL)

⊢ is_model $(mem, indset, ch)$ ⇒
 [] |- Var x Bool === Var x Bool ∧
 ¬([] |- Var x Bool === Var (VARIANT (Var x Bool) x Bool) Bool)

5   From Stateful Back to Stateless

The previous sections have explained the semantics and soundness proof for stateless HOL. Our overall goal is to prove the soundness of a conventional stateful implementation, so we formalise a stateful version of HOL (our rough diamond contains a brief overview [11]) and give it semantics by translation into the stateless version.

The only significant difference between the stateful and stateless versions is that the stateless carries definitions of constants as tags on the terms and types. The translation from the stateful version, which has an explicit global context, simply inlines all the appropriate definitions into the terms and types.

We define this translation from stateful to stateless HOL using inductively defined relations for translation of types and terms. The translation of stateful sequents into stateless sequents is defined as the following relation.

seq_trans $((defs, hs'), c')$ (hs, c) ⟺
 pairwise (term $defs$) hs' hs ∧ term $defs$ c' c

Here $defs$ is the global context in which the stateful theorem sequent has been proved, and **term** is the translation relation for terms. The definition of **term** (and **type** similarly) is straightforward and omitted due to space constraints.

We prove, by induction on the stateful inference system, that any sequent that can be derived in the stateful version can be translated into a provable stateless sequent.

Theorem 4 (Stateful HOL contained in stateless HOL)

⊢ (type_ok $defs$ ty' ⇒ ∃ ty. type $defs$ ty' ty ∧ type_ok ty) ∧
(term_ok $defs$ tm' ⇒ ∃ tm. term $defs$ tm' tm ∧ term_ok tm) ∧
$((defs, hs')$ |- c' ⇒ ∃ hs c. seq_trans $((defs, hs'), c')$ (hs, c) ∧ hs |- $c)$

6   Verifying the Kernel in CakeML

To construct a verified CakeML implementation of the stateful HOL inference rules, we take the implementation of the HOL Light kernel (extended with our constant specification principle) and define each of its functions in HOL4 using a state-and-exception monad. Using previously developed proof automation [10], these monadic functions are automatically turned into deep embeddings (CakeML abstract syntax) that are proved to implement the original monadic functions.

It only remains to show the following connection between the computation performed by the monadic functions, and the inference system for stateful HOL from §5: any computation on good types, terms and theorems will produce good types, terms and theorems according to stateful HOL. A type, term or theorem sequent is "good" if it is good according to `type_ok`, `term_ok`, or (`|-`) from stateful HOL. Here `hol_tm`, `hol_ty` and `hol_defs` translate into the implementation's representations.

⊢ TYPE *defs ty* ⟺ `type_ok` (`hol_defs` *defs*) (`hol_ty` *ty*)
⊢ TERM *defs tm* ⟺ `term_ok` (`hol_defs` *defs*) (`hol_tm` *tm*)
⊢ THM *defs* (Sequent *asl c*) ⟺
    (`hol_defs` *defs*,map `hol_tm` *asl*) |- `hol_tm` *c*

The prover's state *s* implements logical context *defs*, if STATE *s defs* holds. We omit the definition of STATE.

For each monadic function, we prove that good inputs produce good output. For example, for the ASSUME function, we prove that, if the input is a good term and the state is good, then the state will be unchanged on exit and if the function returned something (via `HolRes`) then the return value is a good theorem:

⊢ TERM *defs tm* ∧ STATE *s defs* ∧ ASSUME *tm s* = (*res*,*s'*) ⇒
    s' = *s* ∧ ∀ *th*. *res* = `HolRes` *th* ⇒ THM *defs th*

We prove a similar theorem for each function in the kernel, showing that they implement the stateful inference rules correctly. As another example, take the new rule for constant specification: we prove that if the state is updated then the state is still good and the returned theorem is good.

⊢ THM *defs th* ∧ STATE *s defs* ⇒
  case `new_specification` *th s* **of**
    (HolRes *th*,*s'*) ⇒ ∃ *d*. THM (*d*::*defs*) *th* ∧ STATE *s'* (*d*::*defs*)
  | (HolErr *msg*,*s'*) ⇒ *s'* = *s*

By expanding the definition of THM in these theorems, then applying Theorems 4 and 2, we see that each monadic function implements a valid deduction according to the semantics of HOL. We then compose with the automatically synthesised certificate theorem for the CakeML implementation, to finish the proof about the CakeML implementation of the kernel. The automatically proved certificate theorem for the monadic `new_specification` function is shown below. These certificate theorems are explained in Myreen and Owens [10].

```
⊢ DeclAssum ml_hol_kernel_decls env ⇒
  EvalM env (Var (Short "new_specification"))
    ((PURE HOL_KERNEL_THM_TYPE -M-> HOL_MONAD HOL_KERNEL_THM_TYPE)
       new_specification)
```

7   Related Work

For classical higher-order logic, apart from Harrison's mechanisation [4] of the semantics that we extend here, Krauss and Schropp [6] have also formalised a translation to set theory automatically producing proofs in Isabelle/ZF.

Considering other logics, Barras [3] has formalised a reduced version of the calculus of inductive constructions, the logic used by the Coq proof assistant, giving it a semantics in set theory and formalising a soundness proof in Coq itself. The approach is modular, and Wang and Barras [15] have extended the framework and applied it to the calculus of constructions plus an abstract equational theory.

Myreen and Davis [9] formalised Milawa's ACL2-like first-order logic and proved it sound using HOL4. This soundness proof for Milawa produced a top-level theorem which states that the machine-code which runs the prover will only print theorems that are true according to the semantics of the Milawa logic. Since Milawa's logic is weaker than HOL, it fits naturally inside HOL without encountering any delicate foundational territory such as the assumption on Theorem 1.

Other noteworthy prover verifications include a simple first-order tableau prover by Ridge and Margetson [13] and a SAT solver algorithm with many modern optimizations by Marić [8].

8   Conclusion

CakeML In the context of the CakeML project overall (https://cakeml.org), our verified implementation of the HOL Light kernel is an important milestone: the first verified application other than the CakeML compiler itself. This validates both the methodology of working in HOL4 and using automated synthesis to produce verified programs, and the usefulness of the CakeML language for a substantial application. At this point we have a verified compiler, and a verified application to run on it. What remains to be done is the creation of reasoning tools for CakeML programs that do not fit nicely into the HOL4 logic. In particular, we want to establish that arbitrary – possibly malicious – client code that constructs proofs using the verified HOL Light kernel cannot subvert the module-system enforced abstraction that protects the kernel and create false theorems.

Reflections on Stateless HOL. Our choice to use stateless HOL was motivated by a desire to keep the soundness proof simple and close to Harrison's by avoiding introduction of a context for definitions. We avoided the context, but stateless HOL did introduce some significant complications: the inference system and semantics both become mutually recursive, and care must be taken to avoid terms with no semantics. The dependence of `typeset` on `semantics` is necessary for type definitions, but it could perhaps be factored through a context. Similarly, the move to relations instead of functions seems reasonable given the side-conditions on the definitional rules, but one could instead use a total lookup function to get definitions from a context.

Overall it is not clear that stateless HOL saved us any work and it is clear that it led to some loss of abstraction in our formalisation. Separating the context from the representation of types and terms is closer to the standard approaches adopted in the mathematical logic literature and would help to separate concerns about the conservative extension mechanisms (which we expect to be loosely specified) from concerns about the semantics of types and terms (which we expect to be deterministic functions of the context). After submitting this paper, we experimented with a formalisation of the semantics using a separate context, and would now recommend the context-based approach as simpler and more expressive.

Acknowledgements. We thank Mike Gordon, John Harrison, Roger Jones, Michael Norrish, Konrad Slind, and Freek Wiedijk for useful discussions and feedback. The first author acknowledges support from Gates Cambridge. The third author was funded by the Royal Society, UK.

References

1. Arthan, R.: HOL formalised: Semantics, http://www.lemma-one.com/ProofPower/specs/spc002.pdf
2. Arthan, R.: HOL constant definition done right. In: Klein, G., Gamboa, R. (eds.) ITP 2014. LNCS (LNAI), vol. 8558, pp. 531–536. Springer, Heidelberg (2014)
3. Barras, B.: Sets in Coq, Coq in sets. J. Formalized Reasoning 3(1) (2010)
4. Harrison, J.: Towards self-verification of HOL Light. In: Furbach, U., Shankar, N. (eds.) IJCAR 2006. LNCS (LNAI), vol. 4130, pp. 177–191. Springer, Heidelberg (2006)
5. Harrison, J.: HOL Light: An overview. In: Berghofer, S., Nipkow, T., Urban, C., Wenzel, M. (eds.) TPHOLs 2009. LNCS, vol. 5674, pp. 60–66. Springer, Heidelberg (2009), http://www.cl.cam.ac.uk/~jrh13/hol-light/
6. Krauss, A., Schropp, A.: A mechanized translation from higher-order logic to set theory. In: Kaufmann, M., Paulson, L.C. (eds.) ITP 2010. LNCS, vol. 6172, pp. 323–338. Springer, Heidelberg (2010)
7. Kumar, R., Myreen, M.O., Norrish, M., Owens, S.: CakeML: a verified implementation of ML. In: Principles of Prog. Lang. (POPL). ACM Press (2014)
8. Marić, F.: Formal verification of a modern SAT solver by shallow embedding into Isabelle/HOL. Theor. Comput. Sci. 411(50), 4333–4356 (2010)
9. Myreen, M.O., Davis, J.: The reflective Milawa theorem prover is sound (Down to the machine code that runs it). In: Klein, G., Gamboa, R. (eds.) ITP 2014. LNCS (LNAI), vol. 8558, pp. 421–436. Springer, Heidelberg (2014)

10. Myreen, M.O., Owens, S.: Proof-producing translation of higher-order logic into pure and stateful ML. Journal of Functional Programming FirstView (January 2014)
11. Myreen, M.O., Owens, S., Kumar, R.: Steps towards verified implementations of HOL Light. In: Blazy, S., Paulin-Mohring, C., Pichardie, D. (eds.) ITP 2013. LNCS, vol. 7998, pp. 490–495. Springer, Heidelberg (2013), "Rough Diamond" section
12. Norrish, M., Slind, K., et al.: The HOL System: Logic, 3rd edn., http://hol.sourceforge.net/documentation.html
13. Ridge, T., Margetson, J.: A mechanically verified, sound and complete theorem prover for first order logic. In: Hurd, J., Melham, T. (eds.) TPHOLs 2005. LNCS, vol. 3603, pp. 294–309. Springer, Heidelberg (2005)
14. Slind, K., Norrish, M.: A brief overview of HOL4. In: Mohamed, O.A., Muñoz, C., Tahar, S. (eds.) TPHOLs 2008. LNCS, vol. 5170, pp. 28–32. Springer, Heidelberg (2008)
15. Wang, Q., Barras, B.: Semantics of intensional type theory extended with decidable equational theories. In: CSL. LIPIcs, vol. 23, Schloss Dagstuhl - Leibniz-Zentrum fuer Informatik (2013)
16. Wiedijk, F.: Stateless HOL. In: Types for Proofs and Programs (TYPES). EPTCS, vol. 53 (2009)

Verified Efficient Implementation of Gabow's Strongly Connected Component Algorithm

Peter Lammich

Technische Universität München
lammich@in.tum.de

Abstract. We present an Isabelle/HOL formalization of Gabow's algorithm for finding the strongly connected components of a directed graph. Using data refinement techniques, we extract efficient code that performs comparable to a reference implementation in Java. Our style of formalization allows for reusing large parts of the proofs when defining variants of the algorithm. We demonstrate this by verifying an algorithm for the emptiness check of generalized Büchi automata, reusing most of the existing proofs.

1 Introduction

A strongly connected component (SCC) of a directed graph is a maximal subset of mutually reachable nodes. Finding the SCCs is a standard problem from graph theory with applications in many fields [27, Chap. 4.2].

There are several algorithms to partition the nodes of a graph into SCCs, the main ones being the Kosaraju-Sharir algorithm [28], Tarjan's algorithm [29], and the class of path-based algorithms [25,22,7,4,9].

In this paper, we present the verification of Gabow's path-based SCC algorithm [9] within the theorem prover Isabelle/HOL [24]. Using refinement techniques and efficient verified data structures, we extract Standard ML (SML) [21] code from the formalization. Our verified algorithm has a performance comparable to a reference implementation in Java, taken from Sedgewick and Wayne's textbook on algorithms [27, Chap. 4.2].

Our main interest in SCC algorithms stems from the fact that they can be used for the emptiness check of generalized Büchi automata (GBA), a problem that arises in LTL model checking [30,10,6]. Towards this end, we extend the algorithm to check the emptiness of generalized Büchi automata, reusing many of the proofs from the original verification.

Contributions and Related Work. Up to our knowledge, we present the first mechanically verified SCC algorithm, as well as the first mechanically verified SCC-based emptiness check for GBA. Path-based algorithms have already been regarded for the emptiness check of GBAs [26]. However, we are the first to use the data structure proposed by Gabow [9].[1] Finally, our development is

---

[1] Although called Gabow-based algorithm in [26], a union-find data structure is used to implement collapsing of nodes, while Gabow proposes a different data structure [9, pg. 109].

G. Klein and R. Gamboa (Eds.): ITP 2014, LNAI 8558, pp. 325–340, 2014.

a case study for using the Isabelle/HOL Monadic Refinement and Collection Frameworks [14,19,17,18] to engineer a verified, efficient implementation of a quite complex algorithm, while keeping proofs modular and reusable.

This development is part of the Cava project [8] to produce a fully verified LTL model checker. The current Cava model checker converts GBAs to standard Büchi automata, and uses an emptiness check based on nested depth first search [5,23]. Using GBAs directly typically yields smaller search spaces, thus making tractable bigger models and/or more complex properties [6].

The Isabelle source code of the formalization described in this paper is publicly available [15].

Outline. The rest of this paper is organized as follows: In Section 2, we recall Gabow's SCC algorithm and present our extension for the emptiness check of generalized Büchi automata. Moreover, we briefly introduce the Isabelle/HOL Refinement and Collection Frameworks. Section 3 presents the formalization of the abstract algorithm, Section 4 presents the refinement to Gabow's data structure, and Section 5 presents the refinement to executable code. Finally, Section 6 reports on performance benchmarks and Section 7 contains conclusions and directions of future work.

2   Preliminaries

In this section, we present the preliminaries of our formalization. Subsection 2.1 recalls Gabow's algorithm and GBAs. Subsection 2.2 outlines our verification approach based on the Isabelle/HOL Refinement and Collection Frameworks.

2.1   Path-Based Strongly Connected Component Algorithms

Let $G = (V, E)$ be a finite directed graph over nodes V and edges $E \subseteq V \times V$. A *strongly connected component* (SCC) is a maximal set of nodes $U \subseteq V$, such that all nodes in U are *mutually reachable*, i.e. for all $u, v \in U$, there is a directed path from u to v.

A *path based SCC algorithm* is a depth first search (DFS) through the graph that, whenever it finds a back edge, contracts all nodes in the cycle closed by this edge [9]. To distinguish contracted nodes from nodes of the original graph, we refer to the former ones as *c-nodes*.

The algorithm starts with the original graph and a path that consists of a single arbitrary node. In each step, an edge starting at the end of the path is selected. If it leads back to a c-node on the path, all c-nodes on the cycle formed by this back edge and the path are collapsed. If the edge leads to an unfinished c-node, this node is appended to the path. Otherwise, the edge is ignored. If all edges from the end of the path have already been considered, the last c-node is removed from the path and marked as finished. At this point, the last c-node represents an SCC of the original graph. If the path becomes empty, the algorithm is repeated for another unfinished node, until all nodes have been finished.

Implementation. The problem when implementing this algorithm is to keep track of the collapsed nodes in the graph efficiently. Initially, general set merging algorithms were proposed for identifying the collapsed nodes [25,22]. The idea of [9] (variants of it are also used in [7,4]) is to represent the current path by two stacks: A stack S that contains nodes of the original graph, and a *boundary stack* B that contains indexes into the first stack, which represent the boundaries between the collapsed nodes. For example, to represent the path $[\{a,b\},\{c\},\{d,e\}]$, one uses $S = [a,b,c,d,e]$ and $B = [0,2,3]$. Moreover, the (partial) *index map* I maps each node on S to its index. I is also used to represent nodes that belong to finished c-nodes, by mapping them to a special value (e. g. a negative number).

Collapsing is always due to a back edge (u,v), where u is in the last c-node of the path. Thus, it can be implemented by looking up the index $I\,v$ of v, and then popping elements from B until its topmost element is less than or equal to $I\,v$. Appending a new c-node to the path is implemented by pushing it onto S and its index onto B. Removing the last c-node from the path is implemented by popping elements from S until its length becomes *top B*, and then popping the topmost element from B.

With this data structure, the algorithm runs in time $O(|V| + |E|)$, i. e. linear time in the size of the graph [9].

For our main purpose, i. e. LTL model checking, we generalize the algorithm to only consider the part of the graph that is reachable from a set of start nodes $V_0 \subseteq V$. This is easily achieved by only repeating the algorithm for unfinished nodes from V_0.

Generalized Büchi Automata. Generalized Büchi Automata (GBA) [30] have been introduced as a generalization of Büchi automata (BA) [3] that allows for more efficient automata based LTL model checking [31].

A GBA is a finite directed graph (V, E) with a set of initial nodes $V_0 \subseteq V$, a finite set of *acceptance classes* C, and a map $F : V \to 2^C$. As we are only interested in emptiness, we need not consider an alphabet.

An *accepting run* is an infinite path starting at a node from V_0, such that a node from each acceptance class occurs infinitely often on that path. A GBA is *non-empty*, if it has an accepting run. As the GBA is finite, this is equivalent to having a reachable *accepting cycle*, i. e. a cyclic, finite path with at least one edge that contains nodes from all acceptance classes. Obviously, a graph has a reachable accepting cycle iff it has a reachable non-trivial SCC that contains nodes from all acceptance classes. Here, an SCC is called *non-trivial*, if there is at least one edge between its nodes.

To decide emptiness, we don't need to compute all SCCs first: As the c-nodes on the path are always subsets of SCCs, we can report „non-empty" already if the last c-node on the path contains nodes from all acceptance classes, after being collapsed (i. e. becoming non-trivial). This way, the algorithm reports non-emptiness as soon as it has seen all edges of an accepting cycle.

To implement this check efficiently, we store the set of acceptance classes for each c-node on the path. This information can be added to the B stack, or maintained as a separate stack. On collapsing, the sets belonging to the collapsed c-nodes are joined. This adds a factor of $|C|$ to the run time. However, $|C|$ is typically small, such that the sets can be implemented efficiently, e. g. as bit vectors.

2.2   Refinement Based Program Verification in Isabelle/HOL

Our formalization is done in four main steps, using Isabelle/HOL [24]:

1. Verification of the abstract path-based algorithm.
2. Refinement to Gabow's data structure.
3. Refinement to efficient data structures (e. g. arrays, hash tables).
4. Extraction of Standard ML code.

The key advantage of this approach is that proofs in one step are not influenced by proofs in the other steps, which greatly reduces the complexity of the whole development, and makes more complex developments possible in the first place.

With its refinement calculus [1] that is based on a nondeterminism monad [32], the Monadic Refinement Framework [19,17] provides a concise way to phrase the algorithms and refinements in Steps 1–3. The Isabelle Collection Framework [14,16] contributes the efficient data structures. Moreover, we use the Autoref tool [18] to add some automation in Step 3. Finally, we use the Isabelle/HOL code generator [11,12] in Step 4. In the following, we briefly recall these techniques.

Isabelle/HOL. Isabelle/HOL [24] is a theorem prover for higher order logic. The listings contained in this paper are actual Isabelle/HOL source, sometimes slightly polished for better readability. We quickly review some non-standard syntax used in this paper: Functions are defined by sequential pattern matching, using $\equiv$ as defining equation operator. Theorems are written as $[\![P_1, \ldots, P_n]\!] \implies Q$, which is syntactic sugar for $P_1 \implies \ldots \implies P_n \implies Q$.

Program Refinement. The Monadic Refinement Framework represents programs as a monad over the type *'a nres = res 'a set | fail*. A result **res** X means that the program nondeterministically returns a value from the set X, and the result **fail** means that an assertion failed. The subset ordering is lifted to results:

$$\mathbf{res}\ X \leq \mathbf{res}\ Y \equiv X \subseteq Y \mid \_ \leq \mathbf{fail} \equiv \mathit{True} \mid \_ \leq \_ \equiv \mathit{False}$$

Intuitively, $m \leq m'$ (*m refines m'*) means that all possible values of m are also possible values of m'. Note that this ordering yields a complete lattice on results, with smallest element **res** $\{\}$ and greatest element **fail**. The monad operations are then defined as follows:

$$\mathbf{return}\ x \equiv \mathbf{res}\ \{x\}$$
$$\mathbf{bind}\ (\mathbf{res}\ X)\ f \equiv \mathit{Sup}\ \{f\ x \mid x {\in} X\} \mid \mathbf{bind}\ \mathbf{fail}\ f \equiv \mathbf{fail}$$

Intuitively, **return** x is the result that contains the single value x, and **bind** $m\ f$ is sequential composition: Choose a value from m, and apply f to it.

As a shortcut to specify values satisfying a given predicate Φ, we define **spec** $\Phi \equiv$ **res** $\{x \mid \Phi\ x\}$. Moreover, we use a Haskell-like do-notation, and define a shortcut for assertions:

assert $\Phi \equiv$ **if** Φ **then return** () **else fail**

Recursion is defined by a fixed point:

rec $B\ x \equiv$ **do** $\{$**assert** (*mono B*); *gfp B x*$\}$

As we use the greatest fixed point, a non-terminating recursion causes the result to be **fail**. This matches the notion of total correctness. Note that we assert monotonicity of the recursive function's body B, which typically follows by construction [13]. On top of the **rec** primitive, we define loop constructs like **while** and **foreach**.

In a typical development based on stepwise refinement, one specifies a series of programs $m_1 \leq \ldots \leq m_n$, such that m_n has the form **spec** Φ, where Φ is the specification, and m_1 is the final implementation. In each refinement step (from m_{i+1} to m_i), some aspects of the program are refined.

Example 1. Given a finite set S of sets, the following specifies a set r that contains at least one element from every non-empty set in S:

$$sel_3\ S \equiv \textbf{do}\ \{\textbf{assert}\ (\textit{finite S}); (\textbf{spec}\ r.\ \forall s \in S.\ s \neq \{\} \longrightarrow r \cap s \neq \{\})\}$$

This specification can be implemented by iteration over the outer set, adding an arbitrary element from each non-empty inner set to the result:

```
sel₂ S ≡ do {
    assert (finite S);
    foreach S (λs r.
        if s={} then return r else do {x←spec x. x∈s; return (insert x r)}
    ) {} }
```

Using the verification condition generator (VCG) of the monadic refinement framework, it is straightforward to show that sel_2 is a refinement of sel_3:

lemma $sel_2\ S \leq sel_3\ S$
    unfolding $sel_2\_def\ sel_3\_def$
    by (*refine_rcg foreach_rule*[**where** $I=\lambda it\ r.\ \forall s \in S - it.\ s \neq \{\} \longrightarrow r \cap s \neq \{\}$])
    auto

As constructs used in monadic programs are monotonic, sel_3 can be replaced by sel_2 in a bigger program, yielding a correct refinement.

Data Refinement. In a typical refinement based development, one also wants to refine the representation of data. For example, we need to refine the abstract path by Gabow's data structure. A data refinement is specified by a single-valued *refinement relation* between concrete and abstract values. Equivalently, it can be expressed by an *abstraction function* from concrete to abstract values and an invariant on concrete values. A prototypical example is implementing sets by distinct lists, i. e. lists that contain no duplicate elements. Here, the refinement relation $\langle R \rangle$ $list\_set\_rel$ relates a distinct list to the set of its elements, where the elements are related by R.

Given a refinement relation R, we define the function $\Downarrow_R$ to map results over the abstract type to results over the concrete type:

$$\Downarrow_R (\mathbf{res}\ A) \equiv \mathbf{res}\ \{c \mid \exists a \in A.\ (c,a) \in R\} \quad | \quad \Downarrow_R \mathbf{fail} \equiv \mathbf{fail}$$

Thus, $m_1 \leq \Downarrow_R m_2$ states that m_1 is a refinement of m_2 w. r. t. the refinement relation R, i. e. all concrete values in m_1 correspond to abstract values in m_2.

The Monadic Refinement Framework implements a refinement calculus [1] that is used by the VCG for refinement proofs. Moreover, the Autoref tool [18] can be used to automatically synthesize the concrete program and the refinement proof, guided by user-adjustable heuristics to find suitable implementations of abstract data types. For the algorithm sel_2 from Example 1, Autoref generates the implementation

$sel_1\ Xi \equiv foldl$
$\quad (\lambda\sigma\ x.\ \textbf{if}\ is\_Nil\ x\ \textbf{then}\ \sigma\ \textbf{else let}\ xa = hd\ x\ \textbf{in}\ glist\_insert\ op = xa\ \sigma)\ [\,]\ Xi$

and proves the theorem

$(Xi1,\ X1) \in \langle\langle Id \rangle list\_set\_rel \rangle list\_set\_rel \Longrightarrow$
$\quad \textbf{return}\ (sel_1\ Xi1) \leq \Downarrow_{\langle Id \rangle list\_set\_rel}\ (sel_2\ X1)$

By default, Autoref uses the Isabelle Collection Framework [14,16], which provides a large selection of verified collection data structures.

Code Generation. After the last refinement step, one typically has arrived at a deterministic program inside the executable fragment of Isabelle/HOL. The code generator [11,12] extracts this program to Standard ML code. For example, it generates the following ML function for sel_1:

```
fun sel₁ xi =
  List.foldl (fn sigma ⇒ fn x ⇒
    (if Autoref_Bindings_HOL.is_Nil x then sigma
      else let val xa = List.hd x;
        in Impl_List_Set.glist_insert Arith.equal_nat xa sigma
        end)) [] xi;
```

3 Abstract Algorithm

In this section, we describe our formalization of the abstract path based algorithm for finding SCCs. The goal is to formalize two variants of the algorithm, one for computing a list of SCCs, and another for emptiness check of GBAs, while sharing common parts of the proofs. For this purpose, we first define a skeleton algorithm that maintains the path through the graph, but does not store the found SCCs. This skeleton algorithm helps us finding invariants that

```
skeleton ≡ do {
  let D = {};
  foreach^{outer_invar} V_0 (λv_0 D_0. do {
    if v_0 ∉ D_0 then do {
      let (p,D,pE) = initial v_0 D_0;

      (p,D,pE) ← while^{invar v_0 D_0} (λ(p,D,pE). p ≠ []) (λ(p,D,pE). do {
        (vo,(p,D,pE)) ← select_edge (p,D,pE);
        case vo of
          Some v ⇒ do {
            if v ∈ ⋃set p then return (collapse v (p,D,pE))
            else if v ∉ D then return (push v (p,D,pE))
            else return (p,D,pE)
          }
        | None ⇒ return (pop (p,D,pE))
      }) (p,D,pE);
      return D
    } else
      return D_0
  }) D
}
```

Listing 1.1. Skeleton of a path-based algorithm

hold in all path-based algorithms, and can be used as a starting point for defining the actual algorithms. Listing 1.1 displays the code of the skeleton algorithm.[2] It formalizes the path-based algorithm sketched in Section 2.1: First, we initialize the set D of finished nodes. Then, we iterate over the root nodes V_0 of the graph, and for each unfinished one, we start the inner loop of the search algorithm, which runs until the path becomes empty again. In the inner loop, we additionally keep track of the current path p and a set pE of *pending edges*, i.e. edges that have not yet been explored and start from nodes on the path. For better manageability of the proofs, we have defined constants for the basic operations:

$$initial\ v_0\ D_0 \equiv ([\{v_0\}],\ D_0,\ E \cap \{v_0\} \times UNIV)$$

```
select_edge (p,D,pE) ≡ do {
  e ← select (pE ∩ last p × UNIV);
  case e of
    None ⇒ return (None,(p,D,pE))
  | Some (u,v) ⇒ return (Some v, (p,D,pE − {(u,v)}))
}
```

$$collapse\ v\ (p,D,pE) \equiv let\ i=idx\_of\ p\ v\ in\ (take\ i\ p\ @\ [\bigcup set\ (drop\ i\ p)],D,pE)$$
$$where\ idx\_of\ p\ v \equiv THE\ i.\ i<length\ p \wedge v \in p!i$$
$$push\ v\ (p,\ D,\ pE) \equiv (p\ @\ [\{v\}],\ D,\ pE \cup E \cap \{v\} \times UNIV)$$
$$pop\ (p,\ D,\ pE) \equiv (butlast\ p,\ last\ p \cup D,\ pE)$$

---

[2] Shortened a bit by removing some **assert**-statements.

These constants are defined over the whole state (p, D, pE) of the inner loop, even if they only work on parts of it. This allows for nicer refinement proofs, as operations on the abstract state are refined to operations on the concrete state, without exposing the inner structure of the states, which differs on the concrete and abstract domain. Note that this also results in a more modular correctness proof, as invariant preservation can be shown separately for each operation.

We briefly explain the operations: The *initial* operation initializes the state for the inner loop with a path that consists of the single node v_0. The *select_edge* operation checks if there is a pending edge from the end of the path. If there is no such edge, it returns *None* and does not change the state. Otherwise, it removes the edge from the set of pending edges, and returns its target node. The operation *collapse* first determines the index of the node on the path, and then collapses the corresponding suffix of the path. The operation *push* appends a new node to the path, and the operation *pop* removes the last node from the path.

3.1 Invariants

Correctness of while and foreach loops is proved by establishing a loop invariant. Moreover, we have to show that the body of a while loop transforms states within a well-founded relation, and that the set iterated over by a foreach loop is finite.

We specify invariants for the skeleton algorithm and show that they are preserved by the operations inside the loop. The invariants and the preservation lemmas are then reused for the actual algorithms. In Listing 1.1, the loops are annotated with their invariants, such that the VCG sees them.

The invariant of the outer loop depends on the nodes it still to be iterated over, and on the finished nodes D. It states that (1) we only iterate over start nodes, (2) nodes that we have already iterated over are finished, (3) finished nodes are reachable, and (4) edges from finished nodes lead to finished nodes again. The invariant is formalized using the locale mechanism of Isabelle/HOL [2]:

locale *outer_invar* = *digraph_loc* + **fixes** it **and** D
  assumes *1*: $it \subseteq V_0$ **and** *2*: $V_0 - it \subseteq D$ **and** *3*: $D \subseteq E^*$ "V_0 **and** *4*: E "$D \subseteq D$

Here, *digraph_loc* defines a finite directed graph, represented by its edge relation E and a set of initial nodes V_0. The set V of nodes is implicitly fixed to the universal set *UNIV*, and thus not explicitly mentioned in the formalization. Moreover, r^* denotes the reflexive transitive closure of a relation r, and r "$s = \{y. \ \exists x \in s. \ (x,y) \in r\}$ denotes the image of a set s under a relation r.

The invariant *invar* v_0 D_0 (p, D, pE) of the inner loop is more complex. We only sketch its main idea here, and refer the reader to the actual formalization [15] for more details. The main parts of the inner loop's invariant are:

(1) Edges from finished nodes lead to finished nodes; nodes on the path are not finished; non-pending edges from the path lead either to nodes on the path or to finished nodes.
(2) Only pending edges may go back on the path.
(3) The nodes inside a c-node on the path are mutually reachable.

(1) is a standard invariant for DFS. (2) ensures that all cycles that have already been seen are collapsed, and (3) ensures that the c-nodes on the path are always subsets of SCCs. In particular, when a c-node is popped from the path, it has no pending edges left. Then, it easily follows from the invariant that this c-node is a maximal set of mutually reachable nodes, i.e. an SCC.

We now apply the VCG to the skeleton algorithm, after unfolding the definition of *select_edge*. This leaves us with proof obligations to show invariant preservation for each of the operations in the loop. These are proved as separate lemmas, to be reused later. For example, we prove for the pop operation:

invar_pop: $[\![invar\ v_0\ D_0\ (p,\ D,\ pE);\ p \neq [\,];\ pE \cap last\ p \times UNIV = \{\}]\!]$
$\implies invar\ v_0\ D_0\ (pop\ (p,\ D,\ pE))$

To show termination of the inner loop, we define an edge to be *visited* if it is not pending and starts at a finished node, or at a node on the path. We then show that, in each step, either the set of visited edges grows, or it remains the same and the path length decreases. Technically, we define a well-founded relation over the state of the while loop and show that the operations are compatible with it. These verification conditions are also proved as separate lemmas.

3.2   Computing the SCCs

In a next step, we extend the skeleton algorithm to actually compute a list of SCCs of the graph. We define the algorithm *compute_SCC* by replacing the statement **return** $(pop\ (p,D,pE))$ in Listing 1.1 with **return** $(last\ p\#l,pop\ (p,D,pE))$, and pass the list l through the inner and outer loop, initializing it to the empty list.

In order to specify the intended result, we first define a strongly connected component as a maximal mutually connected set of nodes:

is_scc $E\ U \equiv U \times U \subseteq E^* \wedge (\forall U'.\ U' \supset U \longrightarrow \neg\ (U' \times U' \subseteq E^*))$

Then, we define the intended result as a list that covers all reachable nodes and contains SCCs in (reverse) topological order:

compute_SCC_spec $\equiv$ **spec** l.
$\bigcup set\ l = E^*\ ``V_0 \wedge (\forall U \in set\ l.\ is\_scc\ E\ U)$
$\wedge (\forall i\ j.\ i < j \wedge j < length\ l \longrightarrow l!j \times l!i \cap E^* = \{\})$

Next, we extend the invariant of the skeleton algorithm. The invariant extension is the same for the inner and outer invariant, and states that the list computed so far (1) covers exactly the finished nodes and (2) contains SCCs in (3) reverse topological order. The new invariants can be elegantly defined using the locale mechanism of Isabelle/HOL:

locale *cscc_invar_ext* = *digraph_loc* + **fixes** $l\ D$
  assumes *1:* $\bigcup set\ l = D$ **and** *2:* $\forall U \in set\ l.\ is\_scc\ E\ U$
  and *3:* $\bigwedge i\ j.\ [\![i < j;\ j < length\ l]\!] \implies l!j \times l!i \cap E^* = \{\}$

locale *cscc_outer_invar* = *outer_invar* + *cscc_invar_ext*
locale *cscc_invar* = *invar* + *cscc_invar_ext*

In order to prove the algorithm correct, we have to show that the extended invariant is preserved. We add the following rule to the VCG:

cscc_invarI: $[\![invar\ v_0\ D_0\ s;\ invar\ v_0\ D_0\ s \implies cscc\_invar\_ext\ (l,\ s)]\!]$
$\implies cscc\_invar\ v_0\ D_0\ (l,\ s)$

We also add the analogous rule *cscc_outer_invarI* for the outer loop's invariant. These rules split a proof of the extended invariant into a proof of the original invariant and a proof of the invariant extension.

As we already have proved lemmas for the verification conditions concerning *invar*, we only have to prove lemmas for the invariant extension. For example, for finishing a node, we prove

cscc_invar_pop: $[\![cscc\_invar\ v_0\ D_0\ (l,\ p,\ D,\ pE);\ invar\ v_0\ D_0\ (pop\ (p,\ D,\ pE));$
$p \neq [];\ pE \cap last\ p \times UNIV = \{\}]\!]$
$\implies cscc\_invar\_ext\ (last\ p\ \#\ l,\ pop\ (p,\ D,\ pE))$

The other operations, i. e. *collapse* and *push*, have not been modified at all, and also do not change the parts of the state that the invariant extension depends on. Thus, proving preservation of the invariant extension for these operations is straightforward. Moreover, the termination argument from the skeleton algorithm can be reused. Finally, we prove the theorem *compute_SCC* ≤ *compute_SCC_spec*, which states that the SCC algorithm behaves according to its specification. This is straightforward, using the VCG with the invariant preservation lemmas from the skeleton algorithm together with the new ones for the invariant extension.

While the formalization of the skeleton algorithm and the invariants requires about 1300 lines of proof text, the extension to compute SCCs requires only about 300 lines.

3.3   Emptiness Check for GBA

The extension to check for emptiness of GBAs is more complex, but is formalized in the same way. We sketch the extension here very briefly, and refer the reader to the actual formalization [15] for details.

Starting from the skeleton algorithm, we extend the collapse operation to check whether the collapsed c-node contains nodes from all acceptance classes. If so, we break the loop immediately and return the result for non-emptiness. It contains the two sets $\bigcup butlast\ p$ and *last p*, which can be used to reconstruct the accepting run: The path reaching the accepting cycle only contains nodes from the first set, the accepting cycle itself only contains nodes from the second set.

The invariant for the outer loop is extended to state that there is no accepting cycle within finished nodes. The invariant of the inner loop is extended to state that there is no accepting cycle over visited edges. The extension of the skeleton algorithm to GBA emptiness check requires about 700 lines of proof text.

4   Implementation Using Gabow's Data Structure

In the last section, we described the verification of the abstract path based algorithm. In this section, we describe the refinement to Gabow's data structure, which was already sketched in Section 2.2.

We implement the stack S and the boundary stack B by lists of nodes and natural numbers, respectively. The index map I is implemented as a function from nodes to *node_state option*, where

node_state = DONE | STACK nat

Finished nodes are mapped to *Some DONE*, nodes on the stack are mapped to *Some (STACK j)*, where j is the index of the node in S, and nodes not yet seen are mapped to *None*.

We additionally use the *pending stack P* to store the pending edges. P contains entries of the form *nat × node set*. An entry $(j,succs)$ means that the edges $\{S!j\} \times succs$ are pending. The pending stack only contains entries with non-empty second component, and the entries are always sorted by first component. Thus, the last entry $(j,succs)$ of P contains the pending edges for the last node on S that has pending edges left. By comparing j to *last B*, one efficiently checks whether this node belongs to the last collapsed node on the path. Also, pushing a new node is efficiently implemented by pushing an entry for its successors, if any, onto P. The invariant for Gabow's data structure is, again, formalized as a locale, based on the locale GS, which fixes (S, B, P, I):

locale *GS_invar = GS +*
$(*1*)$ **assumes** *sorted B* **and** *distinct B* **and** *set B ⊆ {0..<length S}*
$(*2*)$ **and** $S{\neq}[] \implies B{\neq}[] \wedge B!0=0$ **and** *distinct S*
$(*3*)$ **and** $(I\ v = Some\ (STACK\ j)) \longleftrightarrow (j{<}length\ S \wedge v = S!j)$
$(*4*)$ **and** *sorted (map fst P)* **and** *distinct (map fst P)*
$(*5*)$ **and** *set P ⊆ {0..<length S}×{x. x{\neq}\{\}}*

Intuitively, Line 1 states that the boundary stack is sorted, distinct, and contains valid indexes into S. Line 2 states that a non-empty stack implies a non-empty boundary stack with the first boundary being 0, and that S is distinct. Line 3 states that the index map is consistent with the stack. Finally, Lines 4 and 5 state that the first elements of the pending edge stack are sorted and distinct, and that the pending edge stack contains valid indexes into the stack and no empty successor sets.

To map concrete to abstract states, we define (in the locale GS):

seg_start i ≡ B!i
seg_end i ≡ **if** $i+1 = length\ B$ **then** *length S* **else** $B!(i+1)$
seg i ≡ {S!j | j. seg_start i ≤ j ∧ j < seg_end i}

$p\_\alpha ≡ map\ seg\ [0..<length\ B]$
$D\_\alpha ≡ \{v.\ I\ v = Some\ DONE\}$
$pE\_\alpha ≡ \{\ (u,v)\ .\ \exists j\ I.\ (j,I){\in}set\ P \wedge u = S!j \wedge v{\in}I\ \}$
$GS\_\alpha ≡ (p\_\alpha,D\_\alpha,pE\_\alpha)$

Here, *GS_α* is the abstraction function, mapping the concrete state (S, B, P, I) (fixed by the locale *GS*) to its corresponding abstract state. Finally, we define *GS_rel* as the refinement relation induced by *GS_α* and *GS_invar*. Similarly, we define *oGS_rel* for the state of the outer loop.

Next, we provide concrete versions of the operations and show that they refine their abstract counterparts. For example, for the pop operation, we define (in *GS*):

pop_impl ≡ **do** {
    $I \leftarrow$ *mark_as_done* (*seg_start* $(|B| - Suc\ 0)$) (*seg_end* $(|B| - Suc\ 0)$) *I*;
    return (*take* (*last B*) *S*, *butlast B, I, P*) }

Here, *mark_as_done l u* marks the nodes in $\{S!i \mid l \leq i \land i < u\}$ as finished. We show the following refinement lemma:

pop_refine: $[\![((S,B,I,P),\ p,\ D,\ pE) \in GS\_rel;\ p \neq [\ ];\ pE \cap last\ p \times UNIV = \{\}]\!]$
    $\implies$ *pop_impl* $(S,B,I,P) \leq \Downarrow_{GS\_rel}$ (**return** (*pop* (*p, D, pE*)))

After having defined the other operations, and shown similar lemmas for them, we finally define *skeleton_impl* and show that it refines *skeleton*. Exploiting the automation provided by the Refinement Framework, this is straightforward:

theorem *skeleton_impl* $\leq \Downarrow_{oGS\_rel}$ *skeleton*
    unfolding *skeleton_impl_def skeleton_def*
    by (*refine_rcg skeleton_refines*)
        (*vc_solve* (*nopre*) *solve: asm_rl I_to_outer simp: skeleton_refine_simps*)

4.1   Refinement of SCC Computation and GBA Emptiness Check

In order to implement the actual algorithm for computing a list of SCCs, the only thing we have to add is a function that builds a set out of the last segment of *S*. This set is added to the output list upon finishing a c-node. This extension is straightforward, and, all together, the formalization requires less than 100 lines. It completely reuses what is already proved for the skeleton algorithm.

The refinement for the GBA emptiness check is more complicated. For better manageability, it is split into two steps: In the first step, the sets of acceptance classes for each c-node on the path are explicitly maintained in a list *A*, and the check after the collapse operation is refined to use *A*. Proving refinement is quite simple as only redundant information is added. In the second step, we refine the algorithm to use Gabow's data structure.

For a clearer structure of the formalization, we decided to define new constants for the operations of the emptiness check algorithm, which use the operations from the skeleton, and add the new functionality for keeping track of *A*. For the collapse operation, we have to compute *idx_of v* twice: Once in the original collapse operation, and a second time for updating *A*. Thus, we add another refinement step to refine this operation to an optimized version that computes *idx_of v* only once. Note that this refinement is limited to the collapse operation, and does not affect the overall proof.

All together, the implementation of the emptiness check requires 900 lines of proof text. Using the refinement framework, we have broken down the formalization into small, manageable steps: First, we introduced the list A, then we introduced Gabow's data structure, and for the collapse operation, we added an additional optimization step. If we would have done all these in one big step, we would have ended up with a complicated formalization, which is hard to maintain or change. In contrast, our approach has already proven its flexibility: In a first version, we had only formalized the inner loop of the algorithms. Thus, we could only handle graphs where all nodes are reachable from a single node v_0. Later, we added the outer loop without any major problems.

5   Refinement to Efficient Standard ML Code

In order to generate efficient SML code, we first have to decide for the data structures used to implement the stacks S, B, and P, and the map I. We resort to the large selection of verified data structures provided by the Isabelle Collection Framework [14,16]: For the stacks, we use arrays with dynamic resizing, which have an amortized constant time per operation. For the index map I, we also use an array, assuming the nodes to be natural numbers.[3] Once we have fixed the refinement relations, the Autoref tool [18] synthesizes and proves correct the refined versions of the algorithms automatically. Finally, the code generator [11,12] is used to extract the SML code.

For computing the SCCs, we encode the output as a list of distinct lists. This is adequate as we only prepend items to the output, which is a constant time operation. Moreover, when extracting an SCC from the path, we know that each node occurs at most once on the path. Thus, building a distinct list of these nodes can be done in linear time. When adding a corresponding assertion to the program, the Autoref tool performs this optimization automatically.

For the GBA emptiness check, we represent the acceptance classes C by natural numbers in the range $\{0..<|C|\}$, and use bitvectors to implement the sets of acceptance classes that are stored on the stack.

6   Benchmarks

We have benchmarked the extracted code against a reference implementation of Gabow's algorithm in Java, taken from Sedgewick and Wayne's book on algorithms [27] and slightly adapted to work with an explicit set of root nodes. To produce the input graphs, we used a random graph generator, also taken from [27], and let it produce graphs with the number of edges $|E|$ in the range from 10^5 to 10^6. Each graph has $|V| = \lfloor 6\sqrt{|E|} \rfloor$ nodes, and $\lfloor \frac{|V|}{10} \rfloor$ SCCs. The results of the benchmark are displayed in Figure 1, using a log-log scale where

---

[3] This choice is adequate for comparison with the Java reference implementation, which also uses an array for I. For model checking, a hash table is more adequate, which can be used by changing only a few lines of the formalization.

the y-axis is the required time to determine the lists of SCCs in milliseconds, and the x-axis is the number of edges. We compiled the extracted code with PolyML 5.5.1 and MLton 20130715, and used Java 7 for the reference implementation. All tests where performed on an x86/64 linux platform.

All implementations scale linearly with the graph size. On first glance, our implementation looks slightly faster than the Java reference implementation. However, performance in Java is very unpredictable, in particular due to just in time compilation taking place in parallel to program execution. Thus, we have modified the Java implementation to allow it to "warm up" with a few graphs, before we started the measurement. This ensures that the JIT compiler has gathered enough statistics about the program and actually finished compilation. The "Java*" - line displays the results for the modified program,

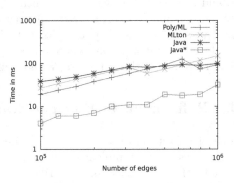

Fig. 1. Benchmarking of extracted code against Java reference Implementation

which are roughly one order of magnitude better, but required some non-obvious modification to the Java program, exploiting intimate knowledge of the JIT compiler.

7   Conclusion

We have presented a verification of two variants of Gabow's algorithm: Computation of the strongly connected components of a graph, and emptiness check of a generalized Büchi automaton. We have extracted efficient code with a performance comparable to a reference implementation in Java.

We have modularized the formalization in two directions: First, we share most of the proofs between the two variants of the algorithm. Second, we use a stepwise refinement approach to separate the algorithmic ideas and the correctness proof from implementation details. Sharing of the proofs reduced the overall effort of developing both algorithms. Using a stepwise refinement approach allowed us to formalize an efficient implementation, without making the correctness proof complex and unmanageable by cluttering it with implementation details.

Our development approach is independent of Gabow's algorithm, and can be reused for the verification of other algorithms.

Current and Future Work. Currently, we are integrating our algorithm into the Cava [8] verified LTL model checker. We expect a considerable improvement in checking speed. Moreover, fine-tuning of the used data structures, e. g. using bit vectors and machine words instead of the currently used arbitrary precision integers may give some performance improvements. The foundations for using those low-level data structures in Isabelle/HOL have recently been laid [20].

Acknowledgement. We thank the anonymous reviewers for their helpful comments.

References

1. Back, R.J., von Wright, J.: Refinement Calculus — A Systematic Introduction. Springer (1998)
2. Ballarin, C.: Interpretation of locales in Isabelle: Theories and proof contexts. In: Borwein, J.M., Farmer, W.M. (eds.) MKM 2006. LNCS (LNAI), vol. 4108, pp. 31–43. Springer, Heidelberg (2006)
3. Büchi, J.R.: On a Decision Method in Restricted Second-Order Arithmetic. In: International Congress on Logic, Methodology, and Philosophy of Science, pp. 1–11. Stanford University Press (1962)
4. Cheriyan, J., Mehlhorn, K.: Algorithms for dense graphs and networks on the random access computer. Algorithmica 15(6), 521–549 (1996)
5. Courcoubetis, C., Vardi, M., Wolper, P., Yannakakis, M.: Memory-efficient algorithms for the verification of temporal properties. Formal Methods in System Design 1(2/3), 275–288 (1992)
6. Couvreur, J.-M., Duret-Lutz, A., Poitrenaud, D.: On-the-fly emptiness checks for generalized Büchi automata. In: Godefroid, P. (ed.) SPIN 2005. LNCS, vol. 3639, pp. 169–184. Springer, Heidelberg (2005)
7. Dijkstra, E.W.: A Discipline of Programming, ch. 25. Prentice Hall (1976)
8. Esparza, J., Lammich, P., Neumann, R., Nipkow, T., Schimpf, A., Smaus, J.-G.: A fully verified executable LTL model checker. In: Sharygina, N., Veith, H. (eds.) CAV 2013. LNCS, vol. 8044, pp. 463–478. Springer, Heidelberg (2013)
9. Gabow, H.N.: Path-based depth-first search for strong and biconnected components. Information Processing Letters 74(3-4), 107–114 (2000)
10. Geldenhuys, J., Valmari, A.: More efficient on-the-fly LTL verification with Tarjan's algorithm. Theor. Comput. Sci. 345(1), 60–82 (2005)
11. Haftmann, F.: Code Generation from Specifications in Higher Order Logic. Ph.D. thesis, Technische Universität München (2009)
12. Haftmann, F., Nipkow, T.: Code generation via higher-order rewrite systems. In: Blume, M., Kobayashi, N., Vidal, G. (eds.) FLOPS 2010. LNCS, vol. 6009, pp. 103–117. Springer, Heidelberg (2010)
13. Krauss, A.: Recursive definitions of monadic functions. In: Proc. of PAR, vol. 43, pp. 1–13 (2010)
14. Lammich, P., Lochbihler, A.: The isabelle collections framework. In: Kaufmann, M., Paulson, L.C. (eds.) ITP 2010. LNCS, vol. 6172, pp. 339–354. Springer, Heidelberg (2010)
15. Lammich, P.: Formalization of Gabow's algorithm, Isabelle Theories, http://www21.in.tum.de/~lammich/isabelle/gabow
16. Lammich, P.: Collections Framework. In: Archive of Formal Proofs formal proof development (December 2009), http://afp.sf.net/entries/Collections.shtml
17. Lammich, P.: Refinement for monadic programs. In: Archive of Formal Proofs formal proof development (2012), http://afp.sf.net/entries/Refine_Monadic.shtml

18. Lammich, P.: Automatic data refinement. In: Blazy, S., Paulin-Mohring, C., Pichardie, D. (eds.) ITP 2013. LNCS, vol. 7998, pp. 84–99. Springer, Heidelberg (2013)

19. Lammich, P., Tuerk, T.: Applying data refinement for monadic programs to Hopcroft's algorithm. In: Beringer, L., Felty, A. (eds.) ITP 2012. LNCS, vol. 7406, pp. 166–182. Springer, Heidelberg (2012)

20. Lochbihler, A.: Native word. Archive of Formal Proofs Formal proof development (September 2013), http://afp.sf.net/entries/Native_Word.shtml

21. Milner, R., Tofte, M., Harper, R., MacQueen, D.: The Definition of Standard ML (Revised). MIT Press (1997)

22. Munro, I.: Efficient determination of the transitive closure of a directed graph. Information Processing Letters 1(2), 56–58 (1971)

23. Neumann, R.: A framework for verified depth-first algorithms. In: McIver, A., Höfner, P. (eds.) Proc. of the Workshop on Automated Theory Exploration (ATX 2012), pp. 36–45. EasyChair (2012)

24. Nipkow, T., Paulson, L.C., Wenzel, M.: Isabelle/HOL. LNCS, vol. 2283. Springer, Heidelberg (2002)

25. Purdom Jr., P.: A transitive closure algorithm. BIT Numerical Mathematics 10(1), 76–94 (1970)

26. Renault, E., Duret-Lutz, A., Kordon, F., Poitrenaud, D.: Three SCC-based emptiness checks for generalized Büchi automata. In: McMillan, K., Middeldorp, A., Voronkov, A. (eds.) LPAR-19. LNCS, vol. 8312, pp. 668–682. Springer, Heidelberg (2013)

27. Sedgewick, R., Wayne, K.: Algorithms, 4th edn. Addison-Wesley Professional (2011)

28. Sharir, M.: A strong-connectivity algorithm and its applications in data flow analysis. Computers & Mathematics with Applications 7(1), 67–72 (1981)

29. Tarjan, R.: Depth-first search and linear graph algorithms. SIAM Journal on Computing 1(2), 146–160 (1972)

30. Vardi, M.Y., Wolper, P.: Reasoning about infinite computations. Information and Computation 115, 1–37 (1994)

31. Vardi, M., Wolper, P.: An automata-theoretic approach to automatic program verification. In: Proceedings of the 1st Symposium on Logic in Computer Science, pp. 322–331 (1986)

32. Wadler, P.: Comprehending monads. In: Mathematical Structures in Computer Science, pp. 61–78 (1992)

Recursive Functions on Lazy Lists via Domains and Topologies

Andreas Lochbihler[1] and Johannes Hölzl[2]

[1] Institute of Information Security, ETH Zurich, Switzerland
andreas.lochbihler@inf.ethz.ch
[2] Institut für Informatik, TU München, Germany
hoelzl@in.tum.de

Abstract. The usual definition facilities in theorem provers cannot handle all recursive functions on lazy lists; the filter function is a prime counterexample. We present two new ways of directly defining functions like filter by exploiting their dual nature as producers and consumers. Borrowing from domain theory and topology, we define them as a least fixpoint (producer view) and as a continuous extension (consumer view). Both constructions yield proof principles that allow elegant proofs. We expect that the approach extends to codatatypes with finite truncations.

1 Introduction

Coinductive datatypes (codatatypes for short) are popular in theorem provers [4,5,8,16,18,20], especially to formalise different forms of computation. Possibly infinite (lazy) lists, the most prominent example, are used to e.g. model traces of finite and infinite executions [17]. Today, Isabelle/HOL has a definitional package to construct codatatypes and define primitively corecursive functions [5].

codatatype α llist $= [] \mid \alpha \cdot \alpha$ llist

Yet, not all functions of interest are primitively corecursive; and the definition facilities based on well-founded recursion [13,21] cannot handle them either, when they produce infinite codatatype values by infinite corecursion. Hence, such functions have to be defined by other means. In this paper, we consider recursive functions that are notoriously hard to define [8], because their recursive specification does not uniquely determine them. In particular, we focus on the best-known example lfilter given by the specification (SPEC).[1]

$$\begin{aligned}
\text{lfilter } P \; [] &= [] \\
\text{lfilter } P \; (x \cdot \overline{xs}) &= (\text{if } P \; x \text{ then } x \cdot \text{lfilter } P \; \overline{xs} \text{ else lfilter } P \; \overline{xs})
\end{aligned} \qquad \text{(SPEC)}$$

Eq. (SPEC) is not primitively corecursive, as no constructor guards the second recursive call. Neither can well-founded recursion handle it, as an infinite list $\overline{xs}$

---

[1] We prefix functions on lazy lists with l to distinguish them from their counterpart on finite lists; variables for lazy lists carry overbars $\overline{xs}$, for finite lists underbars $\underline{xs}$.

G. Klein and R. Gamboa (Eds.): ITP 2014, LNAI 8558, pp. 341–357, 2014.

$$\mathsf{lfilter}\ P\ (\mathsf{lfilter}\ Q\ \overline{xs}) = \mathsf{lfilter}\ (\lambda x.\, P\ x \wedge Q\ x)\ \overline{xs} \qquad \text{(CONJ)}$$

$$\mathsf{lset}\ (\mathsf{lfilter}\ P\ \overline{xs}) = \mathsf{lset}\ \overline{xs} \cap \{\, x \mid P\ x \,\} \qquad \text{(LSET)}$$

$$\mathsf{lfilter}\ P\ \overline{xs} = [] \longleftrightarrow \forall x \in \mathsf{lset}\ \overline{xs}.\, \neg\, P\ x \qquad \text{(NIL)}$$

$$\mathsf{ldistinct}\ \overline{xs} \longrightarrow \mathsf{ldistinct}\ (\mathsf{lfilter}\ P\ \overline{xs}) \qquad \text{(LDISTINCT)}$$

$$\mathsf{lrel}\ R\ \overline{xs}\ \overline{ys} \wedge (\forall x\ y.\, R\ x\ y \longrightarrow (P_1\ x \longleftrightarrow P_2\ y))$$
$$\longrightarrow \mathsf{lrel}\ R\ (\mathsf{lfilter}\ P_1\ \overline{xs})\ (\mathsf{lfilter}\ P_2\ \overline{ys}) \qquad \text{(LREL)}$$

Fig. 1. Proven properties of lfilter

causes infinite recursion. Nor does (SPEC) fully specify lfilter: for $P = (\lambda\_.\, \mathsf{False})$ and $\overline{xs} = x \cdot x \cdot x \cdots$ the infinite repetition of some x e.g. (SPEC) collapses to the vacuous condition lfilter $P\ \overline{xs} = $ lfilter $P\ \overline{xs}$, i.e. lfilter $P\ \overline{xs}$ could be any lazy list.

As HOL functions are total anyway, HOL users often "totalise" a function with a partial specification such that proving becomes easier. Following this tradition, we want to define lfilter such that lfilter $P\ \overline{xs} = []$ whenever $\overline{xs}$ contains no elements satisfying P. This way, equations like (CONJ) and (LSET) in Fig. 1 hold unconditionally, even if $\overline{xs}$ is infinite and none of its elements satisfies P (and Q).

Of course, lfilter can be defined in an ad hoc fashion (see §5), but this has two drawbacks. First, one must come up with another construction for each new function. Second, such constructions typically lack a proof principle. Thus, proofs get cluttered with construction details as the definition must be unfolded.

In this work, we present two approaches to defining functions such as lfilter. They are inspired by two views on specifications like (SPEC). First, we can think of lfilter as corecursively producing a list lazily, i.e. a function of type $\beta \Rightarrow \alpha$ llist for some state type β; when another element is requested, it calls itself with an updated state. Borrowing ideas from domain theory, we turn α llist into a complete partial order, lift it point-wise to $\beta \Rightarrow \alpha$ llist, and take the least fixpoint of the functional associated with (SPEC) for lfilter (§2). Alternatively, we can also view lfilter as recursively consuming a lazy list, i.e. a function of type α llist $\Rightarrow \beta$ for some result type β. In §3, we therefore define lfilter on finite lists by (primitive) recursion and continuously extend it to infinite lists via topological limits.

Clearly, the two approaches require more machinery than ad hoc constructions. But more importantly, both approaches yield proof principles: either structural induction on lazy lists and fixpoint induction (§2.3) or uniqueness of limits and convergence on a closed set (§3.2). They allow elegant proofs with a high degree of automation. To show them in action, we prove the five exemplary properties listed in Fig. 1. Since lfilter both produces and consumes a lazy list, it is a good example to compare the two approaches. We do so in §4.

In this paper, we focus on lfilter, but we have defined more functions on lazy lists this way. Our approach simplifies their (formerly ad hoc) definitions and the proofs in an existing codatatype library [16]. We expect that our approach generalises to a large class of codatatypes (§6) and can be ported to other systems.

2   The Producer View: Least Fixpoints

In this section, we formalise lfilter as the least fixpoint solution to (SPEC). This construction views lfilter as a function that produces a lazy list. First, we define lfilter as a least fixpoint (§2.2) borrowing ideas from domain theory (introduced in §2.1). Next, we set up the infrastructure for the induction proofs (§2.3). Finally, we show how to prove the five properties of lfilter (§2.4) listed in Fig. 1.

2.1   Background on Orders and Fixpoints

In this section, we review some domain theory formalised in plain HOL [14].

An *order* $\leq$ for a given type is a binary relation that is reflexive, transitive, and antisymmetric. Given an order $\leq$, a *chain* Y is a set whose elements are all related in $\leq$ (predicated by chain $(\leq)\ Y$). An order $\leq$ on some type α and a function $\bigvee :: \alpha$ set $\Rightarrow \alpha$ form a *chain-complete partial order* (ccpo) iff $\bigvee Y$ denotes the least upper bound (lub) of every chain Y w.r.t. $\leq$, i.e. for all Y with chain $(\leq)\ Y$, if $x \in Y$, then $x \leq \bigvee Y$, and whenever $x \leq z$ for all $x \in Y$, then $\bigvee Y \leq z$. As the empty set is a chain, every ccpo has a least element bottom $\bot = \bigvee \emptyset$. For example, the type of sets α set ordered by inclusion $\subseteq$ forms a ccpo with lub $\bigcup Y$ and bottom $\emptyset$. An order $\leq$ is lifted pointwise to functions: $f \uparrow \leq g$ denotes $\forall x.\ f\ x \leq g\ x$. Analogously, the lub $\uparrow\bigvee Y$ on a chain Y of functions is determined pointwise: $\uparrow\bigvee Y\ x = \bigvee\{f\ x.f \in Y\}$. If $(\bigvee, \leq)$ is a ccpo, so is $(\uparrow\bigvee, \uparrow\leq)$. Similarly, $\leq\times\leq'$ orders pairs component-wise according to $\leq$ and $\leq'$, resp.; and $(\bigvee\times\bigvee')Y = (\bigvee(\pi_1 \lq Y), \bigvee'(\pi_2 \lq Y))$ computes the lub component-wise. Here, π_1 and π_2 are the projections, and $f \lq A$ denotes the image of the set A under the function f. If $(\bigvee, \leq)$ and $(\bigvee', \leq')$ are ccpos, so is $(\bigvee\times\bigvee', \leq\times\leq')$.

A function f is *monotone* w.r.t. $\leq$ and $\leq'$ (written mono $(\leq)\ (\leq')\ f$) iff $f\ x \leq' f\ y$ for all $x,\ y$ with $x \leq y$. A monotone function f is *(order) continuous* w.r.t. $(\bigvee, \leq)$ and $(\bigvee', \leq')$ iff it preserves lubs of non-empty chains (written mcont $(\bigvee, \leq)\ (\bigvee', \leq')\ f$). Formally, $f\ (\bigvee Y) = \bigvee'(f \lq Y)$ for all Y with chain $(\leq)\ Y$ and $Y \neq \emptyset$. A continuous function f is *strict* iff it propagates $\bot$, i.e. $f\ (\bigvee \emptyset) = \bigvee'\emptyset$. A predicate P is *admissible* (written adm $(\bigvee, \leq)\ P$) iff $P\ (\bigvee Y)$ for all non-empty chains Y such that $P\ x$ for all $x \in Y$. Admissibility is closed under composition with continuous functions, i.e. if adm $(\bigvee, \leq)\ (\lambda x.\ P\ x)$ and mcont $(\bigvee', \leq')\ (\bigvee, \leq)\ f$, then adm $(\bigvee', \leq')\ (\lambda x.\ P\ (f\ x))$.

Let $F :: \alpha \Rightarrow \alpha$ be a monotone function on a ccpo $(\bigvee, \leq)$. Then, by the Knaster-Tarski fixpoint theorem, F has a *least fixpoint* fixp $(\bigvee, \leq)\ F$, which is given by the lub of the transfinite iteration of F starting at $\bot$.

2.2   Definition

As mentioned in §1, we define lfilter P as the least fixpoint of the functional associated to the specification (SPEC); since lfilter passes the predicate P unchanged to the recursive calls, we treat it as a fixed parameter. Thus, we obtain the functional $\mathsf{F}_P :: (\alpha$ llist $\Rightarrow \alpha$ llist$) \Rightarrow (\alpha$ llist $\Rightarrow \alpha$ llist$)$ given by

$$\mathsf{F}_P\ f\ \overline{xs} = (\text{case } \overline{xs} \text{ of } [] \Rightarrow [] \mid x \cdot \overline{xs'} \Rightarrow \text{if } P\ x \text{ then } x \cdot f\ \overline{xs'} \text{ else } f\ \overline{xs'}) \quad (1)$$

For the Knaster-Tarski fixpoint theorem, we need a ccpo on α llist $\Rightarrow \alpha$ llist for which F_P is monotone. It suffices to provide one for α llist and lift it point-wise to functions with codomain α llist. We choose the prefix order $\sqsubseteq$, which (2) defines coinductively. The least upper bound $\bigsqcup :: \alpha$ llist set $\Rightarrow \alpha$ llist is given by primitive corecursion (3). Here, lhd and ltl return the head and tail of a lazy list, resp.; and the definite descriptor $\iota x.\, P\, x$ denotes the unique x such that $P\, x$ if it exists and is unspecified otherwise. We show the ccpo properties for $(\bigsqcup, \sqsubseteq)$ by (rule or structural) coinduction.

$$\frac{}{[] \sqsubseteq \overline{ys}} \qquad \frac{\overline{xs} \sqsubseteq \overline{ys}}{x \cdot \overline{xs} \sqsubseteq x \cdot \overline{ys}} \qquad (2)$$

$$\bigsqcup Y = (\text{if } Y \subseteq \{[]\} \text{ then } [] \\ \qquad\quad \text{else let } Y' = \{\overline{xs} \in Y.\, \overline{xs} \neq []\} \text{ in } (\iota x.\, x \in \mathsf{lhd}\,{}^{\backprime}\, Y') \cdot \bigsqcup(\mathsf{ltl}\,{}^{\backprime}\, Y')) \qquad (3)$$

The prefix order is a natural choice, as it makes the constructor $\_ \cdot \_$ monotone in the recursive argument. Monotonicity is crucial for the existence of the fixpoint (see below). Moreover, the least element $[]$ carries the least information possible. In fact, $\sqsubseteq$ corresponds to the approximation order on the domain of infinite streams, when we interpret $[]$ as "undefined", the additional value that each domain contains. In this view, a finite lazy list represents the set of all its extensions at the end, and this set shrinks when we extend it.

Now, we define lfilter as the least fixpoint of F_P in the ccpo $(\uparrow\bigsqcup, \uparrow\sqsubseteq)$ using the **partial-function** package by Krauss [14]. Given the specification (SPEC) as input, it constructs the functional F_P, proves monotonicity, defines lfilter as the least fixpoint, and derives (SPEC) and a fixpoint induction rule (4) from the definition. The monotonicity proof decomposes the functional syntactically into primitive operations and uses their monotonicity properties. For F_P, we provide the monotonicity theorem for $\cdot$ as a hint, which follows directly from (2).

This completes our first definition of lfilter. After some preparations (§2.3), we prove in §2.4 that lfilter is in fact the desired solution for (SPEC).

2.3   Preparations for Proofs by Induction

Least fixpoints and the ccpo structure on lazy lists provide two induction proof principles, which we review now. Every least fixpoint definition generates an induction rule; the one for lfilter is shown in (4). The second premise requires that the statement Q to be proved holds for the least function $\lambda\_.\, []$ where the fixpoint iteration starts; and in the inductive step (third premise), some underapproximation f replaces the function lfilter P. Admissibility (first premise) ensures that Q is preserved when taking the lub of the iteration for the fixpoint.

$$\frac{\mathsf{adm}\,(\uparrow\bigsqcup, \uparrow\sqsubseteq)\, Q \qquad Q\,(\lambda\_.\, []) \qquad \forall f.\, Q\, f \wedge f \uparrow\sqsubseteq \mathsf{lfilter}\, P \longrightarrow Q\,(\mathsf{F}_P\, f)}{Q\,(\mathsf{lfilter}\, P)} \qquad (4)$$

Fixpoint induction corresponds to the producer view, as it assumes nothing about the parameter; rather, f in the inductive step returns a prefix of lfilter P.

Alternatively, structural induction over a lazy list (5) is available. The inductive cases (second and third premise) yield that the property P holds for all finite lists (predicate lfinite). Admissibility (first premise) ensures that P also holds for the whole list $\overline{xs}$, as all finite prefixes of $\overline{xs}$ form a chain with lub $\overline{xs}$. Clearly, structural induction takes the consumer point of view, because in typical use cases, it acts on a variable $\overline{xs}$ that a function takes as argument.

$$\frac{\mathsf{adm}\ (\bigsqcup, \sqsubseteq)\ P \qquad P\ [] \qquad \forall x\ \overline{xs}.\ \mathsf{lfinite}\ \overline{xs} \wedge P\ \overline{xs} \longrightarrow P\ (x \cdot \overline{xs})}{P\ \overline{xs}} \tag{5}$$

Both induction principles require that the inductive statement is admissible. Müller et al. [19] have already noted in the context of Isabelle's LCF formalisation HOLCF that admissibility is often harder to prove than the inductive steps. Huffman [12] describes the syntax-directed approach to automate these proofs. Proof rules such as (6) first decompose the statement into atoms along the logical connectives. Others then separate each atom into a predicate and its arguments (interpreted as a function of the induction variable). If the arguments are continuous, it suffices to show admissibility of the predicate; HOLCF includes admissibility rules for comparisons and (in)equalities. This approach works well in practice, because all HOLCF functions are continuous by construction.

$$\frac{\mathsf{adm}\ (\bigvee, \leq)\ (\lambda x.\, \neg P\ x) \qquad \mathsf{adm}\ (\bigvee, \leq)\ (\lambda x.\, Q\ x)}{\mathsf{adm}\ (\bigvee, \leq)\ (\lambda x.\, P\ x \longrightarrow Q\ x)} \tag{6}$$

We have proved a similar set of syntax-directed proof rules. They achieve a comparable degree of automation for discharging admissibility conditions. However, they require some manual setup, in particular continuity proofs. We will discuss this now at three examples, namely lfilter, lmap and lset.

First, we prove that lfilter P is continuous. This will allow us later to switch from the producer to the consumer view, i.e. from (4) to (5). As we have defined lfilter P as a least fixpoint, we can leverage the general result that least fixpoints preserve monotonicity and continuity (Thm. 1).

Theorem 1 (Least fixpoints preserve monotonicity and continuity). *Let $(\bigvee, \leq)$ be a ccpo, let $F :: (\beta \Rightarrow \alpha) \Rightarrow (\beta \Rightarrow \alpha)$ satisfy mono $(\uparrow \leq)\ (\uparrow \leq)\ F$. If F preserves monotonicity (continuity), then F's least fixpoint is monotone (continuous). Formally, let g abbreviate fixp $(\uparrow \bigvee, \uparrow \leq)\ F$. If mono $(\leq')\ (\leq)\ (F\ f)$ for all f with mono $(\leq')\ (\leq)\ f$, then mono $(\leq')\ (\leq)\ g$. If mcont $(\bigvee', \leq')\ (\bigvee, \leq)\ (F\ f)$ for all f with mcont $(\bigvee', \leq')\ (\bigvee, \leq)\ f$, then mcont $(\bigvee', \leq')\ (\bigvee, \leq)\ g$.*

Hence, it suffices to show that F_P in (1) is monotone and continuous in $\overline{xs}$ provided that f is so, too. Like for admissibility, we follow a syntax-directed decomposition approach, as continuity is preserved under function composition. We prove rules that decompose the expression into individual functions and then show that they themselves are continuous. Unfortunately, control constructs like case and if are in general neither monotone nor continuous if the branching term depends on the argument. In (1), this is the case for the case combinator.

As we frequently prove functions on α llist strict, we derive a specialised continuity rule (7) (and an analogous monotonicty rule) for an arbitrary ccpo $(\bigvee, \leq)$ with bottom $\bot$. (It cannot handle non-strict functions like $++$ defined in (27).)

$$\frac{\forall x.\, \mathsf{mcont}\ (\bigsqcup, \sqsubseteq)\ (\bigvee, \leq)\ (\lambda \overline{ys}.\, f\ x\ \overline{ys}\ (x \cdot \overline{ys}))}{\mathsf{mcont}\ (\bigsqcup, \sqsubseteq)\ (\bigvee, \leq)\ (\lambda \overline{xs}.\, \mathsf{case}\ \overline{xs}\ \mathsf{of}\ [] \Rightarrow \bot \mid x \cdot \overline{ys} \Rightarrow f\ x\ \overline{ys}\ \overline{xs})} \tag{7}$$

By (7), it suffices to show for all x that $\lambda \overline{ys}.\, \mathsf{if}\ P\ x\ \mathsf{then}\ x \cdot f\ \overline{ys}\ \mathsf{else}\ f\ \overline{ys}$ is monotone and continuous if f already is. Note that the branching condition $P\ x$ no longer depends on the bound variable $\overline{ys}$, so it suffices to prove that the individual branches are monotone and continuous; rule (8) formalises this.

$$\frac{\mathsf{mcont}\ (\bigvee, \leq)\ (\bigvee', \leq')\ f \qquad \mathsf{mcont}\ (\bigvee, \leq)\ (\bigvee', \leq')\ g}{\mathsf{mcont}\ (\bigvee, \leq)\ (\bigvee', \leq')\ (\lambda x.\, \mathsf{if}\ c\ \mathsf{then}\ f\ x\ \mathsf{else}\ g\ x)} \tag{8}$$

Finally, we are left with proving that $\lambda \overline{ys}.\, x \cdot f\ \overline{ys}$ and $\lambda \overline{ys}.\, f\ \overline{ys}$ are monotone and continuous, which follows immediately from $\cdot$ and f being so. Although this proof seems lengthy on paper, it is a one-liner in Isabelle, as its rewriting engine performs the decomposition automatically thanks to the setup outlined above.

The above illustrates how to prove continuity of functions defined in terms of fixp. Other functions are defined by other means, but we want to prove them continuous, too. For example, the codatatype packages defines lmap f, which applies f to all elements of a lazy list (9), and lset, which converts a lazy list to the set of its elements, but not in terms of fixp. The easiest way to prove continuity is to show that they are the least fixpoint of the functionals in (10) and (11), resp. Then, we reuse Thm. 1 and our machinery from above.

$$\mathsf{lmap}\ f\ [] = [] \qquad\qquad \mathsf{lmap}\ f\ (x \cdot \overline{xs}) = f\ x \cdot \mathsf{lmap}\ f\ \overline{xs} \tag{9}$$

$$\mathsf{M}_f\ g\ \overline{xs} = (\mathsf{case}\ \overline{xs}\ \mathsf{of}\ [] \Rightarrow [] \mid x \cdot \overline{ys} \Rightarrow f\ x \cdot g\ \overline{ys}) \tag{10}$$

$$\mathsf{S}\ f\ \overline{xs} = (\mathsf{case}\ \overline{xs}\ \mathsf{of}\ [] \Rightarrow \emptyset \mid x \cdot \overline{ys} \Rightarrow \{x\} \cup f\ \overline{ys}) \tag{11}$$

The proofs for lmap $f = \mathsf{fixp}\ (\uparrow\bigsqcup, \uparrow\sqsubseteq)\ \mathsf{M}_f$ and lset $= \mathsf{fixp}\ (\uparrow\bigcup, \uparrow\subseteq)\ \mathsf{S}$ fall into two parts: (i) monotonicity of M_f and S is shown by **partial-function**'s monotonicity prover and (ii) the actual fixpoint equation by the proof principle associated with the definition (structural coinduction for lmap; lset requires two separate directions with induction on lset and fixpoint induction, resp.). Monotonicity is needed to unfold the fixpoint property in the (co)inductive steps.

2.4   Proving the Properties

With all these preparations in place, we now show how they yield concise proofs for the properties of interest. We start with (NIL), i.e. that the least fixpoint indeed picks the desired solution for (SPEC). First, we illustrate the obvious approach of proving the two directions separately. From right to left, given $\neg P\ x$ for all $x \in \mathsf{lset}\ \overline{xs}$, we must show $\mathsf{lfilter}\ P\ \overline{xs} = []$, or, equivalently, $\mathsf{lfilter}\ P\ \overline{xs} \sqsubseteq []$.

Structural coinduction does not work here, as (SPEC) may recurse forever, but fixpoint induction is good at proving upper bounds, $[]$ in our case. Admissibility of $\lambda f. \forall \overline{xs}. (\forall x \in \mathsf{lset}\ \overline{xs}. \neg P\ x) \longrightarrow f\ \overline{xs} \sqsubseteq []$ follows directly from decomposition and admissibility of comparisons. In contrast, from left to right by contraposition, we have to prove the non-trivial lower bound $[] \sqsubset \mathsf{lfilter}\ P\ \overline{xs}$ under the assumption $P\ x$ for some $x \in \mathsf{lset}\ \overline{xs}$. Fixpoint induction cannot do this, so we resort to other proof principles. Fortunately, induction on $x \in \mathsf{lset}\ \overline{xs}$ is available, and the cases are solved automatically by rewriting.

Alternatively, we can switch to the consumer view and prove (NIL) directly by induction on $\overline{xs}$ using (5). Rewriting solves the inductive cases. Regarding admissibility of $\lambda \overline{xs}. \mathsf{lfilter}\ P\ \overline{xs} = [] \longleftrightarrow \forall x \in \mathsf{lset}\ \overline{xs}. \neg P\ x$, the rules decompose it into four atoms: $\lambda \overline{xs}. \mathsf{lfilter}\ P\ \overline{xs} \neq []$ and $\lambda \overline{xs}. \mathsf{lfilter}\ P\ \overline{xs} = []$ and $\lambda \overline{xs}. \forall x \in \mathsf{lset}\ \overline{xs}. \neg P\ x$ and $\lambda \overline{xs}. \exists x \in \mathsf{lset}\ \overline{xs}. P\ x$. For the (in)equalities and bounded quantifiers, we have admissibility rules, and their arguments $\mathsf{lfilter}\ P$ and lset are continuous by §2.3. Therefore, this proof of (NIL) is automatic.

lemma $\mathsf{lfilter}\ P\ \overline{xs} = [] \longleftrightarrow \forall x \in \mathsf{lset}\ \overline{xs}. \neg P\ x$ **by**(induction $\overline{xs}$) simp-all

Next, we prove property (CONJ) from the introduction. Taking the consumer view, the proof is a one-liner by induction on $\overline{xs}$ plus rewriting, because we have already shown that $\mathsf{lfilter}\ P$ is continuous. Fixpoint induction can also prove (CONJ), but the two directions "$\sqsubseteq$" and "$\sqsupseteq$" must be shown separately. Moreover, we still need continuity of $\mathsf{lfilter}\ P$ for admissibility, because when going from left to right, we have to replace $\mathsf{lfilter}\ Q$ in the context $\forall \overline{xs}. \mathsf{lfilter}\ P\ (\bullet\ \overline{xs}) \sqsubseteq \dots$.

Property (LSET) is similar to (CONJ). We show it by induction on $\overline{xs}$; admissibility requires continuity of lset, $\mathsf{lfilter}$, and $\cap$. Fixpoint induction is also possible.

In the remainer of this section, we prove two more properties with user-defined predicates. The predicate $\mathsf{ldistinct}$ denotes that all elements of a lazy list are distinct, and the relator $\mathsf{lrel}\ R\ \overline{xs}\ \overline{ys}$ lifts a binary relation R point-wise to the lazy lists $\overline{xs}$ and $\overline{ys}$. The rules below define them coinductively.

$$\frac{}{\mathsf{ldistinct}\ []} \qquad \frac{x \notin \mathsf{lset}\ \overline{xs} \qquad \mathsf{ldistinct}\ \overline{xs}}{\mathsf{ldistinct}\ (x \cdot \overline{xs})} \tag{12}$$

$$\frac{}{\mathsf{lrel}\ R\ []\ []} \qquad \frac{R\ x\ y \qquad \mathsf{lrel}\ R\ \overline{xs}\ \overline{ys}}{\mathsf{lrel}\ R\ (x \cdot \overline{xs})\ (y \cdot \overline{ys})} \tag{13}$$

Proofs by induction require admissibility of the statement. As $\mathsf{ldistinct}$ is a new predicate, we prove admissibility directly by unfolding the definition and by coinduction on $\mathsf{ldistinct}$. The proof for lrel is similar. Moreover, we also show that non-distinctness is admissible; this follows from prefixes of distinct lists being distinct. Now, we are ready to show properties (LDISTINCT) and (LREL) from Fig. 1.

Taking the consumer view, we show (LDISTINCT) by induction on $\overline{xs}$; as $\overline{xs}$ occurs in the assumptions, rule (6) requires that the negated assumption, i.e. non-distinctness, be admissible, too (there is no rule for negation). The inductive steps are solved automatically, as we can rewrite $\mathsf{lset}\ (\mathsf{lfilter}\ P\ \overline{xs})$ with (LSET).

Alternatively, we can also take the producer view, i.e. fixpoint induction on $\mathsf{lfilter}$. This demonstrates another limitation of fixpoint induction: recall that fixpoint induction replaces lfilter P by some underapproximation f, i.e. we cannot use (LSET) for rewriting lset ($f\ \overline{xs}$). Fortunately, we get $f\ \uparrow\sqsubseteq$ lfilter P in the inductive step and derive lset ($f\ \overline{xs}$) $\subseteq$ lset $\overline{xs}$ by monotonicity of lset. Otherwise, we would have had to re-prove (LSET) simultaneously in the inductive step. This modularity problem frequently arises with fixpoint induction: all required properties of a function have to be threaded through one big induction, which incurs losses in proof automation and processing speed.

Finally, consider (LREL). Note that the property of not being related in lrel is not admissible. This means that the decomposition rules do not work if the induction variable under lrel in an assumption. Thus, we cannot induct over $\overline{xs}$ (unless we prove admissibility manually, but we would rather not). We use fixpoint induction instead. Yet, the two occurrences of lfilter in (LREL) have different types. As (4) replaces only occurrences of the same type, we resort to parallel fixpoint induction. The general parallel fixpoint induction rule for two ccpos ($\bigvee, \leq$) and ($\bigvee', \leq'$) with least elements $\bot$ and $\bot'$ and two monotone functionals F and G is shown below. Since the projections π_1 and π_2 are monotone and continuous, the parallel fixpoint induction proof becomes fully automatic again.

$$\frac{\mathsf{adm}\ (\bigvee\!\times\!\bigvee', \leq\times\leq')\ (\lambda x.\, P\,(\pi_1\, x)\,(\pi_2\, x))\quad P\,\bot\,\bot'\quad \forall x\, y.\, P\, x\, y \longrightarrow P\,(F\, x)\,(G\, y)}{P\,(\mathsf{fixp}\ (\bigvee, \leq)\ F)\,(\mathsf{fixp}\ (\bigvee', \leq')\ G)}$$

3   The Consumer View: Continuous Extensions

Some proofs about lfilter in §2.4 already took the consumer point of view. Now, we do so also for defining lfilter. In general, we first define a function on finite lists α list and then extend it to lazy lists. For the running example, we first define filter :: ($\alpha \Rightarrow$ bool) $\Rightarrow \alpha$ list $\Rightarrow \alpha$ list on finite lists using primitive recursion (14). Then, we define lfilter as the *continuous extension* of filter (15).

$$\begin{aligned}
\text{filter } P\ [] &= [] \\
\text{filter } P\ (x \cdot \underline{xs}) &= (\text{if } P\ x \text{ then } x \cdot \text{filter } P\ \underline{xs} \text{ else filter } P\ \underline{xs})
\end{aligned} \tag{14}$$

$$\text{lfilter } P\ \overline{xs} = \bigsqcup\{\lceil\text{filter } P\ \lfloor\overline{ys}\rfloor\rceil \mid \overline{ys} \in {\downarrow}\overline{xs}\} \tag{15}$$

where ${\downarrow}\overline{xs} = \{\,\overline{ys} \mid \text{lfinite } \overline{ys} \wedge \overline{ys} \sqsubseteq \overline{xs}\,\}$ denotes the set of finite prefixes of $\overline{xs}$, $\lceil\_\rceil$ embeds finite lists in lazy lists, and $\lfloor\_\rfloor$ is its inverse. This construction yields the same function as the least fixed point in §2.2—see §3.4 for the proofs.

Why do we call this a continuous extension? To generalise this construction method, we introduce a topology on lazy lists with two properties (§§3.2, 3.3). First, every chain of finite lists "converges" towards a lazy (possibly finite) list. Second, every lazy list can be "approximated" by a set of finite lists. Hence, continuous extensions are unique if they exist. So, we extend a function f :: α list $\Rightarrow \beta$ to a function lf :: α llist $\Rightarrow \beta$ by picking the continuous one. This also explains why this is the consumer view: the codatatype is an argument to the function, and the codomain is an arbitrary topology. For unique extensions, the codomain must be a T2 topology.

3.1 Topology in Isabelle/HOL

This section summarises the formalisation of topologies in Isabelle/HOL [11].[2]

A *topology* is specified by the *open sets* (predicate open). In a topology, the whole space must be open (its elements are called *points*), and binary intersection and arbitrary union must preserve openness. A predicate P is a *neighbourhood* of a point x if it holds on an open set which contains x. A *punctured neighbourhood* P of x (written P at x) is a neighbourhood of x which not necessary holds on x. A point x is *discrete* iff $\{x\}$ is open. A topology is called a *T2 space*, if for every two points $x \neq y$ there exists two disjoint neighbourhoods P_x at x and P_y at y.

A function f *converges* on a point x (written $f \xrightarrow{\ x\ } y$) iff for all open sets Y around y the predicate $\lambda x.\ f\ x \in Y$ is a punctured neighbourhood of x. The function f is *continuous* at x iff $f \xrightarrow{\ x\ } f\ x$. Clearly, convergence is meaningless for discrete points x, as $\{x\}$ is open. Also, each f is then continuous at x.

A set is *closed* iff its complement is open, a predicate P is closed iff $\{x \mid P\ x\}$ is closed. Closedness of predicates is preserved under composition with continuous functions. *Convergence on a closed set* (16) is our main proof principle. If the predicate $P \circ f$ is a punctured neighbourhood of x and P is closed (closed $\{x \mid P\ x\}$), then P also holds at the point x itself, unless x is discrete.

$$\frac{\neg\,\mathsf{open}\ \{x\} \qquad f \xrightarrow{\ x\ } y \qquad \mathsf{closed}\ \{x \mid P\ x\} \qquad P \circ f\ \mathsf{at}\ x}{P\ y} \tag{16}$$

3.2 Topology on a Chain-Complete Partial Order

In this section, we introduce a topology for ccpos. In the ccpo topology, an open set is not accessible from outside, i.e. whenever the least upper bound of a non-empty chain is in the open set, then their intersection is not empty (17).

$$\mathsf{open}\ S \longleftrightarrow (\forall C.\ \mathsf{chain}\ C \longrightarrow C \neq \emptyset \longrightarrow \bigvee C \in S \longrightarrow C \cap S \neq \emptyset) \tag{17}$$

It differs from the usual Scott topology only in that open sets need not be upward closed. We omit this condition for two reasons: (i) we need a T2 space, but the Scott topology is not, and (ii) finite lists should be discrete, i.e. open $\{\overline{xs}\}$ if $\overline{xs}$ is finite. Every ccpo topology is a T2 space, since open as defined in (17) fulfills the topology axioms and separation of points.

As mentioned in §3.1, convergence $f \xrightarrow{\ x\ } y$ ignores the value of f at the point x. Thus, if x is discrete, convergence is meaningless at this point. To avoid this issue, we introduce variants of convergence and punctured neighbourhood.[3]

$$f \xrightarrow{\ x\ }' y \longleftrightarrow \text{if open } \{x\} \text{ then } f\ x = y \text{ else } f \xrightarrow{\ x\ } y \tag{18}$$

$$P\ \mathsf{at}'\ x \longleftrightarrow \text{if open } \{x\} \text{ then } P\ x \text{ else } P\ \mathsf{at}\ x$$

---

[2] As the topology formalisation relies on type classes, we now switch to type classes for ccpos, too. Hence, we no longer write the ccpo $(\bigvee, \leq)$ as a parameter for constants like adm and mcont. Instead, it is taken from the type class.

[3] The formalisation of convergence in Isabelle/HOL uses topological filters for the argument, as described in [11]. The punctured neighbourhoods at and at' are topological filters, but for a shorter presentation we avoid their introduction.

For continuity, both limits are equivalent: $f \xrightarrow{\ x\ } f\ x$ iff $f \xrightarrow{\ x\ }' f\ x$. For at' we get a stronger variant of (16) as proof principle: No matter if x is discrete, the closed predicate P holds on x if P is a punctured neighbourhood of x.

$$\frac{\mathsf{closed}\ \{x \mid P\ x\} \qquad P\ \mathsf{at}'\ x}{P\ x} \tag{19}$$

When the convergence limit exists, we select it with definite description: $\mathsf{Lim}\ f\ x = \iota y.\ f \xrightarrow{\ x\ }' y$. As a ccpo topology is a T2 space, the limit is unique.

3.3 Constructing lfilter

As lazy lists are a ccpo, they also form a ccpo topology as described in §3.2. We first observe that the finite lists are dense in this topology, i.e. every lazy list is the limit of a sequence of finite lists. Moreover, a lazy list is discrete iff it is finite: $\mathsf{open}\ \{\overline{xs}\} \longleftrightarrow \mathsf{lfinite}\ \overline{xs}$. This yields a nice characterization of at' (20), from which we easily derive that at' behaves as expected on the constructor $\_\cdot\_$ (21).

$$P\ \mathsf{at}'\ \overline{xs} \longleftrightarrow \exists \overline{ys} \in {\downarrow}\overline{xs}.\ \forall \overline{zs} \in {\downarrow}\overline{xs}.\ \overline{ys} \sqsubseteq \overline{zs} \longrightarrow P\ \overline{zs} \tag{20}$$

$$(\lambda \overline{ys}.\ x \cdot \overline{ys}) \xrightarrow{\ \overline{xs}\ }' x \cdot \overline{xs} \qquad\qquad \frac{(\lambda \overline{zs}.\ f\ (x \cdot \overline{zs})) \xrightarrow{\ \overline{xs}\ }' y}{f \xrightarrow{\ x \cdot \overline{xs}\ }' y} \tag{21}$$

Hence, at' behaves as expected on finite and infinite lists. Thus, we define $\mathsf{lfilter}\ P\ \overline{xs}$ as the limit of $\mathsf{filter}\ P$:

$$\mathsf{lfilter}\ P\ \overline{xs} = \mathsf{Lim}\ (\lambda \overline{ys}.\ \lceil \mathsf{filter}\ P\ \lfloor \overline{ys} \rfloor \rceil)\ \overline{xs} \tag{22}$$

Before proving lfilter's properties, we must prove that it continuously extends filter. Extension (23) shows that they coincide on finite lists. This follows from (18) and uniqueness of limits by unfolding the definitions of lfilter and Lim.

$$\mathsf{lfinite}\ \overline{xs} \longrightarrow \mathsf{lfilter}\ P\ \overline{xs} = \lceil \mathsf{filter}\ P\ \lfloor \overline{xs} \rfloor \rceil \tag{23}$$

Then, we show that lfilter is continuous everywhere (25). It suffices to show that the limit exists, as uniqueness of limits then ensures continuity. To that end, we prove the theorem (24): if a function f is monotone on all finite lazy lists, then it converges on $\overline{xs}$ to the lub of the image of $\overline{xs}$'s finite prefixes under f. This also completes the proof, as filter is monotone. Our initial definition (15) follows from these rules.

$$\frac{\forall \overline{ys}\ \overline{zs}.\ \overline{ys} \sqsubseteq \overline{zs} \wedge \mathsf{lfinite}\ \overline{zs} \longrightarrow f\ \overline{ys} \leq f\ \overline{zs}}{f \xrightarrow{\ \overline{xs}\ }' \bigvee(f\ \text{`}\ {\downarrow}\overline{xs})} \tag{24}$$

$$\mathsf{lfilter}\ P \xrightarrow{\ \overline{xs}\ }' \mathsf{lfilter}\ P\ \overline{xs} \tag{25}$$

3.4 Proving with Topology

In this section, we prove that the definition in (22) satisfies the specification (SPEC) and the properties from Fig. 1. In general, reasoning about lfilter first reduces the property on lazy lists to a property on finite lists. The characterisation of at$'$ on lazy lists (20) yields the following proof principle. It is derived from (19) by taking $\overline{ys} = []$ as witness for the existential quantifier in (20).

$$\frac{\text{closed } \{\overline{xs} \mid P\ \overline{xs}\} \qquad \forall \overline{zs} \in {\downarrow}\overline{xs}.\ P\ \overline{zs}}{P\ \overline{xs}} \tag{26}$$

This proof rule splits a goal $P\ \overline{xs}$ into two subgoals: (i) closed $\{\overline{xs} \mid P\ \overline{xs}\}$ and (ii) $\forall \overline{zs} \in {\downarrow}\overline{xs}.\ P\ \overline{zs}$. Closedness is usually proved automatically in two steps. First, $P\ xs$ is decomposed into an atomic predicate and functions. These are then shown closed and continuous using pre-proven theorems such as closedness of equality in a T2 space (§3.1) and continuity of lfilter (25). In subgoal (ii), ${\downarrow}\overline{xs}$ consists only of finite lists. Hence, we have indeed reduced the statement from arbitrary lazy lists to their finite subset. This goal is proved either by induction on lfinite $\overline{zs}$, or by rewriting with equations like (23) into functions over finite lists. For proving the specification (SPEC) and the properties (CONJ, LSET), this approach suffices. We also use it to show that our two definitions of lfilter from (§2.2) and (22) are equivalent.

Note that the second goal keeps the prefix relation between $\overline{zs}$ and $\overline{xs}$. Crucially, this maintains the relation of subgoal (ii) to further assumptions that are not part of the predicate P. When we prove (LDISTINCT), we operate only on the conclusion ldistinct (lfilter $P\ \overline{xs}$). Closedness (subgoal (i)) follows from ldistinct being closed and lfilter being continuous by preservation of closedness under composition with continuous functions. Subgoal (ii) is

$$\text{ldistinct } \overline{xs} \longrightarrow \forall \overline{zs} \in {\downarrow}\overline{xs}.\ \text{ldistinct (lfilter } P\ \overline{zs}).$$

As prefixes of distinct lists are distinct, it suffices to show the following for all $\overline{zs}$.

$$\text{ldistinct } \overline{zs} \longrightarrow \text{lfinite } \overline{zs} \longrightarrow \text{ldistinct (lfilter } P\ \overline{zs})$$

Existing lemmas about filter and ldistinct suffice to show this, but induction on lfinite $\overline{zs}$ would work, too.

Property (NIL) is more complicated. The statement is not a closed predicate, so we cannot easily reduce it to finite lists. Instead we prove the direction from left to right using (LSET), and the converse using our approach from above.

4 Comparison

In this section, we compare our approaches least fixpoints (§2) and continuous extensions (§3) in five respects: the requirements on the codatatype and on the type of the function, the role of monotonicity, proof principles, and proof elegance.

Ccpo Structure on the Codatatype. Both approaches require a ccpo structure on the codatatype. As monotonicity is crucial for definitions and proofs (see below), functions of interest (and the constructors in particular) should be monotone. For lazy lists, the prefix order with $[]$ as the least element is a natural choice. The extended naturals enat $= 0 \mid$ eSuc enat are a ccpo under the usual ordering $\leq$. Even terminated lazy lists given by (α, β) tllist $=$ TNil $\beta \mid$ TCons α $((\alpha, \beta)$ tllist$)$ form a useful ccpo under the prefix ordering extended with TNil b as least element for any user-specified, but fixed b. Yet, we have not found useful ccpos for co-datatypes without finite values like infinite lists α stream $=$ Stream α $(\alpha$ stream$)$.

Type Restrictions on Function Definitions. For recursive definitions, the two approaches pose different requirements on the function. Least fixpoints need the ccpo on the codomain whereas the domain can be arbitrary. Therefore, this works for functions that produce a codatatype value such as iterate below. In contrast, they cannot handle functions that only consume a codatatype value such as lsum, which sums over a lazy list. Dually, continuous extensions require a ccpo topology on the domain whereas the codomain can be any T2 space. This works for functions that consume a codatatype value such as lsum, but this approach cannot define producers such as iterate.

$$\text{iterate } f \; x = x \cdot \text{iterate } f \; (f \; x)$$
$$\text{sum } [] = 0 \qquad \text{sum } (x \cdot \underline{xs}) = x + \text{sum } \underline{xs} \qquad \text{lsum } \overline{xs} = \text{Lim sum } \overline{xs}$$

Monotonicity. To derive the recursive specification from the definition, we have to show well-definedness for both approaches. For least fixpoints, the associated functional must be monotone, i.e. recursion may only occur in monotone contexts. For example, this approach cannot handle lmirror, because concatenation $++$ is not monotone in its first argument, which contains the recursive call.

$$[] ++ \overline{ys} = \overline{ys} \qquad\qquad (x \cdot \overline{xs}) ++ \overline{ys} = x \cdot (\overline{xs} ++ \overline{ys}) \qquad (27)$$
$$\text{lmirror } [] = [] \qquad\qquad \text{lmirror } (x \cdot \overline{xs}) = x \cdot (\text{lmirror } \overline{xs} ++ [x]) \qquad (28)$$

This shows how the choice of ccpo determines what functions can be defined. The **partial-function** package [14] automates the monotonicity proof and derives the recursive specification. Note that the defined function need not be monotone itself; we can e.g. define $++$ as a least fixpoint for (27).

Continuous extensions need a different form of monotonicity. To derive the recursive equations of the continuous extension, we must show that the limit exists. By (24), it suffices to show that the function (not the functional) is monotone. Thus, we cannot define lmirror as a continuous extension, either. This time, the problem is not with $++$, but rather lmirror, which is not monotone.

Another difference to least fixpoints is that the function need not be continuous at all points, as the continuous extension is defined pointwise. This is essential for functions like lsum that are well-defined only on a subset of its parameters such as the lists of positive real numbers extended with infinity.

Proof Principles. The main advantage of our approaches over ad hoc constructions like in [16] is that they bring their own proof principles: fixpoint induction (4) and structural induction (5), convergence on a closed set (19). They all require admissibility of the induction statement, since closed $\{x \mid P\ x\}$ in a ccpo topology coincides with adm $(\bigvee, \leq)$ P—just unfold the definition of open sets (17) to see this. The two notions of continuity are closely related, too. Monotonicity and order continuity imply convergence in the ccpo topology. The converse does not hold; this reflects difference between the point-wise flavour of continuous extensions and the function-as-a-whole style of least fixpoints.

Convergence on a closed set (26) and structural induction on lazy lists (5) take the consumer view, i.e. they only work for functions that consume a codatatype value. Interestingly, the former generalises the latter. Convergence keeps the bound $\overline{zs} \in \downarrow\overline{xs}$. In comparison, structural induction relaxes the bound $\overline{zs} \in \downarrow\overline{xs}$ to lfinite $\overline{zs}$ and inlines the induction on lfinite $\overline{zs}$. More abstractly, (5) reduces the statement directly to an induction on the finite subset of lazy lists. In contrast, the topological approach translates it to a corresponding statement on the type of finite lists by rewriting with identites such as (23)—the latter is then typically shown by induction.

Fixpoint induction has no counterpart in continuous extensions, as it is a proof principle for producers. It is harder to use than induction on lazy lists, see §2.4 for examples. In particular, fixpoint induction cannot show non-trivial lower bounds. However, it allows to prove properties such as (LREL) where the other principles fail. In fact, we have not yet been able to prove (LREL) by topological means, as we are not yet able to handle general predicates over two variables.

Proof Elegance. As a rough measure of proof elegance, we take the size of proofs for the five properties in Fig. 1. In the fixpoint approach, they all consist of just two steps: (i) the induction method generates the admissibility condition and the inductive cases, and (ii) an automatic proof method solves them immediately. Similarly, the topological approach first applies the proof principle (19) and then solves the subgoals. The level of automation is similar, except when we have to show the statement on finite lists by induction, which does not happen automatically. In summary, proving (CONJ, LSET, NIL, LDISTINCT) takes between 2 and 5 steps with an average of 2.75. For comparison, the former ad hoc construction of lfilter in [16] requires for proving the properties in Fig. 1 between 2 and 35 steps each with an average of 13—not even counting any of the auxiliary lemmas such as (30).

5   Related Work

Functions on Codatatypes. Devillers et al. [8] compare different formalisations of lazy lists that were available in 1997. They note the general difficulty of defining lfilter and lconcat—given by (29)—and proving their properties.

$$\text{lconcat } [] = [] \qquad \text{lconcat } (\overline{xs} \cdot \overline{xss}) = \overline{xs} ++ \text{lconcat } \overline{xss} \qquad (29)$$

In [20], Paulson describes the construction of codatatypes in Isabelle and the primitively corecursive definition of the well-known functions lmap and ++ with coinduction as proof principle for equality. He notes that he did not know of a natural formalisation for lconcat in HOL. Later, he defined lfilter using an inductive search predicate (file LFilter.thy distributed with Isabelle until 2009). Thus, all proofs about lfilter need corresponding lemmas about the search predicate. For example, his 72-line proof of (CONJ) needs seven auxiliary lemmas. For comparison, ours is one line—our preparations are not negligible, but we reuse monotonicity and continuity in many lemmas. In Coq, Bertot [4] relies on a similar search predicate; he transforms non-local properties like sortedness into local ones to simplify proofs.

Matthews [18] presents a framework to define corecursive functions via contractions for converging equivalence relations (CER) over a well-founded relation, Gianantonio and Miculan generalise CERs to complete ordered families of equivalences (COFE) [10]. CERs and COFEs require uniqueness of the specification and therefore yield a proof principle for equality. To prove contraction for lfilter, Matthews needs an inductive search predicate similar to Paulson's, and a search function that returns the first index of an element satisfying P.

Charguéraud [7] formalised the optimal fixpoint (OFP) combinator in Coq. It allows to define a large class of recursive functions, but it cannot pick any particular solution if the specification is not unique. This is arguably closer to the specification, but it complicates proofs: for the OFP of (SPEC), e.g. (LSET) holds only if $\overline{xs}$ is finite or P holds for infinitely many elements of $\overline{xs}$. For proof principles, he relies on a generalisation of COFEs, as the OFP does not provide any.

The Coinductive library [16], developed by the first author, includes functions on lazy lists and lemmas about them. The approach in this paper simplifies the definitions of and proofs about lfilter and similar functions. Previously, their definition was rather involved; lfilter was defined as the corecursive unfolding of ldropWhile; ldropWhile depended on ltakeWhile, llength, and ldrop; and ldrop on further functions. The auxiliary functions have some value of their own, so the overhead was limited. Yet, the theorems about lfilter (like those in Fig. 1) needed other theorems about the auxiliary functions. Thus, definitions and proofs both lacked elegance. The proof of (CONJ) e.g. required the specialised lemma (30).

$$\mathsf{lhd}\ (\mathsf{ldropWhile}\ P\ (\mathsf{lfilter}\ Q\ \overline{xs})) = \mathsf{lhd}\ (\mathsf{ldropWhile}\ (\lambda x.\ P\ x \vee \neg\, Q\ x)\ \overline{xs}) \quad (30)$$

Domain-Theoretic Approaches. Formalisations of domain theory and Scott's logic of computable functions (LCF) exist in HOL [1], Coq [3], and Isabelle/HOL [12]. They provide facilities to define domains and (non-terminating) recursive functions as least fixpoints as well as sophisticated proof automation. They support embedding of ordinary functions into LCF, but not the converse.

Although domains and codatatypes both contain infinite values, they are different, as all domains contain the value "undefined". Coinductive lists e.g. either end with [] or are infinite. In contrast, LCF lists can also end with undefined, e.g. filtering an infinite list whose elements all violate the predicate returns "undefined" instead of []. Thus, coinductive lists are almost isomorphic to infinite streams in HOLCF, except that the domain package additionally requires that the element type α forms a ccpo, too.

Undefinedness plays a central role in LCF: it conceptually represents all values, as monotonicity and continuity permit replacing undefined with a more specific value. This is sensible in modelling functional programs, but also complicates the theorem statements and their proofs (see e.g. [6]). Being based on HOL, our approach need not treat [] specially and can therefore deal with non-continuous functions, too. Our choice of topology reflects this, too. In our ccpo topology, finite values x are discrete, i.e., open $\{x\}$. In contrast, all Scott-open sets S are upward closed, i.e. if S contains x, then S contains all elements greater than x, too. Hence, our topology is finer than the Scott topology, so more functions are continuous, e.g. lsum on lists with a finite number of negative elements.

Two works have applied basic domain theory for defining recursive functions in HOL. First, Agerholm [2] suggested to define arbitrary recursive function as the least fixpoint in a domain by lifting the function's codomain; when termination has been shown, his tool then casts the function back to plain HOL. Hence, our application with infinite recursion is out of scope. Second, Krauss [14] realised that a tail-recursive or monadic function can be defined as a least fixpoint, because its syntactic structure ensures monotonicity. He formalised the relevant concepts in Isabelle and implemented the **partial-function** package. To our knowledge, this has only been used for the option and state-exception monads. We re-use and extend his work to define non-monadic functions on codatatypes.

Topology for Domain Theory. We do not know of any formalisation that defines recursive functions using topology except for Lester [15]. He formalises the Scott topology of a directed complete partial order in PVS and uses it to prove the existence of the fixpoint operator. Friedrich [9] formalises the Scott topology in Isabelle/HOL to characterise liveness and security properties topologically.

6   Beyond lfilter and Lazy Lists

We have described how to use domain theory and topology to define recursive functions on codatatypes. The presentation has focused on the function lfilter, as it illustrates the main ideas well and allows us to compare the approaches. But they are not restricted to it. We have used them with the same ccpo to define lconcat (29), ldropWhile (31), and ldrop (32) and to prove numerous lemmas. These functions pose the same challenge of unbounded, unproductive recursion as lfilter. In addition, the definition of lconcat relies on ++ being monotone (and continuous in the topological approach) in the second argument, which contains the recursive call, and ldrop shows that we handle multiple parameters.

$$
\begin{aligned}
&\text{ldropWhile } P \; [] &&= [] \\
&\text{ldropWhile } P \; (x \cdot \overline{xs}) &&= (\text{if } P \; x \text{ then ldropWhile } P \; \overline{xs} \text{ else } x \cdot \overline{xs})
\end{aligned}
\tag{31}
$$

$$
\text{ldrop } 0 \; \overline{xs} = \overline{xs} \qquad \text{ldrop } n \; [] = [] \qquad \text{ldrop } (\text{eSuc } n) \; (x \cdot \overline{xs}) = \text{ldrop } n \; \overline{xs} \tag{32}
$$

In terms of automating the definitions and proofs, we have used only standard Isabelle tools so far. Hence, we have not yet reached the level of sophisticated packages such as HOLCF [12]. Indeed, our approaches offer more flexibility, as they use the full function space and allow non-continuous functions to some extent. Better automation of the function definitions is left as future work.

It is not yet clear which codatatypes can be turned into useful ccpos. Clearly, it should be possible for codatatypes with finite truncations, i.e. whenever there is a non-recursive constructor. Then, this constructor can cut off a possibly infinite subtree and thus serve as bottom element. Possibly-infinite lists (α llist and (α, β) tllist) and binary trees (α tree $=$ Leaf | Node α (α tree) (α tree)) fall in this class. Conversely, if the codatatype contains only infinite values, e.g. infinite lists (α stream), a general approach seems impossible. Codatatypes with nested recursion such as α rtree $=$ Tree α (α rtree llist) will be more challenging. Working out the precise boundaries of the approach is left as future work. We hope that such insights will lead to automated constructions of ccpos for codatatypes.

Acknowledgements. J.C. Blanchette, J. Breitner, O. Maric, D. Traytel, and the anonymous reviewers suggested many textual improvements. A. Popescu helped generalising our topology on lazy lists to ccpos. Hölzl is supported by DFG grant Ni 491/15-1.

References

1. Agerholm, S.: LCF examples in HOL. In: Melham, T.F., Camilleri, J. (eds.) HUG 1994. LNCS, vol. 859, pp. 1–16. Springer, Heidelberg (1994)
2. Agerholm, S.: Non-primitive recursive function definitions. In: Schubert, E.T., Alves-Foss, J., Windley, P. (eds.) HUG 1995. LNCS, vol. 971, pp. 17–31. Springer, Heidelberg (1995)
3. Benton, N., Kennedy, A., Varming, C.: Some domain theory and denotational semantics in Coq. In: Berghofer, S., Nipkow, T., Urban, C., Wenzel, M. (eds.) TPHOLs 2009. LNCS, vol. 5674, pp. 115–130. Springer, Heidelberg (2009)
4. Bertot, Y.: Filters on coInductive streams, an application to Eratosthenes' sieve. In: Urzyczyn, P. (ed.) TLCA 2005. LNCS, vol. 3461, pp. 102–115. Springer, Heidelberg (2005)
5. Blanchette, J.C., Hölzl, J., Lochbihler, A., Panny, L., Popescu, A., Traytel, D.: Truly modular (co)datatypes for Isabelle/HOL. In: Klein, G., Gamboa, R. (eds.) ITP 2014. LNCS (LNAI), vol. 8558, pp. 93–110. Springer, Heidelberg (2014)
6. Breitner, J., Huffman, B., Mitchell, N., Sternagel, C.: Certified HLints with Isabelle/HOLCF-Prelude. In: Haskell and Rewriting Techniques, HART (2013)
7. Charguéraud, A.: The optimal fixed point combinator. In: Kaufmann, M., Paulson, L.C. (eds.) ITP 2010. LNCS, vol. 6172, pp. 195–210. Springer, Heidelberg (2010)
8. Devillers, M., Griffioen, D., Müller, O.: Possibly infinite sequences in theorem provers: A comparative study. In: Gunter, E.L., Felty, A.P. (eds.) TPHOLs 1997. LNCS, vol. 1275, pp. 89–104. Springer, Heidelberg (1997)
9. Friedrich, S.: Topology. Archive of Formal Proofs, Formal proof development (2004), http://afp.sf.net/entries/Topology.shtml

10. Di Gianantonio, P., Miculan, M.: A unifying approach to recursive and co-recursive definitions. In: Geuvers, H., Wiedijk, F. (eds.) TYPES 2002. LNCS, vol. 2646, pp. 148–161. Springer, Heidelberg (2003)

11. Hölzl, J., Immler, F., Huffman, B.: Type classes and filters for mathematical analysis in Isabelle/HOL. In: Blazy, S., Paulin-Mohring, C., Pichardie, D. (eds.) ITP 2013. LNCS, vol. 7998, pp. 279–294. Springer, Heidelberg (2013)

12. Huffman, B.: HOLCF'11: A Definitional Domain Theory for Verifying Functional Programs. PhD thesis, Portland State University (2012)

13. Krauss, A.: Partial and nested recursive function definitions in higher-order logic. J. Autom. Reasoning 44(4), 303–336 (2010)

14. Krauss, A.: Recursive definitions of monadic functions. In: PAR 2010. EPTCS, vol. 43, pp. 1–13 (2010)

15. Lester, D.R.: Topology in PVS: continuous mathematics with applications. In: AFM 2007, pp. 11–20. ACM (2007)

16. Lochbihler, A.: Coinductive. Archive of Formal Proofs, Formal proof development (2010), http://afp.sf.net/entries/Coinductive.shtml

17. Lochbihler, A.: Making the Java memory model safe. ACM Trans. Program. Lang. Syst. 35(4), 12:1–12:65 (2014)

18. Matthews, J.: Recursive function definition over coinductive types. In: Bertot, Y., Dowek, G., Hirschowitz, A., Paulin, C., Théry, L. (eds.) TPHOLs 1999. LNCS, vol. 1690, pp. 73–90. Springer, Heidelberg (1999)

19. Müller, O., Nipkow, T., von Oheimb, D., Slotosch, O.: HOLCF = HOL + LCF. J. Funct. Program. 9, 191–223 (1999)

20. Paulson, L.C.: Mechanizing coinduction and corecursion in higher-order logic. J. Logic Comput. 7(2), 175–204 (1997)

21. Slind, K.: Function definition in higher-order logic. In: von Wright, J., Harrison, J., Grundy, J. (eds.) TPHOLs 1996. LNCS, vol. 1125, pp. 381–397. Springer, Heidelberg (1996)

Formal Verification of Optical Quantum Flip Gate

Mohamed Yousri Mahmoud[1], Vincent Aravantinos[2], and Sofiène Tahar[1]

[1] Electrical and Computer Engineering Dept., Concordia University,
1455 De Maisonneuve Blvd. W., Montreal, Canada
{mo_solim,tahar}@ece.concordia.ca
http://hvg.ece.concordia.ca
[2] Software and Systems Engineering, Fortiss GmbH,
Gürickestraße 25, 80805, Munich, Germany
aravantinos@fortiss.org
http://www.fortiss.org/en

Abstract. Quantum computers are promising to efficiently solve hard computational problems, especially NP problems. In this paper, we propose to tackle the formal verification of quantum circuits using theorem proving. In particular, we focus on the verification of quantum computing based on coherent light, which is typically light produced by laser sources. We formally verify the behavior of the quantum flip gate in HOL Light: we prove that it can flip a zero-quantum-bit to a one-quantum-bit and vice versa. To this aim, we model two optical devices: the beam splitter and the phase conjugating mirror and prove relevant properties about them. Then by cascading the two elements and utilizing these properties, the complete model of the flip gate is formally verified. This requires the formalization of some fundamental mathematics like exponentiation of linear transformations.

Keywords: Quantum optics, Quantum flip gate, Beam splitter, Phase conjugating mirror, Theorem proving, HOL Light.

1 Introduction

Classical computers (i.e., Turing machines) inefficiently solve hard computational problems, e.g., NP and NP-complete problems. In 1980, Feynman proposed a new machine model which uses quantum mechanics: the quantum computer [4]. This model showed that it can solve some hard problems in polynomial time: a well known example is Shor's algorithm for integer factorization [11]. This result has great consequences on computational theory in general, and security of systems in particular: quantum cryptography became a hot area of research where powerful and secure systems are developed. In addition, limitations are arising in the everlasting quest for more powerful classical computers: power dissipation problems, density limitations, and all their workarounds like multi-core systems. This all shows how important quantum computers could be in the future.

G. Klein and R. Gamboa (Eds.): ITP 2014, LNAI 8558, pp. 358–373, 2014.

The quantum computer model proposed by Feynman consists of a new notion of a bit, called quantum bit (abbreviated as *qbit*), and a set of universal quantum gates, e.g., the flip gate (the quantum counterpart of the classical NOT gate) and the Hadamard gate [17]. A quantum circuit is made of a collection of these gates and qbits. Different means and technologies can be used to implement this model, such as: superconducting circuits [1], ion traps [6], quantum dots [12] and optical circuits [8]. Optical circuits and ion traps are today the most promising ones since they can realize the highest number of bits in laboratory, till now [9]. In this work, we focus on optical circuits which serve as the basis of several implementations of quantum computers, e.g., [19] and [10]. A major task for each of these implementations is to make sure that it satisfies the proposed specifications in the original mathematical model. This verification process is of course very different from its counterpart for classical computers.

For quantum mechanics, and more specifically quantum optics, the available verification methods are lab-simulation and paper-and-pencil, the latter is assisted by numerical methods or computer algebra systems ("CAS"). In lab-simulation, the systems are simulated *physically* in an optical laboratory, i.e., a physical system is set up, whose basic components have properties similar to the ones of the intended system. It is then assumed that this simulation system will behave in a way similar to the actual system to be verified. Note that using computers for the simulation of quantum systems is so complex that it cannot be efficient enough to verify a complete system [4]. In the paper-and-pencil approach, the whole verification process is done by modeling the system and proving– using existing physics knowledge– that the system satisfies its specifications. However, this process is handled by a human and is thus very error-prone, particularly when the system is very large and especially when considering the complex mathematics that one has to deal with in quantum mechanics. Thus, computer methods are used to help the human and decrease the risk of errors: numerical methods (typically Matlab [20]) and Computer Algebra Systems ("CAS", typically Mathematica [3]). Both are used to help the simplification and generation of intermediate mathematical steps. However, these tools are not sufficient: they cannot fully substitute for the paper-and-pencil approach since they cannot mathematically express the whole model of the system. Moreover, they are also error-prone because of the numerical approximations and heuristics used in their computations. This is particularly true for complex computations involved in quantum mechanics. Therefore, we propose to use the theorem proving for the verification of quantum optical computers.

As a first step towards our ultimate goal, in this paper, we focus on the formalization of quantum computers implemented by coherent light (typically laser light). In particular, we formally verify the behavior of one of the universal quantum gates in this implementation, the *flip gate*. To this end, we have to consider the formalization of both physical and mathematical aspects. Mathematically, we implement the quantum operator exponentiation which is similar to exponentiation, but in infinite-dimension linear spaces. We then use this as well as some preliminary work presented in [13] and [14] to develop the theory of coherent light. Coherent light is at the essential basis of two important optical elements: the beam splitter and the phase conjugate mirror, from which the flip gate can finally be built. This development demonstrates the theoretical feasibility of our approach: starting from the formalization of some abstract theory, we progressively build a model for concrete implementation of a practical quantum gate and verify that it has the expected behavior. *This work was completely implemented in HOL Light, the sources are available at [15].*

The rest of the paper is organized as follows: Section 2 gives preliminaries about quantum optics and quantum computers, and recalls the formalization of some of the foundational notions. Section 3 presents the formal development of the exponentiation of quantum operators. Section 4 describes the coherent light formalization and Section 5 deals with the flip gate verification and the formalization of the required devices. Finally, we conclude the paper in Section 6.

2   Preliminaries

In this section, we briefly introduce some notions of quantum computers and quantum optics, in particular optical coherent light. We then give more details about quantum operators that are useful in quantum optics, specifically when implementing a flip gate. We finally give the basic formal mathematical definitions that are used in our formalization.

2.1   Quantum Systems

A quantum system is fully described with a so-called *quantum state*, generally noted $|\psi\rangle$. Mathematically, a quantum state is a square integrable complex-valued function whose square integration is equal to one. Square integrable complex-valued functions form an inner product space whose product $\langle f|g\rangle$ is the integration of the multiplication of f by the conjugate of g.

For every system there is a (finite or infinite) set of quantum states $|\psi_1\rangle$, $|\psi_2\rangle$, ..., called *basis states*, which have the property that every state of the system can be expressed as a linear combination of them, i.e., for every state $|\psi\rangle$ of the system, there are complex numbers c_1, c_2, ... such that:

$$|\psi\rangle = \sum_{i=1,2,\ldots} |c_i| * |\psi_i\rangle \tag{1}$$

where $\sum |c_i|^2 = 1$.

An example of such a system is the basic component of the quantum computer: the quantum bit (or *qbit*). Similar to classical bits, a quantum bit is a quantum system with two basis states $|0\rangle$ and $|1\rangle$. However, contrary to its classical counterpart, the state of a qbit is not only $|0\rangle$ or $|1\rangle$, but can be a mix thereof. Indeed, such a state can be expressed as $|\psi\rangle = \alpha|0\rangle + \beta|1\rangle$, where $|\alpha|^2 + |\beta|^2 = 1$ (according to Equation (1)).

Another example of a quantum system is light: in quantum optics, light is considered as a stream of particles called photons, in contrast to the classical theory that considers light as an electromagnetic wave. As a quantum system, light has an infinite countable set of basis states $|0\rangle$, $|1\rangle$, ..., called *Fock states*. Light in a fock state $|n\rangle$ contains n photons. Light is said to be *coherent* if the number of photons in the light stream (at any time instant) is probabilistically Poisson distributed, i.e., the probability of having n photons is: $P(N = n) = \frac{|\alpha|^n e^{-|\alpha|}}{n!}$ for some complex number α. The modulus of $|\alpha|$ represents the expected number of observed photons. The coherent light is then in the quantum state $|\alpha\rangle$ which can be decomposed according to Equation (1) as follows:

$$|\alpha\rangle = e^{-\frac{|\alpha|^2}{2}} \sum_{n=0} \frac{\alpha^n}{\sqrt{n!}} |n\rangle \tag{2}$$

The essential idea of using quantum optics, and more specifically coherent light, to implement quantum computers is to realize the states $|0\rangle$ and $|1\rangle$ by the states $|0\rangle$ and $|\alpha\rangle$ of light, respectively.

2.2 Quantum Operators

Similar to classical physics, the state of a system can evolve over time. Actually, in the case of quantum physics, it can also evolve just by being observed. In any case, the evolution of a state must be a function mapping the state to another one. Since states are functions themselves, such a function is actually an operator. These operators are even restricted to be linear transformations over the state space.

In order to compute with qbits, one needs operators applied to them. As for classical circuits, this is achieved through *gates*. The quantum computer model is made of nine such gates, which we will not detail here since our focus in this paper is only one: *the quantum flip gate*. The flip gate (or Pauli-X gate) is equivalent to the classical NOT gate: applying it to $|0\rangle$ yields $|1\rangle$ and vice versa. However, due to its quantum nature, it is capable of much more: for any α, β, $\alpha|0\rangle + \beta|1\rangle$ is turned into $\alpha|1\rangle + \beta|0\rangle$.

In the case of optics, there are two basic quantum operators: the *creator* and *annihilator* operators. The creator operator is defined by:

$$\hat{a}^\dagger |n\rangle = \sqrt{n+1}|n+1\rangle \tag{3}$$

and the annihilator by:

$$\hat{a}|n\rangle = \sqrt{n}|n-1\rangle \tag{4}$$

As their names suggest, the annihilator $\hat{a}$ decreases the number of photons by one (i.e., destroys a photon) and the creator $\hat{a}^\dagger$ increases it by one. Note that the resulting quantum state is not exactly the demoted one, since it is scalar-multiplied by $\sqrt{n+1}$ and $\sqrt{n}$, respectively. However, scalar multiplication actually does not change a quantum state behavior. Thereby, the resulting state still has $n-1$ photons.

Solving Equation (3) as a recurrence relation, we obtain a general representation of any fock state $|n\rangle$:

$$|n\rangle = \frac{(\hat{a}^\dagger)^n \, |0\rangle}{\sqrt{n!}} \tag{5}$$

where $|0\rangle$ is called *vacuum* state since it does not contain any photon. Note here that the power notation used in $(\hat{a}^\dagger)^n$ means the application of the creation operator n times (recall that quantum operators are functions).

According to Equations (2) and (5), we can re-express coherent states in terms of the vacuum state and creation operator:

$$|\alpha\rangle = e^{-\frac{|\alpha|^2}{2}} \left(\sum_{n=0} \frac{(\alpha\hat{a}^\dagger)^n}{n!} \right) |0\rangle \tag{6}$$

Note that, for a linear operator $a^\dagger$, $(\alpha\hat{a}^\dagger)^n = \alpha^n(\hat{a}^\dagger)^n$.

This allows us to introduce the *displacement* operator $D(\alpha)$, which is essential for the implementation of the flip gate:

$$|\alpha\rangle = D(\alpha)|0\rangle \tag{7}$$

Here, $D(\alpha) = e^{\alpha\hat{a}^\dagger} \, e^{-\alpha^*\hat{a}} \, e^{[\alpha\hat{a}^\dagger, -\alpha^*\hat{a}]}$, where $*$ denotes the scalar multiplication with quantum operators, $**$ denotes the multiplication between quantum operators, and $[op_1, op_2] = op_1 ** op_2 - op_2 ** op_1$. The proof of Equation (7) can be found in the literature, e.g., in [16]. Note the use of exponentiation *over operators*, which is defined as follows:

$$e^{\hat{O}} = \sum_{i=0} \frac{\hat{O}^i}{i!} \tag{8}$$

Though defined similarly to the classical exponential, its properties are very different.

The importance of the displacement operator is that it can be physically realized by a quantum optical device called a *beam splitter* [18]. Therefore it is an essential ingredient in the implementation of quantum computers using coherent light, as we will see in Section 5.

2.3   Quantum State Space Formalization

After presenting the essential quantum physics notions, we now briefly review the formalization of inner product spaces which was presented in [13].

First, since a quantum state is a complex-valued function, we defined a HOL type for that: $\texttt{cfun} = \texttt{A} \rightarrow \texttt{complex}$, where *cfun* stands for *complex function*. A is a type variable, allowing our formalization to be used to model both finite-dimension systems like quantum computers, and infinite-dimension systems like quantum light.

Additions and scalar multiplications are defined easily as the corresponding point-wise operations, which allows us to characterize the notion of linear subspace as follows:

Definition 1
is_cfun_subspace (spc : cfun → bool) ⇔
  ∀x y. x IN spc ∧ y IN spc ⇒
  x + y IN spc ∧ (∀a. a%x IN spc) ∧ cfun_zero IN spc

where cfun_zero is the constantly null function, and % denotes the scalar multiplication. The notion of inner space is then defined as follows:

Definition 2
is_inner_space ((s, inprod) : (qs → bool) × (cfun → cfun → complex)) ⇔
  is_cfun_subspace s ∧
  ∀x. x ∈ s ⇒
    real (inprod x x) ∧ 0 ≤ real_of_complex (inprod x x) ∧
    (inprod x x = Cx(0) ⇔ x = qs_zero) ∧
    ∀y. y ∈ s ⇒
    cnj (inprod y x) = inprod x y ∧
    (∀a. inprod x (a%y) = a * (inprod x y)) ∧
    ∀z. z ∈ s ⇒
    inprod (x + y) z = inprod x z + inprod y z

where real x states that the complex value x has no imaginary part, and real_of_complex is a function converting a complex number into a real one (if it is real).

Once these bases are set, we can define the notion of operator over an inner space. This is achieved by first defining the type cop = cfun → cfun. A linear operator is then characterized as follows:

Definition 3
is_linear_cop (op : cop) ⇔
  ∀x y. op (x + y) = op x + op y ∧ ∀a. op (a % x) = a % (op x)

In addition, quantum operators must satisfy the property of being self-adjoint:

Definition 4
is_self_adjoint (s, inprod) op ⇔
  is_inner_space (s, inprod) ⇒
    is_linear_cop op ∧
    ∀x y. inprod x (op y) = inprod (op x) y

As seen in the previous section, exponentiation of operators requires their infinite summation. We first define infinite summation over functions:

Definition 5
cfun_sums innerspc f l s ⇔
  cfun_lim innerspc (λn. cfun_sum (s INTER (0..n)) f) l sequentially

which formalizes the fact that $\lim_{n \to \infty} \sum_{i=0}^{n} f_i = l$: INTER is the sets intersection operator, cfun_lim is the notion of limit defined for quantum states, cfun_sum is finite summation over quantum states, and sequentially means that the summation index will be increased sequentially, i.e., 1,2,3,.. More details about implementing infinite summation and related notions are presented in [14].

In practice it is more convenient to actually retrieve the limit in a functional way. To do so we use the Hilbert choice operator @ as follows:

Definition 6

`cfun_infsum innerspc s f = @l. cfun_sums innerspc f l s`

This is useful only at the condition that the sum is convergent, which we express by the following predicate:

Definition 7

`cfun_summable innerspc s f = ∃l. cfun_sums innerspc f l s`

In conjunction with infinite summation, *bounded* operators are of particular importance. Indeed, the application of a bounded operator commutes with infinite summation: i.e., for a bounded operator `cop`:

$$\text{cop (cfun\_infsum f s)} = \text{cun\_infsum } (\lambda n.\ \text{cop (fn))}\ \text{s.}$$

Bounded operators are defined as follows:

Definition 8

`is_bounded (s, inprod) h ⟺ is_inner_space (s, inprod)`
`⇒ is_closed_by s h ∧ ∃B. 0 < B∧`
`(∀x. x IN s ⇒ cfun_norm inprod (h x))) ≤ B * cfun_norm inprod x)))`

where `is_closed_by s h ⟺ ∀x.x IN s ⇒ h x IN s`, and `cfun_norm inprod x =` $\sqrt{\text{real\_of\_complex (inprod x x)}}$. A linear operator `h` is bounded if for all `x` the norm of `h x` is lower or equal to the norm of `x` up to multiplication by a scalar `B`. Note that `B` does not depend on `x`.

3   Quantum Operator Exponentiation

Quantum operator exponentiation is essential for the formalization of the displacement operator. In order to tackle the exponentiation, we have first to consider the infinite summation over quantum operators, which is done simply by using the pointwise infinite summation over complex functions:

Definition 9

`cop_sums (s, inprod) f l set ⟺ ∀x. x IN s ⇒`
`    cfun_sums (s, inprod) (λn.(f n) x) (l x) set`

This definition is an easy adaptation of the `cfun` case: the only differences are the types of `f`, `l`, and `set`, and the fact that the pointwise definition is restricted to the values that belong to the inner space. This latter point is very important since this summation might not exist for some operators, if defined over the complete extension of `cfun`: for instance, many sequences of square-integrable functions do not have a limit that remains square-integrable.

Similarly to `cfun_infsum` and `cfun_summable`, we then define `cop_infsum` and `cop_summable`:

Definition 10
cop_infsum innerspc s f = @l. cop_sums innerspc f l s
cop_summable innerspc s f = ∃l. cop_sums innerspc f l s

Finally, we can use cop_infsum to define quantum operator exponentiation according to Equation (8):

Definition 11
cop_exp innerspc (op : cfun → cfun) ⇔
    cop_infsum innerspc (from 0) ($\lambda n. \frac{1}{!n}$ % (op pow n))

where from 0 denotes the set $\mathbb{N}$. We prove many properties about the exponentiation but we will present in detail the proof of only one of them, and will only mention the end result for others. We start by proving that cop_exp (cop_zero) = I, which is the scalar counterpart of $e^0 = 1$. To do so, we first need to provide the property using the predicate definition, i.e., cop_sums, as follows:

Theorem 1
∀is. is_inner_space is ⇒
    cop_sums innerspc ($\lambda n. \frac{1}{!n}$ % (cop_zero pow n)) I (from 0)

where cop_zero = λx : cfun. cfun_zero is the operator constantly equal to cfun_zero. In addition, we recall that I is the identity operator (to ease the understanding, one can remark that it corresponds to the identity matrix in a finite dimension vector space). We then use the uniqueness of cop_infsum to re-express the property in terms of cop_exp. The unicity theorem is as follows:

Theorem 2
∀s inprod f set l x.
    x IN s ∧ cop_sums (s, inprod) f l set ⇒
        (cop_infsum (s, inprod) set f) x = l x

It states that *if the summation has a limit*, then this limit is unique. Therefore it is also equal to cop_infsum (s,inprod) set f on the considered inner space, since the definition of cop_infsum is precisely to be any of these limits. Note that we cannot ensure that cop_infsum (s,inprod) set f and l are equal since we do not know how they affect elements outside s. This is not a restriction, on the contrary: it ensures that our theory indeed has a non-trivial model. If this was not the case, the inner space of square-integrable functions could not be used with our formalization.

In the end, we obtain the following theorem stating indeed the intended property:

Theorem 3
∀s inprod x.
    x IN s ∧ is_inner_space(s, inprod) ⇒
        cop_exp (s, inprod) cop_zero x = x

Another important property is the commutativity of exponentiation with the scalar multiplication of its argument:

Theorem 4
∀s inprod a x.
    x IN s ∧ is_inner_space(s, inprod) ⇒
        (cop_exp(s, inprod) (λy.a%y)) x = cexp a%x

The scalar counterpart of this theorem is the $e^{(a.1)} = e^a.1$: indeed the identity plays here the role of the unity. Note that this result shows the compatibility of our definitions with the ones defined in HOL Light for infinite dimension linear spaces.

Like for the scalar exponentiation, cop_exp is not a linear function over operators. However, a property which has no counterpart for scalars is the linearity of cop_exp op (which is an operator). This property is essential to the development of the flip gate: indeed, it allows to generalize the effect of the gate on basis states $|0\rangle$ and $|1\rangle$ to any mixed state $c_1 |0\rangle + c_2|1\rangle$. It also helps a lot in the intermediate steps of many proofs, by allowing to move in and out scalar values multiplied by states, i.e., cop_exp op(a % x) = a%(cop_exp op x). The linearity of cop_exp op is however true only on the concerned inner space. Therefore, we need a definition which is relaxed w.r.t. Definition 3:

Definition 12
is_set_linear_cop s (op : cop) ⇔
    ∀x y. x IN s ∧ y IN s ⇒ op (x + y) = op x + op y ∧
    ∀a. op(a % x) = a % (op x)

The linearity of cop_exp op can then be proved, as long as op is itself a linear operator:

Theorem 5
∀s inprod op.
    cop_summable innerspc (from 0) (λn.$\frac{1}{!n}$ % (op pow n)∧is_linear_cop op⇒
        is_set_linear_cop s (cop_exp (s, inprod) op)

This concludes our formalization of operators exponentiation.

4   Coherent Light Formalization

In this section, the formal definition of the coherent state of light is presented, which we then re-express in terms of the displacement operator (according to the presentation of Section 2.2). This is carried out in three steps: 1) quantum light formalization, 2) formalization of fock states (which are the basis of quantum optics states space), and 3) coherent states formalization.

4.1   Single Mode

The basic building block of formalizing light in quantum theory is the formal development of electromagnetic fields [2]: Quantum physics studies a light stream as an electromagnetic field. Such a field can be reduced to the superposition of several single-mode (i.e., single resonance frequency) fields. The formal definition of a single-mode filed is as follows:

Definition 13

is_sm $((sp, cs, H), \omega, vac) \Leftrightarrow$
  is_qsys $(sp, cs, H) \wedge 0 < \omega \wedge \exists q \ p. \ cs = [q; p]$
  $\wedge \ \forall t.$is_observable sp $(p \ t) \wedge$ is_observable sp$(q \ t)$
  $\wedge$ H $t = \frac{\omega^2}{2}\%((q \ t) \ pow \ 2) + \frac{1}{2}\%((p \ t) \ pow \ 2)$
  $\wedge$is_qst sp vac $\wedge$ is_eigen_pair (H t) $(vac, \frac{planck*\omega}{2})$

A single-mode field is characterized by five elements: sp is the quantum states space of the field; cs lists the elementary observables of the mode, p and q are the *canonical coordinates* of the field, out of which we build the creator and annihilator operators; H is expressing the amount of energy inside the field; ω is the resonance frequency; and vac refers to the vacuum state. More details about is_sm can be found in [13].

As explained in Section 2.2, a single-mode field in a fock state (or photon number state) $|n\rangle$ is a light stream containing exactly n photons. These states are crucial because they form the basis of the single-mode quantum states space, and they are widely used in the development of quantum cryptography systems. According to Equation 5, we can formally define a fock state as follows:

Definition 14

let $(((s, inprod), cs, H), \omega, vac) = sm$ in
  fock sm $0 = vac \wedge$ fock sm (SUC n) $=$
    get_qst inprod (creat_of_sm sm (fock sm n)))

where get_qst returns the normalized version of a vector, i.e., the vector divided by its norm. This is to ensure that the norm of the resulting quantum state is equal to one. Using this definition and the infinite summation, a coherent state can be defined as follows:

Definition 15

coherent sm $\alpha =$
  let sm $= ((s, inprod), cs, H), \omega, vac$ in
    $\exp(-\frac{|\alpha|^2}{2})\%$ cfun_infsum $(s, inprod)$ (from 0) $(\lambda n. \frac{\alpha^n}{\sqrt{n!}}\%($fock sm $n))$

where α is the state parameter (recall that the number of photons in a coherent stream is Poisson distributed with expectation $|\alpha|^2$). Note that Definition 15 corresponds to Equation (2).

As usual, we will often need to be able to tell when the sum in the above definition is convergent. We define therefore the predicate coherent_summable:

Definition 16

coherent_summable sm $\alpha \Leftrightarrow$
  let $(((s, inprod), cs, H), \omega, vac) = sm$ in
    cfun_summable $(s, inprod)$ (from 0) $(\lambda n. \frac{\alpha^n}{\sqrt{!n}}\%($fock sm $n))$

We refer the reader to [14] for more details about the formalization of fock and coherent states.

The implementation of quantum coherent computer is based on the idea of expressing coherent beams in terms of the displacement operator, since it can be easily realized using an optical beam splitter. Let us first give the formal definition of the displacement operator:

Definition 17

disp sm α =
let qspc = (qspc_of_sm sm) in
$\underbrace{(\text{cop\_exp qspc } (\alpha \% \text{ creat\_of\_sm sm})}_{1} ** \\ \underbrace{\text{cop\_exp qspc } (-(\text{cnj v}) \% \text{ a\_of\_sm sm})}_{2} ** \\ \underbrace{\text{cop\_exp qspc } ((\text{v} \% \text{ creat\_of\_sm sm}) \text{ com } ((\text{cnj v}) \% \text{ a\_of\_sm sm}))}_{3}$

where op1 com op2 = op1 ** op2 - op2 ** op1 (called the *commutator* of op1 and op2), and creat_of_sm and a_of_sm are functions that return the creator and annihilator operators, respectively.

To express a coherent state in terms of the displacement operator, we study the effect of this operator on the vacuum state: the underlined operator 3 in Definition 17 will collapse to a scalar value because creat_of_sm sm com (a_of_sm sm) = I; and since the two other operators are linear, we can get this scalar outside. The next step is to study the effect of the underlined operator 2 on the vacuum state. The following theorem shows that it actually acts like the identity:

Theorem 6

$\forall$s inprod cs H ω vac.
   let sm = ((s, inprod), cs, H), ω, vac in
   is_sm sm $\wedge$ exp_summable (qspc_of_sm sm) (α % a_of_sm sm)
      $\Rightarrow$ cop_exp (qspc_of_sm sm) (α % a_of_sm sm) vac = vac

where qspc_of_sm returns the corresponding quantum states space of a given field. Thus the resulting state is again vac. It only remains to establish the effect of the underlined operator 1:

Theorem 7

$\forall$s inprod cs H ω vac α.
   let sm = ((s, inprod), cs, H), ω, vac in
   is_sm sm $\wedge$ ($\forall$m. creat_of_sm sm (fock sm m) $\neq$ cfun_zero)
   $\wedge$ exp_summable (qspc_of_sm sm) (α creat_of_sm sm)
   $\wedge$ cfun_summable (s, inprod) (from 0)(λn. $\frac{\alpha \text{ pow } n}{\sqrt{!n}}$ % fock sm n)
      $\Rightarrow$ cop_exp (qspc_of_sm sm) (α % creat_of_sm sm) vac =
      cfun_infsum (s, inprod) (from 0)(λn. $\frac{\text{a pow } n}{\sqrt{!n}}$ % fock sm n)

which corresponds almost to the definition of coherent light (see Definition 15): it differs only by multiplication with a scalar value. One then just needs to combine these results in order to obtain the final expression of coherent light in terms of the displacement operator:

Theorem 8

$\forall$s inprod s H ω vac α.
   let sm = ((s, inprod), cs, H), ω, vac in

$\text{is\_sm sm} \wedge \text{exp\_summable (qspc\_of\_sm sm) (cnj}(-\alpha) \text{ \%a\_of\_sm sm)}$
$\wedge \ (\forall \text{n.creat\_of\_sm sm (fock sm n)} \neq \text{cfun\_zero}))$
$\wedge \ \text{cfun\_summable (s, inprod) (from 0)}(\lambda \text{n.} \frac{\alpha \ \underline{\text{pow}} \ \text{n}}{\sqrt{!\text{n}}} \ \% \ \text{fock sm n})$
$\text{is\_sm sm} \wedge \ \text{exp\_summable (qspc\_of\_sm sm) }(\alpha \ \text{creat\_of\_sm sm)}$
$\ \ \ \ \Rightarrow \text{coherent sm } \alpha = \text{(disp sm } \alpha) \text{ vac}$

In the next section, we will see how this expression of coherent states helps in the development of the quantum flip gate.

5   Quantum Flip Gate Verification

In this section we detail the implementation of the optical flip gate [19], and explain the idea behind it. Recall that $|vac\rangle$ and $|\alpha\rangle$ are meant to implement the qbits $|0\rangle$ and $|1\rangle$, respectively. The specification of a flip gate is that it should turn $c_1 |vac\rangle + c_2 |\alpha\rangle$ into $c_1 |\alpha\rangle + c_2 |vac\rangle$, for all $c_1, c_2 \in \mathbb{C}$. The intended implementation of the gate is represented in Figure 1. First a beam splitter

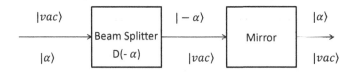

Fig. 1. Flip gate optical implementation

realizes a $-\alpha$ displacement operator. Then a phase conjugating mirror generates a beam identical to the input beam but with a reverse phase, which yields an output of $|-\alpha\rangle$ for an input of $|\alpha\rangle$.

We start by demonstrating the effect of the proposed optical flip gate on each optical qbit separately. Then, we generalize the result to any mixed qbit by using the linearity of the gate.

We start by formalizing the phase conjugating mirror as follows:

Definition 18
$\text{mirror sm} =$
$\ \ \ \ \text{let sm} = ((\text{s, inprod}), \text{cs, H}), \omega, \text{vac in}$
$\ \ \ \ \text{cop\_exp (s, inprod) (i}\pi \ \% \ \text{n\_of\_sm sm)}$

We will see later that applying such quantum operator to a coherent beam result in the same beam in the reverse direction (i.e., the input beam is $|\alpha\rangle$ and the output is $|-\alpha\rangle$). This is exactly what a phase conjugating mirror does. Note that we use again quantum operator exponentiation.

The following property is the key to verify that the mirror implements phase shifting:

Theorem 9

$\forall$s inprod cs H ω vac θ n.

  let sm $=$ ((s, inprod), cs, H), ω, vac in

  is_sm sm $\wedge$ exp_summable (qspc_of_sm sm) ($i\theta$ % n_of_sm sm)

  $\wedge$ creat_of_sm sm (fock sm n) $\neq$ cfun_zero

      $\Rightarrow$ $\underline{\text{cop\_exp (qspc\_of\_sm sm) ($i\theta$ \% n_of_sm sm)}}_1$ (fock sm n) $=$

      $\underline{\text{(cexp ($i\theta$) pow m) \% (fock sm n)}}_2$

The underlined expression 1 is called a *phase shifter operator*. It is a generalization of the behavior of the phase conjugating mirror, except it considers any angle θ instead of just π. Theorem 9 shows the effect of such an operator on fock states: the underlined expression 2 shows that it generates the same state but shifted by θ. By specifying $\theta = \pi$, we can then easily prove the effect of the mirror on coherent states:

Theorem 10

$\forall$s inprod cs H ω vac α.

  let sm $=$ (((s, inprod), cs, H), ω, vac) in

  is_sm sm $\wedge$ cfun_summable (s, inprod) (from0) (λn. $\frac{\alpha^n}{\sqrt{!n}}$ % fock sm n)

  $\wedge$ mirror_summable sm $\wedge$ is_bounded (qspc_of_sm sm) (mirror sm)

  $\wedge$ ($\forall$n.creat_of_sm sm (fock sm n) $\neq$ cfun_zero))

      $\Rightarrow$ mirror sm (coherent sm α) $=$ coherent sm ($-\alpha$)

where mirror_summable is similar to the summable notions defined before: we define a new predicate only for simplicity. The purpose of this predicate is to ensure that the mirror operator exists.

The former theorem proves that the mirror indeed behaves as expected when applied to the qbit $|1\rangle$. We now show that it is also the case for $|0\rangle$, i.e., for the vacuum state vac:

Theorem 11

$\forall$s inprod cs H ω vac.

  let sm $=$ ((s, inprod), cs, H), ω, vac in

  is_sm sm $\wedge$ coherent_summable sm 0

      $\Rightarrow$ coherent sm 0 $=$ vac

Combined with the previous theorem, this confirms that the vac state is unchanged by the mirror.

We now complete the formalization by the beam splitter, which is modeled by the displacement operator. In case that the input to the beam splitter is vac then the output will be coherent sm ($-\alpha$) according to Theorem 8. For coherent sm α input, it results in vac according to the following theorem:

Theorem 12

$\forall$s inprod cs H ω vac α.

  let sm $=$ ((s, inprod), cs, H), ω, vac in

  is_sm sm $\wedge$ ($\forall$b.exp_summable (s, inprod) (b%a_of_sm sm))

  $\wedge$ ($\forall$m. creat_of_sm sm (fock sm m) $\neq$ cfun_zero) $\wedge$ coherent_summable sm α

$\wedge$ exp_summable (qspc_of_sm sm) (α creat_of_sm sm)
$\wedge$ is_bounded (s, inprod) (a_of_sm sm) $\wedge$ (coherent sm $\alpha \neq$ cfun_zero)
$\wedge$ ($\forall$x op. is_linear_cop op $\wedge$ x IN s $\Rightarrow$
   (cop_exp (s, inprod) ($-$op) $**$ cop_exp (s, inprod) (op)) x = x)
   $\Rightarrow$ disp sm ($-\alpha$) (coherent sm α) = vac

The last conjunction in the premises shows an assumed property about exponentiation of quantum operators. Such property requires the proof of a the general theorem of Baker-Campbell-Hausdorff [7] [1]. A major step towards proving Theorem 12 is to evaluate the effect of cop_exp (a_of_sm sm) on coherent beams. The following theorem shows such effect:

Theorem 13
$\forall$s inprod cs H ω vac α.
   let sm = ((s, inprod), cs, H), ω, vac in
   is_sm sm $\wedge$ exp_summable (s, inprod) (cnj α%a_of_sm sm)
$\wedge$ ($\forall$m. creat_of_sm sm (fock sm m)$\neq$cfun_zero) $\wedge$ coherent_summable sm α
$\wedge$ is_bounded (s, inprod) (a_of_sm sm) $\wedge$ (coherent sm $\alpha \neq$ cfun_zero)
   $\Rightarrow$ cop_exp (qspc_of_sm sm) ((cnj α)%a_of_sm sm) (coherent sm α) =
   cexp((norm α)^2)%(coherent sm α)

Now, we have all ingredients to construct the flip gate and verify its behavior. The formal definition of the flip gate is made through the cascading of the mirror and beam splitter elements. This is defined as an operators' multiplication (i.e., function composition):

Definition 19
flip_gate α sm = (mirror sm) $**$ (disp sm ($-\alpha$))

Based on above definition and using Theorems 10-12, we prove the correction of the gate behavior in one single theorem as follows:

Theorem 14
$\forall$s inprod cs H ω vac α.
   let sm = ((s, inprod), cs, H), ω, vac in
   is_sm sm $\wedge$ exp_summable ($\forall$b. (s, inprod) (b%a_of_sm sm)
$\wedge$ ($\forall$m. creat_of_sm sm (fock sm m) $\neq$ cfun_zero)
$\wedge$ ($\forall$b. coherent_summable sm b)
$\wedge$ ($\forall$c. cfun_summable (s, inprod) (from 0) (λn.($\frac{c^n}{\sqrt{!n}}$)%fock sm n))
$\wedge$ ($\forall$d. exp_summable (s, inprod) (%creat_of_sm sm (0)))
$\wedge$ is_bounded (s, inprod) (a_of_sm sm)
$\wedge$ (coherent sm $\alpha \neq$ cfun_zero)$\wedge$
$\wedge$ (cop_exp (s, inprod) ($-$op) $**$ cop_exp (s, inprod) (op)) x = x)
$\wedge$ mirror_summable sm $\wedge$ is_bounded (qspc_of_sm sm) (mirror sm)
   $\Rightarrow$ (flip_gate α sm) (coherent sm α) = vac
   $\wedge$ (flip_gate α sm) vac = coherent sm α

---

[1] The proof of the Baker-Campbell-Hausdorff theorem is very complex and requires a lot of prerequisites that are not available in HOL Light. The formal verification of this theorem in HOL Light is part of our future work.

In a nutshell, Theorem 14 proves that a coherent beam $|\alpha\rangle$ ($|vac\rangle$) passes through a beam splitter, which in turn generates $|vac\rangle$ ($|-\alpha\rangle$), then the beam experiences a mirror which reflects it in the opposite direction to generate $|vac\rangle$ ($|\alpha\rangle$). Hence, we have the realization of the quantum flip gate. Note that given the linearity of the optical elements, this result generalizes for any mixed state $c_1 * |\alpha\rangle + c_2 * |vac\rangle$.

6   Conclusion

Quantum optics explores extremely useful phenomena and properties of light as a stream of photons. However, the analysis of quantum optical systems is complex. Traditional analysis techniques – simulation in optical laboratories, paper-and-pencil, numerical methods, and computer algebra systems – suffer from many problems: safety, cost, lack of expressiveness, human error. We believe that the proposed formalization of quantum optics can contribute to propose an alternative tackling these limitations.

Coherent light (or states) is an essential notion in quantum optics since it eases the analysis of many quantum systems. One of the most interesting applications of coherent light is quantum computers. Coherent states are proposed to model quantum bits [19], by taking $|vac\rangle$ and $|\alpha\rangle$ as $|0\rangle$ and $|1\rangle$, respectively. Many quantum gates were implemented based on this model. In this paper, we considered the quantum flip gate, which converts $\delta|0\rangle + \beta|1\rangle$ into $\beta|0\rangle + \delta|1\rangle$. We verified the behavior of this gate, which requires many formalization tasks. We started by developing the required mathematical foundations, in particular summation over quantum operators and exponentiation of quantum operators. Then we presented the formal definition of coherent beam, and expressed coherent states in terms of the displacement operator, which can be physically implemented as a beam splitter. The gate itself consists of a phase conjugating mirror along with a beam splitter. Therefore, we formalized the mirror and a displacement operator (or, equivalently, a beam splitter) and proved the required theorems to verify the gate behavior.

In the future, we plan to formalize other gates and handle other quantum computer implementations, for example the one based on squeezed states (a special case of coherent light). We also plan to extend our work to multi-mode systems, which are very useful for complicated quantum gates, in particular those that use the phenomena of entailment and teleportation [5].

References

1. Clarke, J., Wilhelm, F.K.: Superconducting quantum bits. Nature 453, 1031–1042 (2008)
2. Dirac, P.A.M.: The fundamental equations of quantum mechanics. Proceedings of the Royal Society A: Mathematical, Physical and Engineering Sciences 109(752), 642–653 (1925)
3. Feagin, J.M.: Quantum Methods with Mathematica. Springer (2002)
4. Feynman, R.: Simulating physics with computers. International Journal of Theoretical Physics 21, 467–488 (1982)

5. Furusawa, A., van Loock, P.: Quantum Teleportation and Entanglement: A Hybrid Approach to Optical Quantum Information Processing. Wiley (2011)
6. Haeffner, H., Roos, C.F., Blatt, R.: Quantum computing with trapped ions. Physics Reports 469(4), 155–203 (2008)
7. Hall, B.: Lie Groups, Lie Algebras, and Representations: An Elementary Introduction. Graduate Texts in Mathematics. Springer (2003)
8. Jennewein, T., Barbieri, M., White, A.G.: Single-photon device requirements for operating linear optics quantum computing outside the post-selection basis. Journal of Modern Optics 58(3-4), 276–287 (2011)
9. Ladd, T.D., Jelezko, F., Laflamme, R., Nakamura, Y., Monroe, C., O'Brien, J.L.: Quantum computers. Nature 464, 45–53 (2010)
10. Li, Y., Browne, D.E., Kwek, L.C., Raussendorf, R., Wei, T.: Thermal states as universal resources for quantum computation with always-on interactions. Physical Review Letter 107, 060501 (2011)
11. Lomonaco, S.J.: Quantum Computation: A Grand Mathematical Challenge for the Twenty-first Century and the Millennium. American Mathematical Society (2002)
12. Loss, D., DiVincenzo, D.P.: Quantum computation with quantum dots. Physical Review A 57, 120–126 (1998)
13. Mahmoud, M.Y., Aravantinos, V., Tahar, S.: Formalization of infinite dimension linear spaces with application to quantum theory. In: Brat, G., Rungta, N., Venet, A. (eds.) NFM 2013. LNCS, vol. 7871, pp. 413–427. Springer, Heidelberg (2013)
14. Mahmoud, M.Y., Tahar, S.: On the quantum formalization of coherent light in HOL. In: Badger, J.M., Rozier, K.Y. (eds.) NFM 2014. LNCS, vol. 8430, pp. 128–142. Springer, Heidelberg (2014)
15. Mahmoud, M.Y., Aravantinos, V.: Formal verification of optical quantum flip gate - HOL Light script,
   http://hvg.ece.concordia.ca/projects/qoptics/flipgate.html
16. Mandel, L., Wolf, E.: Optical Coherence and Quantum Optics. Cambridge University Press (1995)
17. Nielsen, M.A., Chuang, I.L.: Quantum Computation and Quantum Information: 10th Anniversary Edition. Cambridge University Press (2010)
18. Paris, M.G.A.: Displacement operator by beam splitter. Physical Letters A 217(2-3), 78–80 (1996)
19. Ralph, T.C., Gilchrist, A., Milburn, G.J., Munro, W.J., Glancy, S.: Quantum computation with optical coherent states. Physical Review A 68, 042319 (2003)
20. Tan, S.M.: A computational toolbox for quantum and atomic optics. Journal of Optics B: Quantum and Semiclassical Optics 1(4), 424 (1999)

Compositional Computational Reflection

Gregory Malecha[1], Adam Chlipala[2], and Thomas Braibant[3]

[1] Harvard University SEAS, Cambridge, MA, USA
gmalecha@cs.harvard.edu
[2] MIT CSAIL, Cambridge, MA, USA
adamc@csail.mit.edu
[3] INRIA, Rocquencourt, France
thomas.braibant@inria.fr

Abstract. Current work on computational reflection is single-minded; each reflective procedure is written with a specific application or scope in mind. Composition of these reflective procedures is done by a proof-generating tactic language such as Ltac. This composition, however, comes at the cost of both larger proof terms and redundant preprocessing. In this work, we propose a methodology for writing composable reflective procedures that solve many small tasks in a single invocation. The key technical insights are techniques for reasoning semantically about extensible syntax in intensional type theory. Our techniques make it possible to compose sound procedures and write generic procedures parametrized by lemmas mimicking Coq's support for hint databases.

Keywords: Computational reflection, automation, Coq, verification.

1 Introduction

Imperative program verification requires orchestrating many different reasoning procedures. For it to scale to more sophisticated languages and larger programs, these procedures must be efficient. When using a proof assistant, a popular way to achieve good performance is with computational reflection [2], a technique for discharging proof obligations by running verified programs implemented in the proof assistant's logic.

While individually these procedures are fast, composing them relies on non-reflective, proof-generating tactic languages like Ltac [7]. While simple and flexible, this method is expensive. The brunt of the cost of computational reflection is in setting up the procedure and constructing the proof term; the actual computation is often relatively cheap. Composing many small reflective procedures requires paying this price for many moderately sized proof obligations rather than once for the entire goal.

To achieve this composition without returning to the non-reflective world, high-level reflective procedures must support extension to reason about domain-specific problems. Tactic-based languages support patterns such as higher-order tactics and hint databases that allow extending automation after the fact. For example, Coq's `autorewrite` tactic is based on hint databases that package

G. Klein and R. Gamboa (Eds.): ITP 2014, LNAI 8558, pp. 374–389, 2014.

Specification & Implementation	Domain-Specific Hints & Proof

```
Definition bstM : bmodule := {
bfunction "lookup"("s", "k", "tmp") [lookupS]
  "s" ←* "s";;
  [∀ s, ∀ t,
   PRE[V] bst' s t (V "s") * mallocHeap
   POST[R] ⌈ (V "k" ∈ s) \is R ⌉ *
          bst' s t (V "s") * mallocHeap]
  While ("s" ≠ 0) {
    "tmp" ←* "s" + 4;;
    If ("k" = "tmp") {
      Return 1 (* Key matches! *)
    } else {
      If ("k" < "tmp") {
        "s" ←* "s" (* Lower key *)
      } else {
        "s" ←* "s" + 8 (* Higher key *)
      }
    }
  };;
  Return 0
}.
```

quantified invariants (bracket spanning the PRE/POST lines)

```
(* Representation predicate for BSTs *)
Definition bst (s : set) (p : W) :=
  ⌈ freeable p 2 ⌉ * ∃ t, ∃ r, ∃ junk,
    p ↦ r * (p + $4) ↦ junk * bst' s t r.

(* A standard tree refinement hint *)
Theorem nil_fwd : ∀ s t (p : W), p = 0 →
  bst' s t p ⟹ ⌈ s ≃ empty ∧ t = Leaf ⌉.
Proof. destruct t; sepLemma. Qed.
(* ...more hints... *)

(* Combine the hints into a package. *)
Definition hints : HintDatabase.
prepare (nil_fwd, bst_fwd, cons_fwd)
        (nil_bwd, bst_bwd, cons_bwd).
Defined.

(* Prove partial correctness. *)
Theorem bstMOk : moduleOk bstM.
Proof. vcgen; abstract (sep hints; auto). Qed.
                              prove using hints
```

user predicate (bracket) · refinement hints (bracket) · combine hints (bracket)

Fig. 1. Verified implementation of binary search trees implementing finite-set "lookup"

together a collection of rewrites and associated tactics to solve side conditions. These features, however, have not made their way to reflective procedures.

In this work, we focus on building extensible reflective procedures that perform many reasoning steps in a single invocation. Figure 1 demonstrates the degree of automation that we achieve applying our techniques to program verification in Bedrock [4], a Coq [6] library for low-level, imperative programming. Note that the implementation is completely separated (with the exception of loop invariants) from the automated verification. Effective automation for verifying such a program requires simultaneously reasoning about abstract predicates, low-level machine words, and high-level sets. To that end, our automation (sep) is written modularly and composed into large reflective procedures. Reasoning for problem-specific constructs is incorporated via HintDatabases that are constructed completely automatically from both fully verified reflective procedures (similar to Ltac's Hint Extern) and guarded rewriting lemmas (similar to Ltac's Hint Rewrite ... using ...). The latter of these is constructed completely automatically from standard lemmas like nil_fwd above, which drastically lowers the overhead of applying our automation to reason about new abstract predicates.

In the rest of the paper we discuss the techniques that we have developed to support that kind of sophisticated reasoning reflectively. We begin with a short primer on computational reflection (Section 2) before discussing our technical contributions, which correspond to the features of our reflective procedures:

- Our proof procedures reason extensively about two forms of *variables* (Section 3.2): variables introduced by *existential quantifiers* in the goal and *unification variables* introduced by Ltac before our reflective procedures run.
- Our proof procedures *reason semantically about an open-ended set of symbols and types* (Section 3.3). Our approach allows us to build independent

```
Inductive sexpr := (* syntactic separation logic formulas *)
| Star (l r : sexpr) | Opaque (p : nat) | Emp.

Fixpoint sexprD (ps : list hprop) (s : sexpr) : hprop :=
  match s with
  | Star l r ⇒ sexprD ps l * sexprD ps r
  | Opaque f ⇒ nth_with_default (default := ⌈ False ⌉) ps f
  | Emp ⇒ ∅
  end.
Definition check_entailment (l r: sexpr): bool := (* reflective procedure *)

Theorem check_entailment_sound : ∀ ps lhs rhs, (* soundness proof *)
  check_entailment lhs rhs = true →
  sexprD ps lhs ⊢ sexprD ps rhs. (* separation-logic entailment *)
```

Listing 1. A reflective entailment checker for propositional separation logic

procedures for reasoning about different domains, such as lists and bit-vectors, and compose them after the fact.

– Finally, our proof procedures are easy to *customize and extend without knowing about the details of reflection*. One drawback of reflective verification has been the need to write and verify programs in order to extend the automation. Combining the above techniques, we have built a more elementary interface that allows users to construct verified hint databases from Coq theorems completely automatically and pass them to generic reflective automation that applies the theorems (Section 3.4).

After presenting our technical contributions, we evaluate the performance and power of our automation (Section 4) and discuss related work (Section 5). Our techniques are implemented in the MirrorShard library that lays the foundation for the Bedrock automation. Both repositories are available online.

https://github.com/gmalecha/mirror-shard/
https://github.com/gmalecha/bedrock-mirror-shard/

2   Simple Entailment: A Computational Reflection Primer

Before diving into the novel bits, we sketch the high-level approach of computational reflection. We use entailment checking in a toy fragment of propositional separation logic [16] as our running example. Separation logic describes the program state compositionally by splitting it into disjoint pieces using the separating conjunction (notated $*$), which has the empty state (notated $\emptyset$) as its unit. For example, the formula $P * Q * R * \emptyset$ states that the entire program state can be divided into three disjoint parts described respectively by the opaque propositions P, Q, and R.

The first step in using computational reflection is to define a syntactic (called "reified") representation of formulas (sexpr); Listing 1 shows the code. The denotation function (sexprD) formalizes its meaning. Star and Emp represent $*$ and $\emptyset$ respectively; while Opaque n, for some index n, represents an uninterpreted proposition in the ps environment, e.g. P, Q, and R above. This indirection provides a decidable equality on `sexpr`, which allows us to detect (conservatively) when two opaque propositions are equal. When `ps` does not contain a value for an index, our denotation function uses $\lceil$`False`$\rceil$, a contradictory assertion.

Next we write a function (`check_entailment`) that determines whether the entailment is provable. Our simple algorithm erases all $\emptyset$ terms and crosses common terms off both sides of the entailment. If both sides wind up empty, then the entailment is provable. In order to use the procedure to prove an entailment, we prove a derived proof rule (the Coq theorem `check_entailment_sound`). The premise to this inference rule asserts that the function returns `true`, which can be checked efficiently by running the computation. If the result is `true`, the premise is justified by the reflexivity of equality. Notice that arbitrary entailments can be proved using this theorem by (1) reifying their syntax into the `sexpr` type and (2) applying the theorem with the quantifiers instantiated appropriately.

3 Composing Procedures

The entailment checker in the previous section is a good start, but it is not up to the challenges of program verification. Throughout this section we discuss how our technical contributions enable us to take it from a toy decision procedure to an extensible entailment checker capable of proving complex goals.

3.1 Syntax

Before we present our technical contributions, we set the stage with some more conventional elements of our syntax (shown in Listing 2).

The biggest inadequacy of the syntax presented in Section 2 is the representation of predicates. To illustrate the problem, consider the proposition $p \mapsto x$, expressing that the pointer p points to the value x. In the previous syntax this formula might be represented as `Opaque 1`, making it impossible to determine equivalence with $p + 0 \mapsto x$, which would be reified using a different index, e.g. as `Opaque 2`, since the terms are not *syntactically* equal.

To address this problem, we replace `Opaque n` with `Pred n xs` where `xs` is a list of arguments to the nth predicate. Because these arguments are not separation-logic formulas, we introduce a second syntactic category (`expr`) to represent them. We could stop here if all arguments to predicates were e.g. machine words, which would be quite restrictive. To enable `expr` to represent expressions of an open set of types, we introduce a third syntactic category for types (`typ`) and an associated denotation function (`typD`). This denotation function shows up in the return type of `exprD`, which determines the meaning of a syntactic expression at a given (syntactic) type. When the expression does not have the given type, `exprD` returns `None`, signaling a type error.

Our new syntax also supports constants using the `Const` constructor of `expr`. While constants are special cases of 0-ary function symbols, distinguishing them allows our reflective procedures to compute with them. The price that we pay for this flexibility is an additional parameter (`ts`, introduced by the `Variable`

```
Inductive typ := tyProp | tyType (idx : nat). (* type syntax *)
Variable ts : list Type. (* remaining definitions are parametrized by ts *)
Definition typD (t : typ) : Type :=
  match t with
  | tyProp ⇒ Prop
  | tyType i ⇒ nth_with_default (default := Empty_set) ts i
  end.

Inductive expr := (* expression syntax *)
| Func (f : nat) (args : list expr) | Equal (t : typ) (l r : expr)
| Const (t : typ) (val : typD t) | Var (idx : nat) | UVar (idx : nat).

Definition env := list { t : typ & typD t }. (* variable environments *)

Record func := (* syntactic functions *)
{ Args: list typ; Range: typ; Impl: fold_right (→) (typD Range) (map typD Args) }

Definition exprD (fs: fenv) (us vs: env) (e: expr) (t: typ)
: option (typD t) := ...

Inductive sexpr := (* separation logic syntax *)
| Star (l r: sexpr) | Pred (p:nat) (args:list expr) | Emp | Inj (p:expr)
| Exists (t : typ) (s : sexpr).

Record pred := (* syntactic separation logic predicates *)
{ PArgs : list typ ; PImpl : fold_right (→) hprop (map typD PArgs) }.

Definition sexprD (fs : fenv) (ps : penv) (us vs : env) (e : sexpr) : hprop := ...
```

Listing 2. Our three-level, extensible syntax & its denotation.

line) to `expr` and `sexpr` to represent the type environment. Beyond constants, we also support injecting propositions into separation-logic formulas using `Inj` and polymorphic equality using `Equal` in `expr`. The latter is important since our extensible function environment does not support polymorphic definitions, an issue we discuss in more detail in Section 4.3.

The final syntactic forms are for binders and are discussed in the next section.

3.2   Binders and Unification Variables

Existential quantification is common in verification conditions for functional correctness, especially when reasoning about data abstraction. As a result, quantifier support is essential to fully reflective reasoning.

Our syntax supports existential quantifiers in separation-logic formulas using the `Exists` constructor. Syntactically, variables are represented using de Bruijn indices, and the environment (`vs : env`) is encoded as a list of dependent pairs of values and their syntactic types. The denotation of an existential quantifier prepends the quantified value to the variable environment, while the denotation of a variable looks up the value and checks it against the expected type.

The final syntactic form, `UVar`, represents Coq unification variables, which are placeholders for currently unknown terms. Our procedures determine appropriate values for these variables using a reflective unification procedure coded in Gallina. As we will see in Section 3.4, our ability to implement unification reflectively is a powerful feature of our approach.

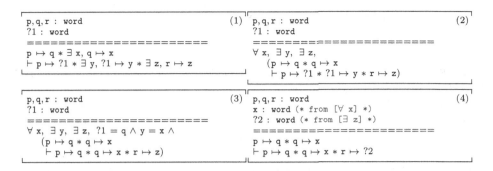

Fig. 2. Representation of quantifiers and unifications as they pass through our verification procedures: (a) initial goal; (b) result of lifting quantifiers; (c) direct output of the unification procedure; (d) after simplification with Ltac

Figure 2 shows how our reflective procedures manipulate quantifiers and unification variables that occur in entailments. Note that while we show each step as an individual goal, all of the steps except the last are performed within a single reflective call.

Box (1) A simple entailment that might be passed to our reflective checker. As in Coq, unification variables are prefixed with question marks. For clarity, we include them explicitly above the line, implicitly representing their contexts as the identifiers that occur above them[1]. For example, the term used to instantiate ?1 can mention any of p, q, and r.

Box (2) The normal form that our procedures use lifts quantifiers to the top. Existentials to the left are introduced as Vars that are universally quantified, while those to the right are represented as UVars and are existentially quantified. Here, the leading quantifiers are represented syntactically as lists of types, and the denotation function interprets them with the appropriate quantifiers.

Box (3) The result of unification is an instantiation of the unification variables. Semantically, this instantiation is a conjunction of equations, each between a unification variable and its instantiation. Here we see that ?1 was unified with q, and the value of the existentially quantified y has been chosen to be the newly introduced x.

Box (4) From here we cannot go any further reflectively, since unification variables only exist at the meta-level and thus cannot be manipulated in Coq's logic Gallina. Post-processing with Ltac cleans up the goal in Box (3) to look like the goal in Box (4). In particular, universally quantified variables are pulled into the context; unification variables are constructed for leading existentials using eexists; and instantiations are side-effected into the proof state by solving leading equations using reflexivity[2].

---

[1] The context of unification variables is not given to our procedures. They make the simplifying assumption that all terms are available in all contexts.

[2] If our reflective procedure instantiates a unification variable using terms outside of its context, reflexivity will fail, leaving the (likely unsolvable) goal to the user.

3.3 Compositional Semantic Reasoning

Only a small subset of operators are explicit in the syntax; the rest are represented by Func and Pred. For example, when we reason about the expression Star x y, the denotation, eliding the environments, is sexprD x * sexprD y. However, when we reason about the expression Func 0 [x;y], the denotation becomes, again eliding the environments and the error-handling code, (getFunc fs 0) (exprD x) (exprD y). To reason about the latter, we must express our environment assumptions as premises to the soundness proof.

To explain our technique for expressing these constraints, we introduce the following simple procedure for reasoning about the commutativity of addition.

```
Definition prove_plus_comm ts (e : expr ts) : bool :=
  match e with
  | Equal 1 (Func 0 [x ; y]) (Func 0 [y' ; x'])  ⇒ expr_eq x x' && expr_eq y y'
  | _ ⇒ false
  end.
```

Already this procedure makes the assumption that nat is at type index 1 and plus is at function index 0. We could prove this procedure sound for the environments [bool;nat] and [plus], but this proof would not be useful for extended environments. To develop reflective procedures independently and link them together after the fact, we need a compositional way to express these assumptions.

Our approach is to use a computational, rather than propositional, constraint formulation. To express constraints computationally, we quantify over an *arbitrary* environment and compute a *derived* environment that manifestly satisfies the constraints and is otherwise exactly the same as the original. The following function derives an environment from e that is guaranteed to satisfy c.

```
Fixpoint applyC (T: Type) (d : T) (c: constraints T) (e: list T) : list T :=
  match c with
  | nil ⇒ e
  | Any :: c' ⇒ hd_with_default (default := d) e :: applyC T d c' (tl e)
  | Exact v :: c' ⇒ v :: applyC T d c' (tl e)
  end.
```

To see applyC in action, we return to our example and declare the type environment constraints for the commutativity prover. Note that by using Any in position 0, we allow other procedures to choose a meaning for tyType 0.

```
Let TC : constraints Type := [Any; Exact nat].
```

Next we state the constraints for the function environment, which requires a syntactic representation of the plus function. Since the type of this syntactic representation depends on the type environment, we apply our technique, parametrizing by an arbitrary environment and retrofitting it with our constraints via applyC. In code:

```
Definition plus_fn (ts : list Type) : func (applyC TC ts) :=
{ Args := [tyType 1; tyType 1] ; Range := tyType 1 ; Impl := plus }.
```

This term type checks because `applyC` and `typD` reduce, making the following equations hold *definitionally.*

$$\mathsf{typD}(\mathsf{applyC\ TC}\ ts)(\mathsf{tyType}\ 1) \equiv \mathsf{typD}\ (\mathsf{hd}\ \mathsf{d}\ ts :: \mathsf{nat} :: \mathsf{tl}\ (\mathsf{tl}\ ts))\ (\mathsf{tyType}\ 1) \equiv \mathsf{nat}$$

The essential enabling property of `applyC` is that when c is a cons cell, the result is *syntactically* a cons cell and is not blocked by a `match` on e.

If we had stated the property propositionally and *proved* the equality, then `Impl` would require an explicit cast, like so.

```
Definition plus_fn_bad ts (pf : TC ⊨ ts) : func ts :=  (* ⊨ is 'holds on' *)
{ Args := [tyType 1; tyType 1] ; Range := tyType 1
; Impl := match compatible_reduces pf in _ = t return t → t → t with
          | eq_refl ⇒ plus end }.
```

Reasoning about casts in intensional type theory is difficult because the discriminee of the `match` must reduce to a constructor before the match can be eliminated. This behavior blocks conversion, making "seemingly equal" terms unequal. Our technique, on the other hand, does not even manifest the cast.

Using these definitions, we can prove the soundness of our simple commutativity prover using the following theorem statement.

```
Let FC ts : constraints (function (applyC TC ts)) := [Exact (plus_fn ts)].
Theorem prove_plus_comm_sound
: ∀ (ts : list Type), let ts' := applyC TC ts in
  ∀ (fs : functions ts'), let fs' := applyC FC fs in
  ∀ e₁ e₂, WellTyped ts' fs' e₁ (tyType 1) → WellTyped ts' fs' e₂ (tyType 1) →
    prove_plus_comm e₁ e₂ = true →
    exprD ts' fs' e₁ (tyType 1) =ₙₐₜ exprD ts' fs' e₁ (tyType 1).
```

With this formulation, `getFunc fs' 0` reduces to `plus`, making the proof follow from the commutativity of `plus`; a simple proof for a simple property.

```
  ∀ fs, let fs' := applyC FC fs in (getFunc fs' 0) x y = (getFunc fs' 0) y x
≡ ∀ fs, x + y = y + x
```

The fact that `applyC c l = l` when $c \models l$ justifies the completeness of the technique. Any theorem that is provable with the propositional formulation is also provable with our computational formulation.

Composition. `applyC`'s computational properties make it well-suited for composition. When two constraints are compatible, i.e. they do not specify different values for any index, `applyC` commutes *definitionally.*

$$\mathsf{applyC}\ C_1\ (\mathsf{applyC}\ C_2\ e) \equiv \mathsf{applyC}\ C_2\ (\mathsf{applyC}\ C_1\ e)$$

We can leverage this property for easy composition of functions and proofs without needing to reason about casts. For example, suppose we have two functions (analogously proofs) phrased using `applyC`, say p_1 and p_2, with different, but compatible, constraints TC_1 and TC_2. Composing each function with the application of the other's constraints gives us the following:

```
(fun ts ⇒ p₁ (applyC TC₂ ts)) : ∀ ts, expr (applyC TC₁ (applyC TC₂ ts)) → bool
(fun ts ⇒ p₂ (applyC TC₁ ts)) : ∀ ts, expr (applyC TC₂ (applyC TC₁ ts)) → bool
```

Since these types are definitionally equal, we can interchange the terms, for example, adding them to the same list or composing them via a simple disjunction without the need for explicit proofs or explicit casts:

```
Definition either ts (p₁ p₂ : expr ts → bool) (e : expr ts) : bool := p₁ e || p₂ e.
```

Packaging. To simplify passing provers around, we package them with their constraints and their soundness proofs using Coq's dependent records[3].

```
Record HintDatabase :=
{ Types : constraints Type
; Funcs : ∀ ts, constraints (func (applyC Types ts))
; Prover : ∀ ts, ProverT (applyC Types ts)
; Prover_sound : ∀ ts fs,
    ProverOk (applyC Types ts) (applyC (Funcs ts) fs) (Prover ts) }.
```

The first two fields represent the constraints for the type and function environments, with the latter phrased using our technique. The `Prover` field would contain the functional prover code, e.g. `prove_sum_comm` wrapped in our prover interface. The final field, `Prover_sound`, would contain the proof that the procedure is sound, derived from `prove_sum_comm_sound`.

To invoke a reflective procedure with a particular hint database, we rely on Ltac to handle the constraints. For example, the top-level soundness lemma for cancellation has the following form:

```
Theorem Apply_cancel_sound ts (fs : fenv ts)
  (prover : ProverT ts) (prover_ok : ProverOk ts fs prover)
: ∀ (lhs rhs : sexpr ts) (us vs : env ts), ...
```

To apply such a theorem, our reification process projects out the constraints from the hint database and uses them as base environments when constructing the syntactic terms. We can then instantiate the prover and its soundness proof to work on the extended environments (distinguished using primes) by simple application.

```
Apply_cancel_sound ts' fs'
  (hints.(Prover) ts' : ProverT ts' (*≡ProverT (applyC hints.(Types) ts')*))
  (hints.(Prover_sound) ts' fs' : ProverOk ts' fs') ...
```

Taking a closer look at the types we notice that the definition of `applyC` justifies the type assertions for the final two arguments. For example, the third argument actually has type `ProverT (applyC hints.(Types) ts')`, but since we construct `ts'` to retain the entries of `hints.(Types)`, this type is definitionally equal to `ProverT ts'`.

---

[3] The `HintDatabase` record in MirrorShard contains an additional field for the constraints on the predicate environment, but our example does not require it.

3.4   Generic Extension with Reified Lemmas

Writing and verifying reflective procedures can be cumbersome. For example, when verifying linked data structures like lists, the automation often requires rewriting by a separation-logic entailment. If each new entailment lemma required writing and verifying a new reflective procedure, users would spend more time building automation than using it. In Ltac, the sort of generic procedure we want is provided by the parametrized tacticals `rewrite` or `autorewrite`. In this section we show how the techniques from the previous two sections allow us to implement a generic, reflective rewriting procedure for separation-logic formulas that is parametrized by a list of lemmas to rewrite with.

As with all reflective tasks, the first hurdle is the representation. The variable-related features from Section 3.2 provide a simple way to represent lemmas.

```
Record lemma (ts : list Type) :=        (* Example:                      *)
{ Foralls : list typ                    (* ∀ x ls,                       *)
; Hyps : list (expr ts)                 (*    x = 0 →                    *)
; Lhs : sexpr ts ; Rhs : sexpr ts }.    (*    llist x ls ⊢ ⌈ ls = nil ⌉ *)

Definition lemmaD ts (fs : fenv ts) (ps : penv ts) (l : lemma) : Prop :=
  forallEach ts l.(Foralls) (fun vs ⇒
    implEach ts fs nil vs l.(Hyps)
      (sexprD ts fs ps nil vs l.(Lhs) ⊢ sexprD ts fs ps nil vs l.(Rhs))).
```

The function `lemmaD` translates a reified lemma statement into a Gallina proposition. Here, `forallEach` introduces universal quantifiers for the given types, packaging the quantified variables into an environment (`vs`) that it passes to the continuation. This environment is then used as the variable environment for the premises (denoted using `implEach`) and the conclusion.

We reduce the problem of determining when a lemma can be used to a unification problem. Our procedure replaces the universally quantified Vars in the lemma statement with UVars, setting up a unification "pattern." This pattern is then passed to the unification procedure that we mentioned in Section 3.2. If unification succeeds, we get a substitution that we can use to instantiate the lemma. Using provers like `prove_plus_comm` from the previous section allows us to discharge the premises. If all of the premises are discharged, we can replace the candidate term completely reflectively. Our rewriting procedure is able to rewrite in the premise and in the conclusion reusing mostly the same code and proofs in both cases.

The most difficult part of the proof lies in justifying the existence of values for the quantified variables. The unification procedure returns the expressions to use but, in order to support finding them incrementally, it only guarantees that they are well-typed in the environment that contains the new unification variables. Justifying that these expressions do not mention the new unification variables requires reasoning about the acyclicity of the instantiation, which is guaranteed by the occurs check in the unification algorithm. While complex, this proof is done once and used whenever we need to type-theoretically strengthen the result of unification.

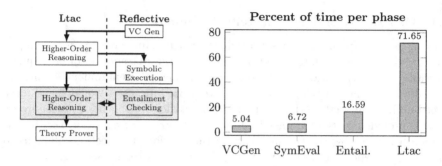

Fig. 3. Verification process and the breakdown of verification time

To make rewriting hints easy to use, we have completely automated the construction of syntactic lemmas and their composition into (an extended version of) the hint databases described in the previous section. These extended hint databases carry two lists of lemmas, one for forward rewriting and the other for backward rewriting, as well as their corresponding proof terms.

4   Evaluation and Discussion

The techniques in Sections 3.2-3.4 form the core insights of the MirrorShard framework. In this section we discuss our results applying this framework to verify programs written in the Bedrock system. We begin with an overview of the end-to-end verification process before justifying our claims from the introduction about the benefits of building broader reflective procedures.

We restrict our evaluation to a collection of data structure libraries including a memory allocator, linked-list operations, sets implemented as unsorted lists and binary trees, and queues. Together, these constitute approximately 355 lines of code that generate 253 verification conditions. The source code to these examples is found in the **examples** directory of our **bedrock-mirror-shard** repository.

4.1   The Verification Procedure

The end-to-end automation that verifies an entire Bedrock module is broken down into three independent, reflective tasks (verification condition generation, symbolic execution, and entailment checking) punctuated by Ltac-based higher-order reasoning. Figure 3 shows the overall process. We focus on the latter two tasks that, combined, apply to a single verification condition. Each verification condition assumes a precondition and that a particular code path is followed. The obligation is to show either that the code runs without errors (progress) or finishes in a state satisfying some postcondition (preservation). We focus on the second case since it is more interesting.

To solve a preservation verification condition, symbolic execution runs to compute the (strongest) postcondition of the path under the precondition. Next, an Ltac tactic runs to determine the postcondition. We use Ltac because we may require non-trivial higher-order reasoning (for instance, if the postcondition comes from the spec of a first-class function being called). This step reduces the goal to a separation-logic entailment that is discharged by our entailment checker. Because higher-order function specifications may involve nested assertions about specifications for other functions, entailment checking and the Ltac for higher-order reasoning run in a loop. Finally, user-defined Ltac runs to discharge any side conditions that could not be solved by our reflective procedures. In practice, these side conditions tend to be the pure parts of specifications, e.g. reasoning about Coq's `length` function when verifying its Bedrock implementation.

Figure 3 shows how the verification time is distributed between the different phases across our data-structure examples. Note that while our reflective procedures end up doing most, if not all, of the heavy lifting, almost three-quarters of our verification time is spent running Ltac, suggesting that while we could further optimize our reflective procedures, the biggest improvements would come from making more of the verification reflective.

4.2 Reflective Performance

Previous work [2,11] has demonstrated the performance and scaling benefits of reflective automation, and our work enjoys similar benefits. More central to our thesis is the benefit of reflective composition and user extension, which we evaluate in the context of symbolic execution.

Consider the following path through the length function for linked lists:

```
assume( *(Sp+4) ≠ 0 );    (* not at the end of the list *)
*(Sp+8) := *(Sp+8) + 1 ;  (* increment the length counter *)
Rv := *(Sp+4) ;           (* get the next pointer *)
*(Sp+4) = *Rv             (* update "current" *)
```

The references from the stack pointer Sp are to local variables. Sp+8 is the location of the length counter, and Sp+4 is the location of the "current" pointer. The first line is the result of knowing that the conditional comparing **current** to **null** returned false, implying that evaluation is not at the end of the list, which justifies the memory dereference on the last line where the code reads the **next** pointer of the current linked-list cell ($**(Sp+4)$).

In order to exploit this information during symbolic execution, our symbolic executor uses the following hint, provided in a hint database:

```
Lemma llist_cons_fwd : ∀ ls (p : W), p ≠ 0
    → llist ls p ⊢ ∃ x, ∃ ls', ⌈ ls = x :: ls' ⌉ * ∃ p', (p ↦ x, p') * llist ls' p'.
```

This lemma is fed to the reflective rewriting framework discussed in Section 3.4, which exposes the ↦ predicate that symbolic execution knows how to interpret[4].

Without this mechanism, we could achieve the same automation by running an Ltac loop bouncing between our reflective symbolic execution and the `autorewrite` tactic to perform this rewriting. In the above example, this loop

---

[4] Not just ↦ but also some other "base" predicates are interpreted by independent, user-defined reflective procedures that plug into our symbolic execution framework.

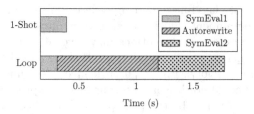

Fig. 4. Performance of 1-shot symbolic execution versus Ltac composition using `autorewrite` on the small example goal

would call symbolic execution, which would get stuck on the final instruction, falling back on `autorewrite` to expose the cons cell, enabling a second call to symbolic evaluation to complete the task.

Figure 4 shows how the loop approach compares to our fully reflective procedure (1-Shot). Using the latter, the entire symbolic execution takes 0.39 seconds, less than half the amount of time (0.89 seconds) taken by `autorewrite` to perform just the rewriting. Overall the reflective composition results in a 4.6x speedup over the Ltac-based composition on this goal, translating into 44 seconds when applied to the entire linked-list module.

While our reflective rewriter is not as powerful as `autorewrite`, it is customizable in the same way. Further, because it is written in Gallina rather than hardcoded inside Coq, we can extend it with smarter unification that, for example, can reason about provable rather than just definitional equality.

4.3   Limitations and Future Work

MirrorShard's success as the core automation for Bedrock is strong evidence for its expressivity. However, the expressivity of Coq's logic limits the power of reflective procedures.

MirrorShard's computational formulation of constraints relies crucially on constants in certain places, for example the indices of types. For example, while it is easy to write a procedure that is sound for any environment where `nat` is located at position 1, it is more difficult to write a procedure parametrized by x that is sound for any environment where `nat` is at position x. While it is possible to manipulate proofs explicitly and achieve the latter degree of parametrization, in this work we have opted for the simpler solution. As we expand the ideas and techniques beyond separation logic, developing more parametrized procedures will likely become more important.

MirrorShard's syntax does not support a general notion of binders, only existential quantifiers in separation-logic formulas. While this limitation has not been problematic for entailment checking and symbolic execution, it prevents us from reasoning about e.g. inline functions and `match` expressions. Supporting a general notion of binder may provide a way to automate reflectively some of the tasks that we currently accomplish in Ltac, increasing the scope of reflection and further improving performance.

While binders should be within our grasp, restrictions of the logic put other features, like general support for polymorphic types, farther out of reach. Type functions can be encoded for special fixed arities, but a general solution allowing arbitrary arities requires universe polymorphism. Universe polymorphism as described by Harper and Pollack [12] is slated for Coq 8.5 and will solve some of these issues.

Finally, general value-dependent types pose an even greater problem. The MirrorShard representation stratifies the type and term languages, but truly dependent types would require these to be unified, making the type of the denotation function mention itself in the style of very dependent functions [13].

5   Related Work

MirrorShard is not the first verified implementation of separation-logic automation, but it is the first to support modular user extension. Marti and Affeldt [15] implemented a verified version of Smallfoot [1], and Stewart et al. [17] verified a more sophisticated heap theorem prover based on paramodulation. Both of these systems are limited to the standard points-to and singly-linked list predicates, and extending either to support user-defined abstract predicates with equations would likely require a considerable overhaul of both the procedure and its proof.

While program verification is our application, our technical contributions are our techniques for phrasing, composing, and extending reflective procedures and their proofs. The applicability of these techniques extends well beyond program verification. Several projects have built large, generic reflective procedures. In his PhD thesis, Lescuyer [14] describes a reflective implementation of an SMT solver. While he also uses an environment-based representation, he is unable to reason about it semantically. As a result, it is not clear how to support first-class hint databases or include additional theories that need to reason semantically about symbols represented using the environment. Our work also supports quantifiers.

Similar to our rewriting engine is the work by Braibant and Pous on reasoning modulo associativity and commutativity [3]. Like Lescuyer, they specialize their procedures for reasoning semantically about a fixed set of symbols (in their case an abstract commutative, associative operator), which removes the need to reason about multiple types or multiple operators. Our techniques support both.

Recent work by Claret et al. [5] on posterior simulation for reflective proofs aims to make it easier to write reflective procedures by supporting side effects and branching proof search efficiently. This goal is complementary to our own work and offers a method of automatic caching for results of the (potentially large) proof searches that our extensible procedures enable. This caching may become essential if reflective procedures begin to rely heavily on speculation.

In the wider sphere of proof automation, Mtac [18] proposes a monadic language for writing Gallina terms that are run during program elaboration. Unlike MirrorShard, Mtac supports dependent and polymorphic types; however, its support for binder manipulation is less sophisticated. For example, it does not appear to be possible to apply lemmas without knowing their types a priori, making it difficult to parametrize by lemmas that are applied automatically.

The Ssreflect tactic library [10] has become a popular alternative to Ltac. Ssreflect provides a higher-level tactic language and support for "small-scale" reflection. The tactics aim to make it easier to refactor proofs and lemmas, but it is still focused on smaller reasoning steps. This approach avoids the need to compose reflective procedures but requires more effort by the user to determine and perform the appropriate reasoning explicitly.

One of the core problems that we overcome in our formulation is the expression problem. Our concrete syntax is similar to that of Garillot and Werner [9], though their work does not suggest any methods for achieving semantic reasoning, which is essential to reasoning about actual terms. Delaware et al. [8] recently proposed another solution to this problem using Church encodings. While useful for reasoning about the metatheoretic properties of programming languages, it is not clear that Church encodings completely solve the issues that arise in computational reflection. In particular, representing terms as functions can make them costly to compute with and type check.

6   Conclusions

In this work we presented three novel techniques for building extensible reflective procedures in Coq. First, we presented a reflected representation of unification variables and existential quantifiers, which we reason about using a verified unification algorithm. Our second technique is a simple encoding of extensible syntax suitable for computational reflection, plus a formulation of constraints that allows reasoning about this representation without any runtime overhead. Our third technique is a method for building first-class, reflected hint databases that can be used by reflective procedures.

These techniques form the core technical insights of MirrorShard, a reusable Coq library for reflective procedures about separation logic. The extensibility of these procedures allows them to reason about broader problems by reflectively orchestrating general and domain-specific reasoning. Our evaluation shows that this approach can provide a significant speedup over performing the extensible reasoning in a hybrid of reflective procedures and Ltac.

Acknowledgments. The authors thank Patrick Hulin and Edward Z. Yang for their contributions to the MirrorShard implementation. We received helpful feedback on this paper from: Andrew W. Appel, Jesper Bengtson, Josiah Dodds, Georges Gonthier, Daniel Huang, Andrew Johnson, Jacques-Henri Jourdan, Scott Moore, Greg Morrisett, and Kenneth Roe. This work has been supported in part by a Facebook Fellowship, an NSF Graduate Research Fellowship, NSF grant CCF-1253229, AFRL under agreement FA8650-10-C-7090, and DARPA under agreement number FA8750-12-2-0293. The U.S. Government is authorized to reproduce and distribute reprints for Governmental purposes notwithstanding any copyright notation thereon. The views and conclusions contained herein are those of the authors and should not be interpreted as necessarily representing the official policies or endorsements, either expressed or implied, of DARPA or the U.S. Government.

References

1. Berdine, J., Calcagno, C., O'Hearn, P.W.: Smallfoot: Modular automatic assertion checking with separation logic. In: de Boer, F.S., Bonsangue, M.M., Graf, S., de Roever, W.-P. (eds.) FMCO 2005. LNCS, vol. 4111, pp. 115–137. Springer, Heidelberg (2006)
2. Boutin, S.: Using reflection to build efficient and certified decision procedures. In: Ito, T., Abadi, M. (eds.) TACS 1997. LNCS, vol. 1281, pp. 515–529. Springer, Heidelberg (1997)
3. Braibant, T., Pous, D.: Tactics for reasoning modulo AC in Coq. In: Jouannaud, J.-P., Shao, Z. (eds.) CPP 2011. LNCS, vol. 7086, pp. 167–182. Springer, Heidelberg (2011)
4. Chlipala, A.: The Bedrock structured programming system: Combining generative metaprogramming and Hoare logic in an extensible program verifier. In: Proc. ICFP, pp. 391–402. ACM (2013)
5. Claret, G., del Carmen González Huesca, L., Régis-Gianas, Y., Ziliani, B.: Lightweight proof by reflection using a posteriori simulation of effectful computation. In: Blazy, S., Paulin-Mohring, C., Pichardie, D. (eds.) ITP 2013. LNCS, vol. 7998, pp. 67–83. Springer, Heidelberg (2013)
6. Coq Development Team. The Coq proof assistant reference manual, version 8.4 (2012)
7. Delahaye, D.: A tactic language for the system Coq. In: Parigot, M., Voronkov, A. (eds.) LPAR 2000. LNCS (LNAI), vol. 1955, pp. 85–95. Springer, Heidelberg (2000)
8. Delaware, B., d.S. Oliveira, B.C., Schrijvers, T.: Meta-theory a la carte. SIGPLAN Not. 48(1), 207–218 (2013)
9. Garillot, F., Werner, B.: Simple types in type theory: Deep and shallow encodings. In: Schneider, K., Brandt, J. (eds.) TPHOLs 2007. LNCS, vol. 4732, pp. 368–382. Springer, Heidelberg (2007)
10. Gonthier, G., Mahboubi, A., Tassi, E.: A Small Scale Reflection Extension for the Coq System. Rapport de recherche RR-6455, INRIA (2008)
11. Grégoire, B., Leroy, X.: A compiled implementation of strong reduction. In: Proc. ICFP (2002)
12. Harper, R., Pollack, R.: Type checking with universes. Theoretical Computer Science 89(1), 107–136 (1991)
13. Hickey, J.J.: Formal objects in type theory using very dependent types. In: Foundations of Object Oriented Languages 3 (1996)
14. Lescuyer, S.: Formalisation et développement d'une tactique réflexive pour la démonstration automatique en Coq. Thèse de doctorat, Université Paris-Sud (January 2011)
15. Marti, N., Affeldt, R.: A certified verifier for a fragment of separation logic. Computer Software 25(3), 135–147 (2008)
16. Reynolds, J.C.: Separation logic: A logic for shared mutable data structures. In: Proc. LICS, pp. 55–74. IEEE Computer Society (2002)
17. Stewart, G., Beringer, L., Appel, A.W.: Verified heap theorem prover by paramodulation. In: Proc. ICFP (2012)
18. Ziliani, B., Dreyer, D., Krishnaswami, N., Nanevski, A., Vafeiadis, V.: Mtac: A monad for typed tactic programming in Coq. In: Proc. ICFP (2013)

An Isabelle Proof Method Language

Daniel Matichuk[1,2], Makarius Wenzel[3], and Toby Murray[1,2]

[1] NICTA, Sydney, Australia*
[2] School of Computer Science and Engineering, UNSW, Sydney, Australia
[3] Univ. Paris-Sud, Laboratoire LRI, UMR8623, Orsay, F-91405, France
CNRS, Orsay, F-91405, France

Abstract. Machine-checked proofs are becoming ever-larger, presenting an increasing maintenance challenge. Isabelle's most popular language interface, Isar, is attractive for new users, and powerful in the hands of experts, but has previously lacked a means to write automated proof procedures. This can lead to more duplication in large proofs than is acceptable. In this paper we present Eisbach, a proof method language for Isabelle, which aims to fill this gap by incorporating Isar language elements, thus making it accessible to existing users. We describe the language and the design principles on which it was developed. We evaluate its effectiveness by implementing some tactics widely-used in the seL4 verification stack, and report on its strengths and limitations.

1 Introduction

Machine-checked proofs, developed using interactive proof assistants, present an increasing maintenance challenge as they become ever larger. For instance, the proofs and specifications that accompany the formally verified seL4 microkernel now comprise 480,000 lines of Isabelle/HOL [9], while Isabelle's Archive of Formal Proofs http://afp.sf.net now comprises over 900,000 lines. Each of these developments is updated to ensure it runs with each new Isabelle release. Large proofs about living software implementations present the additional maintenance challenge of having to be updated as the software to which they apply evolves over time, as is the case with seL4.

The Isabelle proof assistant [15, §6] provides various languages for different purposes, which sometimes overlap and sometimes complement each other. Most commonly used is the Isar language for theory specifications and structured proofs [14]. Isabelle/Isar sits alongside Isabelle/ML, which exposes the full power of system implementation and extension, including the ability to implement new sub-languages of the Isabelle framework. Isar itself is devoid of computation, but it may appeal to arbitrarily complex proof tools from the library: so-called *proof methods*. These are are usually implemented in Isabelle/ML. Isabelle/ML is integrated into the formal context of Isabelle/Isar, and supports referring to logical entities or Isar elements via *antiquotations* [13]. While this makes it reasonably easy to access the full power of ML in proofs, the vast majority of Isabelle theories are written solely in Isar: the AFP comprises just 50 ML files, as compared to 1663 Isar (.thy) files only 6 of which embed ML code.

---

* NICTA is funded by the Australian Government through the Department of Communications and the Australian Research Council through the ICT Centre of Excellence Program.

G. Klein and R. Gamboa (Eds.): ITP 2014, LNAI 8558, pp. 390–405, 2014.

The Isar proof language does not support proof procedure definitions directly, but this hasn't prevented large verifications from being completed: the seL4 proofs rely mainly on two custom tactics. This can be partly explained by the power of existing proof tools in Isabelle/HOL. However, it has arguably led to more duplication in these proofs than is acceptable; managing duplication has been a challenge for the seL4 proofs in [1]. This duplication makes proof maintenance difficult, and highlights the barrier to entry when implementing proof tools in Isabelle/ML. If automation can be expressed at a high level, a wider class of users can maintain and extend domain-specific proof procedures, which are often more maintainable than long proof scripts.

In this paper, we present a proof method language for Isabelle, called Eisbach, that allows writing proof procedures by appealing to existing proof tools with their usual syntax. The new Isar command **method-definition** allows proof methods to be combined, named, and abstracted over terms, facts and other methods. Eisbach is inspired by Coq's Ltac [4], and includes similar features such as matching on facts and the current goal. However, Eisbach's matching behaves differently to Ltac's, especially with respect to backtracking (see Section 3.5). Eisbach continues the Isabelle philosophy of exposing carefully designed features to the user while leaving more sophisticated functionality to Isabelle/ML: small snippets of ML may be easily included on demand. Eisbach benefits from general Isabelle concepts, while easing their exposure to users: pervasive backtracking, the structured proof context with named facts, and attributes to declare hints for proof tools.

The following simple example defines a new proof method which identifies a list in the conclusion of the current subgoal and applies the default induction principle to it with the existing method *induct*. All newly emerging subgoals are solved with *fastforce*, with additional simplification rules given as argument.

method-definition *induct-list* **facts** *simp* =
(**match** *?concl* **in** *?P* (*?x* :: *'a list*) $\Rightarrow$ (*induct ?x* $\mapsto$ *fastforce simp: simp*))

Now *induct-list* can be called as a proof method to prove simple properties about lists.

lemma *length* (*xs* @ *ys*) = *length xs* + *length ys* **by** *induct-list*

The primary goal of Eisbach is to make writing proofs more productive, to avoid duplication, and thereby lower the costs of proof maintenance. Its design principles are:

- To be easy to use for beginners and experts.
- To expose limited functionality, leaving complex functionality to Isabelle/ML.
- Seamless integration with other Isabelle languages.
- To continue Isar's principle of readable proofs, creating readable proof procedures.

We begin in Section 2 by recalling some concepts of Isabelle and Isar. Section 3 then presents Eisbach, via a tour of its features in a tutorial style, concluding with the development of a solver for first order logic. We describe Eisbach's design and implementation in Section 4, before evaluating it in Section 5 by implementing the two most widely-used proof methods of the seL4 verification stack, and comparing them against their original implementations. Section 6 then surveys related work on proof programming languages, to put Eisbach in proper context. In Section 7 we compare Eisbach to Coq's Ltac and Mtac before considering future work and concluding.

2 Some Isabelle Concepts

Isabelle was originally introduced as yet another *Logical Framework* by Paulson [12], to allow rapid prototyping of implementations of inference systems, especially versions of Martin-Löf type theory. Some key concepts of current Isabelle can be traced back to this heritage, although today most applications are done exclusively in the object-logic Isabelle/HOL, and the general system framework has changed much in 25 years.

Isabelle/Pure is a minimal version of higher-order logic, which serves as general framework for Natural Deduction (with arbitrary nesting of rules). There are Pure connectives for universal parameters $\bigwedge x. \; \Box$, premises $A \implies \Box$, and a notion of schematic variables $?x$ (stripped outermost parameters). The Pure connectives outline inference rules declaratively, for example conjunction introduction $A \implies B \implies A \wedge B$ or well-founded induction $wf \, r \implies (\bigwedge x. \; (\bigwedge y. \; (y, x) \in r \implies P \, y) \implies P \, x) \implies P \, a$.

Isabelle/HOL is a rich library of logical theories and tools on top of Isabelle/Pure. It is the main workhorse for big applications, but is subsumed by the general concepts of Isabelle, so w.l.o.g. it is subsequently not explained further.

The logical framework of Isabelle/Pure is augmented by extra-logical infrastructure of Isabelle/Isar, which provides the general setting for structured reasoning. The actual Isar proof language [14] is merely an application of that: it provides particular expressions for human-readable proofs within the generic framework. Some of the underlying concepts of Isabelle architecture are outlined below, as relevant for Eisbach.

Fact. While the inference kernel operates on *thm* entities (as in LCF or HOL), Isabelle users always encounter results as *thm list*, which is called *fact*. This represents the idea of multiple results, without auxiliary conjunctions to encode it within the logic. There is notation to append facts, or to project sub-lists, without any formal reasoning involved.

Goal State. Following [12], the LCF goal state as auxiliary ML data structure is given up, and replaced by a proven theorem that states that the current subgoals imply the main conclusion. Goal refinement means to infer forwards on the negative side of some implication, so it appears like backwards reasoning. The proof starts with the trivial fact $C \implies C$ and concludes with zero subgoals $\implies C$, i.e. C outright. Administrative goal operations, e.g. shuffling of subgoals or restricted subgoal views, work by elementary inferences involving $\implies$ in Isabelle/Pure. While outermost implications represent subgoals, outermost goal parameters correspond to *schematic variables* (or meta-variables), but the latter aspect is subsequently ignored for simplicity.

Tactic. Isabelle tactics due to [12] follow the idea behind LCF tactics, but implement the backwards refinement more directly in the logical framework, without replaying tactic justifications (as still seen in HOL or Coq today). This avoids the brittle concentration of primitive inferences at qed-time. Moreover, backtracking is directly built-in, by producing an unbounded lazy list of results, instead of just zero or one. LCF-style **tacticals** are easily recovered, by composing functions that map a goal state to a sequence of subsequent goal states. Rich varieties of combinators with backtracking are provided, although modern-time proof tools merely use a more focused vocabulary.

Subgoal Structure. An intermediate goal state with n open subgoals has the form $H_1 \implies ... H_n \implies C$, each with its own substructure $H = (\bigwedge x. \; A \, x \implies B \, x)$, for zero or more *goal parameters* (here *x*) and *goal premises* (here *A x*). Following [12], this local context is implicitly taken into account when natural deduction rules are composed by *lifting, higher-order unification*, and *backward chaining*. Isar users encounter this operation frequently in the proof method *rule*, and the rule attributes *OF* or *THEN*.

Other proof tools may prefer direct access to hypothetical terms and premises, when inspecting a subgoal. In Isabelle today the concept of **subgoal focus** achieves that: the proof context is enriched by a fixed term *x* and assumed fact *A x*, and the subgoal restricted to *B x*. After refining that, the result is retrofitted into the original situation.

Proof Context. Motivated by the Isar proof language [14], the structured proof context provides general administrative structure, to complement primitive *thm* values of the inference kernel. The idea is to provide a first-class representation in ML, of open situations with hypothetical terms (fixed variable *x*) and assumptions (hypothetical fact *A*); Hindley-Milner type discipline with schematic polymorphism is covered as well. Proof contexts are not restricted to this logical core, but may contain arbitrary tool-specific **context data**. A typical example is the standard environment of facts (see above), which manages both static and dynamic entries: a statically named fact is interchangeable with its *thm list* as plain value, but a dynamic fact is a function depending on the context.

Attributes. Facts and contexts frequently occur together, and may modify each other by means of attributes (which have their own syntax in Isar). A **rule attribute** modifies a fact depending on the context (e.g. *fact [of t]* to instantiate term variables), and a **declaration attribute** modifies the context depending on a fact (e.g. *fact [simp]* to add Simplifier rules to the context). Such declarations for automated proof tools also work in hypothetical contexts, with fixed *x* and assumed *A x*. There is standard support to maintain named collections of dynamic facts, with attributes to add or delete list entries.

3   Eisbach

3.1   Isar Proof Methods

Eisbach provides the ability to write automated reasoning procedures to non-expert users of Isabelle, specifically users only familiar with the use of Isabelle/Isar [14].

Isar is a document-oriented proof language, focusing on producing and presenting human-readable formal proofs. Such proofs are a structured argument about why a claim is true, with invocations to proof methods to decompose a claim into multiple goals or to solve outstanding proof goals. For the purposes of this paper, proof method invocations come in two forms: structured and unstructured.

The structured form is "**by** *method₁ method₂*", where the initial *method₁* performs the main structural refinement of the goal, and the terminal (optional) *method₂* may solve emerging subgoals; the proof is always closed by implicit *assumption* steps to finish-off trivial subgoals. For example, "**by** (*induct n*) *simp-all*" splits-up a problem by induction and solves it by simplification, or "**by** (*rule impI*)" applies a single rule and expects the remaining goal state to be trivial up to unification.

The unstructured form is "**apply** *method*", which applies the proof method to the goal without insisting the proof be completed; further **apply** commands may follow to continue the proof, until it is eventually concluded by the command **done** (without implicit steps for closing). After one or two **apply** steps, the foreseeable structure of the reasoning is usually lost, and the Isar *proof text* degenerates into a *proof script*: understanding it later typically requires stepping through its intermediate goal states.

The *method* expressions above may combine basic proof methods using Isar's method *combinators*. Unlike former tacticals, there is only a minimalistic repertoire for repeated application, alternative choice, and sequential composition (with backtracking). Such methods are used in-place, to address a particular proof problem in a given situation.

Eisbach allows compound proof methods to be named, and extend the name space of basic methods accordingly. Method definitions may abstract over parameters: terms, facts, or other methods. Additionally, Eisbach provides an expressive matching facility that can be used to manage control flow and perform proof goal analysis via unification.

Subsequently, we will follow the development of a small first order logic solver in Eisbach, gradually increasing its scope and demonstrating the main language elements.

3.2   Combinators and Backtracking

There are four combinators in Isar. Firstly, "," is sequential composition of two methods with implicit backtracking: "*meth1*,*meth2*" applies *meth1*, which may produces a set of possible results (new proof goals), before applying *meth2* to all results produced by *meth1*. Effectively this produces all results in which the application of *meth1* followed by *meth2* is successful.

At the end of each **apply** command, the first successful result from all those produced is retained.

The second Isar combinator is "|", alternative composition: "*meth1*|*meth2*" tries *meth1* and falls through to *meth2* when *meth1* fails (yields no results). The third combinator "?" is a unary combinator that suppresses failure: *meth*? returns the original proof state when *meth* fails, rather than failing. Lastly, "+" is a unary combinator for repeated method application: *meth*+ repeatedly applies *meth* until *meth* fails, at which point it yields the proof state obtained before the final failing invocation of *meth*.

A typical method invocation might look as follows:

lemma $P \wedge Q \longrightarrow P$ **by** ((*rule impI*, (*erule conjE*)?) | *assumption*)+

Which, informally, says: "Apply the implication introduction rule, followed by optionally eliminating any conjunctions in the assumptions. If this fails, solve the goal with an assumption. Repeat this action until it is unsuccessful."

As well as the above lemma, this invocation will prove the correctness a small class of propositional logic tautologies. With the **method-definition** command we can define a proof method that makes the above functionality available generally.

method-definition *prop-solver*$_1$ = ((*rule impI*, (*erule conjE*)?) | *assumption*)+
lemma $P \wedge Q \wedge R \longrightarrow P$ **by** *prop-solver*$_1$

3.3   Abstraction

We can abstract this method over its introduction and elimination rules to make it more generally applicable. The **facts** keyword declares fact parameters for use in the method.

These arguments are provided when the method is invoked, in the form of lists of facts for each, using Isar's standard method-sections syntax. Below we generalise the method above over its *intro* and *elim* rules respectively that it may apply.

> **method-definition** *prop-solver$_2$* **facts** *intro elim* =
> $((rule\ intro,\ (erule\ elim)?)\ |\ assumption)+$
>
> **lemma** $P \wedge Q \longrightarrow P$ **by** $(prop\text{-}solver_2\ intro\text{:}\ impI\ elim\text{:}\ conjE)$

Above, the introduction and elimination rules need to be provided for each method invocation. Traditionally Isabelle proof methods avoid this by using tool-specific data as part of the proof context, which are managed using *attributes* (see Section 2) to add and remove entries. A method invocation retrieves the facts that it needs to know about whenever it is invoked, using the run-time proof context.

Eisbach supports creating new fact collections in the context using a new Isar command **declare-attributes**. A fact parameter [*p*] surrounded by square brackets declares that fact to be backed by the fact collection *p*. It can be augmented further when a method is invoked using the common syntax *meth p: facts*, but can also be managed in the proof context with the Isar command **declare**.

> **declare-attributes** *intro elim*
>
> **method-definition** *prop-solver$_3$* **facts** [*intro*] [*elim*] =
> $((rule\ intro,\ (erule\ elim)?)\ |\ assumption)+$
>
> **declare** *impI* [*intro*] **and** *conjE* [*elim*]
>
> **lemma** $P \wedge Q \longrightarrow P$ **by** *prop-solver$_3$*

Methods can also abstract over terms using the **for** keyword, optionally providing type constraints. For instance, the following proof method *elim-all* takes a term *y* of any type, which it uses to instantiate the *x*-variable of the *allE* (forall elimination) rule before applying that rule as an elimination rule. The instantiation is performed here by Isar's *where* attribute. This has the effect of instantiating a universal quantification $\forall x.$ $P\ x$ in one of the current assumptions by replacing it with the term $P\ y$.

> **method-definition** *elim-all* **for** $Q :: {}'a \Rightarrow bool$ **and** $y :: {}'a =$
> $(erule\ allE\ [where\ P = Q\ and\ x = y\ ])$

The term parameters *y* and *P* can be used arbitrarily inside the method body, as part of attribute applications or arguments to other methods. The expression is type-checked as far as possible when the method is defined, however dynamic type errors can still occur when it is invoked (e.g. when terms are instantiated in a parameterized fact). Actual term arguments are supplied positionally, in the same order as in the method definition.

> **lemma** $\forall x.\ P\ x \Longrightarrow P\ a$ **by** $(elim\text{-}all\ P\ a)$

3.4 Custom Combinators

The four existing combinators in Isar (mentioned above) quickly prove to be too restrictive when writing tactics in Eisbach. A fifth combinator ("$\mapsto$") was added, which takes two methods and, in contrast to ",", invokes the second method on *all* subgoals produced by the first. This is necessary to handle cases where the number of subgoals produced by a method cannot be known statically.

lemma *True* ∧ *True* ∧ *True* **by** (*intro conjI* ↦ *rule TrueI*)

To more usefully exploit Isabelle's backtracking, the explicit requirement that a method solve all produced subgoals is frequently useful. This can easily be written as a *higher-order method* using "↦". The **methods** keyword denotes method parameters that are other proof methods to be invoked by the method being defined.

method-definition *solve* **methods** *m* = (*m* ↦ *fail*)

Given some method-argument *m*, *solve m* applies the method *m* and then fails whenever *m* produces any new unsolved subgoals – i.e. when *m* fails to completely discharge the goal it was applied to.

With these simple features we are ready to write our first non-trivial method. Returning to the first order logic example, the following method definition applies various rules with their canonical methods.

method-definition *prop-solver* **facts** [*intro*] [*dest*] [*elim*] [*subst*] =
    (*assumption*
    | *rule intro* | *drule dest* | *erule elim* | *subst subst* | *subst* (*asm*) *subst* |
    (*erule notE* ↦ *solve prop-solver*))+

The only non-trivial part of this method definition is the final alternative (*erule notE* ↦ *solve prop-solver*). Here, in the case that all other alternatives fail, the method takes one of the assumptions ¬ *P* of the current goal and eliminates it with the rule *notE*, causing the goal to be proved to become *P*. The method then recursively invokes itself on the remaining goals. The job of the recursive call is to demonstrate that there is a contradiction in the original assumptions (i.e. that *P* can be derived from them). Note this recursive invocation is applied with the *solve* method to ensure that a contradiction will indeed be shown. In the case where a contradiction cannot be found, backtracking will occur and a different assumption ¬ *Q* will be chosen for elimination.

After declararing some standard rules to the context, e.g. (*P* ⟹ *False*) ⟹ ¬ *P* as [*intro*] and ¬ ¬ *P* ⟹ *P* as [*dest*], the *prop-solver* becomes capable of solving non-trivial propositional tautologies.

lemma (*A* ∨ *B*) ∧ (*A* ⟶ *C*) ∧ (*B* ⟶ *C*) ⟶ *C* **by** *prop-solver*

3.5   Matching

Matching allows the user to introspect the goal state, and to implement more explicit control flow. When performing a match, the user provides a term or fact collection *ts* to match against, along with a collection of pattern-method pairs (*p*, *m*): roughly speaking, when the pattern *p* matches any member of *ts*, the *inner* method *m* will be executed with schematic variables mentioned in *p* appropriately instantiated. In the case of matching against a fact collection, an optional name may be given for each pattern, which will be bound to the fact that was successfully matched out of term ts. The special term *?concl* is always defined to be the conclusion of the first subgoal, and the special fact *prems* is always defined to be the premises of the first subgoal. Using either in the term *t* allows the user to perform matches against (i.e. to introspect) the current goal-state; doing so causes an implicit *subgoal focus* (see also Section 4) which binds these two names appropriately, creating a local context of local goal parameters (as fixed term variables) and premises (as hypothetical theorems).

In the following example we extract the predicate of an existentially quantified conclusion in the current subgoal and search the current premises for a matching fact. If both matches are successful, we then instantiate the existential introduction rule with both the witness and predicate, solving with the matched premise.

> **method-definition** *solve-ex* =
> (**match** *?concl* **in** $\exists x.\ ?Q\ x \Rightarrow$
>   (**match** *prems* **in** $U: Q\ ?y \Rightarrow (rule\ exI\ [where\ x = y\ \textbf{and}\ P = Q, OF\ U])))$

The first match matches the pattern $\exists x.\ ?Q\ x$ against the current conclusion, binding the pattern $?Q$ to a particular term Q in the inner match. Next the pattern $Q\ ?y$ is matched against all premises of the current subgoal. Once a match is found, the local fact U is bound to the matching premise and the variable y is bound to the matching witness. The existential introduction rule $P\ x \Longrightarrow \exists x.\ P\ x$ is then instantiated with y as the witness and Q as the predicate, with its proof obligation solved by the local fact U (using the Isar attribute *OF*). The following example is a trivial use of this method.

> **lemma** *halts p* $\Longrightarrow \exists x.\ halts\ x$ **by** *solve-ex*

Matching is performed from top to bottom, considering each pattern in turn until a match is found. When attempting to match a pattern, Eisbach tries to match the pattern against all provided terms/facts before moving on to the next pattern. Successful matches serve as cut points for backtracking. Specifically, once a match is made no other patterns will be attempted regardless of the outcome of the inner method *m*. However, all possible unifiers of that pattern will be explored, re-executing the method *m* with different variable bindings when backtracking.

The method *foo* below fails for all goals that are conjunctions. Any such goal will match the first pattern, causing the second pattern (that would otherwise match all goals) to never be considered. If multiple unifiers exist for the pattern $?P \wedge ?Q$ against the current goal, then the failing method *fail* will be (uselessly) tried for all of them.

> **method-definition** *foo* =
> (**match** *?concl* **in** $?P \wedge ?Q \Rightarrow fail \mid ?R \Rightarrow prop\text{-}solver)$

This behaviour is in direct contrast to the backtracking done by Coq's Ltac [4], which will attempt all patterns in a match before failing. This means that the failure of an inner method that is executed after a successful match does not, in Ltac, cause the entire match to fail, whereas it does in Eisbach. In Eisbach the distinction is important due to the pervasive use of backtracking. When a method is used in a combinator chain, its failure becomes significant because it signals previously applied methods to move to the next result. Therefore, it is better for match to not mask such failure in Eisbach. One can always rewrite a match in Eisbach using the combinators *?* and | to have it try subsequent patterns in the case of an inner-method failure. The following proof method, for example, always invokes *prop-solver* for all goals because its first alternative either never matches or (if it does match) always fails.

method-definition foo_1 =
  ((**match** *?concl* **in** *?P* ∧ *?Q* ⇒ *fail*) | (**match** *?concl* **in** *?R* ⇒ *prop-solver*))

Note that matching can be performed against arbitrary terms or facts, with *?concl* and *prems* being special cases. For example, we could match out of a given set of facts to locate rules with matching assumptions and conclusions.

method-definition *match-rules* **facts** *my-facts* =
  (**match** *my-facts* **in** *U*: *?P* ⟹ *?Q* **and** *U'*: *Q* ⟹ *?R*
  ⇒ (*rule U* [*THEN U'*]))

This example demonstrates use of the **and** keyword, which chains patterns linearly. First, a fact matching *?P* ⟹ *?Q* is found and named *U*. Then, having bound *Q* from the pattern, a fact matching *Q* ⟹ *?R* is matched from *my-facts*. If patterns are matched, then *U* and *U'* are bound to local facts and the method body is executed.

lemma
  assumes f_1: *A* ⟹ *B* **and** f_2: *B* ⟹ *C*
  shows *A* ⟹ *C* **by** (*match-rules my-facts*: $f_1 f_2$)

3.6   Example

We complete our tour of the features of Eisbach by extending the propositional logic solver presented earlier to first-order logic. The following method instantiates universally quantified assumptions by simple guessing, relying on backtracking to find the correct instantiation. Specifically, it instantiates assumptions of the form ∀*x*. *?P x* by finding some type-correct term *y* by matching other assumptions against *?H ?y*, using type annotations to ensure that the types match correctly. The use of the previously defined *elim-all* method here ensures that the same assumption that was matched is the one that will be eliminated. The same matching is also performed against the conclusion to find possible instantiations there too.

method-definition *guess-all* =
  (**match** *prems* **in** *U*: ∀*x*. *?P* (*x* :: *'a*) ⇒
  (**match** *prems* **in** *?H* (*?y* :: *'a*) ⇒
  (*elim-all P y*)
  | **match** *?concl* **in** *?H* (*?y* :: *'a*) ⇒
  (*elim-all P y*)))

The higher order pattern *?H ?y* is used to find arbitrary subterms *y* within the premises or conclusion of the current goal. It makes use of Isabelle/Pure's workhorse of higher order unification (although matching involves pattern-matching only). While such a pattern-match need not bind all variables to be valid, to avoid trivial matches, Eisbach considers only those matches that bind all variables mentioned in the pattern.

The inner-match must be duplicated over both the premises and conclusion because of the logical distinction between facts (the premises) and terms (the conclusion). This might look strange to users of Coq's Ltac, where these notions are identified; however, it does not limit the expressivity of Eisbach.

Similar to our previous *solve-ex* method, we introduce a method which attempts to guess at an appropriate witness for an existential proof. In this case, however, the method simply guesses the witness based on terms found in the current premises, again using higher order matching as in the *guess-all* method above.

method-definition *guess-ex* =
(**match** *?concl* **in**
$\exists x.\ ?P\ (x :: \prime a) \Rightarrow$
  (**match** *prems* **in** *?H* $(?x :: \prime a) \Rightarrow$
    (*rule exI* [*where* $x = x$ **and** $P = P$])))

These methods can now be combined into a surprisingly powerful first order solver.

method-definition *fol-solver* =
(($guess\text{-}ex \mid guess\text{-}all \mid prop\text{-}solver) \mapsto solve\ fol\text{-}solver$)

The use of *solve* here on the recursive call to the method ensures that the recursive subgoals are solved. Without it, the recursive call could potentially prematurely terminate and leave the goal in an unsolvable state (due to an incorrect guess for a quantifier instantiation).

After declaring some standard rules in the context, this method is capable of solving various standard problems.

lemma $(\forall x.\ P\ x) \wedge (\forall x.\ Q\ x) \Longrightarrow (\forall x.\ P\ x \wedge Q\ x)$
and $\exists x.\ (P\ x \longrightarrow (\forall x.\ P\ x))$
and $(\exists x.\ \forall y.\ R\ x\ y) \longrightarrow (\forall y.\ \exists x.\ R\ x\ y)$
by *fol-solver*+

4 Design and Implementation

A core design goal of Eisbach is a seamless integration with other Isabelle languages, notably Isar, ML, and object-logics. The primary motivation clearly being to make it accessible to existing Isabelle/Isar users, with a secondary objective of both forward and backward compatibility.

4.1 Static Closure of Concrete Syntax

Isabelle provides a rich selection of powerful proof methods, each with its own parser and invocation style. Additionally, Isabelle's theorem attributes, which perform context and fact transformations, have their own parsers of arbitrary complexity. Rather than re-write these tools to support Eisbach, we exploited an existing feature of the Isabelle parsing framework whereby tokens have values (types, terms and facts) assigned to them implicitly during parsing.

This implicit value assignment mechanism is the main workhorse of Eisbach, allowing it to embed most Isar syntax as uninterpreted token lists. Eisbach then simply serves as an interpretation environment: when a proof method is applied Eisbach instantiates these token values appropriately based on the supplied arguments to the method or results of matching, and then executes the resulting method body.

Although this presents some technical challenges and requires some minor modifications to Isar, this proves to be a very effective solution to performing this kind of language extension. The necessity of this patching will ideally disappear as the design and implementation principles of Eisbach mature, and thus motivate the incorporation of appropriate concepts into core Isabelle.

4.2 Subgoal Focusing

In Isabelle there is a logical distinction between universally quantified parameters (such as x in $\bigwedge x.\ P\ x$) and arbitrary-but-fixed terms (such as x in $P\ x$). A subgoal in the former form does not allow the x to be explicitly referenced (for example, *my-fact* [*where* $y = x$] does not produce a valid theorem). To deal with this, a set of so-called "improper" methods (like *rule-tac*) have traditionally been used, which are aware of this peculiarity.

It is important to note that premises within a subgoal are not local facts. In a structured Isar proof, assumptions are stated explicitly in the text via **assumes** or **assume** and are accessible to attributes etc. In contrast, the local prefix $\bigwedge x.\ A\ x \Longrightarrow \square$ of a subgoal is not accessible to structured reasoning yet.

To allow the user to write methods that can operate directly on subgoal structure, we decided to expose Isabelle's *subgoal focusing* to Eisbach. Focusing creates a new goal out of a given subgoal, but with its parameters lifted into fixed variables and premises into local assumptions. This allows for uniform treatment of the goal state when matching and parameter passing. In Eisbach, focusing is implicitly triggered whenever the special term *?concl* or special fact *prems* are mentioned. Focusing causes these names to be bound to the conclusion and premises of the current subgoal, respectively.

To support its use in Eisbach, the existing subgoal focusing was enriched to be more generally applicable. Premises, while turned into a local fact, still remain part of the goal. This allows methods like *erule* to still remove premises from the goal.

5 Application and Evaluation

To evaluate Eisbach we re-implemented two existing proof methods: *wp* and *wpc*, which are VCGs currently released as part of the AutoCorres framework [8]. They were used extensively in the full functional correctness proof of seL4 [10] for both invariant and refinement proofs. They were originally designed for performing "weakest-precondition" style reasoning against a shallowly embedded monadic Hoare logic [3]. The intelligence of these methods lies in their large collection of stored facts, and have proven to be more generally useful in other projects [11].

Together these two methods comprise 500 lines of Isabelle/ML, and 60 lines of Isabelle/Isar for setup. However, they may be implemented in Eisbach almost trivially.

The Eisbach implementation of *wp* degenerates into the structured application of some dynamic facts: *wp* supplies facts about monadic functions (e.g. Hoare triples), *wp-comb* contains decomposition rules for postconditions, and *wp-split* splits goals across monadic binds.

method-definition *wp* **facts** [*wp*] [*wp-comb*] [*wp-split*] =
  ((*rule wp* | (*rule wp-comb, rule wp*)) | *rule wp-split*)+

This obscures some details from the original implementation, in particular that the collection of *wp* rules grows quite large and relying exclusively on rule resolution to apply it is costly. This suggests potential improvements to Eisbach, such as allowing facts in the context to be explicitly indexed.

The Eisbach implementation of *wpc* is slightly more involved. It makes use of a simple custom attribute *get-split*, defined in Isabelle/ML, to retrieve the *case-split rule* for a given term; such rules are used to decompose case distinctions on datatypes. The *apply-split* method applies the retrieved case-split rule, specialized to the current goal.

> **method-definition** *apply-split* **for** f =
> ⁣ (**match** [[*get-split f*]] **in** U: *?P* **and** *TERM* *?x* ⇒
> ⁣ (**match** *?concl* **in** *?R f* ⇒
> ⁣ (*rule* U [*THEN iffD2, of x R*]))))

We defined another higher-order method *repeat-new* to repeatedly apply a provided method *m* to all produced subgoals.

> **method-definition** *repeat-new* **methods** *meth* = (*meth* ↦ (*repeat-new meth*)?)

This method is then used in conjunction with worker lemmas to produce one subgoal for each constructor.

> **method-definition** *wpc′* **for** f **facts** [*wpc-helper*] =
> ⁣ (*apply-split f*,
> ⁣ *rule wpc-helperI*,
> ⁣ *repeat-new* (*rule wpc-processors*) ↦ (*rule wpc-helper*))

Finally, *wpc* matches the underlying monadic function out of the current Hoare triple subgoal.

> **method-definition** *wpc* =
> ⁣ (**match** *?concl* **in** ⦃*?P*⦄ *?f* ⦃*?Q*⦄ ⇒ (*wpc′ f*) | ⦃*?P*⦄ *?f* ⦃*?Q*⦄,⦃*?E*⦄ ⇒ (*wpc′ f*))

Together, combined with a large body of existing lemmas, these methods calculate weakest-precondition style proof obligations for the monadic Hoare logic of [3]. Additionally, with appropriate lemmas and some additional match conditions for *wpc*, these methods are easily extended to other calculi such as that from [11].

To evaluate the effectiveness of these re-implemented methods, we re-ran the invariant proofs for the seL4 abstract functional specification using them in place of their original implementations. These proofs constitute about 60,000 lines, including whitespace and comments. About 100 lines of Isabelle/ML were required to maintain syntactic compatibility, and an approximately 0.5% change to the proof text itself was required to resolve cases where proofs relied on quirky behaviour of the original methods in very specific situations. The total running time for the proof increased from 8 minutes to 19 minutes (run on an i7 quad-core 2.8Ghz iMac with 8GB of memory), indicating that there is certainly room for optimization, but also that the overhead introduced by Eisbach is not insurmountable.

See https://bitbucket.org/makarius/method_definition/get/6f90e104b1a4.zip for the full sources for these methods, with the implementation of Eisbach, the monadic Hoare logic from AutoCorres and several non-trivial examples.

6   Related Work

The relation of proofs versus programs, proof languages versus programming languages, and ultimately the quest for adequate *proof programming languages* opens a vast space of possibilities that have emerged in the past decades, but the general problem is still not settled satisfactorily. Different interactive provers have their own cultural traditions and approaches, and there is often some confusion about basic notions and terminology. Subsequently we briefly sketch important lines of programmable interactive proof assistants in the LCF tradition, which includes the HOL family, Coq, and Isabelle itself.

The original **LCF** proof assistant [7] has pioneered a notion of *tactics* and *tacticals* (i.e. operators on tactics) that can be still seen in its descendants today. An **LCF tactic** is a proof strategy that reduces a goal state to zero or more subgoals that are sufficient to solve the problem. Tactics work in the opposite direction than inferences of the core logic, which take known facts to derive new ones.

This duality of backward reasoning from goals versus forward reasoning from facts is reconciled by *tactic justifications*: a tactic both performs the goal reduction and records an inference for the inverse direction. At the very end of a tactical proof, all justifications are composed like a proof tree, to produce the final theorem. This could result in a late failure to finish the actual proof, e.g. due to programming errors in the tactic implementation.

ML was invented for LCF as the *Meta Language* to implement tactics and other tools around the core logical engine. Proofs are typically written as ML scripts, but the activity of building up new theory content and associated tactics is often hard to distinguish from mere application of existing tools from some library. The bias towards adhoc proof programming is much stronger than in, for instance, Isabelle theories today.

The **HOL** family [15, §1] continues the LCF tradition with ML as the main integrating platform for all activities of theory and tool development (using Standard ML or OCaml today). Due to the universality of ML, it is of course possible to implement different interface languages on the spot. This has been done as various "Mizar modes" to imitate the mathematical proof language of Mizar [15, §2], or as "SSReflect for HOL Light" that has emerged in the Flyspeck project, inspired by SSReflect for Coq [5].

The HOL family has the advantage that explorations of new possibilities are easy to get started on the bare-bones ML top-level interface. HOL Light is particularly strong in its minimalistic approach. In contrast, Isabelle tools need to take substantial system infrastructure and common conveniences for end-users into account.

Coq [15, §4] started as another branch of the LCF family in 1985, but with quite different answers to old questions of how proofs and programs are related. While the HOL systems have replaced LCF's *Logic of Computable Functions* by simply-typed classical set-theory (retaining the key role of the Meta Language), Coq has internalized computational aspects into its type-theoretic logical environment. Consequently, the OCaml substrate of Coq is mainly seen as the system implementation language, and has become difficult to access for Coq users. Implementing some *Coq plug-in* requires separate compilation of OCaml modules which are then linked with the toplevel application. An alternative is to *drop* into an adhoc OCaml shell interactively, but this only works for the bytecode compiler, not the native compiler (preferred by default).

Since Coq can be understood as a dependently-typed functional programming language in its own right, it is natural to delegate more and more proof tool development into it, to achieve a grand-unified formal system eventually. A well-established approach is to use *computational reflection* in order to turn formally specified and proven proof procedures into inferences that don't leave any trace in the proof object. Recent work on Mtac [16] even incorporates a full tactic programming language into Coq itself.

Ltac is the untyped tactic scripting language for Coq [4], and has been successfully applied in large Coq theory developments [2]. It has familiar functional language elements, such as higher order functions and let-bindings. However, it contains imperative elements as well, namely the implicit passing of the *proof goal* as global state. The main functionality of Ltac is provided by a *match* construct for performing both goal and term analysis. Matching performs *proof search* through implicit backtracking across matches, attempting multiple unifications and falling through to other patterns upon failure. Although syntactically similar to the match keyword in the term language of Coq, Ltac tactics have a different formal status than Coq functions. Although this serves to distinguish logical function application from on-line computation, it can result in obscure type errors that happen dynamically at run-time.

Mtac is a recently developed *typed tactic language* for Coq [16]. It follows an approach of dependently-typed functional programming: the behaviour of *Mtactics* may be characterized within the logical language of the prover. Mtac is notable by taking the existing language and type-system of Coq (including type-inference), and merely adds a minimal collection of monadic operations to represent impure aspects of tactical programming as first-class citizens: unbounded search, exceptions, and matching against logical syntax. Thus the formerly separate aspect of tactical programming in Ltac is incorporated into the logical language of Coq, which is made even more expressive to provide a uniform basis for all developments of theories, proofs, and proof tools. Thanks to strong static typing, Mtac avoids the dynamic type errors of Ltac.

This mono-cultural approach is quite elegant for Coq, but it relies on the inherent qualities of the Coq logic and its built-in computational world-view. In contrast, the greater LCF family has always embraced multiple languages that serve different purposes: classic LCF-style systems are more relaxed about separating logical foundations from computation outside of it (potentially with access to external tools and services). Eisbach continues this philosophy. In Isabelle, the art of integrating different languages into one system (not one logic) is particularly emphasized: standard syntactic devices for quotation and anti-quotation support embedded sub-languages.

SSReflect [5] is the common label for various tools and techniques for proof engineering in Coq that have emerged from large verification projects by G. Gonthier. This includes a sophisticated *proof scripting language* that provides fine-grained control over moves within the logical subgoal structure, and nested contexts for single-step equational reasoning. Actual *small-scale reflection* refers to implementation techniques within Coq, for propositional manipulations that could be done in HOL-based systems by more elementary means; the experimental SSReflect for HOL-Light re-uses the proof scripting language and its name, but without doing any reflection.

SSReflect emphasizes concrete proof scripts for particular problems, not general proof automation. Scripts written by an expert of SSReflect can be understood by the same, without stepping through the sequence of goal states in the proof assistant. General tools may be implemented nonetheless, by going into the Coq logic. The SSReflect toolbox includes specific support for generic theory development based on *canonical structures*. More recent work combines that approach with ideas behind Mtac, to internalize a generic proof programming language into Coq, in analogy to the well-known type-class approach of Haskell, see [6].

7   Conclusion and Future Work

In this paper we have presented Eisbach, a high-level language for writing proof methods in Isabelle/Isar. It supports familiar Isar language elements, such as method combinators and theorem attributes, as well as being compatible with existing Isabelle proof methods. An expressive **match** construct enables the use of higher-order matching against facts and subgoals to provide control flow. We showed that existing methods used in large-scale proofs can be easily implemented in Eisbach. The resulting implementations are far smaller, and easier to understand.

Of the proof programming languages mentioned in Section 6, Eisbach purposefully resembles Coq's Ltac most closely. However, it seamlessly integrates with core Isabelle technologies (fact collections, pervasive backtracking, subgoal focusing) to allow powerful methods to be easily and succinctly written. When building on top of Isabelle/Isar, it made most sense to implement an untyped proof programming language, rather than trying to emulate ideas from languages like Mtac. This is because we wanted Eisbach to be able to invoke existing Isar proof methods, which are untyped. While the absence of typed proof procedures hasn't hindered the development of large-scale proofs, the ability to annotate proof methods with information about how they are expected to transform the proof state is potentially attractive. Although higher order methods can approximate run-time method contracts, we would be free to implement arbitrary contract specification languages because proof methods exist outside the logic of Isabelle/Pure, however this avenue of inquiry remains unexplored.

The evaluation demonstrates that Eisbach can already be effectively used to write real-world proof tools, however it still lacks some important features. Firstly, some debugging features are planned, beyond the current solution of manually printing intermediate goal states. Traces of matches and method applications will be presented, ideally with some level of interaction from the user. Additionally more structured language elements would provide a more natural integration with Isar (e.g. explicit subgoal production and addressing). We would also like Eisbach to support parallel evaluation by default. Method combinators outline a certain structure that should be used as a *parallel skeleton* wherever possible. For example, ↦ could use a parallel version of the underlying tactical THEN_ALL_NEW, analogous to the existing PARALLEL_GOALS tactical of Isabelle/ML. Ultimately we plan to include Eisbach in a future Isabelle release, with the aim of it becoming the primary means of writing proof methods.

Acknowledgements. We would like to thank Gerwin Klein, who was involved in the discussions on the design of Eisbach and who provided early feedback on this paper. Thanks also to Peter Gammie, Magnus Myreen, and Thomas Sewell for feedback on drafts of this paper.

References

[1] Bourke, T., Daum, M., Klein, G., Kolanski, R.: Challenges and experiences in managing large-scale proofs. In: Jeuring, J., Campbell, J.A., Carette, J., Dos Reis, G., Sojka, P., Wenzel, M., Sorge, V. (eds.) CICM 2012. LNCS, vol. 7362, pp. 32–48. Springer, Heidelberg (2012)

[2] Chlipala, A.: Mostly-automated verification of low-level programs in computational separation logic. ACM SIGPLAN Notices 46(6), 234 (2011)

[3] Cock, D., Klein, G., Sewell, T.: Secure microkernels, state monads and scalable refinement. In: Mohamed, O.A., Muñoz, C., Tahar, S. (eds.) TPHOLs 2008. LNCS, vol. 5170, pp. 167–182. Springer, Heidelberg (2008)

[4] Delahaye, D.: A tactic language for the system Coq. In: Parigot, M., Voronkov, A. (eds.) LPAR 2000. LNCS (LNAI), vol. 1955, pp. 85–95. Springer, Heidelberg (2000)

[5] Gonthier, G., Mahboubi, A.: An introduction to small scale reflection in Coq. J. Formalized Reasoning 3(2) (2010)

[6] Gonthier, G., Ziliani, B., Nanevski, A., Dreyer, D.: How to make ad hoc proof automation less ad hoc. J. Funct. Program. 23(4), 357–401 (2013)

[7] Gordon, M.J., Milner, R., Wadsworth, C.P.: Edinburgh LCF. LNCS, vol. 78. Springer, Heidelberg (1979)

[8] Greenaway, D., Andronick, J., Klein, G.: Bridging the gap: Automatic verified abstraction of C. In: Beringer, L., Felty, A. (eds.) ITP 2012. LNCS, vol. 7406, pp. 99–115. Springer, Heidelberg (2012)

[9] Klein, G., Andronick, J., Elphinstone, K., Murray, T., Sewell, T., Kolanski, R., Heiser, G.: Comprehensive formal verification of an OS microkernel. ACM Transactions on Computer Systems (TOCS) (to appear)

[10] Klein, G., Elphinstone, K., Heiser, G., Andronick, J., Cock, D., Derrin, P., Elkaduwe, D., Engelhardt, K., Kolanski, R., Norrish, M., Sewell, T., Tuch, H., Winwood, S.: seL4: Formal verification of an OS kernel. In: SOSP, Big Sky, MT, USA, pp. 207–220. ACM (October 2009)

[11] Murray, T., Matichuk, D., Brassil, M., Gammie, P., Klein, G.: Noninterference for operating system kernels. In: Hawblitzel, C., Miller, D. (eds.) CPP 2012. LNCS, vol. 7679, pp. 126–142. Springer, Heidelberg (2012)

[12] Paulson, L.C.: Isabelle: the next 700 theorem provers. In: Odifreddi, P. (ed.) Logic and Computer Science. Academic Press (1990)

[13] Wenzel, M., Chaieb, A.: SML with antiquotations embedded into Isabelle/Isar. In: Carette, J., Wiedijk, F. (eds.) Workshop on Programming Languages for Mechanized Mathematics (PLMMS 2007), Hagenberg, Austria (June 2007)

[14] Wenzel, M.: Isabelle/Isar—a versatile environment for human-readable formal proof documents. PhD thesis, Technische Universität München (2002)

[15] Wiedijk, F. (ed.): The Seventeen Provers of the World. LNCS (LNAI), vol. 3600. Springer, Heidelberg (2006)

[16] Ziliani, B., Dreyer, D., Krishnaswami, N.R., Nanevski, A., Vafeiadis, V.: Mtac: a monad for typed tactic programming in Coq. In: Morrisett, G., Uustalu, T. (eds.) ICFP. ACM (2013)

Proof Pearl: Proving a Simple Von Neumann Machine Turing Complete

J Strother Moore

Dept. of Computer Science, University of Texas, Austin, TX, USA
moore@cs.utexas.edu
http://www.cs.utexas.edu

Abstract. In this paper we sketch an ACL2-checked proof that a simple but unbounded Von Neumann machine model is Turing Complete, i.e., can do anything a Turing machine can do. The project formally revisits the roots of computer science. It requires re-familiarizing oneself with the definitive model of computation from the 1930s, dealing with a simple "modern" machine model, thinking carefully about the formal statement of an important theorem and the specification of both total and partial programs, writing a verifying compiler, including implementing an X86-like call/return protocol and implementing computed jumps, codifying a code proof strategy, and a little "creative" reasoning about the non-termination of two machines.

Keywords: ACL2, Turing machine, Java Virtual Machine (JVM), verifying compiler.

1 Prelude

I have often taught an undergraduate course at the University of Texas at Austin entitled *A Formal Model of the Java Virtual Machine*. In the course, students are taught how to model sophisticated computing engines and, to a lesser extent, how to prove theorems about such engines and their programs with the ACL2 theorem prover [5]. The course starts with a pedagogical ("toy") JVM-like model which the students elaborate over the semester towards a more realistic model, which is then compared to an accurate JVM model[9]. The pedagogical model is called M1: a stack based machine providing a fixed number of registers (JVM's "locals"), an unbounded operand stack, and an execute-only program providing the following bytecode instructions ILOAD, ISTORE, ICONST, IADD, ISUB, IMUL, IFEQ, GOTO, and HALT, with unbounded arithmetic.

This set of opcodes was chosen to allow students to easily implement and verify some simple M1 programs. On the last class day before Spring Break, 2012, the students complained that it was very hard to program M1; that in fact, it was "probably impossible" to do "real computations" with it because it lacks a "less than" comparator and procedures![1]

---

[1] Such judgements are obviously naive and ill-informed; any machine with a branch-if-0, a little arithmetic, and some accessible infinite resource is Turing Complete.

G. Klein and R. Gamboa (Eds.): ITP 2014, LNAI 8558, pp. 406–420, 2014.

My response was "Well, M1 can do anything a Turing machine do." But on my way home that evening, I felt guilty:

M1 is a pedagogical device, designed to introduce formal modeling to the students and inculcate the idea that expectations on hardware and software can often be formalized and proved. I shouldn't just say it's Turing Complete. I should show them how we can prove it with the tools they're using.

Fortunately, I had Spring Break ahead of me and thus was born this project.

2 Source Files

The complete set of scripts for this project are part of ACL2's Community Books. See the Community Books link on the ACL2 home page [6]. After downloading and installing the books visit `models/jvm/m1/`. See the README file there. References to *.lisp files below are for that directory. If you have a running ACL2 session you could (include-book "models/jvm/m1/find-k!" :dir :system) and (in-package "M1") to inspect everything with ACL2 history commands. This paper is a guide.

3 Related Work

Turing Completeness proofs for various computational models have been a staple of computer science since the time of Turing and Church. Mechanically checked proofs of other important theorems in meta-mathematics (the Church-Rosser theorem, the Cook-Levin theorem, and Gödel's First Incompleteness Theorem) are less common but have been done with several provers. Here I focus on *mechanically checked* formal proofs of the computational completeness of a programming language.

As far as I am aware, the first and only such proof was done in 1984 [2], when Boyer and I proved that Pure Lisp was Turing Complete, using the prover that would become Nqthm. We were asked to prove completeness by a reviewer of [3], in which we proved that the halting problem for Pure Lisp was undecidable; the reviewer objected that we had not proved Pure Lisp Turing Complete.

An important distinction between [2] and the present work is that the "suspect" computational model in the former is the lambda calculus with general recursion (Pure Lisp), whereas here it is a very simple Von Neumann machine (or imperative programming language) similar to the contemporary JVM and its bytecode language [8].

While I'm unaware of other mechanically checked proofs that a given programming language is Turing Complete, this work also involves proofs of properties of low-level assembly code and a verifying compiler. This tradition goes back at least as far as the mechanically checked proof of a compiler by Milner and Weyhrauch in 1972 [10]. Highlights of subsequent systems verification work involving such mechanically checked reasoning include the "CLI verified (hardware/software) stack" of [1], and of course the even more realistic results of the seL4 microkernel [7] and VCC projects [4]. But even with a verified program one must prove that the specification is Turing Complete.

4   Turing Machines

The present work uses the same Turing machine model as [2] (ported from Nqthm to ACL2) which was accepted by the reviewers of that paper. The model is based on Rogers' classic [12] formalization. A Turing machine description, tm, (sometimes called an "action table") is a finite list of 4-tuples or *cells*, $\langle st_{in}, sym, op, st_{out}\rangle$. Rogers represents a tape as a pair of half tapes, each being a (finite but extensible) list of 0s and 1s. The concatenation of these two half tapes corresponds to the intuitive notion of a tape (extensible in both directions) with a read/write head "in the middle." Rogers shows one may start with an extensible finite tape. The read/write head is thought of as positioned on the first symbol on the right half. The interpretation of each cell in description tm is "if, while in state st_{in}, sym is read from the tape, perform operation op on the tape and enter state st_{out}." Here, st_{in} and st_{out} are symbolic state names, sym is 0 or 1, and op is one of four values meaning write a 0, write a 1, shift left, or shift right. The machine halts when the current state and symbol read from the tape do not match any st_{in} and sym in tm.

We define tmi ("Turing machine interpreter") to take a Turing machine state name, tape, and a Turing machine description and a number of steps, n. Tmi returns either nil ("false") meaning the machine did not reach a halted state in n steps, or the final tape produced after n steps. By our choice of tape representation, a tape can never be nil and so the function tmi indicates whether the computation halted in n steps and the final tape if it did halt. See the definition tmi in tmi-reductions.lisp. I will colloquially refer to tmi as our "official" model of Turing machines.

In our official model, Turing machine descriptions and cells are lists constructed with cons, state names are Lisp symbols (e.g., Q1, LP, TEST), "symbols" on the tape are integers 0 or 1, and operations are Lisp objects 0, 1, L, or R. See the definition of *rogers-program* in tmi-reductions.lisp for an example.

5   M1

M1 is defined in a similar style but takes an M1 state and a number of steps. An M1 state contains a program counter ("pc"), a list of integers denoting register values, a stack of integers, and a program; all components of an M1 state are represented with lists, symbols and numbers in the obvious way. The integers are unbounded, the stack may grow without bound. An arbitrary number of registers may be provided but the number of allocated registers never grows larger than the largest register index used in the program. Programs are finite and fixed ("execute only").

Programs are lists of the *instructions* as described below. The notation "$reg[i]$" denotes the contents of register (JVM local variable) i. "$reg[i] \leftarrow v$" denotes assignment to a register; "$pc \leftarrow v$" denotes assignment to the program counter. The notation "$\ldots, x, y, a \Rightarrow \ldots, v$" describes the manipulation of the stack as per [8] and means that three objects, x, y, and a, are popped from the stack (with a being the topmost) and v is pushed in their place. That portion of the stack ("$\ldots$") deeper than x is unaffected. The first six instructions below always increment the pc by 1, i.e., $pc \leftarrow pc + 1$ is implicit.

instruction	stack	description
(ILOAD n) :	$\ldots \Rightarrow \ldots, reg[n]$	
(ISTORE n) :	$\ldots, v \Rightarrow \ldots$	$reg[n] \leftarrow v$
(ICONST k) :	$\ldots \Rightarrow \ldots, k$	
(IADD)	: $\ldots, x, y \Rightarrow \ldots, x + y$	
(ISUB)	: $\ldots, x, y \Rightarrow \ldots, x - y$	
(IMUL)	: $\ldots, x, y \Rightarrow \ldots, x \times y$	
(GOTO d) :	$\ldots \Rightarrow \ldots$	$pc \leftarrow pc + d$
(IFEQ d) :	$\ldots, v \Rightarrow \ldots$	$pc \leftarrow pc + (\textbf{if } v = 0 \textbf{ then } d \textbf{ else } 1)$
(HALT)	: $\ldots \Rightarrow \ldots$	no change to state

Note that by not changing the state, the HALT instruction causes the machine to stop. We consider an M1 state *halted* if the pc points to a HALT instruction.

To *step* an M1 state the instruction at pc in the program is fetched and executed as described above. We define (M1 s n) to step state s n times and return the final state. See m1.lisp for complete details of the M1 model.

An example of an M1 program to compute the factorial of register 0 and leave the result on top of the stack is:

program	pc	pseudo-code
'((ICONST 1)	; 0	
(ISTORE 1)	; 1	$reg[1] \leftarrow 1$
(ILOAD 0)	; 2	
(IFEQ 10)	; 3	if $reg[0] = 0$, then jump to 13
(ILOAD 1)	; 4	
(ILOAD 0)	; 5	
(IMUL)	; 6	
(ISTORE 1)	; 7	$reg[1] \leftarrow reg[1] \times reg[0]$
(ILOAD 0)	; 8	
(ICONST 1)	; 9	
(ISUB)	; 10	
(ISTORE 0)	; 11	$reg[0] \leftarrow reg[0] - 1$
(GOTO -10)	; 12	jump to 2
(ILOAD 1)	; 13	
(HALT))	; 14	halt *with $reg[1]$ on top of stack*

This program runs forever (never reaches the HALT) if $reg[0]$ is negative.

If we require as a precondition that $reg[0]$ is a natural number, a statement of total correctness can be paraphrased as: If s is an M1 state with pc 0, the natural number n in $reg[0]$ and the list above as the program, then there exists a natural number i such that (M1 s i) is a halted state with $n!$ on top of the stack.

To state and prove such a theorem it is convenient to define a witness for the existentially quantified i. This witness is delivered by a user-defined *clock function* that takes n as an argument and returns a natural number.

The ACL2 Community Books directory `models/jvm/m1/` contains many example M1 programs along with machine checked proofs of their correctness via such clock functions and other methods[2].

6   The Correspondence Conventions

To state Turing equivalence I followed the approach of [2]. Paraphrasing it into the M1 setting, I set up a correspondence between official Turing machine representations of certain objects (e.g., machine descriptions and state names) and their M1 representations. The former are composed of lists, symbols, and integers; the latter are strictly numeric.

Consider an arbitrary cell, $\langle st_{in}, sym, op, st_{out} \rangle$, in a Turing machine description tm. Given that tm contains only a finite number of state name symbols, we can allocate a unique natural to each and represent these naturals in binary in a field of width w (which depends on the number of state names in tm). We could represent each of the four possible op as naturals in 2 bits but we allocate 3 bits. Let the numeric encodings of the four elements of *cell* be $st_{in}', sym', op', st_{out}'$, respectively. Then the encoded *cell* is $cell' = st_{in}' + 2^w sym' + 2^{w+1} op' + 2^{w+4} st_{out}'$.

Using this convention we can represent a list of cells, tm, as follows. The empty list is represented as an encoded "cell" of 0s with $op' = 4$ (using the otherwise unnecessary 3rd bit of op'). We call this value *nnil*. A non-empty list whose first cell is represented by *cell'* and whose remaining elements are recursively represented by *tail* is represented by $cell' + 2^{4+2w} tail$.

The tape (which, recall, also encodes the read/write head "in the middle") is represented on M1 as two natural numbers, one specifying (via its binary expansion) the contents of the tape and the other specifying the head position (via the number of bits in the left-half tape). Henceforth I use these conventions:

level	variable	value
Official	*tm*	: a Turing machine description
	st	: a Turing machine state name
	tape	: a Turing machine tape (with encoded head)
M1	*w*	: width of a state symbol encoding req'd by *st* and *tm*
	nnil	: the marked encoded "cell" ($op' = 4$) (wrt w)
	tm'	: the M1 (numeric) representation of *tm* (wrt w and *nnil*)
	st'	: the M1 (numeric) representation of *st*
	tape'	: the M1 (numeric) representation of *tape* contents
	pos'	: the M1 (numeric) representation of the head position
	s_0	: the initial M1 state described below

---

[2] See [11].

The initial M1 state s_0 is an M1 state with program counter 0, thirteen registers set to 0, the stack in which st', $tape'$, pos', tm', w, and $nnil$ have been pushed, and finally, as the program, a certain, fixed list of M1 instructions. That list of M1 instructions, called Ψ and described below, is (allegedly) a Turing machine interpreter in the programming language of M1. Note that s_0 does not specify how long the Turing machine is to run.

The macro with-conventions in theorems-a-and-b.lisp formally defines these conventions. The macro binds the ACL2 variable s_0 (aka s_0) to the value above, in terms of the variables tm, st, and tape. Technically, it binds w, nnil, tm', st', tape', and pos' as specified in terms of tm, st, and tape, and binds s_0 in terms of those auxiliary variables.

7   Theorems Proved

The discussion in [2] requires us to prove:

Theorem A. If tmi runs forever on st, $tape$, and tm then M1 runs forever on s_0. More precisely, we phrase this in the contrapositive and say that if M1 halts on s_0 in i steps then there exists a j such that tmi halts in j steps.

Theorem B. If tmi halts on st, $tape$, and tm in n steps, there exists a k such that M1 halts on s_0 in k steps and computes the corresponding tape.

```
(defthm theorem-A
  (with-conventions
   (implies (natp i)
            (let ((s_f (m1 s_0 i)))
              (implies
               (haltedp s_f)
               (tmi st tape tm (find-j st tape tm i))))))
   :hints ...)

(defthm theorem-B
  (with-conventions
   (implies (and (natp n)
                 (tmi st tape tm n))
            (let ((s_f (M1 s_0 (find-k st tape tm n))))
              (and (haltedp s_f)
                   (equal (decode-tape-and-pos
                           (top (pop (stack s_f)))
                           (top (stack s_f)))
                          (tmi st tape tm n)))))))
```

Note that when the tmi expressions are used as literals (e.g., in the conclusion of theorem-A and the hypothesis of theorem-B) it is equivalent to asserting termination (non-nil returned value) of the tmi run. When tmi is used in the equality, we know the value is a tape and the equality checks the correspondence with what M1 computes.

In formalizing these statements there is an opportunity to subvert our goal by defining a devious sense of correspondence! The correspondence has access to the full power of the logic and *could*, for example, compute the right answer from tm, st, and tape and encode it into s_0. The correspondence above is not "devious."

It remains to explain the fixed M1 program, Ψ, and the witness functions find-j and find-k which constructively establish the existence of the step counts mentioned in the informal statements of the theorems. But first, it is convenient to refine tmi into a function that operates on the kind of data M1 has: numbers.

8   Refinement

We refine the official definition of tmi into a function named tmi3 and verify that it corresponds to tmi modulo the representational issues. The proof is done in several steps which successively implement the change of representations of *tm* and *tape*.

- tmi1 is like tmi but for a renamed *tm* with numeric state names
- tmi2 is like tmi1 but for *tm'*, *w* and *nnil*
- tmi3 is like tmi2 but for *tape'* and *pos'*

This concludes with the theorem tmi3-is-tmi in tmi-reductions.lisp. It is tmi3 we will implement on M1.

9   The M1 Program Ψ

Key to our proof is the definition of an M1 program Ψ for interpreting arbitrary Turing machine descriptions on a given starting state and tape. Ψ either runs forever or HALTs; and when it halts, the representation of the official final tape can be recovered from the M1 state.

Given the limited instruction set of M1, it is necessary to implement some simple arithmetic utilities as M1 programs. Ψ is then the concatenation of all these programs together with "glue code" permitting procedure call and return.

name	stack	description
LESSP :	$\ldots, x, y \Rightarrow \ldots, v$	$v = (\textbf{if } x < y \textbf{ then } 1 \textbf{ else } 0)$
MOD :	$\ldots, x, y \Rightarrow \ldots, (x \bmod y)$	
FLOOR :	$\ldots, x, y, a \Rightarrow \ldots, (a + \lfloor x/y \rfloor)$	
LOG2 :	$\ldots, x, a \Rightarrow \ldots, (a + log_2(x))$	
EXPT :	$\ldots, x, y, a \Rightarrow \ldots, (a + x^y)$	

We underline program names to help the reader; MOD names an M1 program, mod names an ACL2 function[3]. For brevity, the descriptions above do not include the effects of these programs on the pc or registers. In addition, certain obvious

---

[3] ACL2 is case insensitive; formally MOD is 'MOD.

preconditions obtain (e.g., for **FLOOR**, all operands are natural numbers and y is non-0).

With these primitives and subroutine call/return it is not difficult to define slightly higher level **M1** programs for accessing encoded Turing machine descriptions, states, and tapes. The names below are all prefixed with 'n' because these functions are the numeric correspondents of functions in the official model of **tmi**. In the following, $cell'$ is the numeric encoding of some cell $\langle st_{in}, sym, op, st_{out} \rangle$, $st'_{in}, sym', op', st'_{out}$ are the corresponding numeric encodings, tm is assumed non-empty (and so its car is a cell with encoding car' and its cdr is a list of cells with encoding cdr', and tm' is not $nnil$), w is the width of the state symbol encoding, and $nnil$ is the marked cell.

name	:	stack		description
NST-IN	:	$\ldots, cell', w$	$\Rightarrow$	$\ldots, st'_{in}$
NSYM	:	$\ldots, cell', w$	$\Rightarrow$	$\ldots, sym'$
NOP	:	$\ldots, cell', w$	$\Rightarrow$	$\ldots, op'$
NST-OUT	:	$\ldots, cell', w$	$\Rightarrow$	$\ldots, st'_{out}$
NCAR	:	$\ldots, tm', w$	$\Rightarrow$	$\ldots, car'$
NCDR	:	$\ldots, tm', w$	$\Rightarrow$	$\ldots, cdr'$

With these programs we can implement **M1** programs for implementing the numeric version of **tmi**.

NCURRENT-SYM $: \ldots, tape', pos' \Rightarrow \ldots, sym'$
Description: sym' is the symbol at position pos' of $tape'$

NINSTR1 $: \ldots, a, b, tm', w, nnil \Rightarrow \ldots, cell'$
Description: $cell'$ is the first encoded cell in tm with $st'_{in} = a$ and $sym' = b$, if any, or -1 if no such cell exists

NEW-TAPE2 $: \ldots, op', tape', pos' \Rightarrow \ldots, tape'_{nx}, pos'_{nx}$
Description: op' is the encoding of a tape operation; $tape'_{nx}$ and pos'_{nx} are produced by performing that operation on $tape'$ and pos'

TMI3 $: \ldots, st', tape', pos', tm', w, nnil \Rightarrow \ldots, tape'_{nx}, pos'_{nx}$
Description: This is the **M1** program that interprets the Turing machine tm with initial state st and input tape $tape'$ and pos'. The program returns the $tape'_{nx}$ and pos'_{nx} representing the final tape if the machine halts, or runs forever otherwise.

Note that "tmi3" is both the name of an **M1** program and of a function defined in ACL2 as part of our refinement of **tmi** to the **M1** representations. However, the program **TMI3** takes the six arguments listed above, while the function tmi3 takes an additional argument: the number of steps to take, n.[4] The program **TMI3** may run forever. The function tmi3 is total.

MAIN $: \ldots, st', tape', pos', tm', w, nnil \Rightarrow \ldots, tape'_{nx}, pos'_{nx}$

---

[4] Actually, the function tmi3 takes six, not seven, arguments because it does not need $nnil$: it is determined from w.

Description: By convention, our compiler starts execution with the MAIN program and our MAIN just calls TMI3 above.

10 Verifying Compiler

Writing the sixteen programs above is tedious if done directly. Perhaps the main problem is that M1 does not support subroutine call and return: M1 operates on one "flat" program space! Furthermore, the machine does not provide "computed jumps" like the JVM's JSR (which pops the stack into the pc). There is a strict separation of data from pcs. Every GOTO and IFEQ is pc relative, but the distance skipped is always some constant specified in the instruction. Of course, writing the programs is only part of the battle: they must also be verified to implement the ACL2 function tmi3.

I thus decided to write a compiler from a simple "Toy Lisp" subset to M1 code. The compiler takes as input a system description, containing source code and specifications for every subroutine.

The verifying compiler is called defsys (see defsys.lisp). Ψ is generated by the defsys expression in implementation.lisp. Every subroutine to be compiled is given a name, a list of :formals, an :input precondition, an :output specification describing the top of the stack, and the Toy Lisp source :code. It was sufficient and convenient to support only tail-recursive source code functions. As illustrated by main below, a subroutine may return multiple values and provision is made via so-called "ghosts" to model partial programs with total functions. Finally, optional arguments :ld-flg and :edit-commands allow the user to debug and modify the generated events. Inspection of implementation.lisp will reveal that three edit commands were used to augment the automatically generated commands. These generally inserted additional lemmas to prove before certain automatically generated theorems.

```
(defsys :ld-flg nil          ; debugging aid
  :modules
  ((LESSP :formals (x y)
          :input (and (natp x)
                      (natp y))
          :output (if (< x y) 1 0)
          :code (IFEQ Y
                      0
                      (IFEQ X
                            1
                            (LESSP (- X 1) (- Y 1)))))
   (MOD :formals (x y)
        :input (and (natp x)
                    (natp y)
                    (not (equal y 0)))
        :output (mod x y)
        :code (IFEQ (LESSP X Y)
```

```
                    (MOD (- X Y) Y)
                    X))
    ...
    (MAIN :formals (st tape pos tm w nnil)
          :input (and (natp st)
                      (natp tape)
                      (natp pos)
                      (natp tm)
                      (natp w)
                      (equal nnil (nnil w))
                      (< st (expt 2 w)))
          :output (tmi3 st tape pos tm w n)
          :output-arity 4
          :code (TMI3 ST TAPE POS TM W NNIL)
          :ghost-formals (n)
          :ghost-base-value (MV 0 st tape pos)))
     :edit-commands ...)  ; user-added modifications
```

Toy Lisp is just the subset of ACL2 composed of variable symbols, quoted numeric constants, the function symbols +, -, * (primitively supported by M1), the form (MV a_1 ... a_n) for returning multiple values, the form (IFEQ a b c) (which is just ACL2's (if (equal a 0) b c)), and calls of primitive and defined Toy Lisp functions.

In addition to producing the M1 object code, the compiler (a) provides a call/return protocol, (b) links symbolic names to actual pcs (and generates appropriate relative jumps), and (c) produces the ACL2 commands (definitions and theorems) establishing that the object code is correct with respect to the Toy Lisp and that the Toy Lisp implements the :output specification.

If the maximum number of registers required by any subroutine's body is max, the call/return protocol requires $2max + 1$ registers. We divide them into max so-called A-registers, $max + 1$ B-registers. The A-registers are for use by the subroutine body and the B-registers are used by the call/return protocol. For simplicity we assume (and check) that the maximum number of registers used by a subroutine body is equal to the number of input parameters of the subroutine.

Note that of the sixteen programs sketched above, TMI3 has the most parameters: 6. Thus, we need 13 registers.

The basic protocol for calling a subroutine subr of n arguments with arguments $a_1, \ldots, a_n$, is as follows: the caller pushes $a_1, \ldots, a_n$, and the pc to which subr should return. The caller then jumps to the pc of subr. At that pc, a prelude for subr pops $a_1, \ldots, a_n$, and pc into the B-registers. It then protects the caller's environment by pushing the first n A-registers onto the stack, followed by the return pc from the B-registers. Finally, it moves the other B-registers (containing $a_1, \ldots, a_n$) to the first n A-registers[5]. A symmetric postlude supports returning

---

[5] The only way to move a value from one register to another is via the stack; only the topmost item on the stack can be accessed per instruction.

$k \leq n$ values on the stack. At the conclusion of the postlude, the code jumps to the return *pc*.

But how can M1 jump to a pc found on the stack if the ISA firmly separates "data" from "pcs"? The answer is quite tedious: the compiler keeps track of every call of each subroutine; the postlude for each subroutine concludes with a "big switch" which compares the "return *pc*" (data on the stack) to the known pc of each call and then jumps to the appropriate pc.

The compiler works in several passes. The first pass compiles the object code but includes symbolic labels and pseudo-instructions for CALL and RET. The second pass expands the CALL and RET "instructions" into appropriate sequences of M1 code. The last pass removes and replaces labels by relative jumps to the appropriate pcs. The compiler saves the output of the three passes in the ACL2 constants *ccode*, *acode*, and *Psi* respectively. These may be inspected after the compiler is run.

The key to generating the clock functions is just to count instructions in the prelude, loop, and postlude of each subroutine.

Defsys generates certain definitions and theorems for each subroutine, admits the definitions under the logic's definitional principle, and proves the theorems. The important ones are noted below for <u>LESSP</u>[6]. Recall that <u>LESSP</u> takes two arguments, x and y. The :input condition of the module is that both x and y are naturals. The :output condition is that 1 or 0 is on top of the stack, depending on whether $x < y$. The source :code for the module is shown above. When *rpc* is mentioned below it is the return pc from some call of <u>LESSP</u> in Ψ. When s is mentioned it is an M1 state with program Ψ. Toy Lisp translations to ACL2 have names beginning with "!".

Def (!lessp x y): the ACL2 function !lessp is defined

```
(defun !lessp (x y)
  (if (and (natp x) (natp y))   ; :input condition
      (if (equal y 0)           ; translated Toy Lisp
          0
          (if (equal x 0)
              1 (!lessp (- x 1) (- y 1)))))
      nil))
```

Def (lessp-loop-clock x y): defined to compute the number of M1 steps from the loop in <u>LESSP</u> to the postlude

Def (lessp-clock *rpc* x y): defined to compute the number of M1 steps to get from the top of the prelude in <u>LESSP</u> through the return to *rpc*

Thm lessp-loop-is-!lessp: if the pc in s is at the top of the loop in <u>LESSP</u>, with x and y (satisfying the stated :input conditions) in the first two A-registers, then after (lessp-loop-clock x y) steps the pc is at the postlude, all of the A-

---

[6] It is easiest to inspect the results by loading the project into ACL2 (Section 2) typing (pe '*name*), where name is the name of an event mentioned here.

registers except the first two are unchanged, and (!lessp x y) has been pushed on the stack

Thm lessp-is-!lessp: if the pc in s is poised at the pc of <u>LESSP</u> and the stack contains at least three values, x, y, and rpc, where x and y satisfy the :input conditions on lessp and rpc is a known return pc from <u>LESSP</u>, then after (lessp-clock rpc x y) steps the pc is rpc, the A-registers are unchanged, and (!lessp x y) has been pushed onto the stack obtained by popping off x, y, and rpc

Thm !lessp-spec: if x and y satisfy the :input conditions for lessp, then (!lessp x y) is as specified by the :output, i.e., it is 1 or 0 depending on $(x < y)$.

Putting the last two theorems together allows ACL2 to deduce that every jump to <u>LESSP</u> in Ψ just advances the pc to the return pc, pops the arguments and the return pc off the stack, and pushes 1 or 0 according to the specification, without changing the A-registers.

Defsys compiles M1 code and generates and proves analogous definitions and theorems for every module. Thus, it compiles <u>TMI3</u> and proves that running that code produces the results specified by tmi3. The only wrinkle in this story is that tmi3 takes a step-count argument while the program <u>TMI3</u> does not. However, provision is made for this via the user-supplied "ghost" parameters of defsys. The clock function tmi3-clock and the :code function !tmi3 are augmented by an additional formal parameter, the user-supplied :ghost-formal n. In recursion (once per iteration), these functions decrement n and halt if $n = 0$. No such parameter exists in the compiled code. But defsys proves that the code, when run according to tmi3-clock, returns the same result as tmi3 (both wrt n), or else is left "still running" at the top of its loop.

11   Finishing the Proof

From the theorems in tmi-reductions.lisp we get that the official Turing machine interpreter, tmi, is equal to tmi3 modulo the representations, for any Turing step count n.

From implementation.lisp we get theorems about <u>MAIN</u>, its ACL2 analogue, !main and its :output specification function tmi3. In particular the theorem main-is-!main tells us that if invoked appropriately and run for main-clock M1-steps (for exactly n iterations), the result is exactly described by its Lisp analogue !main: If !main reports halting after n iterations, then the final M1 state has as its pc the return pc of the call of <u>MAIN</u> in Ψ, and the stack contains same tape and position computed by !main; and if !main reports that it did not halt (in n iterations) the M1 state is poised at the top of the loop in the <u>TMI3</u> program.

Meanwhile, !main-spec tells us that !main computes the same thing as tmi3.

Since main-clock starts counting from the pc of <u>MAIN</u> and Ψ just pushes the return pc, jumps to <u>MAIN</u>, and HALTs, we define (find-k st $tape$ tm n) to be just 2 more than `main-clock` on the corresponding arguments st', $tape'$, pos', tm', w, $nnil$ and n.

In `theorems-a-and-b.lisp` we combine these results in the `simulation` theorem, which states that an M1 run starting in initial state s_0 and taking (`find-k st tape tm n`) steps is halted precisely if `tmi` halts in n steps, and furthermore, that if `tmi` halts in n steps, then the answer in the final M1 state corresponds to the tape computed by `tmi`.

Now we wish to prove theorem A and B. In fact, `theorem-B` (see page 411) follows easily from the `simulation` theorem.

Theorem A (page 411) requires more work. Recall that it deals with the non-termination of the two machines. Informally, it says that if `tmi` fails to terminate, then so does M1. But we phrased it in the contrapositive: if M1 terminates, then so does `tmi`.

Here we know that M1 halts on s_0 after i steps and we must define `find-j` to return a number of steps sufficient to insure that `tmi` halts. Notice that the previously defined `find-k` counts M1-steps and now we seek to count `tmi` steps.

Two observations are important in defining `find-j`. The first is a theorem called `find-k-monotonic` in `theorems-a-and-b.lisp` which states that if `tmi` has not halted after n steps then (`find-k st tape tm n`) < (`find-k st tape tm n + 1`). This is actually an interesting non-trivial theorem to prove, whose proof involved the only use of traces in the script.

The second observation is an easy one called `m1-stays-halted`: once M1 has halted, it stays halted. Thus, if M1 is halted after i steps it is halted after any greater number of steps.

We can then define `find-j` to find a j at which (`tmi st tape tm j`) is halted given that we know (`M1 s_0 i`) is halted. The definition searches upwards from $j = 0$: if `tmi` is halted at j, return j; if (`find-k st tape tm j`) $\geq i$, return j; else search from $j + 1$.

This is a well-defined function: the recursion terminates because the `find-k` expression is growing monotonically and will therefore eventually reach the fixed i, if the earlier exit is not taken first.

It is easy to see that if M1 is halted at i, then `tmi` is halted at (`find-j st tape tm i`): either `find-j` returns a j (in the first exit) known to be sufficient or else it returns a j such that (`find-k st tape tm j`) $\geq i$. But our second observation above shows that M1 must thus be halted at (`find-k st tape tm j`). And if M1 is halted there, then `tmi` must be halted at j, by the `simulation` theorem.

That completes our proof sketch of theorem A.

12   Efficiency Considerations

M1 is an executable operational model which ACL2 can execute at about 500,000 M1 bytecodes/sec. We can therefore run Ψ to simulate Turing machines. The clock function `find-k` tells us exactly how long we must run it to simulate a given `tmi` run of n steps.

Consider Rogers' Turing machine description for doubling the number on the tape [12]. Suppose the tape starts with Rogers' representation for 4 on the tape. Running tmi experimentally reveals it takes 78 Turing steps to reach termination and compute a tape representing 8. We can use find-k to determine how long it takes M1 to simulate this computation. And the answer is:

103,979,643,405,139,456,340,754,264,791,057,682,257,947,240,629,585,359,596

or slightly more than 10^{56} steps!

The primary reason our implementation is so inefficient is that tapes and Turing machine descriptions are represented as large (bit-packed) integers and must be unpacked on M1 with programs that use <u>LESSP</u>. But the only way to answer the question "is $x < y$?" for two naturals x and y on M1 is to subtract 1 from each until one or the other becomes 0, because the only test M1 programs can perform is equality against 0. Thus it takes exponential time to unpack[7].

The efficiency of our M1 Turing machine interpreter would be much improved if M1 provided the JVM instruction IFLT (branch if negative) or IF_ICMPLT (branch if $x < y$). Further improvement could be made by having IDIV (floor), or bit-packing operations like ISHR (shift right), IAND (bit-wise and), etc., and perhaps arrays (with IALOAD and IASTORE), to represent the tape. Minor further improvements could be had by supporting JSR or even INVOKESTATIC or INVOKEVIRTUAL to make call/return simpler. All of these features are supported on our most complete JVM model, M6[9].

Another obvious approach would have been to compile the Turing machine description tm into an M1 program. Had I done so, the proofs of theorems A and B would have required proving that the Turing machine compiler was correct for all possible Turing machine descriptions. By representing Turing machine descriptions as data to be interpreted, I could limit my compiler's task to proving that its output was correct on the 16 Toy Lisp modules discussed. Put succinctly, it is easier to write a verifying compiler than to verify a compiler.

13 Project History

I developed M1 in 1997 to teach my JVM modeling course, which I subsequently taught about ten times. While the ISA of M1 changed annually to make homeworks harder or easier, programming M1 and proving correctness of my programs became almost second nature to me.

The question of M1's computational power arose in class in March, 2012. I completed the first version of this proof March 10–18, 2012 after coding Ψ by hand in 804 M1 instructions and manually typing the specifications and lemmas. I was helped enormously by the 1984 paper [2] and my experience with M1.

After Spring Break, I gave two talks on the proof: one to the Austin ACL2 research group and one to my undergraduate JVM class. Neither talk went smoothly and I learned a lot about the difficulty of presenting the work. A few

---

[7] And ACL2 would take exponential time evaluating find-k except for the theorems in find-k!.lisp.

weeks later, in early April, 2012, I decided to implement the verifying compiler. The present version of the proof was polished by April 14, 2012.

14   Conclusion

Aside from the satisfaction of formally revisiting the roots of computer science, this work allowed me to go back into class after Spring Break and say:

  M1 *can do anything a Turing machine can do. Here's a proof.*

References

1. Bevier, W., Hunt Jr., W.A., Moore, J.S., Young, W.: Special issue on system verification. Journal of Automated Reasoning 5(4), 409–530 (1989)
2. Boyer, R.S., Moore, J.S.: A mechanical proof of the turing completeness of pure lisp. In: Bledsoe, W.W., Loveland, D.W. (eds.) Contemporary Mathematics: Automated Theorem Proving: After 25 Years, vol. 29, pp. 133–168. American Mathematical Society, Providence (1984)
3. Boyer, R.S., Moore, J.S.: A mechanical proof of the unsolvability of the halting problem. Journal of the Association for Computing Machinery 31(3), 441–458 (1984)
4. Cohen, E., Dahlweid, M., Hillebrand, M., Leinenbach, D., Moskal, M., Santen, T., Schulte, W., Tobies, S.: VCC: A practical system for verifying concurrent C. In: Berghofer, S., Nipkow, T., Urban, C., Wenzel, M. (eds.) TPHOLs 2009. LNCS, vol. 5674, pp. 23–42. Springer, Heidelberg (2009)
5. Kaufmann, M., Manolios, P., Moore, J.S.: Computer-Aided Reasoning: An Approach. Kluwer Academic Press, Boston (2000)
6. Kaufmann, M., Moore, J.S.: The ACL2 home page. Dept. of Computer Sciences, University of Texas at Austin (2014),
   http://www.cs.utexas.edu/users/moore/acl2/
7. Klein, G., Elphinstone, K., Heiser, G., Andronick, J., Cock, D., Derrin, P., Elkaduwe, D., Engelhardt, K., Kolanski, R., Norrish, M., Sewell, T., Tuch, H., Winwood, S.: seL4: Formal verification of an os kernel. In: ACM Symposium on Operating Systems Principles, pp. 207–220 (October 2009)
8. Lindholdm, T., Yellin, F.: The Java Virtual Machine Specification, 2nd edn. Prentice Hall (1999)
9. Liu, H.: Formal Specification and Verification of a JVM and its Bytecode Verifier. PhD thesis, University of Texas at Austin (2006)
10. Milner, R., Weyhrauch, R.: Proving compiler correctness in a mechanized logic. In: Machine Intelligence 7, pp. 51–72. Edinburgh University Press (1972)
11. Ray, S., Moore, J.S.: Proof styles in operational semantics. In: Hu, A.J., Martin, A.K. (eds.) FMCAD 2004. LNCS, vol. 3312, pp. 67–81. Springer, Heidelberg (2004)
12. Rogers, H.: A Theory of Recursive Functions and Effective Commputability. McGraw-Hill (1967)

The Reflective Milawa Theorem Prover Is Sound
(Down to the Machine Code That Runs It)

Magnus O. Myreen[1] and Jared Davis[2]

[1] Computer Laboratory, University of Cambridge, UK
[2] Centaur Technology, Inc., Austin TX, USA

Abstract. Milawa is a theorem prover styled after ACL2 but with a small kernel and a powerful reflection mechanism. We have used the HOL4 theorem prover to formalize the logic of Milawa, prove the logic sound, and prove that the source code for the Milawa kernel (2,000 lines of Lisp) is faithful to the logic. Going further, we have combined these results with our previous verification of an x86 machine-code implementation of a Lisp runtime. Our top-level HOL4 theorem states that when Milawa is run on top of our verified Lisp, it will only print theorem statements that are semantically true. We believe that this top-level theorem is the most comprehensive formal evidence of a theorem prover's soundness to date.

1 Introduction

Theorem provers like HOL4, Coq, and ACL2 are each meant to reason in some particular logic, are each written in a programming language like ML, OCaml, or Lisp, and are each executed by a runtime like Poly/ML, the OCaml system, or Clozure Common Lisp. If we want to make sure that a theorem prover can only prove true statements, we should ideally show that:

A. the logic is sound,

B. the theorem prover's source code is faithful to its logic, and

C. the runtime executes the source code correctly.

In this paper, we explain how we have used the HOL4 theorem prover to establish these three properties about the Milawa theorem prover.

Milawa [2] is a theorem prover inspired by NQTHM and ACL2. Unlike these programs it has a small kernel, somewhat like an LCF-style system. This kernel notably performs reflection and includes a mechanism that modifies the kernel at runtime. High-level tactics like (conditional) rewriting are added into the kernel through a sequence of reflective extensions.

Our proofs of *A* through *C* for the Milawa prover are the key lemmas in a single, top-level HOL4 theorem: when the kernel of the Milawa theorem is run on our verified Lisp runtime, Jitawa [13], it will only ever prove statements that are true with respect to the semantics of Milawa's logic. This theorem means, for instance, that no matter how reflection or any other operation is used, the

G. Klein and R. Gamboa (Eds.): ITP 2014, LNAI 8558, pp. 421–436, 2014.

statement 'true equals false' can never be proved. This top-level theorem relates the semantics of the logic (not just syntactic provability) all the way down to the concrete x86 machine code.

We believe this work provides the most comprehensive formal evidence of a theorem prover's soundness to date, as the combination of these three properties have, to our knowledge, never before been formally proved for any interactive theorem prover.

2   Milawa in a Nutshell

Before delving into the details of our formalizations and soundness results, we start with a high-level description of the Milawa theorem prover.

ACL2-like. Milawa follows the Boyer-Moore tradition of theorem provers. Like NQTHM and ACL2, its logic is essentially a clean subset of first-order Lisp. Also like these systems, its top-level loop processes user-provided *events*. Events steer the prover process. A user can submit events which, for example, cause the prover to define a new function or prove a specific theorem. However, Milawa is simpler than ACL2 in many ways. Milawa is particularly minimalist in its user-interface and debugging output: it really just processes a list of events, aborting if any event is unacceptable.

Small kernel. The most important difference between Milawa and ACL2 is that Milawa has a small logical kernel, somewhat like an LCF-style prover. In contrast, other Boyer-Moore systems have no cordoned off area for soundness-critical code. This design means that the authors of ACL2 must program very carefully to avoid accidentally introducing soundness bugs. But it also means that ACL2 can make greater leaps in reasoning and perform well on large-scale applications; the ACL2 design avoids the LCF-bottleneck where all proofs must *at runtime* boil down to the primitive inferences of the logic.

Reflection. Milawa was designed to show that it is possible to combine the benefits of a small trusted kernel and, at the same time, avoid the LCF-bottleneck. Milawa has approximately 2000 lines of soundness-critical Lisp code. This Lisp code initializes the system and sets up the top-level event handling loop. An important part of this code is the *initial proof checker*. This initial proof checker only accepts proofs that use the primitive inferences of the logic, very much like an LCF-style kernel. In order to allow larger steps in proofs, Milawa supports a special event that replaces the prover's current proof checker with a new, user-supplied proof checker. For this *switch event* to be accepted, we must first prove that the new user-supplied proof checker (which is just a function in the logic of Milawa) can only prove statements that the initial proof checker could have proved. The initial proof checker lives within the Milawa logic. Every function defined in the logic is also defined outside in the underlying Lisp runtime.

Bootstrapping. By (repeatedly) replacing the initial proof checker with new, improved checkers that can make larger leaps in their proofs, we can build a prover that performs ACL2-style proofs where, e.g., conditional rewriting is treated as a single inference step. We call the soundness critical code—the initial 2000-line Lisp program—Milawa's kernel. The Milawa theorem prover is what this kernel morphs into after running through a long list of events (the *bootstrapping* sequence) that ultimately installs a powerful, ACL2-like proof checker. This final proof checker allows for high-level, Boyer-Moore style steps such as rewriting, case splitting, generalization, cross-fertilization, and so forth.

3 Method

To prove the soundness of Milawa in HOL4, we proceeded as follows.

A. We started by formalizing Milawa's logic, following closely the detailed prose description given in Chapter 2 of Davis [2]. We then proved the logic is sound. This part was largely a routine formalization and soundness proof (Section 4), but we did hit some surprises involving the termination obligations Milawa generates (Section 7).

B. Next we turned our attention to the implementation of Milawa's kernel. Our task here was to verify these 2,000 lines of Lisp code with respect to the behavior of Jitawa [13], our verified Lisp runtime. We proved a connection using the following steps (Section 5).

  1. Jitawa's correctness theorem is stated in terms of a read-eval-print loop which reads ASCII input. Using rewriting, we evaluated its parser on the ASCII definition of Milawa's kernel.

  2. Once the ASCII input had been turned into appropriate abstract syntax, we ran a proof-producing tool [12] to translate deeply embedded Lisp programs into their 'obvious' shallowly embedded counterparts.

  3. Given the convenient shallow embeddings, we proved that Milawa's main loop maintains an invariant that implies that all proved theorems are true w.r.t. our semantics of Milawa's logic.

C. We had already verified our Lisp runtime, Jitawa, as described in a previous paper [13]. What remained was to connect the results from *A* and *B* to Jitawa's top-level correctness theorem (Section 6).

The result of combining *A*, *B* and *C* is a top-level theorem (Section 6) that relates logical soundness all the way down to machine-code execution. We found mistakes in Milawa's implementation, but no soundness bugs (Section 7).

4 Milawa's Logic

We start with a formalization and soundness proof of Milawa's logic. Milawa targets a first-order logic of recursive functions with induction up to ϵ_0, similar to the logics of NQTHM and ACL2. The objects of the logic are the natural numbers, symbols, and conses (ordered pairs) of other objects; we call these objects S-expressions. The logic has primitive functions for working with S-expressions like equality checking, addition, cons, car, cdr, etc., whose behavior is given with axioms. Starting from these primitives, we can define recursive functions that look like Lisp programs. An introduction to the logic can be found in Chapter 2 of Davis [2].

The Milawa logic is considerably weaker than popular higher-order logics. Thanks to this, its soundness can be established using higher-order logic as the meta-logic. In this section, we explain how we have used the HOL4 system to formalize the syntax (Section 4.1), semantics (4.3) and rules of inference (4.4) of the Milawa logic, and to mechanically prove the soundness of its inference rules (4.5) and definition principle (4.6). In later sections, we connect these soundness proofs to the theorem prover's implementation.

4.1   Syntax of Terms and Formulas

We formalize the syntax of the Milawa logic as the following datatype:

sexp	::=	Val *num* | Sym *string* | Dot *sexp* *sexp*	S-expression
prim	::=	If | Equal | Not | Symbolp | Symbol_less	
	|	Natp | Add | Sub | Less | Consp | Cons	
	|	Car | Cdr | Rank | Ord_less | Ordp	
func	::=	PrimitiveFun *prim*	primitive functions
	|	Fun *string*	user-defined
term	::=	Const *sexp*	constant S-expression
	|	Var *string*	variable
	|	App *func* (*term* list)	function application
	|	LamApp (*string* list) *term* (*term* list)	λ formals body actuals
formula	::=	¬*formula*	negation
	|	*formula* ∨ *formula*	disjunction
	|	*term* = *term*	term equality

These type definitions are not quite enough to capture correct Milawa syntax. We write separate well-formedness predicates called term_ok and formula_ok to formalize the additional requirements. In particular,

- every function application must have correct arity and refer to a known function with respect to the context (see below), and
- every lambda application must have the same number of formal and actual parameters, must have distinct formal parameters, and its body may not refer to variables besides its formal parameters; these requirements make substitution straightforward.

The term_ok and formula_ok well-formedness predicates depend on a *logical context*, π, which will be explained below.

4.2   Context

The definitions of the syntax, semantics and inference rules all depend on information regarding user-defined functions. To keep the formalization simple, we chose to combine all of this information into a single mapping, which we call the *logical context*. We model the logical context as a finite partial map π from function names, of type *string*, to elements of type:

$$string \; \mathsf{list} \times func\_body \times (sexp \; \mathsf{list} \rightarrow sexp)$$

The first component, *string* list, names the formal parameters for the function. The second component, *func_body*, gives the syntactic definition for the function. This *func_body* is usually either (1) the right-hand side of a definition, for an ordinary function defined by an equation, or (2) a variable name and property, for a witness (Skolem) function. For reasons that will be explained in Section 4.6, we also allow the omission of the function body, i.e., a None alternative.

func_body ::=	Body *term*	concrete term (e.g. recursive function)
	| Witness *term string*	property, element name
	| None	no function body given

Finally, the *sexp* list $\rightarrow$ *sexp* component is an *interpretation function*, which is used in the definition of the semantics. These interpretation functions specify what meaning the semantics is to assign to applications of user-defined functions. In the next section, we will see a well-formedness criteria that relates the interpretation functions with the syntax in *func_body*.

4.3   Semantics

Next, we define a semantics of Milawa's formulas. We present these semantics in a top-down order. Our topmost definition is validity: a Milawa formula p is *valid*, written $\models_\pi p$, if and only if (1) p is syntactically correct w.r.t. the logical context π and (2) p evaluates to true in π for all variable instantiations i.

$$(\models_\pi p) \;\; = \;\; \mathsf{formula\_ok}_\pi \; p \; \wedge \; \forall i. \; \mathsf{eval\_formula} \; i \; \pi \; p$$

We define the evaluation of a formula with respect to a particular variable instantiation i. Our formula evaluator, eval_formula i π, is built on top of a term evaluator, eval_term i π, as follows. The syntax overloading can be confusing in the following definition. On the left-hand side $\neg$, $\vee$ and = are the constructors for the *formula* type, while on the right-hand side $\neg$ and $\vee$ are the usual Boolean connectives and = is HOL's equality predicate.

eval_formula i π $(\neg p)$	=	$\neg$(eval_formula i π p)
eval_formula i π $(p \vee q)$	=	eval_formula i π p $\vee$ eval_formula i π q
eval_formula i π $(x = y)$	=	(eval_term i π x = eval_term i π y)

We define term evaluation with respect to a variable instantiation i. Here $[[v_1, \ldots, v_n] \mapsto [x_1, \ldots, x_n]]$ is a function that maps v_i to x_i, for $1 \leq i \leq$

n, and all other variable names to NIL. Below map is a function such that map $f\ [t_1, t_2, \ldots, t_n] = [f\ t_1, f\ t_2, \ldots, f\ t_n]$.

$$
\begin{aligned}
\text{eval\_term } i\ \pi\ (\text{Const } c) &= c \\
\text{eval\_term } i\ \pi\ (\text{Var } v) &= i(v) \\
\text{eval\_term } i\ \pi\ (\text{App } f\ xs) &= \text{eval\_app } (f, \text{map } (\text{eval\_term } i\ \pi)\ xs, \pi) \\
\text{eval\_term } i\ \pi\ (\text{LambdaApp } vs\ x\ xs) &= \text{let } ys = \text{map } (\text{eval\_term } i\ \pi)\ xs \text{ in} \\
&\qquad \text{eval\_term } [vs \mapsto ys]\ \pi\ x
\end{aligned}
$$

Application of a function to a list of concrete arguments, a list of type *sexp* list, is evaluated according the following eval_app function. This function evaluates primitive functions according to eval_primitive and user-defined functions according to the interpretation function *interp* stored in the logical context. The interpretation functions will be explained further below.

$$
\begin{aligned}
\text{eval\_app } (\text{PrimitiveFun } p, args, \pi) &= \text{eval\_primitive } p\ args \\
\text{eval\_app } (\text{Fun } name, args, \pi) &= \text{let } (\_, \_, interp) = \pi(name) \text{ in} \\
&\qquad interp(args)
\end{aligned}
$$

We omit the definition of eval_primitive, which is lengthy and straightforward, but note that it is a total function. A few example evaluations:

$$
\begin{aligned}
\text{eval\_primitive Add } [\text{Val } 2, \text{Val } 3] &= \text{Val } 5 \\
\text{eval\_primitive Add } [\text{Val } 2, \text{Sym "a"}] &= \text{Val } 2 \\
\text{eval\_primitive Cons } [\text{Val } 2, \text{Sym "a"}] &= \text{Dot } (\text{Val } 2)\ (\text{Sym "a"})
\end{aligned}
$$

The definitions above constitute the semantics of Milawa. Clearly, this semantics is intimately dependent on the interpretation functions stored inside the context π. In order to make sure that these interpretation functions are 'the right ones', i.e., correspond to the syntactic definitions of the user-defined functions, we require that the context is well-formed, i.e., satisfies a predicate we will call context_ok.

For a context to be well-formed, any user-defined functions with an entry of the following form in the logical context π,

$$
\pi(name)\ =\ (formals, \text{Body } body, interp)
$$

must have the *interp* function return the same value as an evaluation of *body* with appropriate instantiations of the formal parameters, i.e., the following *defining equation* must be true:

$$
\forall i.\ interp(\text{map } i\ formals)\ =\ \text{eval\_term } i\ \pi\ body
$$

Note that this is a non-trivial equation since eval_term, which appears on the right-hand side of the equation, can refer to *interp* via eval_app. Indeed, proving soundness of the definition principle requires proving that the termination obligations generated by Milawa imply that our interpetation is total (Section 4.6).

A similar condition applies to witness functions. If,

$$
\pi(name)\ =\ (formals, \text{Witness } prop\ var, interp)
$$

is true then the following implication must hold. This implication states that, if there exists some value v such that property *prop* is true when variable names *var* :: *formals* are substituted for values v :: *args* in *prop*, then *interp*(*args*) returns some such value v. Here the test for 'is true' is Lisp's truth test, i.e., 'not equal to NIL'. These witness functions are explained in Davis [2].

$$\forall args.$$
$$(\exists v.\ \mathsf{eval\_term}\ [var :: formals \mapsto v :: args]\ \pi\ prop \neq \mathtt{NIL}) \Longrightarrow$$
$$\mathsf{eval\_term}\ [var :: formals \mapsto interp(args) :: args]\ \pi\ prop \neq \mathtt{NIL}$$

The well-formedness criteria for contexts puts no restrictions on the *interp* function if the function body is None.

The full definition of the well-formedness criteria for contexts, context_ok, is given below. Here free_vars is a function that computes the list of free variables of a term, and list_to_set converts a list to a set.

$$\mathsf{context\_ok}\ \pi\ =$$
$$(\forall name\ formals\ body\ interp.$$
$$(\pi(name) = (formals, \mathsf{Body}\ body, interp)) \Longrightarrow$$
$$\mathsf{term\_ok}_\pi\ body \wedge \mathsf{all\_distinct}\ formals \wedge$$
$$\mathsf{list\_to\_set}\ (\mathsf{free\_vars}\ body) \subseteq \mathsf{list\_to\_set}\ formals \wedge$$
$$\forall i.\ interp(\mathsf{map}\ i\ formals) = \mathsf{eval\_term}\ i\ \pi\ body) \wedge$$
$$(\forall name\ formals\ prop\ var\ interp.$$
$$(\pi(name) = (formals, \mathsf{Witness}\ prop\ var, interp)) \Longrightarrow$$
$$\mathsf{term\_ok}_\pi\ prop \wedge \mathsf{all\_distinct}\ (var :: formals) \wedge$$
$$\mathsf{list\_to\_set}\ (\mathsf{free\_vars}\ prop) \subseteq \mathsf{list\_to\_set}\ (var :: formals) \wedge$$
$$\forall args.$$
$$(\exists v.\ \mathsf{eval\_term}\ [var :: formals \mapsto v :: args]\ \pi\ prop \neq \mathtt{NIL}) \Longrightarrow$$
$$\mathsf{eval\_term}\ [var :: formals \mapsto interp(args) :: args]\ \pi\ prop \neq \mathtt{NIL})$$

4.4 Inference Rules

Due to space constraints, this section will only sketch a few of Milawa's 13 inference rules. Two of the simplest are shown below. Here milawa_axioms is a set consisting of the 56 axioms from Davis [2]. Most of these are basic facts about the primitive functions, e.g. term equality is reflexive, symmetric and transitive; the Less primitive is irreflexive and transitive, etc.

$$\frac{\vdash_\pi a \vee (b \vee c)}{\vdash_\pi (a \vee b) \vee c}\ \text{(associativity)} \qquad \frac{a \in \mathsf{milawa\_axioms}}{\vdash_\pi a}\ \text{(basic axiom)}$$

The most complicated inference rule allows induction according to a user-defined measure over the ordinals up to ε_0. We omit the presentation of that lengthy inference rule, which Chapter 6 of Kaufmann et al. [6] explains in detail.

Apart from the normal inference rules, we also include rules that allow function definitions to be looked up from the logical context, e.g.

$$\frac{\pi(name) = (formals, \mathsf{Body}\ body, interp)}{\vdash_\pi \mathsf{App}\ (\mathsf{Fun}\ name)\ (\mathsf{map}\ \mathsf{Var}\ formals) = body}$$

4.5 Soundness and Consistency

We state the soundness theorem for Milawa's inference rules as follows:

$$\forall \pi\ p.\ \ \text{context\_ok}\ \pi \wedge (\vdash_\pi p) \implies (\models_\pi p)$$

We have proved this statement by induction over the inference rules $\vdash_\pi$. Proving soundness of the induction rule was the most interesting case: this proof required induction over the ordinals up to ε_0, for which we need to know that less-than over these ordinals is well-founded. Fortunately, Kaufmann and Slind [7] had already formalized this result in HOL4. The soundness of the induction rule follows almost directly from their result.

The soundness theorem from above lets us immediately prove many reassuring corollaries. For instance, since $\models_\pi$ T = NIL is false and $\vdash_\pi$ T = T is true we know that Milawa's inference rules are consistent.

4.6 Soundness Preserved by Function Definitions

As part of our verification of Milawa's kernel (Section 5.4), we have proved that the kernel maintains an invariant which states that (1) the current logical context π is well-formed, context_ok π, and (2) that all theorems the Milawa theorem prover has accepted are provable using the inference rules based on that current context π, i.e., for any formula p accepted by the kernel, we have $\vdash_\pi p$. However, when new definitions are made the logical context is extended. In order to maintain our invariant, we must hence show that properties (1) and (2) carry across context extensions.

Proving that property (1) carries across is straightforward since the syntactic inference rules only make tests for inclusion in the context.

Proving that well-formedness of the context, i.e., property (2), carries across context extensions is less straightforward. The main complication is that we need to find an interpretation for the new function that agrees with the syntax of the new definition. Using HOL's choice operator, we define a function new_interp (definition omitted) which constructs such an interpretation if such an interpretation exists. This reduces the goal to proving that an interpretation exists. For witness functions, this proof is almost immediate. For conventional functions, this proof required showing that a $\vdash_\pi$-proof of the generated termination obligations is sufficient to imply that a suitable interpretation exists. Below definition_ok requires that certain syntactic conditions are true and that the termination obligations can be proved.

$$\forall \pi\ name\ formals\ body.$$
$$\text{context\_ok}\ \pi \wedge \text{definition\_ok}\ (name, formals, body, \pi) \implies$$
$$\text{context\_ok}\ (\pi[name \mapsto (formals, body, \text{new\_interp}\ \pi\ name\ formals\ body)])$$

5 Correctness of Milawa's Implementation

With logical soundness out of the way, our next goal was to show that the source code of the Milawa kernel respects the logic's inference rules.

First, some background: in previous work [13], we introduced the Jitawa Lisp runtime. Jitawa is able to *host* the Milawa theorem prover. By this, we mean that it is able to execute Milawa's kernel all the way through its *bootstrapping process* [2], a long sequence of definitions, proofs and reflective extensions which ultimately extend the kernel with many high-level proof procedures like those of NQTHM and ACL2. As part of the Jitawa work, we developed an operational semantics for the Lisp dialect that Jitawa executes, and proved that the x86 machine code for Jitawa implements this semantics.

Milawa's kernel is about 2,000 lines of Lisp code. In this section, we explain how we have proved that this Lisp code is faithful to Milawa's inference rules w.r.t. the operational semantics that Jitawa has been proved to implement.

5.1   From ASCII Characters to a Shallow Embedding in HOL4

The top-level Jitawa semantics describes how S-expressions are to be parsed from an input stream of ASCII characters and then evaluated. One of the simplest functions in Milawa's kernel is shown below. This function will be used as a running example of how we lift Lisp functions into HOL to make interactive verification manageable.

```
(defun lookup-safe (a x)
  (if (consp x)
      (if (equal a (car (car x)))
          (if (consp (car x))
              (car x)
              (cons (car (car x)) (cdr (car x))))
          (lookup-safe a (cdr x)))
      nil))
```

When Jitawa reads the ASCII definition of `lookup-safe`, it parses the lines above and, as far as its operational semantics is concerned, turns them into a datatype of the form:

App Define [Const (Sym "LOOKUP-SAFE"), Const (...), Const (...)]

We wrote a custom conversion (based mostly on rewriting) in HOL4 which parses the source code for Milawa's 2000-line kernel into abstract datatypes such as the expression above. The evaluation of the parser happens inside the HOL4 logic, so the result is a theorem of the form string_to_prog milawa_kernel_lisp = . . .

When Jitawa evaluates the Define expression from above, a definition for `lookup-safe` is added to its list of functions. The new entry is:

function name: "LOOKUP-SAFE"
parameter list: "A", "X"
function body: If (App (PrimitiveFun Consp) [Var "X"])
                    (If (App (PrimitiveFun Equal) [...])
                       (If (App (PrimitiveFun Consp) [...] (...) (...))
                       (App (Fun "LOOKUP-SAFE") [...]))
                    (Const (Sym "NIL"))

Instead of performing tedious proofs directly over deep embeddings such as that above, we developed a tool that automatically translates these deep embeddings into shallow embeddings and, in the process, proves that the shallow embeddings accurately describe evaluations of the deep embeddings. The details of this tool are the subject of a separate paper [12], but the net effect of using it on lookup-safe is easy to see: we get a simple HOL function,

$$\begin{aligned}
\text{lookup\_safe } a\ x\ =\ \ &\text{if consp } x \text{ then} \\
&\quad \text{if } a = \text{car (car } x) \text{ then} \\
&\qquad \text{if consp (car } x) \text{ then} \\
&\qquad\quad \text{car } x \\
&\qquad \text{else cons (car (car } x)) \text{ (cdr (car } x)) \\
&\quad \text{else lookup\_safe } a \text{ (cdr } x) \\
&\text{else Sym "NIL"}
\end{aligned}$$

and a theorem relating the deep embedding to this shallow embedding, stated in terms of the application relation $\xrightarrow{\text{ap}}$ of Jitawa's semantics:

$$\dots \implies (\text{Fun "LOOKUP-SAFE"}, [a, x], state) \xrightarrow{\text{ap}} (\text{lookup\_safe } a\ x, state)$$

Here *state* is Jitawa's mutable state which has, e.g., the I/O streams and the list of function definitions. The state is not changed by lookup_safe because lookup-safe is a pure function. Extracted impure functions take the state as input and produce a new state as output, e.g. Milawa's admit_defun function returns a (value, new-state) pair:

$$\dots \implies (\text{Fun "ADMIT-DEFUN"}, [cmd, s], state) \xrightarrow{\text{ap}} (\text{admit\_defun } cmd\ s\ state)$$

5.2   Milawa's Proof Checkers and Reflection

The largest and most important pure function in Milawa is its initial proof checker, proofp. This function is given an *appeal* (an alleged proof) to check. It walks through the appeal, checking that each proof step is a valid use of some inference rule. When Milawa starts, it uses proofp to check alleged proofs of theorems and termination obligations. But the kernel can later be told to start using some user-defined function, say new-proofp, to check proofs. Typically new-proofp can accept "higher level" proofs that use new inference rules beyond the "base level" rules available in proofp. The kernel will only switch to new-proofp after establishing its *fidelity claim*: whenever new-proofp accepts a high-level proof of ϕ, there must exist a base-level proof of ϕ that proofp would accept.

We prove that proofp is faithful to the inference rules of the Milawa logic. That is, whenever proofp is given well-formed inputs and it returns something other than NIL, the conclusion of the alleged proof is $\vdash_\pi$-provable. Here *axioms* and *thms* are lists of formulas, and *atbl* is an arity table.

$\forall appeal\ axioms\ thms\ atbl.$
   $\text{appeal\_syntax\_ok } appeal \land \text{atbl\_ok } \pi\ atbl \land$
   $\text{thms\_inv } \pi\ thms \land \text{thms\_inv } \pi\ axioms \land$
   $\text{proofp } appeal\ axioms\ thms\ atbl \neq \text{Sym "NIL"} \implies \vdash_\pi \text{conclusion\_of } appeal$

To accommodate the reflective installation of new proof checkers, the invariant we describe in the next section requires that the property above must always hold for whatever function is the current proof checker. It turns out that Milawa's checks of the fidelity claim are sufficient to show that a new-proofp may only be installed when it satisfies this property.

5.3   Milawa's Invariant

As it executes, Milawa's kernel carries around state with several lists and mappings that must be kept consistent. Its program state consists of:

- a list of axioms and definitions,
- a list of proved theorems,
- an arity table for syntax checks (e.g., are all mentioned functions defined? are they called with the right number of arguments?),
- the name of the current proof checker (proofp, new-proofp, . . .), and
- a function table that lists all the definitions that have been given to the Lisp runtime, and the names of functions that must be avoided since they have a special meaning in the runtime (error, print, define, funcall, . . .).

There is also state specific to the Lisp runtime's semantics:

- its view of how functions have been defined,
- its input and output streams, and
- a special *ok* flag that records whether an error has been raised.

Finally, for our soundness proof, there is also logical (ghost) state:

- a logical context π must also be maintained.

A key part of our proof was to formalize the invariant that relates these state elements. For the most part, the dependencies and relationships between the state components were obvious, e.g. each entry in the function table must have a corresponding entity inside the runtime's function table, and since this is a reflective theorem prover each function in the logic must have an entry in the runtime's function table.

A few details were less straightforward. Each layer has its own abstraction level, e.g. the kernel and runtime allow macros but these are expanded away in the logic, and the function table uses S-expression syntax but the runtime's operational semantics only sees an abstraction of this syntax. There are also some language mismatches: the logic has primitives (e.g. ordp and ord-<) which are not primitive in the runtime, and the runtime has several primitives that are not part of the logic (e.g., funcall, print, error). To further complicate matters, some of these components can lag behind: the function table starts off mentioning functions that have not yet been defined in the logic. Such functions can only be defined using exactly the definition given in the function table, otherwise the defining event, admit-defun or admit-witness, causes a runtime error. We will explain this invariant in more detail in forthcoming journal article and/or extensive technical report.

We proved that each event handling function, e.g. `admit-thm`, `admit-defun`, `admit-switch` etc., maintains the invariant. As a result, the kernel's top-level event-handling loop maintains the invariant.

5.4   Theorem: Milawa Is Faithful to Its Logic

Milawa's kernel reads input, processes it, and then prints output that says whether it has accepted the proofs and definitions it has been given. In order to make it clearer what Milawa claims to have proved, we extended Milawa with a new event, (`admit-print` ϕ), which causes ϕ to be printed if it has already been proved as a theorem, or else fails. For instance, this new event can print:

```
(PRINT (THEOREM (PEQUAL* (+ A B) (+ B A))))
```

We formulate the soundness of Milawa as a guarantee about the possible output: whatever the input, Milawa will only ever print THEOREM lines for formulas that are true w.r.t. the semantics $\models_\pi$ of the logic. More precisely, we first define what an acceptable line of output is w.r.t. a given logical context π:

$$
\begin{aligned}
\mathsf{line\_ok}\ (\pi, l)\ =\ & (l = \texttt{"NIL"})\ \vee \\
& (\exists n.\ (l = \texttt{"(PRINT}\ (n\ \dots\ )\texttt{)"}) \wedge \mathsf{is\_number}\ n)\ \vee \\
& (\exists \phi.\ (l = \texttt{"(PRINT (THEOREM}\ \phi)\texttt{)"}) \wedge \mathsf{context\_ok}\ \pi \wedge \models_\pi \phi)
\end{aligned}
$$

We then prove that Milawa's top-level function, `milawa_main`, only produces output lines that satisfy `line_ok`, assuming that no runtime errors were raised during execution, i.e., that *ok* is true. Here `compute_output` (definition omitted) is a high-level specification of what output lines coupled with their respective logical context the input *cmds* produces.

$$
\begin{aligned}
& \exists ans\ k\ output\ ok. \\
& \quad \mathsf{milawa\_main}\ cmds\ \mathsf{init\_state} = (ans, (k, output, ok))\ \wedge \\
& \quad (ok \implies (ans = \mathsf{Sym}\ \texttt{"SUCCESS"})\ \wedge \\
& \qquad\quad \mathsf{let}\ result = \mathsf{compute\_output}\ cmds\ \mathsf{in} \\
& \qquad\quad \mathsf{every\_line}\ \mathsf{line\_ok}\ result\ \wedge \\
& \qquad\quad output = \mathsf{output\_string}\ result)
\end{aligned}
$$

This approach works in part because Jitawa's print function, though used by Milawa's kernel, is not made available in the Milawa logic. In other words, a user-defined function can't trick us into invalidly printing (`PRINT (THEOREM ...)`).

This soundness theorem can be related back to the operational semantics of Jitawa through the following theorem, which was automatically derived by our tool for lifting deep embeddings into shallow embeddings:

$$
\dots \implies (\mathsf{Fun}\ \texttt{"MILAWA-MAIN"}, [input], state) \xrightarrow{\mathsf{ap}} (\mathsf{milawa\_main}\ input\ state)
$$

6   Top-level Soundness Theorem

Now we are ready to connect the soundness result from above to the top-level correctness theorem for Jitawa, which was proved in previous work [13]. Jitawa's top-level correctness theorem is stated in terms of a machine-code Hoare triple [11] that can informally be read as saying: if Jitawa's implementation is started from a state where enough memory is allocated (init_state) and the input stream of ASCII characters holds *input* for which Jitawa terminates, then either an error message is reported or a final state described by $\xrightarrow{\text{exec}}$ is reached for which *ok* is true and *output* is the final state of the output stream (final_state).

{ init_state *input* * pc *pc* * ⟨terminates_for *input*⟩ }
 pc : code_for_entire_jitawa_implementation
{ error_message ∨ ∃*output*. ⟨([], *input*) $\xrightarrow{\text{exec}}$ (*output*, true)⟩ * final_state *output* }

Roughly speaking, $\xrightarrow{\text{exec}}$ involves parsing some input, evaluating it with $\xrightarrow{\text{ap}}$, and printing the result. By manually unrolling $\xrightarrow{\text{exec}}$ to reveal the $\xrightarrow{\text{ap}}$ relation for the call of `milawa_main`, it was straightforward to prove our top-level theorem relating Milawa's soundness down to the concrete x86 machine code.

This theorem, shown below, can informally be read as follows: if the ASCII input to Jitawa is the code for Milawa's kernel followed by a call to Milawa's main function on any input *input*, then the machine-code implementation for Jitawa will either abort with an error message, or succeed and print line_ok output (according to compute_output) followed by SUCCESS. Here strings are lists of characters, hence the use of list append (++) for strings.

∀*input* *pc*.
  { init_state (milawa_implementation ++ "(milawa-main '*input*)") * pc *pc* }
   pc : code_for_entire_jitawa_implementation
  { error_message ∨ (let *result* = compute_output (parse *input*) in
                 ⟨every_line line_ok *result*⟩ *
                 final_state (output_string *result* ++ "SUCCESS")) }

7   Quirks, Bugs and Other Points of Interest

We ran into some surprises during the proof.

Two minor bugs. No soundness bugs were found during our proof, but two minor bugs were uncovered and fixed. One was a harmless omission in the initial function arity table. The other allowed definitions with malformed parameter lists (not ending with `nil`) to be accepted. We don't see how these bugs could be exploited to derive a false statement, but the latter could probably have lead to undefined behavior when using a Common Lisp runtime, instead of our verified Lisp runtime.

Complication with termination obligations. In its current form, Milawa will only accept user-defined functions when their termination obligations are proven. However, the termination obligations can, in some cases, mention the function that is being defined. For instance, when defining a function like:

$$f(n, k) = \text{if } n = 0 \text{ then } k \text{ else } f(n - 1, f(n - 1, k + 1))$$

Milawa will require that the following termination condition has been proved for some measure function m:

$$n \neq 0 \implies m(n - 1, f(n - 1, k + 1)) <_{\text{ord}} m(n, k) \land$$
$$m(n - 1, k + 1) <_{\text{ord}} m(n, k)$$

But note that this statement mentions function f, i.e., f ought to be part of the logical context π in order for this formula to be well-formed (formula_ok). Milawa's kernel gets around this problem by checking the proof of such termination obligations in a half-way state, where the f is acceptable syntax but the defining equation is not yet available as a theorem. Our formalization of the logic checks the termination obligations in a similar half-way state: the termination obligations are checked in a state where the function's name is available in the context but the function body is set to None (Section 4.2).

Extensions. Once we had completed the full soundness proof, we took the opportunity to step back and consider what part of the system can be made better without complicating the soundness proof.

Evaluation through reflection: The original version of Milawa only used reflection to run the user-defined proof-checkers. However, one can equally well prove theorems by evaluation in the runtime, since all function defined in the logic also have a counter-part in the runtime. We have implemented and proved sound such an event handler (`admit-eval`).

Support for non-terminating functions: Note that our formalization of Milawa's logic only requires that there must exist an interpretation in HOL for each of the functions living in Milawa's logic. This means, e.g., that tail-recursive functions can be admitted without any proof of the termination obligations, because any tail-recursive function can be defined in HOL without a termination proof. We have proved that it is sound to extend a context with any recursive function that passes a simple syntactic check, which tests whether all recursive calls are in tail position. This extension has not been implemented in the Milawa kernel because, if it were there, Milawa might not terminate, which composition with the correctness theorem for our Lisp implementation requires (Section 6).

8   Summary and Related Work

Davis' dissertation [2] describes how the Milawa theorem prover is constructed using self-verification from a small trusted kernel. In this paper, we have explained how we have verified in HOL4 that this kernel is indeed trustworthy.

We have proved that the implementation of the Milawa theorem prover can never prove a statement that is false when it is run on Jitawa, our verified Lisp implementation. This theorem goes from the logic all the way down to the machine code. To the best of our knowledge, this is the most comprehensive formal evidence of a theorem prover's soundness to date.

Related Work. The most closely related work is that of Kumar et al. [8] which aims to verify a similar end-to-end soundness result for a version of the HOL light theorem prover. Kumar et al. have a verified machine-code implementation of ML [9] (the dialect is called CakeML) and have an implementation of the HOL light kernel which has been proved sound w.r.t. a formal semantics of higher-order logic (HOL). At the time of writing, this CakeML project has not yet composed the correctness theorem for the ML implementation with the soundness result for the verified implementation of the HOL light kernel.

Kumar et al. based their semantics of HOL on work by Harrison [5], in which Harrison formalized HOL and proved soundness of its inference rules. Harrison's formalization did not include any definition mechanisms.

A reduced version of the Calculus of Inductive Constructions (CiC), i.e., the logic implemented by the Coq proof assistant, has also been formalized. Barras [1] has given reduced CiC a formal semantics in set theory and formalized a soundness proof in Coq. Recently, Wang and Barras [15] showed that the approach is modular and applied the framework to the Calculus of Constructions plus an abstract equational theory.

Milawa's logic is a simplified variant of the ACL2 logic. The ACL2 logic has previously been modeled in HOL, most impressively by Gordon et al. [3,4]. In this work, ACL2's S-expressions and axioms are formalized as a shallow embedding in HOL. ACL2's axioms are proven to be theorems in HOL, and a mechanism is developed in which proved statements can be transferred between HOL4 and ACL2. Our work is in many ways cleaner, e.g., Milawa's S-expressions do not contain characters, strings or complex rationals, which clutter proofs. As part of our previous work on the verified Jitawa Lisp implementation, we proved that the axioms of Milawa (milawa_axioms from Section 4.4) are compatible with Jitawa's semantics. In the current paper, we went much further: we formalized the logic, proved soundness of all of Milawa's inference rules and proved soundness of the concrete implementation of Milawa w.r.t. Jitawa's semantics.

Other theorem prover implementations have also been verified. Noteworthy verifications include Ridge and Margetson [14]'s soundness and completeness proofs for a simple first-order tableau prover that can be executed in Isabelle/HOL by rewriting, and the verification of a SAT solver with modern optimizations by Marić [10]. Marić suggests that his SAT solver can be used as an automatically Isabelle/HOL-code-generated implementation.

Source Code. Milawa's soundness proof and all auxiliary files are available at http://www.cl.cam.ac.uk/~mom22/jitawa/, and the Milawa theorem prover is available at http://www.cs.utexas.edu/~jared/milawa/Web/.

Acknowledgments. We thank Thomas Sewell for commenting on drafts of this paper. The first author was funded by the Royal Society, UK.

References

1. Barras, B.: Sets in Coq, Coq in sets. J. Formalized Reasoning 3(1) (2010)
2. Davis, J.C.: A Self-Verifying Theorem Prover. PhD thesis, University of Texas at Austin (December 2009)
3. Gordon, M.J.C., Hunt Jr., W.A., Kaufmann, M., Reynolds, J.: An embedding of the ACL2 logic in HOL. In: International Workshop on the ACL2 Theorem Prover and its Applications (ACL2), pp. 40–46. ACM (2006)
4. Gordon, M.J.C., Reynolds, J., Hunt Jr., W.A., Kaufmann, M.: An integration of HOL and ACL2. In: Formal Methods in Computer-Aided Design (FMCAD), pp. 153–160. IEEE Computer Society (2006)
5. Harrison, J.: HOL light: An overview. In: Berghofer, S., Nipkow, T., Urban, C., Wenzel, M. (eds.) TPHOLs 2009. LNCS, vol. 5674, pp. 60–66. Springer, Heidelberg (2009)
6. Kaufmann, M., Manolios, P., Strother Moore, J.: Computer-Aided Reasoning: An Approach. Kluwer Academic Publishers (June 2000)
7. Kaufmann, M., Slind, K.: Proof pearl: Wellfounded induction on the ordinals up to ϵ_0. In: Schneider, K., Brandt, J. (eds.) TPHOLs 2007. LNCS, vol. 4732, pp. 294–301. Springer, Heidelberg (2007)
8. Kumar, R., Arthan, R., Myreen, M.O., Owens, S.: HOL with Definitions: Semantics, Soundness, and a Verified Implementation. In: Klein, G., Gamboa, R. (eds.) ITP 2014. LNCS (LNAI), vol. 8558, pp. 302–317. Springer, Heidelberg (2014)
9. Kumar, R., Myreen, M.O., Norrish, M., Owens, S.: CakeML: a verified implementation of ML. In: Sewell, P. (ed.) Principles of Programming Languages (POPL). ACM (2014)
10. Marić, F.: Formal verification of a modern SAT solver by shallow embedding into Isabelle/HOL. Theor. Comput. Sci. 411(50) (2010)
11. Myreen, M.O.: Verified just-in-time compiler on x86. In: Hermenegildo, M.V., Palsberg, J. (eds.) Principles of Programming Languages (POPL). ACM (2010)
12. Myreen, M.O.: Functional programs: Conversions between deep and shallow embeddings. In: Beringer, L., Felty, A. (eds.) ITP 2012. LNCS, vol. 7406, pp. 412–417. Springer, Heidelberg (2012)
13. Myreen, M.O., Davis, J.: A verified runtime for a verified theorem prover. In: van Eekelen, M., Geuvers, H., Schmaltz, J., Wiedijk, F. (eds.) ITP 2011. LNCS, vol. 6898, pp. 265–280. Springer, Heidelberg (2011)
14. Ridge, T., Margetson, J.: A mechanically verified, sound and complete theorem prover for first order logic. In: Hurd, J., Melham, T. (eds.) TPHOLs 2005. LNCS, vol. 3603, pp. 294–309. Springer, Heidelberg (2005)
15. Wang, Q., Barras, B.: Semantics of intensional type theory extended with decidable equational theories. In: Computer Science Logic (CSL). LIPIcs, vol. 23. Schloss Dagstuhl – Leibniz-Zentrum fuer Informatik (2013)

Balancing Lists: A Proof Pearl

Guyslain Naves[1] and Arnaud Spiwack[2]

[1] Aix Marseille Université, CNRS, LIF UMR 7279, 13288, Marseille, France
guyslain.naves@lif.univ-mrs.fr
[2] Inria Paris-Rocquencourt
ENS, Paris, France
arnaud@spiwack.net

Abstract. Starting with an algorithm to turn lists into full trees which uses non-obvious invariants and partial functions, we progressively encode the invariants in the types of the data, removing most of the burden of a correctness proof.

The invariants are encoded using non-uniform inductive types which parallel numerical representations in a style advertised by Okasaki, and a small amount of dependent types.

1 Introduction

Starting with a list lst, we want to turn it into a binary tree tr of the following form (in Ocaml):

```
type α tree =
  | Node of α tree * α * α tree
  | Leaf
```

With the constraints that lst must be the infix traversal of tr and that tr must be *full*, in the sense that every level except the last are required to be completely filled. Such a function turns, in particular, sorted lists into balanced binary search trees.

There are a number of folklore algorithms to achieve this result in linear time. Here we consider one of these algorithms, presented in Section 2, which repeatedly pairs up trees of height h in a list to form a list of trees of height $h + 1$. Our interest in this algorithm sprouts from the fact that its correctness is not obvious; in particular the invariants are complex: the main loop operates on a list of length $2^k - 1$ whose elements are alternately of two distinct forms.

In Sections 3 and 4 we show refinements of the algorithm where the invariants are pushed into the types, leading to a complete and short proof of correctness in Coq.

2 A Balancing Algorithm

We start by giving a first, simple, implementation of the balancing algorithm. The heart of the algorithm relies on using an alternating list of length $2^k - 1$,

G. Klein and R. Gamboa (Eds.): ITP 2014, LNAI 8558, pp. 437–449, 2014.
© Springer International Publishing Switzerland 2014

where odd-position elements are trees and even-position elements are labels, of type α (indices starting from 1). A full tree of height k can be decomposed into the first $k-1$ levels, containing $2^{k-1}-1$ internal nodes, and the kth level, which contains both nodes and leaves. Thus, the $2^{k-1}-1$ labels in the alternating list will be used to label the internal nodes in the $k-1$ first levels of the balanced tree, while the 2^{k-1} trees, all of height at most one at first, will constitute the level k.

Though we could encode labels as trees of height one in the alternating list, we rather use an appropriate type for the sake of readability:

```
type α tree_or_elt =
  Elt of α
  Tree of α tree
```

We decompose the problem into two parts: computing an alternating list of length 2^k-1 from an arbitrary list of labels, and then transforming this alternating list into a balanced tree. We first show how to solve the second part: turning an alternating list into a full tree.

Given an alternating list lst, by pairing the trees in lst using only one traversal of the list, we obtain an alternating list with exactly half as many trees. Each pairing requires two trees and one label used as a root. In order to build a list that is alternated, we also need a second label, that is kept as a single element. This explains why we consider at each step the four first elements of the list.

A single traversal, encoded by pass : α tree_or_elt list $\rightarrow$ α tree_or_elt_list, reduces an alternating list of length $2^k-1 \geqslant 3$ to an alternating list of length $2^{k-1}-1$. By iterating this process using loop : α tree_or_elt list $\rightarrow$ α tree, we reduce the original list to a list of length one, whose one element is a balanced tree t such that the infix traversal of t is the initial list.

```
let join left node right = Tree (Node (left, node, right))

let rec pass = function
  | Tree left :: Elt root :: Tree right :: Elt e :: others →
      join left root right :: Elt e :: pass others
  | [Tree left; Elt root; Tree right] → [join left root right]
  | _ → assert false

let rec loop = function
  | [] → Leaf
  | [Tree t] → t
  | list → loop (pass list)
```

Notice how the invariant that alternating lists have length 2^k-1 is maintained: this is because, for $k \geqslant 2$, we have $2^k-1 = 4 \times (2^{k-2}-1) + 3$, hence we obtain an alternating list of length $2 \times (2^{k-2}-1) + 1 = 2^{k-1}-1$.

It remains to show how to transform a list of labels of length n into an alternating list of trees and labels. Each of the original trees has height zero or one: they are leaves or contain only one label. Because we want a list of length precisely $2^k - 1$, for $k = 1 + \lfloor \log_2 n \rfloor$, it means we need $2^k - 1 - n$ leaves. This quantity is computed as the variable missing. The function pad computes the alternating list by creating as many leaves as needed, alternating them with elements, and once enough leaves are created, promotes all the odd-position labels into trees.

```
let complete list =
  let n = List.length list in
  let rec pow2 i = if i <= n then pow2 (2*i) else i in
  let missing = (pow2 1) - n - 1 in
  let rec pad missing = function
    | head :: tail when missing <> 0 →
        Tree Leaf :: Elt head :: pad (missing - 1) tail
    | odd :: even :: others → join Leaf odd Leaf :: Elt even :: pad 0 others
    | [single] → [join Leaf single Leaf]
    | [] → []
  in
  pad missing list
```

The balancing algorithm balance: α list $\rightarrow \alpha$ tree is thus given by the composition of loop with complete:

```
let balance list = loop (complete list)
```

As for the complexity of this algorithm, notice that pass and complete are both clearly in linear-time in the length of the lists on which they work, while loop recurses on lists whose length are halved at each recursive step. Hence balance is a linear-time algorithm.

3   Removing Partial Functions

The loop function of Section 2 relies on the invariant that the list argument has length $2^k - 1$. Additionally, all the odd-position values must be of the form Tree t, whereas all the even-position values must be of the form Elt x. If either of these invariants is broken, we would run into the assert false of pass.

It is not immediately apparent that these properties hold. If it does not take a tremendous effort to convince oneself that the balance function of Section 2 is indeed correct, a direct mechanically checked proof would not be very practical.

3.1   Length Invariants

Our goal in this section is to avoid resorting to assert false. In addition to making sure that balance indeed terminates with a value, it will make it considerably simpler to implement the balancing algorithm in Coq in Section 4. To achieve this goal, it is necessary to have more precise types. Let us focus first on the length invariants: we will need to define a type which contains exactly the non-empty lists of length $2^k - 1$.

A data structure which holds $2^k - 1$ elements brings complete binary trees to mind. Even if it is possible – though not necessary convenient – to represent complete binary trees in Ocaml, they are not the appropriate structure. First, because complete binary trees are full trees and are, hence, unlikely to serve as a useful intermediate data structure to build a full tree. Second because there is a simpler – albeit more exotic – alternative.

Indeed, lists can be seen as decorated unary numbers: there is an element at each successor. Different kinds of lists can be obtained, more or less systematically, by varying the numerical representation. This idea goes back to Guibas & al. in [1] and a fairly thorough exploration can be found in Okasaki [2, Chapters 9&10]. In the simplest cases, the analogous list structure corresponds to a structurally recursive exponentiation algorithm. For regular lists, a list of size n whose elements have type a can be recursively defined with the following equations:

$$\begin{cases} a^0 & = 1 \\ a^{n+1} & = a \times a^n \end{cases}$$

Replacing unary numbers with binary numbers, we obtain the binary exponentiation algorithm:

$$\begin{cases} a^{2^0-1} = 1 \\ a^{2n} = (a^2)^n \\ a^{2n+1} = a \times (a^2)^n \end{cases}$$

Okasaki [2, Chapter 10] uses a non-uniform inductive type to encode the latter exponentiation algorithm into a type of lists he calls *binary lists*. We are only interested in lists of length $2^k - 1$, that is a length written only with the digit 1 in binary representation. So following Okasaki, but skipping the second equation above (which corresponds to the digit 0) we define the following non-uniform inductive type, which we call *power lists*:

```
module PowerList = struct

  type α t =
    Zero
    TwicePlusOne of α * (α*α) t

end
```

This type actually appears in Okasaki [2, Chapter 10] as an introduction to non-uniform binary lists. Relatedly, Okasaki [3] leverages a tail-recursive binary exponentiation algorithm to define a type capturing precisely square matrices; on the other hand, Myers [4] introduced a flavour of list based on *skew binary numbers* which are not easily captured as exponentiation.

Although the power lists may look like some sort of trees, it is not a very accurate depiction. The easiest way to picture how power lists works is to see TwicePlusOne as a fancy (::), then the lists with, respectively, 1, 3, 7, and 15 elements are as follows:

- [1]
- [1;(2,3)]
- [1;(2,3);((4,5),(6,7))]
- [1;(2,3);((4,5),(6,7));(((8,9),(10,11)),((12,13),(14,15)))]

Elements appear in order, like in a regular list, but they are packed twice as tightly after each TwicePlusOne.

Just like with regular lists, there is a *map* function for power lists. Due to the non-uniformity it is a little trickier[1] than the regular list map: in the recursive steps, the argument function f needs to process two consecutive elements at a time.

module PowerList = **struct**
   ⋮

  let rec map : α β. $(\alpha{\rightarrow}\beta)$ $\rightarrow$ α t $\rightarrow$ β t = **fun** f $\rightarrow$ **function**
    | Zero $\rightarrow$ Zero
    | TwicePlusOne (elt,lst) $\rightarrow$
      let f' (x,y) = f x , f y **in**
      TwicePlusOne (f elt , map f' lst)

end

3.2 Alternation

In Section 2, labels are separated from trees dynamically. The pass function verifies that trees and labels are interleaved properly, and fails if they are not.

In this section, instead, we consider a variant of α PowerList.t where every odd position contains a tree, and every even position contains an element. More generally, we define a type (ω,η) AlternatingPowerList.t where odd positions have type ω, and even positions have type η. Such a list should have the following pattern:

- [ω;(η,ω);((η,ω),(η,ω))]

After the first element, which must have type ω, there is no difference between even and odd positions: indeed, excluding the first element, we are actually building an $(\eta{*}\omega)$ PowerList.t. Hence the definition:

module AlternatingPowerList = **struct**
  type (ω,η) t =
    | Zero
    | TwicePlusOne of ω * $(\eta{*}\omega)$ PowerList.t

  end

---

[1] The type annotation on PowerList.map informs Ocaml that map is a non-uniform recursive function. Without the type annotation, Ocaml simply assumes that map is uniformly recursive and fails to typecheck since f and f' have different types.

For brevity, let us write PL and APL for PowerList and AlternatingPowerList respectively.

Using these alternating power lists, we can define a version of the pass function free of assert false. Indeed, consider an alternating power list of length at least 3: it is of the form APL.TwicePlusOne (a, PL.TwicePlusOne ((b,c), lst)), where lst has type ((η*ω)*(η*ω)) PowerList.t. The pass function of Section 2, as it happens, manipulates its arguments by groups of four elements: basically, pass is simply a map over lst.

We hence define the function pass which joins the trees in a list of length $2^{k+2} - 1^2$, producing a list of length $2^{k+1} - 1$. The function loop is virtually unchanged from Section 2, except it acts on power lists:

```
let pass left (root,right) apl =
  let join_up ((single,left),(root,right)) =
    single, Node (left,root,right)
  in
  APL.TwicePlusOne ( Node (left,root,right) , PL.map join_up apl)

let rec loop : ε. (ε tree,ε) APL.t → ε tree = function
  | APL.Zero → Leaf
  | APL.TwicePlusOne (tree,PL.Zero) → tree
  | APL.TwicePlusOne (tree,PL.TwicePlusOne (pair,apl)) →
      loop (pass tree pair apl)
```

3.3 Padding

Now that there is no more assert false in the code of loop, we need to change the complete function of Section 2 so that it returns an $(\alpha\ \text{tree}, \alpha)$ APL.t rather than a list. The heart of this section is a function which turns an α list into an $(\alpha*\alpha\ \text{tree})$ PL.t. The final function, which produces an $(\alpha\ \text{tree}, \alpha)$ APL.t is a simple wrapper around the former.

We want to turn a list lst of length $n + 1$ into a pair of its first element, converted into a tree, plus a power list of length $2 \times (2^k - 1) \geqslant n$ representing its tail tail. Each element of the power list is a pair, whose first term is an element, and its second term is a tree of height at most one. In particular, the length of the returned power list is always even, so if tail has odd length, we will need to insert at least a Leaf. This suggests that we may inspect the parity of the length of tail, and insert an extra element precisely when it is odd. This leads to a slightly different padding procedure than that of Section 2, in particular the leaves are not inserted at the same position, but it is inconsequential.

An α list of even length can be turned into an $(\alpha*\alpha)$ list whose length is halved. This turns out to be interesting for our recursion, since it mimics the inductive step of power lists. Also, in the case of even length, we need to distinguish the

---

[2] To ensure that its argument list has at least three elements, pass takes the three first elements as extra arguments. In other words pass t (x,s) l is meant to be read as pass (APL.TwicePlusOne (t , PL.TwicePlusOne ((x,s),l))).

empty case from the non-empty case: the former will be turned into the empty power list APL.Zero while the latter will be turned into a power list of the form APL.TwicePlusOne((x,y),l), where x and y correspond to the two first elements of tail. These different cases are represented in the following view:

```
type α parity =
  | Empty
  | Odd of α * (α*α) list
  | Even of (α*α) * (α*α) list

let pair_up lst =
  let succ elt = function
    | Empty → Odd (elt, [])
    | Odd (b,pairs) → Even ((elt,b), pairs)
    | Even (bc,pairs) → Odd (elt, bc::pairs)
  in
  List.fold_right succ lst Empty
```

The padding function itself, of_list, is at first sight far from intuitive. Let us recall that we want to turn a list of labels of arbitrary length, into a power list of pairs. A label can be thought of as a bit of weight 2^0, while a pair of labels would be a bit of weight 2^1, and so on. At first, all our bits have weight 2^0 and consists in one label each. We can build bits of higher weight by pairing up two bits of the same weight. A bit made up only of labels is called *pure*. We can also double the weight of a bit by interlacing leaves with it (with the function pad), but this gives a bit made of pairs of labels and trees, call them *impure*. Lastly, we can also transmute a pure bit into an impure bit of the same weight (with the function coerce), by replacing odd-position labels by trees of height one.

Each recursive step consists in taking a list of pure bits of the same weight 2^k, and outputting exactly one impure bit of size 2^{k+1}, plus a list of pure bits of weight 2^{k+1}, which is converted recursively. We thus obtain, successively, one bit of each weight from 2^1 to 2^l, for some l, encoding a list of length $2^{l+1} - 2$, as expected.

At any recursive step, suppose first that the number of bits of weight 2^k is odd. As we need to compute only bits of weight 2^{k+1}, one of them impure, we are forced to use pad on one bit, and to pair up the others. Suppose now that the number of bits of weight 2^k is even. In that case, we can pair them all into bits of weight 2^{k+1}, and then use coerce on one of them to make the impure bit.

The last difficulty is that pad and coerce both depend on the current weight of the bits, hence we need to update them at each recursive step. pad must add leaves between every two consecutive labels, in even positions, while coerce must upgrade every even-position label into a tree of height one. This leads to the following definition:

```
module PowerList = struct
    ⋮
    let rec of_list : α β. (α→β) → (α*α→β) → α list → β t =
    fun pad coerce bits →
        let pad' (x,y) = (pad x, pad y) in
        let coerce' (x,y) = (coerce x, coerce y) in
        match pair_up bits with
        | Empty → Zero
        | Odd (a,pures) → TwicePlusOne (pad a, of_list pad' coerce' pures)
        | Even (ab,pures) →
            TwicePlusOne (coerce ab, of_list pad' coerce' pures)
end
```

With that function, we can conclude our implementation. Again writing PL and APL for PowerList and AlternatingPowerList respectively:

```
module AlternatingPowerList = struct
    ⋮
    let of_list leaf up id = function
    | [] → Zero
    | a::l →
        let pad x = id x , leaf in
        let coerce (x,y) = id x , up y in
        TwicePlusOne (up a, PowerList.of_list pad coerce l)
end
let singleton x = Node(Leaf,x,Leaf)
let balance l =
    loop (APL.of_list Leaf singleton (fun e→e) l)
```

The final function, balance : α list → α tree, implements the same algorithm as Section 2 without any partial functions.

What we may have lost in this section, compared to the simple algorithm, is the simplicity of the complexity analysis of the algorithm. The subtleties of the main functions require a finer analysis. Consider first the function PL.map: clearly the number of recursive calls depends only logarithmically on the number of elements in the power list. But each recursive call uses as its first argument a function twice as complex than the previous one. This leads to the following inequation over the complexity $C(n, m, f)$ of map, where n is the number of elements in the power list, m is the size of the elements in the power list, and f is the complexity of the mapped function:

$$C(n, m, f) \leq f(m) + C\left(\frac{n-1}{2}, 2m, k \mapsto 2 \times f(k/2) + O(1)\right) + O(1)$$

From there, it is easy to prove that $C(n, m, f) = n.f(m) + O(n)$, so that PL.map runs indeed in linear-time, and so is loop. Similarly the complexity of PL.of_list can be described by a higher-order recursive inequation (almost the same as above, except that the complexity depends on two functions and the constant term is replaced by a linear term), whose solution gives also a linear-time complexity.

4   Turning to Coq

There is still a property of the algorithm that the implementation of Section 3 does not make obvious: that the algorithm actually does build *full* trees. In this section we shall build into the type of balance that its output is indeed full.

To that effect, we will use Coq rather than Ocaml. Even if it is possible, with some effort, to represent full trees and implement the algorithm in Ocaml – and relatively easy in Haskell – a Coq implementation also gives us termination by construction. Coq forces every recursion to be structural, which will prove to be rather entertaining.

At a superficial level, a visible difference with the Ocaml implementation is that Powerlist.t and AlternatingPowerList.t must be decorated with the k such that the length is $2^k - 1$: it is the structural recursion parameter of the balance_powerlist function. Because it makes the code simpler, we will use a recursive definition rather than an inductive one:

Module PowerList.

   Fixpoint T (A:**Type**) (k:nat) :=
   match k **with**
   | 0 $\Rightarrow$ unit:**Type**
   | S k' $\Rightarrow$ A $*$ T (A$*$A) k'
   end.

End PowerList.

We will also need a version where k can be arbitrary. For that purpose we use Coq's type of dependent pairs { n:nat & F n}. The constructor for dependent pairs is written ⟨ n , x ⟩. The implicit version comes with constructors – tpo stands for "twice plus one":

Module PowerList.
   ⋮

   Definition U (A:**Type**) := { k:nat & T A k }.
   Definition zero {A:**Type**} : U A := ⟨ 0 , tt ⟩.
   Definition tpo {A:**Type**} (a:A) (l:U (A$*$A)) : U A :=
   let '⟨k,l⟩ := l in
   ⟨ S k , (a,l) ⟩.

End PowerList

The definition of AlternatingPowerList.T and AlternatingPowerList.U are similar.

4.1   Full Trees

To code full trees, we index trees by their height, and specify that leaves can happen only at height 0 or 1:

Inductive FullTree (A:**Type**) : nat → **Type** :=
| Leaf$_0$: FullTree A 0
| Leaf$_1$: FullTree A 1
| Node {k:nat} : FullTree A k → A → FullTree A k → FullTree A (S k).

If we omitted the constructor Leaf$_1$, we would have a definition of complete binary trees: both subtrees of a node are complete binary trees of the same height. We allow the full trees to be incomplete by letting Leaf$_1$ take the place of nodes on the last level.

Using the type FullTree A k in place of the type α tree, the functions pass and balance_powerlist are virtually unmodified[3] with respect to Section 3. Only their types change to reflect the extra information:

Definition pass {A k p} : APL.T (FullTree A (S p)) A (S (S k)) →
    APL.T (FullTree A (S (S p))) A (S k).
Fixpoint loop {A k p} : APL.T (FullTree A (S p)) A (S k) →
    FullTree A (plus k (S p)) {struct k}.

The algorithm indeed builds only full trees.

4.2   Structural Initialisation

The padding conversion from lists to power lists, in Section 3, is not structural due to the use of pair_up in the recursive call. To tackle this recursion, we shall make use of another intermediate structure. What we need, essentially, is that all the calls to pair_up are pre-calculated, so the intermediate structure will be like parity except that the calls to $(\alpha * \alpha)$ list are replaced by inductive calls.

As it turns out, this is another non-uniform datatype which corresponds to a numerical representation. Indeed, any natural number can be written in binary with digits 1 and 2 (but not 0). In this system, for example, $8 = 1 \times 2^2 + 1 \times 2^1 + 2 \times 2^0$ is represented as 112. Here is the definition, where tpo reads "twice plus one" and tpt "twice plus two":

Module BinaryList.

Inductive T (A:**Type**) : **Type** :=
| zero
| tpo (a:A) (l:T (A*A))
| tpt (a b: A) (l:T (A*A)).

End BinaryList.

---

[3] In fact, as can be seen from its type, loop only handles non-empty alternating power lists. This is due to a small technicality: the recursive step of loop is the case S (S k), but Coq does not recognise S k as a structural subterm of S (S k), so the definition from Section 3 does not verifies Coq's structural recursion criterion. As a workaround, the empty case is moved to the balance function.

To turn a non-empty list into a BinaryList.T, all we need is a function cons of type A → T A → T A to add an element in front of the list. On the numerical representation side, it corresponds to adding 1. It behaves like adding 1 in the usual binary representation, except that 1-s are turned into 2-s without a carry and 2-s into 1-s while producing a carry:

Module BinaryList.
$$\vdots$$

   Fixpoint cons {A} (a:A) (l:T A) : T A :=
   match l **with**
   | zero ⇒ tpo a zero
   | tpo b l ⇒ tpt a b l
   | tpt b c l ⇒ tpo a (cons (b,c) l)
   end.

   Definition of_list {A} (l:list A) : T A :=
   List.fold_right cons zero l.

End BinaryList.

Note that while cons takes, in the worst case, logarithmic time with respect to the length of the list, building a list by repeatedly using cons is still linear. Indeed, as previously mentioned, cons mimics the successor algorithm for binary numbers, whose amortized complexity is well-known to be constant.

We also need a function which turns a T (A∗A) into a T A. This is effectively multiplication by 2. The lack of 0 among the digits[4] makes this process recursive. A simple presentation of the doubling algorithm consists in adding a 0 at the end of the number, then eliminating the 0 using the following equalities:

$$\begin{cases} 0 = \cdot \\ x20 = x12 \\ x10 = x02 \end{cases}$$

In terms of binary lists:

Module BinaryList.
$$\vdots$$

   Fixpoint twice {A} (l:T (A∗A)) : T A :=
   match l **with**
   | zero ⇒ zero
   | tpo (a,b) l ⇒ tpt a b (twice l)
   | tpt (a,b) cd l ⇒ tpt a b (tpo cd l)
   end.

End BinaryList.

We can now write a structurally recursive padding function, using binary lists as the structural argument. As we do not know in advance the length of

---

[4] The constructor zero represents an empty list of digits.

the produced list, a PowerList.U is returned. We write BL as a shorthand for BinaryList:

Module PowerList.
⋮

Fixpoint of_binary_list {A X} (d:A→X) (f:A∗A→X) (l:BL.T A) : U X :=
match l **with**
| BL.zero ⇒ zero
| BL.tpo a l ⇒
    tpo (d a) (of_binary_list (d×d) (f×f) l)
| BL.tpt a b l ⇒
    tpo (f (a,b)) (of_binary_list (d×d) (f×f) l)
end.

End PowerList.

Where g×f is the function which maps (x,y) to (g x,f y).

The rest follows straightforwardly, and we can define the following functions which conclude the algorithm (BL, PL, and APL stand for BinaryList, PowerList, and AlternatingPowerList respectively):

Module AlternatingPowerList.
⋮

Definition of_binary_list {A Odd Even}
    (d:Odd) (f:A→Odd) (g:A→Even) (l:BL.T A) : U Odd Even :=
match l **with**
| BL.zero ⇒ zero
| BL.tpo a l ⇒
  let d' x := (g x , d) in
  tpo (f a) (PL.of_binary_list d' (g×f) (BL.twice l))
| BL.tpt a b l ⇒
  let d' x := (g x , d) in
  tpo (f a) (PL.of_binary_list d' (g×f) (BL.tpo b l))
end.

Definition of_list {A Odd Even}
    (d:Odd) (f:A→Odd) (g:A→Even) (l:list A) : U Odd Even :=
of_binary_list d f g (BL.of_list l).

End AlternatingPowerList.

Definition singleton {A:**Type**} (x:A) : FullTree A 1 :=
Node Leaf$_0$ x Leaf$_0$.

Definition balance {A:**Type**} (l:list A) : { k:nat & FullTree A k } :=
let '⟨k,l⟩ := APL.of_list Leaf$_1$ singleton (**fun** x⇒x) l **in**
match k **with**
| 0 ⇒ **fun** _ ⇒ ⟨ 0 , Leaf$_0$ ⟩
| S k ⇒ **fun** l ⇒ ⟨ plus k 1 , loop l ⟩
end l.

5   Conclusion

The balance function of Section 4 is, by virtue of its type alone, a total function which turns lists into full binary trees. Yet, to the cost of using intermediary data-structures, it effectively implements the algorithm of Section 2.

The missing piece is to prove that the infix traversal of balance l is indeed l. The infix traversal of a (full) tree is represented in Coq with the functions

Fixpoint list_of_full_tree_n {A n} (t:FullTree A n) : list A :=
  match t **with**
  | Leaf$_0$ $\Rightarrow$ []
  | Leaf$_1$ $\Rightarrow$ []
  | Node _ t$_1$ x t$_2$ $\Rightarrow$
    list_of_full_tree_n t$_1$ ++ [x] ++ list_of_full_tree_n t$_2$
  end.

Definition list_of_full_tree {A} (t:{ k:nat & FullTree A k }) : list A :=
  list_of_full_tree_n (projT2 t).

We can then state the theorem:

Theorem balance_preserves_order A (l:list A) :
  list_of_full_tree (balance l) = l.

The proof is short and straightforward: we define a traversal function for each intermediate structure; and state a variant of balance_preserves_order for each intermediate function. Proving the intermediate lemmas is not difficult and can be mostly automatised: we use a very simple generic automated tactic, which discharges most goals. This theorem concludes our easy formal proof of the balancing algorithm.

References

1. Guibas, L.J., McCreight, E.M., Plass, M.F., Roberts, J.R.: A new representation for linear lists. In: Proceedings of the Ninth Annual ACM Symposium on Theory of Computing, STOC 1977, pp. 49–60. ACM Press, New York (1977)
2. Okasaki, C.: Purely functional data structures. Cambridge University Press (1999)
3. Okasaki, C.: From fast exponentiation to square matrices. In: Proceedings of the Fourth ACM SIGPLAN International Conference on Functional Programming, ICFP 1999, pp. 28–35. ACM Press, New York (1999)
4. Myers, E.W.: An applicative random-access stack. Information Processing Letters 17, 241–248 (1983)

Unified Decision Procedures
for Regular Expression Equivalence

Tobias Nipkow and Dmitriy Traytel

Fakultät für Informatik, Technische Universität München, Germany

Abstract. We formalize a unified framework for verified decision procedures for regular expression equivalence. Five recently published formalizations of such decision procedures (three based on derivatives, two on marked regular expressions) can be obtained as instances of the framework. We discover that the two approaches based on marked regular expressions, which were previously thought to be the same, are different, and we prove a quotient relation between the automata produced by them. The common framework makes it possible to compare the performance of the different decision procedures in a meaningful way.

1 Introduction

Equivalence of regular expressions is a perennial topic in computer science. Recently it has spawned a number of formalized and verified decision procedures for this task in interactive theorem provers [3, 6, 10, 19, 21]. Except for the formalization by Braibant and Pous [6], all these decision procedures operate directly on variations of regular expressions. Although they (implicitly) build automata, the states of the automata are labeled with regular expressions, and there is no global transition table but the next-state function is computable from the regular expressions. The motivation for working with regular expressions is simplicity: regular expressions are a free datatype which proof assistants and their users love because it means induction, recursion and equational reasoning—the core competence of proof assistants and functional programming languages. Yet all these decision procedures based on regular expressions look very different. Of course, the next-state functions all differ, but so do the actual decision procedures and their correctness, completeness and termination proofs. The contributions of our paper are the following:

- A unified framework (Sect. 3) that we instantiate with all the above approaches (Sects. 4 and 5). The framework is a simple reflexive transitive closure computation that enumerates the states of a product automaton.
- Proofs of correctness, completeness and termination that are performed once and for all for the framework based on a few properties of the next-state function.
- A new perspective on partial derivatives that recasts them as Brzozowski derivatives followed by some rewriting (Sect. 4).
- The discovery that Asperti's algorithm is not the one by McNaughton-Yamada [20], as stated by Asperti [3], but a dual construction which apparently had not been considered in the literature and which produces smaller automata (Sect. 5).
- An empirical comparison of the performance of the different approaches (Sect. 6).

The discussion of related work is distributed over the relevant sections of the paper.

G. Klein and R. Gamboa (Eds.): ITP 2014, LNAI 8558, pp. 450–466, 2014.

2   Preliminaries

Isabelle/HOL (see [22, Part I] for a recent introduction) is based on Church's simple type theory. Types τ are built from type variables α, β, etc. via function types, other type constructors are written postfix. The notation $t :: \tau$ means that term t has type τ. Types $\alpha\,set$ and $\alpha\,list$ are the types of sets and lists of elements of type α. They come with the following vocabulary: function set (conversion from lists to sets), $[]$ (empty list), # (list constructor), @ (append), hd (head), tl (tail) and map.

Recursive functions over datatypes are executable, and Isabelle can generate from them code in functional languages [15]. This includes functions on finite sets [14]. Unless stated otherwise all functions in this paper are executable.

Locales [4] are Isabelle's tool for modelling parameterized systems. A locale fixes parameters and states assumptions about them:

locale $A = $ **fixes** x_1 **and** ... **and** x_n **assumes** $n_1 : P_1\,\overline{x}$ **and** ... **and** $n_m : P_m\,\overline{x}$

In the context of the locale A, we can define constants that depend on the parameters x_i and prove properties about those constants using the assumptions P_i (accessed under the name n_i). Parameters can be instantiated: **interpretation** J: A **where** $x_1 = t_1 \ldots x_n = t_n$. The command issues proof obligations $P_i\,\overline{t}$ (that the user must discharge) and exports constants and theorems from the locale with x_i instantiated to t_i. Multiple interpretations of the same locale are possible; the prefix "J." disambiguates different instances.

Regular expressions are defined as a recursive datatype:

datatype $\alpha\,rexp = 0 \mid 1 \mid A\,\alpha \mid \alpha\,rexp + \alpha\,rexp \mid \alpha\,rexp \cdot \alpha\,rexp \mid (\alpha\,rexp)^*$

with the usual (non-executable) semantics $\mathscr{L} :: \alpha\,rexp \to \alpha\,lang$, where $\alpha\,lang$ is short for $(\alpha\,list)\,set$. In concrete regular expressions, we sometimes omit the constructor A for readability. The recursive function nullable $:: \alpha\,rexp \to bool$ satisfies nullable $r \leftrightarrow [] \in \mathscr{L}\,r$. The functions $\Sigma :: \alpha\,rexp \to \alpha\,set$ and atoms $:: \alpha\,rexp \to \alpha\,list$ compute the set and list of atoms (the arguments of constructor A) in a regular expression. The (non-executable) *left quotient* of a language $L :: \alpha\,lang$ w.r.t. some $a :: \alpha$ is defined by $\mathscr{D}\,a\,L = \{w \mid a \# w \in L\}$. The extension of $\mathscr{D}$ from single symbols to words $w :: \alpha\,list$ can be expressed as fold $\mathscr{D}\,w$ where fold $:: (\alpha \to \beta \to \beta) \to \alpha\,list \to \beta \to \beta$.

3   Regular Expression Equivalence Framework

Regular expression (language) equivalence is usually reduced to (language) equivalence of automata. In principle our framework does the same, except that we construct the automata on the fly and replace the traditional transition table by computations on regular-expression-like objects. We start by relating regular expressions and automata.

Left quotients of a regular language L can be understood as states of a deterministic automaton $\mathscr{M}_L$ with the initial state L, the transition function $\mathscr{D}$, and the accepting states being those languages K for which $[] \in K$ holds. This automaton (restricted to reachable states) is finite and minimal by the Myhill-Nerode theorem.[1] The following locale captures this left-quotient-based view of an automaton:

---

[1] Note that the Myhill-Nerode relation $\approx_L$ can be defined as $v \approx_L w \leftrightarrow$ fold $\mathscr{D}\,v\,L = $ fold $\mathscr{D}\,w\,L$. The quotient of $\mathscr{M}_L$ by this relation is isomorphic to $\mathscr{M}_L$; hence $\mathscr{M}_L$ is minimal.

locale *rexpDA* =
fixes $\iota :: \alpha\, rexp \rightarrow \sigma$ **and** L $:: \sigma \rightarrow \alpha\, lang$ **and** $\delta :: \alpha \rightarrow \sigma \rightarrow \sigma$ **and** o $:: \sigma \rightarrow bool$
assumes ιL: L $(\iota\, r) = \mathscr{L}\, r$ **and** δL: L $(\delta\, a\, s) = \mathscr{D}\, a\, (\text{L}\, s)$ **and** oL: o $s \leftrightarrow [] \in \text{L}\, s$

The parameters ι and L formalize what "regular-expression-like" means: $\iota\, r$ embeds the regular expression r into a state of type σ, whereas L gives elements of σ a language semantics, which coincides with the language semantics of regular expressions by the assumption ιL. The function δ is the symbolic computation of left quotients on σ according to δL. It can be regarded as the transition function of an automaton with states in σ and the initial state $\iota\, r$. Accepting states of this automaton are given by o.

Let us develop and verify some algorithms in the context of *rexpDA*. For a start, regular expression matching is easy to define

match $r\, w$ = o (fold $\delta\, w\, (\iota\, r)$)

and prove correct: match $r\, w \leftrightarrow w \in \mathscr{L}\, r$.

Now we tackle the equivalence checker. We follow the well-known product automaton construction where language equivalence means o $s_1 \leftrightarrow$ o s_2 for all states (s_1, s_2) of the product automaton. Alternatively, one can view this procedure as the construction of a bisimulation relation between two automata: language equivalence and existence of a bisimulation coincide for deterministic automata [24]. The set of reachable states of an automaton can be obtained as the reflexive transitive closure of the start state under $\lambda p.$ map $(\lambda a.\, \delta\, a\, p)$ *as* where *as* $:: \alpha\, list$ is the alphabet.

We define a reflexive transitive closure operation

rtc $:: (\alpha \rightarrow bool) \rightarrow (\alpha \rightarrow \alpha\, list) \rightarrow \alpha \rightarrow (\alpha\, list \times \alpha\, set)\, option$

where type $\alpha\, option$ is the datatype None | Some α. It is used to encode whether the closure is finite (Some is returned) or infinite (None is returned). The function rtc is defined using a while combinator and is executable (provided its arguments being executable); the result Some corresponds to a terminating computation [19]. The definition can be found in Isabelle/HOL's library theory While_Combinator under its full name rtrancl_while. The parameters and result of rtc p *next start* have the following meaning: Predicate p is a test that stops the closure computation if an element not satisfying p is found; this is merely an optimization. Function *next* maps an element to a list of successors. Of course *start* is the start element. A result Some (ws, Z) means that the closure computation terminated with a worklist *ws* and a set of reachable elements Z. If *ws* is empty, Z is the set of all elements reachable from *start*; otherwise, the computation was stopped because an element not satisfying p was found. More precisely, we proved

$$\text{rtc } p \text{ } next \text{ } start = \text{Some } (ws, Z) \implies$$
$$\textbf{if } ws = [] \textbf{ then } Z = R \land (\forall z \in Z.\ p\, z) \textbf{ else } \neg p \text{ (hd } ws) \land \text{hd } ws \in R \qquad (1)$$

where $R = \{(x, y) \mid y \in \text{set } (next\, x)\}^* \text{ `` } \{start\}$ and " is infix relation application: $r \text{ `` } \{x\} = \{y \mid (x, y) \in r\}$.

The state space of the product automaton is computed as follows:

closure $:: \alpha\, list \rightarrow \sigma \times \sigma \rightarrow ((\sigma \times \sigma)\, list \times (\sigma \times \sigma)\, set)\, option$
closure *as* = rtc $(\lambda (s, t).\ \text{o } s \leftrightarrow \text{o } t)\ (\lambda (s, t).\ \text{map } (\lambda a.\ (\delta\, a\, s, \delta\, a\, t))\ as)$

The predicate $\lambda(s, t).\ \mathsf{o}\,s \longleftrightarrow \mathsf{o}\,t$ stops the computation as soon as a contradiction to language equality is found. The actual language equivalence checker merely needs to test if the worklist is empty at the end:

eqv :: $\alpha\,rexp \rightarrow \alpha\,rexp \rightarrow bool$
eqv $r\,s =$ **case** closure (atoms r @ atoms s) ($\iota\,r, \iota\,s$) **of**
$\qquad\qquad$ Some ($[], \_$) $\Rightarrow$ True
$\qquad\qquad | \_ \Rightarrow$ False

The alphabet given to closure is the concatenation of the atoms in the two expressions.

Soundness of eqv is an easy consequence of the following property, which in turn follows from (1):

$$\text{closure (atoms } r \text{ @ atoms } s)\,(\iota\,r, \iota\,s) = \text{Some }(ws, Z) \implies \\ ws = [] \longleftrightarrow \mathsf{L}\,r = \mathsf{L}\,s \tag{2}$$

Theorem 1 (in *rexpDA*). eqv $r\,s \implies \mathscr{L}\,r = \mathscr{L}\,s$.

This is a partial correctness statement because it assumes that the call to closure in eqv returns Some, i.e. terminates.

Termination of closure needs finiteness of the underlying automaton. Therefore we extend *rexpDA* with an explicit assumption of finiteness:

locale *rexpDFA* = *rexpDA* +
assumes *fin*: finite $\{\text{fold } \delta\,w\,(\iota\,r) \mid w :: \alpha\,list\}$

In this context the termination lemma for closure is an easy consequence of *fin* and the following termination property of rtc:

finite $(\{(x, y) \mid y \in \text{set }(f\,x)\}^* \text{ '' } \{x\}) \implies \exists y.\ \text{rtc } p\,f\,x = \text{Some } y$

Lemma 2 (in *rexpDFA*). closure *as* ($\iota\,r, \iota\,s$) $\neq$ None.

Together with (2) this implies completeness of eqv:

Theorem 3 (in *rexpDFA*). $\mathscr{L}\,r = \mathscr{L}\,s \implies$ eqv $r\,s$.

This is the end of all considerations about equivalence of regular expressions. The rest of the paper merely needs to focus on various methods for turning regular expressions into finite automata in the sense of *rexpDFA*.

Note that $\mathscr{M}_L$ defined above constitutes a first valid interpretation of *rexpDFA*. The proof of *fin* requires the Myhill-Nerode theorem.

interpretation M : *rexpDFA* **where**
$\quad \iota\,r \quad = \mathscr{L}\,r$
$\quad \delta\,a\,L = \mathscr{D}\,a\,L$
$\quad \mathsf{o}\,L \quad = [] \in L$
$\quad \mathsf{L}\,L \quad = L$

This interpretation is not executable because neither its next-step function $\mathscr{D}$ (being based on infinite sets of words defined by a set comprehension) nor the equality on $\sigma = \alpha\,lang$ (which is needed for the closure computation) is executable.

4 Derivatives

In 1964, Brzozowski [7] showed how to compute left quotients syntactically—as derivatives of regular expressions. Derivatives have been rediscovered in proof assistants by Krauss and Nipkow [19] and Coquand and Siles [10]. Our first executable instantiations of the framework reuse infrastructure from earlier formalizations in Isabelle [19, 26].

A refinement of Brzozowski's approach, partial derivatives, was introduced by Antimirov [2] and formalized by Moreira *et al.* [21] in Coq and by Wu *et al.* [27] in Isabelle. Partial derivatives operate on finite sets of regular expressions. They can be viewed either as a nondeterministic automaton with regular expressions as states or as the corresponding deterministic automaton obtained by the subset construction.

In the following, we integrate the two notions in our framework and show how derivatives can be used to simulate partial derivatives without invoking sets explicitly.

4.1 Brzozowski's Derivatives

Given a letter c and a regular expression r, the *(Brzozowski) derivative* der $:: \alpha \rightarrow \alpha\, rexp \rightarrow \alpha\, rexp$ of r w.r.t. a is defined by primitive recursion:

$$
\begin{aligned}
&\text{der} \_ \mathbf{0} && = \mathbf{0} \\
&\text{der} \_ \mathbf{1} && = \mathbf{0} \\
&\text{der } a \ (\mathsf{A}\ x) && = \text{if } x = a \text{ then } \mathbf{1} \text{ else } \mathbf{0} \\
&\text{der } a \ (r + s) && = \text{der } a\ r + \text{der } a\ s \\
&\text{der } a \ (r \cdot s) && = \text{if nullable } r \text{ then } (\text{der } a\ r \cdot s) + \text{der } a\ s \text{ else der } a\ r \cdot s \\
&\text{der } a \ (r^*) && = \text{der } a\ r \cdot r^*
\end{aligned}
$$

It follows by induction on r that the language of the derivative der $a\ r$ is exactly the left quotient $\mathscr{D}\, a\ (\mathscr{L}\, r)$. This property corresponds exactly to the assumption δL of the locale *rexpDA*. Hence it suggests the following interpretation:

interpretation *rexpDA* **where**
$$\iota\, r = r \qquad \delta\, a\, r = \text{der } a\ r \qquad o\, r = \text{nullable } r \qquad \mathsf{L}\, r = \mathscr{L}\, r$$

Unfortunately, the sound equivalence checker that is produced by this interpretation is useless in practice, because it will rarely terminate. For example, the automaton constructed from the regular expression a^* is infinite, as all derivatives w.r.t. words a^n are distinct: fold der $a^1\ a^* = \mathbf{1} \cdot a^*$; fold der $a^{n+1}\ a^* = \mathbf{0} \cdot a^* + \text{fold der } a^n\ a^*$.

Fortunately, Brzozowski showed that there are finitely many equivalence classes of derivatives modulo associativity, commutativity and idempotence (ACI) of the $+$ constructor. We prove that the number of distinct derivatives of r modulo ACI is finite: finite $\{[\text{fold der } w\ r]_\sim \mid w \in (\Sigma\, r)^*\}$ where $[r]_\sim = \{s \mid r \sim s\}$ denotes the equivalence class of r and the ACI equivalence $\sim$ is defined inductively as follows.

$$
r + (s + t) \sim (r + s) + t \qquad r + s \sim s + r \qquad r + r \sim r
$$

$$
r \sim r \qquad \frac{r \sim s}{s \sim r} \qquad \frac{r \sim s \quad s \sim t}{r \sim t}
$$

$$
\frac{r_1 \sim s_1 \quad r_2 \sim s_2}{r_1 + r_2 \sim s_1 + s_2} \qquad \frac{r_1 \sim s_1 \quad r_2 \sim s_2}{r_1 \cdot r_2 \sim s_1 \cdot s_2} \qquad \frac{r \sim s}{r^* \sim s^*}
$$

ACI-equivalent regular expressions $r \sim s$ have the same atoms and same languages, and their equivalence is preserved by the derivative: der $b\, r \sim$ der $b\, s$ for all $b \in \Sigma r$. This enables the following interpretation that operates on ACI equivalence classes. We obtain a first totally correct and complete equivalence checker $\mathsf{D}_\sim$.eqv in Isabelle/HOL.

interpretation $\mathsf{D}_\sim : \mathit{rexpDFA}$ **where**
$$\iota\, r \quad = [r]_\sim$$
$$\delta\, a\, [r]_\sim = [\text{der } a\, r]_\sim$$
$$\mathsf{o}\, [r]_\sim \quad = \text{nullable } r$$
$$\mathsf{L}\, [r]_\sim \quad = \mathscr{L}\, r$$

Fig. 1. Derivative automaton modulo ACI for $a^* \cdot b$

Technically, the formalization defines a quotient type [18] of "regular expressions modulo ACI" to represent equivalence classes and uses Lifting and Transfer [17] to lift operations on regular expressions to operations on equivalence classes. The above presentation of definitions of the locale parameters by "pattern matching" on equivalence classes resembles the code generated by Isabelle for quotients (a pseudo-constructor [14], $[\_]_\sim$, wraps a concrete representative r), rather than the actual definitions by Lifting.

Since the equivalence checker must compare equivalence classes, the code generation for quotients requires an executable equality (i.e. a decision procedure for $\sim$-equivalence). We achieve this through an ACI normalization function $\langle\_\rangle$ that maps a regular expression r to a canonical representative of $[r]_\sim$ by sorting all summands w.r.t. an arbitrary fixed linear order $\preceq$ while removing duplicates. The definition of $\langle\_\rangle$ employs a smart (simplifying) constructor $\oplus$, whose equations are matched sequentially.

$$\langle 0 \rangle \quad = 0 \qquad\qquad (r + s) \oplus t = r \oplus (s \oplus t)$$
$$\langle 1 \rangle \quad = 1 \qquad\qquad r \oplus (s + t) = \textbf{if } r = s \textbf{ then } s + t$$
$$\langle A\, a \rangle \quad = A\, a \qquad\qquad\qquad\qquad \textbf{else if } r \preceq s \textbf{ then } r + (s + t)$$
$$\langle r + s \rangle = \langle r \rangle \oplus \langle s \rangle \qquad\qquad\qquad\qquad\qquad \textbf{else } s + (r \oplus t)$$
$$\langle r \cdot s \rangle = \langle r \rangle \cdot \langle s \rangle \qquad r \oplus s \quad = \textbf{if } r = s \textbf{ then } r$$
$$\langle r^* \rangle \quad = \langle r \rangle^* \qquad\qquad\qquad\qquad \textbf{else if } r \preceq s \textbf{ then } r + s \textbf{ else } s + r$$

We obtain an executable decision procedure for ACI equivalence: $r \sim s \leftrightarrow \langle r \rangle = \langle s \rangle$. This makes $\mathsf{D}_\sim$.eqv executable, yielding verified code in different functional programming languages via Isabelle's code generator. Yet, the performance of the generated code is disappointing. Fig. 1 shows why: Derivations clutter concrete representatives with duplicated summands. Further derivation steps perform the same computation repeatedly and hence become increasingly expensive. This bottleneck is avoided by taking canonical ACI-normalized representatives as states yielding a second interpretation.

interpretation D: *rexpDFA* **where**

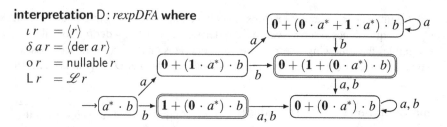

$$\iota\, r\ \ = \langle r \rangle$$
$$\delta\, a\, r = \langle \mathsf{der}\, a\, r \rangle$$
$$o\, r\ \ = \mathsf{nullable}\, r$$
$$\mathsf{L}\, r\ \ = \mathscr{L}\, r$$

Fig. 2. ACI-normalized derivative automaton for $a^* \cdot b$

A few points are worth mentioning here: First, D does not use the quotient type—it operates directly on canonical representatives and therefore can use structural equality for comparison (rather than $\sim$). Second, the interpretations $D_\sim$ and D yield structurally the same automata, although with different labels. Fig. 2 shows the automaton produced by D for $a^* \cdot b$. This observation—which enables us to reuse the technically involved proof of $D_\sim.\mathit{fin}$ to discharge $D.\mathit{fin}$—relies crucially on our normalization function $\langle \_ \rangle$ being idempotent and well-behaved w.r.t. derivatives:

Lemma 4. *We have* $\langle \langle r \rangle \rangle = \langle r \rangle$ *and* $\langle \mathsf{der}\, b\, \langle r \rangle \rangle = \langle \mathsf{der}\, b\, r \rangle$ *for all* $b \in \Sigma\, r$.

The automaton from Fig. 2 shows that the state labels still contain superfluous information, notably in the form of $\mathbf{0}$s and $\mathbf{1}$s. A coarser relation than $\sim$-equivalence, denoted $\approx$, adresses this concern. We omit the straightforward inductive definition of $\approx$, which cancels $\mathbf{0}$s and $\mathbf{1}$s where possible and takes the associativity of concatenation $\cdot$ into account. Coarseness ($[r]_\sim \subseteq [r]_\approx$) together with $D_\sim.\mathit{fin}$ implies finiteness of equivalence classes of derivatives modulo $\approx$: finite $\{[\mathsf{fold}\,\mathsf{der}\,w\,r]_\approx \mid w \in (\Sigma\,r)^*\}$.

As before, to avoid working with equivalence classes, we use a recursively defined $\approx$-normalization function $\langle\!\langle \_ \rangle\!\rangle$ similar to $\langle \_ \rangle$ (it corresponds to the *norm* function from the formalization by Krauss and Nipkow [19]). However, $\langle\!\langle \_ \rangle\!\rangle$ (also $\approx$) is not well-behaved w.r.t. derivatives: for example, $\langle\!\langle \mathsf{der}\, a\, \langle\!\langle ((a+\mathbf{1}) \cdot (a \cdot a)) \cdot b \rangle\!\rangle \rangle\!\rangle \neq \langle\!\langle \mathsf{der}\, a\, (((a+\mathbf{1}) \cdot (a \cdot a)) \cdot b) \rangle\!\rangle$. The normalization would need to take the distributivity of $\cdot$ over $+$ into account to prevent this disequality, but even with this addition a formal proof of well-behavedness seems difficult. Furthermore, our evaluation (Sect. 6) suggests that not too much energy should be invested in finding this proof. Thus, the following interpretation gives only a partial correctness result.

interpretation N: *rexpDA* **where**

$$\iota\, r\ \ = \langle\!\langle r \rangle\!\rangle$$
$$\delta\, a\, r = \langle\!\langle \mathsf{der}\, a\, r \rangle\!\rangle$$
$$o\, r\ \ = \mathsf{nullable}\, r$$
$$\mathsf{L}\, r\ \ = \mathscr{L}\, r$$

Fig. 3. Normalized derivative automaton for $a^* \cdot b$

In practice, we did not find an input for which N would construct an infinite automaton. For the example $a^* \cdot b$ it even yields the minimal automaton shown in Fig. 3.

4.2   Partial Derivatives

Partial derivatives split certain $+$-constructors into sets of regular expressions, thus capturing ACI equivalence directly in the data structure. The automaton construction for a regular expression r starts with the singleton set $\{r\}$. More precisely, partial derivatives pder $:: \alpha \rightarrow \alpha\,rexp \rightarrow (\alpha\,rexp)\,set$ are defined recursively as follows:

$$
\begin{aligned}
\text{pder}\_\mathbf{0}\ \ \ &= \{\} \\
\text{pder}\_\mathbf{1}\ \ \ &= \{\} \\
\text{pder } a\ (\mathsf{A}\ x) &= \textbf{if } x = a \textbf{ then } \{\mathbf{1}\} \textbf{ else } \{\} \\
\text{pder } a\ (r + s) &= \text{pder } a\ r \cup \text{pder } a\ s \\
\text{pder } a\ (r \cdot s) &= \textbf{if } \text{nullable } r \textbf{ then } (\text{pder } a\ r \odot s) \cup \text{pder } a\ s \textbf{ else } \text{pder } a\ r \odot s \\
\text{pder } a\ (r^*) &= \text{pder } a\ r \odot r^*
\end{aligned}
$$

Above, $R \odot s$ is used as a shorthand notation for $\{r \cdot s \mid r \in R\}$. The definition yields the characteristic property of partial derivatives by induction on r:

$$
\mathscr{D}\,a\,(\mathscr{L}\,r) = \bigcup\nolimits_{s \in \text{pder } a\ r} \mathscr{L}\,s
$$

Following this characteristic property, we can interpret the locale *rexpDFA*. The automaton constructed by P for our running example is shown in Fig. 4.

interpretation P: *rexpDFA* **where**

$$
\begin{aligned}
\iota\,r &= \{r\} \\
\delta\,a\,R &= \bigcup\nolimits_{r \in R} \text{pder } a\ r \\
\mathsf{o}\,R &= \exists r \in R.\ \text{nullable } r \\
\mathsf{L}\,R &= \bigcup\nolimits_{r \in R} \mathscr{L}\,r
\end{aligned}
$$

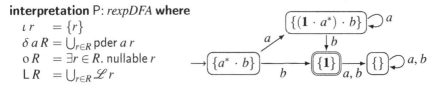

Fig. 4. Partial derivative automaton for $a^* \cdot b$

The assumptions of *rexpDA* (inherited by *rexpDFA*) are easy to discharge. Just as for Brzozowski derivatives, only the proof of finiteness of the reachable state space P.*fin* poses a challenge. We were able to reuse the proof by Wu *et al.* [27] who show finiteness when proving one direction of the Myhill-Nerode theorem. Compared with the proof of D.*fin*, the formal reasoning about partial derivatives appears to be more succinct.

There is a direct connection between pder and der that seems not to have been covered in the literature. It is best expressed in terms of a recursive function pset $:: \alpha\,rexp \rightarrow (\alpha\,rexp)\,set$ that translates derivatives to partial derivatives: pset (der $a\,r$) = pder $a\,r$.

$$
\begin{aligned}
\text{pset } \mathbf{0}\ \ \ &= \{\} & \text{pset } (r + s) &= \text{pset } r \cup \text{pset } s \\
\text{pset } \mathbf{1}\ \ \ &= \{\mathbf{1}\} & \text{pset } (r \cdot s) &= \text{pset } r \odot s \\
\text{pset } (\mathsf{A}\ x) &= \{\mathsf{A}\ x\} & \text{pset } (r^*) &= \{r^*\}
\end{aligned}
$$

A finite set R of regular expressions can be represented uniquely by a single regular expression ΣR, a sum ordered w.r.t. $\preceq$. Hence, we have $\Sigma\,\text{pset (der } a\ r) = \Sigma\,\text{pder } a\ r$, meaning that we can devise a normalization function $\langle\!\langle\!\langle r \rangle\!\rangle\!\rangle = \Sigma\,\text{pset } r$ that allows us to simulate partial derivatives while operating on plain regular expressions. Alternatively, $\langle\!\langle\!\langle \_ \rangle\!\rangle\!\rangle$ can be defined using smart constructors (with sequentially matched equations):

$$\langle\!\langle 0 \rangle\!\rangle = 0 \qquad\qquad 0 \boxplus r = r$$
$$\langle\!\langle 1 \rangle\!\rangle = 1 \qquad\qquad r \boxplus 0 = r$$
$$\langle\!\langle A\,a \rangle\!\rangle = A\,a \qquad\qquad (r+s) \boxplus t = r \boxplus (s \boxplus t)$$
$$\langle\!\langle r+s \rangle\!\rangle = \langle\!\langle r \rangle\!\rangle \boxplus \langle\!\langle s \rangle\!\rangle \qquad r \boxplus (s+t) = \textbf{if } r = s \textbf{ then } s+t$$
$$\langle\!\langle r \cdot s \rangle\!\rangle = \langle\!\langle r \rangle\!\rangle \boxdot s \qquad\qquad\qquad \textbf{else if } r \preceq s \textbf{ then } r + (s+t)$$
$$\langle\!\langle r^* \rangle\!\rangle = r^* \qquad\qquad\qquad\qquad \textbf{else } s + (r \boxplus t)$$
$$0 \boxdot r = 0 \qquad\qquad r \boxplus s = \textbf{if } r = s \textbf{ then } r$$
$$(r+s) \boxdot t = (r \boxdot s) \boxplus (s \boxdot t) \qquad\qquad \textbf{else if } r \preceq s \textbf{ then } r + s$$
$$r \boxdot s = s \cdot t \qquad\qquad\qquad\qquad \textbf{else } s + r$$

This definition allows to contrast the implicit quotienting performed by partial derivatives with the qoutienting modulo ACI equivalence ($\sim$). They turn out to be incomparable: $\langle\!\langle \_ \rangle\!\rangle$ does not simplify the second argument of concatenation $\cdot$ and the argument of iteration *, but erases 0s and uses left distributivity.

Finally, we obtain a last derivative-based interpretation using the characteristic property $\langle\!\langle \text{der } b\,r \rangle\!\rangle = \sum(\text{pder } b\,r)$ and P.$\textit{fin}$ to discharge the finiteness assumption $\textit{fin}$.

interpretation PD: $\textit{rexpDFA}$ **where**
$$\iota\,r = \langle\!\langle r \rangle\!\rangle$$
$$\delta\,a\,r = \langle\!\langle \text{der } a\,r \rangle\!\rangle$$
$$o\,r = \text{nullable } r$$
$$\mathsf{L}\,r = \mathscr{L}\,r$$

Whenever P yields an automaton for r with states labeled with finite sets of regular expressions X_i, PD constructs structurally the same automaton for r labeled with $\sum X_i$.

5 Marked Regular Expressions

One of the oldest methods for converting a regular expression into an automaton is based on the idea of identifying the states of the automaton with positions in the regular expression. Both McNaughton and Yamada [20] and Glushkov [13] mark the atoms in a regular expression with numbers to identify positions uniquely. In this section, we formalize two recent reincarnations of this approach due to Fischer $\textit{et al.}$ [11] and Asperti [3]. They are based on the realization that in a functional programming setting, it is most convenient to represent positions in a regular expression by marking some of its atoms. First we define an infrastructure for working with marked regular expressions. Then we define and relate both reincarnations in terms of this infrastructure.

$\textit{Marked}$ regular expressions are formalized by the following type synonym (where the value True denotes a marked atom)

$$\alpha\,\textit{mrexp} = (\textit{bool} \times \alpha)\,\textit{rexp}$$

We convert easily between $\textit{rexp}$ and $\textit{mrexp}$ with the help of map_rexp, the map function on regular expressions:

```
strip = map_rexp snd
emtpy_mrexp = map_rexp (λr. (False, r))
```

The language $\mathscr{L}_m :: \alpha\ mrexp \to \alpha\ lang$ of a marked regular expression is the set of words that start at some marked atom:

$$
\begin{aligned}
\mathscr{L}_m\,0 &= \{\} \\
\mathscr{L}_m\,1 &= \{\} \\
\mathscr{L}_m\,(\mathsf{A}\,(m,a)) &= \mathbf{if}\ m\ \mathbf{then}\ \{[a]\}\ \mathbf{else}\ \{\} \\
\mathscr{L}_m\,(r+s) &= \mathscr{L}_m\,r \cup \mathscr{L}_m\,s \\
\mathscr{L}_m\,(r \cdot s) &= (\mathscr{L}_m\,r \cdot \mathscr{L}(\mathsf{strip}\ s)) \cup \mathscr{L}_m\,s \\
\mathscr{L}_m\,(r^*) &= \mathscr{L}_m\,r \cdot \mathscr{L}(\mathsf{strip}\ r)^*
\end{aligned}
$$

The function final $:: \alpha\ mrexp \to bool$ tests if some atom at the "end" of a given regular expression is marked:

$$
\begin{aligned}
\mathsf{final}\ \mathbf{0} &= \mathsf{False} \\
\mathsf{final}\ \mathbf{1} &= \mathsf{False} \\
\mathsf{final}\ (\mathsf{A}\,(m,a)) &= m \\
\mathsf{final}\ (r+s) &= (\mathsf{final}\ r \vee \mathsf{final}\ s) \\
\mathsf{final}\ (r \cdot s) &= (\mathsf{final}\ s \vee \mathsf{nullable}\ s \wedge \mathsf{final}\ r) \\
\mathsf{final}\ (r^*) &= \mathsf{final}\ r
\end{aligned}
$$

Marks are moved around a regular expression by two operations. The function read $a\ r$ unmarks all atoms in r except a:

$$
\begin{aligned}
&\mathsf{read} :: \alpha \to \alpha\ mrexp \to \alpha\ mrexp \\
&\mathsf{read}\ a = \mathsf{map\_rexp}\ (\lambda(m,x).\ (m \wedge a = x, x))
\end{aligned}
$$

Its characteristic lemma is that it restricts $\mathscr{L}_m\,r$ to words whose head is a:

$$
\mathscr{L}_m\,(\mathsf{read}\ a\ r) = \{w \in \mathscr{L}_m\,r \mid w \neq [] \wedge \mathsf{hd}\ w = a\}
$$

The function follow $m\ r$ moves all marks in r to the "next" atom, much like an ε-closure; the mark m is pushed in from the left:

$$
\begin{aligned}
&\mathsf{follow} :: bool \to \alpha\ mrexp \to \alpha\ mrexp \\
&\mathsf{follow}\ m\ \mathbf{0} = \mathbf{0} \\
&\mathsf{follow}\ m\ \mathbf{1} = \mathbf{1} \\
&\mathsf{follow}\ m\ (\mathsf{A}\,(\_,a)) = \mathsf{A}\,(m,a) \\
&\mathsf{follow}\ m\ (r+s) = \mathsf{follow}\ m\ r + \mathsf{follow}\ m\ s \\
&\mathsf{follow}\ m\ (r \cdot s) = \mathsf{follow}\ m\ r \cdot \mathsf{follow}\ (\mathsf{final}\ r \vee m \wedge \mathsf{nullable}\ r)\ s \\
&\mathsf{follow}\ m\ (r^*) = (\mathsf{follow}\ (\mathsf{final}\ r \vee m)\ r)^*
\end{aligned}
$$

The characteristic lemma about follow shows that the marks are moved forward, thereby chopping off the first letter (in the generated language), and that the parameter m indicates whether every "first" atom should be marked:

$$
\mathscr{L}_m\,(\mathsf{follow}\ m\ r) = \{\mathsf{tl}\ w \mid w \in \mathscr{L}_m\,r\} \cup (\mathbf{if}\ m\ \mathbf{then}\ \mathscr{L}(\mathsf{strip}\ r)\ \mathbf{else}\ \{\}) - \{[]\}
$$

5.1 Mark After Atom

In the work of McNaughton-Yamada-Glushkov, the mark indicates which atom has just been read, i.e. the mark is located "after" the atom. Therefore the initial state is special because nothing has been read yet. Thus we express the states of the automaton as a pair of a boolean (True means that nothing has been read yet) and a marked regular expression. The boolean can be viewed as a mark in front of the automaton. (Alternatively, one could work with an explicit start symbol in front of the regular expression.) We interpret the locale *rexpDFA* as follows:

interpretation A: *rexpDFA* **where**
$\iota\, r \quad\quad = (\text{True}, \text{emtpy\_mrexp}\, r)$
$\delta\, a\, (m, r) = (\text{False}, \text{read}\, a\, (\text{follow}\, m\, r))$
$\text{o}\, (m, r) \quad = (\text{final}\, r \vee m \wedge \text{nullable}\, r)$
$\text{L}\, (m, r) \quad = \mathscr{L}_m(\text{follow}\, m\, r) \cup (\textbf{if}\, \text{o}\, (m, r)\, \textbf{then}\, \{[]\}\, \textbf{else}\, \{\})$

The definition of δ expresses that we first build the ε-closure starting from the marked atoms (via follow) and then read the next atom. With the characteristic lemmas about read and follow (and a few auxiliary lemmas), the locale assumptions are easily proved. This yields our first version of automata based on marked regular expressions.

Finiteness of the reachable part of the state space is proved via the lemma

$$\text{fold}\, \delta\, w\, (\iota\, r) \in \{\text{True}, \text{False}\} \times \text{mrexps}\, r$$

where mrexps :: $\alpha\, rexp \to (\alpha\, mrexp)\, set$ maps a regular expression to the finite set of all its marked variants, i.e. mrexps $r = \{r' \mid \text{strip}\, r' = r\}$; its actual recursive definition is straightforward and omitted.

Now we take a closer look at the work of Fischer *et al.* [11], which inspired the preceding formalization. They present a number of (not formally verified) matching algorithms on marked regular expressions in Haskell that follow McNaughton-Yamada-Glushkov. This is their basic transition function:

```
shift :: bool → α mrexp → α → α mrexp
shift _ 0 _        = 0
shift _ 1 _        = 1
shift m (A (_, x)) c = A (m ∧ (x = c), x)
shift m (r + s) c  = shift m r c + shift m s c
shift m (r · s) c  = shift m r c · shift (final r ∨ m ∧ nullable r) s c
shift m (r*) c     = (shift (final r ∨ m) r c)*
```

A simple induction proves that their shift is our δ:

$$\text{shift}\, m\, r\, x = \text{read}\, x\, (\text{follow}\, m\, r)$$

Thus we have verified their shift function. Fischer *et al.* optimize shift further, which is still quadratic due to the calls of the recursive functions final and nullable. They simply cache the values of final and nullable at all nodes of a regular expression by adding additional fields to each constructor. We have verified this optimization step as well, yielding another interpretation A_2 (omitted here).

5.2   Mark Before Atom

Instead of imagining the mark to be after an atom, it can also be viewed to be in front of it, i.e. it marks possible next atoms. This is somewhat dual to the McNaughton-Yamada-Glushkov construction. It leads to the following interpretation of the *rexpDA* locale:

interpretation B: *rexpDFA* **where**
$\iota\, r$ $\quad\quad = (\text{follow True } (\text{emtpy\_mrexp } r), \text{nullable } r)$
$\delta\, a\, (r, m) = \textbf{let } r' = \text{read } a\, r \textbf{ in } (\text{follow False } r', \text{final } r')$
$o\, (r, m) \quad = m$
$L\, (r, m) \quad = \mathcal{L}_m\, r \cup (\textbf{if } m \textbf{ then } \{[]\} \textbf{ else } \{\})$

The definition of δ expresses that we first read an atom and then build the ε-closure. The assumptions of *rexpDA* and *rexpDFA* are proved easily just like in the previous interpretation with marked regular expressions.

The interesting point is that this happens to be the algorithm formalized by Asperti [3]. Although he says that he has formalized McNaughton-Yamada, he actually formalized the dual algorithm. This is not easy to see because Asperti's formalization is considerably more involved than ours, with many auxiliary functions. Strictly speaking, his algorithm is a variation of ours that produces the same automata. The complete proof of this fact can be found elsewhere [16]. Because of the size of Asperti's formalization, there is not enough space here to give the detailed equivalence proof. However, we can take a step towards his formulation and merge follow and read into one function move :: $\alpha \to \alpha\, mrexp \to bool \to \alpha\, mrexp$, the analogue of his homonymous function:

$\text{move}\_\, \mathbf{0}\, \_ \quad\quad\quad = \mathbf{0}$
$\text{move}\_\, \mathbf{1}\, \_ \quad\quad\quad = \mathbf{1}$
$\text{move } c\, (\text{A } (\_, x))\, m = \text{A } (m, x)$
$\text{move } c\, (r + s)\, m \quad = \text{move } c\, r\, m + \text{move } c\, s\, m$
$\text{move } c\, (r \cdot s)\, m \quad = \text{move } c\, r\, m \cdot \text{move } c\, s\, (\text{final1 } r\, c \vee m \wedge \text{nullable } r)$
$\text{move } c\, (r^*)\, m \quad = (\text{move } c\, r\, (\text{final1 } r\, c \vee m)))^*$

where final1 is an auxiliary recursive function (not shown here) with the characteristic property that final1 $r\, c = \text{final } (\text{read } c\, r)$. A simple induction proves that move combines follow and read as in δ:

$\text{move } c\, r\, m = \text{follow } m\, (\text{read } c\, r)$

The function move has quadratic complexity for the same reason as shift. Unfortunately, it cannot be made linear with the same ease as for shift. The problem is that we need to cache the value of final1 $r\, c$ in the previous step, before we know c. We solve this by caching the set of all letters c that make final1 $r\, c$ true. In the worst case, the whole alphabet must be stored in certain inner nodes. However, for an alphabet of fixed size this guarantees linear time complexity. This optimization constitutes a last interpretation B_2.

Even for a fixed alphabet, Asperti's move has quadratic complexity when faced with a tower of stars: each recursive call of move can trigger a call of a function eclose, which has linear complexity. Asperti aimed for compact proofs, not maximal efficiency.

5.3 Comparison

The two constructions may look similar, but they do not produce isomorphic automata. Considering our running example, we display the mark by a "•" before or after the atom. The two resulting automata are shown in Fig. 5. There are special states that cannot

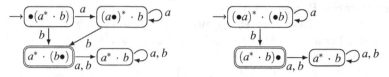

Fig. 5. Marked regular expression automata (A left, B right) for $a^* \cdot b$

be denoted by marking atoms only: $\bullet r$ in A's automaton is the completely unmarked regular expression that is the initial state and $r\bullet$ in B's automaton is a final state.

It turns out that the "before" automaton is a homomorphic image of the "after" automaton. To verify this we specify the homomorphism $\varphi(m, r) = (\text{follow } m\, r, \text{A.o}\,(m, r))$ and prove that it preserves initial states and commutes with the transition function:

$$\varphi(\text{A.}\iota\, r) = \text{B.}\iota\, r \qquad \varphi(\text{A.}\delta\, a\, s) = \text{B.}\delta\, a\,(\varphi\, s) \qquad \varphi(\text{fold A.}\delta\, w\, s) = \text{fold B.}\delta\, w\,(\varphi\, s)$$

A direct consequence is that Asperti's "before" construction always generates automata with at most as many states as the McNaughton-Yamada-Glushkov construction. Formally, in the context of locale *rexpDA* we have defined an executable computation of the reachable state space $\{\text{fold } \delta\, w\,(\iota\, r) \mid w \in (\text{set } as)^*\}$ of the automaton:

$$\text{reachable } as\, r = \text{snd (the (rtc } (\lambda\_.\ \text{True})\,(\lambda s.\ \text{map } (\lambda a.\ \delta\, a\, s)\, as)\,(\iota\, r)))$$

where r is the initial regular expression, as is the alphabet, and the $(\text{Some } x) = x$.

Theorem 5. $|\text{B.reachable } as\, r| \le |\text{A.reachable } as\, r|$ *where* $|\_|$ *is the cardinality of a set.*

In early drafts of this paper, we only conjectured the above statement and unsuccessfully tried to refute it with Isabelle's Quickcheck facility [8]. Later, Helmut Seidl has communicated an informal proof using the above homomorphism to us.

Let us abbreviate the statement of Thm. 5 to $n_b \le n_a$. One may think that n_a is only slightly larger than n_b, but it seems that n_b and n_a are more than a constant summand apart: for a two-element alphabet Quickcheck could refute $n_a \le n_b + k$ even for $k = 100$.

6 Empirical Comparison

We compare the efficiency w.r.t. both matching and deciding equivalence of the Standard ML code generated from eight described interpretations: $\sim$-normalized derivatives (D), $\approx$-normalized derivatives (N), partial derivatives (P), derivatives simulating partial derivatives (PD), mark "after" atom (A), mark "after" atom with caching (A_2), mark "before" atom (B), and mark "before" atom with caching (B_2). The interpretation using the quotient type for derivatives ($D_\sim$) is not in this list, as it is clearly superseded by D. The results of the evaluation, performed on an Intel Core i7-2760QM machine with 8 GB of RAM, are shown in Fig. 6. Solid lines depict the four derivative-based algorithms. Dashed lines are used for the algorithms based on marked regular expressions.

The first two tests, MATCH-R and MATCH-L, measure the time required to match the word a^n against the regular expression $(a + 1)^n \cdot a^n$—a standard benchmark also used by Fischer *et al.* [11]. The difference between the two tests is the definition

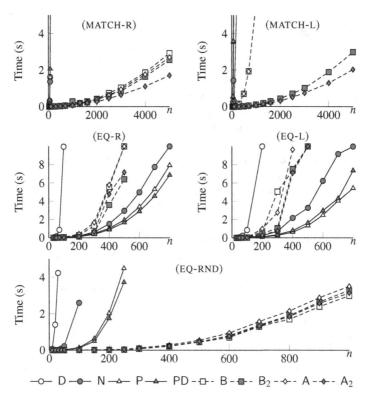

Fig. 6. Evaluation results

of r^n. MATCH-R defines it as the n-fold concatenation associated to the right: $r^4 = r \cdot (r \cdot (r \cdot r))$, whereas MATCH-L associates to the left: $r^4 = ((r \cdot r) \cdot r) \cdot r$. In both tests, marked regular expressions outperform derivatives by far. The normalization performed by the derivative-based approaches (required to obtain a finite number of states for the equivalence check) decelerates the computation of the next state. Marked regular expressions benefit from a fast next state computation. The test MATCH-L exhibits the quadratic nature of the unoptimized matchers A and B (their curves are almost identical and therefore hard to distinguish in Fig. 6). In contrast, A_2 and B_2 perform equally well in both tests, A_2 being approximately 1.5 times faster due to lighter cache annotations.

The next test goes back to Antimirov [1]: We measure the time (with a timeout of ten seconds) for proving the equivalence of a^* and $(a^0 + \ldots + a^{n-1}) \cdot (a^n)^*$. Again two tests, EQ-R and EQ-L, distinguish the associativity of concatenation in r^n. Here, the derivative-based equivalence checkers (except for D) perform better then the ones based on marked regular expressions. In particular, both version of partial derivatives, P and PD, outperform N—since this example was crafted by Antimirov to demonstrate the strength of partial derivatives, this is not wholly unexpected. Comparing EQ-R and EQ-L, the associativity barely influences the runtime.

Finally, to avoid bias towards a particular algorithm, we have devised the randomized test EQ-RND. There we measure the average time (with a timeout of ten seconds) to prove the equivalence of r with itself for 100 randomly generated expressions with n inner nodes (+, ·, or *). Proving $r \equiv r$ is of course a trivial task, but our algorithms do not stop the exploration when the state of the product automaton is a pair of two equal states. This optimization, which is a must for any practical algorithm, is the first step towards the rewarding usage of bisimulation up to equivalence (or even up to congruence) [5]. Without any such optimization, the task of proving $r \equiv r$ amounts to enumerating all derivatives of r, which is exactly what we want to compare. To generate random regular expression we use the infrastructure of SpecCheck [25]—a Quickcheck clone for Isabelle/ML. For computing the average, a timeout counts as 10 second (although the actual computation would likely have taken longer)—an approximation that skews the curves to converge to the margin of 10 seconds. We stopped measuring a method for increasing n when the average approached 5 seconds.

The results of EQ-RND are summarized as follows: $D \gg N \gg P, PD \gg A, A_2, B, B_2$, where $X \gg Y$ means that Y is an order of magnitude faster than X. The algorithm P is noticeably slower than PD—avoiding sets reduces the overhead. Among A, A_2, B, B_2, Asperti's unoptimized algorithm B performs best by a narrow margin. Regular expressions where the caching overhead pays off are rare and therefore not visible in the randomized test results. The same holds for expressions where B produces much smaller automata than A (e.g. the counterexample to $n_a \leq n_b + 100$ from Subsect. 5.3).

Our evaluation shows that A_2 is the best choice for matching. For equivalence checking, the winner is not as clear cut: B (especially when applied to normalized input to avoid quadratic runtime without caching) and PD seem to be the best choices.

7   Extensions

Brzozowski's derivatives are easily extendable to regular expressions intersection and negation—indeed Brzozowski performed the extension right from the start [7]. The number of such extended derivatives is still finite when quotiented modulo ACI.

We [26] have recently further extended derivatives to regular expressions extended with projection, obtaining verified decision procedures for the equivalence of those extended regular expressions and for monadic second-order logics over finite words. The closure computation and its correctness proof follow Krauss and Nipkow [19].

Extending partial derivatives with intersection and negation is more involved [9]. An additional layer of sets must be used for intersections, i.e. the states of our automaton would then be sets of sets of regular expressions. In Sect. 6, we have seen that already one layer of sets incurs some overhead. Hence, the view on partial derivatives as derivatives followed by some normalization is expected to be even more profitable for the extension. The extension of partial derivatives with projection is an easy exercise.

It is unclear how to extend marked regular expressions to handle negation and intersection. The number of possible markings for a regular expression of alphabetic width n is 2^n. However, there exist regular expressions of alphabetic width n using intersection, whose minimal automata have 2^{2^n} states [12].

8 Conclusion

We have shown that all the previously published verified decision procedures for equivalence of regular expressions that operate on regular expressions directly can all be expressed as instances of a generic automaton-inspired framework. The correctness proofs decompose into a generic part that is proved once and for all in the framework and a few specific properties that need to be proved for each instance. The framework caters for a meaningful comparison of the performance of the various instances. Marked regular expressions are superior on average but partial derivatives can outperform them in specific cases. The Isabelle theories are available online [23].

Acknowledgment. We thank Andrea Asperti and Sebastian Fischer for commenting on fine points of their work and Helmut Seidl for contributing an informal proof of Thm. 5. Jasmin Blanchette, Andrei Popescu and three anonymous reviewers helped to improve the presentation through numerous suggestions. The second author is supported by the doctorate program 1480 (PUMA) of the Deutsche Forschungsgemeinschaft (DFG).

References

1. Antimirov, V.: Partial derivatives of regular expressions and finite automata constructions. In: Mayr, E.W., Puech, C. (eds.) STACS 1995. LNCS, vol. 900, pp. 455–466. Springer, Heidelberg (1995)
2. Antimirov, V.: Partial derivatives of regular expressions and finite automaton constructions. Theor. Comput. Sci. 155(2), 291–319 (1996)
3. Asperti, A.: A compact proof of decidability for regular expression equivalence. In: Beringer, L., Felty, A. (eds.) ITP 2012. LNCS, vol. 7406, pp. 283–298. Springer, Heidelberg (2012)
4. Ballarin, C.: Interpretation of locales in Isabelle: Theories and proof contexts. In: Borwein, J.M., Farmer, W.M. (eds.) MKM 2006. LNCS (LNAI), vol. 4108, pp. 31–43. Springer, Heidelberg (2006)
5. Bonchi, F., Pous, D.: Checking NFA equivalence with bisimulations up to congruence. In: Giacobazzi, R., Cousot, R. (eds.) POPL 2013, pp. 457–468. ACM (2013)
6. Braibant, T., Pous, D.: An efficient Coq tactic for deciding kleene algebras. In: Kaufmann, M., Paulson, L.C. (eds.) ITP 2010. LNCS, vol. 6172, pp. 163–178. Springer, Heidelberg (2010)
7. Brzozowski, J.A.: Derivatives of regular expressions. J. ACM 11(4), 481–494 (1964)
8. Bulwahn, L.: The new Quickcheck for Isabelle: Random, exhaustive and symbolic testing under one roof. In: Hawblitzel, C., Miller, D. (eds.) CPP 2012. LNCS, vol. 7679, pp. 92–108. Springer, Heidelberg (2012)
9. Caron, P., Champarnaud, J.-M., Mignot, L.: Partial derivatives of an extended regular expression. In: Dediu, A.-H., Inenaga, S., Martín-Vide, C. (eds.) LATA 2011. LNCS, vol. 6638, pp. 179–191. Springer, Heidelberg (2011)
10. Coquand, T., Siles, V.: A decision procedure for regular expression equivalence in type theory. In: Jouannaud, J.-P., Shao, Z. (eds.) CPP 2011. LNCS, vol. 7086, pp. 119–134. Springer, Heidelberg (2011)
11. Fischer, S., Huch, F., Wilke, T.: A play on regular expressions: functional pearl. In: Hudak, P., Weirich, S. (eds.) ICFP 2010, pp. 357–368. ACM (2010)
12. Gelade, W., Neven, F.: Succinctness of the complement and intersection of regular expressions. ACM Trans. Comput. Log. 13(1), 4:1–4:19 (2012)

13. Glushkov, V.M.: The abstract theory of automata. Russian Math. Surveys 16, 1–53 (1961)
14. Haftmann, F., Krauss, A., Kunčar, O., Nipkow, T.: Data refinement in Isabelle/HOL. In: Blazy, S., Paulin-Mohring, C., Pichardie, D. (eds.) ITP 2013. LNCS, vol. 7998, pp. 100–115. Springer, Heidelberg (2013)
15. Haftmann, F., Nipkow, T.: Code generation via higher-order rewrite systems. In: Blume, M., Kobayashi, N., Vidal, G. (eds.) FLOPS 2010. LNCS, vol. 6009, pp. 103–117. Springer, Heidelberg (2010)
16. Haslbeck, M.: Verified Decision Procedures for the Equivalence of Regular Expressions. B.Sc. thesis, Department of Informatics, Technische Universität München (2013)
17. Huffman, B., Kunčar, O.: Lifting and Transfer: A modular design for quotients in Isabelle/HOL. In: Gonthier, G., Norrish, M. (eds.) CPP 2013. LNCS, vol. 8307, pp. 131–146. Springer, Heidelberg (2013)
18. Kaliszyk, C., Urban, C.: Quotients revisited for Isabelle/HOL. In: Chu, W.C., Wong, W.E., Palakal, M.J., Hung, C.C. (eds.) SAC 2011, pp. 1639–1644. ACM (2011)
19. Krauss, A., Nipkow, T.: Proof pearl: Regular expression equivalence and relation algebra. J. Automated Reasoning 49, 95–106 (2012) (published online March 2011)
20. McNaughton, R., Yamada, H.: Regular expressions and finite state graphs for automata. IRE Trans. on Electronic Comput. EC-9, 38–47 (1960)
21. Moreira, N., Pereira, D., de Sousa, S.M.: Deciding regular expressions (in-)equivalence in Coq. In: Kahl, W., Griffin, T.G. (eds.) RAMiCS 2012. LNCS, vol. 7560, pp. 98–113. Springer, Heidelberg (2012)
22. Nipkow, T., Klein, G.: Concrete Semantics. A Proof Assistant Approach. Springer (to appear), http://www.in.tum.de/~nipkow/Concrete-Semantics
23. Nipkow, T., Traytel, D.: Regular expression equivalence. Archive of Formal Proofs, Formal proof development (2014),
    http://afp.sf.net/entries/Regex_Equivalence.shtml
24. Rutten, J.J.M.M.: Automata and coinduction (an exercise in coalgebra). In: Sangiorgi, D., de Simone, R. (eds.) CONCUR 1998. LNCS, vol. 1466, pp. 194–218. Springer, Heidelberg (1998)
25. Schaffroth, N.: A Specification-based Testing Tool for Isabelle's ML Environment. B.Sc. thesis, Department of Informatics, Technische Universität München (2013)
26. Traytel, D., Nipkow, T.: Verified decision procedures for MSO on words based on derivatives of regular expressions. In: Morrisett, G., Uustalu, T. (eds.) ICFP 2013, pp. 3–12. ACM (2013)
27. Wu, C., Zhang, X., Urban, C.: A formalisation of the Myhill-Nerode theorem based on regular expressions. J. Automated Reasoning 52, 451–480 (2014)

Collaborative Interactive Theorem Proving with Clide

Martin Ring[1] and Christoph Lüth[1,2,*]

[1] Deutsches Forschungszentrum für Künstliche Intelligenz, Bremen, Germany
[2] Universität Bremen, FB 3 — Mathematics and Computer Science, Germany

Abstract. This paper introduces Clide, a collaborative web interface for the Isabelle theorem prover. The interface allows a document-oriented interaction very much like Isabelle's desktop interface. Moreover, it allows users to jointly edit Isabelle proof scripts over the web; editing operations are synchronised in real-time to all users.

The paper describes motivation, user experience, implementation and system architecture of Clide. The implementation is based on the theory of operational transformations; its key concepts have been formalised in Isabelle, its correctness proven and critical parts of the implementation on the server are generated from the formalisation, thus increasing confidence in the system.

1 Introduction

Just like mathematics, interactive theorem proving is at its heart a social activity. Mathematical proof is rarely a solitary activity, it is most often done in collaboration with others. It is thus unfortunate that present theorem prover interfaces have very much been single-user; a *real-time collaborative* user interface, where many users can jointly edit the same proof in the vein of the late Google docs[1] should add much to the user experience, enhance productivity and enable new patterns of interaction between theorem provers and humans. Until now, there have hardly been real-time collaborative user interface for theorem provers, so this hypothesis had to remain untested. This paper presents a first prototype of a real-time collaborative, web-based user interface for a state-of-the-art interactive theorem prover, Isabelle, allowing us to experiment with the collaborative user experience.

As the experience with Google docs shows, collaborative user interfaces thrive when they are available on the web. The web has collaboration built-in, with many users connecting to a single server, and web interfaces offer *eo ipso* a lot of advantages: they are inherently cross-platform, portable and mobile, they require little installation effort (a recent web browser is enough), and only need few resources on the user side. Recent advantages in web technology (collectively and somewhat inaccurately known as 'HTML5') allow the development of

---

* Research supported by BMBF grants 01IW10002 (SHIP), 01IW13001 (SPECifIC).
[1] Now available as Google Drive.

G. Klein and R. Gamboa (Eds.): ITP 2014, LNAI 8558, pp. 467–482, 2014.

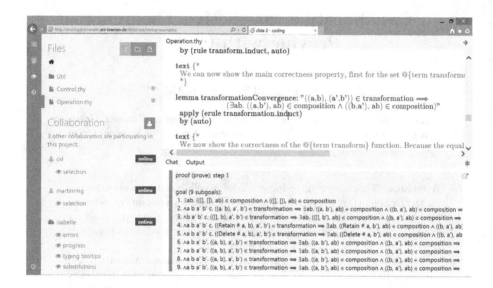

Fig. 1. The Clide user interface: On the left there is a toolbar, on the right a theory has been opened (above). The Isabelle output, here the current proofstate, is shown in a separate part of the window; it can also be inlined.

web interfaces of near-desktop quality. The first version of the Clide system [1] demonstrated a web-based interface for Isabelle; the present work extends this to a truly collaborative setting. This is not a completely trivial exercise; the basic problem is keeping the documents *synchronised* across the different clients (user interfaces), the server and the theorem prover. Fortunately, well explored solutions for this problem exist which we could draw on when implementing our system, namely the theory of operational transformation. We have formalised the basic algorithms of this theory in Isabelle, and generate parts of our implementation from this formalisation.

This paper is structured as follows: we first introduce Clide from the users' perspective, then give the theory and pragmatics of the implementation. We explain the system architecture and the underlying design decisions, and finish with conclusions, where we review related and future work.

2   The User Experience

The interface was designed with an emphasis on typography over superfluous graphics, with a clear arrangement reducing it to the basics such that it does not distract from the main center of attention, the proof script. Pervasive use of HTML5 and JavaScript make the interface very responsive; because the entire user interface is implemented as a single-page application which dynamically changes views through JavaScript the interaction more resembles a desktop application than a web interface of old.

Clide organises the user's work in *projects*, which are collections of files and folders. Projects are the basic unit of granularity for sharing. The Clide user interface has two basic views: the *backstage view* is the starting point, where users can review their projects, create new ones, and select a current one, and the *project view*, where users can create, edit and delete files and folders in the current project.

Fig. 1 shows a screenshot of the project view. On the left, there is a (hideable) tool sidebar, where users can select files, invite collaborators, and access common editing operations such as cut, copy & paste. The tool sidebar further shows the other collaborators and their details. In the center of the view, there is the main file editor, where files are opened in tabs. The editor is based on the CodeMirror editor, and offers a seamless editing of mathematical text in a web-browser, with features such as integration of mathematical symbols, Greek letters, and other Unicode symbols, flexible-width font, super/subscripting and tooltips for text spans. The interaction with Isabelle is very similar to the Isabelle/jEdit interface [2]: users edit the theory while Isabelle processes it asynchronously on the server, sending back the results as they become available. These results can be the prover's state, error or warning messages after executing this particular prover command. The prover messages can be displayed inlined or in a separate window, which is useful for larger proof states. In addition to the messages, the inner syntax of the theory as well as special symbol substitutions are annotated and type information for hovered terms is provided. All these annotations can be deactivated individually if desired. There is also a chat window which allows short text messages to be sent to collaborators. The collaboration is unintrusive, and should be familiar to users of Google docs: each user can see the cursor of other users and their editing operations, taking effect immediately without blocking. Users also see the selection area of other users, which is useful for communication purposes ("Where here is the error?"), and to warn other users that this area of the file is about to be deleted; this feature can be deactivated if it gets too intrusive.

A public evaluation version of the system is online at `http://clide.informatik.uni-bremen.de/`. The evaluation version features public projects, which are open to all users of the system (normally not a desirable state of affairs), which is great to get quickly up and collaborating.

2.1 Use Cases

Collaboration should not be end unto itself. We envisage at least the following use cases for collaborative theorem proving:

- *Scientific collaboration:* two (or more) users are working jointly on a proof, all contributing actively and staying in close contact; collaboration ensures that all participants know the proof, and can continue working on it. In a normal situation, collaborators would be sitting around the same machine; with a collaborative interface this situation can be extended to collaborating across countries and continents (timezone issues notwithstanding).

- *Proof review:* one user is going through the proof, explicating it to others who do not contribute actively, but try to understand what is being formalised. This situation is useful in the classroom, both for lecturers to explain a proof (while the students can interactively explore it), but possibly also for teachers to see how students progress and be able to assist them if needed.
- *Machine-assisted collaboration:* here, other collaborators are software processes. A simple example of this is Clide used as a single-user web-interface: the user still collaborates with Isabelle.

We would be surprised to see massive open online collaboration, where thousands of people work on one single theory. In this situation, an underlying version management and revision control system is needed; the ProofPeer project recently started in Edinburgh is an interesting step in this direction [3].

3   Implementation

Research in the area of *computer-supported cooperative work* (CSCW) goes back to the early eighties of the last century. One of the more challenging concerns has always been real-time activity awareness and coordination [4]. A collaborative system with these properties requires a mechanism to synchronise the distributed document states as quickly as possible across all users without loss of information or diverging documents. This is far from trivial because communication always involves delays (ranging from usual network delay to temporary failure), and thus edits will occur concurrently.

If we only consider the insertion of content the problem has a reasonably easy solution by introducing a partial ordering of concurrently inserted document positions (*e.g.* via vector clocks). Problems come with concurrent insertions and deletions especially if we do not only want state consistency but also basic *intention preservation* [5], which is essential for a usable system. The most popular approach to this problem has been *operational transformation* (OT) [6].

3.1   The Basics of Operational Transformation

The problem of synchronisation is that we may have situations where two operations f and g are applied concurrently to the same document D, and we need to complete the resulting span again with operations f', g' into a common document D' as in (1). Writing *applyOp f D* for the application of an operation f

$$\tag{1}$$

to a document D, the completion of (1) is written as

$$\forall D.\ applyOp\ g'\ (applyOp\ f\ D) = applyOp\ f'\ (applyOp\ g\ D). \tag{2}$$

The basic idea behind OT is to solely consider the operations, and not the documents, and to restrict ourselves to a tractable set of basic operations. Hence, operations are sequences of basic actions, where an actions is: advance one character; insert one character; or delete one character. To apply an operation, we traverse the document and operation simultaneously, and apply the basic actions. Note that this application is partial; we can only apply an operation to a document of the appropriate length. We can then transform these operations against each other, written as $transform\ f\ g = \langle f', g' \rangle$, to obtain the completion (1); applying first f, then g' should be the same as first applying g, then f'.

In order to state (2) point-free, *i.e.* without referring to a document D, we need to define the composition $\circ$ of two operations. Note that this is not simply the concatenation of the two sequences of actions; rather, we merge two sequences into one new sequence which combines the effects of the two operations. We can then drop the document D from the correctness property (2) and state:

$$transform\ f\ g = \langle f', g' \rangle \implies g' \circ f = f' \circ g \qquad (3)$$

The correctness of the composition operation $\circ$ is stated as

$$applyOp\ (g \circ f)\ D = applyOp\ g\ (applyOp\ f\ D) \qquad (4)$$

and together these easily imply (2).

3.2   Formalisation in Isabelle/HOL

We introduce the formalisation of the theory of operational transformation on which our implementation is based. For reasons of space, we do not show the full formalisation; we give enough details to show the actual algorithms, but we elide most lemmas and all Isabelle proofs (most of which are very short anyway).[2]

We start with the basic concepts. For documents, we keep the actual character set as a type parameter, actions are as mentioned above, and operations are then lists of actions:

type_synonym $'char\ document = {}'char\ list$

datatype $'char\ action = Retain \mid Insert\ 'char \mid Delete$

type_synonym $'char\ operation = {}'char\ action\ list$

We can now recursively define the application function:

fun $applyOp :: {}'char\ operation \Rightarrow {}'char\ document \Rightarrow {}'char\ document\ option$
where

$applyOp\ []\ []$	$= Some\ []$
$\mid applyOp\ (Retain\#\ as)\ (b\#\ bs)$	$= Option.map\ (\lambda ds.\ b\#\ ds)\ (applyOp\ as\ bs)$
$\mid applyOp\ (Insert\ c\#\ as)\ bs$	$= Option.map\ (\lambda ds.\ c\#\ ds)\ (applyOp\ as\ bs)$
$\mid applyOp\ (Delete\#\ as)\ (\_\#\ bs)$	$= applyOp\ as\ bs$
$\mid applyOp\ \_\ \_$	$= None$

---

[2] The full theory can be found at http://www.informatik.uni-bremen.de/ ~cxl/papers/itp2014-appendix.pdf for reference.

However, reasoning about this function directly is not straightforward because of its partiality: *applyOp f d* is only defined for a document *d* of a certain *inputLength*, given by the number of *Retain* and *Delete* actions in that operation. A straightforward induction on the definition of *applyOp* would leave us with an induction assumption where it is not immediate that *applyOp* is applicable to its arguments. In order to get around this difficulty, we define the graph of the function as an inductive set:

inductive_set *application* :: ((*'char operation* × *'char document*) × *'char document*) *set* **where**

 empty[intro!]: $(([],[]),[]) \in application$
 | *retain*[intro!]: $((a,d),d') \in application \Longrightarrow ((Retain\#a,c\#d),c\#d') \in application$
 | *delete*[intro!]: $((a,d),d') \in application \Longrightarrow ((Delete\#a,c\#d),d') \in application$
 | *insert*[intro!]: $((a,d),d') \in application \Longrightarrow (((Insert\ c)\#a,d),c\#d') \in application$

We can show that *application* is exactly the graph of *applyOp*:

lemma *applyOpSet*: $((a,d),d') \in application \longleftrightarrow applyOp\ a\ d = Some\ d'$

The composition of two operations traverses through the two operations and combines the actions pointwise. In the following definition, the second argument of the composition is executed after the first one (the other way around as ∘); so *e.g.* a delete action first executed is always kept, because nothing can undo a delete, and an insert action executed second is kept for the same reason. An insert action followed by a retain is just that insert, and insert followed by delete cancel each other out:

fun *compose* :: *'char operation* ⇒ *'char operation* ⇒ *'char operation option* **where**

 compose [] []   = *Some* []
 | *compose* (*Delete*# *as*) *bs*   = *Option.map addDeleteOp* (*compose as bs*)
 | *compose as* (*Insert c*# *bs*) =
   Option.map (*Cons* (*Insert c*)) (*compose as bs*)
 | *compose* (*Retain*# *as*) (*Retain*# *bs*) = *Option.map* (*Cons Retain*) (*compose as bs*)
 | *compose* (*Retain*# *as*) (*Delete*# *bs*) = *Option.map addDeleteOp* (*compose as bs*)
 | *compose* (*Insert c*# *as*) (*Retain*# *bs*) =
   Option.map (*Cons* (*Insert c*)) (*compose as bs*)
 | *compose* (*Insert* _# *as*) (*Delete*# *bs*) = *compose as bs*
 | *compose* _ _   = *None*

The above function uses *addDeleteOp* to insert a delete action. This is an optimisation, where we permute deletes over inserts as much as possible, so we get contiguous sequences of delete and insert actions, which we can later compress for transmission. *addDeleteOp* is defined as follows:

fun *addDeleteOp* :: *'char operation* ⇒ *'char operation* **where**

 addDeleteOp (*Insert c*#*next*) = *Insert c*# *addDeleteOp next*

| $addDeleteOp$ $as = Delete\#as$

The effect of $addDeleteOp$ is to remove the first element of a document:

lemma $addDeleteOpValid$: $applyOp$ $(addDeleteOp$ $a)$ $(c\#d) = applyOp$ a d

Again, in order to be able to show anything about $compose$ we explicitly define its graph as an inductive set.

inductive_set $composition$ :: $(('char\ operation \times\ 'char\ operation) \times\ 'char\ operation)$ set **where** ...

We leave out the lengthy definition; it follows the recursive definition of $compose$ just as $application$ follows the definition of $applyOp$. However, we show that $composition$ is the graph of the $compose$ operation:

lemma $composeSet$: $((a,b),ab) \in composition \longleftrightarrow compose$ a $b = Some$ ab

The first proper result is the correctness of composition (4). It is first shown for the relation $composition$ (omitted), and then for the function $compose$:

theorem $composeCorrect$:
$[\![$ $compose$ a $b = Some$ ab; $applyOp$ a $d = Some$ d'; $applyOp$ b $d' = Some$ d'' $]\!]$
$\implies applyOp$ ab $d = Some$ d''

Finally, we define the $transform$ function, the core algorithm of operational transformation. Recall that the transformation of a and b are two operations a', b' such that a composed with b' is the same as b composed with a'. The transformation is defined recursively, with a lengthy case distinction on the first action of each: e.g., insert actions remain, but cause a retain action to appear in the transformed operation (in order to keep the inserted character); transforming two retain actions results in two retain actions; or a retain and a delete transform to a delete and nothing (reflecting the fact that we either first keep an element, then delete it, or delete it first, without need for a subsequent action):

fun $transform$ :: $'char\ operation \Rightarrow 'char\ operation \Rightarrow ('char\ operation \times\ 'char\ oper$-$ation)\ option$
where
  $transform$ $[]$ $[]$ $= Some$ $([], [])$
$|$ $transform$ $(Insert$ $c\#as)$ bs $=$
            $Option.map$ $(\lambda(at,\ bt).\ (Insert\ c\#\ at,\ Retain\#\ bt))$ $(transform$ as $bs)$
$|$ $transform$ as $(Insert$ $c\#$ $bs) =$
            $Option.map$ $(\lambda(at,\ bt).\ (Retain\#\ at,\ Insert\ c\#\ bt))$ $(transform$ as $bs)$
$|$ $transform$ $(Retain\#$ $as)$ $(Retain\#$ $bs) =$
            $Option.map$ $(\lambda(at,\ bt).\ (Retain\#\ at,\ Retain\#\ bt))$ $(transform$ as $bs)$
$|$ $transform$ $(Delete\#$ $as)$ $(Delete\#$ $bs) = transform$ as bs
$|$ $transform$ $(Retain\#$ $as)$ $(Delete\#$ $bs) =$
            $Option.map$ $(\lambda(at,\ bt).\ (at,\ Delete\#\ bt))$ $(transform$ as $bs)$
$|$ $transform$ $(Delete\#$ $as)$ $(Retain\#$ $bs) =$
            $Option.map$ $(\lambda(at,\ bt).\ (Delete\#\ at,\ bt))$ $(transform$ as $bs)$

| *transform* _ _ $= None$

To our minds, this definition is intricate enough to warrant a formal treatment; at least, we feel more confident about its correctness having done so. We can show that the domain of this function is pairs of operations a and b which have the same input length. To show the main correctness property (3), we define the graph of *transform* as an inductive set (we leave out the lengthy definition):

inductive_set *transformation* :: ((*'c operation* × *'c operation*) × (*'c operation* × *'c operation*)) *set* **where** ...

and show that this is a superset of function graph. With this, we can show the second main result, the correctness of transformation. This is even slightly stronger as (3) as it also states that the composition of a and b' (and implicitly b and a') is defined:

theorem *transformCorrect*: *transform* a b $= Some$ (a', b')
$\implies$ *compose* a $b' \neq None \land$ *compose* a $b' =$ *compose* b a'

Further, we define identity operations *ident*, which consist only of retain actions, and show that they are the left and right unit to the composition operator. Unfortunately because of the optimisation underlying the *addDeleteOp* function, these properties only hold up to normalisation, *i.e.* sorting operations such that a delete action is never followed immediately by an insert action. Moreover, transformation against an identity does not change an operation:

lemma *transformIdL*:
transform (*ident* (*inputLength* b)) $b =$ *Some* (*ident* (*outputLength* b), b)

3.3   Implementing Operational Transformation

The previous section showed the formalisation of the core algorithms of operational transformation. The implementation uses these algorithms to synchronise document states on one server and many clients. Our implementation follows the approach by Google [7] which is a simplification of the original algorithms.

The server keeps a single history $h = \langle a_1, a_2, \ldots, a_n \rangle$ which is a sequence of operations a_i. A revision r_i refers to the state after operation a_i (starting with initial revision r_0). Clients report operations together with a revision number, $\langle b, i \rangle$. On receiving $\langle b, i \rangle$, the operation b is transformed with respect to all operations a_j for $i < j \leq n$, resulting in an operation b', which is appended to the history, and distributed to all other clients as a remote edit. Additionally, the client which sent the operation receives an acknowledgment (see Fig. 2). The correctness property (2) means that all squares in Fig. 2 commute, so the server only needs to append the transformed operation to its history. The server does not need to keep track of the actual document states, it is enough to keep track of the operations.

On the client side, clients need to cater for both local edits (affected by the user) and remote edits (sent from the server). To this end, the client keeps track of which operation $\langle a, r \rangle$ has last been passed on to the server, and waits for an

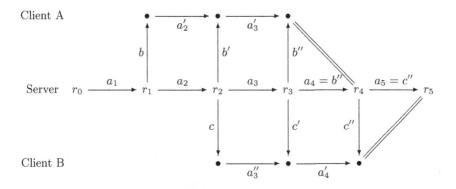

Fig. 2. History management on the server. Here, the current revision is r_3 when client A sends in operation $\langle b, r_1 \rangle$ and client B sends in $\langle c, r_2 \rangle$. The operation b is transformed with respect to $a_3 \circ a_2$ to b'', and appended as a_4 to the history; similarly, the operation c is transformed to c'' with respect to $a_3 \circ a_4$. Client A is sent $a'_3 \circ a'_2$ as the first remote edit, an acknowledgment, and a_5 as a second remote edit, while Client B is sent one remote edit $a'_4 \circ a''_3$, and an acknowledgment. In the end, r_5 becomes the new current revision on the server and both clients A, B. Note that we can resolve the convergence the other way, with first resolving $\langle c, r_2 \rangle$ with respect to a_3 (then a_4 would be c'), and then $\langle b, r_1 \rangle$ with respect to $c' \circ a_3 \circ a_2$ (the result of which would become a_5).

acknowledgment of this operation (we say the operation is *pending*). If further local edits occur while waiting, these are *buffered*. (Note that we only need to buffer one operation, as we can compose multiple edits.) Once the operation $\langle a, r \rangle$ has been acknowledged, the revision is increased, and the buffered operation (if there is any) is passed on and becomes pending. If a remote edit is received before the operation $\langle a, r \rangle$ has been acknowledged, we know it refers to revision r, so the operation is transformed with respect to the pending and buffered operations, applied to the local document, and the revision is increased. In turn, the remote edit operation transforms the pending and buffered operations (see Fig. 3). (Note that the client does not receive its own local edits back as remote edits from the server.)

A huge advantage of our Isabelle formalisation is the ability to generate Scala code [8] for the composition and transformation of operations which we can use in the server application. However, we need an implementation of the same algorithms on the client. Unfortunately, due to the restrictions of the web environment there is no practical alternative to JavaScript as a programming language. This leads to potentially divergent implementations. Experiments showed that in principle we can even go one step further and generate the JavaScript code from Scala with the new *Scala.js* compiler [9], but until this tool moves out of the alpha stage we have to rely on a manual port.

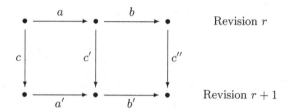

Fig. 3. History management on the client. Operation a is pending, operation b is buffered. If a remote edit c is received, it is transformed with respect to a and b to c'', and applied. a' and b' become the new pending and buffered operations. The revision number r is increased every time a remote edit or acknowledgment is received.

3.4   Further Extensions

In addition to plain text operations we need a mechanism to annotate the document state to integrate information like remote cursors and selections, syntax highlighting, or prover output. The original Google approach was somewhat cumbersome to allow for rich text editing; here, because we do not require annotations to be editable we chose a simpler approach by keeping annotations separate from the operations. In our implementation, annotations consist of two types of actions: $plain(n)$ is equivalent to n retain actions, and $annotate(n, a)$ is equivalent to n retain actions and additionally annotates the document with the annotation a starting at the current position, and ranging n characters.

This means annotations are operations consisting only of retain actions, with the side effect of augmenting the document with additional information. Recall from above that such operations are identities, hence annotations cannot interfere with other operations (see Lemma *transformIdL* in Sect. 3.2). However, other operations have an effect on annotations, so we have to define how delete and insert actions interact with $annotate(a, n)$. We chose to simply extend or shrink the annotation accordingly; this feels like natural behaviour that you would expect if you type inside an annotation of a collaborator.

This leads to a very straightforward integration of annotations. On the client side, annotations are transformed with respect to pending and buffered operations, and only sent on once the pending and buffered operations have been sent on. On the server side, an incoming annotation $\langle a, i \rangle$ is transformed and sent on to the other clients just as with editing operations, except that the sending client receives no acknowledgment and the revision number is not increased. Annotations are identified by their origin and an origin-unique name, and remain until overridden by a subsequent annotation of the same origin and name.

4   System Architecture

Building a modern web application has become an increasingly complex task. The demands placed on such an application have grown rapidly over the last couple of years. To offer an adequate experience to users, the interface must always stay responsive. To be able to handle fluctuating numbers of visitors, the system needs to be scalable. In addition to these universal demands, our application involves very expensive computations on the theorem proving side on one hand, and a highly distributed state due to the real time collaboration between users on the other. The first version of Clide [1] was scalable, responsive and resilient, but to properly integrate collaboration, it soon became obvious that most of the architecture had to be carefully rethought.

For the new iteration of Clide, we chose the *Typesafe Reactive Platform* [10] as a basis because it is event-driven and resilient by design, leading to a responsive and scalable application. The platform includes the *Akka* actor library for concurrency control [11], the functional relational mapper *Slick* as an efficient database integration, and the *Play!* web framework. The uses of *Slick* and *Play!* are obvious, but *Akka* became in fact the most important component for Clide, as it is very well suited for a collaborative architecture.

4.1   Universal Collaboration

To reflect the collaborative nature of the application, instead of offering a specialised API for plug-ins, we utilise the same API for human and non-human users. We call this approach *universal collaboration*. The unification has several advantages: On one hand it simplifies the core system itself, on the other hand it also makes it easier to write plug-ins that involve heavy computations (and thus delays). All the management of distributed asynchronous document states is achieved in the operational transformation framework. This way plug-ins can focus on the important aspects — annotating or otherwise contributing content to documents — in a simple, synchronous manner. Moreover, it is easy for plug-ins to work together without knowing anything about their respective implementations. It is not even required for plug-ins to run on the same machine as the server. Neither is it necessary for the server to know anything about the plug-ins a priori; the plug-ins can actively register with the server via a TCP connection. Users can choose to invite a plug-in into a project just like they would with human collaborators.

4.2   The Clide Infrastructure

The Clide infrastructure consists of modules which are loosely coupled, standalone applications whose actor systems communicate with each other via Akka remoting. The modules themselves can easily be further divided and distributed across different machines which leads to great scalability. Fig. 4 shows an example setup with several modules connected to the `clide-core` module. The modules can be configured to connect to a specific address and are implemented in a way that they retry until they are connected to an instance of `clide-core` and reconnect on network failure or in case the peer is restarted. This way it is very easy to configure a Clide infrastructure because only a couple of configuration files have to be adjusted. The modules do not have to be started in any particular order and individual failures do not propagate.

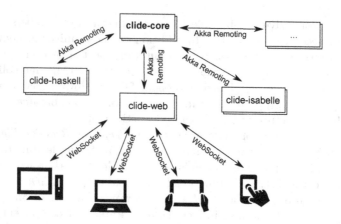

Fig. 4. Overview of the system architecture

The purpose of the `clide-web` module is somewhat special in that it mainly serves as a translator between JSON messages transmitted via WebSockets and the internal message representation. *WebSockets* are a very good fit for the communication between web clients and `clide-web` because their properties are similar to those of message channels connecting actors. WebSockets are full-duplex and thus allow us to send messages from the server to the client without any overhead. `clide-web` directly connects the client with an actor in `clide-core`; that way web clients have the same access to all levels of the API as any other client. The second task of `clide-web` is to provide the resources of the user interface (HTML, CSS and JavaScript files) to the clients. For the user interface we utilised the *angular.js* library which allows for declarative data bindings defined in HTML code. The client side logic and thus also the client side implementation of the operational transformation framework are implemented in CoffeeScript, a language that compiles to JavaScript but compensates for many deficiencies of that language. Because we only used technology from the HTML5 standard, it is possible to use the web interface on any modern, HTML5 compliant web client, i.e. not only on classic computers but also tablets, and given an adapted user interface even smartphones, or television sets.

4.3 The `clide-core` API

Fig. 5 shows a simplified view of the internal actor system with instances of all available types of actors as well as the ownership hierarchy and message flow indicated. The starting point from the outside world is the `UserServer` in the *Global API* which authenticates users, and acts as a message router. It is also possible to sign up a new user here.

Each `UserActor` is responsible for one authenticated user via the *Backstage API*. It allows to create and manage projects and access rights. Peers also get informed about new projects and invitations.

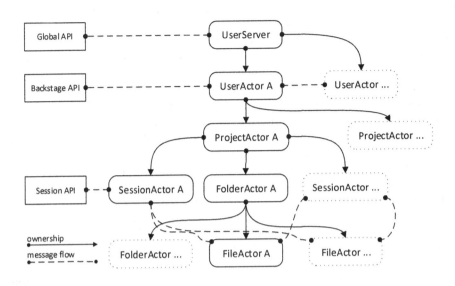

Fig. 5. Actor infrastructure in `clide-core`

Each project is coordinated by a `ProjectActor`. It accepts requests from a `UserActor` to access a project (after checking the rights), and creates a `Session-Actor` for each user accessing the project. The `SessionActor` provides the *Session API* which is the core API in the sense that it provides the editing and annotation operations for one client. The `ProjectActor` coordinates the operations, distributing them to all clients as described in Sect. 3.3. Access to files and folders is given by other dedicated child actors of the `ProjectActor`. Individual `FileActors` manage a single file each and are responsible for the server side operational transformation framework.

4.4   Assistants and Integration with the PIDE Framework

One intriguing aspect of Clide is that collaborators need not be human. We call such non-human collaborators *assistants*, and provide a simplified interface to implement them. It allows easy access to state changes, edit operations and annotations. An assistant only needs to reference the state on which its actions are based, the underlying framework takes care of everything else. This way assistants do not drastically differ from usual plug-ins for IDEs.

There are two ways of building an assistant. The first is to directly use Akka remoting to communicate with the `UserServer` via the messages available in the `clide.actors` package. However, this requires detailed knowledge of the messaging protocol which might still be subject to future modifications and in addition can only be comfortably used from Scala. For this reason there is also a simpler way: Implementing the `AssistantBehavior` interface is all it takes to build an Assistant the easy way. The interface contains a number of abstract methods which will be called as events occur. The implemented interface can then be registered with the generic `AssistantServer` class which will take care of everything else. All method calls within the interface are synchronized. If a calculation takes a long time, subsequent modifications to the document as well as annotation activities (such as cursor movements) will be combined so that the assistant does not lag behind if many changes occur in a short period of time. Communicating back is done via the `AssistantControls` interface which is supplied to the behavior. A short introduction on implementing assistants as well as an API reference can also be found on the GitHub project page.

With this simplified framework, the integration of the PIDE framework [2] (and hence Isabelle) is very straightforward, and a lot of code from Isabelle/jEdit could be reused in the process of implementing the `clide-isabelle` module. When the PIDE framework reports that new information about one of the viewed theories is available, we translate the prover state, type information, inner syntax as well as output messages into Clide annotations and report them back to the assistant framework.

As a case study, we have also implemented an assistant to handle Haskell files (try inviting Haskell as a collaborator on the web site). It calls the Haskell compiler for each source file, parse its error and warning messages and passes them back as annotations. In about two hundred lines of code, we implemented some usable Haskell assistance.

The integration of native processes into a publicly available web interface bear a serious security risk which must be considered when implementing an assistant for Clide. For this reason we start Isabelle in safe mode, disabling all ML integration because that would grant easy access to the server file system.

5   Conclusions

This paper has introduced Clide, a real-time collaborative web interface for the Isabelle theorem prover. It combines modern web technologies with the Isabelle PIDE back-end to offer a user interface which in terms of responsiveness and display of mathematical notation and symbols equals conventional desktop interfaces for Isabelle, such as Isabelle/jEdit or ProofGeneral. The single-user experience resembles Isabelle/jEdit, with the user editing documents which are processed asynchronously on the server in the background, and results appearing as they become available. However, Clide is easier to set up, is mobile, and most of all offers real-time multi-user collaboration. It implements the theory of operational transformations; we have formalised the key concepts in Isabelle, proved its correctness, and derive critical parts of the implementations on the server side from this theory, thus increasing confidence in the system.

5.1   Related Work

There is a lot of related work in the community of computer-supported cooperative work (CSCW) [4,12], including of course the theory of operational transformations [13], but to our knowledge, this is the first fully operational collaborative interface for an interactive theorem prover.

Other web interfaces for theorem provers include Proof Web [14], which is based on older web technologies and hence not as interactive as Clide. There are also a number of mathematical wikis (*e.g.* [15,16,17]) based on interactive provers such as Mizar or Coq, but they resolve collaboration in the typical wiki-style, namely by versioning the edited files or pages; users do not edit the same page at the same time, but instead create different versions.

5.2 Future Work

There are two ways in which this work can be extended. Firstly, there is nothing specific about Isabelle in our framework, except for the fact that the PIDE framework with its Scala API provides a good foundation of our work: it really improves interaction if the theorem prover is multi-threaded and asynchronous, and it helps if everything runs on the same platform. But the system architecture is generic, and could be used to implement a collaborative IDE as well, because all this needs in the first instance is to replace Isabelle with a compiler which analyses the code. There is already a simple assistant for Haskell files; the same approach could be used to integrate batch-based ITPs like HOL4 or HOL-light. Another possible extension would be a 'ProofGeneral module', a generic implementation of script management.

The second main avenue to pursue would be to add more, and richer, assistants. Systems for Proof General such as ML4PG [18], which uses machine learning techniques to help the user find similar proofs, could be adapted to our system, but one could also envisage for example tutoring systems, where the collaborating machine analyses the users errors, and offers helpful advice when certain erroneous patterns occur, or assisted document authoring [19], where for example the collaborating machine suggests an induction scheme based on the current proof state. Apart from actively contributing assistants, there is also a surprising benefit of passively listening collaborators: Clide can be used as a simple API for external applications which can delegate proofs to humans and wait until the theorem prover agrees and then return. The possibilities are endless, and we would like think the generic architecture of our system makes it easy to explore these exciting new avenues of research.

References

1. Lüth, C., Ring, M.: A web interface for Isabelle: The next generation. In: Carette, J., Aspinall, D., Lange, C., Sojka, P., Windsteiger, W. (eds.) CICM 2013. LNCS (LNAI), vol. 7961, pp. 326–329. Springer, Heidelberg (2013)

2. Wenzel, M.: Isabelle/jEdit – A prover IDE within the PIDE framework. In: Jeuring, J., Campbell, J.A., Carette, J., Dos Reis, G., Sojka, P., Wenzel, M., Sorge, V. (eds.) CICM 2012. LNCS (LNAI), vol. 7362, pp. 468–471. Springer, Heidelberg (2012)

3. ProofPeer web page, http://www.proofpeer.net (accessed: January 29, 2014)

4. Dourish, P., Bellotti, V.: Awareness and coordination in shared workspaces. In: Proceedings of the 1992 ACM Conference on Computer-supported Cooperative Work, CSCW 1992, pp. 107–114. ACM (1992)

5. Sun, C., Jia, X., Zhang, Y., Yang, Y., Chen, D.: Achieving convergence, causality preservation, and intention preservation in real-time cooperative editing systems. ACM Trans. Comput.-Hum. Interact. 5(1), 63–108 (1998)

6. Ellis, C.A., Gibbs, S.J.: Concurrency control in groupware systems. SIGMOD Rec. 18(2), 399–407 (1989)

7. Wang, D., Mah, A., Lassen, S.: Google Wave operational transformation, http://www.waveprotocol.org/whitepapers/operational-transform/operational-transform.html (accessed: January 30, 2014)

8. Haftmann, F., Nipkow, T.: A code generator framework for Isabelle/HOL. In: Theorem Proving in Higher Order Logics (TPHOLs 2007), Emerging Trends Proceedings, Dept. of Computer Science, University of Kaiserslautern, pp. 128–143 (2007)

9. Doeraene, S.: Scala.js website, http://www.scala-js.org (accessed: January 29, 2014)

10. Typesafe Inc.: Typesafe reactive platform overview, http://typesafe.com/platform (accessed: January 29, 2014)

11. Wyatt, D.: Akka Concurrency. Artima Press (2013)

12. Greenberg, S., Marwood, D.: Real time groupware as a distributed system: Concurrency control and its effect on the interface. In: Proc. ACM 1994 Conference on Computer Supported Cooperative Work, CSCW 1994, pp. 207–217. ACM (1994)

13. Sun, C., Ellis, C.: Operational transformation in real-time group editors: Issues, algorithms, and achievements. In: Proc. 1998 ACM Conference on Computer Supported Cooperative Work, CSCW 1998, pp. 59–68. ACM (1998)

14. Kaliszyk, C.: Web interfaces for proof assistants. In: Autexier, S., Benzmüller, C. (eds.) Proc. of the Workshop on User Interfaces for Theorem Provers (UITP 2006). ENTCS, vol. 174(2), pp. 49–61 (2007)

15. Urban, J., Alama, J., Rudnicki, P., Geuvers, H.: A wiki for Mizar: Motivation, considerations, and initial prototype. In: Autexier, S., Calmet, J., Delahaye, D., Ion, P.D.F., Rideau, L., Rioboo, R., Sexton, A.P. (eds.) AISC 2010. LNCS, vol. 6167, pp. 455–469. Springer, Heidelberg (2010)

16. Alama, J., Brink, K., Mamane, L., Urban, J.: Large formal wikis: Issues and solutions. In: Davenport, J.H., Farmer, W.M., Urban, J., Rabe, F. (eds.) Calculemus/MKM 2011. LNCS, vol. 6824, pp. 133–148. Springer, Heidelberg (2011)

17. Tankink, C.: Proof in context — web editing with rich, modeless contextual feedback. In: Kaliszyk, C., Lüth, C. (eds.) Proc. 10th International Workshop on User Interfaces for Theorem Provers (UITP 2012). EPTCS, vol. 118, pp. 42–56 (2013)

18. Komendantskaya, E., Heras, J., Grov, G.: Machine learning in Proof General: Interfacing interfaces. In: Kaliszyk, C., Lüth, C. (eds.) Proceedings 10th International Workshop on User Interfaces for Theorem Provers, UITP 2012. EPTCS, vol. 118, pp. 15–41 (2013)

19. Aspinall, D., Lüth, C., Wolff, B.: Assisted proof document authoring. In: Kohlhase, M. (ed.) MKM 2005. LNCS (LNAI), vol. 3863, pp. 65–80. Springer, Heidelberg (2006)

On the Formalization of Z-Transform in HOL

Umair Siddique, Mohamed Yousri Mahmoud, and Sofiène Tahar

Department of Electrical and Computer Engineering,
Concordia University, Montreal, Canada
{muh_sidd,mo_solim,tahar}@ece.concordia.ca

Abstract. System analysis based on difference or recurrence equations is the most fundamental technique to analyze biological, electronic, control and signal processing systems. Z-transform is one of the most popular tool to solve such difference equations. In this paper, we present the formalization of Z-transform to extend the formal linear system analysis capabilities using theorem proving. In particular, we use differential, transcendental and topological theories of multivariate calculus to formally define Z-transform in higher-order logic and reason about the correctness of its properties, such as linearity, time shifting and scaling in z-domain. To illustrate the practical effectiveness of the proposed formalization, we present the formal analysis of an infinite impulse response (IIR) digital signal processing filter.

1 Introduction

In general, dynamics of engineering and physical systems are characterized by differential equations [18] and difference equations [3] in case of continuous-time and discrete-time, respectively. The complexity of these equations varies depending upon the corresponding system architecture (distributed, cascaded, hybrid etc.), nature of input signals and physical constraints. Transformation analysis is one of the most efficient technique to mathematically analyze such complex systems. The main objective of transform method is to reduce complicated system models (i.e., differential or difference equations) into those of algebraic equations. Z-transform [12] provides a mechanism to map discrete-time signals over the complex plane also called z-domain. This transform is a powerful tool to solve linear difference equations (LDE) by transforming them into algebraic operations in z-domain. Moreover, z-domain representation of LDEs is also used for the transfer function analysis of corresponding systems. Due to these distinctive features, Z-transform is one of the main core techniques available in physical and engineering system analysis softwares (e.g., [11,10]) and is widely used in the design and analysis of signal processing filters [12], electronic circuits [3], control systems [4], photonic devices [9] and queueing networks [1].

The main idea of Z-transform can be traced back to Laplace, but it was formally introduced by W. Hurewicz (1947) to solve linear constant coefficient difference equations [7]. Mathematically, Z-transform can be defined as a function

G. Klein and R. Gamboa (Eds.): ITP 2014, LNAI 8558, pp. 483–498, 2014.

series which transforms a discrete time signal $f[n]$ into a function of a complex variable z, as follows:

$$X(z) = \sum_{n=0}^{\infty} f[n]z^{-n} \qquad (1)$$

where $f[n]$ is a complex-valued function ($f : \mathbb{N} \to \mathbb{R}$) and the series is defined for those $z \in \mathbb{C}$ for which the series is convergent.

The first step in analyzing a difference equation (e.g., $x_{n+1} = kx_n(1 - x_n)$) using Z-transform is to apply Z-transform on both sides of a given equation. Next, the corresponding z-domain equation is simplified using various properties of z-transform, such as linearity, scaling and differentiation. The main task is to either solve the difference equation or to find a transfer function which relates the input and output of the corresponding system. Once the transfer function is obtained, it can be used to analyze some important aspects such as stability, frequency response and design optimization to reduce the number of corresponding circuit elements such as multipliers and shift registers.

Traditionally, the analysis of linear systems based on Z-transform has been done using numerical computations and symbolic techniques [11,10]. Both of these approaches, including paper-and-pencil proofs [12] have some known limitations like incompleteness, numerical errors and human-error proneness. In recent years, theorem proving has been actively used for both the formalization of mathematics and the analysis of physical systems. For the latter case, the main task is to identify and formalize the underlying mathematical theories. In practice, four fundamental transformation techniques (i.e., Laplace transform (LT), Z-transform (ZT), Fourier transform (FT), and Discrete Fourier transform (DFT)) are used in the designing and analysis of linear systems. Interestingly, Fourier transform and Discrete Fourier transform can be derived from Laplace transform and Z-transform, respectively. Recently, the formalization of Laplace transform has been reported in [17] using the multivariate analysis libraries of HOL Light [6], with an ultimate goal of reasoning about differential equations and transfer functions of continuous systems. Nowadays, discrete-time linear systems are widely used in the safety and mission critical domains (e.g., digital control of avionics systems and biomedical devices). We believe that there is a dire need of an infrastructure which provides the basis for the formal analysis of discrete-time systems within the sound core of a theorem prover. To the best of our knowledge, so far Z-transform has not been formalized which is an important step towards formal analysis of discrete-time physical and engineering systems.

Our main objective is two-fold: firstly, we aim at extending theorem proving support for linear system analysis. Secondly, we plan to enrich the current foundations of optics formalization [13,15] to reason about futuristic photonic signal processing systems [2,9]. In this paper, we propose Z-transform based system analysis using a higher-order-logic theorem prover. The main idea is to leverage upon the high expressiveness of higher order logic to formalize Equation (1) and use it to verify the classical properties of Z-transform within a theorem prover. These foundations can be built upon to reason about the analytical solutions of difference equations or transfer functions. As a first step towards our ultimate goal, we present in this paper the higher-order logic formalization of Z-transform and its associated region of convergence (ROC). Next, we present the formal verification of its most commonly used properties such as linearity, time delay, time advance and scaling in z-domain. Consequently, we present the formalization of linear constant coefficient difference equation along with the formal verification of its Z-transform by utilizing the above mentioned properties. In order to demonstrate the practical effectiveness of the reported work, we present the formal analysis of an infinite impulse response (IIR) digital signal processing filter.

Formalization reported in this paper has been developed in the HOL Light theorem prover due to its rich multivariate analysis libraries [6]. Another motivation of choosing HOL Light is the existing formalization of Laplace transform and photonic systems which are complementary to achieve our final objective of analyzing linear systems and integrated optics. The source code of our formalization is available for download [14] and can be utilized by other researchers and engineers for further developments and the analysis of more practical systems.

The rest of the paper is organized as follows: Section 2 describes some fundamentals of multivariate analysis libraries of the HOL Light theorem prover. Sections 3 and 4 present our HOL Light formalization of Z-transform and the verification of its properties, respectively. In Section 5, we present the analysis of an IIR filter as illustrative practical application. Finally, Section 6 concludes the paper and highlights some future directions.

2   Preliminaries

In this section, we provide a brief introduction to the HOL Light formalization of some core concepts such as vector summation, summability, complex differentiation and infinite summation [5,6]. Our main intent is to introduce the basic definitions and notations that are going to be used in the rest of the paper.

In the vectors theory formalization, an N-dimensional vector is represented as an $\mathbb{R}^N$ column matrix with individual elements as real numbers. All of the vector operations are then treated as matrix manipulations. Similarly, instead of defining new type, complex numbers ($\mathbb{C}$) can be represented as $\mathbb{R}^2$. Most of the theorems available in multivariate libraries of HOL Light are verified for arbitrary functions with a flexible data-type of ($\mathbb{R}^M \to \mathbb{R}^N$). Next, we present the definitions frequently used in our formalization.

First, generalized summation over arbitrary functions is defined as follows:

Definition 1 (Vector Summation)
⊢ ∀ s f. vsum s f = (lambda i. sum s (λ x. f x$i))

where `vsum` takes two parameters `s : A → bool` which specifies the set over the summation occurs and an arbitrary function `f : (A → ℝ^N)`. The function `sum` is a finite summation over real numbers and accepts `f : (A → ℝ^N)`. For example, $\sum_{i=0}^{K} f(i)$ can be represented as `vsum (0..K) f`.

Next, we present the formal definition of the traditional mathematical expression $\sum_{i=k}^{\infty} f(i) = L$, as follows:

Definition 2 (Sums)
```
⊢ ∀ s f L. (f sums L) s ⇔
            ((λ n. vsum (s ∩ (0..n)) f) → L) sequentially
```

where the types of the parameters are: $(\mathtt{s} : \mathbb{N} \to \mathtt{bool})$, $(\mathtt{f} : \mathbb{N} \to \mathbb{R}^N)$ and $(\mathtt{L} : \mathbb{R}^N)$.

Now, we define the summability of a function $(\mathtt{f} : \mathbb{N} \to \mathbb{R}^N)$, which indeed represents that there exist some $(\mathtt{L} : \mathbb{R}^N)$ such that $\sum_{i=k}^{\infty} f(i) = L$.

Definition 3 (Summability)
```
⊢ ∀ f s. summable s f ⇔ (∃ L. (f sums L) s)
```

The limit of an arbitrary function can be defined as follows:

Definition 4 (Limit)
```
⊢ ∀ f net. lim net f = (@L. (f → L) net)
```

The function lim is defined using the Hilbert choice operator @ in the functional form. It accepts a net with elements of arbitrary data-type A and a function $(\mathtt{f} : \mathtt{A} \to \mathbb{R}^N)$, and returns $(\mathtt{L} : \mathbb{R}^N)$: , i.e., the value to which f converges at the given net. In this paper, we are considering only sequential nets, which describes the sequential evolution of a function, i.e. $f(i), f(i+1), f(i+2), \ldots\ldots$, etc.

Next, we present the definition of an infinite summation which is one the most fundamental requirement in our development.

Definition 5 (Infinite Summation)
```
⊢ ∀ f s. infsum s f = (@L. (f sums L) s)
```

The function infsum is also defined using the Hilbert choice operator @ in the functional form. It accepts a parameter $(\mathtt{s} : \mathtt{num} \to \mathtt{bool})$ which specifies the starting point and a function $(\mathtt{f} : \mathbb{N} \to \mathbb{R}^N)$, and returns $(\mathtt{L} : \mathbb{R}^N)$: , i.e., the value at which infinite summation of f converges from the given s.

In some situations, it is very useful to specify infinite summation as a limit of finite summation (vsum). We proved this equivalence in the following theorem:

Theorem 1 (Infinite Summation in Terms of Sequential Limit)
```
⊢ ∀ s f. infsum s f = lim sequentially (λ k. vsum (s ∩ (0..k)) f))
```

Next, we present the definition of complex differentiation as follows:

Definition 6 (Complex Differentiation)
```
⊢ ∀ f f' net. (f has_complex_derivative f') net ⇔
                (f has_derivative (λx. f' * x)) net
```

The function has_complex_derivative defines the complex derivative in a relational form. Here, $(\mathtt{f} : \mathbb{C} \to \mathbb{C})$ and $\mathtt{f'}:(\mathbb{C})$ represent a given function and the corresponding complex derivative at a given $(\mathtt{net} : (\mathbb{C})\mathtt{net})$, respectively. The function has_derivative is a generalized vector derivative. The above definition can also be described in a functional form as follows:

Definition 7 (Complex Differentiation)
```
⊢ ∀ f x. complex_derivative f x =
        (@f'. (f has_complex_derivative f')) (at x)
```

Note that, the injection from natural numbers to complex numbers can be represented by $\& : \mathbb{N} \to \mathbb{R}$. Similarly, the injection from real to complex numbers is done by $\mathtt{Cx} : \mathbb{R} \to \mathbb{C}$. The real and imaginary parts of a complex number are represented by $\mathtt{Re}$ and $\mathtt{Im}$ both with type $\mathbb{C} \to \mathbb{R}$.

We build upon the above mentioned fundamentals to formalize Z-transform in the next section.

3   Z-Transform Formalization

The unilateral Z-transform [8] of a discrete time function $f[n]$ can be defined as follows:

$$F(z) = \sum_{n=0}^{\infty} f[n]z^{-n} \qquad (2)$$

where f is a function from $\mathbb{N} \to \mathbb{C}$ and z is a complex variable. Here, the definition that we consider has limits of summation from $n = 0$ to $n = \infty$. On the other hand, one can consider these limits from $n = -\infty$ to $n = \infty$ and such a version of Z-transform is called two-sided or bilateral transform. This generalization comes at the cost of some complications such as non-uniqueness, which limits its practicality in engineering systems analysis. On the other hand, unilateral transform can only be applied to *causal* functions, i.e., $f[n] = 0$ for $\forall n.n < 0$. In practice, unilateral Z-transform is sufficient to analyze most of the engineering systems because their designs involve only causal signals [16]. For similar reasons, in [17], the authors formalized the unilateral Laplace transform rather than the bilateral version.

An essential issue of Z-transform of $f[n]$ is whether the $F(z)$ even exists, and under what conditions it exists. It is clear from Equation (2) that Z-transform of a function is an infinite series for each z in the complex plane or z-domain. It is important to distinguish the values of z for which infinite series is convergent and the set of all those values is called the *region of convergence* (ROC). In mathematics and digital signal processing literature, different definitions of ROC are considered. For example, one way is to express z in the polar form ($z = re^{j\omega}$) and then the ROC for $F(z)$ includes only those values of r for which the sequence $f[n]r^{-n}$ is absolutely summable. Unfortunately, to the best of our knowledge, this claim (i.e., absolute summability, e.g., [12,16]) is incorrect for certain functions, for example, $f[n] = \frac{1}{n}u[n-1]$ for which certain values of z result in convergent infinite series, but $x[n]r^{-n}$ is not absolutely summable.

Now, we have two distinct choices for defining ROC: first, z values for which $F(z)$ is finite (or summable) and second, z values for which $x[n]z^{-n}$ is absolutely summable. Most of the textbooks are not rigorous about the choice of ROC and both of these definitions are widely used in the analysis of engineering systems. In this paper, we use the first definition of ROC, which we can define mathematically as follows:

$$ROC = \{z \in \mathbb{C} : \sum_{n=0}^{\infty} f[n]z^{-n} < \infty\} \tag{3}$$

In the above discussion, we mainly highlighted some arbitrary choices of using the definition of Z-transform and its associated ROC. Now, we can formalize Z-transform function (Equation 2) in HOL Light, as follows:

Definition 8 (Z-Transform)
⊢ ∀ f z. z_transform f z = infsum (from 0) (λ n. f n * z⁻ⁿ)

where the z_transform function accepts two parameters: a function $f : \mathbb{N} \to \mathbb{C}$ and a complex variable $z : \mathbb{C}$. It returns a complex number which represents the Z-transform of f according to Equation (2).

Next, we present the formal definition of the ROC as follows:

Definition 9 (Region of Convergence)
⊢ ∀ f. ROC f = {z | summable (from 0) (λ n. f n * z⁻ⁿ)}

Here, ROC accepts a function $f : \mathbb{N} \to \mathbb{C}$ and returns a set of values of variable z for which the Z-transform of $f(n)$ is summable. In order to compute the Z-transform, it is mandatory to specify the associated ROC. Now, we present two basic properties of ROC as follows:

Theorem 2 (ROC Linear Combination)
⊢ ∀ z α β f g. z ∈ ROC f ∧ z ∈ ROC g ⟹
                 z ∈ ROC (λn. α * f n) ∩ ROC (λn. β * g n)

Theorem 3 (ROC Scaling)
⊢ ∀ z α f. z ∈ ROC f ⟹ z ∈ ROC (λn. $\frac{f\,n}{\alpha}$)

where Theorem 2 describes that if z belongs to ROC f and ROC g then it also belongs to the intersection of both ROCs even though the functions f and g are scaled by complex parameters α and β, respectively. Similarly, Theorem 3 shows the scaling with respect to complex division by a complex number α.

4   Z-Transform Properties

In this section, we use Definitions 8 and 9 to formally verify some of the classical properties of Z-transform in HOL Light. The verification of these properties not only ensures the correctness of our definitions but also plays an important role in reducing the time required to analyze practical applications, as described later in Section 5.

4.1   Linearity of Z-Transform

The linearity of the Z-transform is a very useful property while handling systems composed of subsystems with different scaling inputs. Mathematically, it can be defined as:

If $\mathcal{Z}(f[n])\ z = F(z)$ with $ROC = R_f$ and $\mathcal{Z}(g[n])\ z = G(z)$ with $ROC = R_g$, then the following holds:

$$\mathcal{Z}(\alpha * f[n] \pm \beta * g(n))\ z = \alpha * F(z) \pm \beta * G(z) \qquad ROC \supseteq R_f \cap R_g \quad (4)$$

The Z-transform of a linear combination of sequences is the same linear combination of the Z-transforms of the individual sequences. We verify this property as the following theorem:

Theorem 4 (Linearity of Z-Transform)
```
⊢ ∀ z f g α β. z ∈ ROC f ∩ ROC g ⟹
    z_transform (λ n. α * f n + α * g n) z =
            α * z_transform f z + β * z_transform g z
```

where $\alpha : \mathbb{C}$ and $\beta : \mathbb{C}$ are arbitrary constants.

The proof of these theorems are based on the linearity of infinite summation and Theorem 2.

4.2   Shifting Properties

The shifting properties of Z-transform are the most widely used in the analysis of digital systems and in particular in solving difference equations. In fact, there are two kinds of possible shifts: left shift $(f[n + m])$ or time advance and right shift $(f[n - m])$ or time delay. The main idea is to express the transform of the shifted signal $((f[n + m])$ or $(f[n - m]))$ in terms of its Z-transform $(F(Z))$.

Left Shift of a Sequence: If $\mathcal{Z}(f[n])\ z = F(z)$ and m is a positive integer, then the left shift of a sequence can be described as follows:

$$\mathcal{Z}(f[n + m])\ z = z^m F(z) - \sum_{n=0}^{m-1} f[n]z^{-n} \qquad (5)$$

We verify this theorem as follows:

Theorem 5 (Left Shift or Time Advance)
```
⊢ ∀ f z m. z ∈ ROC f ∧ (0 < m) ⟹
    z_transform (λ n. f (n + m)) z =
    z^m * (z_transform f z) - vsum (0..m - 1) (λ n. f n * z^{-n})
```

The verification of this theorem mainly involves properties of complex numbers, summability of shifted functions and splitting an infinite summation into two parts as given by the following lemma:

Lemma 1 (Infsum Splitting)

⊢ ∀ f n m. summable (from m) f ∧ (m < n) ⟹
    infsum (from m) f = vsum (m..n - 1) f + infsum (from n) f

Right Shift of a Sequence: If $\mathcal{Z}(f[n])\, z = F(z)$, and assuming $f(-n) = 0$, $\forall n = 1, 2, .., m$, then the right shift or time delay of a sequence can be described as follows:

$$\mathcal{Z}(f[n-m])\, z = z^{-m} F(z) \tag{6}$$

We formally verify the above property as the following theorem:

Theorem 6 (Right Shift or Time Delay)

⊢ ∀ f z m. z ∈ ROC f ∧ (∀ m. is_causal f m) ⟹
        z_transform (λ n. f (n - m)) z = z⁻ᵐ * (z_transform f z)

Here, is_causal defines the causality of the function f in a relational form to ensure that $f(n-m) = 0$, $\forall n . n < m$. The proof of this theorem also involves properties of complex numbers along with the following two lemmas:

Lemma 2 (Series Negative Offset)

⊢ ∀ f k l. (f sums l) (from 0) ⟹
        ((λ n. f (n - k)) sums l) (from k)

Lemma 3 (Infinite Summation Negative Offset)

⊢ ∀ f k. summable (from 0) f ⟹
        infsum (from 0) (λ n.if k ≤ n then f (n - k) else Cx(&0))
        = infsum (from 0) f

As a direct application of above results, we verify another important property called first-difference, as follows:

Theorem 7 (First Difference)

⊢ ∀ f. z ∈ ROC f ∧ (∀ m. is_causal f m) ⟹
    z_transform (λ n. f (n) - f(n-1)) z = $(1 - z^{-1})$ * (z_transform f z)

4.3   Scaling in Z-Domain

The scaling property of Z-transform plays an important role in the designing of communication systems, such as the response analysis of modulated signals in z-domain. If $\mathcal{Z}(f[n])\, z = F(z)$, then two basic types of scaling can be defined as below:

$$\mathcal{Z}(Z_0^n f[n])\, z = F\left(\frac{z}{Z_0}\right) \tag{7}$$

$$\mathcal{Z}(\omega^{-n} f[n])\, z = F(\omega z) \tag{8}$$

If Z_0 is a positive real number, then it can be interpreted as shrinking or expanding of the z-domain. If Z_0 is a complex with unity magnitude, i.e., $z = e^{j\omega_0}$, then the scaling corresponds to a rotation in the z-plane by an angle of ω_0. Indeed, in communication and signal processing literature, it is interpreted as frequency shift or translation associated with the modulation in the time-domain.

We verify the above theorems in HOL Light as follows:

Theorem 8 (Scaling in z-Domain)
⊢ ∀ f Z_0 z. z_transform (λ n. Z_0^n * f n) z = z_transform f ($\frac{z}{Z_0}$)

Theorem 9 (Scaling in z-Domain (Negative))
⊢ ∀ f ω z. z_transform (λ n. ω^{-n} * f n) z = z_transform f ($\omega * z$)

The verification of above theorems mainly involves the properties of complex power.

4.4 Complex Differentiation

The differentiation property of Z-transform is frequently used together with shifting properties to find the inverse transform. Mathematically, it can be expressed as:

$$\mathcal{Z}(n * f[n])\ z = -z * \left(\sum_{n=0}^{\infty} \frac{d}{dz}(f[n]z^{-n})\right) \tag{9}$$

We prove this property in the following theorem:

Theorem 10 (Complex Differentiation)
⊢ ∀ f z. ≠ Cx(&0) ∧ &0 < Re z ∧ z ∈ (λ n. Cx (&n) * f n)
  ⟹ z_transform (λ n. Cx (&n) * f n) z =
  -z * infsum (from 0) (λ n. complex_derivative (λ z. f n * z^{-n}) z)

The proof of the above theorem requires the properties of complex differentiation, summability and complex arithmetic reasoning.

4.5 Difference Equation

A difference equation characterizes the behavior of a particular phenomena over a period of time. Such equations are widely used to mathematically model complex dynamics of discrete-time systems. Indeed, a difference equation provides a formula to compute the output at a given time, using present and future inputs and past output as given in the following example:

$$y[n] - y[n-1] = \sum_{i=0}^{M} \alpha_i f[n-i] \tag{10}$$

Here, M is called the order of difference equation and α_i represents the list of input coefficients. For a given M^{th} order difference equation in terms of a function $f[n]$, its Z-transform is given as follows:

$$\mathcal{Z}(\sum_{i=0}^{M} \alpha_i f[n-i])\ z = F(z) \sum_{i=0}^{M} \alpha_i z^{-n} \tag{11}$$

We formalize the difference equation as follows:

Definition 10 (Difference Equation)
⊢ ∀ N α_lst f x. difference_eq M α_lst f x =
    vsum (0..M) (λ t. EL t α_lst * f (x - t)* z^{-n}

The function `difference_eq` accepts the order (M) of the difference equation, a list of coefficients α_lst, a causal function f and the variable x. It utilizes the functions `vsum s f` and `EL i L`, which return the vector summation and the ith element of a list L, respectively, to generate the difference equation corresponding to the given parameters.

Next, we verify the Z-transform of the difference equation which is one of the most powerful results of our formalization as will be demonstrated in Section 5.

Theorem 11 (Z-Transform of Difference Equation)
⊢ ∀ M α_lst f x. z ∈ ROC f ∧ z ≠ Cx(&0) ∧
    (∀ m. is_causal f m) ⟹
    z_transform (λx. difference_eq M α_lst f x) z =
    (z_transform f z) * (vsum (0..M) (λ n. EL n α_lst * z^{-n}))

We prove the above theorem by induction and using Theorems 2 and 4 along with the following important lemma about the summability of difference equation:

Lemma 4 (Summability of Difference Equation)
⊢ ∀ M α_lst f x. z ∈ ROC f ∧ (∀ m. is_causal f m)
    ⟹ z ∈ ROC (λx. difference_eq M α_lst f x)

This completes our formalization of the Z-transform and verification of its main properties, which to the best of our knowledge is the first one in higher-order logic. We believe that our formalization can be directly utilized in many applications such as economics, biology, signal processing and control engineering.

5 Application: Formal Analysis of Infinite Impulse Response Filter

In order to illustrate the utilization and effectiveness of the reported formalization, we apply it to analyze a real-world engineering system, i.e., an infinite impulse response filter [12].

Digital filters are fundamental components of almost all signal processing and communication systems. The main functionality of such components are: 1) to limit a signal within a given frequency band; 2) decompose a signal into multiple bands; and 3) model the input-output relation of complicated systems such as mobile communication channels and radar signal processing. The design and analysis of digital filters mainly involves three steps, i.e., the specification of the desired properties of the system, modeling using a causal discrete-time system and realization of overall structure (parallel, cascaded, etc.). Given the filter specifications in terms of frequency response, the first step is to model the filter using constant coefficient difference equations. The next step is to express it in the form of transfer function using the Z-transform properties. Consequently, frequency response analysis and architectural optimization can be performed based on the given specifications.

An impulse response of a system describes its behaviour under an external change (mathematically, this describes the system response when the dirac-delta function is applied as an input [12]). Infinite impulse response (IIR) filters have an impulse response function which is non-zero over an infinite length of time. In practice, IIR filters are implemented using the feedback mechanism, i.e., the present output depends on the present input and all previous input and output samples. Such an architecture requires delay elements due to the discrete nature of input and output signals. The highest delay used in the input and the output function is called the order of the filter.

The time-domain difference equation describing a general M^{th} order IIR filter, with N feed forward stages and M feedback stages, is shown in Figure 1.

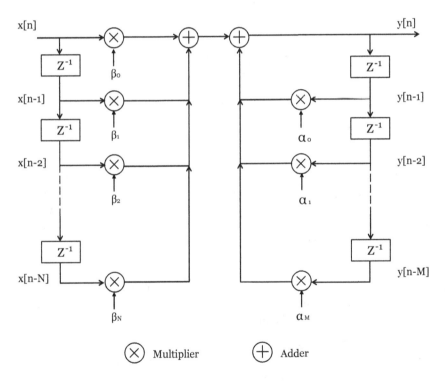

Fig. 1. Generalized Structure of an Mth Order IIR Filter

Mathematically, it can be described as:

$$y[n] = \sum_{i=1}^{M} \alpha_i y[n-i] + \sum_{i=0}^{N} \beta_i x[n-i] \tag{12}$$

where α_i and β_i are input and output coefficients. The output $y[n]$ is a linear combination of the previous N output samples, the present input $x[n]$ and M previous input samples. In case of a time-invariant filter, α_i and β_i are considered constants (either complex ($\mathbb{C}$) or real ($\mathbb{R}$)) to obtain the filter response according to the given specifications.

Our main objective is to formally verify the transfer function and frequency response of an IIR filter which are given as:

$$H(z) = \frac{Y(z)}{X(z)} = \frac{\displaystyle\sum_{i=0}^{N} \beta_i z^{-i}}{1 - \displaystyle\sum_{i=1}^{M} \alpha_i z^{-i}} \tag{13}$$

$$H(\omega) = \frac{\sqrt{(\sum_{i=0}^{N} \beta_i cos(i\omega))^2 + (\sum_{i=0}^{N} \beta_i sin(i\omega))^2}}{\sqrt{(1 - \sum_{i=1}^{M} \alpha_i cos(i\omega))^2 + (\sum_{i=1}^{M} \alpha_i sin(i\omega))^2}} * exp(j * Arg(H(\omega))) \tag{14}$$

where $H(z)$ and $H(\omega)$ represent the filter's transfer function and complex frequency response, respectively. The function $Arg(z)$ represents the argument of a complex number [12]. Equation 14 can be derived from the transfer function $H(z)$ by mapping z on the unit circle, i.e., $z = exp(j * \omega)$. The parameter ω represents the angular frequency.

Based on the above description of the IIR filter, our next move is to conduct its formal analysis, which mainly involves two major steps, i.e., formal description of the model and underlying constraints followed by the formal verification of transfer function and frequency response. As a first step, we build the formal model of the IIR filter using Equation 12.

Definition 11 (IIR Model)
⊢ ∀ x y α_lst β_lst M N n. IIR_MODEL x y α_lst β_lst M N n ⇔
   y n = differen_eq α_lst y M n +
   difference_eq β_lst x N n ∧ HD α_lst = Cx(&0)

The function IIR_MODEL defines the dynamics of the IIR structure in a relational form. It accepts the input and output signals $(x, y : \mathbb{N} \to \mathbb{C})$, a list of input and output coefficients ($\alpha\_lst, \beta\_lst : (\mathbb{C}(list))$), the number of feed forward and feedback stages (N, M) and a variable n, which represents the discrete time.

In order to model $\sum_{i=1}^{M} \alpha_i y[n-i]$ using our definition of difference equation, we added the constraint that the first element (i.e., HD α_lst) of the output coefficients should be 0.

According to the filter specification, we need to ensure that the input and output signals should be causal as described in Section 3. Another important requirement is to ensure that there are no values of z for which denominator is 0, such values are called poles of that transfer function. For the correct operation of the filter, the region of convergence (ROC) should not include any poles. We package these conditions in the following definitions:

Definition 12 (Causality Condition)
⊢ ∀ x y. is_causal_iir x y ⇔
    (∀ k. is_causal x k) ∧ (∀ k. is_causal y k)

Definition 13 (IIR FILTER ROC)
⊢ ∀ x y α_lst M.
    IIR_ROC x y α_lst M =
    z IN (ROC x ∩ ROC y) DIFF
    {z | (Cx(&1) - vsum (1..M) (λ n.EL n α_lst * z^{-n}) = Cx(&0)}

Here, the function is_causal_irr takes two parameters, i.e., input and output, and ensures that both of them are causal. In Definition 13, IIR_ROC specifies the region of convergence of IIR, which is indeed the intersection of ROC x and ROC y, excluding all poles of the transfer function. The function DIFF represents the difference of two sets, i.e., $A \setminus B = \{z : z \in A \wedge x \notin B\}$. Next, we present the formal verification of the transfer function as given in Equation 13.

Theorem 12 (IIR Transfer Function Verification)
⊢ ∀ x y α_lst β_lst M N.
    z ∈ IIR_ROC x y α_lst M ∧
    z ≠ Cx(&0) ∧ is_causal_iir x y ∧
    (∀ n. IIR_MODEL x y α_lst β_lst M N n) ⟹

$$\frac{\text{z\_transform y z}}{\text{z\_transform x z}} = \frac{\text{vsum } (0..N) (\lambda n.\text{EL } n\ \beta\_\text{lst} * z^{-n})}{1 - (\text{vsum } (1..M) (\lambda n.\text{EL } n\ \alpha\_\text{lst} * z^{-n}))}$$

The first and second assumptions describe the region of convergence for the IIR filter. The second assumption ensures the causality of the filter's input and output, and the last assumption gives the time-domain model of the given IIR filter. The proof of this theorem is mainly based on the properties of the Z-transform such as linearity (Theorem 4), time-delay (Theorem 6) and summability of difference equation (Lemma 4). This is a very useful result as it greatly simplifies the reasoning for any given design of IIR. Moreover, this theorem can be used to reason about many important aspects such as stability and architectural optimization. For example, the stability of a given IIR design can be checked by ensuring that all poles of the transfer function lies inside the unit circle (i.e., their magnitude is less than 1).

Next, we verify the frequency response of the filter given in Equation 14 as follows:

Theorem 13 (IIR Frequency Response)

$\vdash \forall$ x y $\alpha\_$lst $\beta\_$lst M N.

    z $\in$ IIR_ROC x y $\alpha\_$lst M $\wedge$

    z = cexp(ii*ω) $\wedge$ is_causal_iir x y $\wedge$

    ($\forall$ n. IIR_MODEL x y $\alpha\_$lst $\beta\_$lst M N n) $\Longrightarrow$

    let H = $\dfrac{\text{z\_transform y z}}{\text{z\_transform x z}}$ and

    num_real = vsum (0..N) (λn.EL n $\beta\_$lst * ccos(n*ω)) and

    num_imag = -vsum (0..N) (λn.EL n $\beta\_$lst * csin(n*ω)) and

    den_real = 1 $-$ (vsum (1..M) (λn.EL n $\alpha\_$lst * ccos(n*ω))) and

    den_imag = vsum (1..M) (λn.EL n $\alpha\_$lst * csin(n*ω)) in

    H = Cx($\dfrac{\text{sqrt}[(\text{num\_real})^2 + (\text{num\_imag})^2]}{\text{sqrt}[(\text{den\_real})^2 + (\text{den\_imag})^2]}$) * cexp(Arg(H))

Where sqrt, cexp and Arg represent the real square root (over reals), complex exponential and argument of a complex number, respectively. The verification of the above theorem is mainly based on Theorem 14 and tedious complex analysis involving complex norms and transcendental functions.

Theorems 12 and 13 provide the generic results due to the universal quantification over the system parameters such as input and output coefficients (α_i and β_k, where $i = 0, 1, 2, \ldots, M$ and $k = 1, 2, \ldots, N$). Next, we utilise these results to formally verify the transfer function and frequency response of a second order low-pass IIR filter as shown in Figure 2. The input and output coefficients are $[0.0605, 0.121, 0.0605]$ and $[1.94, -0.436]$, respectively. We model this structure as follows:

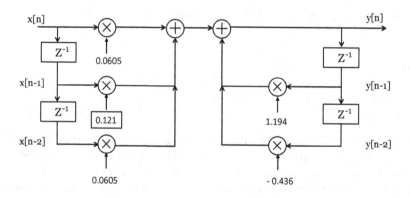

Fig. 2. Second Order Low-Pass IIR Filter

Definition 14 (Second Order IIR Model)

$\vdash \alpha\_$lst = $[\text{Cx}(\&0); \text{Cx}(\frac{\&1194}{\&1000}); -\text{Cx}(\frac{\&436}{\&1000})]$

$\vdash \beta\_$lst = $[\text{Cx}(\frac{\&605}{\&10000}); \text{Cx}(\frac{\&121}{\&1000}); \text{Cx}(\frac{\&605}{\&10000})]$

$\vdash \forall$ x y. SECOND_ORDER_IIR_MODEL x y $\alpha\_$lst $\beta\_$lst $\Leftrightarrow$

    $\forall$ n. y n = differen_eq $\alpha\_$lst y 2 n + difference_eq $\beta\_$lst x 2 n

Here, SECOND_ORDER_IIR_MODEL accepts the input and output signals, a list of input and output coefficients (defined by α_1st, β_1st), and returns the difference equation describing the behaviour of the low-pass IIR filter.

Theorem 14 (Second Order Low-pass IIR Filter Transfer Function)
$\vdash \forall$ x y z. z $\in$ IIR_ROC x y α_1st 2 $\wedge$

   z $\neq$ Cx(&0) $\wedge$ is_causal_iir x y $\wedge$

   (SECOND_ORDER_IIR_MODEL x y α_1st β_1st) $\implies$

$$\frac{\texttt{z\_transform y z}}{\texttt{z\_transform x z}} = \frac{Cx(\frac{\&605}{10000}) + Cx(\frac{\&121}{\&1000}) * z^{-1} + Cx(\frac{\&605}{\&10000}) * z^{-2}}{Cx(\&1) - Cx(\frac{\&1194}{\&1000}) * z^{-1} + Cx(\frac{\&436}{\&1000}) * z^{-2}}$$

The verification of the above theorem is based on Theorem 12.

This completes our formal analysis of a generalized IIR filter which demonstrates the effectiveness of the proposed theorem proving based approach to reason about discrete-time linear systems. The availability of the Z-transform properties greatly simplified the verification of the transfer function and frequency response. The verification of the transfer function and frequency response task took around 150 lines of the HOL Light code and a couple of man-hours each. We believe that reported formalization demonstrates the maturity of interactive theorem provers.

6   Conclusion and Future Directions

In this paper, we reported the formalization of Z-transform which is one of the most widely used transform methods in signal processing and communication theory. We presented the formal definitions of unilateral Z-transform and its associated region of convergence along with the formal verification of some important properties such as linearity, time shifting and difference equations. Finally, in order to demonstrate the effectiveness of the developed formalization, we presented the formal analysis of a generalized infinite impulse repones filter. Consequently, we verified the transfer function and frequency response of a second order low-pass IIR filter.

The utilization of higher-order logic theorem proving in industrial settings (particularly, physical systems) is always questionable due to the huge amount of time required to formalize the underlying theories. Another, important factor is the gap between the theorem proving and engineering communities which limits its usage in industry. For example, it is hard to find engineers with theorem proving background and vice-versa. Our reported work can be considered as a one step towards an ultimate goal of using theorem provers in the design and analysis of systems from different engineering and physical science disciplines (e.g., signal processing, control systems, biology, optical and mechanical engineering).

Our immediate future work is the formalization of the uniqueness theorem of Z-transform [12], which is required to reliably deduce some important properties of difference equations and discrete-time linear systems. The proof of this theorem entails some additional properties of complex differentiation and infinite summations. Another future direction is the formalization of most commonly used inverse transform techniques like power series method, partial fractions and the Cauchy's integral method.

References

1. Alfa, A.S.: Queueing Theory for Telecommunications - Discrete Time Modelling of a Single Node System. Springer (2010)
2. Binh, L.N.: Photonic Signal Processing: Techniques and Applications. Optical Science and Engineering. Taylor & Francis (2010)
3. Elaydi, S.: An Introduction to Difference Equations. Springer (2005)
4. Fadali, S., Visioli, A.: Digital Control Engineering: Analysis and Design. Academic Press (2012)
5. Harrison, J.: Formalizing Basic Complex Analysis. In: From Insight to Proof: Festschrift in Honour of Andrzej Trybulec. Studies in Logic, Grammar and Rhetoric, vol. 10(23), pp. 151–165 (2007)
6. Harrison, J.: The HOL Light Theory of Euclidean Space. Journal of Automated Reasoning 50(2) (2013)
7. Jury, E.I.: Theory and Application of the Z-Transform Method. Wiley (1964)
8. Lathi, B.P.: Linear Systems and Signals. Oxford University Press (2005)
9. Mandal, S., Dasgupta, K., Basak, T.K., Ghosh, S.K.: A Generalized Approach for Modeling and Analysis of Ring-Resonator Performance as Optical Filter. Optics Communications 264(1), 97–104 (2006)
10. Mathematica Guide: Signal Processing Related Functions (2014), http://reference.wolfram.com/mathematica/guide/SignalProcessing.html
11. MathWorks: Signal Processing Toolbox (2014), http://www.mathworks.com/products/signal/
12. Oppenheim, A.V., Schafer, R.W., Buck, J.R.: Discrete-Time Signal Processing. Prentice Hall (1999)
13. Siddique, U., Aravantinos, V., Tahar, S.: Formal Stability Analysis of Optical Resonators. In: Brat, G., Rungta, N., Venet, A. (eds.) NFM 2013. LNCS, vol. 7871, pp. 368–382. Springer, Heidelberg (2013)
14. Siddique, U., Mahmoud, M.Y.: On the Formalization of Z-Transform - HOL Light Script (2014), http://hvg.ece.concordia.ca/projects/signal-processing/z-transform.html
15. Siddique, U., Tahar, S.: Towards the Formal Analysis of Microresonators Based Photonic Systems. In: IEEE/ACM Design Automation and Test in Europe, pp. 1–6 (2014)
16. Sundararajan, D.: A Practical Approach to Signals and Systems. Wiley (2009)
17. Taqdees, S.H., Hasan, O.: Formalization of Laplace Transform Using the Multivariable Calculus Theory of HOL-Light. In: McMillan, K., Middeldorp, A., Voronkov, A. (eds.) LPAR-19. LNCS, vol. 8312, pp. 744–758. Springer, Heidelberg (2013)
18. Yang, X.S.: Mathematical Modeling with Multidisciplinary Applications. John Wiley & Sons (2013)

Universe Polymorphism in Coq

Matthieu Sozeau[1] and Nicolas Tabareau[2]

[1] Inria Paris, PiR2, Univ Paris Diderot, Sorbonne Paris Cité, F-78153 Le Chesnay
[2] Inria Rennes, Laboratoire d'Informatique de Nantes Altantique (LINA)
firstname.surname@inria.fr

Abstract. Universes are used in Type Theory to ensure consistency by checking that definitions are well-stratified according to a certain hierarchy. In the case of the COQ proof assistant, based on the predicative Calculus of Inductive Constructions (pCIC), this hierachy is built from an impredicative sort Prop and an infinite number of predicative $Type_i$ universes. A cumulativity relation represents the inclusion order of universes in the core theory. Originally, universes were thought to be floating levels, and definitions to implicitly constrain these levels in a consistent manner. This works well for most theories, however the globality of levels and constraints precludes *generic* constructions on universes that could work at different levels. Universe polymorphism extends this setup by adding *local* bindings of universes and constraints, supporting generic definitions over universes, reusable at different levels. This provides the same kind of code reuse facilities as ML-style parametric polymorphism. However, the structure and hierarchy of universes is more complex than bare polymorphic type variables. In this paper, we introduce a conservative extension of pCIC supporting universe polymorphism and treating its whole hierarchy. This new design supports typical ambiguity and implicit polymorphic generalization at the same time, keeping it mostly transparent to the user. Benchmarking the implementation as an extension of the COQ proof assistant on real-world examples gives encouraging results.

1 Introduction

Type theories such as the Calculus of Inductive Constructions maintain a universe hierarchy to prevent paradoxes that appear if one is not careful about the sizes of types that are manipulated in the language (e.g. Girard's paradox [1]). To ensure consistency while not troubling the user with this necessary information, systems using typical ambiguity were designed. Typical ambiguity lets users write only anonymous levels (as Type) in the source language, leaving the relationship between different universes to be implicitly inferred by the system. However, the globality of levels and constraints in COQ precludes generic constructions on universes that could work at different levels. Recent developments in homotopy type theory [2] advocate for an extension of the system with universe polymorphic definitions that are parametric on universe levels and instantiated at different

G. Klein and R. Gamboa (Eds.): ITP 2014, LNAI 8558, pp. 499–514, 2014.

ones, just like parametric polymorphism is used to instantiate a definition at different types. This can be interpreted as building fresh instances of the constant that can be handled by the core type checker without polymorphism.

An example of this additional generality is the following. Suppose we define two universes:

```
Definition U2 := Type.
Definition U1 := Type : U2.
```

In the non-polymorphic case but with typical ambiguity, these two definitions are elaborated as $U2 := \text{Type}_u : \text{Type}_{u+1}$ and $U1 := \text{Type}_v : U2$ with a single, global constraint $v < u$.

With polymorphism, U2 can be elaborated as a *polymorphic* constant $U2_u := \text{Type}_u : \text{Type}_{u+1}$ where u is a bound universe variable. A *monomorphic* definition of U1 is elaborated as $U1 := \text{Type}_v : U2_{u'}$ (i.e. $\text{Type}_{u'}$) with a single global constraint $v < u'$ for a fresh u'. In other words, U2's universe is no longer fixed and a fresh level is generated at every occurrence of the constant. Hence, a polymorphic constant can be reused at different, incompatible levels.

Another example is given by the polymorphic identity function, defined as:

$$\text{id}_u := \lambda(A : \text{Type}_u)(a : A), a : \Pi(A : \text{Type}_u), A \to A$$

If we apply id to itself, we elaborate an application:

$$\text{id}_v \ (\Pi(A : \text{Type}_u), A \to A) \ \text{id}_u : \Pi(A : \text{Type}_u), A \to A$$

Type-checking generates a constraint in this case, to ensure that the universe of $\Pi(A : \text{Type}_u), A \to A$, that is $u \sqcup u + 1 = u + 1^1$, is smaller or equal to (the fresh) v. It adds indeed a constraint $u < v$. With a monomorphic id, the generated constraint $u < u$ would raise a universe inconsistency.

In this paper, we present an elaboration from terms using typical ambiguity into explicit terms which also accomodates universe polymorphism, i.e., the ability to write a term once and use it at different universe levels. Elaboration relies on an enhanced type inference algorithm to provide the freedom of typical ambiguity while also supporting polymorphism, in a fashion similar to usual Hindley-Milner polymorphic type inference. This elaboration subsumes the existing universe system of CoQ and has been benchmarked favorably against the previous version. We demonstrate how it provides a solution to a number of formalization issues present in the original system.

To summarize, our contributions are: (i) The design of a universe polymorphic, conservative extension of pCIC, able to handle the whole universe hierarchy and clarifying the inference of universe levels for inductive types. (ii) An elaboration from a source language with typical ambiguity and implicit polymorphism to that core calculus, with a clear specification of the necessary changes in unification. (iii) An efficient implementation of this elaboration as an extension of the CoQ proof assistant. The efficiency claim is backed by benchmarks on real-world developments.

---

[1] $u \sqcup v$ is the least upper bound of two universes.

levels $i, j, le, lt \in \mathbb{N} \cup 0^-$ order $\mathcal{O} ::= \; = \; | < | \leq$

universes u, v $::= i \mid \max(\overrightarrow{le}, \overrightarrow{lt})$ atomic constraint $c ::= i \; \mathcal{O} \; j$

constraints ψ, Θ $::= \epsilon \mid c \wedge \psi$

Fig. 1. Universes

Plan of the Paper. Section 2 introduces the universe structures and a minor variant with constraint checking of the original core pCIC calculus of CoQ. In section 3, we highlight the features of the source language that have to be elaborated and present the enrichment of the core calculus with polymorphic definitions and its conservativity proof. In section 4 we present the elaboration and focus on the subtle issues due to mixing unification and the cumulativity subtyping relation. Finally, we discuss performance benchmarks on real-world examples (§5), and review related and future work (§6).

2   Predicative CIC with Constraint Checking

The current core typechecker implemented in the kernel of CoQ produces a set of universe constraints according to a cumulativity relation while typechecking a term; this set of constraint is then added to the global universe constraints and checked for consistency at the end of typechecking. For a standard presentation of pCIC, see [3]. This design is not suited to support an explicit form of universe polymorphism, where polymorphic definitions carry explicit universe instances.

In this section, we formulate a presentation of the calculus with constraint checking instead of constraint generation. For the purpose of this presentation, we consider a stripped down version of the calculus without Σ and inductive types, however we'll detail how our design extends to these in Section 4.4.

The Core Theory. First we introduce the definitions of the various objects at hand. We start with the term language of a dependent λ-calculus: we have a countable set of variables x, y, z, and the usual typed λ-abstraction $\lambda x : \tau, b$, à la Church, application $t\, u$, the dependent product type $\Pi x : A.B$ and a set of sorts Type_u where u is a universe. Universes are built according to the grammar in Figure 1. We start from the syntactic category of universe *levels* which represent so-called "floating" universe variables i, j, taken from an infinite set of identifiers. Algebraic universes (or universes) are built from levels and formal least upper bound expressions $\max(\overrightarrow{le}, \overrightarrow{lt})$ representing a universe i such that $\overrightarrow{le} \leq i$ and $\overrightarrow{lt} < i$. We identify $\max(i, \epsilon)$ and i. The successor of a universe level i is noted $i + 1$ and encoded as $\max(\epsilon, i)$. For uniformity, we changed the usual set of sorts of CIC to a single sort Type_i and define the propositional sort $\mathrm{Prop} \triangleq \mathrm{Type}_{0^-}$ with the convention that $0^- \leq i$ for all i and $0^- + 1 = 1$. The sort Set is just a synonym for Type_0.

EMPTY

$$\frac{}{\cdot \vdash_\psi}$$

DECL

$$\frac{\Gamma \vdash_\psi T : s \quad x \notin \Gamma}{\Gamma, x : T \vdash_\psi}$$

TYPE

$$\frac{\Gamma \vdash_\psi}{\Gamma \vdash_\psi \text{Type}_i : \text{Type}_{i+1}}$$

VAR

$$\frac{\Gamma \vdash_\psi \quad (x : T) \in \Gamma}{\Gamma \vdash_\psi x : T}$$

APP

$$\frac{\Gamma \vdash_\psi t : \Pi x : A.B \quad \Gamma \vdash_\psi t' : A}{\Gamma \vdash_\psi (t \ t') : B\{t'/x\}}$$

PROD

$$\frac{\Gamma \vdash_\psi A : s \quad \Gamma, x : A \vdash_\psi B : s' \quad (s, s', s'') \in \mathcal{R}}{\Gamma \vdash_\psi \Pi x : A.B : s''}$$

LAM

$$\frac{\Gamma \vdash_\psi \Pi x : A.B : s \quad \Gamma, x : A \vdash_\psi t : B}{\Gamma \vdash_\psi \lambda x : A.t : \Pi x : A.B}$$

CONV

$$\frac{\Gamma \vdash_\psi t : A \quad \Gamma \vdash_\psi B : s \quad A \leq_\psi B}{\Gamma \vdash_\psi t : B}$$

Fig. 2. Typing judgments for $CC_\mathcal{U}$

Universe constraints consist of a conjunction of level (in)equalities which can be seen as a set of atomic constraints. Consistency of a set of constraints is written $\psi \models$, while satisfiability of a constraint is denoted $\psi \models u \, \mathcal{O} \, i$. The kernel only needs to handle constraints of the form $u \, \mathcal{O} \, i$ where i is a level and u an algebraic universe, as shown by Herbelin [4].

Although we do not detail its definition, satisfiability is closed under reflexivity, symmetry and transitivity of $=$, $<$ and $\leq$ where applicable. The satisfiability check is performed using an acyclicity check on the graph generated by the atomic constraints, in an incremental way. Acyclicity ensures that their is an assignment of natural numbers to levels that makes the constraints valid (see [5]). We can hence assume that consistency of constraints is decidable and furthermore that they model a total order:

Lemma 1 (Decidability of Consistency). *If* $C \models$, *then for all levels* i, j *either* $C \wedge i \leq j \models$ *or* $C \wedge j < i \models$.

The typing judgment for this calculus is written $\Gamma \vdash_\psi t : T$ (Fig. 2) where Γ is a context of declarations $\Gamma ::= \cdot \mid \Gamma, x : \tau$, ψ is a set of universe constraints, t and T are terms. If we have a valid derivation of this judgment, we say t has type T in Γ under the constraints ψ. We write $\Gamma \vdash T : s$ as a shorthand for $\Gamma \vdash T : \text{Type}_u$ for some universe u and omit the constraints when they are clear from the context.

The typing rules are standard rules of a PTS with subtyping. The sorting relation $\mathcal{R}$ is defined as:

$$(s, \text{Prop}, \text{Prop}) \qquad \text{impredicative Prop}$$
$$(\text{Type}_u, \text{Type}_v, \text{Type}_{u \sqcup v}) \ \text{predicative Type}$$

We formulate the cumulativity and conversion judgments directly in algorithmic form using a parameterized judgment $T =^R_\psi U$ that can be instantiated with a relation R. Conversion is denoted $T =_\psi U \triangleq T =^=_\psi U$ and cumulativity $T \leq_\psi U \triangleq T =^\leq_\psi U$. The rules related to universes are given in Figure 3. The

R-Type
$$\frac{\psi \models u \; R \; i}{\mathrm{Type}_u =_\psi^R \mathrm{Type}_i}$$

R-Prod
$$\frac{A =_\psi^= A' \qquad B =_\psi^R B'}{\Pi x : A.B =_\psi^R \Pi x : A'.B'}$$

R-Red
$$\frac{A \downarrow_\beta =_\psi^R B \downarrow_\beta}{A =_\psi^R B}$$

Fig. 3. Excerpt of rules for the conversion/cumulativity relation $A =_\psi^R B$

notion of reduction considered here is just the usual β rule, conversion being a congruence for it.

$$(\lambda x : \tau.t) \; u \to_\beta t[u/x]$$

We note $A \downarrow_\beta$ the weak head normal form of A. Note that the R-Red rule applies only if A and B are not in weak head normal form already. The basic metatheory of this system follows straightforwardly. We have validity:

Theorem 1 (Validity). *If $\Gamma \vdash_\psi t : T$ then there exists s such that $\Gamma \vdash_\psi T : s$. If $\Gamma \vdash_\psi$ and $x : T \in \Gamma$ then there exists s such that $\Gamma \vdash_\psi T : s$.*

The subject reduction proof follows the standard proof for ECC[6].

Theorem 2 (Subject Reduction). *If $\Gamma \vdash_\psi t : T$ and $t \to_\beta^* u$ then $\Gamma \vdash_\psi u : T$.*

This system enjoys strong normalization and is relatively consistent to the usual calculus of constructions with universes: we actually have a one-to-one correspondence of derivations between the two systems, only one of them is producing constraints (noted $\Gamma \vdash_{CC} t : T \rhd \psi$) while the other is only checking.

Proposition 1. *Suppose $\psi \models$. If $\Gamma \vdash_{CC} t : T \rhd \psi$ then $\Gamma \vdash_\psi t : T$. If $\Gamma \vdash_\psi t : T$ then $\Gamma \vdash_{CC} t : T \rhd \psi'$ and $\psi \models \psi'$.*

We can freely weaken judgments using larger contexts and constraints.

Proposition 2 (Weakening). *Suppose $\Gamma \vdash_\psi t : T$ and $\psi \models$. If $\Gamma \subset \Delta$ and $\Delta \vdash_\psi$ then $\Delta \vdash_\psi t : T$. If $\psi' \models \psi$ and $\psi' \models$ then $\Gamma \vdash_{\psi'} t : T$.*

It also supports a particular form of substitution principle for universes, as all judgments respect equality of universes. Substitution of universe levels for universe levels is defined in a completely standard way over universes, terms and constraints.

Lemma 2 (Level Substitution). *If $\Gamma \vdash_\psi t : T$, $\psi \models i = j$ then $t =_\psi t[j/i]$.*

3   Predicative CIC with Universe Polymorphic Definitions

To support universe polymorphism in the source language, the core theory defined in Section 2 needs to be extended with the notion of universe polymorphic definition. This section presents this extension and show that it is conservative over the pCIC.

CONSTANT-MONO

$$\frac{\Sigma \vdash_{\Psi}^{d} \qquad \Psi \cup \psi_{c} \models \qquad \Sigma; \cdot \vdash_{\Psi \cup \psi_{c}}^{d} t : \tau \qquad c \notin \Sigma}{\Sigma, (c : \epsilon \vdash t : \tau) \vdash_{\Psi \cup \psi_{c}}^{d}}$$

CONSTANT-POLY

Same Premises

$$\overline{\Sigma, (c : \overrightarrow{i} \vdash_{\psi_{c}}^{d} t : \tau) \vdash_{\Psi}^{d}}$$

Fig. 4. Well-formed global environments

Constant Terms indexed by Universes. Our system is inspired by the design of [7] for the LEGO proof assistant, but we allow arbitrary nesting of polymorphic constants. That is, pCIC is extended by a new term former $c_{\overrightarrow{u}}$ for referring to a constant c defined in a global environment Σ, instantiating its universes at $\overrightarrow{u}$. The typing judgment (denoted $\vdash^{d}$) is made relative to this environment and there is a new introduction rule for constants:

CONSTANT

$$\frac{(c : \overrightarrow{i} \models \psi_{c} \vdash t : \tau) \in \Sigma \qquad \psi \models \psi_{c}\overrightarrow{[u/i]}}{\Sigma; \Gamma \vdash_{\psi}^{d} c_{\overrightarrow{u}} : \tau\overrightarrow{[u/i]}}$$

Universe instances $\overrightarrow{u}$ are simply lists of universe *levels* that instantiate the universes abstracted in definition c. A single constant can hence be instantiated at multiple different levels, giving a form of parametric polymorphism. The constraints associated to these variables are checked against the given constraints for consistency, just as if we were checking the constraints of the instantiated definitions directly. The general principle guiding us is that the use of constants should be *transparent*, in the sense that the system should behave exactly the same when using a constant or its body.

Extending Well-Formedness. Well-formedness of the new global context of constants Σ has to be checked (Fig. 4). As we are adding a global context and want to handle both polymorphic and monomorphic definitions (mentioning global universes), both a global set of constraints Ψ and local universe constraints ψ_{c} for each constant must be handled. When introducing a constant in the global environment, we are given a set of constraints necessary to typecheck the term and its type. In the case of a monomorphic definition (Rule CONSTANT-MONO), we simply check that the local constraints are consistent with the global ones and add them to the global environment. In Rule CONSTANT-POLY, the abstraction of local universes is performed. An additional set of universes $\overrightarrow{i}$ is given, for which the constant is meant to be polymorphic. To support this, the global constraints are not augmented with those of ψ_{c} but are kept locally to the constant definition c. We still check that the union of the global and local constraints is consistent at the point of definition, ensuring that at least one instantiation of the constant can be used in the environment (but not necessarily in all of its extensions).

Extending Conversion. We add a new reduction rule for unfolding constants:

$$c_{\vec{u}} \rightarrow_\delta t[\overrightarrow{u/i}] \qquad \text{when} \qquad (c : \overrightarrow{i} \models \_ \vdash t : \_) \in \Sigma.$$

It is important to notice that conversion must still be a congruence modulo δ. The actual strategy employed in the kernel to check conversion/cumulativity of T and U is to always take the β head normal form of T and U and to do head δ reductions step-by-step (choosing which side to unfold first according to an oracle if necessary), as described by the following rules:

$$\text{R-}\delta\text{-L} \qquad \frac{c_{\overrightarrow{i}} \rightarrow_\delta t \qquad t \; \overrightarrow{a} =^R_\psi u}{c_{\overrightarrow{i}} \; \overrightarrow{a} =^R_\psi u}$$

$$\text{R-}\delta\text{-R} \qquad \frac{c_{\overrightarrow{i}} \rightarrow_\delta u \qquad t =^R_\psi u \; \overrightarrow{a}}{t =^R_\psi c_{\overrightarrow{i}} \; \overrightarrow{a}}$$

This allows to introduce an additional rule for *first-order* unification of constant applications, which poses a number of problems when looking at conversion and unification with universes. The rules for conversion include the following short-cut rule R-FO that avoids unfolding definitions in case both terms start with the same head constant.

$$\text{R-FO} \qquad \frac{\overrightarrow{as} =^R_\psi \overrightarrow{bs}}{c_{\overrightarrow{u}} \; \overrightarrow{as} =^R_\psi c_{\overrightarrow{v}} \; \overrightarrow{bs}}$$

This rule not only has priority over the R-δ rules, but the algorithm *backtracks* on its application if the premise cannot be derived[2].

The question is then, what can be expected on universes? A natural choice is to allow identification if the universe instances are pointwise equal: $\psi \models \overrightarrow{u} = \overrightarrow{v}$. This is certainly a sound choice, if we can show that it does not break the principle of *transparency* of constants. Indeed, due to the cumulativity relation on universes, we might get in a situation where the δ-normal forms of $c_{\overrightarrow{u}} \; \overrightarrow{as}$ and $c_{\overrightarrow{v}} \; \overrightarrow{bs}$ are convertible while $\psi \not\models \overrightarrow{u} = \overrightarrow{v}$. This is where backtracking is useful: if the constraints are not derivable, we backtrack and unfold one of the two sides, ultimately doing conversion on the $\beta\delta$-normal forms if necessary. Note that *equality* of universe instances is forced even if in *cumulativity* mode.

$$\text{R-FO'} \qquad \frac{\overrightarrow{as} =^R_\psi \overrightarrow{bs} \qquad \psi \models \overrightarrow{u} = \overrightarrow{v}}{c_{\overrightarrow{u}} \; \overrightarrow{as} =^R_\psi c_{\overrightarrow{v}} \; \overrightarrow{bs}} \qquad .$$

Conservativity over pCIC. There is a straightforward conservativity result of the calculus with polymorphic definitions over the original one. Below, $T\downarrow^\delta$ denotes the δ-normalization of T, which is terminating as there are no recursive constants. It leaves us with a term with no constants, i.e., a term of pCIC.

---

[2] This might incur an exponential time blowup, nonetheless this is useful in practice.

Theorem 3 (Conservative Extension). $\Sigma; \Gamma \vdash_{\Psi}^{d} t : T \Rightarrow \Gamma{\downarrow}^{\delta} \vdash_{\Psi} t{\downarrow}^{\delta} : T{\downarrow}^{\delta}.$

Proof. The proof goes by mutual induction on the typing, conversion and well-formedness derivations, showing that the three following properties hold:

(1) $\Sigma; \Gamma \vdash_{\Psi}^{d} t : T \Rightarrow \Gamma{\downarrow}^{\delta} \vdash_{\Psi} t{\downarrow}^{\delta} : T{\downarrow}^{\delta}$

(2) $T =_{\Psi}^{R} U \Rightarrow T{\downarrow}^{\delta} =_{\Psi}^{R} U{\downarrow}^{\delta}$

(3) $(c : \overrightarrow{i} \vdash_{\psi_c}^{d} t : \tau) \in \Sigma \Rightarrow$ for all fresh $\overrightarrow{u}$, $\Sigma; \epsilon \vdash_{\psi_c[\overrightarrow{u/i}]} (t[\overrightarrow{u/i}]){\downarrow}^{\delta} : (\tau[\overrightarrow{u/i}]){\downarrow}^{\delta}$ $\qquad\square$

4 Elaboration for Universe Polymorphism

This section presents our elaboration from a source level language with typical ambiguity and universe polymorphism to the conservative extension of the core calculus presented in Section 3.

4.1 Elaboration

Elaboration takes a source level expression and produces a corresponding core term together with its inferred type. In doing so, it might use arbitrary heuristics to fill in the missing parts of the source expression and produce a complete core term. A canonical example of this is the inference of implicit arguments in dependently-typed languages: for example, applications of the id constant defined above do not necessarily need to be annotated with their first argument (the type A at which we want the identity $A \to A$), as it might be inferred from the type of the second argument, or the typing constraint at the point this application occurs. Other examples include the insertion of coercions and the inference of type class dictionaries.

To do so, most elaborations do not go from the source level to the core terms directly, instead they go through an intermediate language that extends the core language with *existential variables*, representing holes to be filled in the term. Existential variables are declared in a context:

$$\Sigma_e ::= \epsilon \mid \Sigma_e \cup (?_n : \Gamma \vdash body : \tau)$$

where *body* is empty or a term t which is then called the *value* of the existential.

In the term, they appear applied to an instance σ of their local context Γ (i.e. an explicit substitution, checked with judgment $\Sigma; \Gamma' \vdash \sigma : \Gamma$), which is written $?_n[\sigma]$. The corresponding typing rule for the intermediate language is:

$$\frac{(?_n : \Gamma \vdash \_ : \tau) \in \Sigma_e \qquad \Sigma_e; \Gamma' \vdash \sigma : \Gamma}{\Sigma_e; \Gamma' \vdash ?_n[\sigma] : \tau[\overrightarrow{\sigma_i/\Gamma(i)}]} \text{EVAR}$$

Elaborating Polymorphic Universes. For polymorphic universes, elaboration keeps track of the new variables, that may be subject to unification, in a *universe context*:

$$\Sigma_u, \Phi ::= \overrightarrow{u_s} \models \mathcal{C}$$

Universe levels are annotated by a flag $s ::= \mathsf{r} \mid \mathsf{f}$ during elaboration, to indicate their rigid or flexible status. Elaboration expands any occurrence of the anonymous Type into a Type_i for a fresh, rigid i and every occurrence of the constant c into a fresh instance c_u ($\overrightarrow{u}$ being all fresh flexible levels). The idea behind this terminology is that rigid universes may not be tampered with during elaboration, they correspond to universes that must appear and possibly be quantified over in the resulting term. The flexible variables, on the other hand, do not appear in the source term and might be instantiated during unification, like existential variables. We will come back to this distinction when we apply minimization to universe contexts. The Σ_u context subsumes the context of constraints Ψ we used during typechecking.

The elaboration judgment is written:

$$\Sigma; \Sigma_e; \Sigma_u; \Gamma \vdash_e t \Leftarrow \tau \rightsquigarrow \Sigma_{e'}; \Sigma_{u'}; \Gamma \vdash t' : \tau$$

It takes the global environment Σ, a set of existentials Σ_e, a universe context Σ_u, a variable context Γ, a source-level term t and a typing constraint τ (in the intermediate language) and produces new existentials and universes along with an (intermediate-level) term whose type is guaranteed to be τ.

Most of the contexts of this judgment are threaded around in the obvious way, so we will not mention them anymore to recover lightweight notations. The important thing to note here is that we work at the intermediate level only, with existential variables, so instead of doing pure conversion we are actually using a unification algorithm when applying the conversion/cumulativity rules.

Typing constraints come from the type annotation (after the :) of a definition, or are inferred from the type of a constant, variable or existential variable declared in the context. If no typing constraint is given, it is generated as a fresh existential variable of type Type_i for a fresh i (i is flexible in that case).

For example, when elaborating an application f t, under a typing constraint τ, we first elaborate the constant f to a term of functional type $f_i : \Pi A : \mathsf{Type}_i.B$, then we elaborate $t \Leftarrow \mathsf{Type}_i \rightsquigarrow t', \Sigma_{u'}$. We check cumulativity $B[t'/A] \leq_{\Sigma_{u'}}$ $\tau \rightsquigarrow \Sigma_{u''}$, generating constraints and returning $\Sigma_{u''} \vdash f_i\ t' : \tau$.

At the end of elaboration, we might apply some more inference to resolve unsolved existential variables. When there are no remaining unsolved existentials, we can simply unfold all existentials to their values in the term and type to produce a well-formed typing derivation of the core calculus, together with its set of universe constraints.

4.2 Unification

Most of the interesting work performed by the elaboration actually happens in the unification algorithm that is used in place of conversion during elaboration.

ELAB-R-TYPE
$$\frac{\psi \cup u \; R \; v \models}{\mathrm{Type}_u \equiv_\psi^R \mathrm{Type}_v \rightsquigarrow \psi \cup u \; R \; v}$$

ELAB-R-PROD
$$\frac{A \equiv_\psi^= A' \rightsquigarrow \psi' \quad B \equiv_\psi^R B' \rightsquigarrow \psi''}{\Pi x : A.B \equiv_\psi^R \Pi x : A'.B' \rightsquigarrow \psi''}$$

ELAB-R-RED
$$\frac{A \downarrow_\beta \equiv_\psi^R B \downarrow_\beta \rightsquigarrow \psi' \qquad A \text{ or } B \text{ not in whnf}}{A \equiv_\psi^R B \rightsquigarrow \psi'}$$

Fig. 5. Conversion/cumulativity inference $\_ \equiv_\_^R \_ \rightsquigarrow \_$

The elaboration rule firing cumulativity is:

SUB
$$\frac{\Sigma; \Sigma_e; \Sigma_u; \Gamma \vdash_e t \rightsquigarrow \Sigma_{e'}; \Sigma_{u'}; \Gamma \vdash t' : \tau' \quad \Sigma_{e'}; \Sigma_{u'} := (\overrightarrow{u_s} \models \psi); \Gamma \vdash \tau' \leq \tau \rightsquigarrow \Sigma_{e''}, \psi'}{\Sigma; \Sigma_e; \Sigma_u; \Gamma \vdash_e t \Leftarrow \tau \rightsquigarrow \Sigma_{e''}; (\overrightarrow{u_s} \models \psi'); \Gamma \vdash t' : \tau}$$

If checking a term t against a typing constraint τ and t is a neutral term (variables, constants and casts), then we infer its type τ' and unify it with the assigned type τ.

In contrast to the conversion judgment $T \leq_\psi U$ which only checks that constraints are implied by ψ, unification and conversion during elaboration (Fig. 5) can additionally *produce* a substitution of existentials and universe constraints, hence we have the judgment $\Sigma_{e'}; \Sigma_{u'} := (\overrightarrow{u_s} \models \psi); \Gamma \vdash T \leq U \rightsquigarrow \Sigma_{e''}, \psi'$ which unifies T and U with subtyping, refining the set of existential variables and universe constraints to $\Sigma_{e''}$ and ψ', so that $T \downarrow^{\Sigma_{e''}} \leq_{\psi'} U \downarrow^{\Sigma_{e''}}$ is derivable[3]. We abbreviate this judgment $T \leq_\psi U \rightsquigarrow \psi'$, the environment of existentials Σ_e, the set of universe variables $\overrightarrow{u_s}$ and the local environment Γ being inessential for our presentation.

The rules related to universes follow the conversion judgment, building up a most general, consistent set of constraints according to the conversion problem. In the algorithm, if we come to a point where the additional constraint would be inconsistent (e.g., in rule ELAB-R-TYPE), we backtrack. For the definition/existential fragment of the intermediate language, things get a bit more involved. Indeed, in general, higher-order unification of terms in the calculus of constructions is undecidable, so we cannot hope for a complete unification algorithm. Barring completeness, we might want to ensure correctness in the sense that a unification problem $t \equiv u$ is solved only if there is a most general unifier σ (a substitution of existentials by terms) such that $t[\sigma] \equiv_{\beta\delta\iota} u[\sigma]$, like the algorithm defined by Abel *et al* [8]. This is however not the case of COQ's unification algorithm, because of the use of the first-order unification heuristic that can return less general unifiers. We now present a generalization of that algorithm to polymorphic universes.

---

[3] $T \downarrow^\Sigma$ denotes the normalization of term T unfolding existentials defined in Σ.

First-Order Unification. Consider unification of polymorphic constants. Suppose we are unifying the same polymorphic constant applied to different universe instances: $c_{\vec{u}} \equiv c_{\vec{v}}$. We would like to avoid having to unfold the constant each time such a unification occurs. What should be the relation on the universe levels then? A simple solution is to force u and v to be equal, as in:

$$\mathrm{id}_j \; \mathrm{Type}_i \equiv \mathrm{id}_m \; ((\lambda A : \mathrm{Type}_l, A) \; \mathrm{Type}_i)$$

The constraints given by typing only are $i < j, l \leq m, i < l$. If we add the constraint $j = m$, then the constraints reduce to $i < m, i < l, l \leq m \Leftrightarrow i < l, l \leq m$. The unification did not add any constraint, so it looks most general. However, if a constant hides an arity, we might be too strict here, for example consider the definition $\mathrm{fib}_{i,j} := \lambda A : \mathrm{Type}_i, A \to \mathrm{Type}_j$ and the unification:

$$\mathrm{fib}_{i,\mathrm{Prop}} \leq \mathrm{fib}_{i',j} \rightsquigarrow i = i' \cup \mathrm{Prop} = j$$

Identifying j and Prop is too restrictive, as unfolding would only add a (trivial) constraint $\mathrm{Prop} \leq j$. The issue also comes up with universes that appear equivariantly. Unifying $\mathrm{id}_i \; t \equiv \mathrm{id}_{i'} \; t'$ should succeed as soon as $t \equiv t'$, as the normal forms $\mathrm{id}_i \; t \to_{\beta\delta}^* t$ and $\mathrm{id}_{i'} \; t' \to_{\beta\delta}^* t'$ are convertible, but i does not have to be equated with i', again due to cumulativity. To ensure that we make the least commitment and generate most general constraints, there are two options. Either we find a static analysis that tells us for each constant which constraints are to be generated for a self-unification with different instances, or we do without that information and restrict ourselves to unifications that add no constraints.

The first option amounts to decide for each universe variable appearing in a term, if it appears at least once only in rigid covariant position (the term is an arity and the universe appears only in its conclusion), in which case adding an inequality between the two instances would reflect exactly the result of unification on the expansions. In general this is expensive as it involves computing (head)-normal forms. Indeed consider the definition $\mathrm{idtype}_{i,k} := \mathrm{id}_k \; \mathrm{Type}_k \; \mathrm{Type}_i$, with associated constraint $i < k$. Deciding that i is used covariantly here requires to take the head normal form of the application, which reduces to Type_i itself. Recursively, this Type_i might come from another substitution, and deciding covariance would amount to do $\beta\delta$-normalization, which defeats the purpose of having definitions in the first place!

The second option—the one that has been implemented—is to restrict first-order unification to avoid arbitrary choices as much as possible. To do so, unification of constant applications is allowed only when their universe instances are themselves unifiable in a restricted sense. The inference rules related to constants are given in Figure 6. The judgment $\psi \models i \equiv j \rightsquigarrow \psi'$ (figure 7) formalizes the unification of universe instances. If the universe levels are already equal according to the constraints, unification succeeds (ELAB-UNIV-EQ). Otherwise, we allow identifying universes if at least one of them is flexible. This might lead to overly restrictive constraints on fresh universes, but this is the price to pay for automatic inference of universe instances.

ELAB-R-FO
$$\dfrac{\overrightarrow{as} \equiv^{=}_{\psi} \overrightarrow{bs} \leadsto \psi' \qquad \psi' \models \overrightarrow{u} \equiv \overrightarrow{v} \leadsto \psi''}{c_{\overrightarrow{u}} \ \overrightarrow{as} \equiv^{R}_{\psi} c_{\overrightarrow{v}} \ \overrightarrow{bs} \leadsto \psi'}$$

ELAB-R-δ-L
$$\dfrac{c_{\overrightarrow{i}} \to_{\delta} t \qquad t \ \overrightarrow{a} \equiv^{R}_{\psi} u \leadsto \psi'}{c_{\overrightarrow{i}} \ \overrightarrow{a} \equiv^{R}_{\psi} u \leadsto \psi'}$$

ELAB-R-δ-R
$$\dfrac{c_{\overrightarrow{i}} \to_{\delta} u \qquad t \equiv^{R}_{\psi} u \ \overrightarrow{a} \leadsto \psi'}{t \equiv^{R}_{\psi} c_{\overrightarrow{i}} \ \overrightarrow{a} \leadsto \psi'}$$

Fig. 6. Unification and constants

ELAB-UNIV-EQ
$$\dfrac{\psi \models i = j}{\psi \models i \equiv j \leadsto \psi}$$

ELAB-UNIV-FLEXIBLE
$$\dfrac{i_f \vee j_f \in \overrightarrow{u_s} \qquad \psi \wedge i = j \models}{\psi \models i \equiv j \leadsto \psi \wedge i = j}$$

Fig. 7. Unification of universe instances

This way of separating the rigid and flexible universe variables allows to do a kind of local type inference [9], restricted to the flexible universes. Elaboration does not generate the most general constraints, but heuristically tries to instantiate the flexible universe variables to sensible values that make the term type-check. Resorting to explicit universes would alleviate this problem by letting the user be completely explicit, if necessary. As explicitly manipulated universes are rigid, the heuristic part of inference does not apply to them. In all practical cases we encountered, no explicitation was needed though.

4.3   Abstraction and Simplification of Constraints

After computing the set of constraints resulting from type-checking a term, we get a set of universe constraints referring to *undefined*, flexible universe variables as well as global, rigid universe variables. The set of flexible variables can grow very quickly and keeping them along with their constraints would result in overly general and unmanageable terms. Hence we heuristically simplify the constraints by instantiating undefined variables to their most precise levels. Again, this might only endanger generality, not consistency. In particular, for level variables that appear only in types of parameters of a definition (a very common case), this does not change anything. Consider for example: id_u Prop True : Prop with constraint Prop $\leq u$. Clearly, identifying u with Prop does not change the type of the application, nor the normal form of the term, hence it is harmless.

We work under the restriction that some undefined variables can be substituted by algebraic universes while others cannot, as they appear in the term as explained in section 3. We also categorize variables according to their global or local status. Global variables are the ones declared through monomorphic definitions in the global universe context Ψ.

Simplification of constraints works in two steps. We first normalize the constraints and then minimize them.

Normalization. Variables are partitioned according to equality constraints. This is a simple application of the Union-Find algorithm. We canonicalize the constraints to be left with only inequality ($<, \leq$) constraints between distinct universes. There is a subtlety here, due to the global/local and rigid/flexible distinctions of variables. We choose the canonical element k in each equivalence class C to be global if possible, if not rigid, and build a canonizing substitution of the form $\overrightarrow{u/k}, u \in C \setminus k$ that is applied to the remaining constraints. We also remove the substituted variables from the flexible set θ.

Minimization. For each remaining flexible variable u, we compute its instance as a combination of the least upper bound (l.u.b.) of the universes below it and the constraints above it. This is done using a recursive, memoized algorithm, denoted lub u, that incrementally builds a substitution σ from levels to universes and a new set of constraints. We start with a consistent set of constraints, which contains no cycle, and rely on this for termination. We can hence start the computation with an arbitrary undefined variable.

We first compute the set of direct lower constraints involving the variable, recursively:

$$\mathsf{L}_u \triangleq \{(\mathsf{lub}\ l, R, u) \mid (l, R, u) \in \Psi\}$$

If L_u is empty, we directly return u. Otherwise, the l.u.b. of the lower universes is computed as:

$$\sqcup_u \triangleq \{x \mid (x, \mathsf{Le}, \_) \in \mathsf{L}_u\} \sqcup \{x + 1 \mid (x, \mathsf{Lt}, \_) \in \mathsf{L}_u\}$$

The l.u.b. represents the minimal level of u, and we can lower u to it. It does not affect the satisfiability of constraints, but it can make them more restrictive. If $\sqcup_u$ is a level j, we update the constraints by setting $u = j$ in Ψ and σ. Otherwise, we check if $\sqcup_u$ has been recorded as the l.u.b. of another flexible universe j in σ, in which case we also set $u = j$ in Ψ and σ. This might seem dangerous if j had different upper constraints than u. However, if j has been set equal to its l.u.b. then by definition $j = \sqcup_u \leq u$ is valid. Otherwise we only remember the equality $u = \sqcup_u$ in σ, leaving Ψ unchanged. The computation continues until we have computed the lower bounds of all variables.

This procedure gives us a substitution σ of the undefined universe variables by (potentially algebraic) universes and a new set of constraints. We then turn the substitution into a well-formed one according to the algebraic status of each undefined variable. If a substituted variable is not algebraic and the substitutend is algebraic or an algebraic level, we remove the pair from the substitution and instead add a constraint of the form $\mathsf{max}(\ldots,\ldots) \leq u$ to Ψ. This ensures that only algebraic universe variables are instantiated with algebraic universes. In the end we get a substitution σ from levels to universes to be applied to the term under consideration and a universe context $\overrightarrow{us'} \models \Psi[\sigma]$ containing the variables that have not been substituted and an associated set of constraints $\Psi[\sigma]$ that are sufficient to typecheck the substituted term. We directly give that information to the kernel, which checks that the constraints are consistent with the global ones and that the term is well-typed.

4.4 Inductive Types

Polymorphic inductive types and their constructors are treated in much the same way as constants. Each occurrence of an inductive or constructor comes with a universe instance used to typecheck them. Conversion and unification for them forces equality of the instances, as there is no unfolding behavior to account for. This setup implies that unification of a polymorphic inductive type instantiated at the same parameters but in two different universes will force their identification, i.e., list_i True $=$ $\mathsf{list}_{\mathrm{Prop}}$ True will force $i = \mathrm{Prop}$, even though i might be strictly higher (in which case it would be inconsistent). These conversions mainly happen when mixing polymorphic and monomorphic code though, and can always be avoided with explicit uses of Type to raise the lowest level using cumulativity. Conservativity over a calculus with monomorphic inductives carries over straightforwardly by making copies of the inductive type.

Inference of Levels. Computing the universe levels of inductive types in the new system is much cleaner than in the previous version, and also simplifies the work done in the kernel. Indeed, the kernel gets a declaration of each inductive type as a record containing the arity and the type of each constructor along with a universe context. Checking the correctness of the level works in two steps. First we compute the *natural* level of the inductive, that is the l.u.b. of the levels of its constructors (optionally considering parameters as well). If the inductive has more than one constructor, we also take the l.u.b with Type_0 because the inductive is then naturally a datastructure in Set. We then take the user-given level and compare it to this natural level. If it is larger or equal then we are done. If the natural level is strictly higher than the user-given level (typically an inductive with multiple constructors declared in Prop), then *squashing* happens:

- If the user-given level is Prop, then we simply restrict eliminations on the inductive to be compatible with the impredicative, proof-irrelevant characterization of Prop.
- If it is Set, then we allow the definition only if in impredicative Set mode: the natural level had to include a large type level. In all other cases, the definition is rejected.

This new way of computing levels results in a clarification of the kernel code.

5   Implementation and Benchmarks

This extension of CoQ (http://github.com/mattam82/coq) will be available in the next major release and supports the formalization of the Homotopy Type

Theory library (HoTT/Coq) from the Univalent Foundations project. It is able to check for example Voevodsky's proof that Univalence implies Functional Extensionality. At the user-level, there is just an additional flag `Polymorphic` that can be used to flag polymorphic definitions. Moving from universe inference to universe checking and adding universe instances on constants required important changes in the tactic and elaboration subsystems to properly keep track of universes. If all definitions are monomorphic, the change is unnoticeable to the user though. As minimization happens as part of elaboration, it sits outside the kernel and does not have to be trusted. There is a performance penalty to the use of polymorphism, which is at least linear in the number of fresh universe variables produced during a proof. On the standard library of CoQ, with all primitive types made polymorphic, we can see a mean 10% increase in time. The main issues come from the redundant annotations on the constructors of polymorphic inductive types (e.g., list) which could be solved by representing type-checked terms using bidirectional judgments, and the choice of the concrete representation of universe constraints during elaboration, which could be improved by working directly on the graph.

6   Related and Future Work

We have introduced a conservative extensions of the predicative calculus of inductive constructions with universe polymorphic definitions. This extension of the system enhances the usability of the original system while retaining its performance and clarifying its implementation.

Other designs for working with universes have been developed in systems based on Martin-Löf type theory. As mentionned previously, we build on the work of Harper and Pollack [7] who were the first to study the addition of universe polymorphic definitions (albeit restricted to no nested definitions, and without considering minimization), and implemented it in LEGO, a system with cumulativity and typical ambiguity.

The Agda programming language provides fully explicit universe polymorphism, making level quantification first-class. This requires explicit quantification and lifting of universes (no cumulativity), but instantiation can often be handled solely by unification. The main difficulty in this setting is that explicit levels can obfuscate definitions and make development and debugging arduous. The system is as expressive as the one presented here though.

The Matita proof assistant based on CIC lets users declare universes and constraints explicitly [10] and its kernel only checks that user-given constraints are sufficient to typecheck terms. It has a notion of polymorphism at the library level only: one can explicitly make copies of a module with fresh universes. In [11], a similar extension of the CoQ system is proposed, with user declarations and a notion of polymorphism at the module level. Our elaboration system could be adapted to handle these modes of use, by restricting inference and allowing users to explicitly declare universes and constraints. This would solve the modularity issues mentionned by Courant, which currently complicate the separate compilation and loading of modules which might have incompatible uses of the same universes. We leave this as future work.

In this study, the impredicative sort Prop is considered a subtype of Type_i by cumulativity. However, this is problematic because we lose precision when typing polymorphic products, e.g., for $empty_i := \forall A : \text{Type}_i, A$. The $empty_i$ definition has type Type_{i+1}, however the instance $empty_{\text{Prop}}$ should be a Prop itself by impredicativity. To handle this, we would need to have conditional constraints that would allow lowering a universe if some other was instantiated with Prop, which would significantly complicate the system. Furthermore, this $\text{Prop} \subset \text{Type}$ rule causes problems for building proof-irrelevant models (e.g., it does not hold in [12], and it is still an open problem to build such a model in a system without judgemental equality, like ECC) and, according to the Homotopy Type Theory interpretation, has computational content. We did not address this issue in this work, but we plan to investigate a new version of the core calculus where this rule would be witnessed as an *explicit* coercion as in [13].

References

1. Coquand, T.: An analysis of Girard's Paradox. In: Proceedings of the First Symposium on Logic in Computer Science, HC, vol. 145. IEEE Comp. Soc. Press (June 1986)
2. The Univalent Foundations Program: Homotopy Type Theory: Univalent Foundations for Mathematics, Institute for Advanced Study (2013)
3. The Coq development team: Coq 8.2 Reference Manual. INRIA (2008)
4. Herbelin, H.: Type Inference with Algebraic Universes in the Calculus of Inductive Constructions (2005) (manuscript)
5. Chan, T.H.: Appendix D. In: An Introduction to the PL/CV2 Programming Logic. LNCS, vol. 135, pp. 227–264. Springer (1982), http://dx.doi.org/10.1007/3-540-11492-0
6. Luo, Z.: An Extended Calculus of Constructions. PhD thesis, Department of Computer Science, University of Edinburgh (June 1990)
7. Harper, R., Pollack, R.: Type Checking with Universes. Theor. Comput. Sci. 89(1), 107–136 (1991)
8. Abel, A., Pientka, B.: Higher-Order Dynamic Pattern Unification for Dependent Types and Records. In: Ong, L. (ed.) TLCA 2001. LNCS, vol. 6690, pp. 10–26. Springer, Heidelberg (2011)
9. Pierce, B.C., Turner, D.N.: Local Type Inference. ACM Transactions on Programming Languages and Systems 22(1), 1–44 (2000)
10. Asperti, A., Ricciotti, W., Coen, C.S., Tassi, E.: A compact kernel for the calculus of inductive constructions. Journal Sadhana 34, 71–144 (2009)
11. Courant, J.: Explicit Universes for the Calculus of Constructions. In: Carreño, V.A., Muñoz, C.A., Tahar, S. (eds.) TPHOLs 2002. LNCS, vol. 2410, pp. 115–130. Springer, Heidelberg (2002)
12. Lee, G., Werner, B.: Proof-irrelevant model of CC with predicative induction and judgmental equality. Logical Methods in Computer Science 7(4) (2011)
13. Herbelin, H., Spiwack, A.: The Rooster and the Syntactic Bracket. CoRR abs/1309.5767 (2013)

Asynchronous User Interaction
and Tool Integration in Isabelle/PIDE

Makarius Wenzel*

Univ. Paris-Sud, Laboratoire LRI, UMR8623, Orsay, F-91405, France
CNRS, Orsay, F-91405, France

Abstract. Historically, the LCF tradition of interactive theorem proving was tied to the read-eval-print loop, with sequential and synchronous evaluation of prover commands given on the command-line. This user-interface technology was adequate when R. Milner introduced his LCF proof assistant in the 1970-ies, but it severely limits the potential of current multicore hardware and advanced IDE front-ends.

Isabelle/PIDE breaks this loop and retrofits the read-eval-print phases into an asynchronous model of document-oriented proof processing. Instead of feeding a sequence of individual commands into the prover process, the primary interface works via edits over a family of document versions. Execution is implicit and managed by the prover on its own account in a timeless and stateless manner. Various aspects of interactive proof checking are scheduled according to requirements determined by the front-end perspective on the proof document, while making adequate use of the CPU resources on multicore hardware on the back-end.

Recent refinements of Isabelle/PIDE provide an explicit concept of asynchronous print functions over existing proof states. This allows to integrate long-running or potentially non-terminating tools into the document-model. Applications range from traditional proof state output (which may consume substantial time in interactive development) to automated provers and dis-provers that report on existing proof document content (e.g. Sledgehammer, Nitpick, Quickcheck in Isabelle/HOL). Moreover, it is possible to integrate query operations via additional GUI panels with separate input and output (e.g. for Sledgehammer or find-theorems). Thus the Prover IDE provides continuous proof processing, augmented by add-on tools that help the user to continue writing proofs.

1   Introduction

Already 10 years ago, multicore hardware has invaded the consumer market, and imposed an ever increasing burden on application developers to keep up with changed rules for *Moore's Law*: continued speedup is no longer for free, but has to be implemented in the application by explicit multi-processing. Isabelle has started to support parallel proof-processing in *batch-mode* already in 2006/2007, and is today routinely using multiple cores, with an absolute speedup factor of the order of 10 (on 16 cores). See also [17] for the situation of Isabelle2013.

* Research supported by Project Paral-ITP (ANR-11-INSE-001).

G. Klein and R. Gamboa (Eds.): ITP 2014, LNAI 8558, pp. 515–530, 2014.
© Springer International Publishing Switzerland 2014

How does parallel processing affect user interaction? After the initial success of parallel batch-mode in Isabelle, it became clear in 2008 that substantial reforms are required in the interaction model, to loosen the brakes that are built into the traditional *read-eval-print* loop. The following aspects are characteristic to asynchronous interaction, in contrast to parallel batch processing:

- demand for real-time reactivity (at the order of 10–100 ms);
- instantaneous rendering of formal content in GUI components;
- continued edits of theory sources, while the prover is processing them;
- treatment of unfinished or failed proof attempts (error recovery);
- cancellation of earlier attempts that have become irrelevant (interrupts);
- orchestration of add-on proof tools that help in the editing process.

The present paper reports on results of more than 5 years towards asynchronous prover interaction, with recent improvements that integrate add-on proof tools via *asynchronous print functions*. All concepts are implemented in the current Isabelle/PIDE generation of Isabelle2013-2 (December 2013)[1]. The PIDE framework is implemented as a combination of Isabelle/ML and Isabelle/Scala, with Isabelle/jEdit the main application and default user-interface. The manual [14] provides further explanations and screenshots; the *Documentation* panel in Isabelle/jEdit includes some examples that help to get started.

The front-end technology of Isabelle/jEdit imitates the classic IDE approach seen in Eclipse, NetBeans, IntelliJ IDEA, MS Visual Studio etc. Fresh users of Isabelle who are familiar with such mainstream IDEs usually manage to get acquainted quickly, without learning about Emacs and the TTY loop first. In contrast, seasoned users of ITP systems may have to spend some efforts to *unlearn* TTY mode and manual scheduling of proof commands.

Subsequently, we assume some basic acquaintance with the look-and-feel of Isabelle/jEdit, but explanations of PIDE concepts are meant to extrapolate beyond this particular combination of prover back-end and editor front-end. Since Isabelle had similar starting conditions as other proof assistants several years ago, like Coq [18, §4], the HOL family [18, §1], PVS [18, §3], or ACL2 [18, §8], there are no fundamental reasons why such seemingly drastic steps from the TTY loop to proper IDE interaction cannot be repeated elsewhere. These explanations of PIDE concepts are meant to help other systems to catch up, although the level of sophistication in Isabelle/PIDE today poses some challenges.

2 PIDE Architecture

PIDE stands for "Prover IDE": it is the common label for efforts towards advanced user-interaction in Isabelle since 2009. The main application of the PIDE framework today is Isabelle/jEdit [13, 14], but there are already some alternative front-ends: Isabelle/Eclipse by A. Velykis, and Clide by C. Lüth and M. Ring [7, 8]. An experiment to connect Coq as alternative back-end is reported in [15].

---

[1] `http://isabelle.in.tum.de/website-Isabelle2013-2`

The general aims of PIDE are to renovate and reform interactive theorem proving for new generations of users, and to catch up with technological shifts (multicore hardware). The PIDE approach is *document-oriented*: all operations by the user, the editor, the prover, and add-on tools are centered around theory sources that are augmented by formal markup produced by proof processing. Document markup for old-school proof assistants is further explained in [11].

The Connectivity Problem. Proof assistants are typically implemented in functional programming languages (like LISP, SML, OCaml, Haskell) that are not immediately connected to the outer world. If built-in interface technology exists, it is typically limited in scope and functionality: e.g. LablGtk for OCaml uses old GTK 2.x instead of GTK 3.x, and GTK is at home only on Linux.

Even if we could assume the ideal multi-platform GUI framework within our prover programming environment, what we really need is a viable text editor or IDE to work with. The Java platform is able to deliver that, e.g. with text editors like jEdit, full IDEs like Eclipse, NetBeans, IntelliJ IDEA, and web frameworks like Play (for remote applications). This observation has lead to the following *bilingual approach* of PIDE with Scala and ML (figure 1).

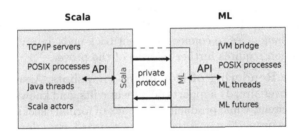

Fig. 1. The bilingual approach of PIDE: Scala and ML connected via private protocol

Here the existing ML prover platform is taken for granted, but its scope extended into the JVM world with the help of Scala [9]. The manner and style of strongly-typed higher-order functional programming in ML is continued with Scala. Both sides happen to provide some tools and libraries for parallel programming with threads, processes, external communication, which serve as starting point for further PIDE functionality. A *private protocol* connects the two worlds: it consists of two independent streams of protocol operations that are a-priori unsynchronized. The conceptual *document-model* that is implemented on both sides is accessible by some *public APIs*, both in Scala and ML.

It is an important PIDE principle to cut software components at these APIs, and not the process boundaries with the protocol. API functions in ML or Scala are statically typed and more abstract than the communication messages of the implementation. APIs are more stable under continuous evolution than a public protocol. The combined Scala and ML sources of Isabelle/PIDE are maintained side-by-side within the same code repository: e.g. `src/Pure/General/pretty.scala` and `src/Pure/General/pretty.ML` for classic pretty-printing in

the style of D.C. Oppen (with support for document markup and font-metrics).
Tools using the PIDE infrastructure may reside in ML (e.g. proof tools that
output document markup), or in Scala (e.g. rendering for particular document
content), or combine both worlds.

PIDE Protocol Layers. Conceptually, the two processes are connected by two
independent streams of *protocol functions*. These streams are essentially symmetric, but input from the editor to the prover is called *protocol command*, and
output from the prover to the editor is called *protocol message*. Syntactically,
a protocol function consists of a name and argument list (arbitrary strings).
Semantically, the stream of protocol functions is applied consecutively to a private *protocol state* on each side; there are extensible tables in Isabelle/Scala and
Isabelle/ML to define the meaning for protocol functions.

The arguments of protocol functions usually consist of algebraic datatypes
(tuples and recursive variants). This well-known ML concept is represented in
Scala by case classes [9, §7.2]. The PIDE implementation starts out with raw
byte streams between the processes, then uses YXML transfer syntax for untyped
XML trees [11, §2.3], and finally adds structured XML/ML data representation
via some combinator library. Further details are explained in [15], including
a full implementation on a few pages of OCaml; the Standard ML version is
part of Isabelle/PIDE. This elementary PIDE protocol stack is easily ported to
other functional languages to connect different back-ends, but actual document-
oriented interaction requires further reforms of the prover.

Approximative Rendering of Document-Snapshots. Assume for the moment that the prover already supports document edits, and knows how to process
partial theory content, while producing feedback of formal checking via messages
(plain text output or markup over the original sources). How does the editor render that continuous flow of information in its single physical instance of the GUI,
without getting blocked by the prover?

The classic approach of Proof General [3] makes a tight loop around each
prover command, and synchronizes a full protocol round-trip for each transaction. This often leads to situations where the editor is non-reactive, not to speak
of the "locked region" of processed text where the user is not allowed to edit.

PIDE avoids blocking by a notion of *document snapshot* and convergence of
content, instead of synchronization. See also figure 2.

The editor and the prover are independent processes that exchange information monotonically: each side uses its present knowledge to proceed, and propagates results to its counterpart. The front-end ultimately needs to render editor
buffers (painting text with colors, squiggly underlines etc.) by interpreting the
source text with its accumulated markup. The flow of information is as follows:

1. editor knows text T, markup M, and edits ΔT (produced by user)
2. apply edits: $T' = T + \Delta T$ (*immediately* in the editor)
3. formal processing of T': ΔM after time Δt (*eventually* in the prover)
4. immediate approximation: $\tilde{M} = revert\ \Delta T$; retrieve M; convert ΔT
5. eventual convergence after time Δt: $M' = M + \Delta M$

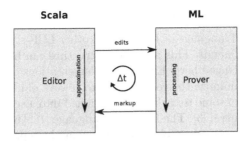

Fig. 2. Approximation and convergence of markup produced by proof processing

This means the editor is streaming edits as document updates towards the prover, which processes them eventually to give feedback via semantic markup. Without waiting for the prover, the document snapshot of the editor uses the edit-distance over the text to stretch or shrink an old version of markup into the space of the new text: *revert* transforms text positions to move before edits, and *convert* to move after edits. The PIDE Scala API allows to make a document snapshot at any time: it remains an immutable value, while other document processing continues in parallel. Thus GUI painting works undisturbed.

A document snapshot is *outdated* if the edit-distance is non-empty or there is a pending "command-exec assignment" by the prover (see also §3). Text shown in an outdated situation is painted in Isabelle/jEdit with grey background. The user typically sees that for brief instances of time, while edits are passed through the PIDE protocol phases. Longer periods of "editor grey-out" (without blocking) may happen in practice, when the prover is unreactive due to heavy load of ML threads or during garbage collection of the ML run-time system.

Decoupling the editor and prover in asynchronous PIDE document operations provides sufficient freedom to schedule heavy-duty proof checking tasks. The prover is enabled to orchestrate parallel proof processing [17] and additional diagnostic tools (see also §5). The concrete implementation requires a fair amount of performance tuning and adjustment of real-time parameters and delays, to make the user-experience smooth on a given range of hardware: in Isabelle2013-2 this is done for high-end laptops or work-stations with 2–8 cores. Continuous proof processing becomes a highly interactive computer game and thus introduces genuinely new challenges to ITP. Even the graphics performance of the underlying OS platform becomes a relevant factor, since many GUI details need to be updated frequently as the editor or prover changes its state.

3   Document Content

The subsequent description of document content refers to data structures managed by a PIDE-compliant prover like Isabelle. This defines declarative outlines and administrative information for eventual processing: part of that is reported to the front-end as "command-exec assignment". Further details of actual execution management are the sole responsibility of the prover (see §4).

3.1   Prover Command Transactions

The theory and proof language of Isabelle and other LCF-style systems consists of a sequence of *commands*. This accidental structure can be explained historically and is *not* challenged here. Existing implementations assume that format, and PIDE aims to minimize the requirement to rework old tools.

A theory consists of some text that is partitioned into a sequence of *command spans* as in Proof General [3]. The Isar proof language [18, §6] demonstrates that linearity is no loss of generality: block structure may be represented by a depth-first traversal of the intended tree, using an explicit stack within the proof state. Note that the superficial linearity of proof documents is in contrast to Mizar articles [18, §2], and most regular programming languages, but PIDE is focused on LCF-style proof assistants.

The internal structure of command transactions with distinctive phases of *read, eval, print* is discussed further in [16]. For PIDE proof documents, these phases are elaborated and specifically managed by the system. At first approximation, a command transaction is a partial function *tr* from some toplevel state st_0 to st_1, with sequential composition of its phases as shown in figure 3.

$$st_0 \xrightarrow{read} \xrightarrow{eval} \xrightarrow{print} st_1 \xrightarrow{read} \xrightarrow{eval} \xrightarrow{print} st_2 \xrightarrow{read} \xrightarrow{eval} \xrightarrow{print} st_3 \cdots$$

Fig. 3. Sequential scheduling of commands: *read, eval, print* loop

Looking more closely, the separate phases may be characterized by their relation to the toplevel state that is manipulated here:

> $tr\ st_0 =$
>   **let** $eval = read\ ()$ **in**   — *read* does not require st_0
>   **let** $st_1 = eval\ st_0$ **in**   — main transition $st_0 \longrightarrow st_1$
>   **let** $() = print\ st_1$ **in** st_1 — *print* does not change st_1

For PIDE, the actual work done in *read, eval, print* does not matter, e.g. commands may put extra syntactic analysis or diagnostic output into *eval*. The key requirement is that all operations are *purely functional* wrt. the toplevel state seen as *immutable value*, optionally with *observable output* via managed message channels (not physical `stdout`). These assumptions are violated by traditional LCF-style provers, including classic Isabelle in the 1990s, so this is an important starting point for reforms of other proof assistants. Command transactions need to be clearly isolated, and operate efficiently in a timeless and stateless manner.

A simple document-model would merely maintain a partially evaluated sequence of command transactions, and interleave its continued editing and execution. This could even work within a sequential prover process, with asynchronous signals for new input, but without multi-threading. On the other hand, explicit threads can simplify the implementation and provide additional potential for performance. In fact, the parallel aspect of proof processing [17] turns out relatively simple, compared to the extra entropy and hazards of user-interaction.

3.2 Document Nodes

Proof documents have additional structure that helps to organize continuous processing efficiently, to provide quick feedback to the user during editing.

The **global structure** is that of a *theory graph*, which happens to be acyclic due to the foundational order of theory content in LCF-style provers. The node *dependencies* are given as a list of imports, cf. the syntax of Isabelle theory headers "**theory** A **imports** $B_1 \ldots B_n$". Parallel traversal of DAGs is a starting point to gain performance and scalability for big theory libraries.

The **local structure** of each document node consists of command *entries*, *perspective* and *overlays*, which are described below.

Node entries are given as a linear sequence of commands (§3.1), but each command span is *interned* and represented by a unique *command-id*. There is a global mapping from *command-id* to the corresponding command transaction, which is updated before applying document edits. This indirection avoids redundant invocation of *read* in incremental processing of evolving document versions, since command positions change more often than the content of command spans.

A *command-id* essentially refers to some function tr on the toplevel state. In different document versions it may be applied in different situations. A particular command application $tr\ st_0$ is called *exec* and identified via some *exec-id*, which serves as a physical transaction identifier of the running command. The *exec-id* identifies both the command execution and its result state $st_1 = tr\ st_0$, including observable output (prover messages are always decorated by the *exec-id*).

For a given document version, the *command-exec assignment* relates each *command-id* to a list of *exec-ids*. An empty list means the command is *unassigned* and the prover will not attempt to execute it. A non-empty list refers to the main *eval* as head, and additional *prints* as tail. *Coincidence* of execs means that in the overall document history, a *command-id* has the same *exec-id* in multiple versions. This re-use of old execution fragments in new versions typically happens, when a shared prefix of commands is unaffected by edits applied elsewhere (see also figure 4). The prover is free to execute commands from different document versions, independently of the one displayed by the editor.

The command-exec assignment is vital for the editor to determine which exec results belong to which command in a particular document version, in order to display the content to the user. Whenever this information is updated on the prover side, the editor needs to be informed about it. Edits that are not yet acknowledged by the corresponding assignment lead to an outdated document snapshot (§2). This intermediate situation is now more often visible in Isabelle/PIDE, because execution is strictly monotonic: while the document is updated the prover continues running undisturbed, so the PIDE protocol thread needs to compete with ML worker threads. In the past, execution was canceled and restarted, but this is in conflict with long-running *eval* and *prints* (cf. §5).

Node perspective specifies *visible* and *required* commands syntactically within the document. The set of visible commands is typically determined by open text windows of the editor. Required commands may be ticked separately by some

GUI panel (Isabelle/jEdit does that only for document nodes, meaning the last command entry of a theory.) Visible commands are particularly interesting for the user and need full execution of *read-eval-print*. Required commands are only needed to get there: *read-eval* is sufficient to produce the subsequent toplevel state. The set of required commands is implicitly completed wrt. the transitive closure of node imports and the precedence relation of command entries.

Commands that are neither visible nor required are left unassigned, and thus remain unevaluated. There is usually a long tail-end of the overall document that is presently unassigned. Likewise, there is a long import chain, where the previous assignment is not changed, because edits are typically local to the visible part. This differentiation of document content by means of the perspective is important for scalability, in order to support continuous processing of hundreds of theory nodes, each with thousands of command entries.

Node overlays assign *print functions* (with arguments) to existing command entries within the document. The idea is to analyze the toplevel state at the point after *eval* via additional *prints*. Document overlays may be added or deleted, without changing the underlying sequence of toplevel states.

The prover also maintains a global table of *implicit print functions* (with empty arguments), which are added automatically to any visible command in the current perspective. This may be understood as a mechanism for default overlays for all commands seen in the document.

Given $st_1 = eval\ st_0$, each print function application $print\ st_1$ is identified by a separate *exec-id*. The observable result of an assigned command is the union of results from the *exec-ids* of its *eval* and all its *prints*. This union is formed by the editor whenever it retrieves information from a document snapshot (§2). It may combine *eval* and *prints* stemming from different document versions due to exec coincidence within the ML process.

3.3   Document Edits

Edits emerge in the editor by inserting or removing intervals of plain text, but these are preprocessed to operate on command entries with corresponding *command-ids* (§3.2). Changes of document node dependencies, perspective, and overlays are represented as edits, too. The PIDE document-model provides one key operation *Document.update* to turn a given document version into a new one, where the edits are syntactically represented as algebraic datatype:

datatype *edit* = *Dependencies* | *Entries* | *Perspective* | *Overlays*
val *Document.update*: *version-id* → *version-id* →
    (*node* × *edit*) *list* → *state* → (*command-id* × *exec-id list*) *list* × *state*

Type *edit* is given in stylized form above: its constructors take arguments, e.g. *Entries* the commands that are inserted or removed. *Document.update* operates on a "big" document *state*, which maintains all accessible versions. This must not be confused with a "small" toplevel state *st* for single commands.

Edits are relative to given document nodes, and can happen simultaneously when the user opens several theory files or the system completes imports transitively. While the editor usually shows a few text windows only, the document-model always works on the whole theory library behind it.

The update assignment (*command-id* × *exec-id list*) *list* is conservative: the list mentions only those *command-ids* that change. The result is reported to the editor to acknowledge the *Document.update*: until that protocol message arrives, the editor re-uses the assignment of the old version, and marks any document snapshots derived from it as outdated (§2).

The prover maintains a command exec assignment for each document version, depending on the visible and required commands of the perspective, and given node overlays: *exec-ids* for *eval* and *prints* are assigned as required. Old assignments are preserved on a common prefix that is not affected by the edits, as illustrated in figure 4. Here st_2 is the last common exec of the old versus new version; subsequent execs are removed and new ones assigned.

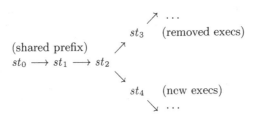

Fig. 4. Update of command-exec assignment, with shared prefix between versions

The precise manner of exec assignment is up to the prover: it can use further information about old versions and more structure of the command language. Earlier observations about the inherent structure of proof documents for parallelization [17, §2.2] apply here as well, but the additional aspect of incremental editing introduces extra complexity. Current Isabelle/PIDE is still based on the simple linear model explained above, with some refinements on how the *read-eval-print* phases of each command transaction is scheduled. This allows internal *forks* of *eval* and independent *prints* as illustrated in figure 5.

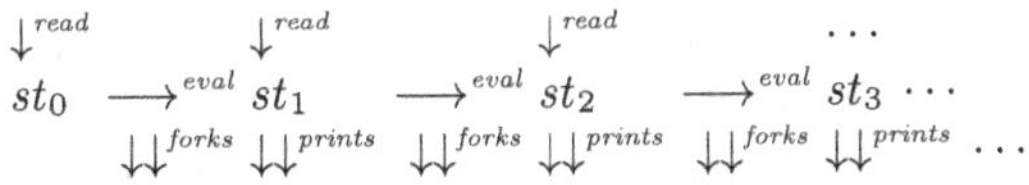

Fig. 5. Parallel scheduling of commands: *read*, *eval*, with multiple *forks* and *prints*

4   Execution Management

There is open-ended potential for sophistication of execution management, to improve parallel performance and reactivity. The subsequent explanations give some ideas about current Isabelle/PIDE, with its recently introduced ML module Execution. Its managed *Execution.fork* now supersedes the earlier approach of goal forks [17, §3.3].

Prerequisite: Future Values in Isabelle/ML. The underlying abstraction for parallel ML programming [17, §3.1] is the polymorphic type α *future* with operations *fork*: $(unit \to \alpha) \to \alpha$ *future* and *join*: α *future* $\to \alpha$ to manage evaluation of functional expressions, with optional *cancel*: α *future* $\to$ *unit*. Moreover, *promise*: *unit* $\to \alpha$ *future* and *fulfill*: α *future* $\to \alpha \to$ *unit* allow to create an open slot for some future result that is closed by external means.

Futures are common folklore in functional programming, but Isabelle/ML implements particular policies that have emerged over several years in pursuit of parallel theorem proving: **strict evaluation** (spontaneous execution via thread-pool), **synchronous exceptions** (propagation within nested task groups), **asynchronous interrupts** (cancellation and signaling of tasks), **nested task groups** (block structure of parallel program), and explicit **dependencies**.

Hypothetical Execution. Each document version is associated with an implicit execution process. After document update, the old execution needs to be turned into a new one, without disturbing active tasks. To this end, Isabelle/PIDE maintains a lazy *execution outline*: chains of commands are composed with their *eval* and *prints* as one big expression, which mathematically determines all prover results beforehand (with corresponding *exec-ids*).

The scheduling diagram of figure 5 illustrates the local structure of this expression: each arrow corresponds to some function application. The global structure has two further dimensions: the DAG of theory nodes and the version history, so many such filaments of *read-eval-print* exist simultaneously.

Since *Document.update* (§3.3) merely performs hypothetical execution, by manipulating a symbolic expression that consists of lazy memo cells, it is able to produce the new assignment quickly and report it back to the editor.

Execution Frontiers. Actual execution is an ongoing process of parallel tasks that force their way through the lazy execution outline. After each document update, the latest document version is associated with a fresh execution, but that needs to coexist with older executions with remaining active tasks.

To prevent conflicting attempts to force these lazy values, the PIDE ML module Execution ensures that at most one execution is formally *running*, in the sense defined below. The module manages a separate notion of *execution-id*, with the following operations:

Execution.start: *unit* $\to$ *execution-id*
Execution.discontinue: *unit* $\to$ *unit*
Execution.running: *execution-id* $\to$ *exec-id* $\to$ *bool*

Execution.start () creates a fresh *execution-id* and makes it the currently running one. *Execution.discontinue* () resets that state: a previously running *execution-id* cannot become running again. *Execution.running execution-id exec-id* requests the exclusive right to explore the given *exec-id*, which is only granted if the *execution-id* is currently running. Moreover, the *exec-id* is registered for management of derived execution forks (see below).

Given a document version, the *execution frontier* is the set of tasks that may explore its execution outline, guarded by invocations of *Execution.running* as shown above. Each PIDE update cycle first invokes *Execution.discontinue*, then updates the document content with its execution outline, and then uses *Execution.start* to obtain a new running *execution-id*. Finally, the exploration tasks are forked as ML futures, with the old execution frontier as tasks dependencies.

Thus the new execution frontier is semantically appended to the old one: the old frontier cannot explore new transactions and finishes eventually (or diverges), afterwards the new execution continues without conflict. This approach enables strictly monotonic execution management: running tasks within the document execution are never canceled; only those tasks are terminated that become inaccessible in the new version (removed commands etc.).

Execution Forks. The running futures of the execution frontier work on command transactions that are presently accessible, guarded by *Execution.running*. This provides a central checkpoint to control access to individual execs within the given execution outline. After having passed *Execution.running* successfully, further future tasks may be managed as follows:

> *Execution.fork*: *exec-id* → (α → *unit*) → α *future*
> *Execution.cancel*: *exec-id* → *unit*

Here the *exec-id* serves as a general handle to arbitrary future forks within that execution context: it is associated with some future task group for cumulative cancellation. Execution management ensures *strict* results: forks need to be joined eventually, and ML exceptions raised in that attempt are accounted to the transaction context. Thus a command transaction may "fail late" due to pending execution forks, even though its *eval* phase has finished superficially, and subsequent commands are already proceeding from its toplevel state.

Execution.fork provides the main programming interface to *forks* of figure 5. The primary application are goal forks in the sense of [17, §3.3], which has been retrofitted into the new execution concept. Note that the Isabelle/PIDE document model still lacks the structural proof forking of batch mode: interactive goal forks are limited to terminal **by** steps (where Isar proofs spend most of the time) or derived definitions with internal proofs like **datatype, inductive, fun**.

Moreover, *Execution.fork* is now used implicitly for diagnostic commands, which are marked syntactically to be state preserving, and can thus be forked immediately in the main evaluation sequence. Such commands are identity functions on the toplevel state, with observable output, and the potential to fail later. Note that **sledgehammer** is such a diagnostic command as well, and several copies put into a theory already causes parallel execution.

5   Asynchronous Print Functions

Diagnostic command output may happen in the main *eval* phase, but this has the disadvantage that linear editing (§3.3) reassigns intermediate execs and thus disrupts the evaluation sequence. PIDE document updates could be made smarter, but it turns out that separate management of *print* phases over existing commands is simpler and more flexible. Further observations indicate that print functions deserve special attention:

- Cumulative *print* operations consume more space and time than *eval*: proof state output is often large and its printing slower than average proof steps.
- Printing depends on document perspective: text that becomes visible requires additional output, but it can be disposed after becoming invisible.
- Printing may fail or diverge, but it needs to be interruptible to enable the system stopping it.
- Different ways of printing may run in parallel, with specific priorities.

These are notable refinements of the former approach [16, §2.3], which was restricted to one *print* as lazy value that was forked eventually; its execution had to terminate relatively quickly, and the result was always stored persistently.

The current notion of asynchronous print functions allows better management of plain proof state output, and more advanced tools to participate in the continuous document processing. The PIDE ML programming interface accepts various declarative parameters to provide hints for execution management:

Startup delay: extra time to wait, after the print becomes active. This latency reduces waste of CPU cycles when the user continues editing and changes already assigned commands again before printing starts.

Time limit: maximum time spent for a potentially diverging print operation.

Task priority: scheduling parameter for the underlying ML future (for task queue management). Note that this is not a thread priority: an already running task of low priority is unaffected by later forks of high priority.

Persistence: keep results produced by *print* (including observable output), or delete them when visibility gets lost.

Application (1): Proof State Output. Printing proof states efficiently is less trivial than it seems. Command-line users do not mind to wait fractions of a second to see the result after each command, but continuous document processing in PIDE means that maybe 10–100 commands become visible when opening or scrolling text windows. If printing requires 10–100 ms for each command, it already causes significant slowdown.

Proof states are now printed asynchronously, with the following scheduling parameters: no startup delay, no time limit, high task priority, no persistence.

The absence of delay and the priority means that the *print* phase runs eagerly whenever possible, after its corresponding *eval* has finished. On multiple cores, the ongoing *eval* sequence proceeds concurrently with corresponding *prints*, resulting in fairly good performance. On a single core, the system adapts its task scheduling to do the interleaving of *eval* versus high priority *prints* sequentially: this is important for the user to proceed, but results in considerable slowdown. The difference can be seen e.g. in the long unstructured proof scripts of `$ISABELLE_HOME/src/HOL/Hoare_Parallel/OG_Hoare.thy`, setting in jEdit *Plugin Options / Isabelle / General / Threads* to 2 versus 1, restarting proof processing via *File / Reload*, scrolling around etc.

Non-persistence is based on the observation that each individual proof state output is reasonably fast, but its result can be big and needs to be stored in the document model (in Scala). For commands that lose visible perspective, the corresponding *print* is unassigned and the document content eventually disposed by garbage collection. Thus we conserve Scala/JVM space, by investing extra ML time to print again later.

Application (2): Automatically Tried Tools. As explained in the manual [14, §2.7], Isabelle/HOL provides a collection of tools that can prove or disprove goals without user intervention: automated methods (*auto*, *simp*, *blast* etc.), **nitpick, quickcheck, sledgehammer, solve-direct**.

In Isabelle Proof General, such tools run synchronously within regular proof state output, and a tight timeout of 0.5s to guarantee reactivity of the command loop. This limits the possibilities of spontaneous feedback by the prover to relatively light-weight tools like **quickcheck** and **solve-direct**, and even that may cause cumbersome delays in sequential command processing.

In Isabelle/PIDE automatically tried tools are asynchronous print functions, with default parameters like this: startup delay = 1s, time limit = 2s, low task priority, persistence.

Thus tools usually run only after some time of inactivity, and do not compete directly with the main *eval* and high-priority *prints*. Persistence is enabled, since tools usually take a long time to produce small output: nothing on failure (or timeout) or a short message on success. In particular, the often unsuccessful applications are retained and not tried again.

Tool output is marked-up as *information message*, which is rendered in Isabelle/jEdit with a blue information icon and blue squiggles for the corresponding goal command. This is non-intrusive information produced in the background, while the user was pondering the text. Cumulatively, automatically tried tools can consume significant CPU resources, though. For high-end work-stations connected to grid power that is rarely a problem, but small mobile devices on batteries should disable extraneous instrumentation.

Application (3): Query Operations. The idea is to support frequently used and potentially long-running diagnostic commands via explicit GUI components in the editor, for example *Sledgehammer* and *Find theorems* as explained in [14, §2.8,§2.9] (with screenshots and minimal examples).

In such situations, Proof General [3] provides a separate command-line to issue state-preserving commands synchronously: the user first needs to move the prover focus to some point in the text and then wait while the query is running.

In Isabelle/PIDE this is now done via asynchronous print functions with explicit document overlays (§3.2). Arguments are provided by some GUI dialog box: input causes a document update that changes the corresponding overlay; the command position is determined from the current focus in the text.

The asynchronous approach allows the user to input the query and start the operation at any time, while the system schedules the print process to run spontaneously after the command that defines its context is evaluated; afterwards it presents query results as they arrive incrementally. There is also a button to cancel the process (notably for *Sledgehammer*). The transitional states of a pending query are visualized by some "spinning disk" icon (with tooltip).

Isabelle/PIDE provides a hybrid module `Query_Operation` in ML and Scala. The ML side accepts a function that takes a toplevel state with arguments and produces output on some private channel; this interface resembles traditional command-line tools. Likewise, the Scala side works with conventional event-based GUI components, without direct exposure to the timeless and stateless PIDE document model. The implementation of the hybrid `Query_Operation` module takes care of the management of different instances for each GUI view, and keeps the connection to running command execs (for cancellation etc.).

This completes the full round-trip of PIDE concepts: from the sequential and synchronous *read-eval-print* loop that connects the user directly to a single command execution, over an intermediate document model that is detached from particular time and space, leading to simple PIDE APIs that recover the appearance of working directly with some command execution that is connected to physical GUI elements. The benefit of this detour is that the system infrastructure is enabled to manage the details of execution efficiently, for many tools on many CPU cores, instead of asking the user to do this sequentially by hand.

6   Conclusion and Related Work

The Isabelle/PIDE approach combines user interaction and tool integration into a uniform document-model. This enables advanced front-end technology in the style of classic IDEs for mainstream programming languages. It also allows us to integrate interactive or automatic theorem proving tools to help the user composing proof documents. The present paper continues earlier explanations of PIDE concepts [12, 11, 13, 15, 16]. The following improvements are newly introduced in the current generation of Isabelle/PIDE (December 2013):

- strictly monotonic document update: avoid cancellation and restart of running command transactions;
- explicit document execution management;
- support for asynchronous print functions, with various execution policies;
- support for document overlays and query operations, with separate GUI components for input and output.

Related Work. Explicit parallelism has been imposed on application developers before, when classic CISC machines became stagnant in the 1990s, and workstation clusters were considered a potential solution. A notable experiment from that time is the *Distributed Larch Prover* [5]: it delegates proof problems to CPU nodes, with a central managing process and some Emacs front-end to organize pending proofs. The report on that early project clearly identities the need to rethink prover front-ends, when the back-end becomes parallel.

Concerning prover front-ends, the main landmark to improve upon raw TTY interaction of proof assistants is *Proof General* by D. Aspinall [3]. It only requires a classic *read-eval-print* loop with annotated prompt and *undo* operation, and thus implements "proof scripting" within the editor. The user can navigate forwards and backwards to move the boundary between the *locked region* of the text that is already checked and the remaining part that is presently edited.

The approach of Proof General was so convincing that it has been duplicated many times, with slightly different technical side-conditions, e.g. in CoqIDE [18, §4] (OCaml/GTK), Matita [2] (OCaml/GTK), Matitaweb [1] (OCaml web server). The great success of Proof General 15 years ago made it difficult to go beyond it. Early attempts by D. Aspinall to formalize its protocol as PGIP and integrate it with Eclipse [4] have never reached a sufficient level of support by proof assistants to become relevant to users. Nonetheless, PGEclipse was an important initiative to point beyond classic TTY and Emacs, into a greater world of IDE frameworks.

Dafny [6] follows a different approach to connect automated theorem proving (Boogie and Z3) with Visual Studio as the IDE. Thus it introduces some genuine user-interaction into a world of automatic SMT solving, bypassing TTY mode. The resulting application resembles Isabelle/jEdit, while the particular proof tools and logical foundations of the proof environment are quite different.

Agora [10] is a recent web-centric approach to document-oriented proof authoring, for various existing back-ends like Coq [18, §4] and Mizar [18, §2]. The main premise of this work is to take the proof assistant *as-is* and to see how much added value can be achieved by wrapping web technology around it. C. Tankink also points beyond classic IDEs, which are in fact already 10–20 years old. More recent movements on IDE design for programming languages integrate old and new ideas of *direct manipulation* of static program text and dynamic execution side-by-side, and a non-linear document-model of source snippets. A notable project is http://www.chris-granger.com/lighttable, which is implemented in Clojure and works for Clojure, Javascript, and Python.

Incidently, interactive proof checking has been based on direct access to proof states from early on, and PIDE already provides substantial support to manage incremental execution and continuous checking of proof-documents. So further alignments with such newer IDE approaches would be a rather obvious continuation of what has been achieved so far, but the ITP community also requires time to get acquainted even with the classic IDE model seen in Isabelle/jEdit.

References

[1] Asperti, A., Ricciotti, W.: A web interface for Matita. In: Jeuring, J., Campbell, J.A., Carette, J., Dos Reis, G., Sojka, P., Wenzel, M., Sorge, V. (eds.) CICM 2012. LNCS (LNAI), vol. 7362, pp. 417–421. Springer, Heidelberg (2012)

[2] Asperti, A., Sacerdoti Coen, C., Tassi, E., Zacchiroli, S.: User interaction with the Matita proof assistant. Journal of Automated Reasoning 39(2) (2007)

[3] Aspinall, D.: Proof General: A generic tool for proof development. In: Graf, S., Schwartzbach, M. (eds.) TACAS 2000. LNCS, vol. 1785, pp. 38–43. Springer, Heidelberg (2000)

[4] Aspinall, D., Lüth, C., Winterstein, D.: A framework for interactive proof. In: Kauers, M., Kerber, M., Miner, R., Windsteiger, W. (eds.) MKM/CALCULEMUS 2007. LNCS (LNAI), vol. 4573, pp. 161–175. Springer, Heidelberg (2007)

[5] Kapur, D., Vandevoorde, M.T.: DLP: A paradigm for parallel interactive theorem proving (1996)

[6] Leino, K.R.M.: Dafny: An automatic program verifier for functional correctness. In: Clarke, E.M., Voronkov, A. (eds.) LPAR-16. LNCS, vol. 6355, pp. 348–370. Springer, Heidelberg (2010)

[7] Lüth, C., Ring, M.: A web interface for Isabelle: The next generation. In: Carette, J., Aspinall, D., Lange, C., Sojka, P., Windsteiger, W. (eds.) CICM 2013. LNCS, vol. 7961, pp. 326–329. Springer, Heidelberg (2013)

[8] Ring, M., Lüth, C.: Collaborative interactive theorem proving with Clide. In: Klein, G., Gamboa, R. (eds.) ITP 2014. LNCS (LNAI), vol. 8558, pp. 467–482. Springer, Heidelberg (2014)

[9] Odersky, M., et al.: An overview of the Scala programming language. Technical Report IC/2004/64, EPF Lausanne (2004)

[10] Tankink, C.: Documentation and Formal Mathematics — Web Technology meets Theorem Proving. PhD thesis, Radboud University Nijmegen (2013)

[11] Wenzel, M.: Isabelle as document-oriented proof assistant. In: Davenport, J.H., Farmer, W.M., Urban, J., Rabe, F. (eds.) Calculemus/MKM 2011. LNCS (LNAI), vol. 6824, pp. 244–259. Springer, Heidelberg (2011)

[12] Wenzel, M.: Asynchronous proof processing with Isabelle/Scala and Isabelle/jEdit. In: Coen, C.S., Aspinall, D. (eds.) User Interfaces for Theorem Provers (UITP 2010). ENTCS (July 2010)

[13] Wenzel, M.: Isabelle/jEdit – A prover IDE within the PIDE framework. In: Jeuring, J., Campbell, J.A., Carette, J., Dos Reis, G., Sojka, P., Wenzel, M., Sorge, V. (eds.) CICM 2012. LNCS (LNAI), vol. 7362, pp. 468–471. Springer, Heidelberg (2012)

[14] Wenzel, M.: Isabelle/jEdit. Part of Isabelle distribution (December 2013), http://isabelle.in.tum.de/website-Isabelle2013-2/dist/Isabelle2013-2/doc/jedit.pdf

[15] Wenzel, M.: PIDE as front-end technology for Coq (2013), http://arxiv.org/abs/1304.6626

[16] Wenzel, M.: READ-EVAL-PRINT in parallel and asynchronous proof-checking. In: Kaliszyk, C., Lüth, C. (eds.) User Interfaces for Theorem Provers (UITP 2012). EPTCS, vol. 118 (2013)

[17] Wenzel, M.: Shared-memory multiprocessing for interactive theorem proving. In: Blazy, S., Paulin-Mohring, C., Pichardie, D. (eds.) ITP 2013. LNCS, vol. 7998, pp. 418–434. Springer, Heidelberg (2013)

[18] Wiedijk, F. (ed.): The Seventeen Provers of the World. LNCS (LNAI), vol. 3600. Springer, Heidelberg (2006)

HOL Constant Definition Done Right

Rob Arthan

Lemma 1 Ltd./ School of Electronic Engineering and Computer Science,
Queen Mary, University of London, UK

Abstract. This note gives a proposal for a simpler and more powerful replacement for the mechanisms currently provided in the various HOL implementations for defining new constants.

1 Introduction

The design of the HOL logic and of its definitional principles [7] evolved in the late 80s and early 90s. Some form of this design has been implemented in HOL4 [8], HOL Light [3], HOL Zero [1], Isabelle/HOL [6] and ProofPower [2]. While the definitional principles have stood the test of time in many practical applications, we believe there is still some room for improvement. This note discusses issues with the mechanisms for introducing new constants and proposes a new and more general mechanism that addresses these issues.

2 The Problem

The original Classic HOL provided a mechanism for defining new constants known as `new_definition`. This worked as follows: given a possibly empty list of variables $x_1, \ldots, x_n$ and a term t whose free variables are contained in the x_i, it introduced a new constant[1] c of the appropriate type and the axiom:

$$\vdash \forall x_1 \ldots x_n \cdot c\, x_1\, \ldots\, x_n = t.$$

This simple mechanism is remarkably powerful but suffered from two significant shortcomings, both pointed out by Roger Jones[2]:

RJ1. The mechanism does not support implicit definitions. As one example, it is pleasant to define the destructors of a data type as the left inverses of the constructors. Thus one wants to define Pre in terms of Suc by:

$$\mathsf{Pre}(\mathsf{Suc}(n)) = n.$$

---

[1] The details of the mechanism for specifying the names of new constants are not important for present purposes.

[2] At various places in this note, I sketch observations made by other people. The wording used is mine and not theirs and any misrepresentation is my responsibility.

G. Klein and R. Gamboa (Eds.): ITP 2014, LNAI 8558, pp. 531–536, 2014.
© Springer International Publishing Switzerland 2014

As another example, the exponential function is naturally defined by a differential equation:

$$\exp(0) = 1$$
$$(\mathsf{D}\exp)(x) = \exp(x).$$

In such cases, the mechanism can be used to define constants having the desired properties, but one has to use the Hilbert choice operator to give witnesses and then derive the implicit definitions as theorems. This results in a loss of abstraction and unintended identities, e.g., the naive way of defining two constants c_1 and c_2 both with the loose defining property $c_i \leq 10$ will result in an extension in which $c_1 = c_2$ is provable.

RJ2. The mechanism is unsound. The condition on the free variables of t is certainly necessary. Without it, we could take t to be a variable, $y : \mathbb{N}$, and define a new constant c satisfying $\vdash \forall y : \mathbb{N} \cdot c = y$. Specialising this in two different ways, we could prove both $c = 1$ and $c = 2$. However, the condition is not sufficient. If $\#$ is a polymorphic function such that $\#X$ is the size of X when X is a finite set, then we can use the mechanism to define a constant $c : \mathbb{N}$ satisfying the axiom $c = \#\{x : \alpha \mid x = x\}$, where α is a type variable. But then if $\mathbf{1}$ and $\mathbf{2}$ denote types with 1 and 2 members respectively, we can instantiate α to prove both $c = \#\{x : \mathbf{1} \mid x = x\} = 1$ and $c = \#\{x : \mathbf{2} \mid x = x\} = 2$.

The fix for **RJ2** was to change `new_definition` so as to check that all type variables appearing anywhere in the term t also appear in the type of the constant c that is being defined. HOL Light, HOL Zero, Isabelle/HOL and ProofPower were all implemented after the problem was known, so they incorporated this solution from scratch. The fix in Classic HOL was carried forward into HOL4.

A new mechanism, `new_specification`, was introduced to address **RJ1**. `new_specification` takes as input a theorem of the form $\vdash \exists v_1 \ldots v_n \cdot p$ and introduces a list of new constants $c_1, \ldots, c_n$ and the axiom

$$\vdash p[c_1/v_1, \ldots, c_n/v_n].$$

`new_specification` requires that the free variables of p be contained in the v_i and that every type variable appearing anywhere in p also appear in the type of each new constant c_i, thus avoiding reintroducing the problem of **RJ2** under a different guise. The result is conservative and hence sound. It also supports a very useful range of implicit definitions. However, there are two issues that I noted during the ProofPower implementation:

RA1. Given `new_specification`, `new_definition` is redundant: what it does can easily be realised by a derived mechanism that given the list of variables $x_1, \ldots, x_n$ and the term t, automatically proves:

$$\vdash \exists y \cdot \forall x_1 \ldots x_n \cdot y \, x_1 \ldots x_n = t$$

and then applies new_specification. Unfortunately, in order to prove existentially quantified statements, one needs a definition of the existential quantifier, and so new_definition seems necessary to avoid a bootstrapping problem. (Since it is only required for bootstrapping, the ProofPower implementation of new_definition only covers the simple case where the axiom has the form $\vdash c = t$.)

RA2. The condition on type variables imposed by new_specification is stronger than one would like. It is natural for certain "concrete" structures to be characterized by more "abstract" properties such as universal mapping properties. For example, data types can be characterized as initial algebras:

$$\forall (z : \alpha)(s : \alpha \to \alpha) \cdot \exists! f : \mathbb{N} \to \alpha \cdot f(0) = z \wedge \forall n \cdot f(\mathsf{Suc}(n)) = s(f(n)).$$

However, the above characterization cannot be used as a defining property for the successor function with new_specification. Characterizing objects by universal properties is endemic in modern mathematics and computer science, so it is irritating to be compelled to resort to circumlocutions.

In HOL4, ProofPower and HOL Zero, new_specification is implemented as a primitive operation. However, in HOL Light, it is derived. I believe this was primarily a consequence of the following design goal for HOL Light:

JH1. The primitive inference system for HOL Light should be defined in terms of language primitives and equality alone and should not depend on the axiomatization of the logical connectives.

A form of new_specification that does not involve existential quantification was implemented in early versions of HOL Light. This took as input a theorem of the form $\vdash p\ t$. Later, to simplify the correctness argument for the system, new_specification was re-implemented as a derived operation that uses the Hilbert choice operator to translate its inputs into a form suitable for new_definition, applies new_definition, then derives the desired axiom to be passed back to the user from the stronger axiom returned by new_definition. Thus HOL Light bypasses **RA1**, but at the price of a certain inelegance, since we have to trust the derived rule to discard the axiom returned by new_definition. This became worse when HOL Light was enhanced to address the following observation of Mark Adams:

MA1. If an LCF style system does not record all the axioms and definitions that have been introduced, the correctness claim for the system has to be defined in terms of a state **and** the sequence of operations which produced that state. This makes it impossible to implement a proof auditing procedure that works by analysing the current state of the system.

As a result of **MA1** axioms and definitions in HOL Light are now recorded. The current HOL Light implementation uses a trick to prevent two constants with the same loose defining property being provably equal. The trick is based on the following idea: to define c_1 and c_2 such that $c_1, c_2 \leq 10$, say, define

$c_1 = (\varepsilon f \cdot \forall n \cdot f(n) \leq 10)\ 1$ and $c_2 = (\varepsilon f \cdot \forall n \cdot f(n) \leq 10)\ 2$; then c_1 and c_2 have the desired property, but $c_1 = c_2$ is not provable. Nonetheless some unintended identities are still provable that would not be provable if new_specification were implemented as a primitive as in HOL4 or ProofPower.

The equivalent of new_specification in Isabelle/HOL is its **specification** command. This is implemented using an equational definition and the choice function, but that definition only exists in a private namespace. Some aspects of the abstraction offered by new_specification are provided by the very popular locale mechanism in Isabelle.

Quantification over type variables as implemented in HOL-Omega [4] obviates many of the problems discussed here. However, our present concern is with improvements that preserve the delightful simplicity of the Classic HOL logic.

3   Proposed Alternative

The proposed alternative is to discard new_definition and to adapt and generalise new_specification so that it does not depend on the meaning of the existential quantifier. The generalised new_specification, which we will call gen_new_specification, takes as input a theorem of the following form

$$v_1 = t_1, \ldots, v_n = t_n \vdash p$$

where the v_i are variables. If all is well, gen_new_specification will introduce new constants $c_1, \ldots, c_n$ and the following axiom:

$$\vdash p[c_1/v_1, \ldots, c_n/v_n].$$

gen_new_specification imposes the following restrictions:

- the v_i must be pairwise distinct;
- the terms t_i must have no free variables;
- the free variables of p must be contained in the v_i;
- any type variable occurring in the type of any subterm of a t_i must occur in the type of the corresponding v_i.

There is no restriction on the type variables appearing in p.

Claim 1. gen_new_specification *is conservative and hence sound.*

Proof: Assume that a sequent $\Gamma \vdash q$ containing no instances of the c_i is provable using the axiom $\vdash p[c_1/v_1, \ldots, c_n/v_n]$ introduced using gen_new_specification. We will show how to transform a proof tree with conclusion $\Gamma \vdash q$ into a proof tree with the same conclusion that does not use the new axiom. First, by simple equality reasoning, derive from the theorem $v_1 = t_1, \ldots, v_n = t_n \vdash p$ that was passed to new_specification, the theorem $\vdash p[t_1/v_1, \ldots, t_n/v_n]$. Now replace each type instance of a c_i in the proof tree with the corresponding type instance of t_i and wherever a type instance of the axiom $\vdash p[c_1/v_1, \ldots, c_n/v_n]$ is used in the proof tree, replace it with the corresponding type instance of a proof tree for $\vdash p[t_1/v_1, \ldots, t_n/v_n]$. By inspection of the primitive inference rules in [3], if one replaces instances of constants in a correct inference by closed terms of the same type in such a way that assumptions or conclusions of the sequents involved that were syntactically identical before the replacement remain syntactically identical, then the result is also a correct inference. As the condition on type variables imposed by gen_new_specification guarantees that two instances of a c_i are syntactically identical iff the corresponding instances of t_i are syntactically identical, we have constructed a correct proof tree whose conclusion is $\Gamma \vdash q$. ∎

Claim 2. gen_new_specification *subsumes* new_definition.

Proof: In the simplest case, to define c with axiom $\vdash c = t$, where t has no free variables and contains no type variables that do not appear in its type, apply gen_new_specification to the axiom $v = t \vdash v = t$. This is all we need to define the logical connectives [3].

For the general case, to define c with axiom $\vdash \forall x_1 \ldots x_n \cdot c \, x_1 \ldots x_n = t$, take the axiom $v = (\lambda x_1 \ldots x_n \cdot t) \vdash v = (\lambda x_1 \ldots x_n \cdot t)$, derive $v = (\lambda x_1 \ldots x_n \cdot t) \vdash \forall x_1 \ldots x_n \cdot v \, x_1 \ldots x_n = t$ from it and then apply gen_new_specification. ∎

Claim 3. gen_new_specification *subsumes* new_specification.

Proof: Given the theorem $\vdash \exists v_1 \ldots v_n \cdot p$, we can derive from it the theorem $v_1 = \varepsilon v_1 \cdot \exists v_2 \ldots v_n \cdot p \vdash \exists v_2 \ldots v_n \cdot p$ and apply gen_new_specification to define a constant c_1 with defining axiom $\vdash \exists v_2 \ldots v_n \cdot p[c_1/v_1]$. Iterating this process we can define $c_2, \ldots, c_n$ such that the defining axiom of c_n is $\vdash p[c_1/v_1, \ldots, c_n/v_n]$. Thus we can achieve the same effect as new_specification at the expense of additional intermediate definitions. This is sufficient to define the constructor and destructors for binary products.

Once we have binary products, we can simulate n-tuples by iterated pairing. This means that given the theorem $\vdash \exists v_1 \ldots v_n \cdot p$, we can derive the theorem $\vdash \exists z \cdot p[\pi_1(z)/v_1, \ldots, \pi_n(z)/v_n]$ in which the n bound variables $v_1, \ldots, v_n$ have been collected into a single n-tuple denoted by the fresh variable z (here π_i denotes the projection onto the i-th factor). Now we can derive from that the theorem $v_1 = t_1, \ldots, v_n = t_n \vdash p$ where t_i is $\pi_i(\varepsilon z \cdot p[\pi_1(z)/v_1, \ldots, \pi_n(z)/v_n])$. Given this theorem as input, gen_new_specification has exactly the same effect as new_specification given the input theorem $\vdash \exists v_1 \ldots, v_n \cdot p$. ∎

4   Conclusion

Let me assess the proposed new definitional mechanism gen_new_specification against the observations that led to it:

RJ1. By claim 3, the support for implicit definitions is at least as good with gen_new_specification as with new_specification. In fact it is better: new_specification cannot define new constants $f : \alpha \to \mathbb{N}$ and $n : \mathbb{N}$ with defining property $\forall x \cdot \neg f \, x = n$, but gen_new_specification can.

RJ2. By claim 1, the proposed alternative is sound. What is more, this proof has been formalised in HOL4: Ramana Kumar, Scott Owens and Magnus

Myreen have recently completed a formal proof of soundness for the HOL logic and its definitional principles including gen_new_specification [5].

RA1. By claim 2, new_definition is no longer required. (As seen in the proof of this claim, the special case needed to define the logical connectives does not involve any reasoning about them, so there is no bootstrapping issue.)

RA2. The restriction on type variables now applies only to the equations that give the witnesses to the consistency of the definition. Defining properties such as initial algebra conditions are supported.

JH1. gen_new_specification is defined solely in terms of equality and primitive language constructs.

MA1. The unintended identities arising as a result of recording definitions in HOL Light will not occur if gen_new_specification is adopted as the primitive mechanism for defining constants.

My conclusion when I wrote the first draft of this note was that the proposal was well worth adopting. It has recently been implemented in HOL4 and Proof-Power. In both cases it is a replacement for new_definition and the existing new_specification has been retained for pragmatic reasons. The ProofPower implementation includes an implementation of the proof of claim 3 above and this completely replaces new_specification in the development of many of the theories supplied with the system, including all the "pervasive" theories such as the theories of pairs and natural numbers that form part of the logical kernel.

Acknowledgments. I would like to thank the ITP 2014 Programme Chairs, the referees, Mark Adams, John Harrison, Roger Jones, Ramana Kumar, Magnus Myreen, Scott Owens, Konrad Slind and Makarius Wenzel for their kind assistance in divers ways in the preparation and publication of this note.

References

1. Adams, M.: HOL Zero, http://www.proof-technologies.com/holzero/
2. Arthan, R., Jones, R.B.: Z in HOL in ProofPower. BCS FACS FACTS (2005-1), http://www.lemma-one.com/ProofPower/index/
3. Harrison, J.: HOL light: An overview. In: Berghofer, S., Nipkow, T., Urban, C., Wenzel, M. (eds.) TPHOLs 2009. LNCS, vol. 5674, pp. 60–66. Springer, Heidelberg (2009), http://www.cl.cam.ac.uk/~jrh13/hol-light/
4. Homeier, P.V.: The HOL-Omega logic. In: Berghofer, S., Nipkow, T., Urban, C., Wenzel, M. (eds.) TPHOLs 2009. LNCS, vol. 5674, pp. 244–259. Springer, Heidelberg (2009)
5. Kumar, R., Arthan, R., Myreen, M.O., Owens, S.: HOL with definitions: Semantics, soundness, and a verified implementation. In: Klein, G., Gamboa, R. (eds.) ITP 2014. LNCS (LNAI), vol. 8558, pp. 308–324. Springer, Heidelberg (2014)
6. Wenzel, M., et al.: The Isabelle/Isar Reference Manual, http://isabelle.in.tum.de/dist/Isabelle2013-2/doc/isar-ref.pdf
7. Norrish, M., et al.: The HOL System: Logic, 3rd edn., http://hol.sourceforge.net/documentation.html
8. Slind, K., Norrish, M.: A brief overview of HOL4. In: Mohamed, O.A., Muñoz, C., Tahar, S. (eds.) TPHOLs 2008. LNCS, vol. 5170, pp. 28–32. Springer, Heidelberg (2008)

Rough Diamond: An Extension
of Equivalence-Based Rewriting

Matt Kaufmann and J Strother Moore

Dept. of Computer Science, University of Texas, Austin, TX, USA
{kaufmann,moore}@cs.utexas.edu
http://www.cs.utexas.edu

Abstract. Previous work by the authors generalized conditional rewriting from the use of equalities to the use of arbitrary equivalence relations. Such (classic) *equivalence-based rewriting* automates the replacement of one subterm by another that may not be strictly equal to it, but is equivalent to it, where this equivalence is determined *automatically* to be sufficient at that subterm occurrence. We extend that capability by introducing *patterned* congruence rules in the ACL2 theorem prover, to provide more control over the occurrences where such a replacement may be made. This extension enables additional automation of the rewriting process, which is important in industrial-scale applications. However, because this feature is so new (introduced January, 2014), we do not yet have industrial applications to verify its utility, so we present a small example that illustrates how it supports scaling to large proof efforts.

Keywords: ACL2, rewriting, congruence, equivalence relation.

1 Introduction

A conditional rewrite rule, $(P_1 \wedge \ldots \wedge P_n \to L = R)$, directs an instance of the term L to be rewritten to the corresponding instance of the term R, provided the corresponding instances hold for hypotheses P_1 through P_n. In previous work [1] we showed how to generalize conditional rewrite rules to allow an arbitrary equivalence relation, $\sim$, in place of $=$, thus: $(P_1 \wedge \ldots \wedge P_n \to L \sim R)$. Key is the use of proved *congruence rules* and *refinement rules*[1] to associate, *automatically*, an equivalence relation with each call of the rewriter, such that it is sound to replace a subterm by one that is equivalent. The above generalized rewrite may then be used when $\sim$ is a refinement of that equivalence relation. This capability is implemented in the ACL2 theorem prover [7,6] and has seen substantial use: as of ACL2 Version 6.4, the community distribution of ACL2 input files [10] contains more than 1800 instances of congruence rules.

We give a preliminary report on a generalization, *patterned congruence rules*, introduced into ACL2 in Version 6.4, January, 2014. At this stage we can only

---

[1] Refinement rules work essentially the same way in this new setting as they did before. We do not mention them further in this paper.

G. Klein and R. Gamboa (Eds.): ITP 2014, LNAI 8558, pp. 537–542, 2014.

guess at uptake of this capability by the ACL2 community, although we do expect it to be used at least by the requester of this feature at Centaur Technology [9].

Our (existing and updated) approach to equivalence-based rewriting differs from approaches based on the use of quotient structures in higher-order logic, for example in HOL [3], Isabelle [4], and Coq [2]. To the best of our knowledge, our approach to first-order equivalence-based rewriting (without quotients) is the only one that automates the tracking of which equivalences are sufficient to preserve in a given context.

We begin in Section 2 by presenting a self-contained example to illustrate our previous work [1]. Section 3 then builds on that example to introduce our extension to patterned congruence rules, followed by a sketch of the relevant algorithm and theory in Section 4. We conclude with a few reflections.

The online ACL2 User's Manual [7] provides user-level introductions to equivalence-based rewriting. See topics EQUIVALENCE, CONGRUENCE, and (for this new work) PATTERNED-CONGRUENCE.

2   Previous Work

The example below uses traditional syntax. Complete ACL2 input is online [5].

The following recursively-defined equivalence relation holds for two binary trees when one can be transformed to the other by some sequence of "flips": switching left and right children.

```
t1 ~ t2 ≜ IF leaf-p(t1) ∨ leaf-p(t2) THEN t1=t2
          ELSE (left(t1) ~ left(t2) ∧ right(t1) ~ right(t2)) ∨
               (left(t1) ~ right(t2) ∧ right(t1) ~ left(t2))
```

When provided a suitable induction scheme, ACL2 automatically proves and stores the theorem that ~ is an equivalence relation. We now define a function that swaps every pair of children in a binary tree (cons is the pairing operation).

```
mirror(tree) ≜ IF leaf-p(tree) THEN tree
               ELSE cons(mirror(right(tree)), mirror(left(tree)))
```

The equivalence-based rewrite rule below directs the replacement of any instance of the term mirror(x) by the corresponding instance of the term x, in contexts for which it suffices to preserve equivalence with respect to ~. Of course, the ordinary rewrite rule mirror(x) = x is not a theorem!

```
REWRITE RULE: tree-equiv-mirror
mirror(x) ~ x
```

The following function returns the product of the numeric elements of the fringe of a tree. It provides an example for sound replacement of mirror(x) by x: ACL2 proves the congruence rule below, stating that the return values are equal for equivalent inputs of the function tree-product. In general, a congruence rule states that the return values of a function call are equal (or more generally, suitably equivalent) when replacing a given argument by one that is equivalent.

```
tree-product(tree) ≜
IF [tree is a number] THEN tree
ELSE IF leaf-p(tree) THEN 1
ELSE tree-product(left(tree)) * tree-product(right(tree))
```

```
CONGRUENCE RULE: tree-equiv-->-equal-tree-product
x ~ y → tree-product(x) = tree-product(y)
```

ACL2 can now prove the following theorem automatically by applying rewrite rule **tree-equiv-mirror** to the term **mirror(x)**. The congruence rule immediately above justifies this rewrite. When that rule is instead a rewrite rule, ACL2 is not able to use either it or **tree-equiv-mirror** to prove the theorem below.

```
THEOREM: tree-product-mirror
tree-product(mirror(y)) = tree-product(y)
```

This particular theorem is easy for ACL2 to prove automatically even without congruence rules or the rewrite rule **tree-equiv-mirror** (though induction would then be required). But to see the scalability of this approach, imagine that there are k_1 functions like **mirror** and k_2 like **tree-product**. If we then prove k_1 rewrite rules like **tree-equiv-mirror** and k_2 congruence rules like **tree-equiv--implies-equal-tree-product**, then these $k_1 + k_2$ rules set us up to perform automatically all $k_1 * k_2$ rewrites like **tree-product-mirror**.

3   Patterned Congruence Rules

A congruence rule, as discussed above, specifies when a given argument of a function call may be replaced by one that is suitably equivalent. A *patterned congruence rule* generalizes this idea by allowing a specified subterm of that call, which is not necessarily a top-level argument, to be replaced by one that is suitably equivalent. The following example is discussed further below.

```
PATTERNED CONGRUENCE RULE: tree-equiv-->-equal-first-tree-data
```
$x \sim y \rightarrow \text{first(tree-data}(x)) = \text{first(tree-data}(y))$

Notice that unlike a "classic" congruence rule, where the replacement of an equivalent subterm is specified at a specific argument of a function call, here x is to be replaced by y at a deeper position: a subterm of a subterm of the call. Indeed, the conclusion of the rule can be an equivalence between complex patterns, for example: $x \sim_1 y \rightarrow f(3, h(u, x), g(u)) \sim_2 f(3, h(u, y), g(u))$. That rule justifies replacement of a term x by a term $y \sim_1 x$, within any term of the form $f(3, h(u, x), g(u))$ that occurs where it suffices to preserve $\sim_2$.

A *patterned congruence rule* is thus a formula of the form $x \sim_{\text{inner}} y \rightarrow L \sim_{\text{outer}} R$, subject to the following requirements. Function symbols $\sim_{\text{inner}}$ and $\sim_{\text{outer}}$ have been proved to be equivalence relations. L and R are function calls such that x occurs in L, y occurs in R, and these are the only occurrences of x and y in the rule. Finally, R is the result of substituting y for x in L.

This rule enables the automatic rewrite of a subterm of L at the position of x to a term that is $\sim_{inner}$-equivalent to x, in any context where it suffices to preserve $\sim_{outer}$. We illustrate this process by continuing the example of the preceding section, this time defining a function that sweeps a tree to collect a *list* of results, whose first element is the product of the numeric leaves (as before). We omit some details; function `combine-tree-data(t1,t2)` returns a list whose first element is the product of the first elements from the recursive calls.

```
tree-data(tr) ≜
IF [tr is a number] THEN [tr, ...]
ELSE IF leaf-p(tr) THEN [1, ...]
ELSE combine-tree-data(tree-data(left(tr)), tree-data(right(tr)))
```

ACL2 can now automatically prove the patterned congruence rule displayed at the start of this section, `tree-equiv-->-equal-first-tree-data`. ACL2 then proves the theorem below as follows, much as it proves Theorem `tree-product--mirror` in the preceding section. First, the patterned congruence rule informs the rewriter that it suffices to preserve $\sim$ when rewriting `mirror(y)`. Hence, the rewrite rule `tree-equiv-mirror` (from the preceding section) is used to replace `mirror(y)` by y. Also as before, this small example suggests the importance of (patterned) congruences for scalability, where $k_1 + k_2$ rules set us up to perform automatically $k_1 * k_2$ different rewrites.

```
THEOREM: first-tree-data-mirror
first(tree-data(mirror(y))) = first(tree-data(y))
```

4 Algorithm Correctness and Patterned Equivalence Relations

In this section we outline briefly the algorithm implemented in ACL2 for using pattern-based congruence rules, and we touch on why it is correct. More details, including discussion of efficiency tricks and addressing of subtle issues (e.g., an example showing that arguments cannot be rewritten in parallel), are provided in a long comment in the ACL2 source code [8]. Of special concern is that ACL2 procedures that manipulate terms must quickly determine the available equivalences on-the-fly and tend to sweep the terms left-to-right, innermost first.

ACL2 implements classic equivalence-based rewriting by maintaining a *generated equivalence relation*, or *geneqv*: a finite list of function symbols that have each been proved to be an equivalence relation, representing the smallest equivalence relation containing them all. Rewriting is inside out, so to rewrite a function call, the rewriter first rewrites each argument of that call. Congruence rules are employed to compute the geneqv for rewriting each argument.

We have incorporated patterned congruence rules into that algorithm without changing its basic structure or efficiency (based on timing the ACL2 regression suite [10]). The key idea is to pass around a list representing so-called *patterned equivalences*, or *pequivs* for short, as defined below. We show how this list is updated as the rewriter dives into subterms, ultimately giving rise to equivalences to add to the current geneqv.

A pequiv is an equivalence relation corresponding to a term L that is a function call, a variable x that occurs uniquely in L, an equivalence relation $\sim$, and a substitution s. The pequiv *based on* L, x, $\sim$, and s is the smallest equivalence relation containing the following relation: $a \approx b$ if and only if there exist substitutions s_1 and s_2 extending s that agree on all variables except perhaps x such that $a = L/s_1$, $b = L/s_2$, and $s_1(x) \sim s_2(x)$.

For a natural number k and function call $C = f(t_1, \ldots, t_k, \ldots, t_n)$, the following notation is useful: $pre(C)$ is the list $(t_1, \ldots, t_{k-1})$, $@(C)$ is t_k, and $post(C)$ is the list $(t_{k+1}, \ldots, t_n)$. Now consider the pequiv based on L, x, $\sim$, and s, and let u be a term $f(u_1, \ldots, u_k, \ldots, u_n)$, where f is the function symbol of L and x occurs in the kth argument of L. We define the *next equiv* as follows when for some substitution s' extending s, $pre(u)$ is $pre(L)/s'$ and $post(u)$ is $post(L)/s'$.[2] Let s' be the minimal such substitution. There are two cases. If x is an argument of L then the next equiv is the equivalence relation, $\sim$. Otherwise the next equiv is the pequiv based on $@(L)$, x, $\sim$, and s'.

The ACL2 rewriter maintains a list of pequivs and a geneqv (list of equivalence relations). Here we outline how those lists change when the rewriter, which is inside-out, calls itself recursively on a subterm. As before [1], classic congruence rules are applied to create a geneqv for the subterm; here we focus on how the list of pequivs contributes to the pequivs and geneqv for the subterm. Consider a pequiv p based on L, x, $\sim$, and s, among the list of pequivs maintained as we are rewriting the term $f(u_1, \ldots, u_k, \ldots, u_n)$, and consider the rewrite of u_k. There are three cases. If the next equiv for p (for position k) is $\sim$, then $\sim$ is added to the geneqv for rewriting u_k. If the next equiv for p is a pequiv p', then p' is added to the list of pequivs for rewriting u_k. Otherwise the next equiv for p does not exist, and p is ignored when rewriting u_k.

The following two theorems (relative to an implicit first-order theory) justify this algorithm. The first explains why a congruence rule justifies the sufficiency of maintaining the corresponding pequiv. The second explains why it suffices to maintain the next pequiv when rewriting a subterm.

Theorem 1. *For a provable patterned congruence rule* $x \sim_{\text{inner}} y \to L \sim_{\text{outer}} R$, *let* $\sim$ *be the pequiv based on* L, x, $\sim_{\text{inner}}$, *and the empty substitution. Then* $\sim$ *refines* $\sim_{\text{outer}}$, *i.e., the following is a theorem:* $x \sim y \to x \sim_{\text{outer}} y$.

Theorem 2. *Let* $\sim_1$ *be a pequiv, let* u *be a term, and assume that the next equiv,* $\sim_2$, *exists for* $\sim_1$ *and* k. *Let* arg *be the* k^{th} *argument of* u, *let* arg' *be a term, and let* u' *be the result of replacing the* k^{th} *argument of* u *by* arg'. *Then the following is a theorem:* $arg \sim_2 arg' \to u \sim_1 u'$.

5 Reflections

ACL2 development began in 1989. Recent years have seen an increase in industrial application, with regular use at Advanced Micro Devices, Centaur Technology, Intel, Oracle, and Rockwell Collins, as well as academia and the U.S.

---

[2] We are simplifying the actual condition here, because the rewriter applies to both a term and a substitution, and this substitution must be applied to $post(u)$.

Government. In order to support these users, we have been continuously improving ACL2; in particular, after the December 2012 release of Version 6.0 through the January 2014 release of Version 6.4, 129 distinct improvements have been reported in RELEASE-NOTES topics of the online ACL2 User's Manual [7].

While some of these improvements may present topics of interest to the ITP community, most are technical and specific to ACL2, as the focus has largely been on direct support for the user community, in particular industrial users. While few of these topics are likely candidates for traditional academic publication, patterned congruence rules seem to us an exception: any modern ITP system might benefit from them, if it is important to perform rewriting efficiently at the scale of industrial projects.

Acknowledgments. We thank the reviewers for helpful remarks. This research was supported by DARPA under Contract No. N66001-10-2-4087 and by ForrestHunt, Inc.

References

1. Brock, B., Kaufmann, M., Moore, J.: Rewriting with equivalence relations in ACL2. Journal of Automated Reasoning 40(4), 293–306 (2008), http://dx.doi.org/10.1007/s10817-007-9095-9
2. Cohen, C.: Pragmatic quotient types in CoQ. In: Blazy, S., Paulin-Mohring, C., Pichardie, D. (eds.) ITP 2013. LNCS, vol. 7998, pp. 213–228. Springer, Heidelberg (2013), http://dx.doi.org/10.1007/978-3-642-39634-2_17
3. Homeier, P.: A design structure for higher order quotients. In: Hurd, J., Melham, T. (eds.) TPHOLs 2005. LNCS, vol. 3603, pp. 130–146. Springer, Heidelberg (2005), http://dx.doi.org/10.1007/11541868_9
4. Huffman, B., Kunčar, O.: Lifting and transfer: A modular design for quotients in Isabelle/HOL. In: Gonthier, G., Norrish, M. (eds.) CPP 2013. LNCS, vol. 8307, pp. 131–146. Springer, Heidelberg (2013), http://dx.doi.org/10.1007/978-3-319-03545-1_9
5. Kaufmann, M.: ACL2 demo of (patterned) congruences, https://acl2-books.googlecode.com/ svn/trunk/demos/patterned-congruences.lisp
6. Kaufmann, M., Manolios, P., Moore, J.S.: Computer-Aided Reasoning: An Approach. Kluwer Academic Publishers, Boston (2000)
7. Kaufmann, M., Moore, J S.: ACL2 home page, http://www.cs.utexas.edu/users/moore/acl2
8. Kaufmann, M., Moore, J S.: Essay on Patterned Congruences and Equivalences, in ACL2 source file rewrite.lisp, https://acl2-devel.googlecode.com/svn/trunk/rewrite.lisp
9. Swords, S.: Personal communication
10. ACL2 Community Books, http://acl2-books.googlecode.com/

Formal C Semantics:
CompCert and the C Standard

Robbert Krebbers[1], Xavier Leroy[2], and Freek Wiedijk[1]

[1] ICIS, Radboud University Nijmegen, The Netherlands
[2] Inria Paris-Rocquencourt, France

Abstract. We discuss the difference between a formal semantics of the C standard, and a formal semantics of an implementation of C that satisfies the C standard. In this context we extend the CompCert semantics with end-of-array pointers and the possibility to byte-wise copy objects. This is a first and necessary step towards proving that the CompCert semantics refines the formal version of the C standard that is being developed in the Formalin project in Nijmegen.

1 Introduction

The C programming language [2] allows for close control over the machine, while being very portable, and allowing for high runtime efficiency. This made C among the most popular programming languages in the world.

However, C is also among the most dangerous programming languages. Due to weak static typing and the absence of runtime checks, it is extremely easy for C programs to have bugs that make the program crash or behave badly in other ways. NULL-pointers can be dereferenced, arrays can be accessed outside their bounds, memory can be used after it is freed, *etc.* Furthermore, C programs can be developed with a too specific interpretation of the language in mind, giving portability and maintenance problems later.

An interesting possibility to remedy these issues is to *reason* about C programs. Static program analysis is a huge step in this direction, but is by nature incomplete. The use of interactive theorem provers reduces the problem of incompleteness, but if the verification conditions involved are just generated by a tool (like Jessie/Why3 [9]), the semantics that applies is implicit. Therefore, the semantics cannot be studied on its own, and is very difficult to get correct. For this reason, to obtain the best environment to reason reliably about C programs, one needs a formal semantics in an interactive theorem prover, like the Cholera [10], CompCert [6], or Formalin [3,4] semantics.

The CompCert semantics has the added benefit that it has been used in the correctness proof of the CompCert compiler. Hence, *if* one uses this compiler, one can be sure that the proved properties will hold for the generated assembly code too. However, verification against the CompCert semantics does not give reliable guarantees when the program is compiled using a different compiler.

The C standard gives compilers considerable freedom in what behaviors to give to a program [2, 3.4]. It uses the following notions of under-specification:

G. Klein and R. Gamboa (Eds.): ITP 2014, LNAI 8558, pp. 543–548, 2014.

- *Unspecified behavior*: two or more behaviors are allowed. For example: order of evaluation in expressions. The choice may vary for each use.
- *Implementation defined behavior*: unspecified behavior, but the compiler has to document its choice. For example: size and endianness of integers.
- *Undefined behavior:* the standard imposes no requirements at all, the program is even allowed to crash. For example: dereferencing a NULL or dangling pointer, signed integer overflow, and a sequent point violation (modifying a memory location more than once between two sequence points).

Under-specification is used extensively to make C portable, and to allow compilers to generate fast code. For example, when dereferencing a pointer, no code has to be generated to check whether the pointer is valid or not. If the pointer is invalid (NULL or a dangling pointer), the compiled program may do something arbitrary instead of having to exit with a nice error message.

Like any compiler, CompCert has to make choices for implementation defined behavior (*e.g.* integer representations). Moreover, due to its intended use for embedded systems, CompCert gives a semantics to various undefined behaviors (such as aliasing violations) and compiles those in a faithful manner.

To verify properties of programs that are being compiled by CompCert, one can make explicit use of the behaviors that are defined by CompCert but not by the C standard. On the contrary, the Formalin semantics intends to be a formal version of the C standard, and therefore should capture the behavior of *any* C compiler. A blog post by Regehr [11] shows some examples of bizarre behavior by widely used compilers due to undefined behavior. Hence, Formalin has to take *all* under-specification seriously (even if that makes the semantics more complex), whereas CompCert may (and even has to) make specific choices.

For widely used compilers like GCC and Clang, Formalin is of course unable to give any formal guarantees that a correctness proof with respect to its semantics ensures correctness when compiled. After all, these compilers do not have a formal semantics. We can only argue that the Formalin semantics makes more things undefined than the C standard, and assuming these compilers "implement the C standard", correctness morally follows.

As a more formal means of validation of the Formalin semantics we intend to prove that CompCert is a refinement of it. That means, if a behavior is defined by the Formalin semantics, then the possible behaviors of CompCert match those of Formalin. As a first step into that direction, we will discuss two necessary modifications to CompCert as displayed in Figure 1. It is important to notice that this work is not about missing features in CompCert, but about *missing behaviors* of features that are in both CompCert and Formalin.

Note that even the Formalin semantics deviates from the C standard. That is because the C standard has two incompatible ways to describe data [1, Defect Report #260]. It uses a low level description of data in terms of bits and bytes called *object representations*, but also describes data *abstractly* in a way that allows various compiler optimizations. For this reason Formalin errs on the side of caution: it makes some behaviors undefined that most people consider to be defined according to the standard.

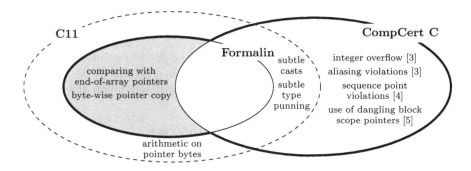

Fig. 1. We extend CompCert C with the behaviors in the shaded area. Each set in this diagram contains the programs that according to the semantics do not have undefined behavior. Since C11 is subject to interpretation, we draw it with a dashed line.

For example, in both Formalin and CompCert, adding 0 to a byte from a pointer object representation is undefined behavior. Both semantics do not just have numeric bytes, but also use symbolic bytes for pointers and uninitialized memory (see the definition **memval** of CompCert in Section 3).

Example. Using CompCert's reference interpreter, we checked that our extensions of CompCert give the correct semantics to:

```
void my_memcpy(void *dest, void *src, int n) {
    unsigned char *p = dest, *q = src, *end = p + n;
    while (p < end) // end may be end-of-array
        *p++ = *q++;
}
int main() {
    struct S { short x; short *r; } s = { 10, &s.x }, s2;
    my_memcpy(&s2, &s, sizeof(struct S));
    return *(s2.r);
}
```

In CompCert 1.12, this program has undefined behavior, for two reasons: the comparison `p < end` that involves an end-of-array pointer, and the byte-wise reads of the pointer `s.r`. Sections 2 and 3 discuss these issues and their resolution.

Sources. Our extension for end-of-array pointers is included in CompCert since version 1.13. The sources for the other extension and the Formalin semantics can be found at **http://github.com/robbertkrebbers**.

2   Pointers in CompCert

CompCert defines its memory as a finite map of blocks, each block consisting of an array of symbolic bytes (and corresponding permissions) [7]. Pointers are pairs (b, i) where b identifies the block, and i the offset into that block.

The C standard's way of dealing with pointer equality is subtle. Consider the following excerpt [2, 6.5.9p6]:

> Two pointers compare equal if and only if [...] or one is a pointer to one past the end of one array object and the other is a pointer to the start of a different array object that happens to immediately follow the first array object in the address space.

End-of-array pointers are somewhat special, as they cannot be dereferenced, but their use is common programming practice when looping through arrays.

```
void inc_array(int *p, int n) {
  int *end = p + n;
  while (p < end) (*p++)++;
}
```

Unfortunately, end-of-array pointers can also be used in a way such that the behavior is not stable under compilation.

```
int x, y;
if (&x + 1 == &y) printf("x and y are allocated adjacently\n");
```

Here, the `printf` is executed only if x and y are allocated adjacently, which may happen as many compilers allocate x and y consecutively on the stack.

In the CompCert semantics, x and y have disjoint block identifiers, and the representations of &x + 1 and &y are thus unequal. Compilation does not preserve this inequality as the blocks of x and y are merged during stack allocation. To ensure preservation of comparisons, the semantics of earlier CompCert versions (1.12 and before) required pointers used in comparisons to be *valid*. A pointer is valid if its offset is strictly within the block bounds. We weakened this restriction on pointer comparisons slightly:

- Comparison of pointers in the same block is defined only if both are *weakly valid*. A pointer is weakly valid if it is valid or end-of-array.
- Comparison of pointers with different block identifiers is defined for valid pointers only.

Our weakened restriction allows common programming practice of using end-of-array pointers when looping through arrays possible, but uses as in the second example above remain undefined. We believe that the above restriction on pointer comparisons is more sensible than the naive reading of the C standard because it is stable under compilation[1].

To adapt the compiler correctness proofs we had to show that all compilation passes preserve weak pointer validity and preserve the new definition of pointer comparisons. Furthermore, we had to modify the definition of memory injections [8] to ensure that also the offsets of weakly valid pointers remain representable by machine integers after each program transformation.

---

[1] Notice that the C standard already makes a distinction between pointers in the same block and pointers in different blocks, for pointer inequalities < and <= [2, 6.5.8p5].

3   Bytes in CompCert

CompCert represents integer and floating point values by sequences of numeric bytes, but pointer values and uninitialized memory by symbolic bytes.

```
Inductive memval: Type :=
  | Undef: memval
  | Byte: byte -> memval
  | Pointer: block -> int -> nat -> memval
  | PointerPad: memval.
```

When storing a pointer (b, i), the sequence Pointer b i 0, ..., Pointer b i 3 is stored, and on allocation of new memory a sequence of Undef bytes is stored (the constructor PointerPad is part of our extension, and is discussed later).

In the version of CompCert that we have extended, it was only possible to read a sequence of Pointer bytes as a pointer value. To make byte-wise reading and writing of pointers possible, we extend values with a constructor Vptrfrag.

```
Inductive val: Type :=
  | Vundef: val
  | Vint: int -> val
  | Vlong: int64 -> val
  | Vfloat: float -> val
  | Vptr: block -> int -> val
  | Vptrfrag: block -> int -> nat -> val.
```

Extending the functions that encode and decode values as memval sequences turned out more subtle than expected. The CompCert compiler back-end must sometimes generate code that stores and later restores the value of an integer register in a stack location. To preserve this value, these memory stores and loads are performed at the widest integer register type, int. For pointer fragments, the top 3 bytes of the in-memory representation are statically unknown, since they can result from the sign-extension of the low byte. Therefore, we abstract these top 3 bytes as the new memval constructor PointerPad.

Arithmetical operations are given undefined behavior on pointer fragments. Reading a pointer byte from memory, adding 0 to it, and writing it back remains undefined behavior. It would be tempting give an ad-hoc semantics to such corner cases, but that will result in a loss of algebraic properties like associativity.

Assignments involve implicit casts, hence char to char casts need to have defined behavior on pointer fragments to make storing these fragments possible. Since the CompCert compiler needs the guarantee that the result of a cast is well-typed (while the CompCert semantics is untyped), neutral casts perform a zero- or sign-extension instead of being the identity. However, since the top 3 bytes of the in-memory representation of pointer fragments are statically unknown, we changed the semantics of a cast from char to char to check whether the operand is well-typed (which vacuously holds for well-typed programs). If not, the behavior of the cast is undefined. This has the desired result that char to char casts can be removed in a later compilation phase.

CompCert 2.2 features a new static analysis that approximates the shapes of values, including points-to information for pointer values. Our `char` values hold more values, and thus this analysis needed some changes. For example, before our extension, the only pointer values that can be read from a given memory location are those that were stored earlier at this exact location using a pointer-wise store. With our extensions, the pointer values thus read can also come from byte-wise pointer fragments that were stored at overlapping locations.

4   Conclusion and Future Work

The two extensions of CompCert described in this paper succeed in giving a semantics to behaviors that were previously undefined. These extensions are a necessary step for cross validation of the CompCert and Formalin semantics. Our treatment of byte-wise copying of objects containing pointers turned out to be more involved than suggested in [7], owing to the nontrivial semantics of casts and changes to the static value analysis. If future versions of CompCert get a type system, our workaround for casts can be removed.

Another behavior that needs attention in future work is CompCert's call-by-reference passing of struct and union values, as discussed in [4]. In this case, as well as in the byte-wise copying case, the approach followed by Norrish [10] (namely, representing values as sequences of bytes identical to their in-memory representations) may provide an alternative solution, but could cause other difficulties with value analysis and compiler correctness proofs.

Acknowledgments. We thank the reviewers for their helpful comments. This work was partially supported by NWO and by ANR (grant ANR-11-INSE-003).

References

1. International Organization for Standardization: WG14 Defect Report Summary (2008), http://www.open-std.org/jtc1/sc22/wg14/www/docs/
2. International Organization for Standardization: ISO/IEC 9899-2011: Programming languages – C. ISO Working Group 14 (2012)
3. Krebbers, R.: Aliasing Restrictions of C11 Formalized in Coq. In: Gonthier, G., Norrish, M. (eds.) CPP 2013. LNCS, vol. 8307, pp. 50–65. Springer, Heidelberg (2013)
4. Krebbers, R.: An Operational and Axiomatic Semantics for Non-determinism and Sequence Points in C. In: POPL, pp. 101–112 (2014)
5. Krebbers, R., Wiedijk, F.: Separation Logic for Non-local Control Flow and Block Scope Variables. In: Pfenning, F. (ed.) FOSSACS 2013. LNCS, vol. 7794, pp. 257–272. Springer, Heidelberg (2013)
6. Leroy, X.: Formal verification of a realistic compiler. CACM 52(7), 107–115 (2009)
7. Leroy, X., Appel, A.W., Blazy, S., Stewart, G.: The CompCert Memory Model, Version 2. Research report RR-7987, INRIA (2012)
8. Leroy, X., Blazy, S.: Formal verification of a C-like memory model and its uses for verifying program transformations. JAR 41(1), 1–31 (2008)
9. Moy, Y., Marché, C.: The Jessie plugin for Deduction Verification in Frama-C, Tutorial and Reference Manual (2011)
10. Norrish, M.: C formalised in HOL. Ph.D. thesis, University of Cambridge (1998)
11. Regehr, J.: (2012), Blog post at http://blog.regehr.org/archives/759

Mechanical Certification of Loop Pipelining Transformations: A Preview

Disha Puri[1], Sandip Ray[2], Kecheng Hao[1], and Fei Xie[1]

[1] Department of Computer Science, Portland State University,
Portland, OR 97207, USA
[2] Strategic CAD Labs, Intel Corporation, Hillsboro, OR 97124, USA

Abstract. We describe our ongoing effort using theorem proving to certify loop pipelining, a critical and complex transformation employed by behavioral synthesis. Our approach is mechanized in the ACL2 theorem prover. We discuss some formalization and proof challenges and our early attempts at addressing them.

Keywords: Behavioral Synthesis, Theorem proving, Electronic System Level Design, Equivalence Checking.

1 Introduction

Behavioral synthesis is the process of compiling an Electronic System Level (ESL) specification of a hardware design into RTL. ESL facilitates fast turnaround time in production of hardware designs by rasing design abstraction; the designer only specifies the design functionality in a high-level language (*e.g.*, SystemC, C, C++, etc.), from which RTL is automatically synthesized. However, its adoption critically depends on our ability to certify that the synthesized design indeed implements the ESL specification. Given the abstraction difference between ESL and RTL, this certification is non-trivial.

Loop pipelining is a critical transformation implemented in most behavioral synthesis tools. Unfortunately, it is also one of the most complex transformations [1]. Furthermore, sequential equivalence checking (SEC) techniques are not directly applicable for certification of synthesized loop pipelines. In this paper, we describe how we are using interactive theorem proving (ACL2) to facilitate this certification. We discuss our early efforts with the proof, some of the challenges and complexities, and the approaches we are exploring to address them.

This work is a part of a project to develop a scalable certification framework for behavioral synthesis. The project has been ongoing for some time, with several mature components; however, the focus of previous work was on automated SEC. The work described here is its first serious "foray" in theorem proving.

2 Background and Context

Behavioral Synthesis and an SEC Framework. Behavioral synthesis transformations are classified into three categories: (1) compiler transformations,

G. Klein and R. Gamboa (Eds.): ITP 2014, LNAI 8558, pp. 549–554, 2014.
© Springer International Publishing Switzerland 2014

(2) scheduling (mapping each operation to a clock cycle), and (3) resource allocation and control synthesis (allocating registers to variables, and generating an FSM to implement the schedule). Loop pipelining is a part of scheduling. Given the abstraction gap between ESL and RTL, there are no obvious mappings between internal variables, rendering SEC ineffective. Applying theorem proving is also challenging: (1) *verifying each synthesized design* requires prohibitive human effort; (2) *verifying a synthesis tool* is infeasible since tool implementations are proprietary (and closely guarded), in addition to being highly complex.

Previous work [2–4] resulted in the following observations. (1) SEC can compare RTL with the intermediate representations (IRs) after compiler and scheduling operations; correspondence between internal variables is preserved, and identified from resource mappings. (2) While transformation *implementations* are proprietary, IRs after successive transformations are available from reports generated during synthesis. (3) IRs are structurally similar across synthesis tools *viz.*, graphs of operations with explicit control/data flow and schedule. Consequently, a formalization called *Clocked Control Data Flow Graph* (CCDFG) was developed for IRs, together with two SEC algorithms, respectively to compare (1) a CCDFG with RTL, and (2) two CCDFGs corresponding to IRs after each successive transformation. However, the latter is effective only if the difference between IRs is small. Loop pipelining substantially changes control/data flow and introduces controls (*e.g.*, to eliminate hazards), making SEC infeasible.

Certifying Loop Pipelining. Our key observation is that it is *not* necessary to verify the implementation of any synthesis tool. Instead, we can (1) develop a *reference* algorithm $\mathcal{A}$ that takes a sequential CCDFG $\mathcal{C}$ and generates a pipelined CCDFG $\mathcal{P}$, (2) use SEC to compare $\mathcal{P}$ with the synthesized RTL $\mathcal{R}$, and (3) prove the correctness of $\mathcal{A}$. The algorithm $\mathcal{A}$ can be much simpler than that of any synthesis tool, since it can use the synthesis tool's report to determine the values of the key parameters (*e.g.*, pipeline interval, number of iterations pipelined, etc.). Viability of this flow was justified previously [5] by developing such an algorithm and using it to compare several synthesized pipelines. However, the algorithm was not verified (indeed, not formalized), rendering the "certification" flow unsound; in fact, we already found errors in that algorithm merely by attempting formalization. Furthermore, since it was not written with reasoning in mind, it is a non-trivial target for mechanical proof. Our current work is a deconstruction of that algorithm, developed from ground up to account for necessary invariants. Note that we are free to choose *any* verifiable implementation without losing the ability to certify designs synthesized by commercial tools.[1]

3   Pipelinable Loop and Correctness Formalization

Pipelinable Loop. A *pipelinable loop* [5] is a loop with (1) no nested structures, (2) one *Entry* and one *Exit* block; and (3) no branching between scheduling

---

[1] One caveat is that we must synthesize pipelines *using the parameters reported by the synthesis tool*; otherwise we may fail to certify correct designs. We have not found this to be a problem in practice.

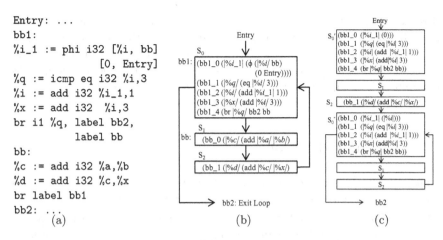

```
Entry: ...
bb1:
%i_1 := phi i32 [%i, bb]
             [0, Entry]
%q := icmp eq i32 %i,3
%i := add i32 %i_1,1
%x := add i32   %i,3
br i1 %q, label bb2,
          label bb
bb:
%c := add i32 %a,%b
%d := add i32 %c,%x
br label bb1
bb2: ...
```

Fig. 1. (a) Loop in LLVM Assembly (b) Fragment of CCDFG corresponding to loop. Scheduling step S_0 has a ϕ-statement (c) ϕ-elimination operation. $\%i\_1$ is assigned 0 in S_0' and $\%i$ in S_0''.

steps. These restrictions are not for simplifying reasoning, but reflect the kind of loops that are actually pipelined, *e.g.*, synthesis tools unroll inner loops (by a compiler transformation) before applying pipelining to the outer loop.

Correctness Statement. Let L be a loop in CCDFG C, and let L_α be the pipelined implementation generated by our algorithm using pipeline parameters α. Let V be the set of variables mentioned in L, and U be the set of all variables in C. Suppose we execute L and L_α from CCDFG states s and s' respectively, such that for each variable $v \in V$, the value of v in s is the same as that in s', and suppose that the state on termination are f and f' respectively. Then (1) for any $v \in V$, the value of v in f is the same as that in f', and (2) for any $v \in (U \backslash V)$, the value of v in f' is the same as that in s'.

Remark: Condition (2) is the *frame rule* which ensures that variables in C that are not part of the loop are not affected by L_α. The algorithm introduces additional variables, eg, *shadow variables* (cf. Section 4). The values of these variables in f' are irrelevant since they are not accessed subsequently.

CCDFG. Formalizing the correctness statement entails defining the semantics of CCDFG. A CCDFG is a control/data flow graph with a schedule. Control flow is broken into basic blocks. Instructions in a basic block are grouped into *microsteps* that are executed concurrently. A schedule is a grouping of microsteps which can be completed within one clock cycle. The instruction language we support is a subset of LLVM [6] which is a front-end for many behavioral synthesis tools [7, 8]; we support assignment, load, store, bounded arithmetic, bit vectors, arrays, and pointer manipulations. As is common with ACL2, we use a state-based operational semantics [9, 10]. Assigning meanings to most instructions is standard; one exception is the ϕ-statement "v := phi [σ bb1] [τ bb2]". If reached from basic block bb1, it is the same as the assignment statement v :=

σ; if reached from bb2, it is the same as v := τ; the meaning is undefined otherwise. Reasoning about ϕ-statement is complex since after its execution from state s, the state reached depends not only on s but previous basic block in the history. We need to handle it since it is used extensively to implement loop tests. A key step in loop pipelining is ϕ-elimination, viz., unrolling the loop once and replacing the ϕ-statement with assignment statements (cf. Fig. 1).

4 Algorithm and Proof

Our algorithm includes (1) ϕ-elimination mentioned above, (2) shadow registers, and (3) superstep construction. Fig. 2 illustrates steps 2 and 3.

Shadow Registers: Consider the CCDFG in Fig. 1. Here %x is written in step S_0 but read in S_2. If the loop is pipelined such that a new iteration is initiated every cycle, then we must ensure that the write from the S_0 step of a subsequent iteration does not overwrite %x before it is read by the S_2 step of the current iteration. This is achieved by introducing a *shadow register* %x_reg that preserves a copy of the "old value of %x" and replacing reads of %x to use %x_reg.

Superstep Construction: We combine scheduling steps of successive iterations into "supersteps" which are scheduling steps for the pipeline. Supersteps account for hazards, viz., if a variable is written in scheduling step S and read subsequently in S' then S' cannot be in a superstep that precedes S. S and S' can be in a single superstep since we implement data forwarding.

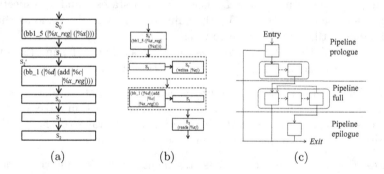

Fig. 2. (a) CCDFG of Fig. 1 after inserting shadow register $\%x\_reg$ for $\%x$. (b) Superstep construction. Horizontal arrows represent data forwarding. (c) Pipelined loop.

Correspondence Relation. Our planned proof involves defining a "correspondence relation" between loops of the sequential and pipelined CCDFGs and proving that it is preserved across loop iterations. The relation is informally paraphrased as follows. "Let S be a sequential loop and G be the pipelined loop generated from our algorithm, constituting prologue G_p, full stage G_l, and epilogue G_e (cf. Fig 2(c)). Let s_l be any state of G poised to execute G_l, and let k be any number such that the loop of G is not exited in k iterations from s_l. Then executing k iterations of G_l from s_l is equivalent to executing k iterations of S together with a collection of "partially completed" iterations of S.[2]

Proof Sketch. The invariant, albeit non-trivial, admits a direct proof of the correctness statement in Section 3. Equivalence of CCDFG states after completing execution of G and S follows from the fact that the epilogue G_e constitutes the incomplete scheduling steps of S. To prove that the relation is invariant across pipeline iterations, note that each new iteration of G_l initiates a new (incomplete) iteration of S, and advances incomplete iterations by one scheduling step; the result follows by rearranging the incomplete iterations, since rearrangement of scheduling steps produces the same computation in the absence of hazards. Thus we need to show that our algorithm generates hazard-free pipelines, which reduces to structural properties of the three components of the algorithm.

5   Current State and Conclusion

As of this writing, we have formalized the correspondence relation, and finished the proof of key lemmas for ϕ-elimination and shadow register. We have also proven an implication chain from the correspondence relation to the correctness statement. Our current ACL2 script has 156 definitions and 300 lemmas, including many lemmas about structural properties of CCDFGs. We admit that the correspondence relation and the proof sketch above, while rather natural on hindsight, are outcomes of lessons learned from several false starts.

Microprocessor pipeline verification is a mature research area [11–13]. Our work, albeit analogous, is different, *e.g.*, we verify an *algorithm* to generate pipelines instead of a specific implementation. Also, recent work on translation validation for software pipelines [1] has parallels to our work. However, their correctness statement is contingent upon the equivalence of a certain symbolic simulation of the two designs, and they do not statically identify data hazards.

Use of theorem proving on industrial flows typically involves either complicated reasoning about (optimized) implementations, or abstracting them significantly to facilitate proof. In contrast, we apply theorem proving on an algorithm that generates reference designs for SEC. This permits adjusting the algorithm (within limits) to suit mechanical reasoning while affording comparison with actual synthesized artifacts. We have made liberal use of this "luxury", *e.g.*, the three components of our algorithm were conceived from a reflection of our invariant and proof sketch. Indeed, we are currently refining the definition of superstep construction to facilitate proof of certain structural lemmas. We believe a similar approach is applicable in other contexts and may provide effective use of theorem proving without exposing confidential intellectual property.

---

[2] The formalization actually characterizes each incomplete iteration, *e.g.*, if the pipeline includes d iterations and successive iterations are introduced in consecutive clock cycles, then the i-th iteration has $i - 1$ incomplete scheduling steps.

References

1. Tristan, J.-B., Leroy, X.: A Simple, Verified Validator for Software Pipelining. In: Hermenegildo, M.V., Palsberg, J. (eds.) Proceedings of the 37th Annual ACM SIGPLAN-SIGACT Symposium on Principles of Programming Languages (POPL 2010), Madrid, Spain, pp. 83–92. ACM (2010)
2. Ray, S., Hao, K., Chen, Y., Xie, F., Yang, J.: Formal Verification for High-Assurance Behavioral Synthesis. In: Liu, Z., Ravn, A.P. (eds.) ATVA 2009. LNCS, vol. 5799, pp. 337–351. Springer, Heidelberg (2009)
3. Hao, K., Xie, F., Ray, S., Yang, J.: Optimizing Equivalence Checking for Behavioral Synthesis. In: Design, Automation and Test in Europe, Dresden, Germany, pp. 1500–1505. IEEE (2010)
4. Yang, Z., Hao, K., Cong, K., Ray, S., Xie, F.: Equivalence Checking for Compiler Transformations in Behavioral Synthesis. In: Byrd, G., Schenider, K., Chang, N., Ozev, S. (eds.) Proceedings of the 31st International Conference on Computer Design (ICCD 2013), Asheville, NC, USA, pp. 491–494. IEEE (2013)
5. Hao, K., Ray, S., Xie, F.: Equivalence Checking for Behaviorally Synthesized Pipelines. In: Groeneveld, G., Sciuto, D., Hassoun, S. (eds.) Proceedings of the 49th International ACM/EDAC/IEEE Design Automation Conference (DAC 2012), San Francisco, CA, USA, pp. 344–349. ACM (2012)
6. Lattner, C., Adve, V.S.: LLVM: A Compilation Framework for Lifelong Program Analysis & Transformation. In: 2nd ACM/IEEE International Symposium on Code Generation and Optimization: Feedback-directed and Runtime Optimization (CGO 2004), San Jose, CA, USA, pp. 75–88. IEEE Computer Society (2004)
7. Cong, J., Liu, B., Neuendorffer, S., Noguera, J., Vissers, K., Zhang, Z.: High-Level Synthesis for FPGAs: From Prototyping to Deployment. IEEE Transactions on CAD of Integrated Circuits and Systems 30, 473–491 (2011)
8. Canis, A., Choi, J., Aldham, M., Zhang, V., Kammoona, A., Anderson, J.H., Brown, S., Czajkowski, T.: LegUp: High-level Synthesis for FPGA-based Processor/Accelerator Systems. In: Wawrzynek, J., Compton, K. (eds.) Proceedings of the 19th ACM/SIGDA International Symposium on Field Programmable Gate Arrays (FPGA 2011), Monterey, CA, USA, pp. 33–36. ACM (2011)
9. Liu, H., Moore, J.S.: Executable JVM Model for Analytical Reasoning: A study. Science of Computer Programming 57, 253–274 (2005)
10. Boyer, R.S., Moore, J.S.: Mechanized Formal Reasoning about Programs and Computing Machines. In: Veroff, R. (ed.) Automated Reasoning and Its Applications: Essays in Honor of Larry Wos, pp. 141–176. MIT Press (1996)
11. Burch, J.R., Dill, D.L.: Automatic Verification of Pipelined Microprocessor Control. In: Dill, D.L. (ed.) CAV 1994. LNCS, vol. 818, pp. 68–80. Springer, Heidelberg (1994)
12. Manolios, P.: Correctness of Pipelined Machines. In: Hunt Jr., W.A., Johnson, S.D. (eds.) FMCAD 2000. LNCS, vol. 1954, pp. 161–178. Springer, Heidelberg (2000)
13. Sawada, J., Hunt Jr., W.A.: Verification of FM9801: An Out-of-Order Microprocessor Model with Speculative Execution, Exceptions, and Program-Modifying Capability. Formal Methods in Systems Design 20, 187–222 (2002)

References

1. Tristan, J.-B., Leroy, X.: A Simple, Verified Validator for Software Pipelining. In: Hermenegildo, M.V., Palsberg, J. (eds.) Proceedings of the 37th Annual ACM SIGPLAN-SIGACT Symposium on Principles of Programming Languages (POPL 2010), Madrid, Spain, pp. 83–92. ACM (2010)

2. Ray, S., Hao, K., Chen, Y., Xie, F., Yang, J.: Formal Verification for High-Assurance Behavioral Synthesis. In: Liu, Z., Ravn, A.P. (eds.) ATVA 2009. LNCS, vol. 5799, pp. 337–351. Springer, Heidelberg (2009)

3. Hao, K., Xie, F., Ray, S., Yang, J.: Optimizing Equivalence Checking for Behavioral Synthesis. In: Design, Automation and Test in Europe, Dresden, Germany, pp. 1500–1505. IEEE (2010)

4. Yang, Z., Hao, K., Cong, K., Ray, S., Xie, F.: Equivalence Checking for Compiler Transformations in Behavioral Synthesis. In: Byrd, G., Schenider, K., Chang, N., Ozev, S. (eds.) Proceedings of the 31st International Conference on Computer Design (ICCD 2013), Asheville, NC, USA, pp. 491–494. IEEE (2013)

5. Hao, K., Ray, S., Xie, F.: Equivalence Checking for Behaviorally Synthesized Pipelines. In: Groeneveld, G., Sciuto, D., Hassoun, S. (eds.) Proceedings of the 49th International ACM/EDAC/IEEE Design Automation Conference (DAC 2012), San Francisco, CA, USA, pp. 344–349. ACM (2012)

6. Lattner, C., Adve, V.S.: LLVM: A Compilation Framework for Lifelong Program Analysis & Transformation. In: 2nd ACM/IEEE International Symposium on Code Generation and Optimization: Feedback-directed and Runtime Optimization (CGO 2004), San Jose, CA, USA, pp. 75–88. IEEE Computer Society (2004)

7. Cong, J., Liu, B., Neuendorffer, S., Noguera, J., Vissers, K., Zhang, Z.: High-Level Synthesis for FPGAs: From Prototyping to Deployment. IEEE Transactions on CAD of Integrated Circuits and Systems 30, 473–491 (2011)

8. Canis, A., Choi, J., Aldham, M., Zhang, V., Kammoona, A., Anderson, J.H., Brown, S., Czajkowski, T.: LegUp: High-level Synthesis for FPGA-based Processor/Accelerator Systems. In: Wawrzynek, J., Compton, K. (eds.) Proceedings of the 19th ACM/SIGDA International Symposium on Field Programmable Gate Arrays (FPGA 2011), Monterey, CA, USA, pp. 33–36. ACM (2011)

9. Liu, H., Moore, J.S.: Executable JVM Model for Analytical Reasoning: A study. Science of Computer Programming 57, 253–274 (2005)

10. Boyer, R.S., Moore, J.S.: Mechanized Formal Reasoning about Programs and Computing Machines. In: Veroff, R. (ed.) Automated Reasoning and Its Applications: Essays in Honor of Larry Wos, pp. 141–176. MIT Press (1996)

11. Burch, J.R., Dill, D.L.: Automatic Verification of Pipelined Microprocessor Control. In: Dill, D.L. (ed.) CAV 1994. LNCS, vol. 818, pp. 68–80. Springer, Heidelberg (1994)

12. Manolios, P.: Correctness of Pipelined Machines. In: Hunt Jr., W.A., Johnson, S.D. (eds.) FMCAD 2000. LNCS, vol. 1954, pp. 161–178. Springer, Heidelberg (2000)

13. Sawada, J., Hunt Jr., W.A.: Verification of FM9801: An Out-of-Order Microprocessor Model with Speculative Execution, Exceptions, and Program-Modifying Capability. Formal Methods in Systems Design 20, 187–222 (2002)

generated from our algorithm, constituting prologue G_p, full stage G_l, and epilogue G_e (cf. Fig 2(c)). Let s_l be any state of G poised to execute G_l, and let k be any number such that the loop of G is not exited in k iterations from s_l. Then executing k iterations of G_l from s_l is equivalent to executing k iterations of S together with a collection of "partially completed" iterations of S.[2]

Proof Sketch. The invariant, albeit non-trivial, admits a direct proof of the correctness statement in Section 3. Equivalence of CCDFG states after completing execution of G and S follows from the fact that the epilogue G_e constitutes the incomplete scheduling steps of S. To prove that the relation is invariant across pipeline iterations, note that each new iteration of G_l initiates a new (incomplete) iteration of S, and advances incomplete iterations by one scheduling step; the result follows by rearranging the incomplete iterations, since rearrangement of scheduling steps produces the same computation in the absence of hazards. Thus we need to show that our algorithm generates hazard-free pipelines, which reduces to structural properties of the three components of the algorithm.

5   Current State and Conclusion

As of this writing, we have formalized the correspondence relation, and finished the proof of key lemmas for ϕ-elimination and shadow register. We have also proven an implication chain from the correspondence relation to the correctness statement. Our current ACL2 script has 156 definitions and 300 lemmas, including many lemmas about structural properties of CCDFGs. We admit that the correspondence relation and the proof sketch above, while rather natural on hindsight, are outcomes of lessons learned from several false starts.

Microprocessor pipeline verification is a mature research area [11–13]. Our work, albeit analogous, is different, e.g., we verify an *algorithm* to generate pipelines instead of a specific implementation. Also, recent work on translation validation for software pipelines [1] has parallels to our work. However, their correctness statement is contingent upon the equivalence of a certain symbolic simulation of the two designs, and they do not statically identify data hazards.

Use of theorem proving on industrial flows typically involves either complicated reasoning about (optimized) implementations, or abstracting them significantly to facilitate proof. In contrast, we apply theorem proving on an algorithm that generates reference designs for SEC. This permits adjusting the algorithm (within limits) to suit mechanical reasoning while affording comparison with actual synthesized artifacts. We have made liberal use of this "luxury", e.g., the three components of our algorithm were conceived from a reflection of our invariant and proof sketch. Indeed, we are currently refining the definition of superstep construction to facilitate proof of certain structural lemmas. We believe a similar approach is applicable in other contexts and may provide effective use of theorem proving without exposing confidential intellectual property.

---

[2] The formalization actually characterizes each incomplete iteration, e.g., if the pipeline includes d iterations and successive iterations are introduced in consecutive clock cycles, then the i-th iteration has $i - 1$ incomplete scheduling steps.

Author Index